Dreaming

Dreaming

A Conceptual Framework for Philosophy of Mind and Empirical Research

Jennifer M. Windt

The MIT Press
Cambridge, Massachusetts
London, England

This book was set in Stone Sans and Stone Serif by Toppan Best-set Premedia Limited.

Library of Congress Cataloging-in-Publication Data

Windt, Jennifer Michelle, 1978–
Dreaming : a conceptual framework for philosophy of mind and empirical research / Jennifer Michelle Windt.
pages cm
Includes bibliographical references and index.
ISBN 978-0-262-02867-7 (hardcover : alk. paper)
ISBN 978-0-262-54991-2 (paperback)
1. Dreams—Philosophy. 2. Sleep—Research. I. Title.
BF1078.W476 2015
154.6′3–dc23
2014034361

To my partner, Stefan Pitz and to my mother, Janice Kaye Windt
In memory of my father, Hans-Joachim Windt

What a singular moment is the first one, when you have hardly begun to recollect yourself, after starting from midnight slumber! By unclosing your eyes so suddenly, you seem to have surprised the personages of your dream in full convocation round your bed, and catch one broad glance at them before they can flit into obscurity. Or, to vary the metaphor, you find yourself, for a single instant, wide awake in that realm of illusions, whither sleep has been the passport, and behold its ghostly inhabitants and wondrous scenery, with a perception of their strangeness, such as you never attain while the dream is undisturbed.

—Nathaniel Hawthorne, *The Haunted Mind* (1837)

Contents

Acknowledgments

This book has been part of my life for a long time, and it has gone through many stages. The most important was the transformation from dissertation to book manuscript. Many people have helped me in various ways throughout this process. First and foremost, I want to thank my supervisors, Thomas Metzinger and Tore Nielsen, whose invaluable comments, questions, criticisms, and advice have made this a much better book than it could ever have been without them. Thomas Metzinger has supported me and my work in many ways throughout the entire project, and I am greatly indebted to him in more ways than I can say. He first kindled my interest in philosophy of mind, long before I ever thought about pursuing this area of research on a professional level, and he first introduced me to the topic of dreaming. He also encouraged me to write this thesis, and then later to turn it into a book. His comments and guidance have profoundly shaped my thinking. I am also extremely grateful to have had Tore Nielsen as a second supervisor. I have admired his work on dreaming since I first became interested in the topic as a student, and I am deeply thankful for his continued interest in my work, his willingness to devote his time to it, and many inspiring conversations. His consistently constructive criticism has truly lifted this book to a new level, and as I was working on the revisions, his comments often opened up new perspectives, enabled insights that I would not have had without him, and provided valuable food for thought. Thanks to Thomas and Tore, this book is different in many ways from the dissertation, and I think that it has greatly improved in this process.

On the institutional side, I am greatly indebted to the Barbara Wengeler Foundation for providing funding in the form, first, of a doctoral grant and then for awarding me with the Barbara Wengeler Prize for the best dissertation in 2012. I also want to thank the Volkswagen Foundation for giving me the opportunity to participate in an interdisciplinary research project on dreaming through the European Platform for Life Sciences, Mind Sciences, and the Humanities. This was a truly formative experience that completely changed my theoretical understanding of dreaming and the methods used in dream research. To the philosophy department of the Johannes Gutenberg University of Mainz, I am thankful for providing an environment in which I was able to grow intellectually and pursue my ideas, and also for giving me the space to develop this project into a book over the last months. And last but certainly not least, I feel extremely lucky to have found a consistently helpful and patient editor in Philip Laughlin at the MIT Press. Judy Feldmann provided invaluable support during a difficult time. I am also grateful to Margarita Encomienda for a beautiful cover design, as well as to Lisa Brown, Christopher Eyer, William G. Henry, Susan Mai, Leonard Rosenbaum, and the entire production team at MIT Press, for their professional and constructive guidance and support. Any mistakes that remain in this book are entirely my own.

In addition, I have received help from many friends and colleagues from Mainz, from the European Platform, and also from philosophy and the fields of dream and consciousness research more generally, and I am thankful to many people for inspiring conversations, constructive criticism, and novel insights. Among them are Adrian Alsmith, David Baßler, Mark Blagrove, Olaf Blanke, Martin Dresler, Daniel Erlacher, Sascha Fink, Dominic Harkness, Jakob Hohwy, Clare Johnson, Ahmed Karim, Nicolas Langlitz, Bigna Lenggenhager, Michael Madary, Levente Móró, Valdas Noreika, Kevin Reuter, Perrine Ruby, Melanie Schädlich, Michaela Schrage-Früh, Michael Schredl, Sophie Schwartz, Eric Schwitzgebel, Mel Slater, Tadas Stumbrys, Katja Valli, Ursula Voss, Markus Werning, Peter Widmer, Yann Wilhelm, Aiha Zemp, and many more. More generally, I would like to thank the members of the MIND Group and of the Theoretical Philosophy Journal Club at Mainz, as well as my students, for many constructive conversations. I also want to thank the authors who contributed to a research topic published in *Frontiers in Psychology* titled "Contrasting Dreaming and Wakefulness: Frontiers in Consciousness Research," which I edited in 2013. Their wonderful contributions necessitated many changes as I was working on the final version of this manuscript, and I am glad that I had the opportunity to make them.

On a more personal level, I want to thank my mother, Janice Kaye Windt, for being there for me. She gave me my love of books, as well as a hobby that has nothing at all to do with books and was an invaluable outlet when the going got tough. I also want to thank her for first teaching me to dream, though in a slightly different sense. And most importantly of all, I want to thank Stefan Pitz for being my love and companion, and for bearing with me through this long and difficult process with endless patience and a healthy dose of humor. He believed in me when I didn't, knew when to distract me and make me laugh, and also when best to leave me alone. I want to thank him for making the past years very happy indeed. I could not have done this without him.

General Introduction: The Conceptualization Problem of Dreaming

Introducing the Conceptualization Problem of Dreaming

Are dreams conscious experiences during sleep? What is it like to dream? Do dreams mimic the phenomenology of standard waking experience, or do they depart from waking perception and imagination, thought, emotion, and bodily experience in subtle and hard-to-grasp ways?

The elusiveness of dreams, so aptly described by Hawthorne, is not just that our ability to recall them begins to fade at the very moment at which, having awakened, we become aware of just having dreamt. The elusiveness of dreams is also of a conceptual nature. From now on, I will call this elusiveness, more technically, the *conceptualization problem of dreaming,* or the problem of how to describe dreaming in terms commonly used for the description of standard wake states. The goal of this book is to lay the groundwork for solving this problem. To do so, it will be necessary to develop a conceptual framework describing not only what it means to say that dreams are conscious experiences but also how to locate dreams relative to concepts such as perception, hallucination, and imagination, as well as thinking, knowledge, belief, deception, and self-consciousness. Along the way, it will frequently become clear that many of these wake-state terms cannot be straightforwardly applied to dreaming and have to be adjusted. This is a challenge, but one that holds promise, namely that the analysis of the concept of dreaming can provide an enrichment and extension of our conceptual repertoire.

A recurring idea is that dreams, conceived of as conscious experiences or phenomenal states during sleep, are an important contrast condition for theories of consciousness and the self. Dreaming arises under different functional and neurophysiological conditions than standard and altered wake states, and as will become clear throughout the book, the phenomenology of dreaming departs from the phenomenology of waking experience in important and sometimes unexpected ways. For this reason, the example of dreaming, properly described, can be informative for existing theories, and it is an important goal for future research to integrate dreaming into an interdisciplinary research program on consciousness and the self.

While there is a wealth of empirical research on sleep and dreaming, the role of dreaming in the interdisciplinary debate on consciousness is still fairly marginalized, and its metatheoretical relevance is often overlooked. On the positive side, several prominent representatives of contemporary philosophy of mind afford an important position to the discussion of dreaming (e.g., Metzinger, 2003, 2009, 2014b; Revonsuo, 2006), and dreaming is also slowly finding its way into textbooks on cognitive neuroscience (Baars & Gage, 2010, pp. 279–282; Stephan & Walter, 2013), as well as into the major online encyclopedia of philosophy (Springett, 2013; Windt, in press). On the negative side, however, both the philosophical and the scientific discussions of dreaming are marked by a profound uncertainty—and in some cases outright controversy—as to how dreams can be described conceptually and how dreaming can be located in the context of general theories of consciousness. The application of wake-state psychological terms such as hallucination and imagination, but also of cognitive terms such as thinking, metacognition, and belief, to dreaming is inherently problematic, and no unified theory of dreaming is currently available. This problem is paralleled by the lack of a commonly accepted definition of dreaming, with researchers defending different definitions ranging from "any mental activity occurring in sleep" to more narrow definitions casting dreams as consistently involving complex, multimodal hallucinations, strong emotions, and specific cognitive deficits (Hobson et al., 2000; Pagel et al., 2001). Underlying this lack of conceptual clarity is a deeper disagreement about the nature of the target phenomenon itself. Consequently, to solve the problem of how to integrate dreaming into the broader theoretical discussion, it is necessary to first solve the problem of how to conceptualize conscious experience in the dream state and to develop a theoretical account of dreaming in its own terms. If we want a theory of consciousness and of the self in general, rather than one that is restricted to the description of standard and altered wake states, the discussion of dreaming can make an invaluable contribution.

In his introduction to *The Concept of Mind*, often considered a turning point in analytic philosophy of mind, Gilbert Ryle compares the work of the philosopher to that of a cartographer. While most people can adequately apply concepts in everyday language, they cannot, according to Ryle, "talk sense about them" (Ryle, 1949/2000, p. 7) or "state the logical relations governing their use" (p. 7). In this, they resemble people who can find their way around their hometown but cannot draw a map of it. Ryle writes that

> to determine the logical geography of concepts is to reveal the logic of propositions in which they are wielded, that is to say, to show with what other propositions they are consistent and inconsistent, what propositions follow from them and from what propositions they follow. (p. 8)

If we extend Ryle's metaphor of philosophy as cartography to dreaming, we might say that dreaming is still a largely uncharted territory in existing theories of the mind, with many parts unknown. Moreover, those who have attempted to chart the conscious experience of dreaming have often reached starkly different conclusions. The situation, then, is much as if one were planning to travel to an unfamiliar country with the help of a collection of maps, some of which point in completely different directions and many of which contain blank spots—much as was the case for the maps used by explorers in earlier centuries. As is the case for historical maps, theoretical accounts of dreaming also contain their share of legends and mythical creatures, and in the course of this book, many of these will be first identified and then adjusted, to appear in a new guise. It is an important aim of this book to bring order into existing maps of the concept of dreaming and fill in regions whose borders so far have only been vaguely sketched.

Ryle's metaphor can also help clarify the importance of the conceptualization problem in another manner. If you want to determine how far a given location is from your home, in what direction it lies, and what route you should take to get there, you must have at least a general idea of your destination. When traveling to a particular region, you will have to know, at least approximately, in which part of the country it is located. But you will also have to know the rough borders both of the country and of its different regions. For the concept of dreaming, this means that to roughly locate it relative to waking consciousness, it will be necessary to determine its borders, delineating the limits of the target phenomenon, and also to accommodate its inherent variability, or describe how different types of dreaming relate to each other.

A first important step, then, is to break the conceptualization problem down into a number of smaller questions. The first concerns the rough outlines of the concept of dreaming and its general characterization as involving conscious experience. While much of the philosophical discussion on dreaming in the twentieth century focused on the question of whether dreams are experiences at all, most authors neglected the further and in many respects more important question of what exactly it means to say that dreams are conscious experiences. This is unfortunate, because to clarify the latter question, we will also have to advance a theory of how (if at all) to apply different wake-state psychological terms to dreaming. Here the relevant question is not so much whether dreams are conscious experiences at all but rather what *kinds of conscious experiences they are.* Answering this question involves charting a rough cartography of dreaming and its regions. To do so, it will not be necessary to determine the exact location of every path, crick, and hill—and indeed such a project would be beyond the scope of this book—but only to chart their rough locations relative to each other. On this basis, it will then become possible to determine the outlines of dreaming in general and to describe the difference between dreaming and standard and altered wake states. Only when such a theory is available will we also have a meaningful answer to the question of what exactly it means to say that dreams are conscious experiences.

The proposed framework will count as successful if it succeeds in rendering the phenomenon of dreaming less, rather than more, mysterious; if it helps us get a conceptual grip on the target phenomenon and thus helps diminish at least some of its original elusiveness; if it is able to contribute to the solution of existing questions and uncertainties; and if, for instance, by proposing points of contact between theoretical and experimental work on dreaming and consciousness research more generally, it can help promote the integration of these different disciplines.

Because dreaming is an inherently interdisciplinary research topic, resolving these questions requires taking not only theoretical and conceptual considerations but also empirical research findings into account. This can again be illustrated with the help of Ryle's metaphor. Even if philosophy is conceptual cartography, a satisfying and fruitful philosophical framework cannot be an exercise in fictional mapmaking only. A map that does not accurately represent actual locations, borders, and distances is for all practical purposes useless; it is not really a map at all. Similarly, though a conceptual framework per se has no real-world consequences, it is not enough to leave it at that. To earn its keep, a conceptual framework should additionally aspire to real-world plausibility, and for the concept of dreaming, which refers to phenomenal experience, this includes phenomenological plausibility. If not, such a framework might be coherent but utterly inept for dealing with the type of phenomena it was originally supposed to describe. Note that we will expect scientific plausibility as well: a conceptual framework of dreaming will be expected not only to take dream reports into account but also to reflect findings from scientific dream research more generally. A successful and constructive theory of dreaming will have to be conceptually sound and coherent as well as empirically informed, and to do so, it will not only have to integrate theoretical insights from philosophy and findings from scientific dream research but also take laypersons' dream reports into account. Moreover, because of its interdisciplinary nature, it will aim not only to inform philosophical theories of dreaming and of consciousness but also to be empirically informative by offering a constructive critique of existing methods in dream research, making suggestions for the interpretation of research results, and formulating questions for future research.

An important secondary aim of the book is to provide an overview of the most important philosophical work on dreaming—both from the history of philosophy and from the contemporary discussion—as well as of the most important research findings. This is crucial, because most readers who are familiar with scientific dream research will not be familiar with the philosophical literature, and vice versa. Moreover, because the philosophical literature on dreaming stems from different historical periods, as well as from different philosophical subdisciplines and traditions, it is important to provide a unified account of how philosophical thinking about dreaming has evolved over the centuries. This approach will at times yield surprising results, showing, for instance, that the Cartesian view of dreaming still resounds in much contemporary work, at least insofar as the phenomenological characterization of dreaming is concerned. Charting the historical development of the concepts of sleep and dreaming also allows me to describe the mutual relationship between the philosophical discussion on dreaming and scientific research, and in the process to shed light on the historical and often contingent roots of current conceptions of sleep and dreaming. By providing an integrated overview of theoretical accounts and scientific findings on sleep and dreaming from different disciplines and different historical periods, this book aims to establish a common ground that can function as a point of departure for future interdisciplinary work on dreaming.

Methodology

Self-reflection is an important part of any philosophical endeavor, and in a philosophical book on the concept of dreaming, several methodological remarks are necessary. Note that the concept of dreaming, as it is investigated here, lies at the intersection of different disciplines, namely, philosophy, psychology, cognitive neuroscience, and scientific sleep and dream research. But it also lies at the intersection of different philosophical subdisciplines such as philosophy of mind and epistemology. The interdisciplinary nature of the target phenomenon presents a challenge for the philosophical analysis of dreaming. Because dreaming is a real-world phenomenon, it is not amenable to an exclusively theoretical approach. Consequently this book departs from the classical analytic project of pure conceptual analysis in that it attempts to combine conceptual and theoretical considerations, mostly from philosophy of mind, with scientific research findings and a critique of the methods employed by scientific sleep and dream research. This interdisciplinary approach is in keeping with the goal of formulating an empirically informed—and informative—theoretical account of dreaming.

This approach differs from existing philosophical approaches in a number of respects. First, philosophers often rely on intuitions and folk psychology, as well as on the use of concepts in ordinary language. There is, however, no singular and unitary folk-psychological or ordinary-language concept of dreaming. Just as dreaming is a heterogeneous phenomenon (Hunt, 1989), so the concept of dreaming is heterogeneous and underdefined, with different uses coexisting not only in theory and research but also in everyday usage. In ordinary language, for instance, dreaming is related to daydreaming, but also has more metaphorical uses such as dream vacations, dream partners, and the like. Consequently, the analysis of ordinary usage will not prove helpful in this context. While a theory of dreaming should not violate the basic tenets of ordinary usage, folk psychology is of little help in solving the more fine-grained aspects of the conceptualization problem of dreaming, and at times it can be outright misleading.

Moreover, as will become clear throughout the book, intuitions are a particularly poor guide to deciding conceptual questions about dreaming. Indeed, many of the existing controversies regarding the concept of dreaming are related to implicit assumptions and unargued-for intuitions about what it is like to dream (or, indeed, what it is like to be awake), often with the effect of introducing a biased and oversimplified view of dreaming. My discussion of the historical literature illustrates the extent to which conceptions of sleep and dreaming have undergone historical changes—often in response to findings from empirical research, which themselves, in turn, are often colored by theoretical preconceptions. The relationship between philosophical theorizing and scientific research is entirely reciprocal in this respect. In addition, several of the existing controversies on dreaming—for instance, regarding the question of whether the phenomenology of dreaming is better likened to waking perception or imagination—hinge on such differing intuitions. And precisely because many conceptual questions about dreaming hinge on the phenomenology of dreaming, and because assumptions about the phenomenology of dreaming are often colored by assumptions about waking consciousness, such an approach is best avoided. But this also means that intuitiveness is not necessarily an indicator of the goodness of a theory of dreaming and thus is not necessarily something that a conceptual framework of dreaming should aspire to. Different aspects of my treatment of dreaming will likely be more or less intuitive to different readers. To the extent that what I say seems radically counterintuitive to some, I would hope that it might be regarded as an exercise in suspending preconceived conceptions about dreaming and assessing the plausibility and explanatory merit of alternative views. Sometimes the best way to become aware of one's own prejudices is to be confronted with those of others, and the consideration of historical bypaths in theorizing about dreaming often provides a particularly instructive contrast to currently popular views and highlights their historical contingency.

My approach also differs from classical analytic approaches in its attempt to integrate empirical research findings with conceptual analysis. This is in keeping with an expanding group of philosophers of mind contributing to the larger project of interdisciplinary consciousness research. However, this approach continues to be controversial and is rejected outright by some. For instance, Bennett and Hacker (2003), in their sweeping and in many ways enlightening book *The Philosophical Foundations of Neuroscience*, restrict the work of philosophers to the disentanglement of conceptual confusions. While my approach affords a central importance to this task as well, I do not share their views that philosophy is inherently and exclusively prescientific and pretheoretical, that philosophy and the neurosciences are separate enterprises treating entirely separate problems, or that neuroscience or empirical research more generally cannot contribute to solving conceptual problems (Bennett & Hacker, 2003, pp. 399–408). Rather, to the extent that concepts refer to real-world phenomena such as dreaming, I think they can—and indeed should—be meaningfully informed by empirical findings.

The relationship between philosophy and the empirical mind sciences, as I view it, is reciprocal, with empirical research results informing conceptual analysis and conceptual and theoretical considerations making valuable suggestions for the interpretation of research results and future research questions.

While my approach shares the interdisciplinary focus of neurophilosophy, as well as its physicalist background assumptions, it cannot properly be considered neurophilosophical, and only a small part of what I have to say directly concerns the neural correlates of dreaming or is informed by neuroscientific research results. This situation owes itself, in part, to the fact that most scientific research on dreaming focuses on the analysis of dream reports, and existing neuroimaging studies of sleep have largely neglected to collect dream reports, meaning that they are better regarded as forming part of sleep research than of dream research proper. However, there is also a deeper reason, namely, that various forms of behavioral evidence—including dream reporting behavior—are, in my view, inextricably linked to the study of dreaming, and in some respects more important for the study of dreaming than neuroscientific evidence. I will also argue, in chapter 8, that purely brain-based approaches are insufficient to explain the phenomenology of dreaming. Instead, to fully understand both the phenomenology of dreaming and its neural correlates, it is necessary to look beyond the brain to the sleeping body. For these reasons, I prefer to describe my approach as *empirically informed and informative*, as this better befits the wide scope of research results that flow into it.

Similarly, while my approach overlaps with experimental philosophy in its rejection of purely a priori conceptual analysis—at least with respect to the concept of dreaming—my approach does not form part of this philosophical subdiscipline, as it does not attempt to resolve theoretical issues by conducting proper experiments.[1] Though I did participate in several interdisciplinary research projects while working on this book, and though I profited immeasurably from them, these results only rarely flow directly into my theoretical argumentation here.

Finally, if it is true that such an interdisciplinary approach continues to be controversial in philosophy of mind, it is even more so with respect to epistemology. Because the epistemological problem of dream skepticism is inseparable from the conceptualization problem, I offer an empirically informed analysis of the background assumptions behind Cartesian dream skepticism and propose, in chapter 10, several real-world variants of dream deception. This approach to the epistemological problem of dream skepticism is unorthodox, and I emphatically intend it not as standing in opposition to existing treatments but as complementing and perhaps extending them.

Overview of the Chapters

For historical and argumentative reasons, the book's first four chapters are devoted to the question of whether dreams are experiences at all. I begin my discussion, in chapter

1, with an analysis of the background assumptions behind Cartesian dream skepticism, both because of the historical importance of this position and because of its theoretical connection to the conceptualization problem. This also allows me to emphasize the connection between the discussion of dreaming in epistemology and philosophy of mind, as well as the connection between Cartesian dream skepticism and skepticism about dreaming and dream reporting. This second kind of skepticism involves the claim, defended most explicitly by Norman Malcolm, that dreams are not conscious experiences for conceptual reasons. This leads to a detailed consideration of the relationship between dreaming and dream reports, which I take to be the foundational epistemological problem for any theoretical account of dreaming, as well as a methodological challenge for scientific dream research. To decide whether dreams are experiences—and thus whether, among other things, they are deceptive experiences, as Descartes claimed—we first have to clarify the relationship between dreaming and dream reporting.

Chapter 2 gives a brief introduction to historical and contemporary dream research. While much that I have to say here will be familiar to those working in the field of scientific dream research, this chapter also charts historical changes in theoretical conceptions of sleep and dreaming and investigates how philosophical and scientific theorizing on these matters have mutually influenced each other.

Chapter 3 is devoted to a detailed discussion of the methodological background assumptions of scientific dream research, ranging from laboratory studies of sleep and dreaming and the analysis of dream reports to the study of sleep behavior in disorders of dreaming and lucid dreams, the losses of dreaming in neurological syndromes, and the possibility of moving beyond dream reports in the study of dreaming. In particular, I confront philosophical skepticism about the trustworthiness of dream reports with a detailed metatheoretical critique of the methods of scientific dream research.

Chapter 4 brings these two strands together to propose an antiskeptical position on the trustworthiness of dream reports and the question of whether dreams are experiences at all.[2] Very briefly, trusting dream reports is, in this view, a methodological necessity and the condition of possibility for scientific dream research. Asking for empirical evidence or scientific proof of the trustworthiness of dream reports is, by contrast, deeply misguided. Moreover, trusting dream reports is theoretically justified because assuming dream reports to be trustworthy with respect to dream experience best explains why we report dreams the way we do. I also indicate how antiskepticism about dream reports is related to the broader debate on the trustworthiness of first-person reports.

Having established the groundwork for my subsequent investigation of dreaming, I then turn, in chapters 5 and 6, to a controversy in the philosophical literature on dreaming, namely, the question of whether dreaming should be modeled on waking perception and hallucination or should rather be described as a form of imaginative experience. While I ultimately reject this question, along with the implicit demand for an account that places dreaming safely on either side of the putative divide between perceiving and imagining, as ungrounded and potentially misleading, an important preliminary conclusion is that dreaming is a test case for theories of perception, hallucination, and imagination. I also tentatively suggest that recent research investigating dreaming and waking mind wandering might suggest a way out of this classical philosophical debate.

Chapters 7 and 8 investigate the relationship between phenomenal selfhood, bodily experience, and the physical body during sleep. A surprising conclusion from chapter 7 is that the phenomenology of embodied selfhood is more variable in dreams than in wakefulness, and dreams occasionally provide examples of phenomenal selfhood, or the experience of being or having a self, even when bodily experience is lost altogether. I introduce different conceptual tools for describing phenomenal embodiment in dreams, ranging from phenomenal indeterminacy and phenomenal disembodiment to strong phenomenal embodiment. Their metatheoretical relevance consists in the fact that the phenomenology of being an embodied self is shown to be unnecessary for the experience of being or having a self.

Chapter 8 then explores the relationship between phenomenal embodiment in dreams and the sleeping physical body. Here I confront the currently popular view of dreams as functionally disembodied states, in which conscious experience is wholly decoupled from sensory input, both from the environment and from the sleeping body, and motor output, with historical theories linking dreaming to illusory own-body perception, as well as scientific research on stimulus incorporation in dreams. The emerging picture from both chapters is that dreams are weakly phenomenally-functionally embodied states, in which bodily experience is attenuated as compared to standard wakefulness but still remains connected in interesting and systematic ways to the sleeping body. This is important, because it means that attempts to model dreaming as a naturally occurring case of cranial envatment, in which conscious experience depends entirely and exclusively on internal brain activity, are oversimplified. Rather, a satisfying explanation of dreaming will have to look beyond the brain and appeal to the sleeping body as well.

Chapters 9 and 10 investigate the relationship between thinking and the phenomenology of being a rational cognitive agent, on the one hand, and genuine insight and deception, on the other. As in chapter 7, chapter 9 begins by developing a conceptual framework for describing instances of thinking, self-reflection, and metacognition in dreams, as well as their relation to lucid dreams and dream control. I also address the thorny question of whether dreams involve beliefs. Again, the emerging picture is that existing hypotheses that cast dreaming as a state of cognitive deficiency in which the phenomenology of being a rational, thinking self is almost completely absent are oversimplified. Instead, dreaming in general, and not just lucid dreaming, involves various forms of conscious thought and metacognitive activity. What is interesting and unique about dreaming is not the absence of conscious thought but the close relationship in which it stands to lower-level, automatically generated imagistic cognition.

The consideration of the problem of dream belief then leads to a reconsideration of the epistemological problem of dream deception. In chapter 10, I argue that a majority of dreams do not fulfill the conditions for counting as deceptive experiences in any interesting sense. While this narrows the scope of the skeptical threat posed by dreaming, my discussion of real-world examples of dream deception increases its bite, showing that dream deception is not just a theoretical possibility but an actual phenomenon, which in some cases can have detrimental consequences for the subject of deception and may even be related to pathological disorders of sleep and dreaming. Aside from retrospective cases of dream deception, in which dream deception extends into wakefulness, instances of cognitive corruption are particularly worrisome, because they suggest that in dreams, the phenomenology of rational thought and genuine insight is profoundly unreliable. I also discuss the analogy between dreams and delusional wake states and offer some suggestions for a nosological analysis of lucid dreaming and its comparison to insight in psychiatric and neurological disorders.

Where chapters 5 through 10 are dedicated to charting both the stereotyped characteristics of a majority of dreams and the inherent variability of the target phenomenon, chapter 11 asks whether beneath this variability it is possible to identify something like a phenomenal core of dreaming, with the goal of developing a unified and distinctive account of dreaming and of how it is distinguished from standard and altered wake states. I propose a new working definition of dreams as *immersive spatiotemporal hallucinations*, according to which dreaming is inextricably connected to the phenomenology of selfhood and minimally involves a sense of self-location within the dream. This view is directly relevant for the current discussion of minimal phenomenal selfhood, or of the minimal conditions for phenomenal selfhood to arise (Blanke & Metzinger, 2009).[3] I also introduce the new technical term *oneiragogia*, or experiences leading into sleep, and show how the conceptual framework proposed by the immersive-spatiotemporal-hallucination model of dreaming can be refined to describe the gradual transition from wakefulness, via sleep-onset imagery, to full-fledged dreams.

Finally, in chapter 12, I take a step back and ask how the conceptual framework for describing dreaming developed over the course of the book can be integrated into a broader framework describing standard and altered wake states. I propose a number of contrast conditions and points of contact between dream research and existing work in interdisciplinary consciousness research that I take to be particularly promising. These are felt presence and full-body illusions; immersive virtual reality and self-movement illusions (or vection); and mind wandering. I also tentatively suggest how the account defended in this book relates to recent work on predictive processing. I close with some reflections on the functions of dreaming. While I take the questions of why we dream and of whether dreaming has an evolutionary function to be problematic for a number of reasons, I argue throughout the book that important progress has been made on understanding the question of why we dream the way we do. By placing this question in a broader theoretical context, we can at least begin to sketch the outlines of an answer, and one that is metatheoretically interesting because it suggests a new perspective on the relationship between perception, mind wandering, and dreaming.

A final note to the reader: This is a long book, and one that aims to cover a broad range of issues from different disciplinary contexts. This necessarily creates redundancy. Moreover, in my attempt to address readers from different disciplinary backgrounds, my discussion of certain issues might seem superficial to some, and overly detailed to others. Clearly, with a project of this scope, it is necessary to find a middle ground between detail and accessibility, and although I have tried, I may not always have succeeded in finding a satisfying compromise. For readers who find that certain sections are too far removed from their own interests, I recommend scanning the introductory and concluding sections of each chapter. While these create additional redundancy, they also, I hope, increase the book's overall readability by providing pointers to the main ground to be covered and the most important results of each chapter.

Also, several of the book's chapters form natural pairs. Although I tend to regard chapters 1 through 4 as a package deal of sorts, they approach similar problems from slightly different perspectives. For those interested in philosophical issues regarding the relationship between dreaming and dream reporting, chapters 1 and 4 will likely be the most interesting. For the more empirically minded, chapters 2 and 3 will be more accessible. From then onward, the chapters proceed in pairs, with chapters 5 and 6, chapters 7 and 8, chapters 9 and 10, and chapters 11 and 12 complementing each other.

1 Dream Skepticism, Skepticism about Dreaming, and the Problem of Dream Experience

1.0 Introduction

The problem of dream experience, understood as a problem for philosophy of mind, is the problem of how to conceptualize conscious experience in the dream state. It is, in other words, the problem of how to situate the concept of dreaming in the logical geography of concepts predominately used to describe standard, altered, and pathological wake states, such as "experience," "perception," "imagination," "belief," "delusion," and "self-consciousness." As discussed in the general introduction, the problem of dream experience, or the *conceptualization problem*, it is not a single problem but a whole bundle of related problems. Many of these raise questions about how to describe conscious experience in the dream state, and to the extent that they do, they presume an affirmative answer to the question of whether dreams are experiences at all, and for logical reasons. Though I will ultimately claim that this latter question is not particularly informative when discussed in isolation, I nonetheless start my inquiry here. This is important because skeptical challenges to dream experience threaten to undercut the rationale for writing a book about dream experience in the first place.

Beginning with the question of whether dreams are conscious experiences at all also makes sense for historical reasons. Although this question is the foundational question for the conceptualization problem, it was not traditionally a topic of philosophical analysis. As will become clear in later chapters, philosophers traditionally endorsed a variety of views on what it is like to dream, but the claim that it is like anything at all mostly had the status of an implicit background assumption. This is particularly obvious in the case of Cartesian dream skepticism, which raises the question of whether we can ever rule out that we are now dreaming rather than awake: if dreams are not experiences, then they are also not deceptive experiences and thus present no cause for epistemological concern. Consequently, skepticism about dream experience undermines (or so it seems) classical dream skepticism, and with it the most popular and widespread reason, historically speaking, for discussing dreaming from a philosophical perspective.

The question of whether dreams are experiences only became a topic of philosophical debate in the second half of the twentieth century. This debate was initiated by Norman Malcolm (1956, 1959/1962), whose central claim was that dreams are not experiences for conceptual reasons. Malcolm (1956) originally developed this position as a response to Cartesian dream skepticism, which he took to exemplify the received or commonsense view of dreaming, but also used this position to challenge the very possibility of scientific dream research, and with it the possibility of an empirically informed philosophical analysis of dream experience. Following Malcolm, the philosophical debate on dream experience in the second half of the twentieth century centered almost entirely on the question of whether dreams are experiences and mostly ignored the challenge of developing an informative and constructive account of what exactly it means to say that they are. Both historically and logically, Malcolm's analysis of dreaming thus lies at the heart of the newer philosophical discussion on dreaming and at the intersection between epistemology and philosophy of mind.

This chapter has three main goals. First, I examine the historical and logical connections between dreaming as an argument for external-world skepticism, as discussed most prominently by Descartes, and internal-world skepticism about dreaming, as proposed by Norman Malcolm and Daniel Dennett. This is important because it suggests that the question of dream experience is relevant not just for philosophy of mind but also for other philosophical subdisciplines, such as epistemology.[1] Second, I expose two central background assumptions behind Cartesian dream skepticism and show how they are both targeted in equal measure by Malcolm's and Dennett's variants of skepticism about dreaming. Importantly, skepticism about dreaming entails skepticism about dream reporting: one cannot be skeptical about dream experience without also being skeptical about the trustworthiness of dream reports. This nearly trivial point highlights the importance of dream reporting as a philosophical problem in its own right. Third, by using this first chapter as a historical and systematic introduction to the philosophical question of whether dreams are experiences, I hope to convince the reader that although skepticism about dreaming (and dream reporting) is systematically flawed, its consideration is an informative and fruitful first step toward developing a more constructive solution to the problem of dream experience. Taking skepticism about dreaming, and the debate about dream experience from the second half of the twentieth century, seriously can provide valuable lessons for a contemporary theory of dreaming. I briefly sketch the outline of such an alternative account in the chapter's final section and develop it in more detail over the following chapters.

1.1 The Background Assumptions behind Cartesian Dream Skepticism

Cartesian dream skepticism is a form of external-world skepticism based on the possibility of dream deception. Its central thesis is that dreaming is subjectively indistinguishable from waking experience and that in virtue of this indistinguishability, we mistakenly believe ourselves to be awake while dreaming. This, in turn, is supposed to show that our belief that we are now awake might be mistaken: at any given moment, it is possible that one is dreaming rather than awake and that all of one's sensory-based beliefs about the external world and even one's own bodily existence are false.

Descartes certainly was not the first to discuss the possibility that we are deceived by our dreams,[2] and in fact skeptical dream arguments were so well-known that in his objections to the *Meditations*, Hobbes remarked that "it is a pity that such a distinguished originator of fresh ideas should come out with this old stuff" (Hobbes & Descartes, 1641/1975–1999, *First Objection*). Though there is some truth in this, it is Descartes's rendition of the dream problem, much more than that of many philosophers before him, that continues to stimulate philosophical discussion to this day.[3] It thus seems fair to say that Descartes appropriated the old problem of dream skepticism. From now on, I refer to it as *Cartesian* dream skepticism.

It is no coincidence that Descartes's project of finding a secure foundation of knowledge starts with the example of dreaming: dreaming is quite possibly the most private of all experiences, in the sense that for dreams, there is perhaps the strongest dissociation between rich inner experience and the absence of externally observable behavior. The privacy of dreaming is reflected in the fleeting nature of dream recall, often rendering the experience of dreaming elusive even to the dreamer. The dream example thus is singularly well poised to illustrate Descartes's epistemic situation at the beginning of the *Meditations*: at the beginning of his inquiries, as in dreams, he can only be certain of his own ongoing experiences, but not about anything they seemingly tell him about the existence of external objects or about his past. The example of dreaming thus aptly illustrates (and perhaps threatens; cf. chap. 10) the rationalist ideal of the lone philosopher reflecting on his beliefs and attempting to discover the foundations of knowledge in himself.

In the *First Meditation*, the dream example appears at a crucial moment. Descartes begins his *Meditations* by critically examining all his beliefs and asking whether he can find any general grounds for calling them into doubt. Pondering the reliability of sensory perception, Descartes briefly considers familiar illusions, for instance, the appearance of small or distant objects (Descartes, 1641/1901, I.4), but he quickly realizes that these cannot justify the more far-reaching and comprehensive doubt he has in mind. Illusions show only that the senses are sometimes deceptive, but there remain enough other cases in which it is "manifestly impossible to doubt" (I.4) the evidence of sensory perception. Consequently, such illusions are insufficient to question the reliability of sensory perception in general. A more general reason for doubt is needed, showing that even in a best-case scenario of sensory perception, which apparently leaves no room for error, the veridicality of sensory-based beliefs is dubitable. An experience of this sort is the following:

> That I am in this place, seated by the fire, clothed in a winter dressing gown, that I hold in my hands this piece of paper, with other intimations of the same nature. But how could I deny that I possess these hands and this body, and withal escape being classed with persons in a state of insanity, whose brains are so disordered and clouded by dark bilious vapors as to cause them pertinaciously to assert that they are monarchs when they are in the greatest poverty; or clothed [in gold] and purple when destitute of any covering; or that their head is made of clay, their body of glass, or that they are gourds? I should certainly be not less insane than they, were I to regulate my procedure according to examples so extravagant. (Descartes, 1641/1901, I.4)

In this type of perceptual situation, it seems that Descartes could only doubt the reliability of his senses if he took himself to be hallucinating in the manner of madmen. As Frankfurt (1970/2008) has argued, this assumption, however, would contradict the Cartesian project of using rational inquiry to find a secure foundation of knowledge. Henceforth Descartes sets aside the possibility that he is mad. Still, dreaming continues to pose a threat:

> Though this be true, I must nevertheless here consider that I am a man, and that, consequently, I am in the habit of sleeping, and representing to myself in dreams those same things, or even sometimes others less probable, which the insane think are presented to them in their waking moments. How often have I dreamt that I was in these familiar circumstances, that I was dressed, and occupied this place by the fire, when I was lying undressed in bed? At the present moment, however, I certainly look upon this paper with eyes wide awake; the head which I now move is not asleep; I extend this hand consciously and with express purpose, and I perceive it; the occurrences in sleep are not so distinct as all this. But I cannot forget that, at other times I have been deceived in sleep by similar illusions; and, attentively considering those cases, I perceive so clearly that there exist no certain marks by which the state of waking can ever be distinguished from sleep, that I feel greatly astonished; and in amazement I almost persuade myself that I am now dreaming. (Descartes, 1641/1901, I.5)

If even one's realistic experience of sitting by the fire could be a dream,[4] then one can never rule out that one is dreaming right now and mistakenly believes that one is awake.[5] Importantly, the dream example provides grounds for doubting sensory-based beliefs in general without undermining the ability of the meditator to engage in rational inquiry. Descartes repeatedly emphasizes this point: "Whether we are asleep or awake, we must only allow ourselves to be convinced by the evidence of reason . . . not by our imagination or our senses" (Descartes, 1637/2008, IV.8); and "although, in truth, I should be dreaming, the rule still holds that all which is clearly presented to my intellect is indisputably true" (Descartes, 1641/1901, V.15). Even in dreams, the evidence of reason, or so Descartes would have it, is trustworthy.

Though limited in scope to sensory-based beliefs, the dream example has far-reaching consequences. From the realization that even a best-case scenario of sensory perception could occur in a dream, Descartes proceeds to apply his radical doubt to all beliefs derived from sensory perception, including his beliefs about the external world, as well as his own body, movements, and sensations.[6] Even if he is only occasionally deceived by his dreams, this justifies the step toward full-blown dream skepticism, according to which all his current sensory-based beliefs are dubitable.

How does this relate to the problem of dream experience? Descartes's initial remark that "the occurrences in sleep are not so distinct as this" (Descartes, 1641/1901, I.5) and his subsequent realization that, as a matter of fact, they are, are telling. Dreams are deceptive because, as he now realizes, they have exactly the same clear and distinct experiential character as waking experience:

> When I considered that the *very same* thoughts (presentations) which we experience when awake may also be experienced when we are asleep, while there is at that time not one of them true, I supposed that all the objects (presentations) that had ever entered into my mind when awake, had in them no more truth than the illusions of my dreams. (Descartes, 1637/2008, IV.1; italics mine)

The failure of the Cartesian dreamer to realize that he is currently dreaming cannot be reduced to simple oversight, lack of attention to detail, or a transient state of cognitive impairment. Instead it is the phenomenal character of dreams—what they feel like to the experiencing subject—that threatens our sensory-based knowledge of the external world. This means that Descartes's version of dream skepticism relies on specific assumptions about the phenomenology of dreaming. The most basic of these is what I call the *experiential hypothesis*.

Experiential hypothesis: Dreams are conscious experiences in the sense that they are phenomenal states, or that it is *like* something to dream.[7]

In fact, Descartes's argument assumes much more. By claiming that dreaming is indistinguishable from waking experience in virtue of the sameness of phenomenal character, he is committed to a particular answer to the question of *what* it is like to dream as well (cf. chap. 5). For now, however, I want to draw attention to another point. Descartes tells us not just that he could conceivably have an ultrarealistic but deceptive dream of sitting by the fire, but that he has had such dreams before, and frequently. He knows this, one might assume, because he has frequently remembered such dreams upon awakening. This commits Descartes to the assumption that dream recall is an adequate source of evidence about what it is frequently (or perhaps even typically) like to dream. As will become clear later, the most charitable and precise formulation is to put this in terms of dream reports: to be trustworthy, general convictions of what it is frequently (or even typically) like to dream will have to be grounded in individual instances of dream recall. And given the fleeting nature of dream recall, the only way to ensure that this is the case is to fixate one's dream in a dream report immediately after awakening. This shift from dream recall to dream reports also allows me to formulate the underlying hypothesis in a way that is compatible with empirical dream research and with the theoretical issues discussed throughout this and the following chapters. Consequently, let us say that Descartes's argument relies on the *transparency hypothesis* of dream reporting:[8]

Transparency hypothesis: Dream reports are epistemically transparent in the sense that they are trustworthy sources of evidence about the occurrence and phenomenal character of experience during sleep.

It remains to be clarified what exactly counts as a dream report and under which conditions such a report should be regarded as transparent, and I return to these questions in the following chapters. At this point, it is important only to note that Descartes's reliance on the experiential hypothesis commits him to the transparency hypothesis as well.

Serious questions can be raised as to whether Descartes's reliance on beliefs about the experiential character of dreaming and the transparency of dream reports is permissible in the context of the *First Meditation*. If Descartes has to rule out that he is now dreaming to trust any of his beliefs about the world, then he cannot base his skeptical argument on any subset of these beliefs. What if, for instance, his beliefs about his previous dreams are themselves just the product of a dream? Moreover, Descartes not only assumes that his own memories of previous dreams are trustworthy, but also implicitly assumes that his description of a deceptive dream will resonate with his readers. If the dream example were a mere autobiographical anecdote about Descartes's own dreams, it would not be philosophically interesting. Rather, its argumentative force depends on what Descartes takes to be commonly accepted folk-phenomenological assumptions about dreaming. All of this requires quite a lot of knowledge that Descartes cannot yet claim to have at this point in his inquiry. Consequently, one might object that Cartesian dream skepticism suffers internal incoherence by relying on exactly the types of beliefs that it intends to undermine.[9]

For now, it is sufficient to note that the Cartesian scenario of dream deception extends beyond the epistemological problem of dream skepticism to questions concerning the experiential character of dreaming and the epistemic status of dream reporting. This makes it susceptible to criticism from philosophy of mind and empirical dream research. At the same time, however, this also makes the dream argument more compelling than versions of external-world skepticism depending on evil demons, lifelong dreams, brains in vats (Putnam, 1981), or *Matrix*-type scenarios (for discussion, see Grau, 2005). Unlike such science-fiction-style thought experiments, Cartesian dream skepticism introduces a nomologically possible form of dream deception. Even more so, it claims that we are *in fact* frequently deceived by our dreams, thus taking the step from mere logical possibility to real-world plausibility.

But did Descartes really take himself to be writing about actual dreams? One might object that for Descartes's purposes, no reference to actual dreams is required, and the mere possibility of wakelike dreams of sitting by the fire would provide sufficient grounds for doubting his sensory-based beliefs. According to Stroud, Descartes "need never have actually dreamt of anything before, and certainly needn't know that he ever has, in order to be worried in the way he is by the thought that he might be dreaming now" (Stroud, 1984, p. 18).

First, note that this is not the same as saying that for the dream example to be worrisome, no knowledge of dreams whatsoever is required. We can grant that a person who has never recalled any dreams of their own might be worried by Descartes's example without thereby granting that the example would be equally worrisome to someone who had no knowledge of dreams whatsoever, or in a world in which no such thing as dreaming existed.[10] Even if Stroud is correct that Descartes's argument does not require firsthand knowledge of his own dreams, this does not show that mere possibility, as opposed to real-world plausibility, is sufficient for the dream example. Of course, deceptive dreams could be logically possible even if real-world dreams were in fact very different from what Descartes describes, if they never shared the phenomenal character of waking experience, or even if they didn't exist at all. But in all these cases, the deceptive dreams described by Descartes would not be dreams in the same sense as the real-world phenomenon of dreaming; calling them dreams at all would be misleading. To the extent that the force of the argument depends at least in part on its psychological effect on the reader, this point is relevant to the version of the dream argument presented by Descartes. Without actual, real-world plausibility, dream skepticism would not have the subject matter that it purports to have: it would not be about actual dreams.

Second, even if the skeptical force of the argument were not diminished by reformulating it in terms of merely possible, rather than actually plausible, dreams, this is not the way Descartes himself presents the argument. Textual evidence strongly suggests that Descartes took himself to be talking about actual dreams and that he based his claims about dreaming on his own firsthand experience of what it is like to dream. Much has been written about several dreams Descartes supposedly experienced as a young man and to which he appears to have attributed considerable importance. Though Descartes's own descriptions of his so-called Olympian dreams have been lost, they were reported by his biographer Adrien Baillet (1691/1946), whom many take to be reliable in this matter (Cole, 1992; Keefer, 1996; Hacking, 2001; but see Clarke, 2006, pp. 58–66). According to Baillet, Descartes experienced the dreams during a period of profound theoretical confusion and uncertainty about his future that culminated, on the day on which he had the dreams, in what he believed to be the discovery of the foundations of a marvelous science (Baillet, 1691/1946; cf. Keefer, 1996). In the *Discourse*, Descartes (1637/2008, II.1) notes that many of the thoughts there described were first formed during his employment in the army, while he was stationed in Neuburg for the winter and spending a quiet day of rest in a heated chamber—on the very day, that is, on which, according to Baillet, he had the dreams. Taken together, this suggests that Descartes himself viewed these dreams—or wanted them viewed (cf. Browne, 1977, p. 273)—as embodying a formative moment in his personal and theoretical development.

Many philosophers have followed suit. Leibniz, one of the earliest commentators on the dreams, remarked that Descartes "decided to reform philosophy after some dreams" (Leibniz, 1880, IV, p. 311; quoted and translated in Browne, 1977, p. 265). Similarly, several contemporary authors have argued that Descartes's theoretical use of the dream example in the *Meditations* was directly influenced by these dreams. For instance, Keefer believes that the dreams "embodied a process of deliberate doubt" (Keefer, 1996, p. 41), and Hacking suggests that they turned dream skepticism into "a *live* skepticism for Descartes, that is, not a mere philosophical position, but genuine doubt" (Hacking, 2001, p. 252). But these issues are mainly of historical interest. What is more important, and what the Olympian dreams—whether real or constructed—show, is Descartes's willingness to pay attention to his dreams and to publicly attribute meaning to them.

Aside from the Olympian dreams, further evidence suggests that Descartes's interest in dreams was not of a purely theoretical nature and that it was sincere. In a letter to Princess Elisabeth of Bohemia, written in 1645, he briefly discusses dreams. Referring to the princess's remark that various illnesses diminish our powers of reasoning and rational thought, he notes:

> Something similar also happens when we sleep, for the most philosophical person in the world does not know how to prevent himself from having bad dreams when his temperament disposes him to them. All the same, experience shows that if one has often had some thought while one has had a free mind, one returns to it often afterward, no matter what indisposition the body has. Thus, I can say that my dreams never represent to me anything upsetting. (Princess Elisabeth of Bohemia & René Descartes, 2007, pp. 106–107)

In sum, even if the reports of the Olympian dreams were fabricated, they were at least fabricated by someone with enough of an interest in dreams to do so in a believable manner. Freud apparently agreed. He declined to interpret the dreams but took the reports to be authentic and even acknowledged their apparent meaningfulness by describing them as "dreams from above" (Freud, 1940/2003, p. 184). And while one may well be skeptical of the latter claim, one does not have to be a fan of psychoanalytic dream theory to think that Freud's judgment as to the authenticity of the dream reports has a certain claim to authority. Similarly, Allan Rechtschaffen, one of the pioneers of laboratory research on dreaming, wrote that "there is nothing in the reported material to preclude its having been dreamed" (Rechtschaffen, personal letter to Cole, February 4, 1986; quoted in Cole, 1992, p. 213). The persistent controversy surrounding the authenticity of the dream reports may also attest to this fact: if fabricated, the dream reports are at least not obviously or unambiguously so.

Descartes's familiarity with his own dreams is also reflected in his solution to the dream problem in the *Sixth Meditation*. Here he takes the step from naive to informed phenomenologist, acknowledging that the dream example given in the *First Meditation* fails to characterize a majority of dreams, and that contrary to his former impression, there is indeed a considerable difference between dreaming and being awake:

> And I ought to reject all the doubts of those bygone days, as hyperbolical and ridiculous, especially the general uncertainty respecting sleep, which I could not distinguish from the waking state: for I now find a very marked difference between the two states, in respect that our memory can never connect our dreams with each other and with the course of life, in the way it is in the habit of doing with events that occur when we are awake. And, in truth, if some one, when I am awake, appeared to me all of a sudden and as suddenly disappeared, as do the images I see in sleep, so that I could not observe either whence he came or whither he went, I should not without reason esteem it either a specter or phantom formed in my brain, rather than a real man. But when I perceive objects with regard to which I can distinctly determine both the place whence they come, and that in which they are, and the time at which they appear to me, and when, without interruption, I can connect the perception I have of them with the whole of the other parts of my life, I am perfectly sure that what I thus perceive occurs while I am awake and not during sleep. And I ought not in the least degree to doubt of the truth of these presentations, if, after having called together all my senses, my memory, and my understanding for the purpose of examining them, no deliverance is given by any one of these faculties which is repugnant to that of any other: for since God is no deceiver, it necessarily follows that I am not herein deceived. But because the necessities of action frequently oblige us to come to a determination before we have had leisure for so careful an examination, it must be confessed that the life of man is frequently obnoxious to error with respect to individual objects; and we must, in conclusion, acknowledge the weakness of our nature. (Descartes, 1641/1901, VI.24)

Descartes can now rest assured, realizing that though he can still potentially be deceived by his dreams, deception is not inevitable. More importantly, having established that God is no deceiver and that clear and distinct ideas are generally trustworthy, the possibility of occasional deception is no longer troubling. Or at least this is the reading Descartes himself suggests; according to Hacking, Descartes's resolution of the dream problem at the end of the *Meditations* only seemingly departs from his genuine doubt about whether he is dreaming or awake and may be no more than "the after-shock giggle of someone who was truly scared" (Hacking, 2001, p. 252).

Be that as it may, even this revised view of dreaming still relies on a number of assumptions about the phenomenology of dreaming, which will again have to be answerable to actual instances of dreaming and hence of dream reporting. I conclude that we may take Descartes's statements about dreams in the *Meditations* as literally referring to actual dream experiences (or at least to ones that are close enough to actual ones to be plausible). For this reason, they are also open to an assessment from the perspectives of scientific dream research and philosophy of mind. Are dreams experiences? More importantly, do dream reports provide evidence for saying that they are? Descartes's example crucially depends on an affirmative answer to both of these questions.

1.2 Norman Malcolm's Denial of Dream Experience

More than three hundred years after Descartes, Norman Malcolm developed his own analysis of dreaming in an attempt to put the recurring nightmare of dream skepticism to rest. Rather than permitting dreams to induce skepticism about our knowledge of the external world, he reversed the problem by introducing a version of internal-world skepticism directed at the conscious experience of dreaming itself. This connection is particularly evident in Malcolm's (1956) article "Dreaming and Skepticism." Here he argued that Cartesian dream skepticism relies on a mistaken conception of dreams as composed of thoughts and sensations during sleep. By denying this claim, which he took to be representative of the received view of dreaming in the psychological and philosophical literature of his time (Malcolm, 1959/1962, p. 4), he aimed to undermine the rationale for Descartes's skeptical argument. Malcolm's major achievement, however, was to expose the vulnerability of the received view itself. This becomes particularly evident in his monograph, first published in 1959 and more simply titled *Dreaming*. Here, he presented a more comprehensive and nuanced version of his argument and focused more strongly on the concept of dreaming itself.

Strictly speaking, Malcolmian skepticism about dreaming is not originally Malcolmian, just as Cartesian dream skepticism is neither originally nor exclusively Cartesian. Several important aspects of Malcolm's argument were foreshadowed by Bouwsma (1977; first published in 1945) and Macdonald (1977; first published in 1953).[11] Still, like Descartes, Malcolm appropriated the problem, and his analysis of dreaming provoked a virtual flood of critical literature. Though his theory, as a whole, lost much of its credibility in the process, the implications of his work are hard to overestimate. Malcolm laid the groundwork for the newer philosophical discussion of dreaming and formulated many of the theoretical problems that are relevant to this day. This impact is reflected in Dunlop's 1977 collection *Philosophical Essays on Dreaming*, most of which, in fact, deal with Malcolmian skepticism about dreaming. In his introduction, Dunlop remarks that "the significance of Malcolm's work can be measured in part by the critical literature that it has engendered; virtually every philosophical article on dreaming published since 1959 addresses some aspect of Malcolm's writings" (Dunlop, 1977, p. 20). This remains true to this day, and entire articles devoted to Malcolm's theses were still appearing forty years after their original publication (e.g., Schröder, 1997; McFee, 1994; for a more recent discussion, see Rosen, 2013, Windt, 2013b).[12]

Crucially, Malcolm's article (1956) and monograph (1959/1962) on dreaming not only drew attention to the problem of dream consciousness but also determined its direction. In particular, Malcolm's denial of dream experience ensured that much of the philosophical work on dreaming focused on this negative claim, rather than developing a positive account of dream experience. This somewhat deplorable tendency to focus on what dreams are not has only been reversed in recent years (see chaps. 5 and 6). For Malcolm, this omission was inherent to his program: "Indeed I am not trying to say what dreaming *is*: I do not understand what it would mean to do that" (Malcolm, 1959/1962, p. 59).

From this, one might conclude that Malcolm's analysis of dreaming is mainly of historical interest and refuting his claims is a mere exercise in historical completeness without advancing the current discussion. But though I follow his critics in rejecting most of his claims, I think this would not do him justice. One of Malcolm's main achievements was to focus the philosophical discussion on a set of entirely new questions about dreaming, some of which were implicit in the older literature but had largely been overlooked. These not only include the questions of whether dreams are experiences and on what grounds this claim can be affirmed or denied, but also concern the relationship between dreams, dream reports, and scientific evidence about dreaming. At least, his work highlighted the urgency of these questions and drew attention to the fact that not just dreaming but dream reporting is a topic of genuine philosophical interest. And, somewhat ironically, Malcolm's adamant rejection of dream research as irrelevant to questions about dream experience also forged a tight link between the philosophical discussion of dreaming and empirical dream research. Even if Malcolm's own answers seem historically outdated, these are still live questions in the sense that premature, misleading, or incomplete answers create theoretical confusion and make theoretical treatments of dreaming vulnerable to counterarguments that could otherwise easily be blocked. Because of this, Malcolm's analysis of dreaming can still function as a valuable corrective in contemporary treatments of dreaming and deserves careful consideration.

Perhaps the best way to approach Malcolm's theoretical treatment of dreaming is to begin with some short remarks of his mentor Ludwig Wittgenstein (for a personal memoir, see Malcolm, 1958).[13] In the *Philosophical Investigations*, Wittgenstein writes:

> People who on waking tell us certain incidents (that they have been in such-and-such places, etc.). Then we teach them the expression "I dreamt," which precedes the narrative. Afterwards I sometimes ask them "did you dream anything last night?" and am answered yes or no, sometimes with an account of a dream, sometimes not. That is the language game. (Wittgenstein, 1953, p. 184)

This shift in perspective from dreaming to dream reporting is crucial for Malcolm's analysis as well:

> Where does the concept of dreaming come from? We are strongly inclined to think of dreaming as an inward state or process of the soul, and to suppose that each of us arrives at the concept

> of dreaming through taking note of the process in himself. But this idea gives rise to insoluble problems. (Malcolm, 1959/1962, p. 54)

Crucially, epistemic access to dreaming is always indirect, twice removed from the actual experience, first through retrospective dream recall, and second—for anyone but the dreamer—through dream reports. An important lesson from Wittgenstein and Malcolm is that our beliefs about dreaming are derived neither from the experience of dreaming nor from the ability to remember dreams, which is again an inner process, but from the *telling* of dreams, or from dream reports: "What we must say, although it seems paradoxical, is that the concept of dreaming is derived, not from dreaming, but from descriptions of dreams, i.e., from the familiar phenomenon that we call 'telling a dream'" (Malcolm, 1959/1962, p. 55).

But Malcolm's argument goes further, claiming that the concept of dreaming is not only historically derived from but logically dependent on dream reporting. The underlying assumption is that concept acquisition fixes reference, that is, that the meaning of a concept is determined by the means by which we acquired it or by which it is taught (for a general introduction to philosophical views on concepts, see Margolis & Laurence, 2011).

A consequence of this view is that the concept of dreaming can be seen as a more elegant and efficient—but also potentially misleading—way of talking about dream reports. This insight requires a reformulation of many older questions about dreaming and reveals them to have been about dream reporting all along. Questions such as "What is it like to dream?" or "Am I deceived by my dreams?" now appear as shorthand for asking what our waking memories of our dreams tell us about what it is like to dream, or whether our waking memories of our dreams provide reasons to think that we are actually deceived during sleep. Crucially, these questions about dreaming can no longer be straightforwardly answered on their own but rather appear contingent on the trustworthiness of dream reports with respect to the dreams they purport to be about. In theorizing about dreams, do we ever refer to dreams, or only to dream reports? And in reporting dreams, do we ever refer to anything at all? The important and lasting lesson from Malcolm's work is that if we do manage to refer to dreams themselves, we do so via dream reports and the memories from which they are derived. Only the added assumptions behind the transparency hypothesis once more turn questions about dream reporting into questions about dreaming. By contrast, Malcolm's central goal was to show that this move is not permissible and that questions about dreaming—insofar as they are understood independently of questions about dream reporting—are nonsensical. Rather than enlarging the scope of philosophical questions about dreaming to include questions about dream reports, he shifted the explanatory target to dream reports alone. In Malcolm's view, questions about dreaming proper were in an important sense off-limits.

Malcolm's strategy for resolving the problematic relationship between dreaming and dream reporting was closely bound up with the criterial view of language and its application to the concepts of sleep and dreaming. Once more, this view is best summed up by Wittgenstein's (1953, §580) dictum that "an inner process stands in need of outward criteria" and the claim that the meaning of mental concepts is determined on the basis of such criteria.[14] Dreaming is an inner process par excellence, but Malcolm additionally assumed that dream reports are the sole criterion of dreaming. If the meaning of the concept of dreaming is wholly determined by the criterion of dream reporting, then the distinction between dreaming and dream reporting itself threatens to collapse. Dream reports are trustworthy in principle precisely because dreaming cannot be sensibly conceived of as anything other than reporting dreams:

> One cannot then permit a question to be raised as to whether those descriptions are in error or not—for this would be to fall back into the original difficulty. One must treat the descriptions as the *criterion* of what the inner occurrences are. (Malcolm, 1959/1962, p. 55)

On its face, it would seem that Malcolm's conceptually motivated skepticism about dream experience is compatible with the transparency hypothesis of dream reporting: after all, in Malcolm's view, the trustworthiness of dream reports is a matter of logical necessity. Note, however, that the kind of unquestionable authority he ascribes to dream reports is very different from that claimed by the transparency hypothesis in three important respects: one, it is not hypothetical but a matter of logical necessity, indeed ruling out the very possibility of misremembering dreams for conceptual reasons; two, dream reports are not transparent, in Malcolm's view, with respect to *experiences* during sleep but are transparent only with respect to dreams; and three, this latter move suggests that the sense in which Malcolm takes dream reports to be transparent with respect to something other than themselves—namely, the separate phenomenon of dreaming—is extremely strained. Later in this section it will become clear that his position on whether dreaming can be conceived of as distinct from dream reporting was indeed complex and not always clear. For now, it is sufficient to note that Malcolm's denial of dream experience resulted from his skepticism about the ability of dream reports to furnish the type of evidence required to regard dreaming as conscious experience during sleep.

Though dreaming was the main target of Malcolm's critique, he mounted his attack against the received view by analyzing the sentences "I am asleep" and "he is asleep" (Malcolm, 1959/1962, pp. 15–18). While the truth of the sentence "he is asleep," to Malcolm, depends on behavioral criteria like closed eyes, rhythmic breathing, and lack of responsiveness (Malcolm, 1956, p. 17), the sentence "I am asleep" is "playful nonsense" (Malcolm, 1959/1962, p. 6). As is the case for claims that one is currently unconscious or dead, "I am asleep" involves a self-contradiction, and its truth can be rejected on purely logical grounds. Malcolm then extends this analysis to all kinds of mental activity during sleep: if one cannot correctly assert that one is currently asleep, neither can one judge, think, or suspect that one is asleep, or indeed have any opinions, doubts, or thoughts in sleep whatsoever (Malcolm, 1956, p. 20). This conclusion is interesting because it makes the meaning of the sentence "I am asleep" and indeed the possibility of entertaining such a thought or belief contingent on the possibility of its verbal expression under circumstances in which it would be true. Muttering the sentence "I am asleep" during an episode of sleep talking is ruled out as a possible counterexample, because this, according to Malcolm, would not qualify as an episode of genuine or sound sleep (Malcolm, 1959/1962, pp. 27–28).

Malcolm's grounds for extending his argument to all kinds of mental activity during sleep are not entirely consistent. In the essay, he unequivocally states that "if a person is in *any* state of consciousness it logically follows that he is not sound asleep" (Malcolm, 1956, p. 21). In the monograph, he defends a weaker and more nuanced version of the argument:

> It could be objected that my argument has shown merely that the *verification* that someone is both asleep and judging is self-contradictory, not that his *being* both asleep and judging is self-contradictory. This is true. The latter notion is not self-contradictory (in the sense of entailing both of a pair of contradictory propositions). But it is senseless in the sense that nothing can count in favour of either its truth or its falsity. (Malcolm, 1959/1962, pp. 36–37)

At times, however, he seems to defend the stronger claim of logical impossibility instead of the weaker verificationist version of his argument in the monograph as well. This becomes especially clear in his analysis of the concept of dreaming.

Dreaming, according to Malcolm, is an exception in that the truth criteria applying to sentences about dreams differ radically from those applying to other first-person, past-tense psychological statements. Dreams by definition occur during sleep, but the only possible criterion for their occurrence is the dreamer's report after awakening. Not only are there no external observational criteria for the occurrence of dreams, but there could be no further evidence to corroborate the dreamer's subsequent testimony of having dreamt, even in principle. Any utterance or movement of the sleeping subject, such as sleepwalking, sleep talking, or tossing and crying during a violent nightmare, would only show that the person was in fact partially awake and in a state merely *resembling* sleep (Malcolm, 1959/1962, pp. 27–28). Consequently they would count as evidence against, rather than for, his subsequent avowal that he had been dreaming, because dreams occur during unqualified sound sleep.

Malcolm concedes that the truth of sentences like "I was afraid" or "I was angry" is also often accepted on the subject's mere testimony (Malcolm, 1959/1962, pp. 91–100), but insists that dream sentences are different in that they could not, even in principle, be expressed by the dreamer in the present tense. Exactly this type of concurrent evidence, according to Malcolm, would be necessary, however, to conceive of the dream as separate from the retrospective report (Malcolm, 1956, p. 33).

The strange consequence of this view is that while there is no contradiction in saying that one knows that one had a certain dream (Malcolm, 1959/1962, p. 64), a truthful dream report does not entail that the sensations, thoughts, or feelings mentioned in the dream report actually occurred:

> If a man had certain thoughts and feelings in a dream it no more follows that he had those thoughts and feelings while asleep, than it follows from his having climbed a mountain in a dream that he climbed a mountain while asleep. (Malcolm, 1959/1962, pp. 51–52)

According to Malcolm, identity of language not only fails to entail identity of experience; for the case of dreaming, conceptual considerations positively rule out the possibility of dream experience altogether: "It is fruitless to argue that this identity of language is due to an identity of experience that lies behind it" (Malcolm, 1959/1962, p. 94). It is only in virtue of the language used to describe dreams that they are said to occur at all; but since there is no criterion for their occurrence over and above the dream report, it would be false to assume that dreams and waking experiences are comparable in any more interesting way. Sentences about dreams and about waking experiences are governed by different modes of discourse, and the "dream-telling" use and the "historical" use of such sentences must not be confused (Malcolm, 1959/1962, p. 95).

A problem that immediately comes to mind is that even if we assume, for the sake of argument, that dreams are not experiences for purely *conceptual* reasons—or even that there is a conceptual contradiction involved in saying one had certain experiences during sleep—this is a conceptual point only and as such has no direct consequences for questions concerning their appropriate *phenomenological* description. The experiential hypothesis, as introduced earlier, is not about the conceptual but about the phenomenological analysis of dreaming, and a point that Malcolm consistently misses is that there is simply no necessary entailment from purely conceptual or analytic arguments to facts about phenomenal experience. Even if the conceptual relationship between dreaming and dream reporting is such that an unreported experience during sleep will not fall under the concept of dreaming, or that seeing something in a dream does not mean the same thing as seeing something in wakefulness, there is still nothing to prevent some phenomenal experience (and perhaps even one that is subjectively indistinguishable from its waking counterpart), call it what you will, from occurring during sleep. Whether or not this is the case simply cannot be resolved on the level of conceptual analysis alone. Even if dreams are not "experiences" (either in Malcolm's sense or in some other technical sense), this is not to say that there is nothing it is like to have them during sleep; even if we lose the term *experience* to refer to dreams, this does not thereby deprive them of their phenomenal character. At best, this tells us something about our use of the terms *experience* and *dream* (or, more likely, about how a particular philosopher suggests using these terms).[15] Importantly, no purely conceptual argument of this type can reveal any necessary truths about the phenomenology of dreaming (or lack thereof).[16] Nagel (1959) puts this point nicely. While he ultimately agrees with Malcolm that dreams are not experiences—the reason being that dreams, unlike illusions, do not occur against a background of nonillusory experience—he argues that this does not demote dreams to mere memory phenomena:

> It is a mistake to invest the demonstration that it is impossible to have experiences while asleep with more import than it has. It is an observation about our use of the word "experience," and no more. It does not imply that nothing goes on in our minds while we dream. True, we cannot have experiences while asleep. But we can have dreams. (Nagel, 1959, p. 114)

From now on, I will call the practice of deriving claims about phenomenal experience from conceptual analysis alone the *fallacy from armchair conceptual analysis*, or more simply: the *armchair fallacy*. As will become clear throughout this book, it is a frequent fallacy in philosophical theorizing about dreaming.[17]

For a similar example of the armchair fallacy that is not related to dreaming, consider the question of whether a conceptual relationship holds between an experience and the subject of experience, or between the occurrence of thoughts and the existence of a thinker. To many, this has seemed to be the case; Descartes's *cogito* argument is a classical example, and the discussion surrounding Shoemaker's principle of immunity to error through misidentification is another (Shoemaker, 1968; Prosser & Recanati, 2012). But does this mean that it is *impossible*, for logical reasons, to *have* an experience (or thought) while failing to *experience* oneself as its subject? Could there be empirical counterexamples? For instance, is the sense of ownership for thoughts lost in schizophrenic thought insertion, or only the sense of agency (for discussion, see Campbell, 1999; Gallagher, 2000; Stephens & Graham, 2000, sec. 6.3)? And can the misattribution of bodily sensations to another in somatoparaphrenia be described as violating Shoemaker's immunity principle (Lane & Liang, 2011; Liang, 2015; Haug & Jung, 2015)? Note that I am not here claiming that these empirical counterexamples succeed but only making the weaker claim that even if it is a conceptual truth that there can be no thought without a thinker or no experience without a subject of experience, this does not entail that for every thought, there is a subject who necessarily *experiences* herself as its thinker or that for every, say, pain experience, the subject of this experience necessarily has the *phenomenology* of ownership for this pain experience. The latter is a phenomenological claim, and its hypothetical (rather than conceptual) nature is revealed by the fact that there is room for meaningful (and controversial) philosophical discussion on this matter. So even if these specific empirical counterexamples fail, it seems that their failure is not a matter of *conceptual* necessity, or at least not obviously so. This would be enough to support my claim that there is no necessary entailment from conceptual arguments to what can or cannot be experienced.

Returning to Malcolm's view, the "sharply different 'grammars'" (Malcolm, 1959/1962, p. 98)—and hence different meanings—of dream reports and reports about past experiences also have important implications for his analysis of dream recall. Remembering a dream, to Malcolm, is not at all the same thing as remembering a past experience (pp. 56–57): remembering a dream means waking up with the impression that something occurred and then realizing the event in question never happened. "To find out one dreamt the incident is to find out that the impression one had on waking is false" (p. 64). Accordingly, Malcolm believes that "statements of the form 'I dreamt so and so' are always inferential in nature" (p. 65): one infers that one has dreamt upon realizing that the events one remembers upon awakening did not in fact occur. In this view, dream reports are profoundly misleading: they may appear to be transparent with respect to experiences during sleep, but this apparent transparency results from the purely superficial similarity of the language used to describe dreams and waking experiences. One of Malcolm's main goals, then, is to show that the language used to describe dreams is in fact inherently deceptive.

For Malcolm, the relationship between dreaming and the waking impression of having dreamt is ambiguous. Dreams are not logically independent of dream reports, because the dreamer's testimony is the only criterion for the dream's occurrence. At the same time, Malcolm falls short of a completely reductive analysis of dreaming because he denies that the concepts of dreaming and dream reporting have the same extension:

> The dream and the waking conviction are not one and the same thing, in the sense that the morning star and the evening star are one and the same. Are they *two* things, numerically different? Let us say so. . . . But not—two logically independent things. (Malcolm, 1959/1962, pp. 59–60)

This view has radical consequences for empirical dream research, which Malcolm flatly rejects as being irrelevant to the concept of dreaming. It is no coincidence that this attack on dreaming came at a time when the discovery of rapid eye movement (REM) sleep, sometimes also called paradoxical sleep, and its correlation with dreaming was revolutionizing the scientific concepts of sleep and dreaming. Suddenly, objective criteria for measuring sleep and dreams were being discovered, and they were threatening to dethrone dream reporting as the sole criterion of dreaming (see chaps. 2 and 3 for details). This situation, to Malcolm, was a call to arms, and his work on dreaming can historically be seen as a direct reaction to the perceived threat from early empirical dream research.

In the monograph, Malcolm quotes extensively from Dement and Kleitman's (1957b) article "The Relation of Eye Movements during Sleep to Dream Activity." In this article, the authors not only correlate the occurrence of dreams—evidenced through dream reports elicited after timed awakenings in the sleep laboratory—with the physiological characteristics of REM sleep but also conjecture about the duration of dreams and their temporal location during sleep (for Malcolm's discussion of these findings, see Malcolm, 1959/1962, pp. 70–82). The importance of this early laboratory work on dreaming can hardly be overestimated: the discovery that dream reports could be elicited after timed awakenings from certain periods of sleep made a systematic, experimentally minded approach to dreaming possible. According to Malcolm, however, Dement and Kleitman's (1957b) conclusion that dreams occur during certain sleep stages and have a measurable duration is based on a misunderstanding of the concept of dreaming. Such a conjecture would only be justified if a "causal or temporal relation between dreams and physical occurrences" (Malcolm, 1959/1962, p. 76) had been directly observed. This, Malcolm continues, is impossible: their findings were based on an observable correlation between retrospective dream reports and previous physical occurrences, not between *dreaming* and concurrent physical occurrences. Because of this, the assumption that the length of a *dream* depends on the duration of a REM phase is senseless. According to Malcolm, the duration of a dream can only refer to one's waking impression or to the length of a dream report. But aside from the common notion that dreams occur during sleep, it makes no sense to speak of the duration of dreams or of their location in physical time (Malcolm, 1959/1962, pp. 77–80). Consequently the distinction between appearance and reality itself does not apply to dreams. There is no objective measurement that could corroborate the subjective dream report, nor could there ever be; the report is all there is:

> If he has a vague impression of his dream then it was a vague dream. If he is not certain whether he dreamt then there is an uncertainty in reality. His impression is the criterion of reality and therefore it cannot be characterized as "subjective." "Subjective" and "objective" are one in the case of dreams. (Malcolm, 1959/1962, pp. 79–80)

Importantly, this means that Malcolm is not so much criticizing the results of a particular research project as denying the possibility of scientific dream research in principle. Note also that his rejection of the appearance–reality distinction with respect to dreaming means that he rejects the question of whether dreams are experiences itself as improper: for Malcolm, there is no separate problem of dreaming, only one of dream reporting, and rejecting the transparency hypothesis suffices for rejecting the experiential hypothesis because the latter fails to represent a *separate* hypothesis about dreaming at all.

According to Malcolm, there is, however, another reason to reject empirical dream research. The adoption of a physiological criterion of dreaming, which Malcolm believes to be implicit in early scientific dream research, would entail a radical change in the everyday concept of dreaming and result in confusion. With the change of criteria, the concept of dreaming would take on a new meaning and would have to be taught differently; we would have to assume that dreams occur independently of, and even in contradiction with, dream reports. But for Malcolm, there is no such thing as misremembering a dream, and the assumption that one could be informed about whether or not or even *what* one had dreamt is simply absurd (Malcolm, 1959/1962,

p. 80). Even if such new criteria of dreaming were adopted, we would no longer be talking about the same phenomenon: the meaning of the concept of dreaming would have undergone too profound a change.[18] Quoting Wittgenstein, Malcolm recommends that we stick to the old concept of dreaming and stop looking for further explanations; dreaming is a proto-phenomenon, part of the language game, and our concept of dreaming is founded on the no less remarkable phenomenon of dream telling, which should be accepted as a given (Malcolm, 1959/1962, p. 87).

While the up-and-coming field of empirical dream research was Malcolm's main opponent in the monograph, the 1956 paper mainly targeted Cartesian dream skepticism. Having rejected the received view of dreaming, Malcolm claimed that the dreams of sound sleep no longer provided a "foothold for philosophical skepticism" (Malcolm, 1956, p. 26), and Cartesian dream skepticism relied on untenable assumptions. Because Malcolm denies that one can form judgments or thoughts during sound sleep and by extension during dreams, the mere notion of trying to determine whether one is awake or asleep is absurd; one cannot doubt whether one is awake or asleep or use a criterion to determine whether one is dreaming unless one is awake. Although you might seemingly (and mistakenly) think you are awake in a dream, this does not mean that you are actually entertaining this thought, but only that you are dreaming that you are. Malcolm does allow that one could be deceived by a dream and even think that one were awake while half asleep (Malcolm, 1956, p. 35). In these cases, however, deception is not inevitable, as there could be means of testing whether one is fully or partially awake (Malcolm, 1959/1962, pp. 116–117). So while dream deception during half sleep remains possible, it does not present the sort of inescapable threat that worried the Descartes of the *First Meditation*: the point of Descartes's dream of sitting by the fire, after all, was that it resisted his attempts at ascertaining whether he was awake.

Just as the exact relation between dreaming and dream reporting is ambiguous in Malcolm's work, so a stronger and a weaker reading of Malcolm's antiskeptical argument is possible. According to the stronger reading, dreaming that one decided or doubted that one was awake would mean nothing more than waking up with the impression of having so decided or doubted. Since one can have no dream experience over and above the waking impression, the skeptic's objection that one could confuse dream thoughts with real thoughts during a dream of sound sleep is invalid: "To a person who is sound asleep, 'dead to the world,' things cannot even seem" (Malcolm, 1956, p. 26).

The weaker version of his antiskeptical argument, which is developed in the monograph (Malcolm, 1959/1962, pp. 101–107), corresponds to the position that the dream is numerically different from the dream report, but not independently verifiable. Although Malcolm still takes the notion of doubting or questioning anything at all during sleep to be senseless, he admits that this "form of reasoning would not, however, remove the perplexity of a determined philosophical sceptic" (p. 117). The skeptic could still rejoin:

> "I don't know whether I am actually in doubt or just dream that I doubt." And so on *ad infinitum*. Nothing can force him to affirm that he *is* perplexed or in doubt and therefore nothing can force him to find in the above reasoning a demonstration that he is awake and not dreaming. (Malcolm, 1959/1962, p. 118)

While this weaker reading of Malcolm's argument admittedly falls short of a convincing refutation of the skeptic, there is a deeper conceptual point: contrary to what we normally assume, one has no way of knowing or observing that one is awake. "There is nothing to decide, no choice to make, nothing to find out. I cannot pass from not knowing whether I am awake or dreaming to knowing that I am awake" (Malcolm, 1959/1962, p. 118).

For Malcolm, the sentence "I am awake" should not be construed as reporting or describing one's current state or condition; like any other assertion, including the nonsensical claim "I am not awake," it is simply a way of *showing* that one is awake. The asymmetry of the first- and third-person, present-tense uses of being asleep also applies to being awake: both are observational terms governed by behavioral criteria when applied from the third-person perspective, and both lose this sense when used in the first person, present tense. During wakefulness, their function is reduced to an inherently meaningless show of behavior, and for sleep and dreaming, the possibility of such an ostentative use of "I am asleep" is ruled out (Malcolm, 1959/1962, pp. 114–120).

Although Malcolm does not explicitly discuss this issue, his analysis of wakefulness, sleep, and dreaming has important conceptual implications for consciousness as well. As Hacking puts it, in a view such as Malcolm's, "mental life becomes parasitic on the world of public affairs" (Hacking, 1975, p. 105). In such a view, the terms *awake* and *asleep* describe different behavioral states. The concept of dreaming, when used as an experiential term to refer to some special kind of sleep experience, is either self-contradictory or unverifiable and senseless. For the same reason, Malcolm's view entails that no positive meaning is attached to the concept of conscious states at all. Consciousness, understood in terms of phenomenal experience or the subjective character of consciousness and as separate from a behavioral state, has no place in such a view, and "the phrases 'mental activity,' 'mental phenomenon,' 'conscious experience,' are so vague that I should not have known what I was asserting" (Malcolm 1959/1962, p. 52). Indeed, nothing in Malcolm's theory requires such a separate description: from the third-person perspective, what is at issue is not consciousness but behavior, and from the first-person perspective, there is nothing to be observed.

Of course, this is exactly why dreaming took on such philosophical importance: it is the most obvious instance of a term referring almost exclusively to first-person conscious experience arising independently of behavioral evidence. Dreaming thus presented a challenge to Malcolm's verificationist view, and it could be dealt with only by making it contingent on dream reports. But this somewhat awkward and uncomfortable solution was threatened by the discovery of REM sleep as an outward criterion of dreaming; the very idea, to Malcolm, seemed preposterous. Verificationism of the sort promoted by Malcolm thus was tied to the idea that mental states have to be open to not just any form of verification but to verification through concurrent behavior. And because he took the link between behavioral criteria and mental states to be one of logical necessity rather than of empirical probability, the possibility of any additional, nonbehavioral (e.g., electrophysiological) criteria was excluded.[19]

Malcolm's analysis of dreaming was thus poised between the budding field of dream research and classical epistemology. And if we assume that dream research is a branch of consciousness research—for dream research, understood as the scientific investigation of conscious experience during sleep, is consciousness research par excellence, though this point is only rarely made explicit[20]—then we can say that Malcolm's analysis was a direct reaction to an early offshoot of consciousness research itself. This is a further indication that something larger than a mere dispute about dreaming was afoot.

1.3 In Defense of Dream Experience: Malcolm's Critics

1.3.1 Objections to Malcolm's Verificationism and His Criterial View of Language

One way to understand the urgency behind Malcolm's critique of the received view of dreaming, as well as the equally impassioned reactions of his critics, is to regard the debate about dream experience as a war of proxy on larger issues. One of these issues concerns the relationship between verification, behavior, and the meaning of concepts. The objections discussed in this section fall into two groups: one, objections to the criterial view of language in general; and two, objections to the criterial analysis of sleep and dreaming in particular. Note that here and in the following sections, my aim is not only to show what is wrong with Malcolm's arguments but also to reconstruct and categorize the main lines of criticism against his theory. Providing an overview of the main arguments of his critics is helpful because it places Malcolm's analysis of dreaming in its proper historical context and highlights some important, and sometimes problematic, changes in philosophical theorizing about dreaming.

To begin with, it is noteworthy that many of Malcolm's specific claims about dreaming, for instance concerning the threat of conceptual change, result directly from the underlying criterial view of language he endorses. By contrast, rejecting this view opens up completely new ways of thinking about dreaming, such as different methods of corroborating dream reports or the possibility of misremembering dreams, which in Malcolm's theory were considered to be the result of conceptual confusion. Accordingly, many of his critics were led to a wholesale rejection of his claims, and indeed his theory of dreaming seemed to many to be a sort of package deal. As Putnam put it, the theory involved "a kind of circularity. If we reject Malcolm's views at one point, we are naturally led to reject them at many others; and if we accept them at one point, we must accept them at many others" (Putnam, 1986, p. 315). Putnam also aptly articulated the general context of Malcolm's analysis of dreaming and the importance of refuting his views not so much for the sake of dreaming but in view of the farther-reaching consequences it entails: "Why is it important to show that Malcolm is wrong? Malcolm's is the sharpest statement of Verificationism in the 1950s. If Malcolm is right, then the 'naive' way of understanding our language and our knowledge is wrong" (Putnam, 1986, p. 306).[21]

One of Putnam's main objections to the "Malcomese idea of language" (Putnam, 1986, p. 324) is that criteria are neither necessary for the intelligibility of sentences nor required for their verification. "Linguistic intelligibility appears to depend on matters quite other than what we can or cannot 'settle with certainty': e.g., on grammar, on ability to occur in coherent and appropriate discourses, on paraphrasability" (p. 309). Putnam offers a similarly deflationary account of the Malcolmian specter of conceptual change, arguing that criteria can change with time, and whether or not a new use of a term is introduced in this process is an open and empirical question. In the case of dreaming, it is not likely to be answered affirmatively: if new physiological criteria were introduced for the occurrence of dreams, Putnam predicts that "virtually a hundred per cent of all hearers will 'pass' these [scientific] discourses without detecting the slightest trace of linguistic oddity" (pp. 314–315). If so, this would suggest that the concept of dreaming had not changed at all. Moreover, proposed correlations between dreaming and rapid eye movements, or hypotheses concerning the duration of dreaming, could be confirmed or disconfirmed depending, for example, on the statistical significance of the observed correlations or how well they fit into an overall explanatory theory of dreaming. Without yielding conclusive verifiability, such kinds of inductive evidence could nonetheless provide sufficient confirmation and are in fact the basis of most scientific theories (p. 317). The duration and temporal location of dreams during sleep—and, with it, the experiential hypothesis—would thus be rendered scientifically investigable.

The dispute between Malcolm and Putnam goes to the heart of philosophical methodology, the relationship between philosophy and the empirical sciences, and the authority of philosophy. According to Putnam, the meaning of a concept such as dreaming, as well as the threat of conceptual change, cannot be determined by a priori conceptual analysis alone; questions about dreaming are open to inductive evidence and empirical predictions. In particular, philosophers conducting armchair conceptual analysis have no privileged access to hidden conceptual truths. Rather, analytic truths (if they exist) should be broadly accessible to laypersons as well. By contrast, Malcolm's analysis of dreaming elevates philosophers to diviners of deep and nonexistent logical connections between concepts:

> The lexicographer would undoubtedly perceive the logical (or semantical) connection between being a pediatrician and being a doctor, but he would miss the *allegedly* "logical" character of the connection between dreams and waking impressions. . . . This "depth grammar" kind of analyticity (or "logical dependence") does not exist. (Putnam, 1986, p. 306)

Malcolm's approach is thus reduced to absurdity: even if Malcolm's mysterious depth grammar existed, it would not be available to ordinary language speakers. The differences between Putnam and Malcolm also hinge on the question of philosophical authority and can be understood, in part, as a dispute about the existence of sharp disciplinary divides between philosophy and the empirical sciences, but also about the ability of philosophers to legislate over the relevance of possible scientific discoveries. Note that in Malcolm's view, empirical dream research is not just irrelevant to the *philosophical* analysis of dreaming. Strictly speaking, philosophical analysis—on threat of conceptual change—shows scientific research on *dreaming* to be impossible and thus prohibits the very existence of an entire field of research.

The broader context of disagreement between Malcolm and his critics is also evident in Chihara and Fodor's (1977) earlier critique. It is noteworthy that the real purpose of their paper was to expose certain implausible conclusions regarding the nature of language and psychology resulting from Wittgenstein's work (p. 174). Malcolm's analysis of dreaming was merely a particularly obvious example of these deeper underlying problems (p. 193). Like Putnam, Chihara and Fodor's main objection to the criterial view of language and verification is that it misconstrues the nature of justification itself:

> It cannot be stressed too often that there exist patterns of justificatory argument which are not happily identified either with appeals to symptoms or with appeals to criteria, and which do not in any obvious way rest upon such appeals. (Chihara & Fodor, 1977, p. 197)

Instead of relying on strictly logical relations between mental predicates and observable criteria, "such justifications depend rather on appeals to the simplicity, plausibility, and predictive adequacy of an explanatory system as a whole" (Chihara & Fodor, 1977, p. 197).

Importantly, without the reliance on criteria for the ascription of psychological terms, Malcolm's arguments for the irrelevance of dream research evaporate. According to Chihara and Fodor, the notion that a dream could occur without the subject's remembering what and even that she had dreamt not only is sensible but could even be confirmed by empirical investigation. They also suggest which conditions such a confirmation process would have to fulfill: if a correlation between a neurological state, dream behavior (such as tossing and crying in sleep), and dream reports had been established, and if another neurological state had been discovered to result in the forgetting of previous experiences, then the observation of both neurological states, accompanied by characteristic dream behavior during sleep, would suggest that the subject had dreamt even if she were unable to report a dream (Chihara & Fodor, 1977, p. 203). This would not change the meaning of the concept of dreaming, which was only loosely correlated with different dream-related behaviors (including dream reporting) in the first place (p. 203). Note that in this view, dream reports not only lose the unique authority they enjoy in Malcolm's theory, but are effectively surpassed by evidence on the neurological and behavioral correlates of dreaming: along with the possibility of misremembering dreams, the possibility of confirming or disconfirming reports of having (or not having) dreamt becomes conceivable. In one respect, then, Malcolm's analysis and Chihara and Fodor's critique are in perfect agreement: both claim that scientific dream research not only introduces new kinds of evidence about dreaming but also diminishes the evidential status of dream reports. With the advent of scientific dream research, dreaming thus appears to become dissociable from dream reports, and the same is true for the study of dreams.

Other critics pursue a more humble but similarly effective strategy. Rather than attacking the criterial view of language in general, they focus on the insufficiencies of Malcolm's analysis of the concepts of sleep and dreaming. Here a first possibility is to claim that behavior alone is insufficient to determine with certainty whether a person is asleep or to learn the meaning of the concept of sleep. For instance, one might argue that the behavioral criterion of sleep cannot rule out nonstandard cases such as feigning sleep or unconsciousness (Caldwell, 1977). Admitting that testimony is needed in such cases as a supplementary criterion is self-defeating, because it shows that one of the main functions of the behavioral criterion—to determine with certainty whether someone is asleep—is not fulfilled (Caldwell, 1977, pp. 165–167). The criterial analysis of sleep is vulnerable because it prohibits the existence of any such counterexamples. Nor can the meaning of the concept of sleep be taught based on the behavioral criterion alone, as one could not learn what sleep is by merely observing someone who was lying down with her eyes closed (p. 168).

Aside from being insufficient, the behavioral criterion may also not be necessary. Canfield (1977) argues that dreams are not unique in lacking a behavioral criterion. Not only do many reports of waking experiences as a matter of fact rely on nothing but testimony, but behavioral criteria cannot be applied to certain sentences even in principle. For instance, the sentence "I am sitting perfectly still" can never be expressed truthfully and consequently lacks a behavioral criterion. Still, it has a clear-cut meaning and is verifiable, and it would be absurd to claim that one cannot think or judge that one is sitting perfectly still because uttering this sentence would result in a performative contradiction. Yet this absurdity would appear to follow from Malcolm's analysis of dreaming. If Canfield's analysis is correct, then the price for demanding a behavioral criterion is too high: it would render sentences such as "I am sitting perfectly still," which appear to be straightforwardly meaningful, senseless.[22] The alternative would be to say that such sentences, like sentences about dreaming, demand a special grammar. To block this somewhat bizarre conclusion, one must admit that sentences about dreaming are no stranger than these other somewhat trivial examples.

Similar claims aimed at dissolving the sharp distinction that Malcolm draws between dream reports and reports of waking experiences are defended by other authors as well. Schröder (1997) notes that as is the case for dream reports, many first-person psychological statements in the past tense will in fact go unchecked. Moreover, in cases in which there is no possibility to check a memory report, "the distinction between the person's remembering correctly and his only being under the impression that he remembers becomes meaningless" (Schröder, 1997, p. 31). Yet such reports are typically not doubted. This suggests that contra Malcolm, "the grammatical affinity between *recollections of thoughts* and *recollections of dreams* is closer than that between *recollections of thoughts* and *observational recollections*" (p. 31).

Siegler (1977) makes a similar point regarding the public verifiability of memories. According to Wittgenstein and Malcolm, what renders dream recall suspicious is its ultimate privacy, whereas memories of waking experiences can be checked against other people's memories. Siegler argues that this public–private distinction, however, is only a matter of degrees, and even public corroboration does not provide absolute certainty in all cases. Therefore the requirement that genuine memories—as opposed to cases of seeming to remember, as Malcolm claims is the case for dreaming—must be capable of public corroboration in every single instance is too strong. Rather than applying to every single memory claim, public verifiability is needed to ensure the credibility of the memory process as a whole. Siegler views memory as a learnable skill that requires the basic ability to express thoughts in the present tense and later report them correctly in the past tense. A person is considered to have mastered this skill if some of his memory claims are supported by independent evidence. The argument rests on the assumption that the reliability of the memory process is independent of the situation, thoughts, or emotions forming the object of the memory report. "Learning to report a dream is not learning a new technique but learning to apply a known technique to a new phenomenon" (Siegler, 1977, p. 274). Consequently, if general evidence indicates that a subject has mastered the ability to give such memory reports, this counts as evidence for the credibility of even those memory reports that in fact cannot be verified, including dream reports.

The objection that immediately comes to mind is that the credibility of memory reports seems to depend on various factors, such as the overall state of mind during which the events, thoughts, or emotions in question were originally experienced, the state in which they are reported, and the time lag between the two. If I am now drunk—or reporting events that happened while I was drunk, or simply a long time ago—my memory reports may not be trustworthy even though I have generally mastered the ability to give more or less accurate memory reports. Consequently it might be objected that Siegler's analysis of dream recall is slightly too permissive. Instead, what is needed is a careful consideration of the conditions under which different types of memory reports in general and dream reports in particular are trustworthy (I return to this issue in chap. 4).

Finally, could the sentence "I am asleep" even be used to express a true proposition? Ayer (1977) argues that it can, suggesting several ways in which it could be corroborated, such as through sleep talking, hypnosis,[23] physiological evidence, or automatic writing during sleep. Though he notes that a majority of dream reports will, as a matter of fact, go untested, he argues that corroborating evidence could be had, at least in principle. Because the same is true for reports about past waking experiences, there is "no reason . . . for putting reports of dreams into a different category" (p. 145) on these grounds. This view corresponds to Ayer's (1946/1952) argument in *Language, Truth, and Logic* that strong or conclusive verifiability in practice is generally too strong a requirement. Rather, a proposition counts as meaningful if it is verifiable in principle—that is, if we know which observations would count as evidence for its truth if we were in a position to make them, and if they would render the proposition probable. Ayer's objection to Malcolm, accordingly, is that dreams can count as experiences during sleep even if we cannot in practice verify many (or even most) dream reports: allowing verifiability in principle—an allowance that seems not even to require science but, if Ayer is correct, could be satisfied by occasional sleep talking—is enough. The connection between dreaming and dream reporting would then no longer be a matter of logical necessity, and the transparency hypothesis of dream reporting could be reconceptualized as an empirical hypothesis.

In sum, based on my analysis of the authors discussed in this section, the criterial view of language seems untenable on several counts.[24] First, the reliance on strict criteria is at odds with widespread views on language use and concept application, as well as with standard scientific justificatory procedure. Second, the criterial view does not justify Malcolm's claim that dream sentences are governed by a different grammar than are sentences referring to past waking experiences: not only does the behavioral criterion of sleep fall short of providing the type of indubitable certainty that Malcolm tends to credit it with, but dreams are also not unique in that they often lack behavioral criteria and only become accessible through retrospective reports. Contra Malcolm, dreaming does not pose a special epistemological problem because of the widespread absence of contemporaneous, publicly observable evidence: certain retrospective reports of waking experiences may be unverifiable, in practice or even in principle, in the same manner as dream reports. At the same time, as kinds of evidence about dreaming other than dream reports become permissible (such as sleep behavior, rapid eye movements, etc.), dream reports lose their status as strict and exclusive criteria, and the possibility of misremembering dreams once more becomes acceptable. Importantly, the experiential and transparency hypotheses now appear to be empirically investigable. The upshot is that contra Malcolm, a deep epistemological and grammatical continuity exists between the dream-telling and historical use of first-person psychological sentences.

I think these lines of criticism suffice to show that Malcolm's reliance on the criterial view of language and his particularly strong brand of verificationism should be rejected. This is all the more true because Malcolm, in his work on dreaming, did not explicitly develop a detailed argument for the criterial view of language (see Chihara, 1977, pp. 253–254). What he did was to apply such a view, which he borrowed from Wittgenstein but failed to make explicit, to the concept of dreaming and show its consequences in this particular case. Nonetheless, without an explicit statement of an overarching theory concerning the criterial view of language, his analysis of dreaming lacks the type of independent support it would require given the foregoing objections.

1.3.2 Phenomenologically Motivated Objections

Several authors argue that the phenomenology of remembering dreams and of awakening provides evidence against Malcolm's theory. Recall that in Malcolm's view, realizing that one dreamt means waking up with the false impression of actually having experienced certain events and then inferring that they only occurred in a dream. Ayer (1977, p. 138) rejects this treatment of dream memories "as though they posthumously masqueraded as waking experiences" as strange and implausible. Aside from young children, who may at times confuse dreams and waking memories, dream memories simply do not act like memories of waking experiences, and no inferential process is needed to identify them as nonveridical. The phenomenology of remembering a dream is quite simply that of waking with the impression of having had certain experiences during sleep (Ayer, 1977, p. 139).

Chihara (1977) notes that Malcolm's account of dream recall is particularly inadequate when applied to cases in which dreams are recalled after a certain time lag. If dream recall were indeed inferential, how would we manage to classify such memories as instances of dream recall hours or even days after the dream occurred? Malcolm's account of dream recall also does not accommodate instances in which a person is unsure of having dreamt at all. By failing to recognize that the distinction between dreaming and dream recall is inherent in our everyday conception of dreaming, "Malcolm completely obscures the place memory has in our common notion of dreaming" (Chihara, 1977, p. 257).

Pears (1977, p. 206) makes the related observation that dream reports are sometimes elaborated or revised over time and that this process involves a sense of effort. This makes the relation between dreaming and dream recall appear almost contingent: if remembering a dream requires time and effort and a dream report can be more or less complete, then it seems equally possible, contra Malcolm, that a dream might not be remembered at all or might be misremembered entirely. Pretheoretically, one's impression upon awakening is not always the ultimate standard of whether, or what, one dreamt (but see chap. 4).[25]

Curley goes a step further, arguing that aside from being phenomenologically implausible and at odds with the pretheoretical concept of dream recall, findings from dream research rule out the alleged dependence of dreaming on dream reporting. According to the hypothesis that dreaming is correlated with REM sleep—supported by the finding that systematic laboratory awakenings result in a high incidence of dream reports in subjects awakened from REM sleep (for details, see chap. 2)—about one-fifth of sleeping time is spent dreaming, but a majority of these dreams are forgotten. This shows that dream reports are not a necessary condition of dreaming (Curley, 1977, pp. 326–327). But there may also be cases in which dream reports provide insufficient evidence of dreaming. A person may be unsure whether an event, such as a knock on the door at night, occurred in a dream, and in trying to decide the question could falsely conclude that it did; in this case, the alleged dream report would not be sufficient evidence for the occurrence of a dream (pp. 336–337). But according to Curley, even a sincere report, in which the subject was entirely certain of having dreamt a particular event, would be insufficient to establish the content of the previous dream. In psychoanalysis, for instance, the therapist might meaningfully wonder whether the patient is unwittingly embellishing or misreporting a dream. Doubting dream reports, then, is not an absurdity but would appear to be a matter of fact, at least in certain contexts.

The possibility of forgetting or misremembering dreams, then, seems to be built into the experience of remembering dreams itself—and with it, quite naturally, comes the assumption that the conscious experience of dreaming is separable from the retrospectively and somewhat contingently remembered and reported dream. This very general observation is supported by dream research, but it is also part of our prescientific discourse on dreaming, thus casting doubt on Malcolm's claim that the notion of misremembering a dream is absurd and would change the concept of dreaming. If this is correct, then Malcolm misconstrues the pretheoretical concepts of dreaming and dream recall, and the alleged depth grammar of dream-related sentences is not only hidden but distinctly at odds with everyday discourse about dreaming. Moreover, with the possibility of misremembering dreams, the experiential and transparency hypotheses become decoupled: saying that dreams are experiences, apparently, is one thing; saying that dream reports reveal this to be the case is another.

Other authors argue that Malcolm misconstrues not only the phenomenology of recalling a dream but also the phenomenology of dreaming itself. Caldwell (1977) argues that the continuous experience of awakening from a dream gives a clear sense to the location of dreams in physical time, a notion that Malcolm had rejected as meaningless. In such cases, experience "does not jump an unbridgeable chasm from the dream to the waking state. One merges into the other around the core of the common event. Thus a fairly precise sense can be given to the expression 'just before'" (Caldwell, 1977, p. 163). Something similar can be said for the transition from wakefulness through hypnagogic imagery to full-fledged dreaming, which can also be continuous (see chap. 11 for details).[26] Why, in the face of such experiences, should we concede Malcolm's point that there is nonetheless a sharp distinction between dreaming and sound sleep, on the one hand, and wakefulness, on the other, such that sentences about these states require sharply different grammars? As Lewis puts it:

> We can, if we like, pedantically insist, in the interest of a theory, that dreaming is one thing, waking another, and that there must be some precise point where experience or process begins after sleep in which there is no mental occurrence at all. But we can only do so by going so sharply against what seem to be familiar incontrovertible facts that only the most overwhelming arguments could justify us. (Lewis, 1969, p. 133)

We have already seen, in the previous section, that Malcolm's arguments for a sharp conceptual or logical distinction between dreaming and wakefulness collapse; now it is becoming increasingly clear that these distinctions are also phenomenologically implausible. They increasingly appear to be conceptual legislation only—and as Lewis notes, of a particularly pedantic type. Importantly, we can now see why the conceptual tools offered by a philosophical theory of dreaming should strive, among other things, toward phenomenological plausibility: failing to do so invites the armchair fallacy, or the practice of deriving (implausible) phenomenological claims from purely conceptual arguments.

A final phenomenological argument defended by several authors refers to lucid dreams, or dreams in which dreamers realize that they are now dreaming and often have some control over the ongoing dream (see chaps. 3 and 9 for details). At least in these cases, as already pointed out by Lewis (1969, p. 132), it seems that people do entertain the thought that they are now dreaming. And as Reed points out, for persons experiencing vivid lucid dreams, "doubting that dreams are experiences will scarcely be easier than doubting past *waking* experience to have occurred" (Reed, 1979, p. 39). Moreover, because lucid and nonlucid dream reports use similar language, not only do lucid dreams seem to provide evidence for saying that some dreams are experiences, but they provide a powerful argument against Malcolm's account of dreaming in general (Reed, 1979, p. 40): denying the experiential character of lucid dreams seems profoundly counterintuitive because of their often vivid phenomenology, and allowing lucid dreams to be experiences but denying that the same is true of nonlucid ones seems arbitrary.

1.3.3 Cartesian Dream Skepticism Revived

Several of Malcolm's critics noted that if Malcolmian skepticism about dreaming is rejected and dreams are indeed experiences, this reintroduces the problem of Cartesian dream skepticism. Ayer emphasizes that despite Malcolm's "valiant efforts" (Ayer, 1977, p. 148) to the contrary, the skeptic's question of whether he is currently awake or dreaming is meaningful. If dreams are experiences, then it is possible to be deceived by one's waking memories about one's dreams, but also by an ongoing dream in which one mistakenly believes oneself to be awake. Ayer does agree with Malcolm on one point, namely, that the coherence test proposed by Descartes at the end of the *Meditations* (see sec. 1.1) is inconclusive. Malcolm, much like Hobbes (Hobbes & Descartes, 1641/1975–1999), had argued that the test was worthless because one could just as well "dream that the right connections hold" (Malcolm, 1959/1962, p. 108). Ayer happily concedes that "Professor Malcolm's argument has this much force, that it is indeed possible to have a dream in which one raises the question whether one is dreaming, applies the coherence test, and wrongly concludes that one is not" (Ayer, 1977, p. 146). According to Ayer, this does not, however, make the test worthless, but merely shows that the conclusion that one is currently awake is always open to future falsification: it could be disproved by the subsequent experience of waking up. The experience of awakening, in turn, could also be deceptive, as in so-called false awakenings or realistic dreams of waking up. This lack of absolute certainty is not, however, a cause of genuine concern:

> In taking it for granted that I am not now dreaming, I am taking it for granted that my present experience is in the main veridical; and the test for this is that further experience confirms it. In making this claim upon the future, I am indeed not betting on a certainty; but this is not to say that the claim has no support at all. (Ayer, 1977, p. 147)

Curley (1977) formulates an empirical argument for the actual existence of deceptive dreams, noting that narcoleptic patients, who are prone to sudden attacks of REM sleep during the daytime, often have extremely realistic dreams of their current environment (see also sec. 10.3.4). Upon awakening, they often do not even realize that they were asleep and mistake their dreams for reality. This, according to Curley, is an actual, real-world example of the type of dream deception described by Descartes (Curley, 1977, pp. 345–346).

Nagel argues that the fan of Cartesian dream skepticism "can accept the linguistic analysis which shows him that dreaming is not a form of experience, while still maintaining the meaningfulness of the expression 'I'm dreaming'" (Nagel, 1959, p. 115). Although this expression is only appropriate in certain strange situations—as might occur in dreams—it is nonetheless meaningful and sometimes even true, namely, when asserted in a dream. So even if Malcolm's rejection of dream experience on conceptual grounds goes through, his original goal of thereby refuting Cartesian dream skepticism fails. Relatedly, Brown notes that even if we accept Malcolm's claim that it is logically impossible to test whether or not one is awake, and even if this test is always only a pseudotest:

> It offers the dreamer no help for him to recall in his dream that if he is awake he can really tell that he is and that if he is asleep he cannot tell (logically) that he is sleeping. Nor can it help him to recall that Malcolm says both the question and the test are absurd. (Brown, 1957, p. 52)

Once more, Malcolm's position is phenomenologically implausible in the light of such examples. As Brown points out, this is especially obvious for cases in which the test is performed in wakefulness. Imagine the case of a person who, having become the victim of a large-scale practical joke, uses the coherence test to determine whether he is dreaming or awake and finds that the events he has just experienced are not systematically related to each other:

> The victim could (and has in the past) come to think that he was dreaming. And in the situation he would not be able to tell whether or not his conclusion was "self-contradictory." . . . In such a case the test would be genuine. (Brown, 1957, p. 52)

Even if performing the test in one's dream was (logically) illegitimate, the illegitimate dream test might be temporarily indistinguishable from the legitimate (and perhaps even inconclusive) daytime test: to the victim's mind, the question of whether he was dreaming or waking would present two genuine alternatives, and this could be the case even if they were not, strictly speaking, logically sound. As Brown points out, the insight that one's doubt is a senseless doubt is not likely to be reassuring in such cases.

1.3.4 A New Kind of Skepticism about Dreaming: Dennett's Cassette Theory

At this point in the debate between the skeptic about dream experience and the proponent of the received view, one might conclude that the received view has not only been saved but also gained additional support from empirical dream research. Rather than changing the folk-psychological concept of dreaming, scientific dream research appears to offer means of corroborating the received view and shedding light on the relationship between dreaming and dream recall. The experiential hypothesis and the transparency hypothesis, it seems, have been turned into empirically investigable hypotheses, open to empirical confirmation and disconfirmation. In this view, not only did Malcolm misconstrue the nature of the question of dream experience by portraying it as a conceptual rather than an empirical question, but his rejection of dream experience is also untenable.

In "Are Dreams Experiences?" (1976), Dennett reopened the case against dreaming and dream reporting from the perspective of scientific dream research. In an essay that is one of the most influential and widely discussed accounts of dreaming following Malcolm's, Dennett apparently comes full circle in his attempt to "undermine the authority of the received view of dreams," albeit with "aims and presuppositions . . . quite antagonistic to Malcolm's" (Dennett, 1976, p. 151). While the title of the essay suggests otherwise, its goal was neither to answer the question of whether dreams are experiences nor to deny that they are, but rather "to treat the question itself as the specimen to be examined" (Dennett, 1979, p. 317; see also Dennett 2014). In this respect, Dennett's discussion of the problem of dream experience is more subtle than Malcolm's, because instead of mounting a full-fledged attack against the received view, Dennett merely wants to expose its vulnerability to empirical disconfirmation (Dennett, 1976, p. 159). Indeed, *not* providing a philosophical answer to the question of dream experience is inherent in Dennett's project. He notes that "to answer it, I claim, one must have a well-confirmed, empirical (psychological, physiological) *theory* of dreams—something no one yet has," emphasizing that his "aim was to wrest the issue out of the hands of philosophers, by showing the systematic inconclusiveness or irrelevance of the sorts of considerations philosophers have traditionally raised" (Dennett, 1979, p. 315).

Aside from rejecting the authority of philosophy to solve the problem of dream experience, Dennett's second goal is to show that subjective testimony is wholly irrelevant to the question of dream experience. In both respects, Dennett's position is diametrically opposed to Malcolm's. In emphasizing the open and empirical nature of the question, Dennett also develops the arguments of many other of Malcolm's critics to their fullest. His analysis of dreaming deserves special attention because "Are Dreams Experiences?" spells out the consequences of regarding the question of dream experience as an empirical question. For this reason, it can be regarded as a test case for an important line of criticism against Malcolm: if Dennett's treatment of dream experience fails, and if its failure is the natural result of viewing the question of dream experience as an empirical question, then the attempt to rehabilitate dream experience by handing it over to scientific dream research fails as well. This is not to say that in such a case, one should revert to Malcolm's view of the problem; the arguments against the criterial analysis of dreaming stand. But it is to say that if Dennett's analysis fails, this would create a vacuum: having ruled out two opposing candidates for saying what kind of question the question of dream experience is, we would be left without any positive account or workable suggestion for its resolution.

In this section, I argue that, at least according to one possible reading of Dennett's argument, this is indeed the case. Constructing the problem of dream experience as an empirical problem leads directly to an inescapable form of skepticism about dreaming and dream reporting. By using Dennettian skepticism about dreaming as a foil for Malcolmian skepticism about dreaming, I hope to show that neither approach is capable of providing a constructive answer to the question of dream experience. My preliminary conclusion is that while Malcolmian skepticism about dreaming can be rejected as relying on mistaken assumptions about meaning, the obvious alternative of viewing the problem of dream experience as an empirical problem is equally flawed, threatening to lead the entire project of dream research *ad absurdum*. If this is correct, then an alternative answer is needed. I provide the outlines of this alternative answer in chapter 4 and briefly hint at my solution in section 1.4. My reconstruction of Dennett's argument takes the first step toward showing why the question of dream experience is not decidable based on empirical evidence alone; chapters 2 and 3 use an extended discussion of the methods and main findings of empirical dream research to further support this view.

At the beginning of what turns into an extended and empirically informed thought experiment, Dennett builds a maximally strong case for empirical dream research. Having established the physiological correlates of dreaming, such as REM periods and wakelike EEG activity, a neurophysiological model of dreaming could make all sorts of predictions about the occurrence, timing, and duration of dreams based solely on physiological measures. Noting that the prospects of a "generalized capacity to predict dream narratives" (Dennett, 1976, p. 153) in any detail are still quite weak, Dennett points out that the capacity to make such predictions is "the natural culmination of the research strategies that are already achieving modest success in 'translating' relatively gross and peripheral nervous-system activity" (p. 153). The plausibility of this claim is enhanced by the fact that Dennett's vision of a future science of dreaming smoothly integrates existing findings from dream research. For instance, he suggests that early findings on the occasional content relativity of REMs—the so-called scanning hypothesis, according to which the eye movements observed in the sleeping subject correspond to subjectively experienced gaze shifts within the dream (e.g., Dement & Kleitman, 1957b)—are a first step toward using physiological measures to predict dream narratives (Dennett, 1976, pp. 152–153). Dennett's key claim is that if dream research were to reach an even higher level of sophistication, such predictions would be valid independently of, and even in contradiction with, subjects' own reports. In this view, dreaming is not only wholly separable from dream recall on the level of conceptual analysis, but the study of dreaming is also *methodologically* independent of the study of dream reports. Dream reports, accordingly, would fail to have any methodological relevance for the study of dreaming at all.

Given this dual separation between dreaming and dream reporting, Dennett points out that a sophisticated theory of dreaming might eventually posit the existence of three separate processes: a presentation process, which would be the physiological correlate of the ongoing dream, a memory-loading process, responsible for the subject's ability to recall the dream upon awakening (Dennett, 1976, p. 154), and a composition process, responsible for the composition of dream narratives (p. 156). Once they had been identified, these three processes could be tampered with—possibly to the point that veridical dream memories could be obliterated and substituted by undreamed but nonetheless recalled dream narratives (p. 156).

> The story told so far does not, I take it, exhibit the conceptual chaos Malcolm imagines; strange as it is, I do not think it would evoke in the layman, our custodian of ordinary concepts, the nausea of incomprehension. As a premise for a science-fiction novel it would be almost pedestrian in its lack of conceptual horizon-bending. (Dennett, 1976, p. 157)

Next Dennett suggests a more parsimonious view requiring only two of these processes. According to the cassette theory of dreaming, previously composed dream cassettes trigger the instantaneous appearance of dream memories upon awakening.[27] One could even imagine an entire, possibly innate, "library of undreamed dream cassettes ready for appropriate insertion in the playback mechanism" (Dennett, 1976, p. 160). Alternatively, an unconscious composition process could be discovered to occur during REM sleep. Dennett emphasizes that the cassette theory could accommodate all physiological data on REM sleep, as well as the occasional incorporation of external stimuli into dreams and the apparent content relativity of REMs. He even suggests how lucid dreams could be explained by the cassette theory: lucidity, which he defines as self-conscious dreaming or the impression of "tinkering" with an ongoing dream, would ensue if the unconscious composition process inserted "traces of itself into the recording via the literary conceit of a dream within a dream" (Dennett, 1976, p. 161; see also Dennett, 1979).

Dennett (1976, p. 161) points out that compared to the received view of dreaming, the cassette theory provides a simpler explanation of so-called precognitive dreams, or dreams in which a complex narrative seemingly builds up toward a climax, which upon awakening is identified with an external stimulus, such as a loud noise in the bedroom. It is also more parsimonious, in that it does not posit an additional presentation process during sleep. Referring to Nagel's (1974) dictum of states being conscious when it is *like something* to be in these states, he remarks:

> on the cassette theory it is not like anything to dream, although it is like something *to have dreamed*. On the cassette theory, dreams are not experiences we have during sleep; where we had thought there were dreams, there is only an unconscious composition process and an equally unconscious memory-loading process. (Dennett 1976, p. 161)

The point is that everyday experience could provide no evidence either for or against this theory: unless dream reports were posited as criteria for the occurrence of dreams during sleep from the outset, the subject would be incapable of distinguishing genuine instances of dream recall from instantaneously inserted ones. The thought experiment is supposed to show that the impression of having experienced a dream does not entail that one actually had the respective experience; as illustrated by the cassette theory, this impression might be profoundly mistaken. By applying the "distinction between remembering and seeming to remember . . . to dream recollections, the concept of dreaming is cast adrift from any criterial anchoring to first-person reports, and becomes (or is revealed to be) a theoretical concept" (Dennett, 1976, p. 165).

In fact, the thought experiment does even more: it denies that dream reports have any evidential status with respect to previous dream experience whatsoever. Note that this is the case whether or not the cassette theory turns out to be true: even if the received view were vindicated and most dream reports turned out to be trustworthy, dream reports would not, indeed could not, play any evidential role in this process, and this could only be established on the basis of independent evidence (cf. Dennett, 1979). This fits with the broader aim of Dennett's thought experiment, which is not to show that the cassette theory is true but rather to use it as a foil for the received view (Dennett, 1979). Indeed, the cassette theory is chosen precisely because of its counterintuitivity (Dennett, 1976, p. 159): it is designed to show that we cannot even rule out such a radically different theory of dreaming based on subjective dream recall. Seen in this context, Dennett's argument is not so much a version of skepticism about dreaming as it is a form of skepticism about the epistemic status of dream reports, suggesting that one might reject the former while adhering to the latter.

While Dennett's position on dreaming is less destructive than Malcolm's in that he does not categorically exclude dreams from the realm of conscious experience in general, it is much broader in scope. Readers familiar with Dennett's work on consciousness and qualia will easily recognize the contrast between the cassette theory of dreaming and the received view as an early formulation of the views elaborated in the *multiple-drafts model* (for details and further references, see Dennett, 1991; Dennett & Akins, 2008). After all, the rejection of the Cartesian theater, or of the view that there is a single place and time where consciousness comes together and at which a presentation process occurs, is inherent to the multiple-drafts model. This broader context is already evident in Dennett's (1976) remark that the cassette theory could be modified to include waking experience as well. In *Consciousness Explained*, Dennett (1991) develops a similar thought experiment, focusing on the contrast between preexperiential memory production and postexperiential memory revision. Again, his point is that it is impossible to distinguish between the two on the basis of subjective testimony (Dennett, 1991, pp. 115–126; see also Dennett & Kinsbourne, 1992). Referring to George Orwell's *1984* scenario of history being rewritten by the Ministry of Truth, Dennett identifies "Orwellian revision" with postexperiential memory revision. He contrasts this with the "Stalinesque method" of staging show trials based on false confessions and fictive evidence, that is, of simulating the presentation process. While the aim of both Orwellian revision and the Stalinesque method consists in the production of misleading archives, there seems to be a clear-cut difference between the retrospective fabrication of historical accounts and the actual staging of simulated events. He notes, however, that it is an illusion to believe that this distinction can be applied "all the way in," and that we can distinguish between hallucinated memories and hallucinated experiences on a very small timescale:

> Here the distinction between perceptual revisions and memory revisions that works so crisply at other scales is no longer guaranteed to make sense. We have moved into the foggy area in which the subject's own point of view is spatially and temporally smeared, and the question *Orwellian or Stalinesque?* loses its force. (Dennett, 1991, p. 119)

As is the case for dreaming in the 1976 paper, the subjective indistinguishability between the two theories finally leads Dennett to identify conscious experience with the production of memory traces: "The Multiple Drafts Model makes 'writing it down' in memory criterial for consciousness. . . . There is no reality of conscious experience independent of the effects of various vehicles of content on subsequent action (and hence, of course, on memory)" (Dennett, 1991, p. 132). Conscious experience is probe dependent: it cannot be conceived of independently of reporting. It is interesting to note that at this point, something resembling the transparency hypothesis does once more play a central role in Dennett's theory. However, it does so only with respect to a drastically altered conception of experience, according to which experience is contingent on retrospective report, rather than conceived of as separate from it. Once more, memory has turned into the *criterion* of previous experience. But this means that Dennett no longer can regard questions about the transparency of dream reports as genuine hypotheses open to empirical investigation. His claims on the probe dependence of conscious experience rule out such a view.

One consequence would seem to be that the problems raised by dreaming are no different from those presented by waking experience. Dreams are particularly vulnerable to skeptical objections because we can only report them retrospectively after awakening (Dennett, 1976, p. 166)—but they are not uniquely vulnerable.[28] Once again, the philosophical debate on dreaming appears to be a proxy war for broader questions concerning the relationship between conscious experience and memory.[29]

While the problem for dream experience generalizes to experience in general, there is still, however, a residual problem for dreaming. Dennett is quick to point out that even if something like the cassette theory turned out to be true for conscious experience in general—if, that is, waking experience could be identified with memory production—this would not automatically rehabilitate dreaming as a form of conscious experience, even in the reduced, presentation-free sense defended in the multiple-drafts model. Dennett points out that even if we accept that memory is a necessary condition for experiencing and that the concept of experience would be hopelessly impoverished without it, it cannot be considered as sufficient unless all nonstandard cases of hallucinated, inserted, or manipulated memories can be ruled out (Dennett, 1976, pp. 167–169). Here the question of dream experience hinges not on the question of whether a presentation process occurs during sleep but on the question of whether the production of dream memories is similar to normal memory-production processes (whatever those may be). Only if deviant cases can be ruled out can dreams be considered to fall within the boundaries of experience.[30] Note that this narrows the scope of the problem considerably: the question is no longer whether dreams are experiences during sleep in the sense of whether it is like something to dream, but whether dream recall results from a normal memory-production process. Dennett even specifies what kind of scientific discoveries could convince him to count dreams as experiences, albeit in this limited sense:

If it turns out that sleep, or at least that portion of sleep during which dreaming occurs, is a state of more or less peripheral paralysis or inactivity; if it turns out that most of the functional areas that are critical to the governance of our wide awake activity are in operation, then there will be good reason for drawing the lines around experience so that dreams are included. If not, there will be good reason to deny that dreams are experiences. (Dennett, 1976, p. 169)

Note that these seem, once more, to be straightforward empirical questions. According to this reading, one might quite simply say that yes, given the findings from modern dream research—and most of the conditions specified by Dennett, as will become clear later (cf. chaps. 2 and 3), were already known to be fulfilled in 1976—dreams are, indeed, experiences, at least in the same sense as waking experiences. Especially with hindsight, and taking into account Dennett's endorsement of something very much like the cassette theory in the multiple-drafts model, it becomes clear, however, that a stronger reading is possible. This stronger reading contrasts with Dennett's explicit but weaker claim that the question of dream experience is a straightforward empirical question; as we have seen, empirical research could only determine whether dream recall involves a standard process of memory production, but would leave open the question of whether dreaming involves a presentation process during sleep. Though Dennett focuses on the weaker reading, even in 1976 there were indications that things are not quite so simple. For instance, Dennett notes that even if the empirical conditions for dream experience stated earlier were fulfilled, and even if the activation, for instance, of the visual processing system during REM sleep were empirically confirmed, this would provide strong but not "overwhelming grounds for deciding that dreams are experiences" (Dennett, 1976, p. 170). The same considerations would apply if certain kinds of dreams, such as recurring dreams, were discovered to have a function, for example, for emotional and cognitive processing. In all of these cases, it seems that any empirical evidence that initially might appear to support the received view could, on closer inspection, also be compatible with an unconscious composition process taking place during REM sleep. Indeed, Dennett emphasizes that the predictive power of the cassette theory is exactly the same as that of the received view. If the composition process occurred during REM sleep, "one might even be able to 'translate' the composition process—that is, predict dream recollections from data about the composition process. This theory looks suspiciously like the elaboration of the received theory, except for lacking the presentation process" (Dennett, 1976, p. 160).

Moreover, note that this problem is systematic. The supposed empirical plausibility of the cassette theory was, after all, the condition for turning it into a powerful rival to the received view in the first place. It seems, however, that once this has been done, the demand for further empirical evidence on the basis of which one could adjudicate between the two theories has no natural end point. Given its initial ability to accommodate the findings even of a highly sophisticated future science of dreaming, why would the cassette theory fail given any particular piece of empirical evidence? If this stronger reading of Dennett is correct, then it is not clear that the difference between the cassette theory and the received view can be investigated empirically at all. The cassette theory would then give way to an inescapable and principled form of skepticism, according to which the cassette theory could be a relevant alternative independently of how much empirical evidence was available.[31]

If this is correct, then Dennett's attempt to portray the problem of dream experience as an empirical question threatens to lead the very project *ad absurdum*. Moreover, this problem is not specific to Dennett's analysis of dream experience—rather, it quite naturally results from regarding the question of dream experience as an empirical question to be settled independently of dream reports. Determining whether this reading is plausible will require taking a careful look at the methods and findings of contemporary dream research, but also at changing theoretical conceptions of sleep and dreaming. Initial support, however, comes from the fact that the same is true for Dennett's rejection of the distinction between Orwellian and Stalinesque memory production in the multiple-drafts model: though in a sense analogous to his dream argument, the distinction between these two models is not, as we have seen, based on empirical considerations but rather results from an inability of the empirical evidence to decide the question.[32]

1.4 Taking Stock: Toward an Alternative Account of the Problem of Dream Experience

If the problem of dream experience is neither conceptual/philosophical nor empirical, then—absent further alternatives—a creeping suspicion arises that it might be a pseudoproblem, an artifact of philosophical debate. Perhaps construing the problems of dream experience and of the transparency of dream reports as genuine questions or hypotheses was misleading all along. Throughout the following chapters, I argue that this is indeed the case. But this is not to say that we can learn nothing constructive from the philosophical discussion of these problems. The positive upshot of the unsatisfying—because inconclusive—philosophical discussion of dream experience is that we can extract from it a more convincing rationale for laying to rest the nagging specter of skepticism about dreaming and dream reporting. For this reason, the discussion of skepticism about dreaming is not merely an exercise in refuting an old and outdated question; it is enormously useful for investigating what kind of problem the problem of dream experience actually is and for solving it in a constructive manner.

Throughout the following chapters, I use examples from dream research to show that the experiential hypothesis and the transparency hypothesis are the very conditions of possibility for scientific dream research, but also for constructing an empirically informed philosophical framework of the concept of dreaming. Without assuming the truth of both hypotheses, by contrast, progress becomes impossible. This, in turn, suggests that many questions about dreaming and dream reporting require reformulation. Rather than asking, "Are dreams experiences?" and "Do dream reports reveal what it is like to dream (or that it is like anything at all)?" we should ask more constructive questions, such as "What does science reveal about dream experience?" and "Which role do—and should—dream reports play in investigating dream experience?" Assuming dreams to be experiences and dream reports to transparently reveal this fact is, I claim, neither a conceptual truth nor based on (or capable of) empirical verification, but a matter of methodological necessity. Although there is a sense in which dream reports are intimately linked to dream experience, and in which the link between dreaming and dream reporting is unbreakable, this is a methodological sense, and one that demands careful qualification.

Finally, if this is correct, can anything positive by said in favor of the experiential hypothesis and the transparency hypothesis? In chapter 4, I argue that inference to the best explanation provides theoretical justification for what was assumed to be true all along, by default: that dreams are experiences, and dream reports transparently show this to be the case. The intuitive plausibility of the experiential and transparency hypotheses is confirmed by explanatory considerations. Saying that these hypotheses better explain dream reporting behavior than alternative hypotheses, in turn, enables dream reports under the *assumption* of transparency *themselves* to turn into evidence about dreaming. As Putnam puts it, "The use of the term is based on the *supposition* that there is something—a 'natural kind,' so to speak—for which our 'criteria' are *good*, but not *perfect* indicators" (Putnam, 1986, p. 311; italics mine). Moreover:

> *Assuming* that dreams take place "in physical time"—i.e., that they start and stop at some time or other—various things *become* inductive evidence that correlations hold: correlations between the things we do with our eyes, muscles, vocal cords, as we sleep and the dream events; and correlations between the neural processes that normally go with "seeing" certain things and dream events. (Putnam, 1986, p. 317; italics mine)

Making certain default assumptions about dreams, then, is the condition for investigating dreaming as a natural phenomenon and for turning physiological processes during sleep into evidence about dreaming in the first place. Given these two assumptions, dream research is extremely fruitful; without them, it could not properly be considered as *dream* research, understood as the investigation of phenomenal experience during sleep, at all.

The strategy of regarding the experiential hypothesis as the best explanation of dream reporting and as further strengthened by being placed in the context of ever-more-elaborate theories of dreaming is not a glamorous solution, nor a new one. It is a compromise between two opposing philosophical treatments of dreaming and dream reporting—of which Malcolm's and Dennett's are the extreme points—that were each in their own way taken too far. But for the same reason, it attempts to combine the best of both approaches: from Malcolm's treatment of dreaming (as well as from Dennett's later remarks on the probe dependence of conscious experience), it inherits the view that the relationship between dreaming and dream reporting is in a certain sense unbreakable, and that this is true pretheoretically and independently of empirical evidence. It also preserves the special status of dream reports as compared to other forms of evidence about dreaming. But from Dennett, and from many of Malcolm's other critics, it takes an openness toward empirical research and a more modest and careful position regarding what can and cannot be settled by means of philosophical argumentation alone. Malcolm's critics were right that it does not take conceptual analysis to show that dreams are experiences; but the question of dream experience does not thereby become an open and empirical question. Assuming that it does, and that it is the task of empirical dream research to *prove* the experiential and transparency hypotheses, renders the question of dream experience unanswerable and makes dream research appear strangely inept. Given pretheoretical qualms about the trustworthiness of dream reports, it seems natural, even advisable, to ask for empirical confirmation. But insofar as this demand comes at the cost of slipping into verificationism, this temptation is better avoided. In this sense, the question of dream experience is indeed a pseudoquestion: no straightforwardly empirical answer is forthcoming. At the same time—and this is another lesson learned from Malcolm's critics—a plausible account of dreaming and dream reporting will have to accommodate the commonsense assumption that dreams can be misremembered or forgotten.

Reconstructing the relationship between dreaming and dream reporting will require a detailed look at the methods and findings of empirical dream research; again, this question is not wholly resolvable from the armchair but can be solved only by taking actual empirical research into account. So far I have only surveyed the theoretical side of the argument and pointed out certain insufficiencies arising from the two main types of philosophical skepticism about dreaming and dream reporting. The task of the next two chapters is to analyze examples from empirical dream research. In particular, can these studies be interpreted as providing empirical evidence for saying that dreams are experiences? Is there any reason to think that dream reports can, as a matter of fact, be verified by independent means? Or does my claim that the experiential and transparency hypotheses have the status of default assumptions better explain the way dream research is actually conducted? In other words, aside from giving an overview of existing theories and methods for the study of dreams, the following chapters can also be seen as an extended plausibility check of the claims sketched in this section. Based on this plausibility check, I then develop a more detailed defense of my theoretical argument in chapter 4.

1.5 Conclusions

One of the most enduring and enigmatic problems in the philosophy of dreaming is the epistemological problem of dream skepticism. Descartes's exposition of the problem, in particular, secured a prominent position for dream-related arguments in the epistemological debate on external-world skepticism. More than three hundred years later, Norman Malcolm responded to Cartesian dream skepticism by introducing his own version of skepticism about dreaming. Since then, Malcolmian skepticism about dreaming and Cartesian dream skepticism have existed alongside each other and continue to be a matter of philosophical discussion.

An important lesson is that these two types of skepticism rely on diametrically opposed background assumptions about dreaming: where Cartesian dream skepticism relied on the experiential and transparency hypotheses as implicit background assumptions, vaguely grounded in folk-phenomenological beliefs about dreaming as well as possibly in Descartes's first-personal familiarity with his own dreams, Malcolm rejected both on conceptual grounds. To him, their falsity was a matter of logical necessity. This point is all the more interesting because Malcolmian skepticism about dreaming can be seen as a direct response to the beginning development of scientific dream research, brought on by the discovery of REM sleep and its supposed correlation with dreaming.

Many of Malcolm's critics, by contrast, argued that the experiential character of dreaming is scientifically investigable. This position was once more made possible, in part, by early scientific dream research, which for the first time suggested kinds of evidence about the occurrence of dreams other than, and possibly even in opposition to, dream reports. It was also at this point that the experiential and transparency hypotheses first became decoupled. Both had been implicitly assumed by Descartes, and more generally by the received view of dreaming, and both had been rejected as conceptually flawed by Malcolm. But now, with the possibility of misremembering dreams and of confirming or disconfirming dream reports, it seemed possible to have evidence supporting the experiential hypothesis while disconfirming the transparency hypothesis. An important lesson here is that a conceptual framework of dreaming and dream reporting that strives toward phenomenological and empirical plausibility should accommodate the familiar phenomenon of misremembering or forgetting dreams. In particular, if the transparency hypothesis is in fact a methodologically necessary default assumption, how can this view account for the fleeting and unstable nature of dream recall?

Finally, whereas Malcolm rejected dream research as irrelevant to the concept of dreaming, Dennett used findings from dream research to construct a new, empirically motivated version of skepticism about dreaming. Like Malcolm's, Dennett's discussion of dreaming shifted the explanatory burden from dreaming to dream reporting, and thus from the experiential hypothesis to the transparency hypothesis. Unlike Malcolm, however, Dennett at times suggests that the two hypotheses might be separable: whether or not dreams are experiences, he believes dream reports to be wholly irrelevant to this question. Again, this raises the challenge of clarifying the exact relationship between the two hypotheses in more detail.

Contrasting Malcolm's and Dennett's versions of skepticism about dreaming is also important for another reason. Just as Malcolmian skepticism about dreaming seems to be a package deal of sorts—insofar as accepting his argument at one point makes it nearly impossible to reject his position at another—Dennett's cassette theory, once acknowledged as a meaningful challenge for empirical research to sort out, quickly outgrows this function. By more or less automatically incorporating almost any imaginable empirical evidence, the cassette theory quickly moves beyond empirical refutation. The upshot is that a fresh assessment of the experiential and transparency hypotheses is required: they are neither conceptual truths nor empirical hypotheses, but rather default assumptions that are methodologically necessary in an empirically informed philosophical analysis of dreaming, but also for empirical dream research itself. From now on, I will speak not of the experiential and transparency *hypotheses* but of the experiential and transparency *assumptions*.

An interesting consequence of this view is that in the end, it turns out that the experiential and transparency assumptions have much the same status that they did in Cartesian dream skepticism: that of default assumptions. The difference, however, is that thanks to Malcolm, the problem of dream reporting is no longer invisible but rather can be made explicit. It now remains to be seen where exactly the concept of dream reporting should be located on the logical map of concepts referring to dreaming. This requires looking critically at how the concepts of sleep and dreaming have changed throughout the history of scientific dream research, which role they play in different theories of dreaming, and which methodological approaches to dreaming and dream reports are currently used.

2 A Short Introduction to Empirical Dream Research: History, Methodology, and Changing Theoretical Conceptions

2.0 Introduction

What is the status of the experiential and transparency assumptions in empirical dream research? If it is true that they have the status of default assumptions and that the problem of dream experience is, in a sense, a pseudoproblem, as suggested at the end of chapter 1, then this should be reflected not only in the methods employed in contemporary dream research but also in changing theoretical conceptions of sleep and dreaming throughout the history of dream research.

This chapter has three main aims. First, by providing a brief overview of the major advances of empirical dream research, it complements my discussion of changes in philosophical theorizing about dreaming from chapter 1. This broad historical perspective places these changes in context and illustrates the way philosophical and scientific theories of dreaming have mutually influenced each other. Second, and for the same reason, this also enables a broader perspective on different methodological approaches and existing controversies in contemporary dream research. The contrast between historical and contemporary conceptions of, and methodological approaches to, sleep and dreaming highlights the historical contingencies and theory dependence not just of past but also of contemporary theories of dreaming, thus setting the stage for a more neutral and unbiased approach. Because these points can best be brought into focus by taking a fairly general and coarse-grained perspective on them, the present chapter does not aim to be comprehensive or complete but only to bring some initial clarity into the discussion. This approach is also in keeping with the chapter's third goal, which is to provide a general introduction for readers who are unfamiliar with empirical dream research. This provides a good starting point for my more detailed and systematic analysis of the methods of contemporary dream research in chapter 3.

2.1 Changing Conceptions of Sleep and Dreaming

The history of theorizing about dreaming is marked by the long-standing assumption that dreaming is not open to third-person observation. In antiquity, dream interpretation and dream incubation (Van de Castle, 1994; Hughes, 2000; Gehring, 2008; cf. sec. 8.2.1) played an important role in medical and religious practice. Dreams were exclusively dealt with through dream reports, but sleep, the "externally observable counterpart to dreaming" (Kroker, 2007, p. 21), played virtually no role in this. Indeed, before the nineteenth century, both sleep dreams and waking hallucinations could be referred to as apparitions or visions (Aleman & Larøi, 2008). Both were classified according to their supposed meaningfulness, whereas their occurrence in sleep or in wakefulness seemed largely irrelevant to their understanding. Kroker (2007) notes that in the absence of any standardized or widely accepted methods for the study of sleep and its relation to dreaming, "knowledge about dreaming amounted to a public accumulation of private experience; there were no attempts to render the experience itself public. There were no empirical codes, save one: the dreamer was the final authority" (p. 20). The analysis of dream reports was the only method for studying dreams, and their transparency was accepted as given.

Aristotle's (350 BCE/1997) treatises on sleep, dreaming, and dream interpretation can be considered as the first attempt to formulate a naturalistic account of sleep and dreaming. They strongly influenced philosophical and psychological theorizing for centuries (Barbera, 2008; Noreika & Windt, 2008). Importantly, however, they also strengthened the view that dreams are unobservable from the third-person perspective.

In *On Dreams*, Aristotle attempts to explain dreaming in physiological terms, linking dream imagery to the residual movements of the sensory organs during sleep. His central argument is that dreams are brought about by the same process as waking perception, but in the absence of appropriate external stimuli. Consequently he thinks that dreams are not mere products of judgment but rather can be described as quasi-perceptual, hallucinatory experiences (cf. sec. 5.1.1). Regarding the external observation of dreams, the crucial point is that, according to Aristotle, sleep itself is defined in negative terms. Aristotle views wakefulness and sleep as contradictory states that depend on the same criterion, that is, the presence or absence of perception (Aristotle 350 BCE/1997, p. 101). Sleep results when what Aristotle calls the primary or central perceptual organ (which he contrasts with the perceptual organs related to the different sensory modalities, i.e., eyes, ears, etc.) temporarily ceases to function; in this respect, sleep is similar to unconsciousness or coma. Sleep, in this view, is a necessary period of rest that often follows eating and physical exertion, and its final cause is to ensure wakefulness, that is, the ability to perceive and think. For Aristotle, sleep is, quite simply, the absence of wakefulness.

In general, defining a phenomenon in purely negative terms is not apt to encourage its systematic observation. But in addition, perception, which for Aristotle is the criterion of wakefulness, is also not directly observable from the third-person perspective. In one sense, perception is a success word, and the correspondence between the phenomenal character of experience and the objects of perception can only be assessed from the third-person perspective: what subjectively seems like a perception could always turn out to be a hallucination or illusion. However, in another sense, perception also involves subjective awareness, and if we want to determine whether someone *heard* a sound or only reacted to it without awareness—that is, if we want to say that perception involves the presence of phenomenal experience—this can only be inferred indirectly through first-person reports and the observation of appropriate behavioral responses to sensory stimulation.

Accordingly, in Aristotle's view, sleep was detected through the absence not of perception itself but of behavioral evidence for perception, that is, through the observation of closed eyelids, motionlessness, lacking responsiveness to sensory stimulation, and so on (for discussion, see Kroker, 2007, p. 31). This dichotomous view of sleep and wakefulness and the resulting negative definition of sleep characterized theoretical approaches to sleep for centuries. The similarity even to Malcolm's view of sound sleep as a uniform state of unconsciousness and of the concept of dreaming as being governed by exclusively first-person criteria is striking.

Aristotle's insistence that animals dream—based on evidence such as dogs barking in their sleep (Kroker, 2007, p. 32)—is contradictory in this context, as such behavior would seem to disrupt the quiescence of sleep. Also, it would seem to suggest that there is a behavioral, third-personally observable criterion of dreaming, after all. In his *History of Animals*, he writes:

> With regard to the sleeping and waking of animals, all creatures that are red-blooded and provided with legs give sensible proof that they go to sleep and that they waken up from sleep; for, as a matter of fact, all animals that are furnished with eyelids shut them up when they go to sleep. Furthermore, it would appear that not only do men dream, but horses also, and dogs, and oxen; aye, and sheep, and goats, and all viviparous quadrupeds; and dogs show their dreaming by barking in their sleep. (Aristotle, 350 BCE/1994–2009c, part 10)

Strangely enough, Aristotle did not believe sleep behavior such as sleep talking to be helpful for the study of human dreaming. In his treatise *On Dreams*, he explicitly states that instances in which persons respond to questions during sleep are not instances of dreaming: they occur in a hybrid state between dreaming and wakefulness, where responsiveness to external stimuli is still possible. Consequently they could still count as actual instances of perception. Dreams, by contrast, occur during sleep in the strict sense of the term and involve the absence of actual instances of perception. Again, the parallel to Malcolm's distinction between wakefulness, sound sleep, and half sleep, where half sleep is a hybrid state between sound sleep and wakefulness, is striking.

The experimental investigation of dreaming was further discouraged by Aristotle's remarks on prophetic dreams. Though generally critical of his contemporaries' belief in prophetic dreams, he did not reject them completely but allowed that they may arise from a residual perception of distant events during sleep. Here the idea was that sleep can enhance a person's sensitivity to subtle or distant stimuli, thus explaining the occurrence of veridical dreams of distant events. He is quick to add, however, that such dreams occur predominately in those whose mind resembles an empty desert (Aristotle 350 BCE/1997, p. 135), thus rendering them more susceptible to subtle movements of the perceptual organs during sleep. As Kroker remarks:

> The social and empirical implications of such a thesis are as unmistakable as they are paradoxical: because those people who pay greatest attention to their dreams are likely dullards, the deranged, or melancholics, their testimony cannot serve as the foundation of any systematic study of dreaming. Yet such a claim also implied that any self-observation of dreams was doomed to failure. A natural history of dreaming was impossible, because such a pursuit could be conducted only by making the observer more like the cowards who experienced fearful dreams or the fools who claimed to see the future. (Kroker, 2007, p. 37)

In sum, as important as Aristotle's work on sleep and dreaming is for the claim that dreams can be understood in naturalistic terms, it also, and somewhat paradoxically, strengthened the view of dreaming as a private, unobservable phenomenon and effectively discouraged attempts to study it scientifically.

The pivotal moment in empirical dream research came in the 1950s with the discovery of the physiological features of REM sleep and their correlation with dreaming (for a general introduction to the history of sleep research, see Pelayo & Guilleminault, 2009). In 1953, Eugene Aserinsky and Nathaniel Kleitman used the electroencephalogram (EEG) and the electrooculogram (EOG) to study attention spans in children and somewhat serendipitously obtained measures of heightened, wakelike brain activity and rapid eye movements during sleep (Aserinsky & Kleitman, 1953, 1955). When they measured similar periods of brain activation and rapid eye movements during adult sleep and were able to elicit dream reports from their participants upon awakening, they hypothesized that rapid eye movement sleep, abbreviated REM sleep, was correlated with dreaming. Their findings were replicated by William Dement and in 1957 led to the joint publication of "The Relation of Eye Movements during Sleep to Dream Activity" (Dement & Kleitman, 1957b). Just two years later, this important research article became the main target of Malcolm's critique in his monograph *Dreaming*. Malcolm's work on dreaming can thus be viewed not just as a direct reaction to these empirical findings but also as an attempt to return to the prescientific assumption, traceable to Aristotle, that dreams are principally unobservable phenomena.

The third cardinal feature of REM sleep, muscular atonia, was discovered in 1959 by French researchers Michel Jouvet and François Michel. REM sleep was now shown to be distinguished from non-REM (NREM) sleep, first, by the occurrence of heightened, wakelike brain activity; second, by the occurrence of rapid eye movements; and third, by muscular atonia or near paralysis. It was also Jouvet who, in 1962, suggested that the periodic occurrence of REM sleep is triggered by the brain stem (for details on the discovery of REM sleep, see Hobson, 1988, pp. 139–154; for a very personal account, see Dement, 1999/2001, pp. 27–50, as well as Jouvet, 1999).

Note how the discovery of REM sleep both introduced a positive criterion for this stage of sleep and shifted the negative Aristotelian conception of sleep to the stages of NREM sleep, which now came to be characterized by the absence of the characteristic features of REM sleep. A predictable result was the almost exclusive focus of scientific sleep research on REM sleep and a comparative neglect of the stages of NREM sleep. Wilse Webb, one of the few early sleep researchers to devote his research efforts to NREM sleep, even founded the Society for the Prevention of Cruelty to Non-REM Sleep in response to this development (see Hobson, 1988, p. 142; see also Webb, 1975).

The discovery of REM sleep was crucial for the evolving understanding of sleep architecture (for details, see Hobson, 1988; Pace-Schott, 2009). Sleep was now seen to be a periodic phenomenon characterized by the cyclic alteration of different sleep stages. The trajectory through these different sleep stages and their respective duration are highly stable across different subjects. NREM or slow-wave sleep, which makes up the larger proportion of sleep, is itself heterogeneous and is commonly divided into four stages. These stages are marked by a progressive set of changes in EEG measures of brain activity, during which the fast, low-voltage waves measured during wakefulness give way to slower, higher-voltage activity. While stage 1 NREM sleep, preceded by drowsiness, is relatively similar to wakefulness in terms of EEG measures, the EEG patterns of stages 2 and 3 are marked by slower and more regular synchronized activity, as well as by the appearance of sleep spindles, a stereotyped waveform associated with the loss of consciousness. Stage 4 is characterized by the predominance of slow-wave, high-voltage activity. The electromyogram (EMG) registers a decrease, but not a total loss, of muscle tone, and the EOG, measuring slow eye movements, is comparatively inactivated. Having reached stage 4, the activation pattern rapidly ascends back through stages 3 and 2 and only then gives way to the characteristic activation pattern of REM sleep (see fig. 2.1).[1]

While this brief description departs considerably from the earlier view that either sleep as a whole or particular sleep stages are uniform periods of quiescence, it is important to note that even this more differentiated classification of sleep stages is both extremely crude and highly idealized.[2] To give a single example, Tanaka et al. (1996) identified nine different EEG stages characterizing the transition from wakefulness to sleep during the hypnagogic (or sleep onset) period, spanning an average of seventeen minutes in their study. According to their scoring system, stages 1–2 correspond to what, according to standard criteria, is classified as wakefulness; EEG stages 3–8 correspond to sleep stage 1; and EEG stage 9 corresponds to sleep stage 2. In other words, the sleep onset period they investigated reaches from wakefulness into sleep stage 2, with sleep stage 1 at its center. They note:

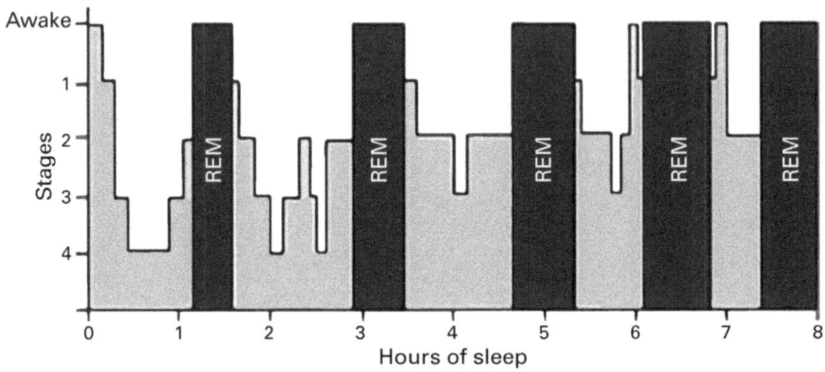

Figure 2.1

All-night sleep cycle: a typical hypnogram from a young, healthy adult. Light-gray areas represent non–rapid eye movement (NREM) sleep. Source: National Institutes of Health, "Sleep, Sleep Disorders, and Biological Rhythms," http://science.education.nih.gov/supplements/nih3/sleep/guide/info-sleep.htm. © 2003 by BSCS. All rights reserved. Used with permission.

> The hypnagogic state is neurophysiologically and psychophysiologically complex, and even paradoxical, and . . . the standard sleep stage criteria, especially for sleep stage 1, are too vague to define when the convergence of behavioral, subjective, and polygraphical measures in the hypnagogic state is being studied. (Tanaka et al., 1996, p. 732)

I discuss the relationship between these nine EEG stages and hypnagogic imagery in more detail in chapter 11; for now, I only want to draw attention to the fact that sleep-stage scoring systems can operate on different levels of granularity and that the relatively coarse-grained distinction of REM and NREM sleep does not necessarily map onto cleanly distinguished, but internally homogeneous sleep stages. Things are further complicated by the fact that behavioral, EEG, and subjective criteria of sleep can also come apart. While longer response times and intermittent response failures place the beginning of "behavioral" sleep at sleep stage 1, sleep spindles as an "objective" marker of sleep characterize sleep stage 2 and even in this state may fail to correspond with subjective judgments of sleep (Tanaka et al., 1996, p. 731). In one study, 60 percent of subjects denied having slept during sleep stage 1; 10 percent denied having slept during sleep stage 2 (with 5 percent claiming they didn't know); and 5 percent denied that they had slept, or said they didn't know, respectively, during sleep stages 3 and 4 (for details, see Webb, 1975). So another issue is how more or less fine-grained EEG scoring systems map onto behavioral criteria and subjective judgments of sleep.[3] This brings me to an important metatheoretical observation. These research findings make it adamantly clear that not only folk concepts such as sound and half sleep but also standard sleep stage scoring criteria are highly idealized tools that should not be confused with the actual phenomena under investigation—even though, ideally, they will reflect them as closely as possible. Obviously, introducing straightforward scoring criteria for sleep and sleep stages is helpful, but it may also obscure the fact that both sleep–wake transitions and transitions between sleep stages are in fact much more complex and in many ways continuous.

Returning to standard sleep-stage scoring, during REM sleep, the EEG once more shows high-frequency, low-voltage activity similar to that seen during wakefulness. The EMG, however, registers a near-complete loss of muscle tone and shows a flat line, and the EOG shows the increased activity associated with the characteristic rapid eye movements. Further characteristics of REM sleep include increased blood pressure, respiratory rate, and pupil diameter, as well as irregular heart rate. Within REM periods, it is common to differentiate between *tonic* or persistent physiological features, such as EEG measures of heightened cortical activation and muscle atonia, and *phasic* or episodic signs, such as rapid eye movements and muscle twitches registered on the EMG. Throughout the REM period, the sleeping position, due to muscular atonia, typically remains stable, but the beginning and the end of REM sleep are often marked by a change in sleeping position. REM sleep may be followed by a brief awakening or a direct return to NREM sleep.

In humans, the duration of each sleep cycle (consisting of the descent from wakefulness through the stages of NREM sleep and back to REM sleep) is relatively stable at about ninety minutes, but the proportionate length of the different sleep stages per cycle varies throughout the night. Specifically, the duration of REM periods, which make up approximately 25 percent of total sleep time in adults, increases throughout the night, ranging from just a few minutes during the first REM period to thirty to fifty minutes in early morning sleep (Siegel, 2002, p. 108). By contrast, the duration of stage 4 NREM sleep decreases throughout the night and disappears entirely from later sleep cycles (Hobson, 1988, p. 148).

Both the total amount of sleep and the proportion of REM sleep change with age. In newborns, REM sleep makes up about 50 percent of total sleep time, amounting to approximately eight hours of REM sleep per day. Adults spend approximately 1.5 to 2 hours per night in REM sleep, and both total sleep time and the proportion of REM sleep continue to decrease with age (Hobson, 2002, pp. 77–79),[4] though the decrease of REM sleep is minimal in adults (Floyd et al., 2007). In older adults, changes in circadian rhythm increase the likelihood of insomnia, and a number of other sleep disorders also appear predominately in older adults (Rissling & Ancoli-Israel, 2009). Dream recall may be modulated, in part, by circadian influences, and these might partly explain age-related changes in dream recall (Wamsley et al., 2007; Chellappa et al., 2009; Chellappa & Cajochen, 2013). Adolescence is also marked by a number of profound changes in sleep architecture, such as the reduction of the quantity of NREM sleep stages 3 and 4 and the timing of sleep (Tarokh & Carskadon, 2009).

Experimental awakenings in the sleep laboratory generally show a relatively close quantitative and qualitative correlation between reports of previous mentation and the physiological measures of the different sleep stages. While awakenings from stage 1 NREM sleep often yield dreamlike reports, reports from the other stages of NREM sleep often describe more thoughtlike, confused, nonprogressive, and repetitive mentation. Awakenings from REM sleep, on the other hand, tend to yield more elaborate, vivid, and emotionally intense reports. Typically, REM reports include multimodal imagery and tend to be both longer and more complex in storylike organization (see Hobson, Pace-Schott, & Stickgold, 2000, pp. 799–803; Antrobus & Wamsley, 2009). The existence of qualitative differences between reports from REM and NREM sleep has been confirmed by a number of studies (e.g., Hobson et al., 2000, p. 800; McNamara et al., 2010). Moreover, whereas 81.9 percent of awakenings from REM sleep yield mentation reports, the incidence of reports following NREM awakenings lies at only 43 percent (Nielsen, 2000a, p. 855). Though not exhaustive, both the quantitative and the qualitative differences between more thoughtlike reports following NREM awakenings and more dreamlike reports from REM sleep make dreaming appear "physically diagnosable" (Hobson, 1988, p. 154). "While recording the state of the brain—by means of brain wave, eye movement, and muscle tone—does not reveal the state of the mind, it can predict mental state with high statistical confidence" (p. 143).

It should be noted, however, that the relationship between dreaming and REM sleep, on the one hand, and less frequent reports of more thoughtlike mentation from NREM sleep, on the other, remains controversial. While most contemporary dream researchers allow that dreaming can also occur during NREM sleep, many hold that REM dreams are typically more complex and perceptually and emotionally more vivid than NREM dreams (e.g., Hobson et al., 2000; Gottesmann, 2006). Others allow that even narratively complex and vivid dreams occur during NREM sleep and object to any strong or exclusive association between dreaming and REM sleep (e.g., Antrobus, 1990; Foulkes, 1993b; Solms, 1997, 2000; Domhoff, 2003). A concessive approach is proposed by Nielsen (2000a; see sec. 2.2 for details). He points out that the controversy as to whether and to what extent dreaming occurs outside REM sleep is in part conceptual: there is currently no standardized and widely accepted definition of dreaming (Pagel et al., 2001), and researchers who regard dreams as involving any conscious mental activity in sleep will tend to find more instances of dreaming during NREM sleep than those reserving the term for specific types of experience during sleep. As Nielsen writes, "Differences in definitions of 'cognitive activity' and/or 'dreaming' presumably account for much of the variability in levels of mentation recall from REM and NREM sleep that has been observed in previous studies" (Nielsen, 2000a, p. 853). However, over and above this lack of conceptual consistency, there is also a controversy as to which types of experience are supported by the different stages of sleep. Consequently the putative correlation between dreaming and REM sleep should be treated with caution.[5]

While the discovery of REM sleep is commonly identified as the beginning of systematic scientific dream research, it is interesting to note that many of the physiological characteristics of dreaming were known long before the discovery of REM sleep, and attempts to systematically study dreams were well under way in the nineteenth century. Though the influence of psychoanalysis eventually pointed popular dream theory in a different direction, the development of experimental psychology in the late nineteenth century led to a surge of physiologically minded dream theories. By the beginning of the twentieth century, dreaming had already been observed to occur in certain sleep periods, and tentative observations suggested a link with periodic body movements and changes in heart rate and respiration (Hobson, 1988, pp. 70–73).

Most nineteenth-century researchers believed dreams to be caused by peripheral bodily sensations or stimuli in the sleeper's environment (see chap. 8 for details; for details on the history of dream research, see Hobson, 1988). However, there were exceptions. Sante de Sanctis (1901), otherwise best known for his theory on animal dreaming (Hobson, 1988), speculated that there might be an internal source of dreaming, as the "dream image . . . must correspond to a heightened and particularly rich level of cerebral activity" (de Sanctis, 1901, p. 148; translation mine).[6] Wilbrand and Charcot observed the cessation of dreaming in general and of visual dreaming in particular to be related to lesions in the occipito-temporal area (Wilbrand, 1887; Charcot, 1883; see also Solms, 1997, 2011; for details on the cessation of dreaming, see sec. 3.6). Several early authors (Raehlmann & Witkowski, 1877, 1878; Ladd, 1892; for discussion, see Gottesmann, 2007) hypothesized that dreaming might be related to periods of eye movement activity in sleep. In sum, many aspects of contemporary dream research, such as its focus on the statistical analysis of subjective dream content, the systematic collection of dream reports, and the interest in the physiological basis of dreaming, were already anticipated by nineteenth-century researchers (for details and discussion, see Schwartz, 2000). Methodologically, most of these early theories were based on the observation of sleeping adults and children, as well as on collections of dream reports elicited by awakening participants throughout the night.

At the same time, systematic self-observation, often including tentative experiments concerning the influence of external stimuli on dreaming, became an important and increasingly popular method of dream research and led to the publication of various dream collections and journals. These publications often stemmed from a scientific interest in dreams, and their authors specialized in the fields of medicine, psychology, and even philosophy. Prominent examples are Havelock Ellis (1911/1931), Alfred Maury (1861), Ernst Meumann (1907), and John Mourly Vold (1910/1912). However, there were also many laymen like the French aristocrat and sinologist Hervey de Saint-Denys (1867) and laywomen like Mary Arnold-Forster (1921) and the German poet Isolde Kurz (1919), whose interest in dreaming was of a more personal or even literary nature and who considered the study of dreams mainly as a pastime. These dream collections owe much of their credibility, and much of their charm, to the open-mindedness with which their authors approached the topic, and many of these early dream researchers explicitly distanced themselves from popular contemporary theories such as psychoanalysis. Moreover, these collections present a rich reservoir of interesting source material.

Importantly, many of these pioneers of dream research went beyond merely chronicling spontaneously recalled dreams. Alfred Maury collected dream reports following systematic awakenings to investigate the influence of external stimuli on dreams (Maury, 1861). Mary Calkins (1893) and Weed et al. (1896) conducted a systematic analysis of dream reports to determine the frequency of imagery in the different sensory modalities in dreams, and the results turned out to be quite similar to those of more recent studies (McCarley & Hoffman, 1981; Strauch & Meier, 1996; for discussion, see Schwartz, 2000). Several oneironauts developed the ability to induce lucid dreams and even to control the content of their dreams at will, allowing them to perform "experiments" during the dream state (e.g., Hervey de Saint-Denys, 1867; Mary Arnold-Forster, 1921; Frederik van Eeden, 1913; see sec. 9.3 for details). The term *lucid dreaming* itself was introduced by Frederik van Eeden (1913) in a paper proposing a taxonomy of different types of dreams.

The similarities between nineteenth-century and contemporary dream research prompted Sophie Schwartz (2000) to postulate a loop of one hundred years in dream research. She notes that the development of nineteenth-century dream research turned in a completely different direction after the publication of Freud's (1899/2003) *Interpretation of Dreams*, namely, toward the view that dreaming is the result of unconscious and disguised wishes. This altered view of the origins of dreaming also led to a methodological focus on free association as a method of dream interpretation. This, as well as the advent of behaviorism, meant that the project of a scientific investigation of dreaming was largely neglected until Aserinsky and Kleitman's (1953) discovery of the correlation between dreaming and REM sleep. Only then did a majority of dream researchers return to more scientifically minded approaches to the physiological correlates of dreaming.

This development notwithstanding, several researchers in the 1920s, 1930s, and 1940s obtained measurements of what later came to be considered as REM sleep in humans and animals (for detailed discussion and further references, see Gottesmann, 2007; Kroker, 2007). The periodic occurrence of wakelike EEG patterns of brain activity throughout the night was already described by Hans Berger in 1929. He even observed that the duration of these activation periods increased throughout the night, which is consonant with the lengthening of REM periods after several hours of sleep. However, he believed these measurements to be associated with intermittent periods of wakefulness, and because he did not observe his participants' sleep directly, the truth was not discovered for another thirty years (Siegel, 2002, p. 107). In keeping with the contemporary view of genuine sleep as a more or less uniform period of reduced brain activity, researchers in the 1930s and 1940s concentrated on what later came to be called NREM sleep and thus failed to discover the relationship between dreaming and REM sleep. As Davis, one of the early EEG researchers, later wrote:

> We let a big one get away that time, although the sheets from the chronograph, laid out on the ping-pong table in the game room, revealed at a glance the periods of slow-wave sleep and the periods of low-voltage sleep. We simply set the latter aside as uninteresting,—a return to light sleep. It was more fun to work with floating ["lapses" of consciousness before sleep onset] and K complexes, which could be obtained very reliably on the drum and try to make something happen in the late low-voltage stage. . . . We saw REM but didn't identify it as a separate state. We worked where the action was. (quoted in Kroker, 2007, p. 304)

And the action, somewhat paradoxically, was believed to occur in deep, slow-wave sleep. In fact, until the discovery of REM sleep in the 1950s, EEG measures of fast, low-voltage brain activity remained so strongly associated with wakefulness that dreaming was initially believed to occur during light sleep or during the transitional state between sleep and wakefulness. Indeed, this association was so strong that it prevented researchers from seeing, quite literally, what their own measurements indicated. Consequently, as dreams became associated with heightened periods of brain activation, their putative location in sleep shifted from deep sleep, as in Aristotelian dream theory, to half sleep and sleep–wake transitions. In this latter view, dreaming is no longer a uniquely sleep-related phenomenon. Although Jacobson (1930), in what Gottesmann calls "the first objective result related to dreaming obtained by using modern methodology" (Gottesmann, 2007, p. 4), had even recorded eye movements during sleep and tentatively related them to dreaming, the general invisibility of REMs persisted (see Kroker, 2007, pp. 303–306; for details on Jacobson, see Kroker, 2007, pp. 237–254). At the same time, the "passive depiction of sleep left dreaming as unworthy of study and ultimately extended the view that sleep was as it was experienced—as nothing. No amount of eye movements observed in the night could change this" (Kroker, 2007, p. 256).

Note that Malcolm's view of dreaming in fact presents a compromise between the Aristotelian view and this newer understanding of dreams: as in Aristotelian dream theory, Malcolm believes dreaming proper to be associated with deep or sound sleep, which in turn is the opposite of wakefulness. For Malcolm, unlike for Aristotle, however, this also meant that dreaming was the opposite of conscious experience: dreams, though located in deep sleep, were precisely *not* conscious experiences. At the same time, Malcolm's view of dreaming shares the view of early dream researchers that such experiences as do occur during sleep occur during half sleep. The difference is merely terminological: for Malcolm, such half-sleep experiences were not proper dreams, whereas for early dream researchers, they were. Nonetheless it is fascinating to note how this particular aspect of Malcolm's view of dreaming parallels, rather than runs contrary to, the predominant view of early dream researchers.

Be that as it may, the "paradigm . . . of dreaming as half-waking, half-sleeping, persisted" (Jouvet, 1999, p. 5) for some time after Aserinsky and Kleitman's seminal work on dreaming. Only gradually did researchers come to embrace the conceptualization of REM sleep as a genuine and unique sleep stage rather than a hybrid state between sleep and wakefulness.[7] This development once more led to a profound change in theoretical thinking about sleep, or what Hobson describes as a "classical example of a paradigm shift" (Hobson, 1988, p. 150). Jouvet writes that the findings from the late 1950s showed that REM sleep, or what he calls paradoxical sleep, was "neither sleeping nor waking. It was obviously a *third state* of the brain, as different from sleep as sleep is from wakefulness" (Jouvet, 1999, p. 5). The paradoxical nature of this discovery lies not only in the fact that sleep could no longer be considered as a unitary phenomenon. Jouvet's theory that the reticular formation, a set of neurons in the brain stem, was the control mechanism for REM sleep also proved that sleep was not, as previously thought, simply a state of passivity. Instead sleep was shown to be an active phenomenon, dependent on the self-activation of the brain during sleep (p. 6).

Because the reticular formation was known to govern cortical activation and posture (via the spinal cord), its activation was linked to awareness and behavioral arousal during wakefulness. Consequently the reticular formation was initially believed to be inhibited during sleep (Hobson, 1988, pp. 127–131). Dennett still explicitly emphasized the connection between reticular activation and awareness in his 1976 article on dreaming. However, he falsely assumed the reticular system to be inhibited throughout sleep and argued that this presented evidence *against* the received view of dreams as experiences:

> One routinely recognized condition for having an experience is that one be conscious, or awake, and dreamers are not. A well-confirmed physiological condition for this is that one's reticular activating system must be "on," which it is not during sleep. The fact that one is in a sound sleep goes a long way to confirming that one is *not* having experiences, as *ordinarily* understood. . . . Lack of reticular system activity strongly suggests that nothing is being experienced during REMs. (Dennett, 1976, p. 163)

However, Jouvet's findings had already suggested the contrary, long before Dennett formulated the cassette theory: not only are the psychophysiological changes characteristic of REM sleep mediated by the brain stem, but the reticular system is actually disinhibited during REM sleep. If the activation of the reticular system is a physiological condition for the occurrence of conscious experience, as suggested by Dennett, then the finding that the same control mechanism governs arousal during wakefulness and REM sleep strengthens, rather than undermines, the received view of dreaming. Likewise, Jouvet suggests that activation of the reticular system in REM sleep, in conjunction with muscular atonia, helps solve the paradox of REM sleep because it provides a possible physiological explanation for the occurrence of conscious experience in a state of outward behavioral passivity and lack of responsiveness (Jouvet, 1999, chap. 1; for details on the role of the reticular activating system in sleep, see Fuller & Lu, 2009b; Garcia-Rill, 2009). The idea is that the activation of the reticular formation in sleep may play a role in providing the internal conditions for the occurrence of conscious experience, while muscular atonia explains why the outward signs of conscious experience are absent during REM sleep/dreaming.

Importantly, the attempt to map the occurrence of dreams onto the physiological characteristics of REM sleep suggests, at least superficially, a view according to which dreaming can be identified with REM sleep. In their initial publication on the correlation between this newly discovered period of active sleep and dreaming, Aserinsky and Kleitman optimistically proclaimed that their method of eliciting dream reports following timed awakenings in the sleep laboratory furnished "the means of determining the incidence and duration of periods of dreaming" (Aserinsky and Kleitman, 1953, p. 274). Similarly, Dement and Kleitman noted:

> An objective measurement of dreaming may be accomplished by recording REM's during sleep. This stands in marked contrast to the forgetting, distortion, and other factors that are involved in the reliance on the subjective recall of dreams. It thus becomes possible to objectively study the effect on dreaming of environmental changes, psychological stress, drug administration, and a variety of other factors and influences. (Dement & Kleitman, 1957b, p. 346)

By finding an objective method for the study of dreaming, researchers were apparently now able to surpass the presumably unreliable method of collecting dream reports and move on to the truly scientific study of dreaming. The situation, then, really is extremely close to Malcolm's scenario of researchers hoping to inform people that they have been dreaming—but, thanks to the close correlation between physiological measures of sleep and subjective reports of dreaming, it seemed, initially, to be based on compelling evidence.

I will return to a more careful analysis of why the notion of conducting dream research independently of dream reports is, in fact, deeply problematic. For now, it is more important to note how the concepts of sleep and dreaming changed in response to the budding field of dream research. In particular, it seems that, to be recognized as an empirically investigable and objectively measurable phenomenon, dreaming first had to become decoupled from deep sleep. Researchers' views of sleep were such that the occurrence of heightened periods of brain activation and their correlation with dreaming initially seemed to speak against, rather than for, the occurrence of dreams during genuine sleep. As in Malcolm's theory, experiential status hinged on proximity to wakefulness. Only when dreaming had been assigned its own sleep stage, namely, REM sleep, and thus had been restricted to certain portions of sleep, was dreaming once more conceived of both as a genuine sleep phenomenon and as one that was objectively measurable. For this latter step to occur, however, the concept of sleep itself had to undergo a profound change, and the former conception of sleep as a unitary state of passivity and rest had to be replaced by a more complex view of sleep as involving different stages of sleep. The continued debate about dreams and NREM sleep suggests that the concept of dreaming may once more be set adrift from its confinement to particular sleep stages, and that with this development, the occurrence of experience in all sleep stages is becoming acceptable. As Schwartz and colleagues aptly put it, "REM sleep is not a necessary, but a facilitating condition for dreaming to occur. Conversely, there is little doubt that dreaming was a necessary condition for REM sleep to become famous" (Schwartz et al., 2005, p. 431). Today, attempts are under way to identify the neural correlates of dreaming independently of sleep stages (Noreika et al. 2009; Siclari et al. 2013, 2014). Here, the basic idea is that progress might be made by identifying a sleep stage with roughly equal proportions of dreaming and nondreaming (as evidenced by retrospective recall). For this particular project, NREM sleep (and in particular stages 2 and 3) are more viable than REM sleep, with its high prevalence of dream reports. Ironically, it thus seems that NREM sleep may be taking center stage in the attempt to identify the neural correlates of dreaming. And even more ironically, because this method enables a no-task, within-state contrast between conscious and unconscious states, identifying the correlates of dreaming may be a step towards identifying the neural correlates of consciousness itself (Siclari et al., 2014; Noreika, 2015). If so, then long-neglected NREM sleep, in yet another interesting historical twist, might yet play a central role in consciousness research.

Finally, note that Aserinsky and Kleitman's initial findings were at least as important for the development of the scientific investigation of sleep as they were for dreaming. The beginning of scientific dream research coincides with the beginning of scientific sleep research. Kroker states:

> Dreaming has in fact played a pivotal and diverse role in sleep's history, particularly regarding the issue of sleep's function. For it was through the problem of dreaming, not of sleep, that psychologists and psychiatrists first took an interest in understanding night-life in terms of function. (Kroker, 2007, p. 6)

In recent years, however, scientific interest in dreaming has been surpassed by a much larger body of research devoted to sleep and its relation to cognitive processes such as memory consolidation, as well as sleep disorders (Kroker, 2007). This comparative marginalization of dream research may in part result from a frustration of early optimism regarding the scientific investigability of dreams via the study of REM sleep.[8] However, overcoming this paradigm, as noted above, also holds promise, and it is possible that real progress will require yet another profound relocation of dreaming within sleep.

2.2 The Refinement of Methods and Proliferation of Rival Theories

The early EEG studies of REM sleep initiated a rapid expansion of empirical dream research in the 1960s and 1970s (Kroker, 2007, p. 327). This initial enthusiasm did not, however, persist. No significant new discoveries ensued, and several initially promising lines of research did not yield the hoped-for results. These included the attempt to correlate reports of visual imagery in dreams with changes in the direction of eye movements during REM sleep, the failure to explain REM rebound following REM deprivation in terms of an increased pressure for dreaming,[9] and the increased blurring of the distinction between dreamlike REM mentation and more thoughtlike reports from NREM sleep (Kroker, 2007, pp. 326–328). The failure to find a sleep-stage or neural correlate that would enable researchers to detect the occurrence of dreaming independently of dream reports continues to present a challenge for dream research and may explain why dream research has once more been marginalized. Thus, after a brief period in which the investigation of dreaming and REM sleep had been united, "by the mid 1970s, they had begun to separate again, as the sleep clinic began to take shape and biomedical interest in establishing sleep as a public-health problem trumped engagement with the idea of using laboratories to comprehend dreaming" (Kroker, 2007, p. 7).

Today, though the disciplines of dream and sleep research continue to overlap, they are largely separate fields, with separate associations (i.e., the International Association for the Study of Dreaming and various national and international associations and societies devoted to sleep and sleep disorders) and separate journals (e.g., *Dreaming* and the *International Journal of Dream Research*, on the one hand, and *Sleep*, *Sleep Medicine*, and many more, on the other). It is common for researchers to study dreaming outside the laboratory and based solely on dream reports—a practice that cannot straightforwardly be considered as sleep research proper (cf. sec. 3.3). Even more frequently, sleep research, including research on the functions of sleep and its relationship to memory processing, as well as neuroimaging studies of sleep stages, is conducted without eliciting dream reports and investigating the relevance of results for dreaming (cf. secs. 3.2, 6.2.2).

Nonetheless, important progress has been made in dream research. The publication of Hall and Van de Castle's *The Content Analysis of Dreams* in 1966 introduced a classification system for the content analysis of dream reports. Hall and Van de Castle proposed nine different categories for scoring the content of dream reports: (1) setting and objects; (2) characters; (3) aggressive, friendly, and sexual interactions; (4) activities; (5) success, failure, misfortune, and good fortune; (6) emotions; (7) modifiers (defined as "any adjective, adverb, or phrase that is used for descriptive elaboration," such as *color*, *size*, or *age* [Hall & Van de Castle, 1966, p. 116]); (8) temporal (references to time units), negative (use of negative words such as *no*, *not*, or *none*), and oral references; and (9) castration. To this day, the Hall and Van de Castle scoring system is widely used and is considered the standard method of content analysis (Domhoff, 1996). In a field otherwise characterized by its controversies and lack of consistent methods and definitions, this is an important exception and ensures the comparability of research results. The original book is out of print, but the central ideas and methods were once more made available to the public in Domhoff's (1996) *Finding Meaning in Dreams*, which also gives an overview of the main studies of dream reports conducted since the publication of Hall and Van de Castle's book in 1966 (see also Domhoff, 2003). Aside from developing their own classification system, Hall and Van de Castle discussed alternative scoring methods, thereby providing an introduction to the main results and methodologies in the field. Such an overview can also be found in Winget and Kramer's (1979) extensive documentation of existing scales and research results. Taken together, these three books provide a comprehensive introduction to dream content analysis, discussing both methodological questions and making hard-to-access research results available.

Hobson's formal analysis is an alternative approach. Dream form is distinguished from dream content by targeting the supposedly universal aspects of dream experience rather than the specific and more idiosyncratic contents of individual reports. According to Hobson (1988; Hobson et al., 2000, p. 799), the major formal features of dreaming are (1) its sensory aspects, marked by the predominance of vision over the other sensory modalities; (2) the frequency of motor imagery; (3) disorientation and discontinuity of persons, times, and places; (4) changes in mnemonic functioning both during and after dreaming, including familiar dream amnesia after awakening; (5) delusional belief in the reality of dream events and loss of critical insight; (6) the thematic unity or narrative structure of dreams; and (7) the intensely emotional character of dreams. To give an example, rather than categorizing the types of characters and objects encountered in dreams, as done in content analysis, the formal approach asks whether these characters and objects were merely seen or also perceived through other sensory modalities, whether they were continuously present throughout the dream or suddenly changed, and so on (for an example of formal analysis using several of his own dreams, see Hobson, 2005).

In Hobson's theory, the formal analysis of dreaming is related to three important assumptions. The first is that there are systematic phenomenological differences between wakefulness, NREM sleep, and REM sleep on the levels of multimodal imagery, cognition, and emotion rather than thematic content (Hobson, 1988, pp. 229–231; Hobson et al., 2000, p. 799). The second is that the formal aspects of dreaming correspond to specific patterns of brain activity that are unique to REM sleep:

> For us, the state-specific changes in brain function virtually guarantee concomitant changes in mental function. . . . With respect, we suggest that the failure to demonstrate psychological differences concomitant with physiological ones must be laid at the door of inadequate psychological methodology. (Hobson et al., 2000, p. 804)

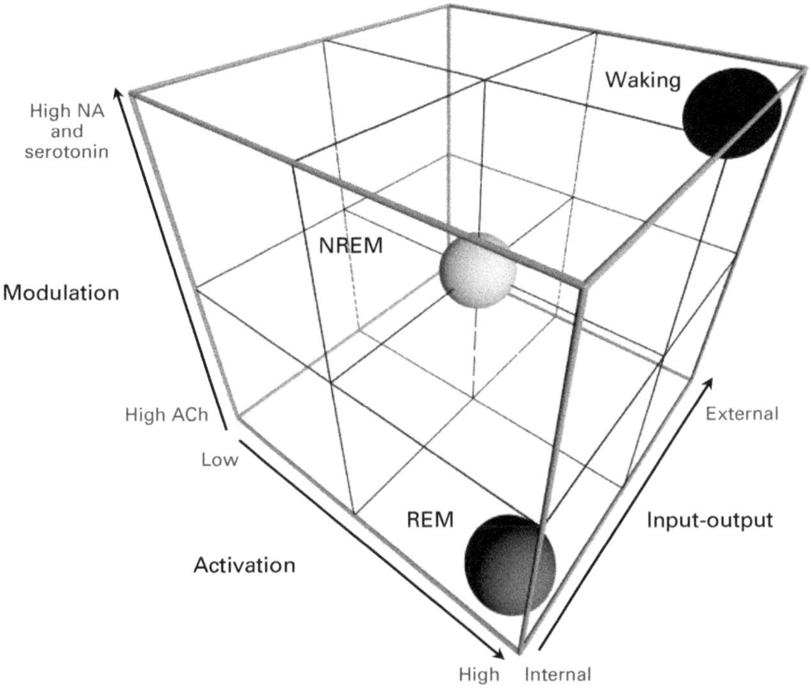

Figure 2.2
The AIM state space model: the transition from waking to NREM and REM sleep. Figure by David Baßler; reprinted from Windt et al. (2013), with friendly permission from Metzler; adapted from Hobson (2009), with friendly permission from Macmillan Publishers Ltd.; see also Hobson et al. (2000).

In keeping with both of these assumptions, Hobson suggests that the differences between waking, REM sleep, and NREM sleep can be explained by mapping them onto a three-dimensional state space. The *AIM model* proposes three parameters—activation (A), information flow (I), and the mode of information processing (M)—for describing state-specific changes accompanying the trajectory from wakefulness through sleep on the phenomenological and neurophysiological levels of description (Hobson et al., 2000, p. 794; see fig. 2.2; see also Hobson, 2009, Hobson 2014). The AIM model is not only the most widely discussed and best-known model of dream generation but is also important because it proposes an integrated theory of consciousness across sleep–wake states.[10] Given that dream research continues to be a fairly marginalized field, this in itself is an important contribution.

The AIM model is an elaborated and updated version of the "activation synthesis" model proposed by Hobson and McCarley in 1977. This earlier model viewed dreaming as the product of the auto-activation of the "brain-mind" during REM sleep, initiated by the random, bottom-up, brain-stem-driven generation of internal signals related to the so-called PGO waves. PGO waves, which designate a characteristic EEG pattern, are bursts of phasic activity commonly associated with heightened levels of recall and particularly intense dream reports (for details, see Steriade, 2009). Automatic signals generated in the pontine brain stem (P) are conducted to the lateral geniculate nucleus of the thalamus (G) and the occipital cortex (O), resulting in patterns of heightened activity in the oculomotor, vestibular, and visual regions of the brain that are believed to underlie the multimodal hallucinatory and emotional quality of dream imagery (see sec. 5.2.2 for discussion). As Steriade (2009, p. 28) puts it, the PGO waves may "represent 'the stuff dreams are made of.'"

More specifically, according to Hobson, the multimodal quality of dreaming arises because during REM sleep, the brain processes internally generated, brain-stem driven signals as if they were the result of external sensory input. These randomly generated signals are subsequently synthesized into a dream narrative by the forebrain. The phenomenally experienced dream is the product of a synthetic, "meaning-added process" (Hobson, 1988, p. 218), resulting from the ongoing attempts of cortical structures to impose coherence on random, brain-stem-driven information input. Ultimately, the process of dream construction is a type of "physiological Rorschach test," resulting from the combination of a "chaotic, bottom-up activation process and its resultant semi-coherent, top-down synthetic process" (Hobson et al., 2000, p. 823). This stands in stark contrast to Freudian dream theory, with its emphasis on a psychologically meaningful process of dream generation. It should be noted, however, that Hobson does not deny that dreams can have meaning and can reflect the personality and concerns of the dreamer. He just thinks that their meaning is transparent and immediately obvious to the dreamer, rather than requiring an elaborate process of interpretation guided by an expert (Hobson, 1988; 2002; 2005).

The AIM model preserves the emphasis of the earlier activation-synthesis model on the factors of brain activation and the internal source of signal generation but adds the dimension of neuromodulation. One of its key claims is that wakelike brain activation, in itself, is insufficient to explain the phenomenology of dreaming. In particular, the characteristic bizarreness of dreams results not simply from the lack of stabilizing sensory input but from a change in the way this information is processed. As overall measures of brain activation approach waking values during REM sleep, the quantity of information input, having reached a minimal level during NREM sleep, increases and shifts to an internal information source. At the same time, a shift in neuromodulation takes place. Whereas wakefulness is characterized by a clear predominance of aminergic (noradrenergic and serotonergic) activity, the ratio shifts to cholinergic predominance and a cessation of aminergic activity during REM sleep (for details, see Jones, 2009a, 2009b). The *reciprocal interaction hypothesis* states that the cholinergic mechanisms that are essential to the generation of REM sleep cannot unfold unless the serotonergic and noradrenergic systems are inhibited (Hobson et al., 2000, p. 816). On the cognitive level, aminergic activity during wakefulness is associated with deliberate attention and volition, whereas "cholinergic activation without aminergic restraint leads to the fictive, illogical, and episodically emotionally charged life of dreams" (Kahn et al., 1997, p. 25). The characteristic cognitive deficiency of dreaming (including the lack of attentiveness, self-reflective awareness, and deficient short-term memory) thus is attributed to a change in neuromodulatory balance (Hobson et al., 2000, pp. 832–833; Hobson & Pace-Schott, 2002).

This latter assumption turns the familiar analogy between dreaming and insanity into a viable scientific hypothesis. Indeed, the third major assumption of Hobson's theory is that dreaming, in terms of both its formal features and the underlying changes in neurochemical functioning, is "a model of psychopathology, a normal functional psychosis whose understanding could provide a key to major psychiatric dysfunctions" (Hobson, 1988, p. 229). In *Dreaming as Delirium*, Hobson (1999) explores the relationship between dreaming and organic mental syndromes in detail and concludes that research on the physiological correlates of dreaming has direct implications for understanding the genesis of pathological psychosis: "dreaming . . . is not *like* delirium. It *is* delirium. Dreaming is not a *model* of a psychosis. It *is* a psychosis. It's just a healthy one" (Hobson, 1999, p. 44). The analogy between the "normal delirium" of dreaming and abnormal or pathological delirium goes beyond their similar formal features, predicting that these states "always have . . . the same kind of cause. . . . The common cause of normal and abnormal delirium is a sudden shift in the balance of brain chemicals" (Hobson, 1999, p. 62; for a more detailed discussion of the relationship between dreaming and delusional wake states, see chap. 10; for a critique of the view that dreaming can be used as a model of psychosis, see Windt & Noreika, 2011).

Due to its concentration on the central differences between conscious states, the AIM model focuses on the stereotypical formal features of dreaming. But it also integrates a number of dissociated or hybrid states, such as hypnagogic imagery experienced at sleep onset, the enactment of dream behavior in RBD (REM sleep behavior disorder; cf. sec. 3.4), and lucid dreaming (Hobson et al., 2000, pp. 838–841).[11] Hobson and colleagues concede that it "may be more useful to envisage psychophysiological continua, manifested at the levels of both the brain and the mind, whose various combinations define not only commonly experienced states of the brain-mind but uncommon ones as well" (p. 842). A general worry, however, is that the model's reliance on the fairly schematic differentiation of three canonical conscious states may be too coarse-grained to capture many of the subtle differences within these states, as well as the gradual transitions between them.

Dreaming, as argued perhaps most emphatically in Hunt's (1989) monograph *The Multiplicity of Dreaming*, is a highly variable process. A general, stereotyped account of the formal features of dreaming is useful to highlight the differences between dreaming and wakefulness, but a more fine-grained description might ultimately be necessary to enable a deeper understanding of the target phenomenon. This is particularly true for dreams that depart from the stereotyped profile suggested by the AIM model—either by occurring outside REM sleep or by lacking bizarre, emotion-laden dream content or cognitive and mnemonic deficiencies. As will become clear in subsequent chapters, an idealized model of dreaming of the type implicit in the AIM model is theoretically fruitful, but only to a point. In particular, it brings the danger of distracting attention from subtle variations in the phenomenology of dreaming and inviting an oversimplified theoretical account (for discussion, see Windt & Noreika, 2011; and see chap. 11).

Mark Solms's (2000) theory of dream generation is probably the most important rival theory to Hobson's and contradicts several of its key claims (for a general discussion, see Domhoff, 2005). In particular, Solms objects to the claim that there is a tight link between the formal features of dreaming and the physiological and neurochemical characteristics of REM sleep, as well as to the resulting assumption that REM sleep dreams can be distinguished qualitatively from more thoughtlike mentation arising during NREM sleep. Solms argues that there is in fact a double dissociation between dreaming and REM sleep, with hallucinatory and narratively complex dreams sometimes being reported following NREM sleep awakenings, and REM sleep awakenings sometimes failing to yield such reports. Though REM sleep awakenings may also fail to elicit dream reports in healthy subjects (Pagel, 2003), this dissociation is particularly pronounced following lesions in the focal forebrain areas, which may lead to a complete loss of dream recall even though REM sleep is preserved. This selective loss of dreaming in general or of visual dreaming in particular is referred to as the Charcot-Wilbrand syndrome (cf. sec. 3.6).

Solms's key claim is that dreaming is independent of the physiological conditions of REM sleep and is triggered by a further mechanism involving the dopaminergic circuits of the ventromesial forebrain (Solms, 2000, p. 849; 2011; for details on the role of the dopaminergic system for sleep regulation, see Fuller & Lu, 2009a). Based on an extensive review of the historical literature on the loss of dream recall following lesions, Solms argues that dreaming is generated by a specific network of forebrain mechanisms and preserved following pontine brain stem lesions. He cites various independent studies reporting a cessation of dreaming following focal forebrain lesions and extensively discusses the early findings by Charcot and Wilbrand on the global or partial cessation of dreaming. Critically, the brain stem was not affected by these lesions, nor was the loss of dreaming accompanied by alterations in the sleep cycle (Solms, 2000, p. 846).[12] While this suggests that REM sleep can occur without dreaming, Solms also claims that forebrain mechanisms can actively generate dreaming outside REM sleep. For instance, vividly experienced recurring nightmares may be triggered by epileptic forebrain seizures during NREM sleep (p. 847). According to Solms, the high prevalence of dreaming in REM sleep "merely reflects the fact that it [REM sleep] is a regular and persistent source of cerebral activation during sleep" (p. 849).

Solms sees dreaming as the product of highly specific patterns of regional forebrain activity, arguing that the parieto-temporo-occipital (PTO) junction, associated with mental imagery in wakefulness, may be responsible for the generation of multimodal hallucinations during sleep. This would indicate that dreaming is linked to visuospatial cognition rather than perception (see sec. 6.2.3) and that "the generation of dream imagery . . . is actively constructed through complex cognitive processes" (Solms, 2000, p. 848). This hypothesis is supported by other studies showing that patients who suffered a selective loss of imagery abilities in wakefulness following neurological damage also reported an impairment or loss of dreaming (Kerr, 1993).

In sum, a central difference between Solms's and Hobson's theories of dream generation is that Solms denies the central role of the brain stem as a dream-state generator, instead claiming that dreaming depends on complex patterns of forebrain activity, particularly at the PTO junction. He also claims that the mesocortical-mesolimbic dopamine system (Solms, 1997, 2000, 2002), rather than the shift from aminergic predominance in wakefulness to cholinergic predominance in REM sleep, is central to dreaming. In doing so, Solms not only questions the central claims of Hobson's theory but also tries to show that dreaming and REM sleep depend on different mechanisms and can be dissociated from each other. At the same time, Solms's project of formulating a neuropsychoanalytical theory of dreaming attempts to integrate neuroscientific findings based on a review of historical case studies of patients with brain lesions with the Freudian view of dreaming as a form of wish fulfillment (see Solms, 1997, 2000, 2011). This interest in preserving certain elements of Freudian dream theory has contributed considerably to the continued controversy between Hobson and Solms, which was expressed in a particularly vocal and heated manner in the "Dream Debate" in Tucson in April 2006, thus exposing the fault lines in contemporary dream research. The most recent twist in this heated debate is Solms's (2014) claim that Hobson's theory of dreaming in fact deeply resembles Freud's.

An intermediate position on dream generation and its relation to REM sleep is Tore Nielsen's covert REM sleep model (2000a; see also Nielsen et al., 2005; Oudiette et al., 2012; Nielsen, 2014). Based on an extensive review of studies on the sleep-stage correlates of dreaming, Nielsen concludes that vivid dream reports, which may approximate or even surpass REM reports in length and complexity, can also be elicited from the stages of NREM sleep, most notably from sleep onset (SO) (see Nielsen, 2000a). He suggests that such dreams are best explained through the occurrence of REM-related processes or so-called *covert REM sleep* occurring during NREM sleep. Nielsen's claim that dreamlike activity, regardless of the sleep stage in which it occurs, may be related to the same underlying neurophysiological processes has the important advantage of acknowledging vivid dream reports from NREM sleep while preserving insights from older studies on the neurophysiological basis of dreaming in REM sleep. Thus he agrees partially with Solms (2000) in that he promotes a more inclusive approach to dreaming beyond the borders of REM sleep, but he agrees with Hobson in emphasizing the importance of REM-like processes for explaining the distinctive quality of dreaming.

There are, however, more subtle issues underlying the differences between these three theories. This is particularly evident in Tore Nielsen's covert REM sleep model. In the previous section, I discussed the example of the hypnagogic period, consisting of nine distinct EEG stages reaching from wakefulness into sleep, and its failure to map cleanly onto NREM sleep stages as commonly scored. I noted that scoring systems can be more or less fine grained, and the questions of what constitutes a sleep stage and of which criteria (behavioral, EEG, or subjective) distinguish sleep stages both from each other and from wakefulness depend in part on the resolution of a given scoring system. If the same reasoning is applied to REM sleep, it suggests that the question of whether REM sleep is necessary and sufficient for dreaming, which is central to the debate on the sleep-stage correlates of dreaming, might be too superficial to be informative: the notion of sleep stages itself is an idealization. Tore Nielsen's (2000a) covert REM sleep model questions the dichotomous distinction between REM and NREM sleep and draws attention to the fact that many of the polysomnographic measures characterizing REM sleep occur during NREM sleep as well but fail to be identified as such because of standard scoring criteria. Nielsen supposes that the use of polysomnographic sleep-stage scoring criteria itself "has helped promote an artificially categorical view of sleep and the unproven notion that sleep stages are *mutually exclusive states separated by abrupt transitions*" (Nielsen, 2014, p. 175). If so, then this oversimplified theoretical conception of sleep may be an artifact of what has arguably been the major methodological advancement in sleep and dream research. By softening the distinction between REM and NREM sleep, the covert REM sleep model lays the groundwork for a more continuous approach, both on the levels of neurophysiology and phenomenology. Though in a very different manner than in the 1950s, dreaming may thus once more play a formative role in emerging conceptions of the different sleep stages: the process of reassigning dreams to all stages of sleep may lead to reconceptualizing the stages of REM and NREM sleep themselves. It now becomes clear that the armchair fallacy of deriving real-world, phenomenological consequences from purely conceptual distinctions is complemented by something like the *scoring fallacy* in sleep and dream research, or the tendency to mistake artificial scoring criteria for real-world distinctions between sleep stages. Tore Nielsen's covert REM sleep model can also be regarded as an attempt to question the blind acceptance of REM and NREM sleep as distinct, nonoverlapping states.[13] Perhaps the covert REM sleep model, then, is a first step toward overcoming conventional sleep-stage distinctions. And perhaps, with time, the question of whether REM sleep can be identified with dreaming will not so much be resolved as disappear as more fine-grained scoring systems are developed.

Moving beyond the REM–NREM controversy, important progress in the understanding of sleep, dreaming, and dreamless sleep has also been enabled by the development of more sophisticated means of measurement and advanced data analysis. The latter include the analysis of spectral power and inter- and intrahemispheric coherence (Palagini et al., 2004; Nielsen & Chénier, 1999; Esposito et al., 2004). The former include imaging studies of sleep stages, which have primarily been conducted with PET (positron emission tomography; for a review of neuroimaging studies of REM sleep and dreaming, see Dang-Vu et al., 2007; see sec. 3.2 for discussion) but have also used the combination of EEG and fMRI (Duyn, 2012); the use of high-density EEG, which offers a much better spatial resolution than classical EEG and so far has mostly been used in NREM sleep (Lustenberger & Huber, 2012); and transcranial magnetic stimulation (TMS), especially during NREM sleep.

The latter method of stimulating the sleeping brain, in combination with EEG, is importantly connected to what is today one of the most widely discussed theories of consciousness: Giulio Tononi's (2004, 2008; for a popular account, see Tononi, 2012) integrated information theory (IIT). The theory assumes that consciousness is a gradual property that can be lost or regained by degrees; it starts out from the phenomenological observation that consciousness "is what vanishes every night when we fall into dreamless sleep and reappears when we wake up or when we dream" (Tononi, 2008, p. 216). Importantly, it offers a precise mathematical model as well as experimental paradigms for investigating changes in information integration across sleep–wake states and links them to changes in the quantity and quality of conscious experience. In doing so, it makes two main assumptions. At any given moment, "(i) the quantity of consciousness corresponds to the amount of integrated information generated by a complex of elements; (ii) the quality of experience is specified by the set of informational relationships generated within that complex" (p. 216). According to IIT, it is not sufficient to have a large repertoire of possible states (or information) to generate consciousness; rather, a conscious physical system must also be sufficiently unified, rather than decomposed into separate parts (p. 219).

In IIT, dreamless sleep has so far received more attention than dreaming. Indeed, in an approach quite contrary to Revonsuo's (2006; cf. sec. 5.1.3) claim that dreaming might be used as a model of consciousness, Massimini and colleagues (2012, p. 46) have proposed that "sleep represents a physiological model of unconsciousness" and that the investigation of dreamless sleep might help identify the necessary and sufficient conditions for the brain to generate conscious experience (Tononi & Massimini, 2008, p. 330). Massimini and colleagues (2012) note that IIT provides a possible explanation for why consciousness is frequently lost in NREM sleep—even though the thalamocortical system remains active rather than being completely shut down, and even though a certain level of responsiveness to external stimuli is preserved (cf. sec. 5.2.3). In particular, the changes in brain activation induced by TMS suggest that during NREM stages 3 and 4, reactivity is preserved in the cortical areas directly targeted by stimulation, but fails to be propagated to other brain regions, or only in a nonspecific manner. This suggests that the loss of consciousness in deep sleep is related to a breakdown of cortical connectivity (Massimini et al., 2005). Tononi and Massimini conclude that taken together, "these TMS–EEG measurements suggest that the sleeping brain, despite being active and reactive, loses its ability of entering states that are both integrated and differentiated; it either breaks down in causally independent modules or it bursts into an explosive and unspecific response" (Tononi & Massimini, 2008, p. 333).

Data from fMRI confirm the shift from global connectivity in wakefulness to local connectivity in NREM sleep, characterized by a disintegration of the frontoparietal network (Spoormaker et al., 2012). By contrast, REM sleep is marked by a cortical response to TMS that is similar to that seen in wakefulness, namely, a partial recovery of cortical connectivity (Massimini et al., 2010; for discussion, see Tononi, 2014). Recently, Casali et al. (2013) proposed that TMS experiments can be used to develop a theory-based index of the level of conscious experience, which they call the perturbational complexity index. They suggest that the level of consciousness can be determined much more precisely by systematically investigating the exact pattern of electrocortical responses following TMS in healthy participants across the sleep–wake cycle, but also during anesthesia or coma. While these developments are extremely promising and potentially have important clinical implications, there is also, however, reason for caution. It should be noted that methodologically, these studies do not form part of dream research proper, not only because they focus mainly on dreamless sleep but also because they typically do not involve the collection of mentation reports following systematic awakenings (cf. sec. 3.2). Yet investigating the loss of consciousness in dreamless sleep may, somewhat paradoxically, contribute to a theoretical understanding of conscious experience itself. This is especially true as this line of research is now beginning to integrate the collection and analysis of mentation reports, thus enabling a within-state contrast between dreaming and nondreaming during NREM sleep (Siclari et al., 2013, 2014).

Finally, without going into details, I want to point out a particularly interesting discovery arising from this type of research. Evidence from human participants and rats now suggests the existence of *local sleep*: slow-wave activity can selectively be induced in circumscribed cortical regions, suggesting that sleep is regulated locally rather than globally (Huber et al., 2004). In sleep-deprived rats, local sleep can even be measured during wakefulness and is associated with reduced task performance (Vyazovskiy et al., 2011). Vyazovskiy and colleagues (2011, p. 445) explain that during sleep deprivation, "subsets of neurons may enter an OFF period in one cortical area but not in another, and even within the same cortical area some neurons may be OFF while others are ON, seemingly in a piecemeal fashion."

Aside from the relevance for research on sleep loss and learning, this discovery could have important conceptual implications as well. So far, saying that someone is half asleep has been related to the fact, roughly, that sleep–wake transitions, as a function of whole-brain changes in activation levels, can be gradual. In this view, sleep–wake transitions can be understood in a purely quantitative fashion, by saying that the brain as a whole is more or less active. Now, however, the concept of local sleep suggests that certain *parts* of the brain can be asleep while others are awake. If so, then sleep is not, apparently, a property of the whole person, or even just of the whole brain, resulting from changing activation levels in a global and highly orchestrated manner, but something that can be detected on the level of local neuronal populations. Of course, some might claim that attributing the concept of sleep to parts of the brain is to commit the mereological fallacy (Bennett & Hacker, 2003): in this view, sleep is a property of the entire person, not of individual body parts, and certainly not of parts of the brain. Yet to exclude the possibility of local sleep might well be to commit the armchair fallacy: if the empirical research is robust, then perhaps an empirically plausible conceptual framework of sleep will have to be adjusted to accommodate the concept of local sleep. If it is indeed methodologically fruitful to say that parts of the brain can be asleep while the person or animal as a whole is awake, making such initially counterintuitive conceptual adjustments may be worthwhile.

Preliminary evidence indicates that this is the case: not only do sleep disorders present in different brain regions, but sleep depth appears to be unevenly distributed, suggesting that understanding sleep–wake transitions may involve acknowledging local rather than global changes (Nobili et al., 2011, 2012). For instance, during sleep onset, thalamic deactivation precedes changes in cortical activation by several minutes (Nobili et al., 2012). Similarly, the combination of wakelike and sleeplike cortical activation patterns might turn out to be characteristic of healthy sleep. If so, the description of NREM sleep parasomnias as *dissociated* states may have to be reconsidered: perhaps parasomnias only result from an imbalance of an otherwise entirely normal coexistence of sleep- and wakelike cortical activity (Nobili et al., 2011). If these findings on local sleep were further substantiated, the received view of sleep as a centrally orchestrated, global whole-brain phenomenon may turn out to be empirically false. If the concept of local sleep leads to a better theoretical understanding of sleep, sleep–wake transitions, and conscious experience during sleep, then making appropriate adjustments to the folk-psychological concept of sleep seems to be warranted, and philosophical conceptual analysis should follow suit.

Returning to dream research proper, and leaving the field of brain research behind for now, another area of progress concerns interindividual differences in dreaming (for a general review, see Domhoff & Schneider, 2008a; Schredl, 2010a; cf. sec. 6.2.2). Early on, researchers were particularly interested in detecting changes in dreaming following sleep deprivation (e.g., Dement, 1960) and during depression (for discussion and further references, see Cartwright, 2010), because they hoped that this might help explain the contribution of dreaming to emotional processing. The dreams of schizophrenic patients have also been the focus of empirical dream research from the beginning in the 1950s and 1960s (e.g., Dement, 1955; for discussion and further references, see Gottesmann, 2006), and the idea that dreaming might be used as a model of psychosis continues to be one of the hot topics discussed today (as in an ESF meeting in Aosta in May 2009, where researchers from psychology, neuroscience, and philosophy were brought together to discuss the feasibility of dreaming as a model of psychosis; see also *From Dreams to Psychosis: A European Science Foundation Exploratory Workshop*, a special issue of *Consciousness and Cognition*, vol. 20, no. 4 [2011]; and Windt & Noreika, 2011, for a critique of this idea; see also chaps. 5, 10). Epilepsy is a further area of interest, and Schönhammer (2004) has suggested that dreams of flying or falling in healthy subjects might be caused by patterns of heightened neuronal excitation—so-called epileptic equivalents—similar to those found in migraine and epilepsy patients, who also report experiencing vestibular hallucinations (Schönhammer, 2004, pp. 165–166). Generally, psychopathological conditions such as depression (Cartwright, 1990, 2001, 2010), schizophrenia (Hunt, 1989, pp. 136–137; Scarone et al., 2007; Limosani et al., 2011a, Limosani et al., 2011b), frontotemporal dysfunction as in epilepsy and Parkinson's disease (Paiva et al., 2011), migraine (Schönhammer, 2004), and sleep disorders (Schredl, 2009) may be accompanied by changes in the intensity, storylike organization, and content of dream reports, as well as by changes in dream recall frequency (for a general review of disturbed dreaming in different medical conditions, see Levin & Nielsen, 2007). These cases are particularly fascinating, because they suggest the possibility of correlating the qualitative differences in dreaming with physiological features such as subclinical migraine attacks, epileptic seizures (Schönhammer, 2004, pp. 209–210), or changes in REM sleep latency and eye movement density in depression (Cartwright, 2010). This latter line of research might suggest that dreaming, after all, fulfills a function in mood regulation (for a recent review, see Perogamvros et al., 2013) and that dreams may even serve as "a screening tool useful in determining a patient's psychological resiliency" (Cartwright, 1990, p. 71), for instance, following major life changes such as divorce.

While these findings suggest that varying rates of dream recall are related to individual differences in the intensity of dreaming, attempts to relate them to personality traits have not yielded straightforward results (for a review, see Blagrove & Pace-Schott, 2010). Hartmann (1998) found that participants with thin boundaries, characterized by open-mindedness, sensitivity, vulnerability, creativity, and artistic ability, had more frequent dream recall and that their dreams were more vivid, emotional, and nightmarelike than those of subjects with thick boundaries. In a study of home dream recall in college students, Watson (2003) found that subjects who are prone to absorption, imaginativeness, daydreaming, and fantasizing had comparatively high rates of dream recall. Other factors, such as total sleep time and the quality of sleep, were not associated with individual differences in dream recall. He interprets these results as indirect evidence for the *salience hypothesis*, or the claim that subjects who are prone to imagination may not only remember more dreams but also dream more intensely and more vividly. However, in a similar study, Schredl, Wittmann, Ciric, and Götz (2003) found that the combined factors of personality traits, creativity, frequency of nocturnal awakenings, and individual attitude toward dreams accounted for only 8.4 percent of variance in individual dream recall, indicating that personality factors exert a much weaker influence on dream recall frequency than is commonly assumed. In his detailed review of studies investigating the relationship between personality traits and dreaming, Blagrove (2007) concludes that while abundant evidence supports the claim that dreams incorporate current concerns, the evidence for a systematic relationship between personality traits and dreaming is mixed and inconclusive.

Subjects experiencing frequent nightmares may be an exception, as trait anxiety, coping styles, and nightmare distress have been linked to frequent nightmares and unpleasant dreams (Nielsen, Stenstrom, & Levin, 2006; Levin & Nielsen, 2007; Germain & Zadra, 2009; Nielsen & Levin, 2009). Nielsen (2010d) found evidence for a gender-specific association between evening (as opposed to morning) types and nightmare frequency such that female, but not male, subjects prone to eveningness—characterized by later sleeping times and a preference for evening activities—report more frequent nightmares.

Gender differences in dreaming have also been studied extensively, mostly through the use of questionnaires (rather than through the collection of dream diaries or laboratory awakenings; see Schredl, 2007, p. 29). While women recall both dreams and nightmares more frequently than men (Nielsen et al., 2006), it is not clear whether this reflects an actual difference in dreaming or gender-specific differences in socialization in dream reporting (Schredl & Reinhard, 2008). Men and women also report dreaming about different topics, with men more frequently reporting dreams about physical aggression, sexuality, and work, and women more frequently reporting dreams about people and clothing (Schredl, 2007). Dream recall frequency and content may also undergo specific changes during the menstrual cycle, pregnancy, and menopause (Garfield, 1988; Lara-Carrasco et al., 2013).

Finally, an important controversy concerns the matter of whether small children and infants dream. David Foulkes, who conducted the most extensive laboratory study of children's dreams (Foulkes, 1999), argues that dreaming is a cognitive achievement, which is gradually developed throughout childhood. He suggests that dreaming is "continuous with our waking reflective ability to think in images" (p. 15) and is associated with the development of visuospatial skills. However, Hobson et al. (2000) and Resnick et al. (1994) argue that while the incidence of dream bizarreness may increase from ages three to eight, recall rates remain stable, and even small children predominately report dreams characterized by active self-participation.

2.3 Conclusions

In sum, while considerable progress has been made in refining the methods used in dream research—for example, through the development of different methods of collecting and analyzing dream reports, but also of measuring brain activity during sleep, as well as advanced methods of EEG analysis—the initial optimism associated with the discovery of REM sleep was followed by a fragmentation of empirical dream research and the proliferation of differing and at times outright contradictory neuroscientific theories of dream generation. Likewise, the psychological mechanisms involved in dreaming, as well as the relationship between dream recall and personality traits and gender, the development of dream recall in children and adolescents, and the relationship between dreaming and psychological well-being, are yet incompletely understood. In this process, the target phenomenon has become ever more complex: sleep has been subdivided into different stages and most recently may be shifting from being regarded as a global, whole-brain phenomenon to a local phenomenon, and different types of dreams and interindividual differences in dreaming have been investigated.

While none of these theoretical controversies, for instance, concerning the sleep-stage and neurophysiological correlates of dreaming, can be resolved in a philosophical book on dreaming, the foregoing analysis highlights the historical contingency of changing theoretical conceptions and the extent to which they influence—and are influenced by—empirical observations. Historically, it is noteworthy that the reassignment of dreaming to genuine rather than half sleep has recently been followed by the recognition that dreaming may occur in all stages of sleep—and possibly, as suggested by the covert REM sleep model, by attempts to overcome the REM–NREM dichotomy altogether. After early optimism concerning the identification of REM sleep and dreaming, researchers are now increasingly recognizing that dreaming and REM sleep are separate phenomena, and that as yet, no sleep-stage correlate of dreaming is available. Even Hobson, otherwise a strong supporter of the REM-dream hypothesis, remarks that "dreaming could not be conceived as confined to the REM state no matter what definition one applied" (Hobson, 1988, p. 143). This means that the concept of dreaming is once more being set adrift from its close association with a particular stage of sleep.

It also seems that changing conceptions of sleep and dreaming are closely related to changing scoring systems. Just as the concept of dreaming is sorely underdefined in current research, so the contemporary scoring criteria for sleep stages might also be too vague in some respects (i.e., NREM sleep stages, as suggested in Tanaka et al., 1996) and too strict in others (thus promoting the REM–NREM dichotomy). If the failure to identify a sleep-stage correlate of dreaming really is, at least in part, an artifact of overly coarse-grained (or overly restrictive) scoring systems, and if the study of dreams occurring in all sleep stages can play some part in overcoming the REM–NREM dichotomy, then perhaps dreaming might once more be rehabilitated as making an important contribution to sleep research proper. In any case, these reflections suggest that the conceptual issues related to dreaming are paralleled by similar issues in sleep-stage scoring, and thus that philosophical-conceptual and methodological approaches may complement each other.

My survey of the historical sources of early dream research shows that in response to new findings on sleep, conceptions of sleep and dreaming changed considerably: from being considered as unobservable phenomena occurring during deep sleep, as in Aristotle's dream theory, dreams came to be viewed as observable, at least indirectly, through dream behavior and dream reports in the nineteenth and early twentieth centuries. However, at the same time, in virtue of their association with heightened periods of brain activity in sleep, dreams were relocated to light sleep and thus partially decoupled from genuine deep sleep. Empirical researchers excluded, or completely ignored, the possibility that such heightened brain activity might occur during sleep. Similarly, Malcolm developed his philosophical argument rejecting the possibility of dream experience during deep sleep. In doing so, he was not simply reacting to the emerging field of scientific dream research but was influenced, at least implicitly, by older conceptions of sleep dating back to Aristotle, namely, the view that sleep is the opposite of wakefulness and that dreams are a phenomenon of deep sleep. In this respect, Malcolm's view also paralleled, rather than ran counter to, early researchers' disbelief that high levels of brain activity could occur during genuine sleep. But this also shows that the relationship between conceptual change and empirical research findings is reciprocal: preexisting conceptions influence the interpretation of research results, and these, in turn, modify existing theoretical and pretheoretical or folk-psychological conceptions. Thus a reassignment of dreams to genuine sleep only became possible when the distinction between two different types of sleep—REM sleep and NREM sleep—became available and when the concept of sleep as a uniform period of passivity and rest came to be abandoned. Now, paradoxically, understanding the neurophysiological mechanisms underlying dreams may once more require a departure from currently accepted sleep-stage scoring systems.

For this reason, this chapter can also be read as a plea against relying too strongly on currently popular theoretical conceptions and unquestioned intuitions about the concepts of sleep and dreaming. It has become clear that these concepts are theory laden and prone to change in response to novel discoveries, and it is even possible that differing conceptions of these phenomena can coexist at the same time. I think this in itself suggests a historically motivated argument against using armchair conceptual analysis to define the concepts of sleep and dreaming: if there is no single, clear, and invariant concept of sleep and dreaming, then the project of purely armchair conceptual analysis and of using intuitions as a guide to theorizing about sleep and dreaming seems increasingly unattractive. Moreover, if there is indeed a close relationship between changing philosophical conceptions of sleep and dreaming on the one hand and empirical research results on the other hand, as I have been suggesting, then this underscores the importance of interdisciplinary dialogue and of an empirically informed and informative philosophical approach.

The next important step is to look more closely at the different methods used in empirical dream research and to analyze the role of the experiential and transparency assumptions in specific empirical research projects.

3 The Methodological Background Assumptions of Scientific Dream Research

3.0 Introduction

This chapter has the double aims of functioning as an extended plausibility check of the arguments developed in chapter 1 and of providing an overview of the main methodological approaches to the study of dreaming from scientific dream research. How do researchers construct the relationship between dreaming, dream reporting, and other types of evidence about dreaming? On which background assumptions do existing methods for the study of dreaming rely? In particular, does my claim that the experiential character of dreams and the transparency of dream reports have the status of methodologically necessary default assumptions retain its plausibility in the face of actual examples from dream research? Or are there any cases in which dream research can be said to yield straightforward empirical evidence confirming the experiential and transparency assumptions, thus once more turning them into genuine hypotheses?

I will argue that the latter is not the case: scientific dream research is best regarded not as confirming the experiential and transparency assumptions but as assuming them, and as a matter of methodological necessity. This, however, raises two further questions: What, if any, are the consequences of this essentially metatheoretical observation for the interpretation of empirical research results? Put differently, if I am right to claim that questions about the experiential character of dreaming and the transparency of dream reports are pseudoquestions in the sense that they are artifacts of a misconceived philosophical problem, then why should dream researchers worry about them at all? And second, what consequences does my discussion of the methods of empirical dream research have for the philosophical discussion on dream experience and dream reporting?

To answer these questions, I discuss the following methods, which I take to be both representative of the main methodologies used in empirical dream research and the ones that are the most important for the experiential and transparency assumptions. These are laboratory studies using a combined methodology of polysomnography and dream reports following timed awakenings from sleep; neuroimaging studies of REM sleep; exclusively report-based studies of spontaneously recalled dreams; sleep behavior as manifested in REM-sleep behavior disorder (RBD) and in lucid dreaming; the global or partial cessation of dreaming and the comparison between dreaming and neuropsychological syndromes; and the possibility of moving beyond dream reports by decoding mental states during sleep from brain activity.

In doing so, I hope not only to show that my reconstruction of the experiential and transparency assumptions as methodologically necessary default assumptions is plausible in view of the methods of scientific dream research, but also to convince readers that this position is both empirically and philosophically informative. This chapter is intended, in part, as an example of how a philosophically motivated metatheoretical critique can support the methodologies used in current dream research, clear up persistent uncertainties, and perhaps even suggest perspectives for future research. At the same time, I hope to show that a philosophical account of dreaming and dream reporting can profit immensely from this type of analysis. In particular, my discussion of the methods of empirical dream research deepens the theoretical understanding of the transparency assumption, shows that it is inextricably linked to the assumption that dreams are reportable experiences (called the *reportability assumption*), and suggests the conditions under which both assumptions apply, and why they have to be restricted in certain respects. I also propose a working definition of what it means to report a dream, and discuss some consequences of this view for the interpretation of results from sleep behavior.

3.1 Studying Dreaming in the Sleep Laboratory: Polysomnography and Timed Awakenings

Sleep laboratory studies using a combined methodology of polysomnographic measurements and timed awakenings from different sleep stages are closely connected to the philosophical discussion on dreaming for historical and methodological reasons. As discussed in chapter 2, they marked the beginning of scientific dream research and led to a paradigm change in the theoretical understanding of sleep and dreaming. The newly emerging view of sleep as a heterogeneous and active process and of dreaming as closely related to the wakelike EEG activity of the brain during REM sleep introduced an additional or, according to Malcolm, rival type of evidence for the occurrence of dreams. Consequently the combination of polysomnography and the collection of dream reports following timed awakenings in the sleep laboratory was likely the most important methodological innovation in the study of dreaming.

Typically, the setup will be as follows: The participant is wired up in a lengthy and somewhat tedious procedure, in which electrodes for EEG measurements are attached to the scalp, around the eyes for EOG measurements of eye movements, and to the chin, leg, or wrist for EMG recordings of muscular activity; heart and respiration rate are also frequently measured. Once this is done, the participant is asked to go to sleep. In addition to the electrodes and cables needed for polysomnographic recordings, there will often be a microphone dangling above the participant's head and a night-vision video camera allowing researchers to monitor sounds and movements throughout the night. This is important in case the participant exhibits sleep behavior, which may be indicative of sleep disorders, but also to enable communication throughout the night. In an adjacent room, one or two researchers use these different information channels to carefully monitor the participant's trajectory from wakefulness through sleep onset and the early stages of NREM sleep into deep, stages 3 and 4 NREM sleep and back "up" into REM sleep—and so on throughout the night. The researchers score the sleep stages in real time, as they are occurring, to determine the ideal time of awakening, which depends on the sleep stage being targeted in the respective study. They then wake the participant up with the help of a beeping tone or voice message via intercom; this is followed by the (often tape-recorded) request to relate one's dream, either by giving a free dream report or answering specific questions—the exact procedure and wording vary. There may also be a computer monitor next to the bed, allowing the participant to type in answers to a questionnaire. Afterward the participant is allowed to go back to sleep, only to be awakened again later in the night.[1] Using this procedure, approximately three to five reports from REM sleep can be collected throughout the night (Kramer, 2007, p. 36). This number can be greatly increased through the use of serial awakenings paradigms.[2]

Many methodological issues are involved. These include the speed of awakening,[3] the situation encountered in the lab, the wording of questions,[4] the gender of the interviewer and the interaction between interviewer and participant, and many more (for details, see Hall & Van de Castle, 1966; Antrobus et al., 1991; Domhoff, 1996, 2003; Winget & Kramer, 1979; Kramer, 2007; Cartwright, 2010). There is a long discussion on the so-called laboratory effect, regarding the extent to which sleeping in the laboratory influences the quantity and quality, as well as the content, of reported dreams (see Hobson et al., 2000; Kramer, 2007; Schredl, 2008a). The experimental situation is frequently incorporated into dreams; experimental awakenings and the attachment to measurement devices cause discomfort; and in prolonged studies, participants may experience motivational problems known as "laboratory fatigue" (Hobson et al., 2000, p. 803). In a review study, Schredl (2008a) found that nearly 20 percent of dream reports gathered in the laboratory contain direct, and nearly 40 percent contain indirect, references to the laboratory situation. An alternative but costly solution is the use of the Nightcap, which enables researchers to gather physiological measurements of sleep phases and conduct timed awakenings while participants sleep at home in their own bedrooms (Hobson et al., 2000, p. 803; Hobson, 2009).

Importantly, the reports thus elicited are taken to reflect, as accurately and in as much detail as possible, whatever was experienced in the immediately preceding sleep stage. For instance, awakening a participant several minutes into a REM phase is believed to yield a report that reflects conscious experience during this REM phase, whereas a report following an NREM stage awakening is taken to reflect whatever was (or was not) experienced during NREM sleep. The assumption that dream reports reflect experiences from the immediately preceding sleep stage may sound trivial, but is important: without it, reports following timed awakenings would not enable researchers to relate dream activity to specific sleep stages at all. Nielsen notes that early dream researchers, committed to the supposed identification of dreaming and REM sleep, attributed "mentation reported from NREM sleep . . . to purportedly confounding factors, for example, recall of mentation from previous REM episodes or subjects' waking confabulations" (Nielsen, 2000a, p. 851). Increasing evidence for the occurrence of dreams during NREM sleep—for instance, from studies investigating mentation reports from the first sleep cycle, when subjects have not yet reached REM sleep (e.g., Suzuki et al., 2004)—has slowly changed this interpretation (Nielsen, 2000a). This is yet another example of the complex relationship between pretheoretical assumptions and changing interpretations of empirical findings: mentation reports following awakenings from NREM sleep only turn into evidence about experiences during preceding NREM phases once they are taken to be transparent.

This suggests two important modifications to the transparency assumption as discussed so far. The first is that in sleep laboratory studies of dreaming, a positive report of having dreamt is afforded as much prima facie credibility as a report of not having experienced anything at all. The transparency assumption, as formulated so far, concerns the trustworthiness of *dream* reports with respect to previous experiences during sleep. If negative reports of nondreaming are regarded as equally trustworthy, this means that the transparency assumption is extended to a further type of report, namely, that of not having dreamt (or of not having experienced anything during sleep, depending on the wording used in a given study). Taken together, the equal trustworthiness of positive and negative reports of dreaming thus gives rise to the further assumption that dreams, at least following timed awakenings, are *reportable* experiences: if something had been experienced during sleep, it would (or at least could) have been reported following awakening (for further discussion, see sec. 4.4).[5] Call this the *reportability assumption*:

Reportability assumption: Dreams are reportable experiences in that the occurrence of experience during sleep can be reported following timed awakenings in the sleep laboratory.

Only under this assumption can reports of nondreaming be considered trustworthy and the putative correlation between REM sleep and dreaming be established in the first place: studying the sleep-stage and neural correlates of dreaming requires contrasting dreaming and nondreaming, and this, in turn, is only possible if both reports of dreaming and of nondreaming are assumed to be equally trustworthy.

The second modification is that in sleep laboratory studies, dream reports are assumed to be transparent only with respect to immediately preceding sleep experiences. Without this assumption, it would be impossible to rule out carryover from earlier sleep stages, and thus mentation reports could not be mapped onto the immediately preceding sleep stages in the second half of the night. This would, once more, undercut the rationale for conducting timed awakenings throughout the night. The transparency and reportability assumptions, then, can be restricted to reports given under the condition of temporal proximity.[6]

We can see both points at work in the original study reported by Aserinsky and Kleitman (1953). In that study, the correlation between dreaming and REM sleep was established not only because participants typically reported dreams following REM sleep awakenings, but also because of their frequent failure to do so following NREM sleep awakenings. In their original publication, Aserinsky and Kleitman observed:

> Of 27 interrogations during ocular motility, 20 *revealed detailed dreams* usually involving visual imagery; the replies to the remaining 7 queries included complete failure of recall, or else, "the feeling of having dreamed," but with inability to recollect any detail of the dream. Of 23 interrogations during ocular inactivity, 19 disclosed complete failure of recall, while the remaining 4 were evenly divided into the other 2 categories. (Aserinsky & Kleitman, 1953, pp. 273–274; italics mine)

Had the authors not taken reports both of dreaming and of nondreaming to be trustworthy with respect to experiences (or their absence) during the preceding sleep stage,[7] the correlation between dreaming and periods of eye motility, later called REM sleep, could not have been established.[8] In particular, their conclusion that "this method furnishes the means of determining the incidence and duration of periods of *dreaming*" (Aserinsky & Kleitman, 1953, p. 274; italics mine) only makes sense given the experiential, transparency, and reportability assumptions.

Other remarks suggest that the authors took a more modest view. For instance, Aserinsky and Kleitman remark that "the ability to *recall* dreams is significantly associated with the presence of the eye movements noted" (Aserinsky & Kleitman, 1953, p. 274; italics mine). However, this more modest view contradicts the self-proclaimed goal of the article, which was, after all, to study dreaming, not dream recall. Though Aserinsky and Kleitman apparently did not consistently subscribe to the transparency and reportability assumptions in their article, my point nonetheless stands that in view of their conclusions, they should have.

This reading is supported by other early studies. In an article titled "The Relation of Eye Movements during Sleep to Dream Activity: An Objective Method for the Study of Dreaming," William Dement and Nathaniel Kleitman stated that "the results of these experiments indicate that *dreaming* accompanied by REMs and a low-voltage electroencephalogram occurred periodically in discrete episodes during the course of a night's sleep" (Dement & Kleitman, 1957b; p. 345; italics mine). Again, this suggests that early dream researchers very naturally relied on the experiential and transparency assumptions: this reading best explains the shift from associating dream *recall* with *previous* REMs to associating *dreaming* with *concomitant* REMs. By now, it should also be clear that these two examples are not special in this respect. Rather, this analysis more generally applies to studies investigating the occurrence of dreams based on the observed correlation between dream reports and preceding sleep stages (for a general review of such studies, see Nielsen, 2000a). Indeed, the transparency and reportability assumptions underlie the very rationale for claiming that *dream* research, as opposed research on dream *reports*, is possible.

Looking at these early examples of laboratory studies is also interesting for another reason. These early researchers not only relied on dream reports to investigate the occurrence of dreams in different sleep stages but also made claims about the duration of dreaming. As we have seen, the duration of dreams is central to the philosophical debate on dream experience because if dreams are not temporally extended experiences, it hardly makes sense to say that they are experiences at all. In his monograph, Malcolm (1959/1962, pp. 70–82) devoted an entire chapter to showing that dreams have no location and duration in physical time, and the central point of Dennett's (1976) cassette theory is that dreams are instantaneously inserted into consciousness at the moment of awakening. The problem of dream duration is also not an exclusively philosophical problem: the view that dreams are instantaneous memory insertions was already discussed by nineteenth-century dream researchers (Maury, 1861; Binz, 1878; Goblot, 1896; for discussion, see also Hall, 1981; Kramer, 2007, pp. 22–24).

There is also a methodological reason for this interest, namely, the project of mapping the timing and duration of dream events onto the timing and duration of brain events. Here we encounter the problem that, in many studies, the duration of the analyzed polysomnographic measurements is much shorter than the presumed duration of the reported dream. Note that most studies analyze EEG recordings from a relatively short period of time, ranging from ten seconds (Wackermann et al., 2002) to five minutes (Bertolo et al., 2003).[9] Even if one assumes that the participant was actually dreaming throughout the analyzed period (e.g., beginning with the onset of REM sleep and up until the moment of awakening), it is not possible to exactly identify the portion of the dream report that corresponds to the measurement period. The only way of doing this would be to ask for subjective estimates of roughly the same time period as that which was analyzed, and thus to fall back on dream reports. This would also, it seems, be required to ensure that the subject was dreaming throughout the analyzed period at all (for a more precise way of timing dream events to map them onto simultaneous changes in brain activity, see sec. 3.5).[10]

In view of these considerations, it is not surprising that claims about the duration of dreaming figured prominently in the interpretation of research results from the beginning. Aserinsky and Kleitman concluded their 1953 paper by stating that their method furnished the means of investigating not just the occurrence but also the duration of periods of dreaming. Crucially, they did not ask subjects about the duration of their dreams at all; instead, having ascertained that REM sleep awakenings yielded dream reports, Aserinsky and Kleitman inferred the duration of dreams from the duration of uninterrupted REM periods, apparently this time without asking their participants whether they had dreamt at all. In doing so, they attempted to move beyond dream reports altogether, thus equivocating between dreaming and REM sleep. This presumes, however, that there is, indeed, a direct correspondence between the subjectively experienced duration of dreams and the objective duration of the respective periods of sleep, such that dream experience begins at the onset of a REM period and continues throughout it, without interruption. It also presumes that the subjective duration of dreams is stable, rather than variable across different types of dreams.

Dement and Kleitman's (1957b) early follow-up study on dreaming and REM sleep tried to rule out the possibility of dreams being instantaneously inserted memories in a more straightforwardly empirical manner. Participants were awakened either five or fifteen minutes after the onset of rapid eye movement activity and asked to estimate the duration of their dream. The authors write that with one exception, "all Ss were able to choose *the correct dream duration* with high accuracy" (Dement & Kleitman, 1957b, p. 343; italics mine) and concluded that "there was nothing in the experiments reported in this paper to indicate that the dream occurred instantaneously, or with great rapidity, as some have supposed" (p. 346). Their remark that participants were able to choose the duration of their dream *correctly*, however, shows that the authors supposed dreams both to be temporally extended experiences and to occur throughout periods of REM sleep from the beginning—had they not, their claim for having an objective measure of the actual length of dreaming would have been ungrounded. Here Malcolm's critique that they were taking the observed eye movements as the "*criterion* of the occurrence, temporal location and duration of dreams" (Malcolm, 1959/1962, p. 74) seems warranted. The subjective estimate was something to be confirmed or disconfirmed by the duration of REMs, not evidence in its own right.

The problem that I want to point out, however, is different from Malcolm's objection that REM periods were taken to furnish a new criterion of dreaming. Specifically, note that the argument concerning the duration of dreams is characterized by a crucial switch in perspective. In stating that their participants were able to correctly estimate the duration of their dreams, Dement and Kleitman (1957b) appear to make the credibility of the collected dream reports contingent on the duration of REM phases. In other words, they assume that the actual duration of the REM phase provides a means of checking the accuracy of the subjectively estimated duration of the dream. So here the transparency of the dream report is made contingent on the sleep-stage correlate it helped identify, in the same study. The obvious objection is that you cannot have it both ways: if you first presume dream reports to be transparent, you can't later confirm their transparency by way of the sleep-stage correlate previously identified on their basis. The argument is circular. A better interpretation is to say that dream reports, under the assumption of transparency, were first used to locate dreaming in REM sleep and then, by the same reasoning, to investigate the duration of dreams.

Aside from asking subjects to estimate the duration of their dreams, which is rarely done, indirect measures of dream duration may be obtained via the length of the dream report, either through word count or by discerning different narrative units, with more units indicating a longer and more complex dream (Kramer, 2007, pp. 56–57). For instance, it is known that the length of dream reports increases throughout the night (Domhoff, 1996, 2003; Kramer, 2007), which fits in well with the increasing duration of REM periods in later sleep cycles. At the same time, this represents an indirect measure of dream duration at best. A number of studies suggest that the perceived duration of a dream increases together with its complexity, with REM dreams being experienced as taking longer than NREM dreams even when the duration of sleep is controlled for (Suzuki et al., 2004; Noreika, Windt, Arstila, Falter, Kiverstein, & Revonsuo, 2010) and participants overestimating the amount of time they actually spent asleep for more complex, but not for simple, dreams (Moiseeva, 1975). Consequently, two separate questions can be raised about the subjective duration of dreaming: One, does the duration of dreaming correspond to the duration of a given period, for example, of REM sleep? And two, does the subjective duration of dreams (how long the dream seemed to last) correspond to the actual time the subject spent dreaming? Progress with these questions, it would seem, can only be made by asking detailed questions about the subjective duration of dreams, in combination with timed awakenings. But if this is correct, then any study making progress on these questions will have to rely heavily on subjective estimates of dream duration.

In sum, it seems that early laboratory studies of dreaming are best reconstructed as relying on the transparency and reportability assumptions. This claim is based not only on textual evidence but also on the finding that when researchers did try to move beyond dream reports or find independent evidence for their corroboration, they failed, and for principled reasons. Importantly, the same is true for all studies investigating the correlation between dreaming and sleep stages more generally: whether or not participants report dreams following REM sleep awakenings remains a test case for the strength of the correlation between dreaming and REM sleep, and in fact, theoretical views on this correlation shifted once enough evidence indicated that dreams can occur outside REM sleep. Trusting dream reports was a crucial step toward shifting theoretical opinion on this matter.

I am not here defending a Malcolmian-style argument according to which dream reports are the sole criterion of dreaming; nonetheless, compared to other types of evidence such as polysomnographic recordings, dream reports take a privileged position. They do so, first, in virtue of their *historical* priority in dream research and, second and more importantly, by imposing a *methodological* constraint on dream research and the predictive power of theories on the sleep-stage and neural correlates of dreaming. Thus one may conclude that the type of experimental approach to dreaming discussed here is constrained, for methodological reasons, by dream reports under the assumption of transparency and reportability. This is both a descriptive claim about how laboratory research on dreaming is typically conducted and a normative claim about how best to interpret research results. Moreover, because future progress hinges on the willingness to take dream reports seriously, an important insight is that rather than questioning or trying to verify the trustworthiness of dream reports, a more promising and likely more productive project is to focus on improving reporting conditions in the laboratory, but also at home.

Before moving on to the next section, I want to consider a possible objection to this claim. In particular, it might seem that the experiential assumption does gain some independent support from these early laboratory studies. Specifically, the observation that the REMing brain is as highly activated as the waking brain, and that its reactivation after the relative quiescence of NREM sleep is related to the reticular activating system, shifts the evidence in favor of the experiential assumption (cf. sec. 1.3.4). REM sleep paralysis, explaining the outward passivity of the subject during REM sleep, as well as the fact that the different physiological characteristics of REM sleep—eye movements, heightened cortical activity, muscular atonia—do not occur randomly but are strongly related to each other, further thickens the plot. Taken together with the high likelihood of dream reports following REM sleep awakenings, this would seem to enhance the empirical plausibility of the experiential assumption considerably. As a historical note, it is worth pointing out that this type of argument—call it the *argument from brain activation*—was actually used by early researchers.

Initially, Dement and Kleitman (1957b) suggested that the presence and duration of dreaming were indicated by rapid eye movements during sleep. In an article published later that same year, however, they rescinded their earlier claim that "eye movements should be the criterion of the presence or absence of dreaming" (Dement & Kleitman, 1957a, p. 685), pointing out that "the EEG is actually a better criterion" (p. 685): "while the amount of eye movement may be large or small depending on the actual imagery of the dream" (p. 685), nonspindling EEG activity was more consistently associated with dreaming. This was also an attractive move:

> With dreaming and hypnagogic reverie assigned to a definite EEG stage, there exists at least a fairly consistent relation between the EEG and levels of consciousness and it seems possible to undertake the second step of considering the neural origin of these patterns as representing the basis of consciousness itself. (Dement & Kleitman, 1957a, p. 687)

Rather than assuming that dreaming could occur throughout sleep and that no particular relation connected brain activation levels with conscious experience, this suggested both a positive and negative marker of consciousness, with spindling activity indicating an absence of dreaming and thus of conscious experience (Dement & Kleitman, 1957a, p. 687).

Though early researchers clearly felt the pull of the argument from brain activation, it has not historically turned out to be quite as attractive as the experiential and transparency assumptions: as we have seen, researchers now accept that dreams occur in all stages of sleep—thus portraying a willingness to rely on dream reports as indicators of previous dream experience even when reports are not preceded by wakelike brain activation. As it happens, sleep spindles and K-complexes, the defining characteristics of stage 2 NREM sleep, are fairly often followed by dream reports (Nielsen, 2000a). Indeed, contra Dement and Kleitman's prediction, the consensus today is that the minimal levels of brain activation required for conscious experience are not known.[11] Though the argument from brain activation may, as a matter of fact, have eased the way toward accepting dreams as experiences during sleep, this is a matter of historical contingency.

To further illustrate this claim, we might imagine a scenario for early dream research in which things had gone slightly differently. Let us say that our imaginary dream researchers found that awakenings from the deep stages of NREM sleep were frequently followed by vivid dream reports. This would have been in keeping with the Aristotelian (and Malcolmian) view that dreams occur in deep sleep. They might, then, easily have attempted to confine dreaming proper to deep sleep, perhaps even suggesting that the presence of wakelike brain activation during sleep indicated an absence of dream experience. After further research, they might, much like real-world dream researchers, have gradually come to accept that dreaming could occur in all sleep stages—and thus was not incompatible with wakelike brain activation during sleep after all, as they had initially believed.

Thought experiments, of course, are notoriously prone to elicit different intuitions in different readers, but this one, I think, is empirically plausible: it really just tells the history of scientific dream research, albeit with a twist, and it builds up to a situation that is no different from the situation of contemporary, real-world dream researchers. As thought experiments go, it is actually pretty unimaginative—and this, I hope, is why it works. Its point is to show that the slight twist at the beginning—the putative finding that vivid dreams are frequently reported from deep sleep rather than from REM sleep—has no necessary consequences for the problem of dream experience whatsoever. It would even have fit the then-popular view of dreaming as a phenomenon of deep sleep. But if this is correct, then the argument from brain activation is not necessary for the endorsement of the experiential assumption, and this will hold true not only for our imaginary but also for real-world dream researchers. As we have seen, this is actually the case: it is supported by the willingness of contemporary, real-world researchers to accept that dreams occur in all sleep stages.

Incidentally, it might even turn out that the study of NREM sleep will be more informative for investigating the neural correlates of dreaming than the study of REM sleep. Recall that Noreika et al. (2009) found that awakenings from NREM stage 3 yielded almost equal numbers of reports of dreaming and dreamless sleep, suggesting that this might be a particularly promising sleep stage for contrasting the presence and absence of phenomenal experience during sleep, and thus for determining the neural correlates of dreaming within, rather than across, sleep stages (see also Siclari, 2013, 2014).

3.2 Studying Sleep without Studying Dreaming: Neuroimaging during Sleep

The development of neuroimaging techniques has changed the face of consciousness research, and the possibility of applying these techniques to the study of dreaming is promising. Despite considerable methodological challenges—in particular, the problem of getting participants to sleep within the noisy and claustrophobic confines of the scanner, on their backs, without moving their heads[12]—several imaging studies, mainly using PET, have been conducted (for excellent reviews connecting the results to dreaming, see Dang-Vu et al., 2005, 2007, 2009; Desseilles et al., 2011; Dresler et al., 2014).[13] Several studies (Czisch et al., 2004; Wehrle et al., 2005, 2007; for discussion and further references, see Desseilles et al., 2011) have shown the feasibility of using fMRI for the study of REM sleep but failed to obtain dream reports. The use of fMRI in dream studies would, however, be desirable. Though noisier than PET and more sensitive to movement artifacts, fMRI does not require the application of a contrast agent and does not involve exposure to ionizing radiation. The cost of fMRI is also much lower than PET, and fMRI has a higher spatial resolution (Baars & Gage, 2010, pp. 573–579; for a general review of the advantages and limitations of fMRI, see Logothetis, 2008). Yet the neuroimaging studies of REM sleep whose results have been related to dreaming have typically involved PET rather than fMRI, and the following remarks refer mostly to PET studies.

Early PET studies showed that the global level of cerebral energy requirements (Maquet et al., 1990) and cerebral blood flow during REM sleep is similar to wakefulness (cf. Madsen & Vorstrup, 1991), thus confirming findings from EEG studies. Later PET studies showed, however, that REM sleep is characterized by a shift in regional patterns of activity compared to wakefulness and NREM sleep (Braun et al., 1997; Maquet et al., 1996, 2000, 2005; Nofzinger et al., 1997). Specifically, the pons, thalamus, temporo-occipital, motor, and limbic and paralimbic areas (including the amygdala) are highly activated during REM sleep, while the dorsolateral prefrontal and inferior parietal cortices are comparatively quiescent. This pattern is telling, as it fits nicely with the formal features of dreaming, that is, the predominance of visual and motor imagery; the frequency of intense, often negative emotions; and cognitive deficits such as a loss of self-awareness, mnemonic deficits, and the delusional belief in the reality of dream events (see fig. 3.1).

At first sight, then, neuroimaging studies of REM sleep seem to be a major step toward fulfilling the dream dreamt by dream researchers since the discovery of REM sleep: they seem to allow researchers, quite literally, to watch the dreaming brain and to identify the neural correlates of specific types of dream experiences, a finding that would go far beyond the establishment of a sleep-stage correlate of dreaming. Though these studies are compelling, let us be clear about what type of correlation this is, exactly: with only a few notable exceptions (Madsen et al., 1991; Dresler et al., 2012; see sec. 3.7 for discussion), neuroimaging studies of REM sleep have not elicited dream reports from participants (see Horikawa et al., 2013a; Horikawa et al., 2013b, discussed in sec. 3.7). So the hypothesized correlation between, say, the predominance of visual imagery in dreams and activation in the visual areas during REM sleep is not founded on a one-to-one mapping between a retrospective dream report and analyzed measurements obtained from the respective REM period. In fact, for most existing neuroimaging studies, it is not known whether the participants in these studies would actually have reported dreams had they been asked to do so. The assumption that they dreamt at all during the recorded REM phases, though likely, is based on the twofold generalization of the finding that most participants in other studies have reported dreams following most awakenings from REM sleep.

In addition, as pointed out earlier, neuroimaging studies are interesting because they go beyond the sleep-stage correlates of dreaming by revealing potential neural correlates of specific types of dream experience. In practice, this is done by correlating the results from neuroimaging studies of REM sleep in one set of participants with the results from the formal analysis of dream reports from different studies, conducted with different participants and under starkly different conditions (i.e., by analyzing dream reports from spontaneous awakenings at home or obtained during EEG studies in the laboratory). But, of course, just as not all participants report dreams following all REM sleep awakenings, so not all dream reports are equal in their formal features and content. So even if it is true that a majority of dreams are characterized by visual and motor imagery, strong emotions, and cognitive deficiencies (Hobson et al., 2000), this does not apply to all dreams and is not necessarily true for the particular REM phases during which neuroimaging data were obtained in a given study. This is not to deny that such generalizations play an important role in dream research. But even so, to make real progress on the neural correlates of dreaming, it would be crucial to have a one-to-one correlation between a dream report and measurements obtained from the corresponding sleep stage.

In their 2002 paper, Schwartz and Maquet address this problem and propose that the combination of imaging of REM sleep with the neuropsychological assessment of

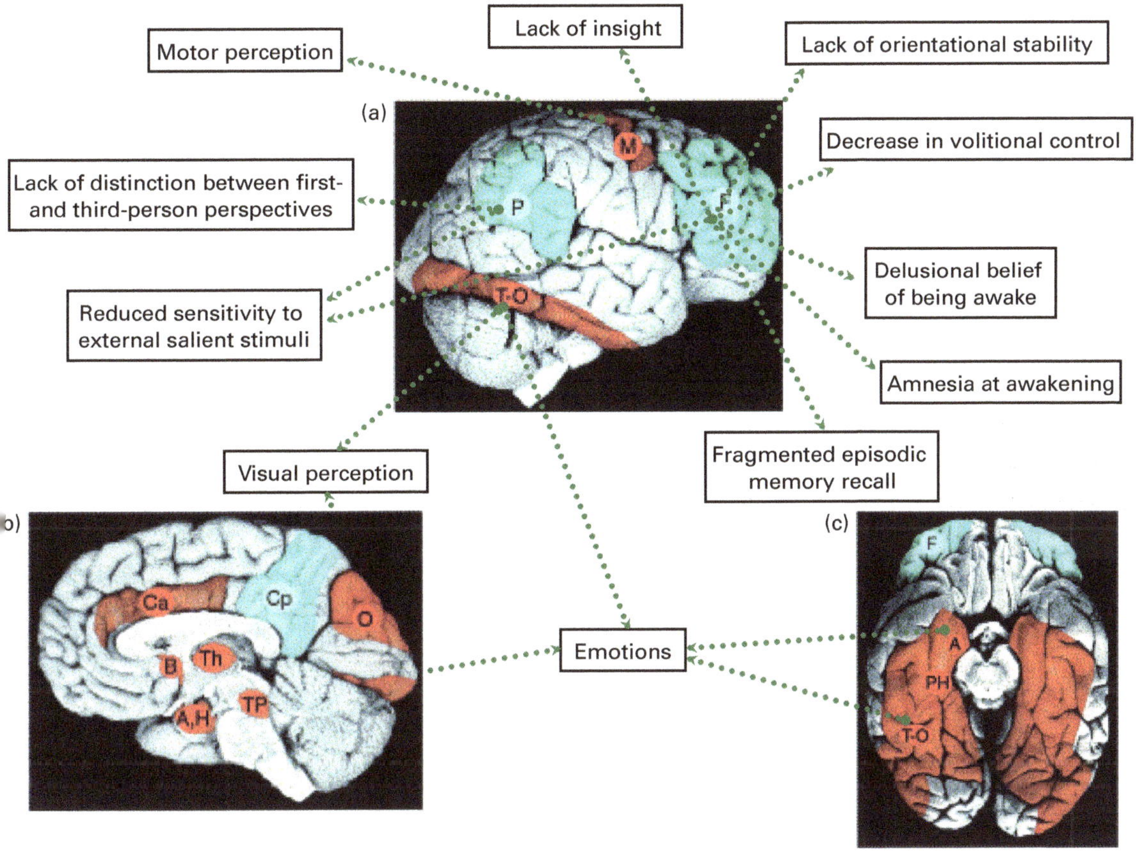

Figure 3.1
Neuroimaging of REM sleep dreaming. Figure 3.1 shows the relative increases (red areas) and decreases (blue areas) in neural activity during REM sleep; green arrows indicate their putative relationships with dream experience. (a) lateral view; (b) medial view; (c) ventral view. A = amygdala; B = basal forebrain; Ca = anterior cingulate gyrus; Cp = posterior cingulate gyrus and precuneus; F = prefrontal cortex (middle, inferior, and orbito-frontal cortices); H = hypothalamus; M = motor cortex; O = occipital-lateral cortex; P = parietal cortex (inferior parietal lobule); PH = parahippocampal gyrus; Th = thalamus; T-O = temporo-occipital extrastriate cortex; TP = pontine tegmentum. Reprinted from Dang-Vu et al. (2005), with friendly permission from EMH Schweizerischer Ärzteverlag AG; adapted from Schwartz and Maquet (2002), with friendly permission from Elsevier.

dream reports could present a solution. Indeed, the importance of using a combined methodology of studying dream reports and neuroimaging data from the same sleep stage as a necessary next step is made explicit in all the major reviews of neuroimaging studies of REM sleep and its relation to dreaming (e.g., Dang-Vu et al., 2005, 2007, 2009; Desseilles et al., 2011).[14] So far, however, this has mostly remained no more than a promising daydream.

Finally, to the extent that existing studies make any claims about the relationship between neuroimaging data and dream experience, they do not investigate the experiential, transparency, and reportability assumptions but assume them. And they do so in a much more general manner than is the case for EEG studies, because they do not deal with dream reports directly. Moreover, they build strongly on the presumed identity between REM sleep and dreaming—an assumption that no longer seems attractive in view of findings on NREM sleep dreams (Desseilles et al. [2011], for instance, explicitly address this problem). To the best of my knowledge, neuroimaging data have not been systematically applied to dream reports from NREM sleep (cf. Kussé et al., 2010).

This is an important omission. Let us assume, for the sake of argument, that neuroimaging studies had established a one-to-one correlation between, say, reports of visual dreaming following REM sleep awakenings and wakelike activation of the visual areas during the preceding REM periods. Even so, the problem would still be that vivid, visual dreams don't only occur during REM sleep but also occur during NREM sleep, which is typically, according to existing studies, characterized by a very different activation pattern in the visual areas. There would now be two possibilities: One, visual imagery in dreams might not require the type of activation seen in REM sleep, after all; in NREM dreams, a weaker (or perhaps even different) type of activation might be sufficient. Two, researchers might find that even visual NREM dreams correlate with wake- (and REM-)like activation of the visual areas; but then, existing neuroimaging findings on the *differences* between REM and NREM sleep would only apply to dreamless NREM sleep (or nonvisual NREM dreams). Though challenging to study, these are, it seems, fairly straightforward empirical alternatives. They raise questions about the neural correlates of specific types of dream experiences that can clearly only be decided, first, by relating dream reports (as well as, preferably, reports of nondreaming) to neuroimaging measurements obtained from the preceding sleep phases; and, second, by contrasting both reports of dreaming/nondreaming and neuroimaging data from REM sleep with those obtained from the stages of NREM sleep. That said, note that the strategy of contrasting measurements from REM and NREM sleep, which underlies such neuroimaging studies, is also problematic for another reason: if current sleep-stage scoring turns out to be too coarse-grained, this type of paradigm might be seen to rely on an overly sharp, dichotomous distinction of REM and NREM sleep.

In sum, existing attempts to relate results from neuroimaging studies of REM sleep to dreaming should be considered as tentative and highly preliminary. Indeed, considering existing neuroimaging studies of REM sleep as *dream* studies would be misguided: studies that do not themselves make any attempt to gather or investigate dream reports are better regarded as sleep than as dream studies. At least currently, studying sleep without studying dream reports amounts, for all practical purposes, to studying sleep without studying dreaming. Indeed, this best describes the available imaging studies; currently, connections to dreaming are primarily drawn in review articles. That said, the applicability of results from neuroimaging studies of sleep to dreaming suggests that neuroimaging studies of dreaming (or "dream imaging"; see Dresler et al., 2014) are indeed feasible—and thus may be seen as a preview of particularly promising and intriguing methodological innovations for the future study of dreaming.

Before moving on to an approach that complements this one—the practice of studying dreaming without studying sleep—I want to point out another way in which neuroimaging studies may lead to a profound rethinking of both sleep and dreaming. In a recent paper, Dang-Vu (2012) reviewed fMRI studies of sleep transcending sleep stages as conventionally defined. While most existing neuroimaging studies have focused on contrasting REM and NREM sleep, the studies reviewed by Dang-Vu investigated phasic activation patterns across sleep stages; and importantly, this altered methodological approach led to new insights that might have been missed by more conventional approaches. In particular, these studies suggest that "even during the deepest stages of sleep, neuronal network activities are sustained and organized by spontaneous brain oscillations of sleep" (Dang-Vu, 2012, p. 154). He concludes:

> While increases in brain activity during REM sleep were expected, given previous PET data, fMRI studies have demonstrated increases of brain activity in phase with specific NREM sleep oscillations, that is, spindles and slow waves. *NREM sleep is not confined anymore to a mere state of brain rest and disconnection, but now appears as an active state organized by phasic neuronal activities.* As suggested by electrophysiological data (Destexhe et al., 2007), these *transient surges in neural activity during NREM sleep might be comparable to "micro-wake fragments"* facilitating neuronal interactions. (Dang-Vu, 2012, p. 164; italics mine)

Perhaps, then, this new generation of neuroimaging studies of sleep beyond the confines of conventional sleep stages is a further indication of a profound rethinking of the concept of sleep and the methodologies involved in its study. And perhaps the reassignment of dreams to all sleep stages is indeed being paralleled by a blurring of existing sleep-stage distinctions and a departure from conventional scoring systems (cf. chap. 2). In another sense, however, this is a case of history repeating itself: while the discovery of REM sleep showed that sleep was not a uniform and passive state of rest, these new fMRI studies now seem to be doing the same for NREM sleep. Dang-Vu's conclusion is also reminiscent of the assumption of early researchers that heightened levels of brain activation are associated with wakefulness or wakelike activity. If the history of sleep and dream research is any indication, then perhaps a next step will be to broaden the conception of NREM sleep proper to include such phasic activation patterns, thus leading to a refinement not only of the concept of sleep but also of wakefulness. Dang-Vu's conclusion, down to its exact wording, thus offers an intriguing glimpse of the future while simultaneously echoing the past.

3.3 Studying Dreaming without Studying Sleep: Exclusively Report-Based Dream Research

Probably the most widespread method used in dream research is the study of dream reports gathered without additional polysomnographic or neuroimaging measurements. In this section, I discuss different methods used for the collection and analysis of dream reports, offer a working definition of dream reporting and a discussion of different types of dream reports, and discuss some challenges for the transparency and reportability assumptions arising from exclusively report-based research.

Dream reports can take different forms, such as free (written or oral) reports, questionnaires, or affirmative probes, in which subjects are instructed to report specific aspects of their dream that might otherwise go unreported (for details and further references, see Winget & Kramer, 1979; Domhoff, 1996, 2003; Kramer, 2007; Schredl, 2010a; cf. chap. 2). The latter type of report has been used successfully in the study of dream emotions. Merritt et al. (1994) found that the frequency of reported dream emotions increased tenfold when subjects were asked to report their emotions on a line-by-line basis compared to the results from the content analysis of free dream reports according to the Hall and Van de Castle (1966) scoring system. This finding suggests that the prevalence of emotion was underestimated in many earlier studies and that dreams are actually "hyperemotional," with emotions being mentioned in 95 percent of dream reports and the average dream report containing several different types of emotions. In a study comparing external and self-ratings of emotions in dream reports, Sikka et al. (2014) found that external ratings underestimate not only the frequency but also the type of emotions experienced in dreams. A particularly surprising result, which contradicts the commonplace assumption that dream emotions are predominately negative (Hobson et al., 2000), was that self-ratings showed positive dream emotions to be six times as frequent as negative ones.

The specificity of questions may also influence the reported frequency of different aspects of dreaming. In a study investigating the frequency of dream-enactment behavior in healthy subjects, Nielsen, Svob, and Kuiken (2009) found that nonspecific questions regarding such episodes produced a much lower prevalence rate (approximately 36 percent) than more specific questions. Adding examples of such behaviors produced a prevalence rate of more than 76 percent, and when an elaborate list with several specific questions was used, the prevalence rate reached 98 percent. This underscores the importance of the precise wording of instructions and of questions.

Aside from dream reports and questionnaires, researchers may also use alternative methods of reporting, such as asking participants to produce dream drawings,[15] or they may use supplementary visual material such as photographs with varying degrees of color saturation, brightness, and clarity and ask participants which of them best compares to visual imagery in their dreams (Rechtschaffen, 1983; Rechtschaffen & Buchignani, 1992). Both methods can help participants describe aspects of dreaming that are hard to capture verbally and may help circumvent the ineffability of certain qualitative aspects of dreaming.

Studies also differ regarding the number of dream reports elicited from individual participants; some studies only collect one dream report per participant, whereas others collect series of reports from the same participant, ranging from two dream reports to longitudinal dream diaries kept over years. Dream sets either can be collected from certain subject groups (e.g., men, women, patient groups) or can focus on certain dream types (e.g., typical dreams, nightmares, etc.). The ideal size of dream samples necessary to replicate the results from normative studies is also crucial (for details, see Domhoff, 1996, 2003).

Results can vary for different types of reports; for instance, verbal reports tend to be longer and more poorly organized than written ones, and report modality (e.g., written versus spoken) may affect the types of experiences described (Casagrande & Cortini, 2008). Other factors include setting (i.e., home versus laboratory reports) and the interpersonal situation (e.g., familiarity with the dream collector, but also gender; see Kramer, 2007, pp. 48–50). The scoring method is crucial (e.g., content or formal analysis; see sec. 2.2), as are the issues of how to ensure interrater reliability (Domhoff, 1996, pp. 28–51), how to control for varying report length, whether to exclude additional remarks about the dream or repetitions from the actual dream report, and so on and so forth (Kramer, 2007, 51–58).

Questionnaires or interviews may refer to specific dreams or to dreaming in general, for instance, by asking participants how frequently they recall dreams, nightmares, or certain dream themes. This difference between collecting reports of specific dreams and asking general questions about dreams is closely related to the issue of temporal proximity. For instance, general questionnaires may ask participants to assess how frequently they experience certain dream themes—such as dreams of being chased—in a typical month, at what age they first experienced such a dream (e.g., Nielsen et al., 2003; Yu, 2008), or how often they dream at all (Schredl & Fulda, 2005).

In addition, even studies aiming to collect reports of specific dreams do not always strive to minimize temporal distance from the reported dreams. For instance, Domhoff (1996, pp. 67–68) proposes the collection of most recent dreams as a reliable and more effective method for dream research. Rather than collecting small samples of dreams from individual participants, this method involves asking large groups of participants—for instance, easily accessible university students—to report the most recent dream they are able to remember and indicate whether it occurred last night, last week, or last month. Based on this information, dreams not recalled within the past few days are ruled out to ensure that the dream reports refer to fairly recent dreams. Yet it seems likely that the type of dream reported in such a situation is quite different from a dream reported immediately upon awakening and that all but the most memorable dreams require attention immediately after awakening to have even a reasonably good chance of being remembered a few days later. Dreams that can be remembered in any detail for hours, let alone days, after their occurrence are likely to form a highly selective subset of particularly memorable dreams, and furthermore one can expect such delayed dream memories to have undergone a considerable amount of editing and restructuring compared to the dream initially remembered upon awakening. The same may be true for typical dream themes, which likely refer to a subset of memorable dreams and are not representative of dreaming in general (Schönhammer, 2004, pp. 291–292).

Similar methodological problems are raised by Bulkeley et al.'s (2005) study of the content of earliest remembered dreams. Such dreams are often nightmares, and 48 percent (for women; for males it is only 20 percent) of them involve bodily misfortunes. This type of study, in which participants are asked to report the earliest dreams they remember, is not so much indicative of actual childhood dreams as suggestive that certain dreams are particularly memorable, even over years. Negative emotions and bodily misfortunes appear to be linked here, and possibly the intensity of these dreams can explain their memorability. Though interesting, these findings may be more informative about the long-term memorability and psychological impact of dreams than about dreaming per se.

How do these methodological points relate to philosophical concerns about the experiential character of dreaming, the transparency of dream reports, and the reportability of dream experience? Clearly, for studies relying exclusively on dream reports, the possibility of finding independent evidence for the corroboration of dream reports is irrelevant. And just as clearly, exclusively report-based studies of dreaming cannot provide any principled support for the transparency and reportability assumptions, nor can they raise any principled doubts about their validity. While exclusively report-based studies of dreaming do investigate, according to the transparency assumption, experience during sleep, they do not form part of sleep research proper in that they do not directly address any of the questions relevant to this field, for instance, questions regarding different sleep stages and their relation to dreaming. Accordingly, any inferences drawn from exclusively report-based studies about the sleep stages in which the respective dreams occurred are highly speculative. That said, exclusively report-based studies can help enrich the theoretical understanding of the transparency and reportability assumptions considerably: the refinement of methods for collecting dream reports is, after all, what this branch of dream research is all about.

An important insight, then, is that to be empirically plausible, a theoretical account of dream reporting will have to be able to accommodate a variety of different types of reports. Based on my earlier remarks, a nonexclusive list will have to include verbal and written reports, as well as answers to specific questionnaires or affirmative probes, but also dream drawings. Perhaps even the behavioral reenactment of specific dreams (such as using one's voice and gestures to illustrate the behavior of a particular dream character) could usefully be classified as dream reporting, and the same seems to be true of pointing at photographs to determine, for instance, the color saturation of dreams. Dream reporting seems to comprise a fairly broad set of different behaviors united by the fact that they are conducted with the *sincere intent of conveying or recording certain relevant information about a specific dream.*

This rough pass at a definition addresses some important points. One, dream reporting is a type of intentional, nonautomatic behavior; it is conducted with the intent of conveying information about previous experience during sleep to another or of recording this information for future reference. It might also be used as a means of jogging one's memory and aiding the process of dream recall: in the process of writing down a dream in my dream diary, I have often recalled parts I had initially forgotten. Not all forms of dream reporting, it is important to note, end up as scientifically investigable dream reports; this requires some additional recording activity, which may be conducted by the researcher making an audiotape of the spoken dream report or by the participant giving a written report of a dream. So it seems that while fixated dream reports of some type or another are crucial for dream research, dream reporting is the more basic and broader category.

Two, the intent of conveying or recording information in a dream report is sincere; a willfully fabricated dream report is not a "real" dream report in this sense, though the two might be hard to tell apart. As always in consciousness research, honesty and trusting the subject are important issues (for a general introduction on first-person reports in consciousness research, see Varela & Shear, 2000; Jack & Roepstorff, 2003; cf. sec. 4.5). Several studies have tried to assess whether participants are honest in reporting their dreams, but it is unclear whether judges can distinguish artificial, that is, deliberately constructed, dream reports from "real" or sincere ones (for details and further references, see Domhoff, 1996, pp. 47–48). One study suggested that the percentage of intentionally fabricated dream reports in studies asking participants to voluntarily report dreams is only 3.8 percent (Tonay, 1990–1991). By contrast, requiring participants to report dreams, for instance, when participating in a class on dreaming, may increase the frequency of fabricated reports considerably. Aside from completely fabricated reports, censorship is also an issue, for instance, when parts of the dream the participant is not comfortable reporting are omitted—which might explain the low instance of sexual dream themes in laboratory studies (Hobson, 1988, p. 245). Investigating this issue in more detail and creating situations that minimize such intentional fabrication and censorship are important—perhaps even asking subjects whether a given dream report was censored could help (for discussion of willful fabrication from a philosophical perspective, see Rosen, 2013).[16]

Three, reports are necessarily selective; no report can convey all possible aspects of an experience at the same time, if only for practical reasons. Reports can vary with respect to their *detail* and *specificity*. For instance, if I report last night's dream by saying that I dreamt of my mother, this report is not at all detailed, but specific in that it reports a certain kind of dream content. For a study investigating the gender of dream characters and their social relationship to the dreamer, this might be enough. By contrast, a very detailed account of what my mother said and did in the dream would not be at all helpful for a study investigating, say, bodily experiences in dreams; in this case, detail may even get in the way of specificity. This is why different types of dream reports may be better suited to different types of questions—and why using affirmative probes to target aspects that may otherwise be underreported can be helpful (such as emotions; cf. Merritt et al., 1994). Relatedly, free dream reports, though probably the most commonly used type of report and the one that is most central for content analysis, may be of very limited use for addressing certain questions about the qualitative character of experience: though detailed, they simply lack the relevant kind of specificity, for instance about the formal aspects of emotional dream experience. An important goal, then, is to ensure that the level of detail and specificity conveyed in dream reports is suited to the respective research question.

Finally, there is another aspect, which might be called *expressive granularity*.[17] Say I am asked to give a detailed report specifically of visual imagery in my dream; it seems that I could do so in various ways, for instance, by telling the dream or producing a dream drawing. Although both the spoken report and the dream drawing might be highly detailed and specific, they might still be poorly suited to questions concerning, for instance, the color saturation, brightness, and clarity of visual dream imagery (Rechtschaffen & Buchignani, 1992). Different types of reports have different degrees of expressive granularity—briefly impersonating a dream character might convey completely different aspects than the most detailed verbal description or the most meticulous drawing. And, of course, the expressive granularity of a given type of report can also be enhanced by giving subjects specific categories with which to describe their dreams—or even developing such categories together with them. In a study investigating variations in subjective experience during the presentation of visual images, Lutz et al. (2002) successfully used experiential categories that had been developed together with their subjects and mapped them onto differences in EEG activity (for a discussion of such neurophenomenological approaches to experience, see also Thompson, 2007, chap. 11). Similarly, guiding dream reports in an interview situation, with a researcher asking specific questions in response to the ongoing report, might also be a helpful tool, especially in explorative studies. Conceivably, this could even be combined with a diary study in the home environment, with subjects giving a written dream report upon awakening and scheduling an interview later in the day (as in descriptive experience sampling; see Hurlburt & Schwitzgebel, 2007; cf. sec. 4.5). These are just two brief examples of how the study of dream reports might profit from methods used in the study of first-person reports of waking experience.

Four, all the examples of reporting behaviors used so far refer to reports of specific dreams. Dream reports do not describe what it is generally like to dream, but describe what it was like to have a particular dream: the dream recalled upon awakening. This claim is in keeping with the principle of temporal proximity: though exclusively report-based studies, typically investigating spontaneous dream recall at home, can only approximate the temporal proximity between report and experience achieved in laboratory studies, it is nonetheless important to counteract dream forgetting by reducing this temporal gap as much as possible. This is why most studies ask subjects to report their dreams immediately upon awakening.

An important consequence is that responses to general questions of the type "how often do you dream" or "do you typically dream in color" (cf. chap. 4), in this view, are not instances of dream reporting at all. An obvious concern is that answers to general questions might be more informative about people's opinions about their dreams than about their actual dreams.[18] One can have a number of opinions about one's dream life—or one's waking life, for that matter—and find these to be false. Casual judgments, concerning either one's dreams or one's waking experiences, should carefully be distinguished from claims based on the observation or recall of specific experiences. As long as this difference is made explicit, it does not create a problem, but it does change the target of the relevant research projects: general questionnaires about dreaming target opinions or general convictions about dreams but are not, strictly speaking, part of *dream* research at all (see secs. 4.3 and 4.4 for further discussion).

A more cautious approach is to say that general opinions about dreaming cannot straightforwardly be applied to actual dreaming unless there is independent (in this case: report-based) evidence for their trustworthiness. Whether they form part of dream research proper would then be an empirical question[19]—and, indeed, studies addressing this question exist. For instance, Zadra and Robert (2012) found that subjects underestimated their dream recall frequency when asked for a general assessment of how many dreams, on average, they remembered per week. When subjects were asked to keep a dream log over a two- to five-week period, it turned out that they remembered more dreams. This was only true, however, for participants who reported dreams on a checklist log, in which they were only asked to report the number of recalled dreams and provide a brief title for each remembered dream. By contrast, subjects who were asked to keep a narrative log, that is, to report all remembered dreams in detail, did not show an increase in dream recall frequency compared to their general estimates. This finding is surprising, because keeping a dream journal is typically thought to improve dream recall. This study, however, suggests that keeping a dream diary, possibly due to motivational factors and the high demand on participants, is not the optimal method for studying dream recall frequency.

Either way, following this study, it is more important to note that at least some opinions about dreaming—namely, retrospective assessments of dream recall frequency—are indeed inaccurate when compared to dream recall frequency as measured by a checklist log. This suggests that transparency cannot be applied in equal measure to all types of reports—as well as a methodology for comparing different types of reports to determine their respective transparency. While a checklist log, in Zadra and Robert's (2012) study, appeared to be best suited for the assessment of (spontaneous) dream recall frequency, such a log would not reveal anything interesting about the content of the specific dreams—for this, the detail and specificity of a narrative log would be needed. Moreover, results may vary depending on which types of reports are used as the baseline. Schredl and Fulda (2005) found that their participants' retrospective estimations of dream recall frequency were indeed similar to the number of dreams they were able to report when keeping a diary for two weeks, which seemed to support the reliability of general opinions about dream recall frequency (see also Schredl, 2004). This conclusion now seems to be called into question by Zadra and Robert's (2012) study—and given future studies, even the comparatively more accurate checklist logs might not turn out to be the ultimate standard for assessing dream recall frequency. The lesson from all of this is that different types of reports can be afforded different degrees of transparency with respect to different aspects of dreaming. Determining which type of report is most trustworthy in light of a specific research question requires the systematic comparison of different types of reports. Understood in this sense, the relative transparency of different types of dream reports is no longer a principled and theoretical problem or a default assumption but a target for empirical dream research.

Finally, an obvious concern regarding exclusively report-based studies is that they target only a subgroup of dreams. Researchers commonly agree that spontaneously recalled dreams make up only a small fraction of actual dreaming. While most subjects can recall four to six dreams per night in the sleep laboratory, they only occasionally remember their dreams spontaneously at home. Spontaneously recalled dreams probably make up no more than 5 to 10 percent of dreams in general (Domhoff, 1996, p. 41). Consequently, exclusively report-based studies only target a subgroup of dreams. Goodenough (1991) has suggested that spontaneously recalled dreams may be more salient and emotionally intense than forgotten dreams, which may be more mundane and less bizarre. If this is correct, then spontaneously recalled dreams, though crucial for our waking convictions and opinions about our dreams, might actually be atypical dreams. It is, however, a matter of controversy whether saliency really is the driving factor behind dream recall. Based on a review of studies comparing dreams reported after timed awakenings with morning dream recall, Domhoff (2003, pp. 42–46), argues that duration and recency are more important than saliency for dream recall.

A careful conclusion is that exclusively report-based studies can strive to enhance the space of actually reported dreams by using different reporting techniques, asking subjects to report their dreams immediately upon awakening, increasing the expressive granularity of reports, and so on—but still, it would not make sense to apply the reportability assumption to dreams reported in these types of studies: clearly, spontaneously recalled dreams are an extremely poor indicator of the number of dreams subjects can report in the laboratory. Spontaneously recalled dreams are not, as modern researchers are well aware, to be conflated with dreaming per se. Whether they can be used as a proxy for the types of dreams reported following timed awakenings continues to be an open question. Yet, such reports are transparent, or so one would hope, with respect to spontaneously recalled dreams. For many research questions, this type of limited or restricted transparency is enough.

3.4 Studying Dreaming through Sleep Behavior I: Sleep Disorders

As several philosophers discussing Malcolm's analysis of dreaming pointed out (cf. sec. 1.3), one possibility for corroborating dream reports is by observing different types of sleep behavior such as sleep talking or sleepwalking. The question, then, is what is known about the association between sleep behavior and conscious experience during sleep.

Many well-known types of sleep behavior, such as sleepwalking, sleep talking, confusional arousals, and sleep terrors, as well as less well-known phenomena such as sleep eating, sleep cooking, and sleep sex, occur during NREM sleep (for a general introduction, see Schenck, 2005, 2007). The specificity and seemingly goal-directed nature of these behaviors is as fascinating as it is potentially dangerous. In one highly publicized case, a man drove fourteen miles to the house of his parents-in-law, where he stabbed his wife's mother and almost killed her father, all while he was asleep. He turned himself in to the police when he began to realize what had happened. He was pardoned—setting a precedent in Canadian legal history (for a detailed account, see Callwood, 1990; see also Cartwright, 2010). Indeed, around 2 percent of the general adult population report having exhibited violent behaviors arising from sleep (Mahowald et al., 2005). Although I am primarily interested in the theoretical questions raised by sleep behavior and its relation to dreaming here, I do want to emphasize at the outset that violent sleep behavior raises pressing medical and legal questions as well (Mahowald et al., 2005; Siclari et al., 2010).

Despite its specificity, sleep behavior is not consistently associated with recall of preceding experience. Children typically have complete amnesia for sleepwalking episodes, but adults can often recall dreamlike experiences, many of which have to do with their being threatened and in imminent danger (Schenck, 2005). Violent sleep behavior, in particular, often appears to be a reaction to a dreamed provocation, but also to noise, touch, or close proximity to another (Siclari et al., 2010). In a study investigating the occurrence of dreamlike mentation during episodes of sleepwalking and sleep terrors, Oudiette, Leu, et al. (2009) found evidence for a variable relationship between dreaming and NREM sleep behavior. Though patients and controls underwent an overnight video-polysomnogram in the sleep laboratory, dream reports were collected in an interview asking patients about *any* dreamlike experience associated with episodes of sleep behavior throughout their lifetime. Seventy-one percent of these patients reported at least one dreamlike mentation associated with such an episode; most of these were short, consisting of a single visual scene, and they were frequently unpleasant, involving aggression, misfortune, or apprehension. However, because of the methodology used, it is not clear whether these reports accurately represent the incidence of dreaming during sleepwalking and night terrors (see sec. 3.3). In general, the association between NREM parasomnias and dreaming therefore needs more systematic study, and not all instances of NREM sleep behavior associated with phenomenal experience should be regarded as *dream*-enactment behavior.

This variable relationship to dreamlike experience is borne out by polysomnographic measurements obtained during episodes of NREM sleep behavior, with EEG recordings during episodes of sleep talking and sleepwalking showing the persistence of sleep, the admixture of sleep and wakefulness, or even complete wakefulness (Schenck, 2007). So far, a single brain imaging study has been conducted immediately after an episode of sleepwalking, finding evidence for a dissociation between motor arousal and sleep (Bassetti et al., 2000). It thus seems possible that the specificity of certain sleep behaviors might be explained by saying that only a subset of sleep behaviors associated with conscious experience involve dreaming, with others involving residual processing of external sensory stimuli.

Could there be cases in which sleep behavior arises independently of conscious experience altogether? Revonsuo (2006) argues that if this were the case, subjects exhibiting sleep behavior could be regarded as temporary zombies, thus casting doubt on the causal relevance of phenomenal experience for behavioral control. He concludes, however, that the evidence at best supports a weak form of zombiehood, noting that in most cases involving complex sleep behavior such as driving, subjects are later able to report at least some fleeting images or impressions, thus ruling out that they were completely unconscious.

Some cases of moderately complex behavior coupled with dense amnesia for the events may, however, indicate nearly complete absence of subjective experience during the episode. Even if in those cases the patients can be regarded as totally nonconscious zombies, their behavior is far from the flexible and integrated types of behavior guided by a consciously apprehended model of the world. (Revonsuo, 2006, p. 391)

If this is correct, it supports the claim that there is indeed a systematic relationship between the complexity of behavior and conscious experience. The next question, then, is whether one could use sleep behavior to corroborate retrospective dream reports, perhaps even informing subjects what they had dreamt based on the behavior they previously exhibited during sleep. Note, however, that Revonsuo's argument is in fact at odds with this line of reasoning. By taking amnesia and reports of nondreaming to be indicative of the absence of conscious experience during the preceding sleep period, he implicitly subscribes not only to the transparency but also to the reportability assumption. Indeed, he notes that the existence of true zombiehood would only be supported if retrospective reports describing a loss of consciousness, including "how phenomenal contents disappeared and a blackout, a total phenomenal void, commenced" (Revonsuo, 2006, p. 386), as well as the reappearance of phenomenal content afterward, were available. But this also means that sleep behavior cannot, as suggested by Malcolm's critics, be used to corroborate dream reports. The opposite is true: retrospective reports are needed to determine whether a given instance of sleep behavior was indeed associated with experience.

To this, one might object that while not at all instances of sleep behavior are associated with experience during sleep or even with dreaming, well-documented cases of *dream*-enactment behaviors exist and at least in these cases, sleep behavior could be used to corroborate dream reports after all.

The strongest examples come from RBD patients who, due to a loss of muscular atonia during REM sleep, exhibit a wide range of complex behaviors during sleep. These include violent behaviors such as kicking or punching, but also nonviolent behaviors, including sexual behavior, mimicking eating, drinking, or smoking, as well as laughing, singing, dancing, whistling, clapping, or gesturing (Oudiette, de Cock, et al., 2009). The frequency of enactment dreams is highly variable, ranging from several per night to once a month (Boeve, 2010). RBD is often associated with or even predates neurodegenerative diseases such as Parkinson's disease or Lewy body dementia and typically arises in older men (Schenck & Mahowald, 1996, 2002; Mahowald & Schenck, 1999; Fantini, Ferini-Strambi, & Montplaisir, 2005; Schenck, 2005; Boeve, 2010; Arnulf, 2012; Postuma et al., 2012).[20]

For the present context, it is noteworthy that the dreams reported by RBD patients often correspond quite closely to the observed behaviors (Nielsen, 2010b). In addition, researchers have long assumed that dreaming undergoes characteristic changes in these patients. In one study using the most recent dream method, RBD patients described more aggressive, action-filled, and stereotyped dreams (often involving animals) than healthy controls (Fantini, Corona, et al., 2005). This change in dreaming, which stands in marked contrast to the placid and mild-mannered daytime temperament of most of these patients (Schenck, 2005; Boeve, 2010), may also lead to an increase in dream recall frequency (Fantini, Corona, et al., 2005).[21] Anecdotal evidence also suggests that recall for RBD enactment dreams is more stable than for ordinary nightmares, with patients sometimes vividly recalling their dreams after days, weeks, or even years (Boeve, 2010).

Does RBD, or dream-enactment behavior more generally, present empirical evidence for the experiential and transparency assumptions? And could dream-enactment behavior even be used as evidence for the experiential character of dreaming independently of dream reports? If so, this would contradict my claim that the experiential and transparency assumptions are default assumptions incapable of empirical corroboration.

On the first view, this claim seems to be supported by anecdotal evidence. As Boeve (2010) writes:

> The vocalizations and behaviors that are exhibited are strikingly consistent with the content of the dreams later reported by the patient—the behaviors mirror the dream content. Bedpartners are often very accurate at predicting what the patient had just dreamed about based on their observations of the dream enactment behavior. One vivid example involved a man who held his wife's head in a headlock while moving his legs as if running while both were attempting to sleep in bed, then exclaimed, "I'm gonna make that touchdown!" and then attempted to forcefully throw her head down toward the foot of the bed. When awakened, he recalled a dream in which he was running for a touchdown, and he spiked the "football" in the end zone. His wife knew precisely what he had been dreaming about. (Boeve, 2010, p. 18)

In a recent study, Leclair-Visonneau and colleagues (2010) aimed to investigate the scanning hypothesis (see sec. 1.3.3; for details, see Schredl & Erlacher, 2011a) by correlating the sleep behavior observed in RBD patients with their eye movements as measured by EOG. Their findings were impressive, suggesting that in more than 80 percent of cases in which apparently goal-directed behavior (such as strangling an attacker, talking to someone and kissing them, climbing up a ladder, and waving and grabbing invisible objects) was observed, the eyes of their patients moved in concert with these movements. However, such goal-directed behaviors were generally rare, and out of 56 patients observed during 77 nights, only 19 apparently goal-directed behaviors were observed in 9 patients. The authors conclude that their results "indicate that the limbs, hands, head and eyes move together in a coherent, directional manner during RBD. The final gaze is placed in the direction and plane of the fictive target" (Leclair-Visonneau et al., 2010, p. 1742), suggesting that eyes and limbs move in response to dream imagery.

Though these findings seem compelling, note that in this study, patients were only interviewed about possible dream experiences after spending a night in the sleep laboratory, and these dream reports were not used in the analysis. The conclusions just summarized were based first on a comparison between observed eye movements and sleep behavior, and second on the comparison between these coordinated sequences of eye, head, and limb movements with those observed in healthy controls instructed to reproduce such behaviors, with eyes closed and eyes open. Yet the authors take their results to be informative about the correlation between eye movements and dream imagery as well:

> The main addition of this study to preceding experiments on the scanning hypothesis (based on dream-recall and a posteriori correlation) is the presence of an external, dreamer-independent, "on line" check of the dream scenario. This is key to avoid the numerous biases associated with the a posteriori recall method (total or partial amnesia, censure, reconstruction of a scenario, clarity of recall, simplicity of the narrative, intervention of an external investigator). (Leclair-Visonneau et al., 2010, p. 1742)

This latter assumption, however, is problematic, as other studies (Scaglione et al., 2005; Oudiette et al., 2011) found that RBD patients may be unable to recall dreams following apparent enactment episodes during sleep. It at least seems to be possible, then, that the occasional failure to report dreams is related to the absence of prior dream experience—this would follow from the reportability assumption, but also from NREM sleep behavior in the absence of dream recall. By contrast, the authors write:

> The major strength of the RBD model is that, in cases with scenic behaviours, the dream report from the dreamer is not necessary, as the behaviour is so demonstrative (especially when speeches accompany the behaviour) that any observer having the same cultural background will be able to determine its nature: smoking, kissing, gesturing "thumbs up," clapping, inspecting the army, giving a lesson. (Leclair-Visonneau et al., 2010, p. 1745)

And because the correlation between apparently goal-directed behavior and eye movements in RBD patients renders a correlation of both with dream imagery probable,

> It is also probable that REMs in normal subjects are directed towards the targeted dream images. In conclusion, our results support a tough link between the dream action (when goal-oriented) during RBD and the direction of the gaze, which can be extended by analogy to normal dreamers in REM sleep. If REMs are always temporally synchronized with mental imagery during REM sleep, they could be a reliable marker of gaze direction during dreams. (Leclair-Visonneau et al., 2010, p. 1745)

In this view, the scanning hypothesis can not only be confirmed independently of dream reports in RBD patients but also be generalized to the dreams of healthy subjects independently of observed behavior. Readers will not be surprised that I would once more plead for caution; rather than proving the scanning hypothesis, I think this study raises fascinating questions for future research and presents a viable method for their investigation: namely, correlating eye movements with apparently goal-directed sleep behaviors and investigating their relation to dream experience, as reported following timed awakenings in the laboratory. By contrast, simply assuming that the proposed methodology could be used to confirm the scanning hypothesis independently of dream reports once more falls prey to the familiar circularity: the behavior observed in RBD can only be considered as *dream*-enactment behavior insofar as its association with dream imagery is supported by retrospective dream reports. This same behavior cannot then be used, however, as evidence for dream-enactment behavior occurring independently of, and even in contradiction to, dream reports: these cases could equally well fail to be *dream*-related in the relevant sense at all.

That said, we have a more basic cause for concern—which once more points to an important desideratum for future research. Though the correlation between dream reports and RBD sleep behavior is indeed impressive, it mostly relies on anecdotal evidence, rather than on a systematic mapping of dream reports onto previously observed sleep behavior. The study by Fantini, Corona, et al. (2005) cited earlier presents an exception in that it systematically compared the dreams of RBD patients with those of a control group, thus presenting the first attempt to study the content of the dreams of RBD patients in detail. However, because this study used the most recent dream method, the differences the authors found in dream recall frequency and dream content between RBD patients and healthy controls are temporally distant from the actual dreams and possibly more error prone than reports given immediately upon awakening. Consequently, it is also unclear whether the dream reports given by RBD patients in their study referred to dreams involving behavioral enactment or not. Without this information, their conclusions can only be regarded as preliminary. Moreover, a recent study investigating dream diaries from RBD patients and healthy controls failed to find a difference in aggressive dream content (D'Agostino et al., 2012). Whether these results are related to the fact that a majority of patients were undergoing medical treatment at the time of the study is unclear. The correlation between changes in dream content and enactment behavior in RBD would have to be investigated in more detail to ensure that it is indeed robust. All the existing evidence suggests that this would be a tremendously promising research project; but once more, it should be clear that to turn this work into a truly vigorous, scientific project, a more cautious look at dream reports following timed awakenings would be needed.

Let us now turn to the claim that evidence from dream enactment in RBD patients can be generalized to the dreams of healthy subjects. A similar analogy between dream-enactment in RBD and standard dreams is defended by Revonsuo (2006), who argues that the conjunction of complex sleep behavior and retrospective dream reports in RBD patients presents evidence against the Dennettian scenario of dreams as instantaneous memory insertions. Revonsuo argues that this is true for cases in which an RBD patient, upon awakening, reports

> a dream in which he underwent exactly the same actions. Thus, the complex behavior that was objectively observed while the sleeper was still in REM sleep can only be explained by referring to the sequence of dream events and dream actions that the dreamer experienced during REM sleep without atonia. There is no reason to suppose that the removal of atonia as such should some-

how cause dream experiences; thus, the most reasonable hypothesis is that dream experiences are brought about in the brain whether or not muscular atonia is present. (Revonsuo, 2006, p. 78)

Importantly, his claim is not that sleep behavior provides independent evidence for the trustworthiness of dream reports; instead, he defends the more cautious claim that RBD dream-enactment behavior only presents evidence for the experiential character of dreaming given its correspondence with the retrospective dream report. In a second step, however, he additionally assumes that because RBD presents evidence that *some* dream reports refer to phenomenal experience during sleep, dream reports can *generally* be regarded as describing experiences in sleep, even in the absence of corroborating behavior. Once the experiential assumption has been established for a limited number of RBD dreams, it can, in this view, be expanded even to cases where the only evidence of previous dreaming consists in a dream report (Revonsuo, 2006, pp. 78–84).

The issue here is not, I should emphasize, whether the dream reports themselves should be trusted in such cases. Rather, it is whether their trustworthiness is justified by the analogy between standard dreams and RBD enactment dreams. As mentioned earlier, the dreams of RBD patients are thought to differ in characteristic ways from the dreams of healthy subjects, and it is not known whether this is only true of their enactment dreams or of their dreams more generally—or how exactly this is related to changes in muscle tone. As Boeve (2010) notes, overt enactment episodes are rarely observed in the laboratory, how to differentiate between pathological and nonpathological increases in EMG remains unclear, and we do not know why the frequency of sleep behavior differs in RBD patients. One possibility, it would seem, is that RBD patients always have more vivid, action-packed dreams but only enact some of them due to a loss of muscular atonia; another is that RBD patients enact only dreams involving particularly vivid motor imagery (or perhaps these dreams are also more vivid and more memorable precisely because they are enacted). However, if enactment is specific to certain types of dream content, then generalizing from these examples to dreams lacking such content does not seem wise.

Moreover, dream-enactment behaviors may also be more prevalent among healthy subjects than previously believed, raising questions about the exclusiveness of this symptom to RBD. Dream-enactment behaviors, for instance, are prevalent among healthy undergraduate students, taking various forms such as speaking, laughing, motor activity, and emotional expressions such as crying, anger, or fear (Nielsen et al., 2009). They also occur especially often postpartum, and in a study conducted by Nielsen and Paquette (2007), approximately 60 percent of recent mothers reported at least one dream-enactment episode. These were often associated with stereotyped nightmares of searching for their baby in the bed and could involve motor activity such as searching behavior, speaking out loud, or crying. While similar types of behavior are often seen in RBD patients, the frequency and severity are greatly reduced in healthy subjects, who report an average of fewer than six episodes per year (Nielsen et al., 2009). Still, this raises the question of whether RBD is simply a more severe expression of nonpathological dream-enactment behavior and of whether the latter might predict future RBD symptoms.

And finally, even if one accepts that dream-enactment behavior in RBD patients supports general conclusions about the experiential character of dreams (Revonsuo, 2006), it is not clear that the same holds true for more specific claims, for instance, regarding the coordination between gaze shifts performed within the dream, real eye movements, and dream body movements (Leclair-Visonneau et al., 2010; Arnulf, 2011). To generalize such findings to standard dreams in healthy subjects, it would seem that questions about the exact relation between changes in dreaming and sleep behavior would first have to be answered. For the time being, however, it seems that we should treat such specific generalizations with caution.

In sum, sleep behavior raises fascinating and potentially fruitful questions about its relation to dreaming, but it does not seem to provide independent evidence of the experiential and transparency assumptions or a means of moving beyond the study of dream reports altogether. Importantly, even if RBD enactment dreams make the experiential assumption seem particularly compelling, they do so for the same reason, essentially, that would also lead to its endorsement for standard dreams: the assumption that dream reports are transparent.

One might still wonder, however, whether the opposite might be the case. That is, could the *lack* of correspondence between sleep behavior and retrospective reports be used to *disconfirm* the transparency assumption? This possibility seems particularly compelling for a more familiar type of sleep behavior: sleep talking. In a series of studies, Arkin and colleagues investigated the correspondence between sleep talking and retrospective dream recall. They found that generally, 70 to 80 percent of instances of sleep talking, or somniloquy, arise from NREM sleep, and only between 20 and 25 percent from REM sleep; interestingly, however, their most prolific sleep talker talked predominately during REM sleep (Arkin et al., 1970b). The level of correspondence between incidences of sleep talking and retrospective mentation reports varied highly across sleep stages. In another laboratory study, Arkin and colleagues (1970a) matched 166 sleep speeches from 28 chronic sleep talkers to retrospective mentation reports; they found that sleep speeches from REM sleep had the highest correspondence with retrospective mentation reports (80 percent), while this was only the case for 45 percent from stage 2 NREM sleep and 21 percent from stages 3 and 4 NREM sleep. Note that these numbers fit in well with the frequency of dream reports following REM (82 percent) and NREM (43 percent) sleep awakenings found in other studies (Nielsen, 2000a). An initially attractive assumption, then, would be that in cases demonstrating no correspondence, this might be due not to forgetting but to an absence of preceding experience: sleep talking might then be better regarded as automatic behavior arising independently of dream experience.

In a study using posthypnotic suggestion, Arkin and colleagues instructed a single chronic sleep talker to describe his dreams while asleep; this increased the overall frequency of sleep talking, as well as the overall degree of correspondence with REM sleep reports (87.5 percent versus 20 percent from NREM sleep). It also diminished dream recall, both at home and in the laboratory, throughout the experiment (Arkin et al., 1966)—an effect that might have been caused, one might speculate, by laboratory fatigue. At least in some cases, sleep speech was more specific than, or even differed from, the retrospective report.

Sleep speech: "Mm—mmm—how about this hat (?) from India—how about—how about—hat from India where—where it was made by camel driver—damn thing [pause] pretty little bugger."

Waking report: "Uh—uhm—uh—gee—I—something about—yeah—a hat—belonging to you know—one of those old sheep herder hats—I don't remember what it was all about though." (Arkin et al., 1966, p. 305)[22]

Rosen (2013) argues that such discordance (e.g. between a sheep herder hat and a camel driver hat) suggests that parts of the dream were forgotten or even fabricated upon awakening. She also worries that dreamers may knowingly distort their reports to cover up embarrassing details or compensate for memory loss (p. 6), for instance to satisfy experimental demands. To this, I would respond as follows. First, while instances of *willful* fabrication are a genuine cause for concern, this is a methodological problem, but does not present a principled objection to the transparency assumption, which only applies to dream reports given with the *sincere* intent of accurately conveying information about preceding experience (cf. sec. 3.3). Second, I agree with her observation that timed awakenings do not *guarantee* a complete and accurate report; however this observation does not depend on discordance between sleep speech and retrospective report, but instead, the dream report itself draws attention to this fact: the various "uhs," "uhms," and "gees" suggest, quite explicitly, that the dreamer is uncertain or hedging. Indeed, Arkin et al. (1970a) point out that in their study, dreamers often described the feeling of having experienced more than they could recall. Thus Rosen's assertion that the dream report is incomplete does not contradict the dream report's trustworthiness but rather assumes it. Third, let us assume, for the sake of argument, that the dreamer had been quite certain, in this case, that he dreamed of a sheep herder's hat rather than a camel driver's, thus directly contradicting his previous sleep speech. Would this thereby be a sufficient reason for doubting the dream report?

I think it would not, though several interpretations are available. Perhaps the dreamer, after awakening, realized that what he took to be an Indian camel driver's hat in his dream was, in fact, a sheep herder's hat (I am not exactly sure what the difference between the two might be, but let's assume there is one and that it is known to the dreamer). Further questioning might even reveal what led the dreamer to this erroneous conclusion and thus produced the misleading sleep speech. Perhaps, the apparent conflict between the sleep speech and the report could be resolved in this manner. But things might turn out otherwise. Perhaps reporting conditions, in this case, were not optimal, thus casting doubt on the transparency of this particular dream report. Even so, this would not be a cause for grave concern: recall that the transparency assumption is a default assumption driven by methodological and pragmatic considerations; in particular, saying that the assumed correspondence between dream reports and dreaming *generally* best explains dream reporting behavior does not exclude the possibility that alternative explanations, such as misremembering, willful fabrication, or suboptimal reporting conditions might *better* explain individual instances of dream reporting. This is always a possibility, as Rosen rightly points out; I merely disagree with her conclusion that this warrants a more principled skepticism about the trustworthiness of dream reporting in general. Such cases, in my view, may even be a valuable way of making progress on identifying ideal reporting conditions and restricting the transparency and reportability assumptions to more convincing cases.

But this latter possibility raises a further and deeper question: is sleep talking a type of evidence about dreaming that potentially trumps retrospective reports? Rosen seems to think, at least for cases that exhibit no clear correspondence between retrospective report and sleep speech, that it might be. As Arkin's (1966) review of the literature on sleep talking shows, however, there is a long-standing dispute about whether sleep talking is, as Adriani (1892) beautifully puts it, a spoken dream reflecting ongoing dream experience (quoted in Arkin, 1966, p. 111) or an instance of automatic sleep behavior unrelated to dream activity. Only in the former case, however, could sleep speech even potentially meet the conditions for being considered as *reporting* behavior—and only then would the transparency assumption apply to sleep speech. Whether (a given instance of) sleep speech can indeed be regarded as a *report* of ongoing experience, however, is a question that would have to be confirmed by the retrospective report; it would depend on whether the sleep speech was accompanied by ongoing experience and on whether it was intended as a sincere report of this experience. Thus, as is the case for other nonverbal forms of sleep behavior, retrospective dream reports (under the assumption of transparency) are a means of investigating the correspondence of sleep speech with dream experience—and hence sleep speech cannot, on pains of circularity, be used to prove or disprove the trustworthiness of dream reports.

As things stand, then, even for sleep speech arising from REM sleep, the lack of a corresponding retrospective report might indicate an absence of preceding experience rather than forgetting or (willful or unwitting) fabrication. It is heartening to see that according to Arkin and colleagues' studies, this rarely occurs for REM sleep; but even if things had turned out otherwise, this would not change my general point. Moreover, in many cases, sleep speech will be so truncated that the retrospective report is needed to make any sense of its relation, if any, to previous experience. Take the following example from stage 2 NREM sleep, where the subject utters a single word: "No-O." The retrospective report reads: "I was thinking that—you told me—to stop dreaming and take a walk, and I said, 'No' because I was hooked up to the bed—and then there were about 6 people in the room unhooking me" (Arkin et al., 1966, p. 387). Here the "no" apparently cannot itself be regarded as an intentional report, and while the retrospective report suggests that it indeed corresponded to ongoing dream experience, this correspondence is only made visible by the retrospective report. Sleep speech, in the absence of retrospective confirmation, has no claims to transparency of its own.

We have seen, then, that even hypothetical intentional instances of reporting while dreaming would not override retrospective dream reports, and for principled reasons. Yet one may wonder whether any such cases exist. The example of lucid dreaming seems to be the most obvious candidate.

3.5 Studying Dreaming through Sleep Behavior II: Lucid Dreams

Lucid dreaming is interesting because it offers the promise of an entirely new type of evidence about dreaming: dream reports obtained directly from the dream state, while the subject is still asleep. For the same reason, it also offers a much more precise methodology for timing the occurrence of particular dream events. The term *lucid dreaming* was coined by the Dutch psychiatrist Frederik van Eeden in 1913.

> In these lucid dreams the reintegration of the psychic functions is so complete that the sleeper remembers day-life and his own condition, reaches a state of perfect awareness, and is able to direct his attention, and to attempt different acts of volition. (van Eeden, 1913)

The phenomenon of becoming aware, while dreaming, of the dream state is much older and has been known throughout history. In his treatise *On Dreams*, Aristotle observed that "often when one is asleep, there is something in consciousness which declares that what then presents itself is but a dream" (Aristotle, 350 BCE/1994–2009a, Pt. 3). One of Descartes's famous Olympian dreams may have involved such lucid insight as well, as he reportedly interpreted his dream while still asleep (Cole, 1992, pp. 149–170; see sec. 1.1). Leibniz was also familiar with the phenomenon and apparently experienced such dreams repeatedly:

> It is to be noted, however, that now and then the dreamer himself observes that he is dreaming, yet the dream continues. Here he must be thought of as if he were awake for a brief interval of time, and then, once more oppressed by sleep, returned to the previous state. It is also to be noted that some men can wake themselves up, and it is a familiar experience of mine that, when some pleasing vision presents itself, I notice that I am dreaming and try my eyes and pull them open with my fingers to admit the light. (Leibniz, 1880; quoted and translated in Hacking, 2001, pp. 252–253)

Unlike Leibniz's description of the phenomenon, as well as many other historical references to lucid dreaming (for details and further references, see LaBerge, 1988a), van

Eeden does not simply identify lucidity with insight into the fact that one is dreaming. Rather, he describes a unique type of dream characterized by both full intellectual clarity and the ability to control the dream. The question of how to define lucid dreaming remains problematic, with some contemporary authors accepting this stronger type of definition (LaBerge, 1990, p. 109; Tholey & Utecht, 1995/2000, pp. 29–30), and others defining lucid dreaming only in terms of insight into the fact that one is currently dreaming (see, e.g., Green, 1968; for a critical discussion, see Windt & Metzinger, 2007).

Because dreaming is often regarded as a state of widespread cognitive deficiency (e.g., Hobson et al., 2000; see chap. 9 for details), the notion of lucid insight in sleep has been met with considerable skepticism. An important objection is that the degree of critical reflection required for lucidity is simply incompatible with dreaming (see sec. 9.2 for further examples of this view).[23] To give an example, Havelock Ellis asserted:

> [We] can never go behind the fantastic universe of our dreams. The validity of that universe is for dreaming consciousness unassailable. . . . The phenomenon occurs; the person who says to himself that he is dreaming believes that he is still dreaming, but one may be permitted to doubt that he is. It seems far more probable that he has for a moment, without realising it, emerged at the waking surface of consciousness. (Ellis, 1911/1931, p. 64)

The objection that dream lucidity is related to a period of actual (rather than metaphorical, as claimed by Leibniz) wakefulness during sleep was targeted directly by two researchers, Keith Hearne and Stephen LaBerge, who independently worked on lucid dreaming as part of their PhD projects in the late 1970s. Both provided evidence for the occurrence of lucid dreams in unequivocal REM sleep by using an ingenious experimental setup (Hearne, 1978; LaBerge, 1988b; Schatzman et al., 1988). Building on the scanning hypothesis, according to which the gaze shifts performed by the dreamer within the dream correspond to the actual eye movements of the sleeping subject as recorded on the EOG, trained lucid dreamers executed prearranged patterns of eye movements while dreaming. The signals for the onset of lucidity and wakefulness—two and four pairs of eye movements, respectively—showed up clearly on the EOG (LaBerge, 1988b, p. 141; see fig. 3.2).[24] Moreover, since subjects confirmed, after awakening, having executed these signals during their dreams, there was a close correspondence between objective polysomnographic measures and subjective reports. This experiment not only proves that critical awareness and deliberate action control are compatible with the dream state but also that dreamers can manipulate one of the physiological features of REM sleep, namely, the eye saccades, while dreaming. Aside from intentionally directed eye movements, LaBerge (1990) has also suggested that muscle twitches registered on the EMG—associated with intentional fist clenching within the dream—could be used in this type of experiment. While a number of studies have targeted the psychophysiological correlation between bodily actions performed within lucid dreams and changes in the sleeping body, eye movement patterns are still typically used in most studies of lucid dreams (for a general discussion, see Erlacher & Schredl, 2008b; and sec. 8.2.2 for details).

These experimental studies of lucid dreaming are important for several reasons. First, they prove that lucid dreams do in fact occur during REM sleep rather than wakefulness (LaBerge, 1990)—or at least this was the standard interpretation so far (see hereafter)—and thereby contradict the widespread notion of REM sleep as a period of state-dependent cognitive deficiency.

Second, several philosophers have argued that the example of lucid dreaming provides empirical evidence against Malcolm's rejection of dream experience and of the possibility of judging during sleep, as well as against Dennett's cassette theory of dreaming (cf. Bernhard, 2005; Revonsuo, 2006, pp. 77–78). Kramer, who is one of the few contemporary dream researchers to explicitly address philosophical skepticism about dreaming, explicitly considers laboratory studies on signal-verified lucid dreaming "a direct empirical contradiction of Malcolm's description of what sleep is like" (Kramer, 2007, p. 19).

Third, to the extent that these deliberately generated eye movement signals are a means of communication, they are not simply behavioral artifacts but real actions. Control over volition and attention enables lucid dreamers to act, as it were, within the dream state. This, in itself, is unique, because nonlucid dreamers, at least to the extent that they do not reflect on their own behavior within the dream, are not agents in a strong sense (for details, see Windt & Metzinger, 2007).[25] The signaling experiment, however, shows that the dreamer's functional isolation from the waking world is not as impenetrable as commonly thought. Though the degree of action control is strongly limited by REM sleep paralysis, lucid dreamers are able to perform goal-directed actions not only within the dream but also, via eye movement signals, in the real world, and these actions have objectively measurable effects. It is because of their intentionality—retrospectively attested to by the dreamer—that these eye movements should be regarded not merely as behavioral artifacts or even behavioral evidence but as communicative actions. In particular, the eye movement pattern is executed with the aim of conveying a certain message, such as "I have just noticed that I am dreaming," "I am awake," or "I am now beginning to execute a predetermined task in my lucid dream" (cf. fig. 3.2). Though simple, these eye movement patterns are clearly used as meaningful signals. For this reason, I think they are not just an example of concurrent behavioral evidence of dreaming—as is the case for RBD enactment dreams—but should count as concurrent dream reports, that is, as behaviors conducted with the sincere intent of conveying or recording certain relevant information about a specific dream (cf. sec. 3.3), even though the expressive granularity of such reports is poor.

This reading is shared by other philosophers working on lucid dreams. Metzinger notes that the ability to submit signals during an ongoing lucid dream can be considered a type of "transreality communication" (Metzinger, 2003, p. 536), and Revonsuo allows that "the lucid dreamer can in principle report in real time to external observers in a sleep laboratory that he is just *now* enjoying dream experiences" (Revonsuo, 2006, p. 78). Contrary to Malcolm's and Dennett's argument, dream reports are not necessarily retrospective—and the existence of concurrent dream reports is an empirical discovery.

That said, my previous point still stands: neither concurrent behavioral evidence nor concurrent dream reporting behavior can provide stand-alone evidence of dream experience: to rule out the danger of false positives, retrospective confirmation is necessary. This is important, because both the successful induction of lucid dreams in the laboratory and the ability to gain sufficient control to execute prearranged gaze shifts within the dream require practice and the waking intention to become lucid and conduct these gaze shifts within one's dream (LaBerge, 1990). And because such waking intentions have a way of cropping up in dreams, it seems conceivable that a dreamer could make these very eye movements within a nonlucid dream, as well. Brooks and Vogelsong (1999) note that they sometimes became lucid only after beginning to conduct previously planned experiments during their dreams, rather than vice versa. Similarly, Stumbrys and Erlacher (2012) report that one of their subjects was not able to confirm having been lucid after being awakened two and one-half hours after the supposed eye movement signals were detected. And Voss et al. (2009) introduced a more complex signal after initially detecting false positives (for a similar suggestion, see Stumbrys, Erlacher, Schädlich, & Schredl, 2012). In most studies, dreams are only scored as lucid when lucidity is confirmed by the retrospective report. Both theoretically and in practice, then, the status of eye movement patterns as concurrent reports depends on retrospective confirmation; and once more, retrospective reports are treated as trustworthy by default.

Concurrent reports of lucid dreaming raise a number of more specific methodological questions. For instance, a more sophisticated way of ensuring that lucidity was continuous rather than lost after initial signaling is to ask participants to repeatedly signal lucidity throughout their dream, rather than only once (Stumbrys et al., 2013). Researchers may also use multiple signals, allowing dreamers to signal not only that they are dreaming but also, for instance, when they think they are awake (Stumbrys et al., 2013). Stumbrys et al. (2012) also emphasize that subjects should be instructed to signal lucidity only when they are certain that they are dreaming, and note that some researchers encourage participants to signal lucidity even when they merely suspect they are dreaming. This introduces a confound between lucid and prelucid dreams and might hinder the identification of the neural correlates of dream lucidity: An interesting perspective might be to use a separate signal for cases in which participants merely suspect they are dreaming and those in which they feel certain that they are. More generally, researchers might consider different ways of increasing the expressive granularity of ongoing reports of lucidity, and perhaps an even more sophisticated vocabulary of eye movement signals could be developed.

Note that these sophisticated methods of signaling ongoing lucidity also, however, potentially threaten the generalizability of signal-verified lucid dreams to lucid dreams reported in diary studies, where researchers rely exclusively on their subjects' own description of a dream as lucid or on lucidity as scored by blind judges (Voss et al., 2009; Stumbrys et al., 2012). A dream scored as lucid in a diary study thus might not fulfill the conditions for lucidity in a laboratory study using a sophisticated signaling method. For instance, Stumbrys et al. (2013) note that their participants frequently forgot to signal lucidity—a shortcoming that would not have showed up in a diary study at all. The consequence is not that diary reports of lucidity are less trustworthy but only that it is unclear whether they involve the more demanding type of lucidity required for laboratory studies.

Lucidity is a learnable skill, and different induction methods, ranging from cognitive techniques (e.g., autosuggestion, forming the intention to remember or recognize that one is dreaming, or reality testing; cf. Paulsson & Parker, 2006) to sensory stimulation during sleep or drug applications, are discussed in the literature. In a review of thirty-five studies, Stumbrys et al. (2012) found, however, that none of these techniques reliably and consistently induce lucidity, though cognitive techniques, alone or in combination with light sensory stimulation, seemed to be particularly promising. The prevalence of lucidity in laboratory dream reports is estimated at only between 1 percent and 3 percent (Moffitt et al., 1988, p. 430; Cicogna & Bosinelli, 2001, p. 29).

In general, lucid dreams are more likely to occur in the second half of the night (LaBerge, 1988b, p. 144) and are associated with high levels of physical activity and affective arousal during the daytime (Price & Cohen, 1988, p. 109). Morning naps appear to be particularly conducive to lucid dreaming, and LaBerge et al. (1994) found that lucid dreams were ten times more likely to occur during naps than in preceding nighttime sleep.

The prevalence of lucid dreaming is difficult to ascertain, but conservative estimates indicate that 50 to 60 percent of the population experience a lucid dream at least once in their lives, and 21 percent report lucid dreams at least once a month (Snyder & Gackenbach, 1988, p. 230; Schredl & Erlacher, 2011b), though the prevalence may be even higher (Mota-Rolim et al. 2013). Lucidity may be especially pronounced in young children (Voss et al., 2012), with lucid insight and dream control peaking in adolescence. This has led Voss and Hobson (2015) to formulate the "brain maturation hypothesis" of lucid dreaming, according to which cerebral diversification, especially in the frontal lobes, leads to a confounding of sleep-dreams and wakefulness in sleep and facilitates the occurrence of lucid dreams.

Studies trying to establish a link between individual differences and the propensity for lucid dreaming have not yielded clear-cut results. In one study, about 45 percent of children and adolescents reported having experienced at least one lucid dream, and lucidity was associated with a preference for reading fantasy and science fiction (Schredl, Henley-Einion, & Blagrove, 2012). Generally, lucid dreamers tend to be more introverted than socially oriented, but more specific hypotheses linking lucidity to visualization, imagery skills, absorption, or physical balance have not been confirmed (Snyder & Gackenbach, 1988; Schredl & Erlacher, 2004). One study found preliminary evidence for an association between lucidity and enhanced performance on cognitive tasks involving frontal areas (Neider et al., 2011). However, another recent study suggests that self-reported dream awareness and lucidity are associated with liberal acceptance bias, or a tendency toward memory errors and false feelings of familiarity (Cortlett et al., 2014).

Assessing the phenomenology of lucid dreams and the aims for cultivating lucidity seems to be more straightforward. Based on the responses of 571 lucid dreamers to an online questionnaire, Stumbrys et al. (2014) found that most had experienced their first lucid dream spontaneously in adolescence. The average duration of lucid dreams was estimated at around fourteen minutes, which is much longer than typical lucid dreams in the laboratory, which last only around two minutes. Subjects were also more likely than not to try to actively influence their dream and execute particular actions in their lucid dreams. Several studies suggest that for most people, the primary motivation for having lucid dreams is recreational: flying, communication with dream characters, and sex are the most frequently planned activities, but lucid dreams are also frequently used for changing nightmares into pleasant dreams, problem solving, creativity, or practicing skill (Schädlich & Erlacher, 2012; Stumbrys et al., 2014).

Early studies suggested that lucid dreams are physiologically correlated with phasic periods of REM sleep, particularly with heightened levels of eye movement density and skin potential, and the onset of lucidity is most likely to occur when levels of physiological activation reach their peak about five to seven minutes into REM sleep (for details, see LaBerge, 1988b, pp. 138–140). Early findings on the correlation between lucid dreaming and specific EEG patterns were, however, controversial and inconclusive (Hunt & Ogilvie, 1988; Ogilvie et al., 1991; Holzinger, 1997; Holzinger et al., 2006). Yet, the general observation of heightened levels of brain activation fits in nicely with anecdotal reports describing the onset of lucidity as involving an increase in the perceptual vividness and emotional intensity of the dream (see, e.g., LaBerge, 1985; LaBerge & Rheingold, 1990; Tholey & Utecht, 1995/2000). However, some reports have also shown lucid dreams to occur during NREM sleep (Cicogna & Bosinelli, 2001; Stumbrys & Erlacher, 2012), and it seems that not all lucid dreams have this vivid and emotional quality (Voss et al., 2013).

Allan Hobson and colleagues (2000, p. 837) hypothesized that the change in cognitive functioning accompanying the transition from nonlucid to lucid dreaming might be related to a selective reactivation of the dorsolateral prefrontal cortex (DLPFC), which is typically deactivated during REM sleep. This hypothesis has gained some support in several recent studies. The first, conducted by Voss et al. (2009), was an EEG study in which participants had been especially trained to induce and signal lucidity in the laboratory. Despite the small sample size—a problem that characterizes earlier studies of lucid dreaming, as well—the results are telling: the authors found that power in the lower gamma band around 40 Hz, especially in the frontal and fronto-lateral regions, was related to the onset of lucidity (for a discussion of the gamma band hypothesis, see Voss & Hobson, 2015). Coherence levels were significantly higher than in nonlucid REM sleep and resembled those seen in wakefulness throughout the entire frequency spectrum, but they were especially pronounced in these regions. The authors tentatively speculate that lucidity results from a wakelike activation of the frontal lobes in association with coherence in posterior structures.

This is an important step toward better understanding the brain basis of lucidity, but it also raises a familiar conceptual issue. Voss et al. (2009; see also Voss & Hobson, 2015) propose that their findings, in particular regarding frontal activity during lucid dreams, provide evidence that lucid dreaming is a hybrid state combining distinctive features of REM sleep and wakefulness. This interpretation would seem to suggest that the very frontal activity that presumably underlies dream lucidity also locates lucid dreams on the borders between REM sleep and wakefulness. In this view, the metaphorical description of lucidity as "waking up in one's dreams" would be complemented by regional shifts in activation and coherence patterns toward wakefulness. Thus the case for saying that lucid dreaming is a genuine REM sleep phenomenon (or even a genuine sleep phenomenon) seems to be weakened by this new evidence. From the perspective of the history of ideas, it is interesting to note how lucidity once more appears to be challenging both conventional conceptions of dreaming and conventional distinctions between (REM) sleep and wakefulness. Indeed, Voss and Voss's (2014; see also Voss & Hobson, 2015) space of consciousness model suggests a more continuous and dynamic view that questions common boundaries between conscious states. Within this model, lucidity involves a transient integration of the typically discrete states of wakefulness and REM sleep.

The second study, conducted by Dresler et al. (2008, 2009, 2012) used a combined methodology of EEG and fMRI to investigate the neural correlates of signal-verified lucid dreams compared to nonlucid REM sleep dreams. They found increased activation in a range of neocortical regions, including the bilateral precuneus, cuneus, and parietal, prefrontal, and occipito-temporal cortices during lucid dreaming. In keeping with the Voss et al. (2009) study, Dresler et al. (2012) also found increased activity in the right DLPFC. They relate their findings to the supposed contribution of self-referential processing (precuneus and fronto-polar activation) and working memory (fronto-parietal network activation) to lucid dreaming (see secs. 9.3 and 9.4 for further discussion). They also suggest that heightened activation in the bilateral cuneus and occipito-temporal cortices, involved in visual perception, may be related to the increase in brightness and clarity of visual imagery often reported to accompany the onset of lucidity.

Dresler et al.'s study is important because it presents the first-ever imaging data from lucid dreaming. As noted earlier (sec. 3.2), participants in most imaging studies of REM sleep are not explicitly asked about their dreams; for this reason, Dresler et al.'s study is valuable from the dual perspectives of lucid and nonlucid dreaming. Accordingly, it might seem tempting to generalize Dresler et al.'s findings to nonlucid dreams as well and to think that signal-verified lucid dreams afford a possibility for studying not just the correlates of specific forms of lucid dream content but of dreaming in general. While Dresler et al. (2008) suggest as much in an earlier publication, they later (in Dresler et al. 2011) more carefully acknowledge that their results may not be generalizable to nonlucid dreams, after all, noting that "it cannot be excluded that neural activity associated with nonlucid dreams differs from that of dream content deliberately induced by a lucid dreamer" (Dresler et al., 2011, p. 1836).

I am not so much interested in the empirical issues here (that is, in questions concerning similarity and differences in activation patterns between lucid and nonlucid dreams) as in the conceptual implications. In particular, note that endorsing a dichotomous conception of lucid and nonlucid dreaming, and one that perhaps even pushes lucidity away from sleep dreaming proper and toward wakefulness, thereby also challenges the generalizability of results from lucid dream studies to dreaming in general. If, on the other hand, one endorses a gradual distinction between lucid and nonlucid dreams, along with a conception of dreaming that is open to many different kinds of dreaming and not restricted to stereotyped characteristics of nonlucid REM sleep dreams (cf. chap. 2), then results from lucid dreaming may be applicable to dreaming proper. The question of whether "the combination of lucid dreaming with neuroimaging and polysomnography is a promising technique to also transfer more sophisticated 'brain reading' tasks . . . to the dreaming state" (Dresler et al., 2011, p. 1836) thus has both methodological and conceptual implications. One could, of course, legislate an answer, but this would hardly do justice to the dynamics of conceptual change in theorizing about sleep, dreaming, and wakefulness.

The metaphorical description of lucidity as waking up in your dreams also points to a practical obstacle many lucid dreamers attest to: the danger of actually waking up following the onset of lucidity. In the literature, lucidity is often described as a balancing act between involvement and detachment, with the double threat of either returning to a nonlucid dream or waking up (e.g., LaBerge, 1985; Brooks & Vogelsong, 1999). In fact, dreamers can even use lucidity to explore the borders of dreaming consciousness. By maintaining awareness while falling asleep, lucid dreamers can consciously observe how isolated hypnagogic imagery gradually unfolds into a multimodal dream (on "wake-induced lucid dreams," see LaBerge, 1985; see also LaBerge & DeGracia,

2000). Likewise, some dreamers report being able to observe how dream imagery fades upon awakening and claim to be able to experience the transition between dreaming and wakefulness without a break in consciousness. Van Eeden (1913) reports a particularly fascinating lucid dream in which he was lying on his stomach in the garden and simultaneously experienced his sleeping body lying in a supine position in bed. In this borderline state, he reportedly was able to experience his dream body slowly slipping into his sleeping body. The phenomenal realism of this experience remained impenetrable even after awakening, and he describes the double recollection of two bodies as "so indubitable that it leads almost unavoidably to the conception of a dream-body" (van Eeden, 1913).

Aside from empirical questions about sleep–wake transitions, methodological questions related to scoring, and conceptual questions concerning the relation between lucidity, dreaming, sleep, and wakefulness, reports such as van Eeden's suggest that dreaming, sleep, and wakefulness are also not as clearly distinguished as commonly assumed on the phenomenological level of description. If one can be aware of gradually entering the dream state and can consciously experience the interplay between dream imagery and waking perception upon awakening, or if dream experience can carry over from sleep into wakefulness, then one might not be able to determine to what extent one is awake or asleep at a given moment. In these cases, one might even maintain lucid awareness *without* being able to clearly identify one's current conscious state.

Saying that the concepts of sleep, dreaming, and wakefulness are subject to change, and that findings on the neurophysiology and phenomenology of lucid dreaming might push researchers toward either a dichotomous or a gradual distinction between lucid and nonlucid dreaming, is one thing; still, the question arises of how, given the current situation, these concepts are best used. To begin with, note that even if we assume that EEG activation levels indicate that lucidity is a hybrid state between *sleep* and wakefulness, it does not follow that lucidity is a hybrid state between *dreaming* and waking consciousness. Consequently, one way of reading the results from the Voss et al. (2009) study is to say that they challenge the association between a particular type of dreaming and sleep. For the time being, I think the most cautious view is to take a permissive approach, according to which lucid dreams are just what the term indicates: a special and in some ways distinctive type of dream. The neural correlates of dreaming are not yet known, and the link between REM sleep (as commonly scored) and dreaming is not, it seems, strong or exclusive enough to warrant the move of making dreaming proper contingent on REM sleep. Dreaming itself is a heterogeneous phenomenon, and if lucid dreaming merely adds to a variability already seen in nonlucid dreaming, then this does not require setting lucid dreams apart from nonlucid ones. If both lucid and nonlucid dreams were found to converge on a single phenomenal core, this would justify classifying lucid dreams as genuine dreams (cf. chap. 11). This would be further strengthened if the distinction between nonlucid and lucid dreaming were found to be gradual. By contrast, if lucid dreaming forms a more or less homogeneous category of its own, and if it is clearly and sharply distinguished from nonlucid dreaming, on both the phenomenological and the neurophysiological levels of description, then it may even eventually make more sense to no longer apply the concept of dreaming to lucid dreaming, and to rather speak of lucidity.

Before moving on, I want to suggest that several recent studies investigating either the transition, on both the phenomenological and the neurophysiological levels of description, between lucid and nonlucid dreams or comparing lucid dreaming and standard waking experience suggest that lucidity is indeed best regarded as continuous with nonlucid dreaming. Though preliminary, these studies also exemplify how the careful analysis of dream reports under the assumption of transparency can enrich both scientific and philosophical theories of dream experience.

The first study (Erlacher et al., 2014) returned to the familiar question concerning the duration of dreams. Rather than relying on subjective estimates of dream duration, however, this study devised a method for measuring the objective duration, using a stopwatch, of performing various tasks in lucid dreams and compared it to their duration in wakefulness. Figure 3.2 shows polysomnographic measurements while subjects counted to 10, 20, and 30 while dreaming, marking the beginning and end of each part of the task with eye movement signals. Retrospective reports were used to confirm the successful completion of the task, as well as to ascertain that lucidity was maintained throughout the dream.

The other two tasks involved walking 10, 20, and 30 steps at a regular pace and performing a gymnastic routine consisting of 6 consecutive parts. Results indicated that all three tasks took longer in dreaming than in wakefulness, with the effect being most pronounced for walking (52.5 percent) and least pronounced but still significant for gymnastics (23.2 percent). The increase for counting (27 percent) was statistically insignificant. Task length and task complexity did not appear to affect the increased duration of actions performed during lucid dreams.[26] This partially contradicts previous studies that had suggested that the time required for counting in lucid dreams was the same as in wakefulness (LaBerge, 1985) and that only the duration of motor activities (in this case, performing squats) was increased in lucid dreams (Erlacher & Schredl, 2004).

Clearly, findings on the duration of lucid dream actions compared to wakefulness are still preliminary. Though Erlacher et al. (2014) explicitly address older theories according to which dreams might be instantaneous memory insertions at the moment of awakening and lack duration altogether (Maury, 1861; Goblot, 1896), their study would likely not convince the determined skeptic about dream experience (cf. chap. 1). Instead, they investigate a more productive question, namely: what is the duration of dream actions? My point here is merely that lucid dreaming presents the opportunity

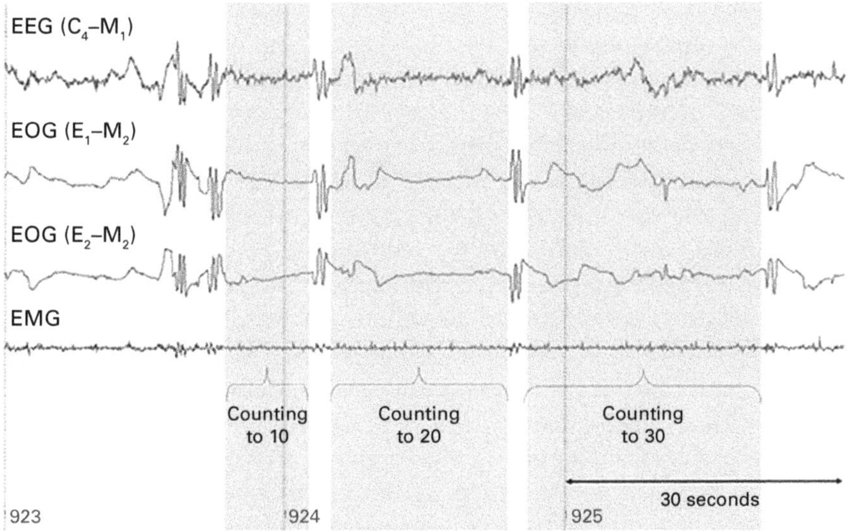

Figure 3.2
A signal-verified lucid dream involving a counting task: five left-right-left-right eye signals mark the onset of lucidity and the beginning and end of the counting intervals (gray area). Reprinted from Erlacher et al. (2014), with friendly permission from *Frontiers in Psychology*.

for turning questions about dream duration into empirically investigable questions, and dream reports play a crucial role in this process. Can these findings be generalized, then, to nonlucid dreams? Erlacher et al. (2014) are duly cautious; however, if lucid and nonlucid dreams are indeed continuous, this would carefully point toward an affirmative answer.

The second study attempted to investigate the DLPFC hypothesis of dream lucidity more directly than existing studies by moving beyond mere correlation to the investigation of the causal contribution of DLPFC activity to lucidity (Stumbrys et al., 2013). In particular, the authors attempted to test whether transcranial direct current stimulation (tDCS) over the DLPFC would increase lucidity. Nineteen participants spent three nights in the sleep laboratory; the first was an adaptation night, the other two nights were used for sham and real stimulation, respectively. The effect on lucidity was weak and only found in the self-reports of frequent lucid dreamers, which the authors attribute to methodological problems. For instance, stimulation frequently disrupted sleep, and though more than one hundred REM awakenings were conducted, only a single signal-verified lucid dream was recorded. Yet this may have occurred because participants simply forgot to signal lucidity, as several dreams were scored as lucid by external judges and were reported to have been lucid by the subjects themselves. However, stimulation did have a more general effect on dreams. For instance, stimulation was associated with an increase in deliberate decisions in nonlucid dreams, and frequent lucid dreamers reported increased awareness that their physical body was asleep and dream characters and dream objects were not real following stimulation. Dream reports were also longer on stimulation nights and were self-rated as being more lucid than following sham stimulation. As the authors point out, the finding that DLPFC stimulation enhanced decision making and working memory fits in well with what is known about the function of the DLPFC in wakefulness. They also note that the stimulation site may have been too specific, as the study by Dresler and colleagues discussed earlier suggests that a wider network beyond frontal regions is associated with lucidity.

Be that as it may, if lucidity is continuous with nonlucid dreaming, and if DLPFC activation is not specific to lucid insight into the fact that one is dreaming but rather is related to a number of cognitive abilities such as decision making, working memory, self-observation, and attention (cf. Muzur et al., 2002; Dang-Vu et al., 2005; Pace-Schott, 2007; Desseilles et al., 2011), then the general effects observed by Stumbrys et al. (2013) would be expected. Moreover, in this view, it would not be surprising that lucidity was only enhanced in frequent lucid dreamers. Perhaps nonlucid dreamers were pushed, metaphorically speaking, in the direction of lucidity, but not enough to realize that they were dreaming. Might future researchers still find more specific correlates of dream lucidity that can be turned on and off, as it were, by brain stimulation? Nothing in the current research results excludes this possibility, and this is an open empirical question. Yet whatever the answer turns out to be, the study conducted by Stumbrys et al. (2013) still suggests an interesting methodology for charting the transitions between nonlucid and lucid dreams.

This interpretation is supported by the third study I wanted to discuss here, which investigated the phenomenological similarities and differences between lucid and nonlucid dreams. In an interdisciplinary project in which I participated together with Ursula Voss, Karin Schermelleh-Engel, Clemens Frenzel, and Allan Hobson, we aimed to investigate the transitions between lucid and nonlucid dreaming in more detail than in previous studies. The project was interdisciplinary in that both empirical/psychological and conceptual/philosophical considerations entered into the study's design, and we considered the relevance of our results not just for lucidity but for the debate on consciousness and dreaming more generally. That said, I do want to emphasize that all the truly hard work was done by my coauthors: my main contribution was to assist in the construction of the initial survey and add some paragraphs to the final paper. In particular, we identified a number of dimensions commonly taken to distinguish lucid and nonlucid dreams in the literature and used these to develop the Lucidity and Consciousness in Dreams (LuCiD) scale. The dimensions we identified were Insight, Control, Thought, Realism, Memory, Dissociation, Negative Emotion, and

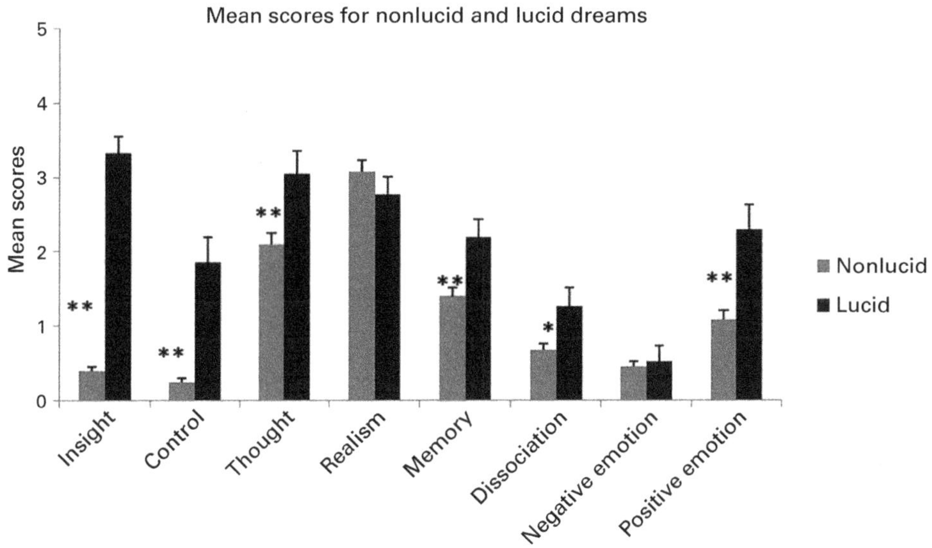

Figure 3.3
The LuCiD scale: mean scores for nonlucid versus lucid dreams. Reprinted from Voss et al. (2013), with friendly permission from Elsevier.

Positive Emotion. The scale was constructed in three steps: (1) a survey involving fifty items; (2) exploratory factor analysis selecting the twenty-eight items that most clearly distinguished lucid and nonlucid dreams; and (3) a confirmatory factor analysis and validation of the scale. We used a combined methodology, asking subjects to report a single dream and respond to a questionnaire with detailed questions about the reported dream. For the confirmatory factor analysis, home reports (paper and pencil), online reports, and reports following timed awakenings in the laboratory were analyzed. The results from laboratory reports were less pronounced than those from home diaries: Control, Thought, Memory, and Positive Emotions were significantly more frequent in home reports, suggesting that these better represent standard lucid dreams than laboratory reports (incidentally, the lower incidence of Control in the laboratory might also explain why lucid dreamers are frequently unable to signal lucidity in the laboratory). The final LuCiD scale can be found in Voss et al. (2013, p. 12). Figure 3.3 gives an overview of the main results.

Not surprisingly, lucid insight most clearly distinguished lucid and nonlucid dreams; after all, realizing that one is dreaming is, by definition, a characteristic of lucid dreams. Yet certain aspects even of this factor could occur in nonlucid dreams as well, so we asked subjects not just whether they realized that they were dreaming, but whether they wondered whether they were, whether they were aware that other dream characters were unreal, and so on. Similarly, based on the literature, a correlation between control and insight was to be expected, though control was more pronounced in dreams reported at home than in the laboratory. It is also not surprising that we found Thought, Memory, and Positive Emotions to be significantly more frequent in lucid dreams than in nonlucid ones, as all of these factors have long been associated with lucidity.

Several other findings, however, are surprising. First, note that despite the significant difference between nonlucid and lucid dreams, thoughts and memories were not completely absent in nonlucid dreams but actually more frequent than has traditionally been assumed (cf. chap. 9 for discussion). Here it seems that a continuous model distinguishing different degrees of memory and reflectiveness in dreams might be more helpful than a dichotomous model sharply distinguishing lucid and nonlucid dreams. Second, note that the factor of Realism did not significantly distinguish lucid and nonlucid dreams. When asked whether the events, thoughts, and emotions experienced in their dreams were realistic by waking standards, our participants answered similarly for nonlucid and lucid dreams. Perhaps, then, lucid insight has to be understood as arising independently of the realistic or bizarre quality of the dream. Indeed, lucidity only rarely arises from the recognition of bizarre elements in dreams, and perhaps such reflection is not necessary to obtain lucid insight (Gackenbach, 1988). Moreover, the presence of memory and thought is not, it seems, sufficient: if it were, it could not occur as frequently as it does in nonlucid dreams. Third, several authors have suggested, as noted earlier, that lucidity is accompanied by a change in the overall experiential quality of the dream, with lucid dreams often having a surreal, dreamlike quality (Brooks & Vogelsong, 1999; LaBerge, 1985; Tholey & Utecht, 1995/2000). Our study, however, suggests that this is not the rule: rather than being an all-pervasive experiential phenomenon, a majority of lucid dreams, it seems, feel very much like standard wakefulness and nonlucid dreams. In a certain sense, then, this study offers a more differentiated and enriched view of lucidity and a more critical assessment of the claim that lucidity is a hybrid state between sleep and wakefulness: again, this leads to the observation that newer and more detailed studies may lead to a changing conception of which kinds of cognitive activity are considered as "wakelike," and perhaps, with a deepening understanding of the transitions between lucid and nonlucid dreams, also as "dreamlike."

What I want to suggest here, however, is even more basic: namely, that our study may suggest how lucidity can contribute to the understanding of dream experience independently of our specific results. Throughout this section, it has become clear that signal-verified lucid dreams cannot, as is the case for other types of dream-enactment behavior, provide independent empirical evidence for the experiential and transparency assumptions. Our study suggests that lucid dreaming might be more interesting for another reason: by providing an interesting contrast condition for the study of nonlucid dreams, and by allowing researchers to chart the continuities between nonlucid and lucid dreams. The method of applying a careful, statistically controlled approach to dream reports, of combining findings from the literature with findings from the analysis of free dream reports and specific questionnaires, both at home and in the laboratory, and of discussing the relevance of results not just for dream research but for the interdisciplinary field of consciousness research may offer new insights about dream experience. More specifically, our study, as we optimistically note, suggests an "empirically-based approach to measure and define consciousness in different types of dreaming" (Voss et al., 2013, p. 10), and thus a methodology for using dream reports to chart the relationship between dreaming, lucid dreaming, and ultimately perhaps also wakefulness.

Finally, a central motivation for endorsing the transparency assumption of dream reporting is pragmatic: the merits of a particular method used for the study of dreaming via dream reports depend on the extent to which this method enables new perspectives for future research. A related prediction is that as new and more precise tools for investigating subjective dream reports are developed, this should facilitate the identification of associated changes in brain activation as well. My last example is a study in which the LuCiD scale was combined with the attempt to induce lucidity via transcranial alternating current stimulation (tACS) over fronto-temporal areas (Voss et al., 2014; see also Voss & Hobson, 2015). Participants were not experienced lucid dreamers, and they were not asked to attempt to become lucid in the laboratory; rather, they were told that the goal of the study was to investigate the effect of tACS on dream content more generally. Twenty-seven healthy subjects spent up to four consecutive nights in the laboratory. Different stimulation frequencies were applied and contrasted with sham stimulation during REM sleep. Participants were then awakened and asked to report any dreams and provide ratings for the items of the LuCiD scale. Effects were specific to 25 Hz and 40 Hz stimulation: these were the only frequencies to influence brain activity (possibly also explaining why a similar effect was not found for the weaker currency used in tDCS studies), and they also had the clearest effect on LuCiD ratings. The latter were most pronounced for Dissociation, or dreams in which participants described experiencing events from the third-person, observer perspective. Insight also received significantly higher ratings during 25 Hz and 40 Hz stimulation, while the results for Control were strongest for 25 Hz. So although stimulation increased the frequency of Insight as the core measure of lucidity, the effect was more general, suggesting, once more, that participants were pushed in the direction of lucidity even when they did not realize that they were dreaming. For now, this leads to the important metatheoretical observation that the efficiency of a method, for instance for experimentally manipulating dreams by inducing changes in brain activation during sleep, depends in part on whether it is combined with a careful and fine-grained way of measuring changes in subjective dream experience as well. I am not here claiming that either tACS or the

LuCiD scale are optimal in this respect; but perhaps, they are a step in the right direction.

3.6 Studying Dreaming through Its Losses: Lesion Studies

Perhaps paradoxically, important information about the neural correlates of dreaming is provided by cases in which dreaming as a whole or certain types of dreaming are lost. By linking changes in the phenomenology of dreaming to brain lesions, researchers can infer which brain regions contribute to normal dreams in healthy subjects (for a review, see Schwartz et al., 2005).

The systematic cessation of dreaming was first described in the nineteenth century. In 1883, Charcot described the case of a patient who lost visual memory, the ability to deliberately invoke mental images, as well as visual dreaming, though he still experienced dreams lacking visual imagery (e.g., dreams involving speech). Four years later, in 1887, Wilbrand described a patient who reported a complete cessation of dreaming, as well as a number of cognitive deficits such as visual irreminiscence, prosopagnosia, and topographagnosia, as well as attention and memory deficits (Charcot, 1883; Wilbrand, 1887; for details, see Solms, 1997). The cessation of dreaming or specifically of visual dreaming is often referred to as the Charcot-Wilbrand syndrome.

Contrary to Charcot's and Wilbrand's observations, evidence suggests that a cessation of dreaming is not always accompanied by further cognitive deficits. Bischof and Bassetti (2004) describe the case of a patient who experienced a complete cessation of dreaming following a stroke, but without exhibiting any specific cognitive deficits during standard neuropsychological testing. The patient's sleep architecture, including the overall amount, density, and latency of REM sleep, was normal, and no sleep disorders such as abnormal breathing or periodic limb movements were observed. Also, the patient's initial inability to remember dreams following spontaneous awakenings was confirmed following REM sleep awakenings. At least some dream recall did return eventually, but the frequency, intensity, and vividness of the patient's dreams were still reduced a year later. The authors conclude that a complete cessation of dreaming is a distinct neuropsychological dysfunction, occurring independently of cognitive deficiencies and in the absence of REM sleep changes.[27] They suggest that the minimal lesion extension is deep bilateral occipital lobe damage including the right inferior lingual gyrus.

In an extensive study, Solms (1997; see also Solms, 2011) investigated changes in and cessation of dreaming in 361 neurological patients, whom he interviewed in detail about their dreams and sleep quality. He distinguished four main disorders of dreaming: (1) loss of visual dreaming (visual anoneira) associated with occipito-temporal lesions and often co-occurring with impairments of visual perception and imagery in wakefulness; some patients also reported that only one aspect of visual imagery (e.g., face, color, or kinematic perception) was lacking in their dreams; (2) the global or complete (though sometimes temporary) cessation of dreaming (global anoneira), associated with parietal and posterior temporo-occipital lesions; (3) dream–reality confusion (anoneirognosis), often accompanied by a quantitative increase in dream frequency and associated with frontal-limbic lesions; and (4) recurring nightmares, often occurring in temporal-limbic seizure activity. He also found further changes in dreaming to be associated with cortical lesions; for instance, reduced complexity and emotional intensity of dreaming were associated with deep bifrontal lesions, while lesions in the anterior parts of the limbic system were associated with an increase in the rate and vividness of dreaming.[28]

Solms (1997) also reviewed 111 cases of a loss of dreaming reported in the literature. Ninety-four patients had lesions in the posterior convexity of the hemispheres, which is in or near the so-called parieto-temporo-occipital (PTO) junction, which he had also found to be associated with a loss of dreaming in his own patients (see also Solms, 2000). Out of 83 unilateral lesions, 48 were localized in the left and 35 in the right hemisphere. Similarly, Doricchi and Violani (1992) found that dream cessation was as frequently associated with unilateral left or right damage as with (often larger) bilateral damage. Both studies thus provided evidence against the older hypothesis that dreaming is related to asymmetrical hemispherical activation during sleep (for discussion, see Schwartz et al., 2005).

Importantly, Solms (1997) found that while the symptoms of the classical Charcot-Wilbrand syndrome (irreminiscence, prosopagnosia, and topographical agnosia or amnesia; for details, see Solms, 1997, pp. 22–25) were typically accompanied by a cessation of visual dreaming or by a global cessation of dreaming, the opposite was not the case. Indeed, most patients reporting a global cessation of dreaming did not suffer from any of the classical Charcot-Wilbrand symptoms. This finding also fits in well with the case study reported by Bischof and Bassetti (2004). Because the same was not true for patients reporting a cessation of visual dreaming, Solms proposes that nonvisual dreaming and global cessation of dreaming should be regarded as two distinct disorders (Solms, 1997, p. 159). Hence, contrary to the classification of Charcot-Wilbrand syndrome as a single disorder, this terminology may in fact conflate two distinct syndromes. Solms proposes the new term of *anoneira*, derived from the Greek noun Ονειρα (dreams), to describe the loss of dreaming, with visual anoneira referring to the loss of visual dreaming described by Charcot, and global anoneira referring to the global cessation of dreaming described by Wilbrand. Typically, global anoneira is an acute syndrome, disappearing within a year, but it may also persist (Solms, 1997, p. 235).

The central question, in the present context, is whether subjects reporting a loss of dreaming are actually experiencing a cessation of dreaming or have merely become unable to remember and report their dreams following awakening. This question is directly relevant to the reportability assumption, and Solms addresses it explicitly. He acknowledges that "subjective reports of global cessation of dreaming might well be artifacts of general amnesia" (Solms, 1997, p. 35), but he argues that amnesia is not necessary for a loss of dreaming to occur, because "although many amnesiacs apparently do not dream, very few non-dreamers are amnesiacs" (p. 35). In his study, only 25 percent of nondreamers suffered from long-term memory disorder, and if the subjective cessation of dreaming were related to a cessation of dream recall rather than of dreaming, one would expect related mnemonic deficits in wakefulness (pp. 33–34). Solms also points out that most amnesic patients in his study were unsure whether or not they dreamed, rather than denying that they dreamed altogether, as was the case for patients suffering from global anoneira. He concludes that these findings "strongly support the view that neurogenic loss of dreaming is a disorder of dreaming rather than a disorder of memory" (p. 35) and hypothesizes that the "global cessation or reduction of dreaming is a primary phenomenon, independent of any disturbance of memory" (p. 36).

Amnesia may not even be sufficient for a reported loss of dreaming: most amnesiacs can recall what happened a few seconds ago and so might even be able to report brief episodes of dreaming immediately following REM sleep awakenings. This is supported by a study investigating the effect of playing Tetris on hypnagogic imagery reported by novice and expert players, as well as by amnesic patients (Stickgold et al., 2000). Although the patients' performance improved only minimally over seven hours, and although they later could not recall having played Tetris, three out of five patients reported Tetris-related imagery after being awakened from sleep onset. The report rate was similar to that seen in novice players and higher than that seen in experts. The type of imagery—falling Tetris pieces—was also similar in all three groups. By contrast, there was only a single report of thinking about Tetris (without related imagery) from an amnesia patient, whereas such thoughts were much more frequently reported by healthy novices and experts. The authors conclude that declarative memory is not required for hypnagogic imagery or reports of such imagery.

One might, of course, still argue that the loss of dream recall is caused by a category-specific type of amnesia (Solms, 1997, p. 33) presenting independently of waking-memory disturbances and selectively leading to the cessation of dream recall, with dream experience being preserved. This possibility would threaten the reportability assumption. I think, however, there are at least three reasons not to be too concerned about this type of objection. The first is that in the absence of straightforward empirical evidence to the contrary, it seems preferable to adopt the simpler and more parsimonious view and assume that the same memory processes are involved in recalling dreams and waking experiences. As long as one supposes that remembering a dream is analogous to remembering waking experiences rather than involving different mnemonic processes, it would seem that a cessation of dream recall refers to an actual cessation of dreaming in cases in which it presents without memory deficits observed in wakefulness.

The objection would be strengthened, however, if empirical evidence for a sleep-specific memorization process were available. Indeed, a study investigating dream recall in twelve patients with medial temporal lobe epilepsy (Fell et al., 2006) might suggest just that (but see Marzano et al., 2011; de Gennaro et al., 2012). Patients were awakened from REM sleep and immediately asked to give a free mentation report and respond to a questionnaire; measurements from scalp EEG and multicontact depth electrodes implanted in the medial temporal lobe from the preceding REM period were analyzed. Subjects were divided into two groups of high and low recallers, with the high-recall group recalling a dream after approximately 80 percent and the low-recall group after approximately 7 percent of awakenings. The low incidence of dream recall in this study may in part be explained by the fact that only reports fulfilling specific criteria (storylike development, self-participation, vivid or bizarre perceptual episodes) were scored as dream reports. Indeed, some of the patients classified as low recallers did recall nondreaming mentation, though overall recall rates were still lower than for high recallers. Importantly, the study found no difference between the two groups on verbal and figurative memory tests. High recallers, however, showed a significantly higher coherence level between the rhinal cortex and the anterior hippocampus during REM sleep, suggesting, according to Fell and colleagues, that dream memorization is linked to connectivity between these two areas. The authors conclude that this supports the view that "dream recall variations are probably largely determined by memory formation during sleep, rather than by daytime memory capability" (Fell et al., 2006, p. 108). As in Solms's study, this suggests that a loss of dream recall cannot be reduced to deficient waking memory. However, while Solms takes a loss of dream recall to reflect a loss of dreaming, Fell and colleagues' findings might be interpreted as suggesting that the problem lies with sleep-specific memory formation rather than with dream production. If this reading turned out to be correct, such findings would, it might seem, pose a more substantial threat to the reportability assumption.

Or would they? I don't think this follows. To see why, let's assume there was indeed empirical evidence for a sleep-specific process of memory formation that is independent of waking memory and strongly predicts subsequent dream recall. In this case, it would still be unclear whether this process contributes to dream recall or dream formation. In view of the important role that waking-memory sources play for the generation of dream content (see sec. 6.2), one could argue that memory is needed not only to remember dreams upon awakening but to experience them in the first place. Indeed, as Solms points out, Charcot (1883) attributed the cessation of visual dream imagery to defective visual memory. Anecdotal evidence also points in this direction. Cole relates the touching story of a patient who, after experiencing a gradual loss of vision over many years and finally losing sight altogether, described being particularly distressed when he realized he was beginning to forget what is wife and children looked like. According to Cole, dreams in which sight was occasionally still preserved became particularly important in this stage. One dream report read as follows:

> I had got out of bed. . . . This toddler came padding in and I could see her quite clearly in the dim light. . . . The first time I had been able to see her. I stared, full of wonder, taking in every detail of her face as she stood there, wreathed in smiles. "So this is her, this is the smile they all talk about." I had a wonderful sense of renewal of contact. . . . Then the dream faded. (quoted in Cole, 1998, p. 30)

Although fading visual memory and imagery in wakefulness seemed to be associated with nonvisual dreams, occasional visual dreams, apparently, accessed memory sources no longer available during wakefulness. A similar experience is related by Rainville, who describes such seeing and reminiscent dreams even after more than twenty-five years of blindness. He notes that his visual dreams are more stable and perhaps even facilitate waking visual imagery, which for him requires deliberate attention, effort, and concentration. For instance, he writes: "Once I dream of her [his daughter] in her new hairdo, that is, once I have *seen* it, she will appear to me in it pretty much consistently from that time on" (Rainville, 1994). This also fits in well with the commonly accepted view that congenitally blind subjects do not experience visual dream content, but visual dreaming is preserved at least occasionally if sight is lost after the age of five to seven years (for discussion and further references, see Kerr & Domhoff, 2004). Dreams, then, may enable vivid access to memories of visual imagery that are no longer readily available in wakefulness, and perhaps they even do so in virtue of some special memorization process related to sleep; after all, one need look no farther than Freud to find descriptions of hypermnesic dreams of long-forgotten events (Freud, 1899/2003, pp. 31). In a sense, the view that dream generation, not just dream recall, depends on memory sources follows quite directly from empiricist intuitions: dreams are at least dependent on waking memories to the extent that the formal features of dreams, though perhaps not their specific content, will ultimately be derived in some way or another from waking impressions (cf. sec. 5.1). So the second reason not to be too worried is that even if dream recall was found to be predicted by a special sleep-related memorization process, this would actually fit in well with scientific and anecdotal evidence for saying that dream production itself is constrained by waking-memory sources. Rather than threatening the reportability assumption, the detected process could contribute to dream formation itself.

Third, let's assume that an even stronger challenge to the reportability assumption was available. Let's assume that the speculative view on the dependence of dream production on memory sources turns out to be empirically invalid. Let's assume, then, that it turns out that in certain patients, perhaps following closely circumscribed lesions, sleep-related dream recall is selectively disturbed, while dream formation itself is unaffected. In these patients, then, the inability to report dreams would no longer indicate the cessation of dream experience. Though clearly relevant, such a finding would not, however, amount to a wholesale refutation of the reportability assumption but only question its applicability in certain special cases: the conditions under which dream experiences are assumed to be reportable, and hence under which reports of dreaming and of nondreaming are assumed to be equally trustworthy, would have to be restricted. That dreams are not readily reportable under just any conditions is, however, no news at all—clearly a majority of dreams do go unreported, and clearly many situations are not conducive to the ability to report dreams: a failure to recall several dreams after an undisturbed night of sleep at home does not, after all, indicate that no dreams occurred. So the damage to the reportability assumption could, it seems, be limited. But we have yet a deeper reason not to be too concerned about this possibility. It is that restricting reportability to certain conditions would require showing that in these new conditions, dreams could once more be reported. Only then would it make sense to assume that they were still occurring, even in the absence of dream recall in other, less-than-optimal conditions. By contrast, the attempt to argue that dreaming is preserved in the absence of any sort of dream recall runs the risk of being methodologically self-defeating.

In sum, findings on the cessation of dreaming, so far, at best are compatible with the reportability assumption and even assume that the cessation of dream recall indicates an underlying cessation of dreaming. At worst, they require the qualification of the reportability assumption, but not its outright rejection. But if this is correct, then studying dreaming and its losses—in particular the global cessation of dreaming—is not so much informative about dream recall as it may help identify the neural basis of dreaming itself.

That said, aside from questions concerning the relationship between dreaming and dream recall, on the one hand, and the cessation of dreaming and the loss of dream recall, on the other, lesion studies also raise the problem of how to distinguish between the loss of dream recall and the inability to report dreams. To give an example, there is some controversy as to whether aphasia is associated with the ability to generate or to report dreams (for discussion, see Solms, 1997). Doricchi and Violani (1992, p. 114) review evidence suggesting that language and semantic functions "are functionally involved both in the decoding of a dream and in its later recall." By contrast, Solms found a high incidence of preserved dreaming among his aphasic patients, confirming results from earlier studies (Solms, 1997, p. 161) and suggesting that the ability to generate language is not necessary for dream generation. It is also, apparently, not necessary for dream reporting: Schanfald et al. (1985) found that stroke patients suffering from aphasia claimed to still experience dreams and were able to draw dream scenes from memory. This once more underscores the importance of improving reporting conditions tailored to the needs and abilities of specific subject groups and to develop a more differentiated view of what it means to report a dream.

Lesion studies are also interesting beyond the investigation of specific changes in, or losses of, dreaming. Schwartz and Maquet (2002) suggest how imaging data from REM sleep, matched with dream reports from the same sleep phase, could be used to develop a neuropsychological approach to dreaming and to identify the correlates of specific types of dream experiences. As a first step, they suggest systematically comparing different types of dream bizarreness to neuropsychological syndromes. For instance, a dream in which I identified, say, George Clooney as my grandmother would be reminiscent of the Fregoli syndrome, in which subjects misidentify strangers as close friends or family members, though they share no obvious physical resemblance. Schwartz and Maquet hypothesize that the occurrence of Fregoli-like phenomena in dreams is related to regional changes in activation patterns similar to the lesion topography that causes the disorder in wakefulness. By matching instances of dream bizarreness to neuropsychological syndromes, one might even be able to predict the pattern of brain activation underlying the occurrence of a bizarre dream feature. They write:

> In this context, the neuropsychological assessment of dream content would provide hypotheses about the distribution of regional activity in these areas. The categorization of dream reports based on the presence of specific bizarre features in dreams might thus offer new constraints to the analysis of REM sleep data that would be acquired immediately before the dream reports. For example, we predict that the activation of V4 will be lower in scans that are followed by dream reports containing achromatopsia than those without. Likewise, the correlation in activity between the fusiform face area and infero-temporal cortex should be weaker in REM sleep scans that are followed by dream reports in subjects with Frégoli syndrome than in those without. . . . Overall, this approach of dream analysis might reveal, for the first time, genuine cerebral correlates of dreaming activity. (Schwartz & Maquet, 2002, p. 29)

Similar interpretations may also be available for the misidentification of places in dreams, the multiplication of visual percepts in time or space (palinopsia and polyopia), or seeing things as too big or too small (micropsia, macropsia).

More generally, this approach to dream reports assumes that "dream reports contain valuable information about the cognitive processes and brain functions that participate in their generation" (Schwartz et al., 2005, p. 427).[29] In particular, the frequency of certain types of dream experience (such as visual imagery or emotion) in large samples suggests that "common underlying patterns of cerebral activity might occur iteratively during sleep" (p. 427), thus predicting a match with findings from neuroimaging studies of sleep, as discussed in section 3.2. "Like circumscribed neuropsychological impairments after focal brain damages, functional dissociations in the sleeping mind can reveal cognitive 'modules' that would otherwise remain undetected since they are normally integrated during wakefulness to form our coherent everyday-life experiences" (p. 433). Clearly this type of approach, in which dream reports play a prominent role, would be a first step toward *dream* imaging, as opposed to REM sleep imaging. It would also present the possibility of extending Dresler et al.'s (2012, 2014) innovative work on lucid dream imaging to nonlucid dreams. And finally, it might also be a way, as Schwartz et al. (2005) suggest, of integrating findings from dream research with consciousness research more generally. Here the analysis of standard dreaming in healthy subjects might, together with lesion studies, be used as a contrast condition for standard waking consciousness.

In sum, lesion studies—including those investigating the losses of dreaming—may be most valuable, paradoxically, for understanding the neural basis of normal dreaming; lesion studies conducted independently of changes in dreaming can act as a contrast condition for familiar types of dream bizarreness; and normal dreaming, taken together with lesion studies, might shed new light on consciousness more generally. The latter two of these strategies assume that dream reports allow researchers to predict underlying changes in brain activation, or, metaphorically speaking, to read the brain via dream reports. The next section discusses recent developments suggesting that the opposite approach might soon be more than science fiction: researchers are increasingly using changes in brain activation to predict changes in dreaming, thereby taking the step from neuroimaging sleep to genuine dream reading. The tantalizing question is whether this step might eventually enable dream research to move beyond the study of dream reports altogether.

3.7 Studying Dreaming without Studying Dream Reports? From Brain Reading to Dream Reading

Since the beginning of scientific dream research, researchers have been dreaming the dream of verifying dream reports or even of studying dreams without dream reports altogether. Science-fictional thought experiments describing such scenarios have been developed for decades. They stand in a broader context of popular reports of dream machines claiming to produce dreamlike imagery in wakefulness or allow their owners to selectively manipulate their sleep dreams, but also of literary or cinematic devices allowing their owners to observe, control, or even enter the dreams of others. Dreams, it would seem, are the ultimate test of our ability to observe and control the minds of others, and dreams' elusive, unpredictable, and private nature makes the challenge of doing so all the more alluring. For this reason, philosophers and researchers alike have repeatedly felt the pull of science-fictional dream-reading devices.[30] This is not the place to attempt anything like an exhaustive treatment of this topic, but here is an exhibit of three examples that I particularly like.

In 1966, in *The Content Analysis of Dreams*—to this day a classic of scientific dream research—Hall and Van de Castle wrote that "conceivably, advances in instrumentation will eventually permit dreams to be 'televized' as they are being dreamed. If this time ever comes, we will no longer have to ask people to tell us their dreams because we can record and study the dream exactly as it was experienced" (Hall & Van de Castle, 1966, pp. 18–19).

Forty years later, a highly similar scenario was proposed by Antti Revonsuo, who, to my knowledge, has done more detailed work on dreaming from the perspective of philosophy of mind than any other author. In a science-fictional thought experiment involving the so-called dream catcher test, Revonsuo (2006, pp. 300–303) proposes criteria for what would count as the discovery of the phenomenal level of organization in the brain. The test would involve two teams of researchers monitoring the same participant: the brain team, collecting data from the human brain (but with no access to information about the representational content of brain states), and the dream team, assembling a database about the phenomenology of the participant's dream based exclusively on dream reports gathered following experimental awakenings in the laboratory. Once they had completed data collection, both teams would construct a 3-D model of their participant's dream. The test would be successful if independent judges succeeded in matching the model based on the first-person dream report to the model based on brain data from the same sleep period with better-than-chance accuracy. Though Revonsuo emphasizes that methods today are probably insufficient to allow researchers to pass the dream catcher test, he takes this to be not just science fiction but a goal for future research. He also uses the test to make a specific prediction:

> Passing the test means that pessimist philosophers who maintain that *no* amount of data can tell *anything* about consciousness have been decisively defeated. Moreover, philosophers who maintain that the contents of consciousness are externally based and therefore not to be found in the brain will also have to reconsider their views. (Revonsuo, 2006, pp. 302–303)

In this way, the dream catcher test—as Revonsuo (2006, p. 302) puts it, "the Turing test of consciousness science"—would be a way of operationalizing his own preferred view of biological realism, according to which phenomenal consciousness is a real level of organization in the brain, and thus of empirically validating, if successful, a philosophical theory of phenomenal experience.

The third example in my cursory exhibit unites an example from literary science fiction from the 1940s with scientific dream research, recent work on virtual reality and brain reading, philosophical considerations, and a discussion of consequences for consciousness research. Emphasizing the importance of new methodologies for gathering first-person data and integrating them with neurophysiological measures, Perogamvros (2013) proposes a new methodology for future research. In this, he is inspired by Adolfo Bioy Casares's 1940 novel *The Invention of Morel*, which describes an apparatus able to "capture and store 'waves' and 'vibrations' from all human senses" (Perogamvros, 2013, pp. 1–2) and project them holographically. From this, Perogamvros extracts two conditions: a multimodal, virtual reality component, which he takes to be exemplified in current virtual reality research, and a brain-reading component, which he takes to be exemplified by current efforts to decode visual experience from fMRI data. By combining these two techniques, that is, virtual environments coupled with fMRI decoders, one might, he suggests, end up with something very similar to Casares's consciousness device. As a more feasible alternative today, however, he suggests that dreaming might be regarded as an "'endogenous' Morel's machine" (p. 5): "Studying an internal state incorporating multimodal integration and reality simulation, such as dreaming, could further elucidate the nature of conscious states" (p. 7) and thus reflect back on current theoretical understanding of consciousness in general. Incidentally, this proposal closely echoes Revonsuo's (2006, pp. 347–351) suggestion of using dreaming as a model system for consciousness research and of combining brain-reading techniques with virtual presentation processes to develop a novel "dream presenting" technology, enabling researchers to "phenomenalize" their data in a multimodal, immersive, and interactive 3-D model. In the seven years since Revonsuo's thought experiment, however, some important developments have taken place, and it seems that Revonsuo's and Perogamvros's visions are slowly inching toward technical feasibility. The rapidity of developments also leads me to expect that this section could well be outdated, with respect to the empirical examples discussed, by the time this book goes to press. Still, I want to complement the three science-fictional thought experiments discussed so far with an equally cursory exhibit of recent empirical examples.

Before I begin, note that a general challenge for decoding mental states from neuroimaging data is that it requires training in the individual participant, as well as a stable and hence predictable association between a particular type of brain activation and a particular type of mental state. Moreover, this association, so far, must be highly stereotyped, enabling researchers to distinguish, for instance, whether a participant is seeing a face or a house (based on differences in activation in the fusiform face area and the parahippocampal place area; for a general introduction and further references, see Haynes & Rees, 2006; Haynes, 2015; for a critique of the background assumptions involved in "reverse inference," or the practice of using imaging data to make inferences about mental states, see Levin & Aharon, 2011). While Haynes and Rees (2006) note that this approach might even be used to decode unconscious or covert mental states (like racial bias or deception), they also point out that generalizations over time, across different instances of the same mental state, and across different individuals are, at present, still highly problematic. One might assume, then, that such generalizations from brain activity underlying perceptual states to dreaming may be particularly challenging, as the neural basis of dreaming and of particular types of dream experiences continues to be unknown.

The first example in my exhibit, though far from resolving these problems, makes some important progress, suggesting that brain-reading technology is becoming increasingly sophisticated. Miyawaki et al. (2008) attempted to overcome the constraints imposed by the fact that today's decoding models are trained on closely circumscribed candidate image sets and consequently are poorly suited to the complexity of perceptual experience.

> We show that once our model is trained with several hundred random images, it can accurately reconstruct arbitrary images (2^{100} possible images), including geometric and alphabet shapes, on a single trial . . . basis, without any prior information about the image. (Miyawaki et al., 2008, p. 916)

This is an important step, greatly increasing the real-world feasibility of decoding mental states from brain states. In their paper's final paragraph, Miyawaki and colleagues offer a speculative outlook, among other things, on dreams, noting that "attempts to reconstruct subjective states that are elicited without sensory stimulation, such as visual imagery, illusions, and dreams," might be a particularly interesting goal for future research (Miyawaki et al., 2008, p. 926) and might help clarify whether the neural basis of these states overlaps with that of stimulus-evoked percepts. Moreover, "the combination of elemental decoders could even reveal subjective states that have never been experienced with sensory stimulation" (p. 926), thus opening up, one might imagine, an entirely new field of investigation for brain-reading technology.

My second example is a recent study, widely discussed in the press, that for the first time achieved the reconstruction of dynamic visual images based on imaging data from brain activity obtained during the presentation of YouTube movies (Nishimoto et al., 2011). This is an important step, because previous studies involving so-called brain-reading techniques focused on the reconstruction of static images from neuroimaging data. Importantly, the movies used to train and test the models were different, suggesting, once more, that this method can "decode" images not previously seen. And again, we find the suggestion that their "framework could potentially be used to decode involuntary subjective mental states (e.g., dreaming or hallucination)" (p. 1645). This would mean, in other words, that their modeling framework, involving training sessions in which blood-oxygen-level-dependent (BOLD) signals obtained while participants repeatedly watch YouTube movies, could be applied to BOLD signals from REM sleep, as well. Hall and Van de Castle's vision of *televising* dream experience by decoding dynamic visual images from brain activity may then actually be possible. At the same time, it seems that dreams may provide a litmus test for assessing the success of novel brain-reading technologies.

Assuming that this is technologically feasible, why not simply apply these novel technologies to dreaming? My third and final example describes a case in which this (or something close enough) has been done.[31] Horikawa et al. (2013a, 2013b) gathered approximately two hundred reports from sleep onset from three subjects sleeping in an fMRI scanner. Words describing visual scenes or objects were extracted from the reports and mapped to a lexical database comprising broader object categories (or synsets; see

Table 3.1
Example of a dream report used in the construction of a "dream-reading machine" (Horikawa et al., 2013b, p. 25).

Report	Word	Base synset
Well, what was that? Two *male persons*, well, what was that? I cannot remember very well, but there were e-mail *texts*. There were also *characters*. E-mail address? Yes, there were a lot of e-mail addresses. And two *male persons* existed.	male person text character	male (male person) writing (written material, piece of writing) character (grapheme, graphic symbol)

table 3.1). Then, in a second step, Internet photos corresponding to the visual object categories identified in the dream reports were shown to the waking participants while they lay in the scanner, to train a computer program to associate the object categories to specific patterns of activity in the visual cortex. Based on a further round of sleeping in the scanner and dream reporting, the researchers then used this program to predict with around 60 percent accuracy the visual objects described in the dream reports.

This, of course, is a single and highly preliminary study. Still, let us assume, for the sake of argument, that it had successfully been replicated with larger subject groups, with reports not only from sleep onset but from all sleep stages, with even higher accuracy and perhaps even with more sophisticated reporting methods. Would this suggest that researchers might indeed move *beyond* dream reports in the future? This is not the place to engage in science-fictional predictions as to whether "dream reading" independently of dream reporting might someday be scientific practice. I merely want to point out that in developing such technologies, dream reports play an indispensable role both during training sessions and in establishing the required stability and accuracy of predictions. At present, predictions count as successful only if they are borne out by dream reports. Far from corroborating the trustworthiness of dream reports, dream reports, under the assumption of transparency, corroborate the accuracy of the predictions and thus are instrumental in developing "dream-reading" technologies in the first place. What this single study shows, then, is that it is simply not enough to use a decoder trained in waking perception to predict dream experience—instead, dream reports play an indispensable role in applying brain-reading technologies to dreaming.

This is in keeping with the only other study, to my knowledge, to have mapped imaging data to specific dream contents, namely, Dresler and colleagues' study of lucid dreaming. While they note that in the future "this technique could be used to inversely infer specific dream content from its underlying neural activity, allowing for true 'dream reading'" (Dresler et al., 2011, p. 1836), they are quick to point out that in their study, data from lucid dreams were gathered under conditions in which both concurrent and retrospective reports were available and indeed methodologically crucial (cf. Dresler et al., 2012). By contrast, applying insights from the successful decoding of images obtained during wakefulness to dreams *without* previously collecting and analyzing dream reports is not feasible. This is not a trivial point: while Horikawa and colleagues themselves are explicit about the role that dream reports played in their study, the press coverage occasionally portrays their results as verifying the accuracy of dream reports and even as suggesting the possibility of investigating forgotten aspects of dream experience (cf. Underwood, 2013).

I think there are two reasons for caution. The first is practical. As the example in table 3.1 shows, the decoder used by Horikawa and colleagues was, in fact, quite limited, in that it did not capture strictly all aspects of dream experience or even all object categories (several more reports are available in the supplement to Horikawa et al.'s article).

Of course, more sophisticated methods could be developed; but still, the number of categories in which the decoder is trained cannot, if only for practical reasons, be endless, meaning that the ability even of a more sophisticated decoder to predict dream experience would be narrower than what participants were able to report. The expressive granularity of dream reports, it seems, outruns that of the decoder.

The second concern is of a more principled, theoretical nature. Let us assume that future scientists really do manage to build a more comprehensive dream-reading machine based on Horikawa and colleagues' findings, as well as on similar, more sophisticated studies. But even then, the dream-reading capacities of this future machine would still be constrained by dream reports, including occasionally incomplete dream recall—and this would even be the case if researchers as a matter of fact ceased to collect dream reports in future studies. The accuracy of their predictions would still be constrained by the reported dreams originally used for the development and testing of this technology and thus would still depend on the ability of their model to predict reportable dreams. It seems, then, that at least in this respect, the methodological background assumptions of scientific dream research have not changed much since the discovery of REM sleep in the 1950s: even future dream-reading technology would be constrained by the space of reportable dreams.

Moving from feasible real-world examples back to science fiction, note that the same is true not just for current attempts to engage in dream reading but also for the thought experiments discussed at the beginning of this section. Perogamvros rightly notes that using dreaming as an endogenous Morel's machine is constrained not just by present technical limitations but also by the problem of how to gather maximally accurate dream reports. In Revonsuo's dream catcher test, dream reports also play a crucial role: note that the model of dream experience based exclusively on findings from brain activation patterns would count as successful only if it could be matched to a model based exclusively on dream reports. In this setup, the accuracy of the dream reports themselves is not questioned but presumed. And returning to the 1960s, Hall and Van de Castle note that in the absence of dream-televising technology, "a dream may be operationally defined as that which a person reports when he is asked to relate a dream. . . . When we speak of a dream we shall always mean a reported dream" (Hall & Van de Castle, 1966, pp. 18–19).

Before concluding this chapter, I want to note that far from moving beyond dream reports, dream-reading technologies might rather present a novel and more accurate way of using dream reports to investigate the neural basis of dream experience. As Horikawa and colleagues (2013, p. 640) point out, the fact that they were able to successfully read out visual dream experiences from cortical activity patterns observed during visual stimulus presentation in wakefulness in the first place suggests a "principle of perceptual equivalence," according to which perception and dream imagery share a common neural substrate. Assuming that they accurately predicted not just dream reports but visual dream experience, this result would thus be a large step toward enriching an account of which types of experiences occur in dreams, and of how to relate dream experience and its neural underpinnings to standard wake states. Applying brain-reading technologies primarily developed for conscious wake states to dreams, then, may be an important step toward an integrative approach to the neural correlates of conscious experience across sleep–wake states.

3.8 Conclusions

Throughout this chapter, it has become clear that the experiential and transparency assumptions are inseparable from the methods of scientific dream research. In fact, as many examples have shown, the careful study of dream reports appears to be the condition for scientific *dream* (as opposed to sleep) research in the first place. This finding thus enhances the plausibility of my claim that the experiential and transparency assumptions are methodologically necessary default assumptions incapable of empirical verification.

Yet we have also seen that the unpredictable outcome of gathering dream reports and the instability of dream recall are often seen as obstacles to be overcome. Researchers occasionally regard the methodological dependence of their work on dream reports as a limitation and, since the beginning of scientific dream research in the 1950s, have been discussing ways of corroborating or verifying dream reports. If my analysis is correct, however, then this is not only impossible but also misguided. It is also unfortunate, because this attempt invites a particularly damaging and unproductive type of skepticism about dream reporting, namely, one that is unresolvable, as argued in chapter 1, by empirical means. In a sense, Malcolm's analysis of dreaming (and with it the whole philosophical debate on dream experience and dream reporting in the second half of the twentieth century) was provoked by what he took to be the problematic attempt to verify dream reports in Dement and Kleitman's early study. To the extent that researchers occasionally do portray their results in this manner, this practice indeed deserves critique; however, and this is my deeper point, the resulting skepticism about the trustworthiness of dream reports misses the deeper logic of scientific dream research, which depended on dream reports rather than questioning or corroborating them all along. Philosophical skepticism about dream reporting, then, targets a practice that, though dream researchers do occasionally appear to say otherwise, actually has very little to do with the way scientific dream research is actually conducted.

Consequently, though the discussion on dream experience and dream reporting is essentially a philosophical problem, it is not exclusively a philosopher's problem. Here is a final example that I hope will illustrate my claim that the metatheoretical critique presented here may be of actual value to dream research. In a chapter on the methodological problems involved in the study of dreaming and the measurement of dream content, Winget and Kramer offer the following remarks:

> *Definition of the dream protocol.*—The *object of our interest is the dream experience.* We *presume,* with the support of some inferential evidence . . . *that the dream experience does indeed exist. Unfortunately,* however, the *dream experience is subjective and,* even with the current electrophysiological techniques, is *not directly available for public scrutiny.* Therefore, *the object of our measurement cannot be the dream as experienced, but must be the report of that experience.* Again, with some indirect evidence, we *assume a significant correspondence between the dream as experienced and the dream as reported.* (Winget & Kramer, 1979, p. 13; italics mine)

They continue, two pages later:

> To say simply that the dream is what the subject reports (Malcolm, 1959) is not specific enough for the scientific exploration of dream content. The final dream protocol that is to be measured may be a rather distilled product of the verbal report. . . . In any event, the exact process used to achieve this distillation should be clearly set forth. (Winget & Kramer, 1979, p. 15)

As has become clear throughout this chapter, two separate problems are at play here: One, the philosophical (and, I would claim, largely futile) question of whether dreams are experiences and whether dream reports transparently show this to be the case; here, note that Winget and Kramer, entirely in keeping with my terminology, portray this as an assumption, not as a fact. They also, however, express regret about the subjective nature of dreaming and thus portray the inherent limitation of their approach as a shortcoming. This latter worry, however, is misguided: the subjective and hence unobservable nature of dreaming is not a contingent fact about dreaming but inherent in saying that dreaming is a form of subjective experience. It is also not unique to dreaming but characterizes the study of conscious experience more generally.

Two, there is the methodological and empirically investigable problem of how to distill a measurable and, in my terminology, maximally transparent or trustworthy product from the verbal report. Here Winget and Kramer are entirely correct to point out that this problem is commonly overlooked by philosophers. But just as philosophers have typically taken an oversimplified view of the nature of dream reports and dream reporting, so researchers have occasionally mistakenly assumed the former philosophical problem to be a real problem for dream research—and thus unwittingly perpetuated misguided forms of skepticism about dreaming and dream reporting. It is true that empirical dream research cannot prove dream reports to be trustworthy; however, this is due neither to a shortcoming of particular research findings, nor to the poor trustworthiness of dream reports, but to the fact that the demand for empirical corroboration itself is misconceived. Overlooking this point, however, perpetuates worries about the trustworthiness of dream reports, thus making report-based dream research appear more vulnerable than it really is and fueling the ultimately paradoxical desire for conducting dream research independently of dream reports. Such efforts, however, would better be invested in more realistic endeavors. The reliance on dream reports is not a limitation of the field of scientific dream research but an inherent part of the project itself, and portraying this as a limitation damages the reputation of an entire field of research by making it appear unscientific and built on a shaky foundation.

One way in which I hope that my alternative account of the experiential and transparency assumptions may prove to be helpful, then, is by freeing empirical dream research from the unsolvable because misconceived challenge of proving the trustworthiness of dream reports. Rather than investigating *whether* dreams are experiences—which is not a genuine question at all—researchers should (and typically do) focus on how best to *use* dream reports to *investigate* dream experience. Throughout this chapter, we have seen a number of ways in which this can be done, ranging from polysomnography and timed awakenings in the sleep laboratory, the promise of neuroimaging studies of dreaming (and not just of sleep), and lessons from exclusively report-based research, to the investigation of sleep behavior, its relation to dream experience, and its potential use as *concurrent* dream reports (as in dream lucidity), to lessons from lesion studies and the (partial) cessation of dreaming, to, finally, novel attempts to predict dream experience based on the known correlation between brain activation and waking stimulus presentation. Particularly promising perspectives include not only contrasting different types of dreams (e.g., lucid and nonlucid) with each other but also contrasting dreams with standard wake states and neuropsychological syndromes and thus aiming at an integrative approach to the study of conscious experience across sleep–wake states (for a detailed discussion of this point, see Windt & Noreika, 2011). By contrast, misplaced skepticism about dream reporting cannot, and should never be, an excuse for neglecting the careful and disciplined study of dream reports. All the existing evidence suggests that the analysis of dream reports is a tremendously promising project—but also that truly vigorous and scientific dream research cannot be achieved without an equally vigorous approach to dream reports.

Aside from these almost programmatic claims about dream research, there are also a number of more specific benefits for the philosophical discussion of dreaming and dream reporting. In particular, the foregoing analysis of the methods of scientific dream research has enriched the theoretical account of the relationship between dreaming and dream reporting considerably. One important lesson from this chapter is that the transparency assumption is more complex than assumed so far: to provide a truly informative basis for empirical research, it has to be complemented by the further assumption that dreams are reportable experiences, and hence that reports of nondreaming, at least under certain conditions, have the same prima facie credibility as reports of dreaming. I call this the reportability assumption. Only if reports both of dreaming and of nondreaming are taken to be equally trustworthy can they be used as indicators not only for the occurrence of dreams but also for their absence. Thus both the transparency and the reportability assumptions are inherent in the project of investigating the neural correlates of dreaming. By contrast, if one admits the possibility of unreportable dreams, this undermines the rationale for contrasting dreams with dreamless sleep, and with it one of the central goals of empirical dream research.

We also now have a working definition of what it means to report a dream: dream reporting is a special type of behavior conducted with the sincere intent of conveying or recording certain relevant information about a specific dream. While the example of lucid dreaming shows that even concurrent forms of dream reporting during sleep are feasible, their status as reports (rather than as automatic sleep or even dream-enactment behavior) depends on retrospective confirmation. Thus retrospective dream reports are indispensable for identifying concurrent neural and behavioral evidence, but also concurrent reports of dreaming. But we have also seen that the transparency and reportability assumptions are most helpful and most informative when they are qualified. A consistent assumption seems to be that temporal proximity is crucial for ensuring the validity of both assumptions; however, a number of further qualifications may be helpful depending on the specific research question asked in a given study. For instance, different types of reports might be transparent with respect to different aspects of dreaming (e.g., checklist logs with respect to dream recall frequency versus verbal reports and affirmative probes with respect to the content and formal features of a specific dream); reports can be tweaked as to their specificity, level of detail, and expressive granularity (e.g., by suggesting specific categories, using questionnaires, or even conducting interviews about specific dream experiences, but also by using different types of reports such as verbal, written, or behavioral reports or dream drawings), and so on. Clearly, a plausible account of dream reporting should adhere to Winget and Kramer's warning that equating the dream with whatever the subject happens to report, or equating the dream report with the actual data used by researchers, is oversimplified and naive. While a philosophical account should clearly acknowledge these issues, it would seem that at this point, the problem of dream reporting—or, better, the multiple problems involved in the collection and assessment of dream reports, the improvement of reporting conditions, and scoring methods—is no longer a principled problem for philosophy but a problem for empirical dream research. And, of course, it is one that researchers have been addressing for decades. It cannot be the goal of a philosophical book on dreaming to contribute to this project; at best, a philosopher can hope to systematize the types of problems raised by dreaming and dream reporting, suggest which ones are productive and which are misguided, and perhaps point out parallels with other debates (such as the debate on first-person reports in consciousness research).

Returning to these types of general questions is the task of the next chapter. There I will also complete my account of the basic problem of whether dreams are experiences at all, or rather: why we should say that they are. In particular, I suggest that this is not just an act of desperation resulting from a methodological necessity, but also theoretically justified.

4 Antiskepticism about Dreaming and Dream Reporting: From Default Assumption to Theoretical Justification

4.0 Introduction

The preceding chapters have established the central role of the experiential, transparency, and reportability assumptions as methodologically necessary default assumptions for scientific dream research. This means, first, that if these assumptions turned out to be false, then dream research—understood as the scientific investigation of conscious experience during sleep—would not exist: it would be deprived of either its target or the primary type of evidence used for its investigation. Second, it has become clear that the truth of these assumptions cannot itself be the object of scientific investigation: attempts to corroborate or verify the trustworthiness of dream reports (or the reportability of dreams) by independent means are either circular or demonstrate their own insufficiency by inviting inescapable skepticism about dreaming and dream reporting.

The next step is to offer a theoretical account showing that the experiential, transparency, and reportability assumptions are indeed justified. Without this additional step, their endorsement might seem like a simple act of desperation, resulting from the insight that their truth is methodologically necessary for scientific dream research. This step is also needed to put skeptical worries about dream reporting to rest. In particular, the determined skeptic might be willing to bite the bullet and accept that if no empirical corroboration of the transparency and reportability assumptions is forthcoming, then scientific *dream* research is impossible. Granted, I have shown that this is a sizable bullet; however, the whole point of the skeptical endeavor is not to endorse a comfortable and attractive theory about dream reporting but to defend one that is true.

The task of this chapter, then, is to develop a better and more powerful response to the skeptic. At the end of chapter 1, I already hinted at how this might be done: inference to the best explanation shows that the transparency and reportability assumptions are not only methodologically necessary for scientific dream research but also theoretically justified because they provide a *better explanation* of dream reporting than skeptical alternatives. In this chapter, I develop this idea in more detail (for a concise version of some of the main ideas discussed here, see Windt, 2013b). Admittedly, this will not amount to an outright refutation of the skeptic (indeed, if I am correct, this is impossible), but it will tip the balance toward antiskepticism about dream reporting. For present purposes, this is enough.

One might wonder why I develop this account only now, after a prolonged detour through the history and methods of scientific dream research. The first reason is that chapters 2 and 3 amounted to an extended plausibility check of my claim that the experiential and transparency assumptions are methodologically necessary default assumptions, thus establishing the basis of the argument presented here. A second reason, however, is that my examination of the methods of scientific dream research has also led to a deepened theoretical understanding of these assumptions and the conditions in which they apply. Consequently, by grounding the theoretical discussion of dreaming and dream reporting in real-world examples from dream research, this extended detour has strengthened my argument against the skeptic considerably. Skepticism about dream reporting is a philosophical and highly abstract problem, but it is also of immediate concern to a real branch of research. Thus a balanced assessment of skepticism about dream reporting should be informed by the actual scientific practice that it targets.

The present chapter, then, can be regarded as a reverse plausibility check, testing my new, empirically enriched account of the problems of dreaming and dream reporting against both familiar and new versions of philosophical skepticism about dream reporting. My goal is not simply to refute the skeptic, but rather to profit from these skeptical arguments and use what is good about them to strengthen my own account. In calling this account "antiskeptical," I want to emphasize both its central claim and its heritage: if it is successful, its success is greatly indebted to the versions of skepticism it ultimately rejects. I also, in the concluding section of this chapter, position my account in the broader context of the current debate on first-person reports and suggest a number of particularly promising projects for the future.

4.1 Inference to the Best Explanation as a Response to the Skeptic

Inference to the best explanation (IBE) plays an important role in scientific investigation but has also often been suggested as a response to different variants of philosophical skepticism. Skeptical arguments typically function by casting doubt either on the possibility of knowledge in general or on particular kinds of knowledge. Specifically, skeptics will often claim that knowledge in a given domain can only be had if certain relevant possibilities threatening this knowledge can be ruled out (for a general introduction to philosophical skepticism, see Stroud, 1984; Landesman, 2002). Typically they cannot. Hence the skeptic has succeeded in undermining the respective knowledge claims.

IBE offers a model of why knowledge claims should be regarded as trustworthy even if skeptical alternatives cannot conclusively be ruled out. Importantly, IBE does not require direct evidence confirming or disconfirming a particular knowledge claim but rather considers an inference as justified if it provides a better or more powerful explanation of the available evidence. As Lipton (2000) puts it, the "governing idea is that explanatory considerations are a guide to inference, that scientists infer from the available evidence to the hypothesis which would, if correct, best explain that evidence" (p. 184; see also Lipton, 2004, 2007). Similarly, Harman (1965) writes:

> In making this inference one infers, from the fact that a certain hypothesis would explain the evidence, to the truth of that hypothesis. In general, there will be several hypotheses which might explain the evidence, so one must be able to reject all such alternative hypotheses before one is warranted in making the inference. Thus one infers, from the premise that a given hypothesis would provide a "better" explanation for the evidence than would any other hypothesis, to the conclusion that the given hypothesis is true. (Harman, 1965, p. 89)

The central motivation for IBE is pragmatic: alternative hypotheses are not falsified empirically but ruled out based on their comparatively weaker ability to explain the available evidence. In keeping with philosophical pragmatism, IBE provides not only methodological justification for treating a given hypothesis as true but also a rationale for saying that it actually is true. As William James wrote, pragmatism is "first, a method; and second, a genetic theory of what is meant by truth" (James, 1907/1981, p. 33). Consequently IBE provides just what is needed in the debate on dreaming and dream reporting: a rationale supporting not only the normative claim that dream reports should be treated as transparent given the methods of scientific dream research (and conditional on the assumption that dreams actually exist) but also the claim that they actually are transparent and that dreams actually are experiences during sleep.

Many authors consider IBE as analogous to abduction (Harman, 1965; Lipton, 2000; Douven, 2011; but see Mackonis, 2013), a term used by Charles Sanders Peirce to refer to a type of inference common in both everyday and scientific reasoning that is distinguished from both deduction and induction (for general discussion, see Burks, 1946; Frankfurt, 1958; Douven, 2011). According to Peirce, pragmatism "is nothing else than the question of the logic of abduction" (Peirce, 1958, 5.196). This is so because "abduction is the process of forming an explanatory hypothesis. It is the only logical operation which introduces any new idea; for induction does nothing but determine a value, and deduction merely evolves the necessary consequences of a pure hypothesis" (5.171). Peirce viewed abduction not just as involving the methodological process of discovering new hypotheses but also as an evidencing process. According to Peirce, our preference for simpler explanations is essentially instinctive, and abduction is "really an appeal to instinct" (1.630). When confronted with a surprising fact, we instinctively embrace the simplest explanation that would, if true, explain why the fact is a matter of course. An explanation counts as successful if it helps avoid future surprise. In this view, inquiry is the process of resolving doubt and arriving at stable belief habits, and one of its aims is to establish belief habits that "lead to the avoidance of all surprise and to the establishment of a habit of positive expectation that shall not be disappointed" (5.197). The truth of a hypothesis thus depends to a considerable extent on its explanatory and predictive power.

Others argue that an additional step is needed to move from abduction to IBE: the assumption not only that a given hypothesis explains why an initially surprising fact is a matter of course—hence providing evidence for its truth—but that this explanation is also better than explanations offered by competing hypotheses (e.g., Mackonis, 2013). As theoretical questions about the relationship between abduction and IBE are not directly relevant to my argument, I will leave them to the side here. The main point, for now, is that from the beginning, abduction was viewed both as a method of discovery and as an evidencing process, with simplicity and intuitive plausibility being regarded, at least initially, as indicators of explanatory goodness.

I will adopt the following reconstruction of the standard form of IBE from Mackonis (2013, p. 977) because it nicely emphasizes the comparative aspect needed for ruling out alternative hypotheses:[1]

(1) The surprising fact, *C*, is observed.
(2) But if *A* were true, *C* would be a matter of course.
(3) No available competing hypothesis can explain *C* as well as *A* does.
(4) Hence, *A* is true.

What, then, makes one explanation better than another? First, note that abduction or IBE should not necessarily be construed as an inference to the likeliest explanation. While the best explanation will typically also be regarded as the likeliest one, a deeper question is how this estimation of likelihood is reached. As Lipton (2000) points out:

> At first glance, it may appear that likeliness is the notion Inference to the Best Explanation ought to employ, since scientists presumably only infer the likeliest of the competing hypotheses they consider. This is, however, probably the wrong choice, since it would severely reduce the interest of the model by pushing it towards triviality. Scientists do infer what they judge to be the likeliest hypothesis, but the main point of a model of inference is precisely to say how these judgements are reached, to give what scientists take to be the symptoms of likeliness. To say that scientists infer the likeliest explanations is perilously similar to saying that great chefs prepare the tastiest meals: true perhaps, but not very informative if one wants to know the secrets of their success. (Lipton, 2000, p. 187)

Instead Lipton suggests that IBE should be construed as inference to the *loveliest* explanation. The "central claim is that scientists take loveliness as a guide to likeliness, that the explanation that would, if correct, provide the most understanding, is the explanation that is judged likeliest to be correct" (Lipton, 2000, p. 187). What exactly the explanatory virtues contributing to loveliness consist in raises a number of theoretical issues, and there is a large debate as to how exactly to spell them out.[2] Based on the literature, however, is seems that scope, precision, simplicity, and unification are plausible candidates. From now on, I will assume that these ingredients indeed contribute to the loveliness of an explanation, and this loveliness can be seen as a symptom of likeliness, and hence as an indicator of truth (for details, see especially Lipton, 2000, 2004; see also Harman, 1965; Douven, 2011; Mackonis, 2013).

Roughly, the loveliness of an explanation depends on

- the range of phenomena to which it can be applied (*scope*), as well as on
- its ability to generate specific predictions (including those about underlying mechanisms; *precision*)
- in a parsimonious manner (*simplicity*)
- that is consistent with existing theories or background assumptions (*unification*).

Relatedly, one explanation will count as lovelier than another if meets these requirements to a fuller extent.

In addition, from the historical literature, it seems that instinctive or intuitive loveliness may contribute to the overall explanatory loveliness of a given hypothesis. I return to this issue in section 4.2.

Before I consider the application of IBE to skepticism about dreaming and dream reporting, note that IBE has long been thought to be of particular value in refuting skeptical arguments.[3] A familiar example is Cartesian dream skepticism. Here, the central claim is that there are two rival hypotheses explaining ongoing sensory experience, namely, one, I am really sitting in my study and looking at the piece of paper in my hands (or rather at the computer monitor, feeling the keyboard, etc.); and two, I am only dreaming that I am (cf. sec. 1.1). And because Descartes applies the argument to a best-case scenario of sensory perception (Stroud, 1984), I cannot use further sensory evidence to decide which hypothesis is correct: looking at the piece of paper (or monitor, keyboard, or what have you) more closely will not help me decide whether I am really seeing it or only dreaming that I am. Hence I cannot conclusively rule out that I am now dreaming.

Arguments from abduction or IBE grant that the rival hypothesis—in this case, the dream hypothesis—presents a possible explanation, but deny that it provides a better or even a particularly good one. Such arguments concentrate on showing that the commonsense, naive realistic view that I am really seeing (rather than merely dreaming that I am seeing) the piece of paper does a better job of explaining the available sensory evidence. This, in turn, justifies the conclusion that it is true.

A classic example is Bertrand Russell's use of IBE as a response to Cartesian dream skepticism in *The Problems of Philosophy*. Russell describes the dream hypothesis as "an uncomfortable possibility; but although it cannot be strictly proved to be false, there is not the slightest reason to suppose that it is true" (Russell, 1912/1999, p. 9). First he notes that the belief that our sensory impressions, or sense-data, represent mind-independent physical objects comes naturally to us; it is a matter of instinctive belief and of common sense. Then he points out that though logically possible, the dream hypothesis "is, in fact, a less simple hypothesis, viewed as a means of accounting for the facts of our own life, than the common-sense hypothesis that there really are objects independent of us, whose action on us causes our sensations" (p. 13). In particular, the dream hypothesis does a worse job at explaining the spatiotemporal continuity of sensory experiences: after all, when I return to my study from a break, my computer monitor is always in the same position that I last saw it in—and when it is not, I can explain this by its having been moved by someone else. Moreover, even if the dream hypothesis could explain isolated instances of regularity in my sensory impressions, it could not as smoothly explain the general continuity of my sensory impressions, nor could it explain why my entire system of beliefs, which is, after all, tailored to the existence of an external, mind-independent world, typically yields successful predictions. Russell concludes that "every principle of simplicity urges us to adopt the natural view, that there really are objects other than ourselves and our sense-data which have an existence not dependent upon our perceiving them" (p. 14).

Russell also notes that our instinctive, commonsense belief in the mind-independent existence of physical objects not only fails to lead to any difficulties "but on the contrary tends to simplify and systematize our account of our experiences" (Russell, 1912/1999, p. 14). The instinctive belief thus has the double advantage of failing to lead to the difficulties suggested by the rival hypothesis and of bringing order into our existing belief system:

> Hence, by organizing our instinctive beliefs and their consequences, by considering which among them is most possible, if necessary, to modify or abandon, we can arrive, on the basis of accepting as our sole data what we instinctively believe, at an orderly systematic organization of our knowledge, in which, though the *possibility* of error remains, its likelihood is diminished by the interrelation of the parts and by the critical scrutiny which has preceded acquiescence. (Russell, 1912/1999, p. 15)

In other words, our commonsense beliefs, thus systematized, lend support to each other, and according to Russell, providing such systematization is one of the functions of philosophy itself (Russell, 1912/1999, p. 16). This means that for Russell, philosophy is, in part, in the business of determining the best explanation of our impressions and systematizing our beliefs accordingly.[4]

In sum, given the stubborn intractability of skeptical arguments, it may well be, as Briesen (2008) puts it, that "the most promising candidate for a successful refutation is the antiskeptical strategy provided by the inference to the best explanation" (Briesen, 2008, p. 24). He takes simplicity to be the main symptom of explanatory goodness and argues that since the skeptical hypothesis is more complex than the antiskeptical alternative, we have a priori justification to regard it as less plausible. Similarly, Vogel (1990) notes that explanatory considerations favor the commonsense view because it offers a simpler, less ad hoc, and better integrated explanation than the dream hypothesis. The next step, then, is to show that the experiential and transparency assumptions have the same explanatory merits.

4.2 Inference to the Best Explanation at Work: Toward an Antiskeptical Account

A first step is to apply the standard format of IBE introduced in the previous section to dreaming and formulate a master argument for saying that dreams are experiences and that dream reports transparently show this to be the case. The initially surprising fact—this is one of the important lessons from Malcolm's and Dennett's versions of skepticism about dreaming and dream reporting—is not the occurrence of conscious experience during sleep but the fact that people often engage in dream reporting behavior upon awakening:[5] they describe having had sometimes vivid experiences during sleep, and they do so using the same language used to describe waking experiences. Accordingly, the master argument for dreaming and dream reporting goes as follows:

(1) The surprising fact, *C*, is observed: *upon awakening, people often report having had experiences during sleep!*
(2) But if *A* (*the transparency assumption*)[6] were true, *C* would be a matter of course.
(3) No available competing hypothesis can explain *C* as well as *A* does.
(4) Hence *A* is true: *dreams are experiences during sleep, and dream reports transparently show this to be the case.*

Note that this master argument does not aspire to psychological plausibility, for instance, by claiming to describe a conscious reasoning process. The initial endorsement of the transparency assumption is not contingent on a foregoing assessment of its explanatory loveliness as compared to rival assumptions. Rather, a psychologically more realistic description, as we have seen, may be to say that the transparency assumption is initially accepted by default: it is instinctively or intuitively taken to be true. At least, assuming this to be so would explain why the received view is so received as to typically not require any special theoretical justification at all: its truth, at least before Malcolm's skeptical argument, was assumed implicitly rather than argued for (cf. chap. 1). What the master argument is supposed to show, then, is why this default assumption is also theoretically justified and can be maintained even in the face of skeptical arguments to the contrary.[7]

Based on this observation, it would seem that intuitive plausibility determines the original endorsement of the transparency assumption but does not contribute to its theoretical justification. Yet we have already seen that instinctiveness or intuitiveness is regarded as an explanatory virtue both within the framework of pragmatism introduced by Peirce and James and by those using IBE as a response to the skeptic. Intuitive plausibility might also be intimately connected to the explanatory virtues of simplicity and unification. Recall that according to Peirce, we instinctively embrace the simplest explanation and prefer it over less parsimonious ones. Moreover, a hypothesis that can easily and effortlessly be integrated with background knowledge and assumptions will also be experienced as intuitively plausible, whereas a counterintuitive hypothesis may require a revision of one's belief system to restore cognitive equilibrium (see Gendler, 2010; Metzinger & Windt, 2014, 2015). Thus it would seem that intuitive plausibility may, all other things being equal, enhance the attractiveness or loveliness of one hypothesis compared to another. Perhaps, then, the intuitive loveliness of a hypothesis is a guide to its explanatory loveliness, after all.

A worry, however, is that intuitive loveliness cannot provide a satisfying response to the skeptic: if it could, there would be no need or room for further theoretical justification of the transparency assumption at all. Indeed, proponents of skepticism about dream experience can happily agree that the received view has intuitive plausibility on its side—they merely contend that in the case of dreaming, this is a superficial and psychological kind of plausibility only. If Malcolm is right, after all, then we are profoundly misled by the fact that we use the same language to describe our dreams and our past waking experiences (cf. sec. 1.2). Relatedly, Dennett notes that he has chosen the cassette theory precisely because it "runs most strongly against our pretheoretical convictions" (Dennett, 1976, p. 159). In assessing the explanatory loveliness of the transparency assumption, then, intuitive loveliness is better left to the side.

Instead the question is whether intuitive loveliness of the kind that leads to the initial endorsement of a hypothesis maps onto explanatory loveliness of the more sophisticated type required for theoretical justification. Metaphorically speaking, intuitive plausibility or loveliness is a promise, and its fulfillment depends on its ability to predict explanatory loveliness as determined by the greater scope, precision, simplicity, and unification of one hypothesis compared to another. This kind of explanatory loveliness will ideally be stable or even increase over time and be capable of broad intersubjective assent. One reason I particularly like Lipton's suggestion of using loveliness as a guide to explanatory goodness is that it invites a dual reading from the perspectives of subjective or intuitive loveliness, on the one hand, and explanatory loveliness in a more objective sense, on the other. Consequently the concept of loveliness seems particularly well tailored to the historical view of abduction and IBE both as a method of discovery and as an evidencing process. The task for IBE, then, is to show that the intuitive loveliness of the transparency assumption corresponds to the more robust and sophisticated kind of explanatory loveliness defined earlier.[8]

The next step, then, is to assess the explanatory goodness of the transparency assumption as compared to a specific rival hypothesis. It makes sense to start with the cassette theory of dreaming: it is already familiar to readers (cf. sec. 1.3.4) and presents a single specific alternative. Moreover, several of Dennett's own remarks suggest that he explicitly took explanatory considerations into account.[9] He notes that with the cassette theory,

> We have a challenge to the received view worth reckoning with. It apparently accounts for all the data of the REM researchers as well as the received view does, so there is no reason for sober investigators not to adopt the cassette theory forthwith if it has any advantages over the received view. And it seems that it does: it has a simple explanation of precognitive dreams (if there are any) and it posits one less process by eliminating a presentation process whose point begins to be lost. (Dennett, 1976, p. 161)

The first central advantage of the cassette theory, in this view, is that it is simpler than the received view in that it does not require three separate processes, namely, a composition process, a presentation process, and a memory-loading process, but only the first and last of these.

There are at least two ways in which one might deny that this is indeed an advantage for the cassette theory. The first is to accept that the cassette theory is simpler than the received view in this respect but to point out that this simplicity is counterbalanced by the other explanatory virtues of the received view. Dennett's argument for preferring the cassette theory because of its greater simplicity, after all, hinges on the claim that the presentation process posited by the received view is explanatorily inert: it does not contribute to the theoretical understanding of dream formation and can thus be dropped without detracting from its explanatory loveliness in any way. By contrast, if the assumption of an additional presentation process led to more precise predictions about the composition of dream narratives or allowed a theory of dream formation to be more unified with other existing theories, this might be a worthwhile trade-off for endorsing a less parsimonious view.

Second, a stronger claim would be to deny that the received view requires a separation between presentation and composition process. The proponent of the received view might accept, for instance, that dream formation (or, as Dennett calls it, composition) involves largely unconscious processes, but yet deny that dream experience (or Dennett's presentation process) is wholly separable from dream formation. Indeed, from the perspective of the received view, the separation between the presentation and composition processes may seem somewhat artificial and even inherently dualistic: once one grants that dreams are experiences, it is not clear why one should conceive of dream formation as entirely separate from dreaming. Instead the received view could posit a single process of dream formation, and although this process would surely differ from the one posited by the cassette theory in that it gave a role to conscious experience, it would not necessarily be less parsimonious. According to this view, the ongoing experience of dreaming would be partly constitutive of the processes involved in dream formation; for a proponent of the received view, this would seem to be a natural position. This is not the place to develop a full-fledged theory of dream formation, and I will leave this second possibility to the side for now (but see chap. 12). Here I will only defend the weaker claim that even if composition and presentation are conceived of as separate, we may still need to consider the latter to understand the former.

Incidentally, this claim seems to be closely related to Dennett's (1991) brief discussion of dreaming and dream formation in *Consciousness Explained*. He begins by somewhat playfully describing the party game of psychoanalysis. One person, the dupe, is asked to leave the room and is told that during his absence, one member of the party will share a recent dream with everyone else. When the dupe returns, he is to use yes/no questions to reconstruct the dream narrative and identify the dreamer based on psychoanalytic considerations. Unbeknownst to the dupe, there is, however, no dream to be discussed. Instead, yes/no answers are given according to three simple rules: if the question ends with a word whose last letter is from the first half of the alphabet, the answer is yes, if it does not, the answer is no; the only exception is when adherence to this rule would contradict the answer to an earlier question. As Dennett notes, a "usual outcome is for a bizarre and often obscene story of ludicrous misadventure to unfold" (Dennett, 1991, p. 11). Yet the dream narrative has no author; it results, as Dennett notes, from "a process of narrative production, of detail accumulation, with no authorial intentions or plans at all—an illusion with no illusionist" (p. 11).

Dennett then suggests that the party game of psychoanalysis fits in well not only with "generate-and-test theories of perception" (Dennett, 1991, p. 12)—that is, with models emphasizing perception as involving expectation-driven hypothesis testing (predictive processing is a recent and much-discussed example; cf. Clark, 2013b; Hohwy, 2013b; and sec. 12.3)—but also with the way actual (and not just party game) dreams are composed. His key idea is that dreams and hallucinations "assuage epistemic hunger" or satisfy curiosity (Dennett, 1991, p. 16). Dream narratives, like hallucinations and even perceptions, are not ready-made for consumption or completely determined by bottom-up processes but responsive to attention, anxieties, interests, or expectations and result from the exploration of the ongoing situation. Importantly, as Dennett points out, this theory would explain not only bizarre transformations in dreams but also why certain themes, objects, or characters can sometimes persist in dreams or perhaps even recur:

> This volatility is just what we would expect from an active but insufficiently skeptical question-asker confronted by a random sample of yeses and noes. At the same time, the persistence of some themes and objects in dreams, their refusal to metamorphose or disappear, can also be tidily explained by our model. (Dennett, 1991, p. 15)

For instance, the model would predict that daytime preoccupation with a certain topic would lead to its incorporation in dreams (as suggested by the continuity hypothesis; cf. sec. 6.2.2), perhaps even in a stereotyped manner ("Here I am taking my final exam again, and once more I am completely unprepared"; on typical dream themes and recurrent dreams, see Nielsen et al., 2003; Zadra, 2001). At the same time, the interplay between epistemic hunger and the gullible, uncritical attitude of the dreamer nicely explains dream bizarreness: one may be curious enough to check whether the ghost is still there, but not enough to notice that its color has changed from blue to green (to use Dennett's own example).

An advantage of this account, as Dennett notes, is that it eliminates the playwright in favor of rule-abiding question answerers; no ready-made library of dream cassettes is needed. However, the question asker—and with him, Dennett notes, "the problem of our consciousness of dreams and hallucinations" (Dennett, 1991, p. 15)—is still left unexplained. Indeed, it would seem that if one takes the party game analogy seriously, the questions will be asked in an intelligent and selective manner, in response to the unfolding narrative as presented to the dupe (i.e., to the dreamer). Clearly this should not lead one to introduce an intelligent, question-asking homunculus to the process of dream formation—this would certainly be a mistake—nor should one assume that the dreamer is consciously reflecting on current dream events to formulate further questions, as is the case for the dupe in the party game; instead, one might say, somewhat more neutrally, that ongoing dream experience evokes certain expectations or emotional reactions, and these, in turn, partly constrain the subsequent events of the dream. Perhaps, then, the game of psychoanalysis can be seen as a first step toward answering a problem that Dennett formulated fifteen years earlier:

> In dreaming there is also a third process that is distinguished both in the layman's version of the received view and in fancier theories, and that is the *composition* of what is presented and recorded. In various ways this process exhibits intelligence: dream stories are usually coherent and realistic (even surrealism has a realistic background), and are often gripping, complex, and of course loaded with symbolism. Dream composition utilizes the dreamer's general and particular knowledge, her recent and distant experience, and is guided in familiar ways by her fears and desires, covert and overt. (Dennett, 1976, p. 156)

This raises the problem of how dreams manage to be responsive to the dreamer's ongoing thoughts and expectations, especially if these thoughts and expectations themselves play no role in shaping the unfolding narrative. At least in 1976, Dennett offered no specific proposal about how the apparent intelligence of dreams could be explained by the cassette theory. Whereas taking the presentation process seriously at least hints at a possible solution, the cassette theory renders the process of dream formation more, rather than less, mysterious. Though the cassette theory provides a simpler explanation, it does so only at the price of leaving out exactly the feature that could begin to provide a positive account of dream formation: namely, the ongoing stream of dream experience itself. But if this is correct, then positing a separate presentation process, though less parsimonious, may nonetheless be crucial for understanding the process of dream composition. As we have seen, it may even lead to specific predictions about dream bizarreness, but also about persistent or even recurrent dream themes, and it appears to enable a unified account of dreaming, (waking) hallucination, and perception. An account of dream formation might even, as hinted at earlier, ultimately do away with the separation between composition and presentation altogether. In either case, the putative advantage gained by the cassette theory's elimination of the presentation process appears to be a dubious asset at best.

What about the second proposed advantage of the cassette theory? According to Dennett (1976), this is its superior ability to explain precognitive dreams, suggesting that the theory is broader in scope. The term is often used more broadly in the popular literature to refer to dreams involving a premonition of the future or extrasensory or telepathic perception of distant events. Dennett, however, uses it only to designate dreams that seemingly build up to a narrative climax—which, upon awakening, is associated with a stimulus in the dreamer's actual environment. He even gives an example from his own dreams:

> In a recent dream of mine I searched long and far for a neighbor's goat; when at last I found her she bleated baa-a-a—and I awoke to find her bleat merging perfectly with the buzz of an electric alarm clock I had not used or heard for months. (Dennett, 1976, p. 157)

Dennett argues that the cassette theory offers a more convincing and parsimonious explanation of such phenomena: the cassette theory does not require a presentation process during sleep that gradually builds up to an external stimulus that has not yet occurred, and thus more elegantly explains precognitive dreams. In the cassette theory, but not in the received view, precognitive dreams no longer appear surprising but rather, or so it may seem, would be expected to result from instantaneous composition/memory insertion at the moment of awakening.[10]

But does the received view really have to accept that precognitive dreams of the type described by Dennett involve an utterly mysterious anticipation of future events? I do not see why this would be the case. After all, if dream formation is partly constrained by ongoing dream experience, the ability to fluidly integrate external stimuli into the unfolding narrative would be expected. One need not assume any mysterious anticipation of a future stimulus to do so—Dennett's own example of his dream goat bleating in unison with the alarm clock may attest to this fact. I think examples of ringing alarm clocks being integrated into dreams in various different and occasionally ad hoc ways provide sufficient support for this claim (for details on stimulus integration in dreams, see sec. 5.2.3). Even if the goat's bleating merged well with the actual sound of the alarm clock, the same stimulus might, with slightly different distortions, have been associated with some other type of dream sound. Rather than saying that the dream was tailored to fit (or even anticipate) the stimulus, the opposite might be true: the exact pattern of stimulus integration might be tailored to fit the dream. Also, Dennett's dream goat may well have bleated without any external support from the alarm clock whatsoever. Most of what is heard—and almost all of what is seen—in dreams occurs without a salient external stimulus source. Moreover, if the alarm clock's sound really was unequivocally goatlike, perhaps the dream did build up to its customary and expected bleating ring. Even if, as was the case for Dennett, the alarm clock had not been used for a long time, the timing and sound of its bleating ring might still be familiar and anticipated. I was once given an alarm clock as a joke that did not ring but belted out an extremely shrill song in a screeching electronic voice—and though I am anything but an early riser, and typically only reluctantly respond to alarm clocks after repeated ringing, I quickly developed the ability to wake up at the very last moment before this one went off—just in time to turn it off, avoid the offensive stimulus, and go back to sleep. Given the sound it produced, I do not think my waking up involved any mysterious powers of precognition, but pure survival instinct. (Unfortunately I can report no cases of integrating its sound into dreams; but perhaps, had it been less shrill, I would indeed have had dreams perfectly timed to coincide with its ring, allowing me to sleep just a few moments longer.) This is not to categorically exclude the possibility that more mysterious and hard-to-accommodate cases might exist (for a discussion of laboratory research on supposedly precognitive and telepathic dreams, cf. Ullman et al., 1973/2002; Radin, 2006); but whatever the verdict turns out to be on such cases, the examples given by Dennett are more mundane and more easily explained, and this is sufficient for present purposes (for a similar analysis of Dennettian precognitive dreams, see Springett, 2013; for a contradictory view, see Rosen, 2013).

Dennett's own arguments for the greater simplicity and scope of the cassette theory, then, fail. If it really turns out that dream formation can fruitfully be explained by taking dream experience into account—and perhaps even by saying that dream formation is at least partly constrained by ongoing dream experience—then this would even seem to tip the balance in favor of the received view. I use the remainder of this section to argue that the explanatory virtues of the received view really do outweigh those of the cassette theory. If this attempt succeeds, its result will be more than a naive-realistic, intuitive endorsement of the transparency assumption: we will have taken a crucial step toward an informed defense of the received view against skeptical objections, and thus toward an antiskeptical account of dream reporting.

First, note that the received view offers a simpler and more unifying account of dream recall than the cassette theory. Recall that two readings of Dennett's thought experiment are possible. In the first reading, dream recall, unlike waking memory, is the product of instantaneous memory insertion at the moment of awakening. In the second reading—which is better in keeping with Dennett's later endorsement of the multiple-drafts model—conscious experience is generally regarded as dependent on retrospective recall. Even in this view, however, "We are still not out of the woods on dreaming . . . for we must define normal memory-entry in such a way as to admit ordinary experience and exclude tampering and other odd cases" (Dennett, 1976, p. 168). Either way, the question of dream experience hinges on whether dream recall is the product of a standard or nonstandard memory production process, with the cassette theory soundly committed to the latter view. Clearly, positing a single process explaining dream recall and the recall of waking experiences is preferable, all things being equal, in virtue of its greater simplicity and higher degree of unification with theories of waking memory.

This view may also have empirical plausibility on its side. One recent study found evidence for the claim that memory consolidation processes are state independent, remaining the same across the sleep–wake cycle. The authors conclude that "the current EEG results, in suggesting that the mechanisms involved in encoding and recall of episodic memories across wakefulness and sleep are the same, undoubtedly strengthen the general notion of a continuity between waking and sleep" (Marzano et al., 2011, p. 6681; cf. de Gennaro et al., 2012). Of course, this is an empirical question, and research results could also point in a different direction. As discussed in section 3.6, researchers might find evidence for a sleep-specific memorization process and may even, as suggested by Fell and colleagues (2006), already have done so. Though this would require a less parsimonious account of dream recall, it could still be plausibly accommodated by the received view, for instance, by saying that sleep-specific memory processes are necessary not only for dream recall but also for dream formation itself. Admittedly, this means that the received view, in being open to both types of theories, generates less precise predictions about the exact process of dream memory formation than the cassette theory. However, for the same reason, it also seems better suited to accommodate existing research findings, whichever way they point. Given the present status of research, this might be an asset rather than a weakness (for a review of the largely contradictory findings on the relation between specific EEG patterns and dream recall, see Chellappa & Cajochen, 2013; Ruby et al., 2013).

Perhaps, however, the same could be said of the cassette theory, after all. Perhaps the cassette theory could likewise accommodate the claim that dreaming is the product of standard memory insertion—the same type of memory insertion, that is, that characterizes waking memories as well. But by forging such a tight link between dreaming and memory—after all, in this generalized version of the cassette theory, memory insertion is all there is to dreaming and waking experience alike—it would seem to predict a closer association between the loss of dream recall and disturbances in waking-memory formation than is supported by the empirical evidence. We have seen, for instance, that the cessation of dreaming can present independently of amnesia (Solms, 1997), and amnesia patients can even report hypnagogic imagery (Stickgold et al., 2000). This suggests that the cessation of dream recall cannot straightforwardly be regarded as a problem of dream forgetting but is better explained by a cessation of dreaming proper (cf. sec. 3.6). While, in the received view, the cessation of dreaming independently of mnemonic disturbances is a matter of course because dreaming is not reducible to a process of memory insertion, the cassette theory would render this observation more, rather than less, surprising: if remembering is all there is to dreaming, why should dream reporting be selectively lost even if mnemonic functioning remains otherwise undisturbed? At this point, the generalized version of the cassette theory would have to collapse, once more, into a version according to which dream recall is the product of dream-specific memory insertion.

The received view also does a better job of explaining why dream reports describing roughly the same types of experiences as occur during wakefulness are preceded by roughly the same types of brain activity as are correlated with such experiences in wakefulness. If dream reports describing wakelike experiences are transparent, this similarity would be expected; as we have seen, the comparison between dream bizarreness and neuropsychological syndromes may even be used to predict specific patterns of brain activation during REM sleep (cf. sec. 3.6; Schwartz & Maquet, 2002). Generally, neuroimaging studies from REM sleep offer preliminary support for the claim that the stereotyped formal features of dreaming—such as the predominance of visual and motor imagery, strong emotions, and cognitive and mnemonic deficits—are associated with activation in the same brain areas that are associated with these types of experiences in wakefulness (cf. sec. 3.2). Further support comes from Horikawa and colleagues' (2013a; cf. sec. 3.7) recent demonstration that it is possible to predict specific categories of visual imagery (such as specific objects) described in sleep-onset reports by decoding brain activity known to be associated with perceiving such objects in wakefulness. According to the cassette theory, these findings would be surprising; according to the transparency assumption, they are to be expected. Consequently the transparency assumption potentially allows for a more unifying account of dreaming and waking experience; and by the same token, it allows for more precise predictions about the formal features of dreaming and underlying brain activation patterns. I am saying not that these predictions should be regarded as confirmed—in the current situation, and given the lack of studies mapping dream reports to neuroimaging data from the preceding sleep stage, such a claim would be premature—but only that the received view at least makes specific predictions about how to investigate the neural underpinnings of dreaming in the first place. By contrast, the cassette theory cannot, ex hypothesi, use the presumed similarity between dreaming and waking experience and their respective correlates to develop a research strategy on dreaming.

Relatedly, it has been suggested that dream-enactment behavior in RBD and signal-verified lucid dreams may support a particularly strong argument against the cassette theory (e.g., Revonsuo, 2006, p. 77; cf. secs. 3.4, 3.5). The transparency assumption proposes a straightforward interpretation of these findings: if observed sleep behavior is retrospectively confirmed in the subjective dream report, then it counts as dream-enactment behavior or even, in the case of signal-verified lucid dreams, as a concurrent dream report. But I have also claimed that the transparency of dream reports cannot be verified in this manner but has to be assumed. So how could the cassette theory account for these findings? Here the proponent of the cassette theory would, it seems, have two options. One, as tentatively suggested by Dennett, lucidity could be explained by saying that the unconscious composition process inserted "traces of itself into the recording via the literary conceit of a dream within a dream" (Dennett, 1976, p. 161), thus resulting in the impression, after awakening, of having been able to tinker with or control the ongoing dream. The eye movement signals submitted by lucid dreamers in the laboratory and the dream-enactment behavior observed in RBD patients might be explained in a similar manner. In the cassette theory, all these cases of sleep behavior would be artifacts of the unconscious composition process during sleep.

This option, however, appears increasingly contrived and ad hoc. For instance, lucid dreamers sometimes use external stimuli as lucidity cues (Stumbrys et al., 2012; cf. sec. 5.2.3), and thus could presumably respond to these cues while dreaming with prearranged eye movement signals. Though external stimuli are apparently not a reliable method for the induction of lucid dreams, there is no principled reason why they could not be used to engage in a two-way communication between researcher and lucid dreamer, perhaps even with the dreamer using different patterns of eye movements to selectively respond to different types of stimuli (auditory versus visual, or what have you). In fact, preliminary evidence suggests that this can actually be done: Strelen (2006) showed that three lucid dreamers were able to discriminate high and low tones ("beeps" and "boops") while dreaming and to selectively respond to them with different eye movement patterns. Applying the cassette theory to this kind of scenario is not impossible; but it would, it seems, strain the limits of credibility.

Two, the proponent of the cassette theory could grant that lucid dreams and perhaps even dreams associated with enactment behavior are experiences during sleep, but deny that the same is true for nonlucid REM dreams. And although one should treat such generalizations from lucid to nonlucid dreams (or dream-enactment to standard dreams) with caution, research does seem to suggest that lucid and nonlucid dreams, for instance, are better described on a continuum than as completely distinct phenomena (cf. Voss et al., 2013). At the least, this seems to be plausible for very basic aspects—such as whether dreams are experiences or not. It would, for instance, be strange to assume that the onset of lucidity—and apparently most lucid dreams arise from previously nonlucid ones (Stumbrys et al., 2012)—is marked by a sharp transition from nonconscious dream composition to lucid dream experience.

Even if these cases could in principle be explained by the cassette theory, the received view is broader in scope in that it applies to standard, lucid, and enactment dreams alike, and it is more unifying, because it allows researchers to place dreaming in the context of what is more generally known about the association between brain activation patterns, behavior, and subjective experience. In this process, existing conceptions of sleep, dreaming, and the association between consciousness, brain activation, and behavior are apt to change—but at least, the received view suggests a unified research strategy for approaching these topics in the first place. The greater scope of the received view also enables it to make more precise predictions. After all, if dreams really do involve vivid motor imagery, then one would expect a loss of REM sleep paralysis to lead to the enactment of dreams. Relatedly, if eye movements during REM sleep really do scan dream imagery, then one would expect lucid dreamers to be able to signal lucidity by making prearranged eye movements in their dreams. This, after all, is how researchers came up with the idea of signal-verified lucid dreams in the first place (cf. LaBerge, 1985, p. 68). On all these counts, then, the received view has explanatory loveliness on its side.

In defending the explanatory loveliness of the received view, it might seem that I am claiming that the transparency and experiential assumptions can, after all, be verified. But as we have seen in chapter 3, this is not the case: no independent evidence for the trustworthiness of dream reports is forthcoming. Yet, part of the reason for thinking that IBE is itself a particularly lovely strategy for defending the received view is that it can account for the increasing loveliness of the transparency assumption without requiring independent evidence in its favor. If we regard familiar findings from neuroimaging studies, RBD enactment dreams, lucid dreams, or the cessation of dreaming as part of an integrated research program on dream experience and evaluate them from the perspective of IBE—rather than asking whether they can verify dream reports—their significance changes entirely. What does or does not count as evidence for a given theory changes with the type of question asked. In particular, while the demand for empirical verification invites inescapable skepticism, an assessment of the same assumptions and the same kinds of evidence in terms of their explanatory loveliness leads to an antiskeptical account of dreaming and dream reporting.

In this context, note that an interesting property of the received view is that it allows researchers to turn findings about brain activation, for instance, during REM sleep, into evidence about dreaming in the first place. As pointed out by Putnam (1986; cf. sec. 1.4), this can only be done by initially assuming dream reports to be transparent and dreams to be experiences during sleep—and once this is done, it turns into a powerful tool for investigating the neural basis of dreaming itself. By contrast, the cassette theory would seem to require a dissociation between brain activation and subjective experience: in the special case of dreaming, wakelike levels of brain activation followed by reports of vivid experience would not count as evidence for the occurrence of conscious experience, nor would high levels of activation in the visual areas count as evidence for visual experience, and so on and so forth. Although findings on NREM dreams suggest that brain activation levels alone provide insufficient evidence of dreaming, denying that there is any meaningful correlation between brain activation and experience in sleep would mean losing the best candidate for a neural correlate of dreaming that is available to date—surely not an attractive possibility. Moreover, at least part of the association between brain activity and conscious experience might be saved by relating NREM dreams to covert REM sleep processes (as suggested by Nielsen, 2000a; see sec. 2.2). Modifying existing theories on the neural correlates of dreaming to incorporate NREM dreams is surely preferable, all things being equal, to giving up on them completely.

But this also points to a property of the received view that can be considered both a strength and potential weakness, depending on one's perspective. Note that in defending the explanatory loveliness of the received view against that of the cassette theory, I have been using empirical findings that were not even available when Dennett first formulated the cassette theory in 1976. Though I suspect that the received view had explanatory loveliness on its side even then,[11] this suggests that the explanatory loveliness of a given theory is not static but partly dependent on research findings. As van Fraassen (1980) points out, IBE is always an inference to the best explanation given to date, and it is always possible that what now appears as the best explanation is merely the best out of a set of bad hypotheses.[12] Clearly, it is always possible that even if a hypothesis is good, a better one might later come along. Explanatory loveliness is a revocable property. In a sense, considering a particular hypothesis as the best explanation is always a prediction: the explanatory loveliness of a hypothesis depends not only on its present ability to satisfy the proposed criteria to a fuller extent than alternative hypotheses, but also on its ability to do so in the future.

A prediction for future dream research would be that because the transparency assumption makes new forms of evidence about dreaming and dream reporting available, its explanatory loveliness should further increase over time. Indeed, a case can be made for saying that the loveliness of the received view has already increased, while for the cassette theory, the opposite seems to be true. At least initially, the received view, while rendering the practice of dream reporting less surprising, may even have rendered other facts about dreaming more so. Specifically, the application of the concept of experience to dreaming suggested a dissociation between conscious experience and behavioral responsiveness that ran strongly contrary to behaviorist intuitions—which, as we have seen in Malcolm's analysis of dreaming, were still strong in the 1950s. Indeed, the popular view of sleep as a state of passivity, taken together with a behaviorist approach to psychology, seems to have prevented early researchers from discovering REM sleep and its association with dreaming, even though the relevant empirical data were available (see chap. 2 for discussion). It would seem that the required revision of the concepts of experience, sleep, and dreaming were indeed too costly to be embraced. The initial, pretheoretical endorsement of the received view, precisely because it required a dissociation between experience and behavioral responsiveness, as well as a profound change in the conceptions of sleep and dreaming, initially produced a surprising new fact. It required a leap of faith, and in this respect it provided a less-than-optimal explanation of dream reporting. As Flanagan notes:

> Dreams might have gone off the consciousness list (or changed status on the list) as inquiry proceeded. . . . But the present discussion shows that, in fact, things are going in the other direction—science is helping to secure the place of dreams as experiences, not to make us remove dreams from the class of experiences. It could have worked the other way (it still might), but it has not thus far. So our commonsense concept of consciousness is open to revision, to being "re-drawn." How consciousness is conceived depends almost completely on how the science of the mind progresses. (Flanagan, 1996, pp. 24–25)

Indeed, there is as yet no final, uncontroversial theory of dreaming, either from the perspective of empirical dream research or from philosophy. But there is certainly an ever-expanding system of interconnected theoretical assumptions about dreaming, and the transparency assumption has been crucial for their development. Dream research and theoretical work on dreaming have provided a much more elaborate theoretical context than was available in 1976, and certainly than when REM sleep was discovered in the 1950s. So far, then, it seems that the promise given by the intuitive loveliness of the received view has been kept: adopting this view has enabled dreaming to become an object of scientific investigation, and it has become clear that dreaming deserves a place on the map of conscious states investigated by the interdisciplinary mind sciences. Where exactly to place dreaming on this map continues to be a matter of controversy, but it is clear that a broader theory of consciousness will ultimately have to account for dreaming, as well.

This in itself, however, suggests that the transparency assumption is an extremely productive explanation: treating dream reports as transparent made the advances of dream research possible in the first place, and this advantage may even outweigh aspects of the received view that initially seemed less than lovely. Had researchers remained neutral as to the trustworthiness of dream reports or even outright denied their trustworthiness, they would have been deprived of the rationale underlying their discoveries. The transparency assumption has also stood its ground against skeptical accounts: though it is not, as the philosophical discussion on this matter shows, the only game in town, I hope to have convinced readers that it is a particularly lovely one.[13] Consequently, the initially instinctive and perhaps even naive trust in dream reports has turned into an informed endorsement of an antiskeptical account of dream reporting. This, I think, is a good reason to be confident not only that the transparency assumption will retain its loveliness in the face of future skeptical attacks, but that its loveliness will also further increase. Seen from this perspective, treating the current explanatory loveliness of the transparency assumption as a prediction of its future success (rather than as a dogmatic and unalterable truth) is not a weakness at all: it suggests that the transparency assumption may, after all, be a well-behaved citizen of scientific approaches to consciousness.

4.3 Lessons from the Debate on Dream Color: From Moderate Distrust to the Ideal Conditions of Dream Reporting

Saying that the explanatory loveliness of the transparency assumption is revocable and a prediction of future success is prone to invite further skeptical objections. Even assuming I have succeeded in making a case for saying that dream reports transparently reveal the *occurrence* of conscious experience during sleep, this is not the same as saying that dream reports are transparent with respect to all aspects of dream experience. Moreover, part of the explanatory loveliness of the transparency assumption is, as I have claimed, that it allows for a unified account of dreaming and waking experience. But then it stands to reason that it will also have to allow for a unified account of reports of dreaming and of waking experience and be able to respond to skeptical objections to both.

In several publications, Eric Schwitzgebel has defended arguments intending to cast doubt on the trustworthiness of first-person reports (e.g., Schwitzgebel, 2002b, 2002c, 2011c, 2013b, 2013c). His position is relevant to the discussion on dream reports for several reasons: One, it is broader in scope in that it challenges the trustworthiness not only of dream reports but of first-person reports in general. Two, it is more specific than the versions of skepticism about dream reporting discussed so far: while accepting that dreams are experiences, Schwitzgebel's main aim is to cast doubt on the trustworthiness of dream reports with respect to certain aspects of dream experience (namely, our ability to report colored dreaming). Three, he challenges the trustworthiness of first-person reports even under ideal reporting conditions—a claim that would, if substantiated, be particularly damaging to my defense of the transparency assumption. I propose, however, a more moderate and constructive reading, according to which Schwitzgebel's skeptical arguments and the ensuing debate on dream color can be regarded as a tool for restricting the transparency assumption to the ideal conditions of dream reporting. Reinterpreting Schwitzgebel's argument as warranting moderate distrust in dream reports is, I claim, an important step toward an antiskeptical account (for similarly constructive replies to Schwitzgebel's argument, see Kriegel, 2013; Spener, 2013; Bayne, 2015).

Schwitzgebel is explicit about the radicalism of his claims. He repeatedly contrasts his view with Descartes's external-world skepticism, claiming that unlike Descartes, he endorses "blanket skepticism" (Hurlburt & Schwitzgebel, 2007, p. 234) about our knowledge of our inner world (for a debate on the consequences for Cartesian epistemology, see Smithies, 2013). Just as Descartes used the dream argument to claim that even in a best-case scenario of sensory perception, we can never rule out that we are now dreaming (cf. sec. 1.1), so Schwitzgebel claims that introspective error occurs even in best-case scenarios of first-person reporting:

> There are major lacunae in our self-knowledge that are not easily filled in, and we make *gross, enduring mistakes* about even the most *basic* features of our *currently ongoing* conscious experience, even in *favorable circumstances of careful reflection*, with remarkable *regularity*. (Schwitzgebel, 2011c, p. 119; italics mine)

If Schwitzgebel is right, then introspective error is not an exception but the rule; it is practically unavoidable and perhaps even undetectable. As damaging as this would be, note that this also places high demands on Schwitzgebel's argument: he would have to show that we are not only occasionally and under certain conditions, but regularly and systematically prone to error, even under ideal reporting conditions.

For present purposes, the main question is whether Schwitzgebel's argument challenges the trustworthiness of dream reports given under ideal conditions: if so, then it threatens to undermine the transparency assumption; if not, then it would only show that the transparency assumption has to be restricted. While the latter reading is more constructive, it is not entirely new: that the trustworthiness of dream reports varies with reporting conditions (e.g., spontaneous dream recall versus timed awakenings in the laboratory) has been known for decades, and large parts of exclusively report-based dream research are dedicated to investigating this question (cf. chap. 3).

Based on a review of historical studies on color in dreams, Schwitzgebel (2002c, 2011c) found evidence for "an arc of opinion: before scientific psychology, a consensus or assumption that dreams are colored; divided opinions into the early twentieth century; a consensus that dreams typically have little color from about 1930 to 1960; and then a sudden overturning of that consensus in the 1960s" (Schwitzgebel, 2011c, p. 5). Schwitzgebel notes that this initially surprising finding might be explained by the widespread availability first of black-and-white and then of color photography and television. He suggests three possible interpretations. The first is that the dreams themselves, influenced by the rise first of black-and-white and then of color television, changed from colored to black-and-white and then back to colored. If this view turned out to be correct, there would be no threat to the transparency assumption. Second, the dreams themselves may have stayed the same, that is, either black-and-white or colored, and media exposure may have influenced only the ability to recall color in dreams. In this view, the dream reports either of those subjects reporting colored dreams or of those reporting black-and-white dreams would be systematically mistaken. The third possibility is that dreams are indeterminate with respect to color. Again, the change from black-and-white to color media may have influenced only the way dreams are reported. In this view, both groups are systematically mistaken about their dreams, as dreams are neither black-and-white nor colored.

How, then, to adjudicate between the three possibilities? Schwitzgebel repeatedly (2002c, 2011c; Schwitzgebel et al., 2006) describes the possibility of a change in dreaming as unlikely and backs this up with two arguments. The first is that one of the few studies (cf. Hall & Van de Castle, 1966) to collect dream reports (rather than use general questionnaires) and analyze them with respect to dream color during the 1940s found a similar frequency of color terms as did studies from the late twentieth century (for details and further references, see Schwitzgebel, 2011c, p. 7). He notes that "if Hall and Van de Castle's respondents were dreaming in less color than people 50 years later, that fact is not reflected in their use of color terms when describing those dreams" (p. 7). Consequently it would seem that people's opinions about colored dreaming (as expressed in their responses to general questionnaires) were not borne out by their actual dreams (as described in individual dream reports).

Second, he notes that a questionnaire study he helped design that asked about colored dreaming found a strong relationship to group-level (but not individual) exposure to black-and-white versus colored media (Schwitzgebel, 2011c, p. 7; see also Schwitzgebel et al., 2006). This again would make it seem more likely that people's opinions and cultural attitudes about colored dreaming change in response to media exposure; if the dreams themselves were changing, one would expect individual media exposure to be the driving factor. Based on these two arguments, Schwitzgebel favors the view that dreams are either richly colored or indeterminate with respect to color, suggesting that at least during certain periods, opinions about colored dreaming have been mistaken. He concludes:

> Although many mornings I remember a dream or two—and sometimes they seem to have been quite vivid—I can't tell you whether those dreams are in color. The historical swings in opinion about black-and-white vs. color dreaming suggest that I am not singularly inept, and that incompetence in assessing the coloration, or lack of it, of our dream life is fairly widespread, despite the considerable confidence people often exhibit when questioned on the matter. We don't know the phenomenology of dreaming nearly as well as we think we do. (Schwitzgebel, 2011c, pp. 14–15)

I think there are at least two reasons for being cautious about this skeptical conclusion. First, note that there is an inherent ambiguity in Schwitzgebel's claim about our knowledge of the phenomenology of dreaming, possibly even pointing toward an equivocation. There are at least two different ways of knowing the phenomenology of dreaming at play in his argument, where one refers to knowledge of *particular dreams*, as they are remembered and reported upon awakening, and the other to *general opinions about dreaming*, reached independently of individual dream recall, as expressed in response to general questionnaires.[14] Schwitzgebel's conclusion suggests that he is skeptical about both.[15]

Second and relatedly, while Schwitzgebel has made a convincing case for doubting the trustworthiness of *opinions* about colored dreaming, I do not think that the same is true for individual dream reports in general, and laboratory reports of the type commonly thought to approximate ideal reporting conditions in particular. Schwitzgebel's research is meticulous, and his arguments are cautious. He explicitly acknowledges the possibility that

> the change in research methodology—from mostly questionnaires in the 1940s and the 1950s to mostly REM awakenings in the 1960s—is partly explanatory. Maybe, for example, REM awakening is a more accurate method of assessing dream content. Then perhaps the questionnaire studies assessed only casual opinion about dreams, which was influenced by media analogies, whereas the REM-awakening studies assessed the dreams themselves—dreams which contained color and always had—and once scientists had this better method in hand, their opinions changed rapidly. (Schwitzgebel, 2011c, p. 10)

Yet he does not find this possibility persuasive, arguing that we have insufficient evidence for saying that dream reports, even under ideal reporting conditions, are trustworthy with respect to dream color. What he lacks, however—or so I would claim—is a specific reason for this persistent distrust.

Perhaps Schwitzgebel simply places higher demands on a theoretical explanation of dream reporting than I do. Whereas my endorsement of the transparency assumption as the loveliest explanation of dream reporting suggests that dream reports are assumed to be transparent by default and that the burden of proof is on the skeptic, Schwitzgebel perhaps thinks that the opposite is the case: skepticism is warranted unless the antiskeptic can produce evidence to the contrary. I suspect, however, that the disagreement between our views cannot wholly be explained in this manner. After all, if Schwitzgebel thought that distrust was the default position about dream reports, then his project of defending an empirically motivated form of skepticism about dream reporting would be largely superfluous. Indeed, he makes it clear that skeptical arguments are needed to undermine people's "cavalier confidence" about their knowledge of subjective experience (Schwitzgebel, 2013c, p. 1163), suggesting that antiskepticism is, once more, the default position. But if this is correct, then he owes an argument for distrusting individual dream reports. Moreover, I suspect that unlike his moderate skepticism about *opinions* about dreaming, he could not supply any specific arguments for distrusting dream reports in general. If so, the problem is not Schwitzgebel's at all but shows that empirically motivated skepticism about dream reports itself is inherently and systematically flawed, at least when it is taken to undermine the *principled* trustworthiness of dream reports (and not restricted to the *relative* trustworthiness of certain kinds of reports).

This is so for two reasons. First, recall that an important reason for thinking that the change in *opinions* about colored dreaming was not accompanied by a change in colored *dreaming* was that this interpretation was suggested by the only study to analyze individual reports of dream color from this period. Consequently, Schwitzgebel's skepticism about opinions about colored dreaming, if successful, is grounded in assuming (at least a subset of) individual reports of colored dreaming to be transparent.

Second and more generally, it seems that any attempt to use empirical evidence to cast doubt on the trustworthiness of dream reports *independently* of reporting conditions is misguided. The untrustworthiness of reports given under *less*-than-ideal conditions does not show that dream reports are systematically untrustworthy even under *ideal* reporting conditions; any discrepancy between different types of reports could only warrant moderate distrust about a particular type of report (e.g., spontaneously recalled dreams versus laboratory awakenings; opinions about dreaming versus individual dream reports, etc.), but not about the trustworthiness of dream reports in general.[16] The trustworthiness of dream reports also cannot be undermined by independent evidence: just as independent evidence cannot verify the trustworthiness of dream reports, so it also cannot be used to falsify them. Rather, trusting dream reports is the condition for turning evidence, for instance, about neural activation patterns during REM sleep, into evidence about *dreaming* in the first place.

That said, I agree with Schwitzgebel that there is some uncertainty about the question of colored dreaming. To begin with, though the Hall and Van de Castle (1966, see esp. chap. 14) study was based on a normative sample of one thousand dream reports, they were not gathered following timed awakenings in the sleep laboratory and can thus be treated at best as a rough proxy of ideal reporting conditions. Consequently, as Schwitzgebel notes, it cannot be ruled out that reports following REM sleep awakenings from this period might have provided evidence of black-and-white dreaming, after all. Were this to be the case, it would not, however, provide evidence for calling such reports into question: rather, it would, I think, count as evidence for a change in dreaming.[17] Such a conflict between reports from REM awakenings and spontaneously recalled dreams would suggest moderate skepticism about the latter. Yet, by the same token, the case for skepticism about *opinions* about colored dreaming would be eroded: opinions about colored dreaming from this period would now apparently be as reliable as reports following REM awakenings. But either way, such findings would not undermine reports given under ideal reporting conditions.

Incidentally, note that this reading provides yet a further sense in which we can know—or fail to know—the phenomenology of dreaming. While the two readings introduced earlier focus on the capacity of individual persons to know the phenomenology either of particular dreams (based on individual instances of dream recall) or by forming general opinions about their own dreams, one can also ask whether researchers have theoretical knowledge about the general incidence of colored dreaming and its stability across times and cultures, but also across and within different individuals. Such a theory, optimally, will be based on the analysis of large data sets, and judging from Schwitzgebel's analysis, it does seem that we still lack the type of detailed and comparative studies needed to form such an overarching theory of dream color. But even if we do not know the phenomenology of colored dreaming in this general theoretical sense, it does not follow that individual dream reports are untrustworthy.

A first important lesson is that the question of whether we dream in color is an empirically tractable question: the answer will depend on reports of dream color given immediately upon awakening, under ideal reporting conditions, and under the assumption of transparency. Given the paucity of studies of individual reports of dream color from the 1930s to the 1960s, one should likely withhold judgment as to whether dreams were colored, black-and-white, or indeterminate during this period. Perhaps dream coloration varies across time or even inter- and intraindividually (cf. Murzyn, 2008)—and perhaps it does not. No a priori reason exists to assume that the question of dream coloration will have a tidy and homogeneous, yes/no answer; variation in experience cannot be ruled out (cf. Hohwy, 2011; Alsmith, 2013). Our failure to know which answer is correct, however, points not to a deep philosophical problem but only to the lack of studies we could use to answer this question.

The second important lesson is that moderate skepticism about certain types of dream reports (or about opinions about dreaming) is extremely helpful in restricting the transparency and reportability assumptions to the ideal conditions of dream reporting. Again we find that skepticism about dream reporting is unhelpful and even misguided when regarded as a principled position, but it is a fruitful and constructive tool for moving forward when read in a moderate, empirically motivated manner. This claim fits in well, I think, with a number of follow-up studies about dream color conducted by different research groups. In particular, the literature on colored dreaming reflects a widespread consensus that there is a hierarchy in the trustworthiness of dream reports ranging from laboratory awakenings to spontaneously recalled dreams, where the latter can at best be used as a proxy for the former, to opinions about dreaming in response to general questionnaires.

Hoss (2010) analyzed tens of thousands of dream reports from an Internet database and from long-term dream journals to determine the frequency of reports of dream color. He was mainly interested in factors determining reports of dream color (such as waking visual experience, personal preferences, or emotional responses to color); for present purposes, however, his methodological conclusions are more relevant. He writes:

> Research designs which capture the dream and color content in laboratory settings, immediately upon awakening from the REM state, appear to capture more dream and color content (in the 80% to 100% range), and thus may provide results closer to that of what colors actually populated the dream. Designs that rely on spontaneous recall and journaling, tend to show color recall falling off to about 25% (typically within an 11% to 46% range). Therefore it would follow that attempting to capture the actual color originating within a dream may rely on REM awakenings. Immediate reporting upon awakening, with a strict questionnaire that focuses on describing the color of each element in the dream, also appears to capture a great deal of color present in the dream. (Hoss, 2010, p. 89)

The contrast between what I call opinions about dreaming and spontaneously recalled dreams was investigated by Schredl et al. (2008). Their aim was to investigate whether reports of black-and-white dreaming are the result of memory problems by comparing subjective estimates of color in dreams in response to a general questionnaire to dream reports given immediately upon awakening, in which subjects were explicitly instructed to report all colored objects in their dream. They found that 10 percent of dreams and dream elements were rated as black-and-white in response to the general questionnaire, but this percentage decreased to 2.7 percent in dream reports given immediately after awakening. They conclude that "it seems very plausible that memory processes affect the retrospective estimates elicited with the questionnaire" (Schredl et al., 2008, p. 179). By contrast, Murzyn (2008) failed to find differences in reports of color in response to questionnaires and dream diaries. She concluded:

> This result can bolster the confidence in the early 20th century questionnaire studies of dreaming, by showing that people can provide reliable information in questionnaires. On the other hand, there is a possibility that the diary method is not immediate enough to counteract memory and selective recall issues (especially in the older population), and that this convergence of methods is an artefact of a confirmation bias. (Murzyn, 2008, p. 1235)

She adds that "without a REM awakening study that would probe the colour of people's dreams more immediately than a dream diary it seems impossible to decide between the two explanations" (Murzyn, 2008, p. 1236). A careful conclusion is that the trustworthiness of opinions about colored dreaming remains poorly understood. Likewise, while the association between age, media experience, and opinions about colored dreaming remains unclear, some evidence suggests that reports of colored dreaming are associated with imagery abilities, but only poorly with color memory (Murzyn, 2012; for a general discussion, see Murzyn, 2010; for Schwitzgebel's own research on media experience and dream color, see Schwitzgebel, 2003; Schwitzgebel et al., 2006). Clearly, investigating the factors affecting opinions about colored dreaming continues to be an important project in its own right.

In sum, while important questions can be raised about the *relative* trustworthiness of different types of reports, it seems that they can only be investigated by defining certain types of dream reports as the gold standard against which others can be tested. For the same reason, the debate about dream color suggests that moderate distrust about certain types of reports and opinions about dreaming is a fruitful strategy for determining the ideal conditions of dream reporting. Perhaps this process might even lead to a better understanding of what ideal reporting conditions consist in and related methodological improvements. But even so, this is not the same as defending principled skepticism about dream reporting under *any* conditions.

Importantly, this position of moderate distrust, combined with the view that first-person reports are indispensable for consciousness research, is compatible with some of Schwitzgebel's own remarks:

> Despite its untrustworthiness, introspection must be given a central role in the study of consciousness. Without introspection, we might not even know that we *are* conscious in the relevant sense—that a stream of phenomenology accompanies our outwardly visible behavior. Behavior

> and physiological measures alone tell us nothing about consciousness unless it is established that those measures correlate with conscious experience; and introspection is the most straightforward way to establish such correlations. (Hurlburt & Schwitzgebel, 2007, p. 53)

This suggests that Schwitzgebel—despite occasional claims to the contrary—is best understood as endorsing a moderate view, according to which "we have *some* sort of attunement to our ongoing conscious experience, and we impoverish ourselves if we try to do without it" (Schwitzgebel, 2011c, p. 118). Despite his skepticism about the trustworthiness of first-person reports, Schwitzgebel nonetheless regards them as methodologically necessary for conducting consciousness research. In fact, by applying his skeptical approach to existing controversies regarding conscious experience (see esp. Schwitzgebel, 2011c, chaps. 1, 2, 4, 6), he constructively contributes to their elucidation.

In keeping with these brief remarks, we have no reason to think that the arguments presented in this section apply only to dream reports. At least it would seem that principled skepticism about first-person reports cannot be constructed in any of the ways discussed here. In particular, skepticism about the trustworthiness of first-person reports cannot be achieved by distrusting first-person reports altogether but rather must contrast different types of reports and weight them with respect to their *relative* transparency. An interesting consequence is that moderate and constructive skepticism necessarily unfolds before the background of the transparency assumption.[18] Understood in this manner, moderate distrust about dream reporting might lead to antiskepticism not only about dream reporting but about first-person reports in general.

4.4 Transparency and Reportability Restricted: Antiskepticism, the Ideal Conditions of Dream Reporting, and the Conservative Approach

We are now ready to bring the results of the preceding chapters together to formulate a restricted version of the transparency and reportability assumptions and summarize the main tenets of antiskepticism about dream reporting. My extended discussion of the transparency assumption and of its status in scientific dream research and in the philosophical debate on dreaming has led to several important preliminary results. First, scientific dream research assumes that at least under certain conditions, dream reports are transparent with respect to preceding dream experience. This background assumption is methodologically necessary for scientific dream research; yet it is very different from the undifferentiated and mostly unarticulated endorsement of the transparency assumption that characterizes the construal of the received view of dreaming that first became the target of philosophical criticism in the 1950s (cf. chap. 1). In particular, a careful consideration of the scientific literature has shown that we need a much more differentiated understanding of different types of reports to make sense of the contribution of dream reports to scientific dream research. To the extent that philosophical skepticism about dream reporting neglects this distinction, it targets a practice that does not exist. Second, we have seen that the transparency assumption of dream reporting is intimately connected to the assumption that dreams, at least under certain conditions, are *reportable* experiences. Laboratory research on dreaming, in particular, depends on the assumption that reports of dreaming and of nondreaming are equally trustworthy (cf. sec. 3.1). Moreover, we have seen that the most constructive reading of skepticism about dream reporting is to regard it as a tool for identifying ideal reporting conditions. Distrusting certain types of dream reports is not a purely theoretical exercise but can be regarded as a research strategy for determining the *relative* transparency of different types of reports, and thus of narrowing down the conditions under which dream reports can and should be trusted.

Taken together, the transparency and reportability assumptions (at least once appropriate restrictions are in place) demarcate the scope and limits of scientific dream research. First, dream reports, under the assumption of transparency, define the maximum database of scientific dream research. This does not mean that no other types of evidence can be used for the study of dreaming—clearly, this would lead the project of scientific dream research *ad absurdum*. What it does mean is that other types of evidence about dreaming, such as polysomnographic measurements, neuroimaging data, or sleep behavior, only turn into evidence about *dreaming* before a general background of trusting dream reports (chap. 3). Second, this means that dream research is constrained by the space of reportable dreams in the sense that studies moving beyond the study of dream reports altogether no longer form part of *dream* research proper. This raises a number of open questions, which I discuss in turn.

- *If the transparency and reportability assumptions have to be restricted to the ideal conditions of dream reporting, does this mean that similar restrictions apply to the experiential assumption as well?*

This is perhaps the easiest question on my list: the answer is clearly no. Saying that dream research is constrained, for methodological reasons, by the space of *reportable* dreams is one thing; saying that only dreams that are *in fact* reported are experiences is another. Only a small percentage of dreams are ever recalled and reported in the first place, and the number of dreams reported in a scientific study necessarily is even more limited. Dreams are not experiences *because* they are reported; it would be absurd to think that the largely contingent fact of whether or not a particular dream is reported could make any difference to whether or not it was experienced. Experiential character per se is not contingent on reportability. Yet *saying* that dreams are experiences is theoretically justified because the experiential assumption is implicit in the transparency assumption, and because the transparency assumption best explains dream reporting. The transparency assumption offers a rationale for answering questions about dream experience, and for present purposes, this is enough; but to think that such theoretical considerations have any consequences for facts about phenomenal experience would be akin to committing the armchair fallacy. Antiskepticism about dream reporting does not claim that dreams should be *identified* with dream reports. It only says that dream experience cannot be *studied* independently of dream reports, and, more generally, that the question of dream experience can only be answered before a background of principled trust in dream reports.

This might seem trivial, but it is not. To see why, recall that questions concerning the relationship between dreaming, dream reporting, and scientific dream research mark a deep divide in philosophical theorizing about dreaming and dream reporting. In particular, Malcolmian skepticism about dreaming says that there is a deep, unbreakable conceptual connection between dreaming and dream reporting, but argues that this very fact speaks against, rather than for, a naturalistically minded approach (Malcolm, 1956, 1959/1962; cf. sec. 1.2). More generally, philosophers who have assumed that a conceptual link exists between dreaming and dream reporting such that grammatical considerations about dream reporting determine whether or not dreams are experiences (and such that decoupling dream experience from dream reporting is ruled out) have typically thought that this undermines the project of identifying physiological correlates of dream experience (e.g., Squires, 1973, 1995; McFee, 1994; Schröder, 1997).[19] At the same time, philosophers regarding the question of dream experience as an empirical question to be handed over to scientific dream research have generally tended to decouple dreaming from dream reporting, resulting in the view that truly scientific dream research will either have to be conducted independently of dream reports or at least have to provide independent evidence for their trustworthiness (Dennett, 1976; Schwitzgebel, 2002c, 2011c; Rosen, 2013).[20] Either way, the strategy of using dream reports for the *scientific* investigation of conscious experience during sleep is threatened. In the former view, scientific dream research is impossible; in the latter, it is only scientific to the extent that it does *not* rely on dream reports. Antiskepticism about dream reporting is attractive precisely because it suggests that using dream reports (under the assumption of transparency) is not only reconcilable with, but the condition for, scientific dream research. Consequently antiskepticism about dream reporting is the only way to acknowledge and indeed make sense of the central contribution of dream reports to the scientific study of dreaming (see table 4.1).

- *What exactly are ideal reporting conditions?*

In the preceding sections, it has become clear that a variety of issues play into this question, including the sleeping environment, timing and method of awakening, the temporal proximity between dreaming and dream reporting, the wording of questions, the types of reports used (e.g., verbal, written, drawings), and so on (cf. sec. 3.3). The most careful conclusion is that reporting conditions are sensitive to all these factors, and probably to many more. It would seem to follow that speaking of ideal reporting

Table 4.1
Skepticism and antiskepticism about dream reporting

	Are dreams experiences?	Is scientific dream research possible?	Can dreams be investigated independently of dream reports?	Can trusting dream reports be reconciled with scientific dream research?
Malcolmian skepticism about dream reporting	**No** The very idea is a conceptual contradiction!	**No** Dream reports are the sole criterion of dreaming; any attempt to investigate dreaming empirically changes the concept of dreaming.		
Dennettian skepticism about dream reporting	**Maybe** The problem of dream experience is an open empirical question . . .	**Yes** . . . to be decided independently of dream reports.		**No** (Or only if independent evidence for trusting dream reports were to become available.)
Empirically motivated skepticism about dream reporting (e.g. Schwitzgebel, Rosen)	**Yes** Empirical evidence shows that dreams are experiences . . .	**Yes** . . . but empirical evidence also suggests that dream reports are not trustworthy.		
Antiskepticism about dream reporting	**Yes** The experiential assumption is implicit in the transparency assumption.	**Yes** The transparency assumption is a methodologically necessary default assumption for scientific dream research and theoretically justified by inference to the best explanation.	**No** Scientific dream research is constrained by the space of reportable dreams.	**Yes** Trusting dream reports (given under ideal reporting conditions) is the condition of possibility for scientific *dream* research!

conditions is an oversimplification, and very likely there is no single master formula. One type of report may be ideal for assessing spontaneous dream recall frequency (e.g., checklist logs, as suggested by Zadra & Robert, 2012), but insufficient for saying anything about the experiential character or content of the recalled dreams. Certain aspects of dream experience may systematically be underreported unless they are explicitly asked for (Merritt et al., 1994). Moreover, if a study is interested only, say, in colored dreaming, then asking for a *detailed* description of all aspects of the dream might be irrelevant and even distracting, thus standing in the way of the required level of specificity. In addition, different types of reports also differ as to their expressive granularity. Even detailed and specific verbal reports may be inappropriate to convey the exact quality of visual imagery; here, asking for drawings or using photographs may help (Rechtschaffen & Buchignani, 1992). Very generally, ideal reporting conditions will require a large degree of tweaking of the required level of detail, specificity, and expressive granularity and will likely have to be determined on a case-by-case basis and in the context of a given study.

Over and above these general remarks, the one factor most researchers apparently agree on—and the one that also lies at the heart of laboratory studies—is *temporal proximity*. While laboratory reports are often preferred over home diaries because conducting timed awakenings is the best way of minimizing the time lag between dreaming and dream reporting (and certainly more easily controlled than simply asking subjects to report their dream immediately upon awakening), this is at best a necessary and not a sufficient condition, and the precise method of reporting—given a verbal or written report, answering a questionnaire, and so on—can still be more or less suited to the questions asked.

Finally, an important lesson from Schwitzgebel is that general questionnaires tapping into casually and unsystematically formed *opinions* either about one's own dreams or about dreaming in general are best not regarded as dream *reports* at all: they lack any direct grounding in, and may even be contradicted by, individual dream reports. Consequently considerations of transparency are not directly applicable to such general questionnaires. The most radical conclusion would be to say that responses to general questionnaires involving "self-reports on dream content are worthless at best and often misleading" (Domhoff, 2003, p. 41). A more charitable interpretation would be to say that investigating folk-phenomenological opinions and beliefs about dreaming has its own interest—though it does not form part of dream research proper.

- *Isn't it profoundly dissatisfying to place so much weight on ideal reporting conditions and then fail to give a single, clear, and comprehensive definition of what this actually means?*

From a certain perspective, it might indeed seem that my answer to the previous question is inappropriately sketchy or even outright evasive. And while it would certainly be nice to have a simpler and more clear-cut answer, I think that there is good reason to resist this temptation.

To begin with, note that an interesting asymmetry characterizes the dialectic between the skeptic about dream reporting and scientific dream research. Very generally, philosophical skepticism is interested not in refuting particular knowledge claims but in casting doubt on our ability to obtain knowledge, either in general or in a certain domain. Skepticism about dream reporting is a local form of skepticism aiming to undermine dream reports as a source of evidence about dreaming. If it succeeds, then we can no longer use dream reports to support knowledge about dreaming or certain aspects of dreaming (such as dreaming in color). By contrast, scientific dream research is not interested in the *general* trustworthiness of dream reports at all: it is interested in the *specific* design of a given study, the adequacy of methods used for the investigation of a particular question, and so on. This asymmetry is a first indication that an empirically plausible account of dream reporting should not be expected to give any general answer to questions about the trustworthiness of dream reports. Demanding one would be to ignore the fact that insofar as the trustworthiness of dream reports is concerned, the philosophical and scientific debates are conducted on two different and perhaps incompatible levels.

Moreover, my analysis so far suggests that while it seems reasonable to assume that the transparency and reportability assumptions will have to be restricted to certain ideal conditions of dream reporting, what exactly these are is an empirical, not a theoretical, question and thus is subject to change in response to novel research findings. If this is correct, then questions about the (relative) trustworthiness of dream reports can now be handed over to scientific dream research. Philosophical analysis can investigate what kinds of questions can meaningfully be asked about the trustworthiness of dream reports, suggest how they might be answered and whether (and why) claims about the trustworthiness of dream reports are theoretically justified—but in this process, the contribution of philosophy is limited to metatheoretical observations. Certainly it would not make sense to make any specific prescriptions about what exactly the ideal conditions of dream reporting are. Legislating a general definition of ideal reporting conditions would be to misconstrue the nature of the problem.

- *Doesn't restricting transparency to ideal reporting conditions rule out the use of dream reports by placing such high demands on transparency as to make it practically unattainable?*

An obvious worry is that the restricted version of the transparency assumption suggested here might be, for all practical purposes, too demanding. Even if, in an ideal world, one might want to restrict one's claims about dreaming to reports gathered under ideal conditions, such a restriction would drastically diminish the available database about dreaming. Given that most studies conducted today are based on dream reports gathered outside the laboratory, this would come close to a practical rejection of most of contemporary dream research. Clearly this option is neither attractive nor constructive.

Recall, however, that an important insight from the debate on dream color is that questions about the trustworthiness of dream reports are best regarded as empirical questions concerning the *relative* trustworthiness of one type of dream report compared to another. Indeed, large portions of dream research can be described as using dream reports not only to discover other types of evidence about dreaming (e.g., polysomnographic evidence, sleep behavior, etc.) but also to investigate the *relative transparency* of different types of dream reports (e.g., laboratory versus spontaneously recalled dreams, etc.). While this is conducted before a background of principled trust in at least *some* dream reports, trust and distrust complement each other in this process. Very generally, we can say that the *relative transparency* of a given (type of) dream report is determined by the extent to which the conditions under which it was gathered *approximate ideal reporting conditions* (see table 4.2). Seen in this manner, transparency is not an all-or-nothing affair but comes in degrees.

To illustrate this with the help of a metaphor, note that the same is true for literal transparency: A window can be clean or smudged; its glass can be clear or colored, and even a stained-glass window in a church will be translucent to a certain degree. It won't thereby automatically be rendered useless for seeing what (if anything) is on the other side, but depending on what one wants to see, its use will be limited. A window also constrains one's view by binding it to a perspective, keeping the viewer at a certain distance from the thing observed, and presenting only part of the scenery. Reducing spatial (or, in the case of dream reports, temporal) distance, but also adjusting the window's position and size (e.g., by tweaking the level of detail, specificity, and expressive granularity), can improve one's view of what one is interested in. Deciding which type of glass (or report) to use will be equally relevant, as will researchers' ability to keep it clean (i.e., avoid unwanted disturbances). In this way, *relative* transparency directly complements my account of ideal reporting conditions as dependent on multiple, case-sensitive factors: transparency is not a static but a highly dynamic property. The concept of relative transparency emphasizes this point and may thus offer a way to reconcile skeptical worries with practical feasibility.

For instance, if laboratory dream reports can be used as a proxy for reports gathered under ideal conditions, then they, in turn, can be used as a baseline for assessing the relative trustworthiness of reports gathered under less-than-ideal conditions. Existing studies comparing spontaneously recalled dreams to laboratory reports suggest that for many research questions, settling for the relative transparency of the former, and thereby ensuring a larger database, might be a worthwhile trade-off. Based on a review of studies comparing laboratory reports to spontaneously recalled dreams, Domhoff notes that "dream reports collected in the sleep laboratory provide a reasonably representative sample of dream life" (Domhoff, 2003, p. 43). He adds that if one takes laboratory reports as a "starting point for assessing the adequacy of home dream reports"

(p. 46), one might cautiously conclude that the same is largely true for spontaneous dream recall at home.[21] The concept of relative transparency thus suggests a specific research strategy for investigating this question and for operationalizing skeptical worries about dream reports in future studies.

In addition, note that regarding *relative* transparency as a dynamic property is interestingly related to the discussion of IBE in section 4.2. There I noted that treating a hypothesis as the best explanation of a phenomenon always takes the form of a prediction of its future success. Now we can see that the same is true for ideal reporting conditions. Even if laboratory reports are currently considered the gold standard of dream reporting (cf. Schredl, 2010a)—and thus the standard by which the relative transparency of other types of reports (e.g., home diaries) is determined—a better standard might come along.[22] Perhaps improved reporting conditions will someday offer an even better explanation of current reporting behavior. Perhaps the transparency assumption for dream reports given under ideal reporting conditions is best regarded as a hypothetical maximum: in practice, most dream reports will fall short, but it can still be used as a theoretical measure of future improvement.

- *Does the reportability assumption strictly rule out the existence of unreportable experiences during sleep?*

One might worry that claiming that dreams are reportable experiences during sleep and that scientific dream research is constrained by the space of reportable dreams invites something very close to the armchair fallacy (cf. sec. 1.2): even if one admits that dream research endorses the reportability assumption for methodological reasons, this is very different from saying that unreportable experiences during sleep do not exist. Methodological constraints have no direct consequences for facts about phenomenal experience during sleep.

This is an important point. Note, however, that the reportability assumption does not say that all dreams are *in fact* reported—it only says that they *could* be reported given (sufficiently) ideal reporting conditions. In addition, if experiences during sleep exist that are principally unreportable—for instance, by being below the threshold even of short-term memory, or by being so subtle that they are completely ineffable and subjects can give no indication even of their occurrence (more subtle, that is, even than white dreams, where subjects can still report the *impression* of having dreamt, though they are unable to remember any details; cf. sec. 3.1)—then these experiences are not the object of scientific dream research and do not count as dreams. An important aim for dream research is to reduce the gap between dreams that are *in fact* reported and *reportable* dreams as much as possible; the underlying rationale is that dreams reported under ideal (or nearly ideal) reporting conditions can then be used as a proxy for reportable experiences during sleep. This is a conceptual and methodological point only: the reportability assumption says nothing about the existence of unreportable experiences during sleep; it only says that if they exist, they do not form part of dream research. Making any assertions about unreportable experiences beyond this point is purely speculative and unscientific.

That said, it does seem plausible that certain types of sleep experiences are beyond the grasp of certain *types* of reports, even in principle. For verbal reports, this is very likely true. It has long been known that subjects can discriminate many more hues of color and many more sounds than they can reliably reidentify, suggesting that certain subtleties of conscious experience are ineffable. If we cannot reliably reidentify the experience associated with a particular hue, for example, then we cannot conceptualize or report it retrospectively. As Raffman (1993, 1995) has argued, this may impose certain limitations on the scientific study of phenomenal experience. For waking experience, this suggests that concurrent reports involving discrimination tasks are more fine grained than retrospective verbal reports. Relatedly, this might suggest that the expressive granularity of dream reports—even if it is improved—falls short of the detail of dream experience, even in principle; and one might speculate not only that this is true for verbal reports but that the resolution of any type of report is limited. We may have to accept, then, that there is something like a maximum resolution of dream phenomenology beyond which asking for more detailed or specific reports simply is not feasible.

Yet a prediction will be that just as transparency is a relative property, so reportability will also vary by degrees. In particular, assuming that certain experiences exist during sleep that are below the threshold of reportability and so elusive as to be beyond the grasp of recall does not contradict the reportability assumption outright: subjects occasionally do report that they are unable to describe certain aspects of their dream, or even that they had a dream but are unable to describe it in any detail. Saying that such white dreams exist is grounded in *reports* of white dreaming; as such, the frequency of white dream reports can become the object of scientific investigation (e.g., Norcika et al., 2009) and may even help determine the limits of reportable experience during sleep. Perhaps investigating the exact wording of white dream reports would be a promising question for future research. Do different types of white dream reports exist? For instance, do subjects report having had detailed experiences that they are unable to describe? Or do they describe experiences that were themselves so subtle and indeterminate as to evade any more detailed descriptions? Charting these transitions between dreaming and nondreaming might be an interesting project for future research—especially as it would seem that a gradual transition is predicted by leading theories of consciousness, such as Tononi's integrated information theory (Tononi, 2004, 2008, 2012; see sec. 2.2). Perhaps white dreams, or a subgroup thereof, are indeed an in-between state, where information integration is sufficient to enable only the most general of retrospective reports. And perhaps in the future, they can be described in mathematically precise terms.

For present purposes, the most important point is that relative reportability requires a distinction between *detailed experiences*, *detailed reports*, and *reports of detailed experiences*. In particular, one can report the occurrence of a detailed experience without thereby being able to give an equally detailed report (and, indeed, demanding that the report replicates the experience in all its detail would be hopelessly unrealistic); and one can report having had an experience that is too vague or indeterminate to be put into words. But just as improving and tweaking the methods of report-based research and thereby perhaps increasing the space of reportable dreams is an important goal for the future, so ignoring dream reports altogether and postulating experiences that even in principle are beyond the grasp of reportability is unscientific.

- *What about using ideal reporters rather than focusing only on ideal reporting conditions?*

A question that is importantly related to the debate on ideal reporting conditions is which subject groups to use in dream research. Here the main question is whether the dreams of subjects who have high levels of spontaneous dream recall can be assumed to be representative of the general population; if so, then all other things being equal, using high recallers would be preferable for practical reasons, because it could maximize the number of dreams reported in a given study. Indeed, given the poor association between dream recall frequency and personality traits, there seems to be a case for maximizing efficiency in this way (Domhoff, 2003, p. 52). However, it is also possible that different people have different types of dreams. For instance, Ruby et al. (2013) recently found evidence for neurophysiological trait differences between high and low recallers, suggesting that this strategy should be assessed with caution.

A related possibility, suggested by Rosen (2013), is to distinguish subjects not by the quantity but by the accuracy of reported dreams. She notes that subjects who are particularly prone to hypnotic suggestion and imagination inflation—that is, mistaking imagined episodes for actual memories—might be more likely to fabricate dream reports as well. She suggests that if certain subjects were found to be particularly prone to mistaking presleep fantasies for dreams, screening could be used to identify them and thus to indirectly improve the accuracy of dream reports. However, a problem with this approach would seem to be that subjects prone to imagination inflation might simply be more likely to integrate presleep fantasies into their dreams as well; hence this would not provide straightforward evidence for saying that their reports were less accurate in virtue of being more closely related to their presleep fantasies.

Finally, one way of improving reporting conditions might be to train participants to be better dream reporters, for instance by teaching them to report exactly those aspects of dream experience that researchers are most interested in in the context of a given study. Relatedly, one might also want to focus on participant groups who have already accumulated a certain degree of expertise in observing and describing their experiences. Meditators are a much discussed example, and it has been suggested that they may be able to give more precise and accurate reports than the average population (Varela, 1996; Lutz et al., 2002; Lutz & Thompson, 2003). One hypothesis is that meditation, but also dance, for instance, might increase sensitivity to interoceptive cues (Sze et al., 2010; Fox et al., 2013; but see Khalsa et al., 2008).

Assuming that certain subject groups indeed enjoy a form of expertise in observing and describing certain types of dream experiences, what exactly would data gathered from their reports tell us? In particular, could the results from such studies be generalized to the dreams of the general public? Here, there is, I think, reason for caution. Developing expertise, even in retrospectively reporting one's dreams, might color one's dreams: even the intention of writing down one's dreams in a diary has a way of finding its way into one's dreams. It is quite possible that training could change the target phenomenon in subtle and hard-to-detect ways. The same might be true for expert groups: even assuming that expert meditators or dancers enjoy greater interoceptive sensitivity, and assuming this would enable them to give more detailed reports of certain bodily experiences in dreams, it is possible that it would be not so much their reports that would be more accurate, but rather that their dreams themselves would be richer in the relevant kind of experience. Much as is the case for the dreams of experienced lucid dreamers or for enactment dreams of RBD patients, there is a genuine concern about the generalizability of such results. As Solomonova and colleagues (2014) point out, however, there are other reasons for being interested in such reports, for instance because they complement reports from untrained participants. Whereas ideal reporting conditions might be most suitable "to uncover what is 'typical' in dreams of a certain socio cultural population (the 'breadth' of dreaming)," the reports of expert groups might give some indication of "what is 'possible' in the dream state (the 'depth' of dreaming)" (Solomonova et al., 2014, p. 1). While it would be oversimplified to expect the dreams of expert groups to reflect the breadth of dreaming, it would be shortsighted, as they rightly point out, to deny the relevance of expertise in probing the depth of dream experience. Lucid control dreams might be particularly useful for this latter purpose, because experienced lucid dreamers can selectively manipulate and alter their ongoing dreams. Again, both types of approaches will be most powerful when they complement each other, and Solomonova and colleagues' distinction between probing the breadth and the depth of dreaming nicely highlights this point.[23]

- *What is the relationship between types and tokens of dream reports? Should all dream reports of the same type be assumed to be equally trustworthy?*

So far, I have been arguing for a hierarchy in the trustworthiness of types of dream reports but have said little about tokens, or individual instances of dream reports. The main reason for this omission is that scientific dream research is not, for the most part, interested in individual reports; it is interested in data sets extracted from large numbers of dream reports. In a typical study, these dream reports will be gathered under identical or at least highly similar conditions (studies comparing different types of reports are an obvious exception) and will be treated as being equally trustworthy.

Of course, particular reasons can always arise for distrusting an individual dream report: perhaps reporting conditions were not, in this case, the same as in the others; perhaps the preceding sleep period (or the process of dream reporting) was disrupted, or perhaps the subject has a parasomnia or is taking medication suspected to have an influence on dreaming, and so on. Screening for these types of cases is standard procedure. It is equally clear that the decision to distrust individual reports cannot be based solely on their failure to converge with the findings supported by a majority of reports. Even if most dream reports describe colored imagery, individual reports describing grayscale dreams are, all other things being equal, equally trustworthy. And even if most people report dreams following REM sleep awakenings, their occasional failure to do so is still to be taken seriously and may even point to a selective loss of dreaming (cf. sec. 3.6). Dream research is not a popularity contest. This is not a trivial point: it means that a certain degree of variability in research results will have to be accounted for by a theory of dreaming. Sometimes, trusting nonstandard and unexpected reports may even enable genuine progress: had researchers not eventually trusted comparatively rare reports of dreaming from NREM sleep, they would not have moved beyond the identification of dreaming and REM sleep. Nonstandard dream reports can be used as a corrective for commonly accepted theoretical assumptions about dreaming. Although I am focusing here mainly on the trustworthiness of types of dream reports, individual dream reports may be theoretically fruitful and, all other things being equal, should be regarded as having the same degree of relative transparency as other reports of the same type, regardless of their content.

Departing from this idea, I want to use the remainder of this section to introduce what I call the *conservative approach* to dream reports. It says that that the transparency assumption is not only restricted to ideal (or approximately ideal) reporting conditions but also limited to the aspects of a dream that were explicitly described in the respective dream report, and only in the absence of evidence to the contrary (e.g., later recantations, uncertainty, etc.; in such cases, it is probably advisable to suspend judgment). This may again sound trivial, but in the following chapters, it will become clear that adhering to the conservative approach raises a number of commonly overlooked questions about the phenomenology of dreaming, with occasionally surprising results. Here I want to give two reasons for thinking that it is a valuable approach. Both of them relate directly to the debate on dream color.

The first has to do with Schwitzgebel's suggestion that dreams might be indeterminate with respect to color. If color indeterminacy is a genuine option at least for some dreams, then it should be reportable. And if it is at least occasionally described in reports, then henceforth this option cannot be excluded unless subjects explicitly describe the contrary. In particular, saying one saw something in a dream does not mean that one's visual experience was either black-and-white or colored: indeterminacy has to be considered as a further option. Generally, so as not to exclude the possibility of nonstandard, unexpected types of dream imagery, which might go unreported unless subjects are explicitly asked about them, careful reporting instructions are needed.

This is not a purely theoretical point. For instance, Schredl et al. (2008) found that the percentage of reports describing black-and-white dreams dropped not only for dreams reported immediately upon awakening but also when subjects were presented with the option that they may have forgotten dream color. Relatedly, Murzyn (2008) asked her subjects to answer yes/no questions about dream color immediately upon awakening and used their answers to define five categories: colored dreams, grayscale dreams, mixed (partially colored, partially grayscale) dreams, dreaming neither in color nor in grayscale, and unrecalled color. Reports of "neither/nor," or indeterminacy, were infrequent, but 9 percent of subjects under twenty-five and 22 percent of subjects over fifty-five reported mixed dreams—a finding that, according to Murzyn, was wholly unexpected. Schwartz and Maquet (2002) also cite evidence of subjects occasionally having achromatic or even hemiachromatic dreams. Reports of mixed dreams do not lend themselves to the interpretation that the subjects reporting them have poor dream recall, or that they are imposing waking assumptions on their dream reports. If they were, within-subject and especially within-dream variability would be unexpected: one would not, it seems, selectively forget or misremember half of the color of one's dream or expect dreams to be partially colored and partially grayscale based on waking experience. Similarly, Domhoff notes that while self-reports in response to general questions (or opinions) about dreaming often cohere with culturally stereotypic distortions, the same is not true for dream reports:

> The dream experience is so direct, immediate, and compelling that a report on it is not likely to be distorted by filtering it through cultural beliefs. This point is especially strong for awakenings in the sleep laboratory while the person is in the process of dreaming. (Domhoff, 2003, p. 41)

We find a similar, though more poetic, formulation in Ellis's *The World of Dreams*. He notes that "we only catch our dreams for a moment by the light, as it were, of the open door as we are emerging from sleep" (Ellis, 1911/1931, p. 3). Even though in reporting dreams, we are, to Ellis, reporting "another kind of consciousness" (p. 3), he cannot bring himself to distrust dream reports given immediately upon awakening:

> It is impossible to believe that the witness of waking consciousness to the nature of the visions it has caught at the threshold between sleeping and waking life is false, and the most convincing evidence of this is the utter unlikeness of these visions to the data of ordinary waking life. (Ellis, 1911/1931, p. 5)

Results may also vary depending on the type of reports used. Rechtschaffen and Buchignani (1992) found that when asked to compare their dreams to photographs with different degrees of color saturation, about 20 percent of participants chose achromatic photographs. Again, we find that increasing the granularity of reports—in this case by adding specific questions or using different types of reports—changes theoretical views of dream color by rendering new aspects of dream experience reportable.

This insight is not new. In 1905, Monroe published a study on what he called the mental elements of dreams in the *Journal of Philosophy, Psychology, and Scientific Methods*. He asked participants to report dreams on six successive mornings, emphasizing that "it was desirable to know the nature of the imagery of the dreams (and the distinction of dreaming *about* a type of imagery and dreaming of *experiencing* that imagery was pointed out)" (Monroe, 1905, p. 650). Seven out of 287 dreams were of a "purely intellectual" character; subjects reported dreaming of their mother but not seeing her, dreaming of a fire alarm but not hearing it, and so on. I like this example because it focuses so explicitly on the distinction between the content of a dream and the way it is experienced. Introducing such explicit distinctions may be necessary to make the subtleties of dream experience visible. In investigating dream experience, overanalogizing dream and waking experience is an implicit danger, and adhering strictly to what is described in reports, plus developing fine-grained questionnaires about the experiential character of dreaming (or the formal features of dreaming; cf. Hobson, 1988; and sec. 2.2.), not just about dream content, might offer an effective way of counteracting researchers' own expectations.[24]

My second reason for endorsing the conservative approach has to do with the variability of dream experience. Recall that one of Schwitzgebel's reasons for distrusting opinions about colored dreaming is that he takes an underlying change in dream experience to be implausible. In this view, variability in beliefs about dreaming can be explained away and fails to reflect changes in dream experience. The argument from variability is a recurring theme in his work on the unreliability of introspection. He summarizes it as follows:

> People often differ greatly in their judgments about their stream of experience (across cultures, between individuals within the same culture, or within the same individual over time). Sometimes, in such cases, it seems unlikely that their actual underlying experiences vary correspondingly. Consequently, some of their judgments—we don't necessarily know which ones—are probably wrong. (Schwitzgebel, 2011c, p. xi; cf. Schwitzgebel, 2013b)[25]

According to the conservative approach, variability in dream reports, all other things being equal, indicates underlying variability in dream experience. Indeed, studies investigating the continuity hypothesis of dreaming have found that dream recall frequency, but also specific aspects of dreaming (content, emotional tone), can vary inter- and intraindividually (cf. sec. 6.2.2). Hunt's (1989) work on the multiplicity of dreaming impressively documents the variability of dreaming, and Hacking (2001) has suggested that cultural differences may have a strong influence on different forms of dreaming. Evidence also points to cross-cultural continuity across dreams. Typical dream themes are fairly stable, with the chase dream being the most frequent dream theme both in historical studies and in different cultures (Nielsen et al., 2003; Yu, 2008). Similarly, sleep paralysis nightmares often have a stereotyped form, and again evidence demonstrates historical and cross-cultural stability (Cheyne, 2003, 2005; Cheyne et al., 1999; cf. sec. 8.2.2 for discussion). For the time being, treating genuine variability in dream experience as a scientifically investigable research question and trying to determine both the factors contributing to variability and aspects that are stable across dreams seems to be the most productive and most cautious option.

In fact, rare and unexpected reports deviating from the stereotyped features of a majority of dreams and from commonly accepted folk-phenomenological or theoretical assumptions about dreaming may even be particularly interesting, because they can counteract premature overgeneralizations and deeply ingrained convictions about dreaming. Returning to the metaphor of literal transparency, note that even clear and perfectly positioned windows can mirror one's own reflection; in this way, dream reports, unless special steps are taken to make room for and perhaps even facilitate unexpected and nonstandard reports, may turn into self-fulfilling prophecies, reflecting only what researchers expected to find all along. Here the conservative approach can be a helpful corrective for antiskepticism about dream reporting. While general characterizations of dreams in terms of their stereotyped features can be helpful, they can also be deceptively simple, obscuring the inherent multiplicity of the target phenomenon. This point will become clearer in the next chapters. For now, we can note that an important goal for dream research is to assemble a catalog of reportable dream experiences and render it as complete as possible. Attempting to render unreported types of dream experiences reportable by designing specific questionnaires or employing interview techniques and timed awakenings is an important goal for this project. Yet the conservative approach is also plea for caution: in descriptions of dream experience, even nonstandard and unexpected types of imagery are never completely off the table, and inferences beyond what is explicitly described in dream reports are risky at best and quite possibly misleading.

Finally, note that this may be true not only for dreaming but for conscious experience in general. In keeping with the conservative approach, and in direct response to Schwitzgebel, Hohwy suggests that the variability of waking experience reports is largely reducible to genuine variability in experience and only partly caused by suboptimal reporting conditions:

> Global unreliability of introspection is, I think, a somewhat less good explanation of the evidence than my explanation in terms of variability of phenomenology and unreliability in some testing

> conditions. (1) As they stand, the variability/limited unreliability hypothesis draws on a wider range of evidence including neurocomputation and neuroimaging. Such integration is normally considered a best-maker for explanation. (Hohwy, 2011, p. 281)

He also notes that the argument from variability is often plagued by an inherent circularity, depending on introspective reports to defeat the reliability of such reports. The same circularity is found in Schwitzgebel's argument on dreaming, where doubts about the reliability of dream reports are generated, in part, by relying once more on dream reports. This suggests that an integrated approach to dream and first-person reports might be a valuable next step: as indicated earlier, the explanatory loveliness of the transparency assumption would be further increased if it allowed for a unified account not only of dreaming and waking experience but also of dreaming and waking experience reports. Moreover, given that the philosophical debate on dream reporting can largely be regarded as a proxy war on the trustworthiness of first-person reports, it stands to reason that an antiskeptical account of dream reporting might be equally applicable to first-person reports in general. The next section takes a first step in this direction.

4.5 Antiskepticism about Dream Reporting and the Debate on First-Person Reports in Consciousness Research

Now that an antiskeptical account of dream reporting is in place, a natural next step is to ask how this account relates to the general debate on first-person reports in interdisciplinary consciousness research. Rather than attempting to give an overview of the general debate, I want to use this section to highlight some interesting points of contact between the theoretical debates on dream reports and first-person reports in general, as well as offer some general remarks on how to situate my account of dream reporting in the broader debate on first-person reports. I will close by sketching some particularly interesting experimental approaches, as well as suggestions for future research.

To begin with, note that early dream researchers not only conceived of themselves as investigating dream experience but explicitly pointed out the relevance of their work for determining the neural basis of consciousness itself. Dement and Kleitman write:

> Since dreaming in the past was said to occur during all the EEG levels of sleep, one had to conclude that at least in sleep there was no particular relation between brain wave patterns and consciousness. . . . That this lack of relation between brain waves and consciousness, as represented by dreaming, is not the true picture seems clearly indicated by this study. (Dement & Kleitman, 1957a, p. 687)

This suggests that the investigation of dreaming is an important step toward the investigation of conscious experience. Yet aside from isolated attempts to integrate the two fields (Allan Hobson and Antti Revonsuo are perhaps the most prominent counterexamples; cf. Hobson et al., 2000; Hobson 2009, 2014; Revonsuo, 1995, 1999, 2006), dream and consciousness research remain largely separate. As Wamsley suggests, this may be related to continued skepticism about first-person reports: "Although cognitive neuroscientists have become increasingly comfortable with using introspective self-report to study wakefulness, conscious experience during dreaming has continued to be treated as a 'special case' placed outside the purview of scientific investigation" (Wamsley, 2013, p. 2).

Changing this view of dreaming is an important goal of this book, and dispelling skeptical worries about dream reports is a first step in this direction. For now, note that differing views on the status of dream reports and their relation to dream experience have been related to differing views on conscious experience from the beginning. To illustrate this point, I want to give three examples from the history of philosophy and show how they are paralleled by contemporary thinking about conscious experience and first-person reports. All three center on the problem of consciousness during deep sleep: do we dream throughout sleep, and is a failure to recall dreams upon awakening evidence for saying that we were unconscious in the preceding sleep stage (for an excellent analysis of the problem of consciousness during deep sleep in Indian philosophy, see Thompson, 2014, 2015; for discussion of this issue, see Windt, 2015)?

The first example takes us back to our starting point in chapter 1, namely, to Descartes's treatment of dreaming in the *Meditations*. Note that Descartes's *cogito* argument, according to which the self is essentially a thinking thing, makes existence contingent on ongoing conscious experience:

> And, let him [the most powerful and cunning being] deceive me as he may, he can never bring it about that I am nothing, so long as I shall be conscious that I am something. So that it must, in fine, be maintained, all things being maturely and carefully considered, that this proposition (pronunciatum) *I am, I exist*, is necessarily true each time it is expressed by me, or conceived in my mind. (Descartes, 1641/1901, II.3)

This view commits him to the view that the soul is conscious throughout dreamless sleep and to what Hill (2004b) calls the *retirement view*, according to which conscious thinking during sleep is only rarely remembered upon awakening. In his replies to Gassendi's objections to the *Second Meditation*, Descartes writes:

> So long as the mind is joined to the body, then in order for it to remember thoughts which it had in the past, it is necessary for some traces of them to be imprinted on the brain; it is by turning to these . . . that the mind remembers. So is it really surprising if the brain of . . . a man in a deep sleep, is unsuited to receive these traces? (Descartes, 1984, II.247)

For Descartes, because the mind is essentially a thinking thing, the failure to recall dreams upon awakening does not indicate an absence of consciousness. Instead the absence of dream recall is best explained by saying that in deep sleep, the soul has retired from the body, thus failing to leave any traces of its activity in the brain.

The second example is John Locke's endorsement of what Hill calls the *blackout view*. Locke (1689/1997, II.I) objected to Descartes's argument in the strongest terms, arguing that the assumption that we dream throughout sleep or indeed have any conscious experiences that we are unable to recall is purposeless and urged in vain (II.I.15). For Locke, consciousness is "the perception of what passes in a man's own mind" (II.I.19); and while he allows that we have experiences in sleep while dreaming, he portrays waking memory as the measure of their occurrence:

> No man's knowledge here, can go beyond his experience. Wake a man out of a sound sleep, and ask him, what he was that moment thinking on? If he himself be conscious of nothing he then thought on, he must be a notable diviner of thoughts, that can assure him, that he was thinking: may he not with more reason assure him, he was not asleep? This is something beyond philosophy; and it cannot be less than revelation, that discovers to another, thoughts in my mind, when I can find none there myself. (Locke 1689/1997, II.I.19)

With characteristic irony, Locke points out a further problem for Descartes's view:

> This I would willingly be satisfied in, whether the soul, when it thinks thus apart, and as it were separate from the body, acts less rationally than when conjointly with it, or no: if its separate thoughts be less rational, then these men must say, that the soul owes the perfection of rational thinking to the body; if it does not, 'tis a wonder that our dreams should be, for the most part, so frivolous and irrational; and that the soul should retain none of its more rational soliloquies and meditations. (Locke 1689/1997, II.I.16)

For Locke, the problem of unconscious thinking throughout sleep is related to the question of personal identity. We can imagine, according to Locke, two men sharing one continuously thinking soul, with one man being asleep while the other is awake and vice versa (Locke 1689/1997, II.I.12). If one man retained no memory of the soul's thoughts and perceptions while it was linked to the other's body, and vice versa, they would be distinct persons—hence refuting Descartes's claim that identity is constituted by continuous though forgotten conscious experience. In this view, identity is guaranteed neither by continuous thinking nor by physical continuity but by psychological continuity or memory.[26]

Finally, a compromise between Descartes's retirement view and Locke's blackout view is the *confusion view* promoted by Leibniz (Hill, 2004b). Responding to both Descartes and Locke, Leibniz (1704/2004) argues that while thinking persists throughout sleep, it is too confused and fragmented to be remembered upon awakening. He has Theophilus say:

> Something remains of all our past thoughts, none of which can ever be entirely wiped out. When we are in dreamless sleep, or when we are dazed by some blow or a fall or a symptom of an

illness or other mishap, an infinity of small, confused sensations occur in us. (Leibniz, 1704/2004, II.1.11)

Later, Theophilus explains:

Not only is it easy to conceive, but something like it can be observed every day of our waking lives! There are always objects that affect our eyes and ears, and therefore touch our souls as well, without our attending to them. Our attention is held by other objects, until a given object becomes powerful enough to attract our attention its way, either by acting more strongly on us or in some other way. It is as though we had been selectively asleep with regard to that object; and when we withdraw our attention from everything all at once the sleep becomes general. It is also a way of getting to sleep—dividing one's attention so as to weaken it. (Leibniz, 1704/2004, II.1.14)

As Hill notes, this compromise between continuous but partly unconscious thinking throughout sleep might allow for a gradual distinction between minute but unconscious sensations and conscious ones, but also between wakefulness and unconscious sleep. There is, however, an inherent ambiguity in this view, regarding what exactly it means for small or confused sensations to be unattended and unconscious. A strong reading would say that in dreamless sleep, these confused sensations are so subtle as to occur below the thresholds of attention, memory, and reportability, even in principle. Textual evidence, however, supports a weaker reading. In the preface to *New Essays on Human Understanding*, Leibniz writes that typically, minute sensations—such as hearing a faint sound, or one that one is accustomed to, such as a waterfall—are not forceful enough to attract attention; they may, however, be remembered if we are alerted to their occurrence immediately afterward. This reading is supported by Leibniz's remark that "these tiny perceptions . . . are more effective in their results than has been recognized. They constitute that *je ne sais quoi*, those flavours, those images of sensible qualities, vivid in the aggregate but confused as to the parts" (Leibniz, 1794/2005, p. 6). This suggests that Leibniz did not mean to divorce these minute sensations from function, or from reportability, but only meant that typically we fail to remember and hence report them.

These three historical positions have certain analogues in the contemporary literature. For instance, the assumption that "mental activity is continuous during sleep" (Winget & Kramer, 1979, p. 14), that "the mind is never 'empty' . . . and that mental activity is continuously present during waking and sleep" (Cicogna et al., 2000, p. 70), roughly analogous to the retirement view, persists in contemporary dream research. To the extent that this is presented as a possibility that cannot be ruled out based on timed awakenings in the laboratory, and to the extent that it is taken to refer to conscious mental activity, it contradicts, as we have seen, the transparency and reportability assumptions. As noted by Locke (though put in slightly less dramatic terms), the assumption of conscious experience independently of and perhaps even in contradiction to reports (where these are given under sufficiently ideal conditions) is mere speculation and transcends the limits of scientific dream research.

The confusion view is also sometimes echoed in current work on dreaming. For instance, Rosen (2013) suggests that dream reports may be especially prone to confabulation because of memory loss, but also because of the bizarre character and the widespread absence of rational thought in dreams. The underlying idea is that dream experience is simply too bizarre to be reported accurately: upon awakening, we naturally rationalize our dreams so as to downplay their dissimilarity from waking experience.

Finally, the essentially Lockean assumption that dreamless sleep is a state of unconsciousness and can be contrasted both with conscious experience during standard wake states and with dreams is an important motivation for integrating dreams into the debate on conscious experience in the first place (Revonsuo, 2006; Metzinger, 2003, 2009; Tononi, 2012).[27] If, however, it should turn out that there is a well-defined and empirically supported case for the occurrence of conscious experience during dreamless sleep, this assumption might well have to be adjusted (Thompson, 2014, 2015; Windt, 2015).

Moving beyond dreams and sleep, we find an interesting parallel to the debate on the reportability of waking experience. There is a long-standing controversy over the relationship between phenomenal and access consciousness, where the former refers to the subjective experiential character of experience and the latter to the ability to access phenomenal states through cognitive functions such as memory, attention, and decision making. Importantly, cognitive access is also the precondition for reporting phenomenal states, either verbally or via behavioral reports (e.g., button press or pointing). Block, in particular, has argued that "phenomenology overflows access" (Block, 2007, p. 487), suggesting that the space of cognitively accessible and hence reportable experience is more narrow than the space of phenomenal experience. In this view, it is impossible, and perhaps even misguided, to attempt to study phenomenal (as opposed to access) consciousness through verbal reports, because this approach introduces a confound between two distinct forms of consciousness. The debate hinges on whether there are—or indeed could be—convincing reasons for thinking that phenomenal consciousness outruns access, even in principle.[28]

The opposite side of the debate is exemplified by Cohen and Dennett's (2011; see also Dennett, 2015) rejection of the distinction between phenomenal and access consciousness. In a Lockean manner, they argue that any theory based on this distinction is not amenable to empirical confirmation or disconfirmation and hence lies outside the scope of science for principled reasons. The emphasize that it is only

> the products of cognitive functions (i.e. verbal report, button pressing etc.) that allow consciousness to be studied at all. A proper neurobiological theory of consciousness must utilize these functions in order to accurately identify which particular neural activations correlate with conscious awareness. (Cohen & Dennett, 2011, p. 358)

According to Cohen and Dennett (2011), this approach renders the "hard problem" of consciousness (Chalmers, 2000), which is an impossible problem insofar as it rests on the distinction between phenomenal and access consciousness, empirically tractable.

Much of the debate has centered on diverging interpretations of Sperling-type experiments (Sperling, 1960; for details and further references, see Block, 2007, 2011; Cohen & Dennett, 2011). In these experiments, participants are briefly shown an array of letters (e.g., three lines of four letters each). When asked to report what they saw, they can only report about four letters. Yet when they are cued as to which line of letters to describe after stimulus presentation, they are able to do so, suggesting that they could have reported four letters from any of the lines. They also report having seen all the letters, even though they are not able to name all of them individually. In one sense, then, it seems that not all aspects of participants' experience are reportable—the capacity of working memory seems to impose certain limits on how many letters can be reported at a time, in any particular trial. However, given that subjects can report letters from any line that is cued with high accuracy, one might also assume that all of the seen letters are indeed reportable—just not all at once, and only in principle, but not in practice. Principled reportability is compatible with the existence of a factual limit to how much can, in fact, be reported in a given trial. This is also in keeping with participants' reports of having seen all the letters in detail. Consequently, one might say that while Block (2007) is correct in claiming that phenomenal consciousness is richer than what, in fact, can be cognitively accessed and reported, this does not show that phenomenal consciousness overflows cognitive accessibility and reportability in a more principled sense. Indeed, in more recent writings, Block has defended a more humble claim, arguing that "the overflow argument does not claim that any of the items in the array are cognitively inaccessible, but rather that necessarily most are unaccessed" (p. 567; for a similar conclusion, see Tye, 2011, chap. 7) and no longer portraying his position as "a vote for inaccessible consciousness" (Block, 2011, p. 574) in any principled sense. Importantly, this means that rather than slipping into the retirement view, Block defends something more akin to the confusion view, according to which the details of conscious experience typically go unreported but do not thereby violate the reportability assumption. Incidentally, note that in this weaker reading, the difference between Block and Cohen and Dennett mostly evaporates: how much is experienced and whether reportability is the measure of experience once more becomes an empirical question.

Rather than pursuing this debate in more detail, I merely want to indicate how antiskepticism about first-person reports, combined with the conservative approach, can help make sense of Sperling-type experiments. The crucial thing to see is that in the Sperling experiment, reporting conditions are improved by cueing subjects to a certain line; because the cue occurs after stimulus presentation, it is not reasonable to assume that subjects just happened to focus on the line that was cued. Rather, subjects saw all the letters in the array; their ability to report no more than four letters at a time is a limitation in working memory, but it does not suggest that conscious experience is divorced from reportability. Recognizing that an experience report is necessarily more limited and less detailed than the original experience is important, but not particularly surprising: any description is limited by taking a certain perspective, and reports are necessarily selective. Sperling-type experiments provide a particularly clear illustration of the distinction between detailed experience, detailed reports, and reports of detailed experiences. Apparently there are systematic limits to how much detail can be given in a report; yet reports can describe the occurrence of detailed experiences (i.e. of seeing more letters than they now can name) even if they fail to reproduce the experience in all its original detail.[29] For the reportability assumption, this is enough (cf. my discussion of white dreaming in sec. 4.4). This insight could also have consequences for dream research: for instance, it may be possible to counteract memory loss for dreaming by asking subjects to report only certain aspects of their experience in detail. Perhaps asking subjects specific questions about their experiences can be regarded as fulfilling a similar function as the cues in Sperling-experiments, thus rendering aspects of dream experience reportable that would otherwise likely have been forgotten. It also suggests that asking participants to report their dream in as much detail as they can may be counterproductive by inviting memory loss for exactly the kinds of rich detail researchers might be interested in.

More generally, antiskepticism, combined with the conservative approach, can now be related directly to classical and contemporary questions about the relationship between subjective experience and first-person reports. First, the retirement view, or the claim that phenomenal experience outruns reportability in principle, is ruled out as transcending the limits of scientific research. Second, the contrast between the blackout and the confusion view can be given an empirical reading. Given that reports of dreaming and nondreaming made under the same conditions are equally trustworthy, reports of an absence of experience before awakening support the blackout view. One might even raise a similar question about wakefulness. While it is commonly assumed that conscious experience and thought are continuous throughout (and perhaps even the hallmark of) wakefulness, recent research suggests that "mind blanking" may be a robust and widespread phenomenon, distinct from mind wandering and occurring frequently in standard wakefulness (Ward & Wegner, 2013). If this is correct, and to avoid lapsing into the retirement view of principally unreportable experience, these findings suggest that even wakefulness includes frequent gaps in conscious experience—and ones that typically go unnoticed. Rather than categorically excluding such a possibility and thereby committing the armchair fallacy, antiskepticism about first-person reports will take such findings seriously.

At the same time, a weak version of the confusion view is strengthened by the finding that subjects can report having had detailed experiences without describing them in detail (i.e., without giving a detailed report); clearly the detail of experience may outrun the detail of reports. Thus a careful compromise between the confusion view—or the finding that the detail of experience can be elusive—and the blackout view—or the finding that experience may not be continuous throughout sleep, or perhaps even throughout wakefulness—may be the best solution. Importantly, as suggested by my discussion of Sperling-type experiments, but also by research on mind blanking, special reporting methods may be required to make both subtle features of experience and the occasional absence of experience visible. Finally, note that according to antiskepticism, the rejection of the distinction between phenomenal and access consciousness should be understood not as a metaphysical claim regarding the impossibility of phenomenal consciousness independent of access consciousness, but rather as a methodological claim regarding the impossibility of studying such states (cf. my discussion of unreportable dream experience in sec. 4.4).

In this, I think antiskepticism is compatible with many current approaches (for an overview, see Varela & Shear, 2000; Overgaard et al., 2008; Irvine, 2013).[30] Overgaard et al. (2008) note that substantial overlap characterizes different methods for examining first-person reports. These include the centrality of first-person reports for consciousness research, the importance of training and unbiased observation on the part of subjects and experimenters, and the use of postexperiment interviews examining aspects of experience that were not investigated during the actual experiment. While I think the applicability of training to dream research is largely limited to training subjects to be better reporters (cf. sec. 4.4), a number of other aspects are more interesting. For instance, Overgaard et al. (2008) note that postexperiment "interviews can inspire a re-grouping of data and new ways to look at data using quantitative methods" (p. 19). Relatedly, starting out with free reports and then developing specific experience categories on their basis, and perhaps even giving an active role to participants in this process (as in Lutz et al., 2002), could be fruitfully used in dream research as well.[31]

Antiskepticism also shares the core intuition behind Froese and colleagues' (2011) approach to the validation and calibration of first-person methods in consciousness research:

> One effective way to evaluate the scientific validity of the phenomenological results of first- and second-person methods is to see whether they help us to do better science or not. In other words, it is not necessarily important that we can validate phenomenological reports by having recourse to our own experience, it is sufficient that we can do better work as a result of having them. (Froese et al., 2011, p. 39)

This careful combination of moderate distrust and acknowledging the centrality of first-person reports, which lies at the heart of antiskepticism, is widely endorsed in philosophical theorizing about first-person reports as well. It fits in well, I think, with Piccinini's (2003, 2009) argument that neither agnosticism nor blind faith about introspection is feasible; rather, he regards subjects giving first-person reports as self-measuring instruments and argues that first-person data culled from such reports are compatible with scientific methodology. These key assumptions of antiskepticism are also compatible with Bayne and Spener's recommendation of introspective humility about first-person reports. They note that "radical scepticism about introspection is untenable: denying that we have any first-person knowledge of consciousness is not a serious possibility" (Bayne & Spener, 2010, p. 2):

> When it comes to introspection we should be neither radical sceptics nor uncritical fideists. Instead, we should adopt an attitude of introspective humility. Introspection is not to be rejected outright, but its use in the study of consciousness should be accompanied by a great deal more caution than is currently the case. (Bayne & Spener, 2010, p. 22)

In recent work, Bayne (2015) continues to emphasize the importance of a differentiated approach to the trustworthiness of introspection, which he contrasts with both radical pessimism and undiscerning optimism.

What about Dennett's own preferred position on the relevance and use of first-person reports? Given his defense of skepticism about first-person reports in general and dream reports in particular, it may seem surprising that antiskepticism about dream reporting actually overlaps with Dennett's heterophenomenology in a number of ways. Dennett originally introduced heterophenomenology as a neutral, scientific approach to phenomenology from a third-person perspective and presented it as descriptive of the way consciousness research is actually conducted (rather than as a primarily normative theory; cf. Dennett, 1991, 2003, 2007, 2013, pp. 340–346). The goal of heterophenomenology is to assemble a catalog of people's beliefs about experience (including, for instance, reports about ineffable experience); a complete catalog of such beliefs would constitute the subject's heterophenomenological world:

> The total set of details of heterophenomenology, plus all the data we can gather about concurrent events in the brains of subjects and in the surrounding environment, comprise the total data set for a theory of human consciousness. It leaves out no objective phenomena and no subjective phenomena of consciousness. (Dennett, 2003, p. 20)

This is the analogue of the reportability assumption: even if unreportable experiences exist, they are not the object of dream or consciousness research; rather, scientific dream (and consciousness) research is constrained by the space of reportable experiences. How, then, if at all, does antiskepticism differ from heterophenomenology? An important controversy about heterophenomenology is related to its alleged neutrality or agnosticism about first-person reports and to Dennett's claim that in a certain sense, first-person reports are best regarded as fictions. Dennett notes that heterophenomenology

> is a cautious, controlled way of taking subjects seriously, as seriously as they could possibly be taken without granting them something akin to papal infallibility, while maintaining (contrary

to everyday interpersonal communicative practice) a deliberate bracketing of the issue of whether what they are saying is literally true, metaphorically true, true under-an-imposed-interpretation, or systematically-false-in-a-way-we-must-explain. (Dennett, 2007, p. 252)

Yet he also insists that first-person reports are akin to a theorist's fiction:

If a human subject says she is rotating a figure in her mind's eye . . . she is speaking at best metaphorically and at worst creating an unwitting fiction, for we can be quite sure that the process going on in her head that inspires and guides her report (to put it as neutrally as possible) is not a process of actual image rotation. (Dennett, 2007, pp. 257–258)

Is this reconcilable with the kind of moderate distrust about dream reports advocated by antiskepticism? To begin with, one might think that Dennett's view that first-person reports are fictitious is related to his occasionally eliminativist views on conscious experience and his claim that there is no such thing as actual phenomenology (Dennett, 1991, p. 365). Dennett (2007) makes it clear, however, that while eliminativism is his destination, heterophenomenology is his neutral starting point. By contrast, antiskepticism is not neutral about the trustworthiness of dream reports with respect to dream experience: antiskepticism says that in reporting dreams, at least under ideal reporting conditions, we are reporting experiences during sleep. In this way, by combining moderate distrust with a general background of trust, antiskepticism suggests a compromise between granting papal infallibility to the subject and withholding judgment.

In another reading—and this is the one suggested by Dennett's remarks on image rotation—heterophenomenology is not so much neutral as to whether subjects are accurately reporting the phenomenal character of their experience but rather says that in reporting their experience, they are creating unwitting metaphors about what is really going on inside their brains. In this reading, first-person reports are fictitious not with respect to experience but with respect to its neural underpinnings. Thompson (2007, pp. 304–305) argues that assessing the truth or falsity of first-person reports based on their ability to match what is going on in the brain is misguided (but see Dennett, 2007). Experience reports carry no particular commitments about brain processes; similarly, one might argue that dream reports carry no commitments about sleep stages—they are only committed to describing previous experience. In reporting a dream of being chased by a monster, I am clearly not *intentionally* trying to convey information about the underlying brain processes involved in this dream or the sleep stage in which it occurred (just as I am not claiming that I was really chased by a monster, that my dream monster really exists, and so on); all my report commits me to is the claim that my experience, in sleep, was that of being chased by a monster, that I had the thoughts and emotions described in the dream report, and so on. Antiskepticism only applies the transparency assumption to the latter type of claim; whether this is (or is not) compatible with heterophenomenology will depend on how exactly one reads Dennett here.

Dennett's (2003, 2007) claim that beliefs about experience are the primary data of consciousness research and that, in reporting experience, we are merely expressing beliefs about experience has also been criticized. Thompson, for instance, notes that this move "collapses the crucial distinction between conscious experience and belief about conscious experience" (Thompson, 2007, p. 307) and overintellectualizes experience reports. In my view, dream experience is indeed distinct from dream reports; however, because dream research is constrained by the space of reportable dreams, its maximum data set comprises dreams reported under (approximately or sufficiently) ideal reporting conditions and under the assumption of transparency. I would not go so far as to say that reports are beliefs about experiences; in my view, this would invite an equivocation between different types of reports and opinions about experiences, which, because they lack any grounding in individual dream recall, are best not considered as reports at all.

What, then, about neurophenomenology? Because it places much weight on first-person reports, it is plausible to assume that it is closely related to antiskepticism. Neurophenomenology aims at developing a common structural level of description that captures the moment-to-moment temporal dynamics of changes in conscious experience and large-scale neural processes (for discussion, see Thompson, 2007, chap. 11). According to Varela, "Phenomenological accounts of the structure of experience and their counterparts in cognitive science relate to each other through reciprocal constraints" (Varela, 1996, p. 343; see Lutz & Thompson, 2003; Thompson, 2007), suggesting that phenomenological analysis guides scientific research and vice versa. While antiskepticism shares the emphasis on first-person reports, I think it differs from neurophenomenology in several respects. First, neurophenomenology builds on the assumption that first-person methods are disciplined practices involving "the systematic training of attention, mindfulness, and self-regulation of emotion" (Thompson, 2007, p. 339). Through training, subjects are supposed to "gain access to aspects of experience that otherwise would remain unnoticed and hence unavailable for verbal report and description" (p. 339), thus rendering their reports more accurate. Second, this approach frequently works within the framework of classical (Husserlian) phenomenology and contemplative traditions. Third, neurophenomenology is primarily interested in relating changes in conscious experience to underlying neural dynamics. Fourth, neurophenomenology often focuses on the structural invariants of experience, such as the intentional structure of experience, or the threefold structure of time consciousness (primal impression, retention, protention).

I think antiskepticism is generally compatible with this approach, but more neutral in all four respects. First, I do not think that training in the sense of cultivating a disciplined *observation* of experience is directly applicable to dream research. Reporting dreams surely requires discipline, and people can be trained to report their dreams in the ways required for a given study. But at least for nonlucid dreams, contemplative techniques have no direct application. Moreover, training is just one way of improving reporting conditions, and one idea behind antiskepticism is that by tweaking, among other things, the timing and method of awakening, the questions asked of subjects, and the interaction with experimenters, the transparency of reports even from untrained participants can be improved. Training, as discussed earlier, is particularly helpful for investigating the depth of dream experience, whereas reports from untrained participants may be most useful for investigating the breadth of dreaming (see Solomonova et al., 2014). A next step, then, would be to conduct studies comparing both types of reports. Second, dream research is not committed to applying concepts from classical phenomenology to dreaming and thus does not share the historical heritage of neurophenomenology, but is broader in this respect. Third, the focus of neurophenomenology on neural dynamics is certainly relevant to dream research insofar as it uses polysomnographic and imaging data, but cannot be generalized to dream research in general. Large parts of contemporary dream research are not directly related to the investigation of underlying neural activation patterns. Could there be neurophenomenological studies of dreaming in this narrower sense as well? Perhaps there could, for instance, by asking trained lucid dreamers to signal changes in experience on a small timescale, enabling researchers to relate them to the underlying neural dynamics (as done, though without the phenomenological background, by Dresler et al., 2011, 2012). Sleep onset experience might be another example (see chap. 11). Fourth, much of what I have to say in this book focuses on the variability of dreaming. My goal is to develop a framework describing the multiplicity of dream experience and to investigate, as far as possible, its breadth; only near the end, in chapter 11, do I return to questions about the invariant phenomenal core of dreaming. In all these respects, I think that my goals are much more basic and coarse-grained than those of neurophenomenology.

Finally, I think that the closest methodological relative of dream research is immediate retrospection of the type used in Hurlburt's descriptive experience sampling (for details and further references, see Hurlburt & Akhter, 2006; Hurlburt & Schwitzgebel, 2007; for critical discussion, see Froese et al., 2011). In descriptive experience sampling, the subject is a given a beeper, which goes off at random times during the day. The subject is instructed to remember in as much detail as possible any conscious experiences at the moment preceding the beep and to take notes immediately after the beep. The final report emerges in a longer conversation with the experimenter, conducted within twenty-four hours of the sampled experience, and can be seen as the result of a collaborative effort rather than a finished statement provided by the subject.

The central advantage of descriptive experience sampling is that it investigates conscious experience in a natural rather than a laboratory setting. While the retrospective nature of reports invites the objection of possible forgetting it also, together with the random, unexpected timing of beeps, ensures that conscious experience is undisturbed by special attention and observation, as is the case in methods involving ongoing introspection (see Lyons, 1986; Overgaard et al., 2008). At the same time, descriptive experience sampling is distinct from casual armchair introspection in that subjects undergo a period of training, with reports from the first sampling day typically only being used to familiarize them with the experimental procedure. Again, as in dream research, subjects are not so much trained in observing experience as in reporting it. A disadvantage of descriptive experience sampling is that such a snapshot approach, in which the focus is only on the last undisturbed moment before the beep, cannot capture the longer-term temporal dynamics of conscious experience (Hurlburt & Schwitzgebel, 2007, pp. 22–23).[32]

Note that the brief description of the ideal conditions of dream reporting from section 4.4 is importantly similar to descriptive experience sampling. Both demand retrospective reports of an experience immediately preceding a beep or tone (though dream reports typically are not limited to the moment preceding awakening but span a larger time frame). And both may combine a free report with an interview involving detailed questions. Consequently, using descriptive experience sampling to gather reports from sleep and wakefulness would be a wonderful opportunity to compare dreaming and waking experience reports; and as we will see in a few moments, some existing studies have already begun to do just that.

In sum, it seems that antiskepticism fits in nicely with a number of prominent positions on first-person reports and may even suggest a particular strategy for dealing with ongoing debates, for instance, on the relationship between phenomenal and access consciousness and Sperling-type experiments. In closing, I want to suggest a more speculative claim. So far, it has become clear that dream reports are often regarded as a particularly easy target for skepticism about first-person reports. I have suggested that this view is misguided: principled skepticism about dream reports fails, and for systematic reasons. Now I want to tentatively suggest that dream reports may actually be more trustworthy, in certain respects, than reports of waking experience. If so, then antiskepticism about dream reporting may not only generalize to antiskepticism about first-person reports but might also be a particularly strong case or indeed a paradigm example for trusting first-person reports.

In particular, note that while dream recall is often thought to be more vulnerable to forgetting than the recall of waking experiences, dreaming may be more impervious to external influences, bias, and expectation. Immediate retrospection of waking experience, as in descriptive experience sampling, minimizes this problem but perhaps does not obliterate it. In particular, there is still an effect of training and possibly of active reconstruction and selection in assembling the retrospective report. By contrast, the risk of altering dreams, especially nonlucid dreams, through concurrent introspection is virtually nonexistent. This is attested to by the difficulty of learning to induce lucid dreams and dream control. Merely intending to become aware of the dream state is not enough, and often even trained lucid dreamers cannot become lucid at will or can do so only after several attempts (Worsley, 1988; Brooks & Vogelsong, 1999; Stumbrys et al., 2012). It is true that dreams change in response to waking concerns—keeping a dream diary almost inevitably leads to dreams of the diary, just as sleeping in a sleep laboratory often leads to dreams about the laboratory and experimenters—but still, dreams are relatively impervious to deliberate ongoing introspection and waking expectations and intentions, and such incorporations are usually unintended and spontaneous. The attenuation of reflective thought, memory, and control over attention and volition often thought to characterize nonlucid dreams (see chap. 9) may thus protect them from many of the disturbances and pitfalls of ongoing introspection in wakefulness. Thus, if studying undisturbed conscious experience is a goal for scientific research, dreaming may offer the optimal conditions for doing so.

Moreover, the cognitive state upon awakening does not lend itself to deliberately censuring and constructing the dream report: at least when participants are asked to report their dreams immediately upon awakening, they are often in a state of stupor or sleep inertia, with little capacity for deliberate rescripting. As Kerouac writes in the introduction to his collection of his dreams:

> *Book of Dreams* was the easiest book to write—When I woke up from my sleep I just lay there looking at the pictures that were fading slowly like in a movie fadeout into the recesses of my subconscious mind—As soon (one minute or so) as I had assembled them together with any earlier dreams of the evening I could catch, like fish in a deep pool, I got my weary bones out of bed & through eyes swollen with sleep swiftly scribbled in pencil in my little dream notebook till I had exhausted every rememberable item—I wrote non-stop so that the subconscious could speak for itself in its own form, that is, uninterruptedly flowing & rippling—Being half awake I hardly knew what I was doing let alone writing. (Kerouac, 1961/2001, p. xvii)

Of course, this description of Kerouac's writing process may itself be not so much an accurate confession as a beautifully scripted marketing strategy. Yet, there would seem to be at least a kernel of truth in it, and the comparative rawness of dream reports can be regarded as a merit, as an indication of their sincerity and authenticity. If dream reports are indeed less amenable than waking experience reports to willful fabrication, this might even counterbalance their major disadvantage, namely, amnesia. It is also possible that the problem of dream forgetting itself is overrated. Rechtschaffen, one of the pioneers of dream research, notes:

> When laboratory subjects are awakened from the rapid eye movement (REM) stage of sleep, they generally have little difficulty in giving a fairly long, detailed report of dreams, with transcripts sometimes running to several typed pages. Indeed, it is often easier to get detailed, articulate re-

ports of ASCs [altered states of consciousness] than detailed reports of normal waking consciousness. When I catch my friends off guard and ask them what had been going through their minds, they generally acknowledge that something had been going on more or less continuously, but the description most often consists of a few relatively noncommunicative, stumbling phrases. (Rechtschaffen, 1975, p. 143)

As it stands, this is no more than an anecdotal observation. Yet, it suggests that a systematic comparison of experience reports across the sleep–wake cycle might be particularly fruitful. In fact, a beeper method has already been used to investigate the prevalence of thinking and hallucinations across the sleep–wake cycle (Fosse et al., 2001). And in a number of studies, Horton (2011a, 2011b) investigated how participants recall and recognize reports of dream and waking events from their own diaries, as well as the effect of rehearsing dream and waking reports on subsequent recall. Another promising possibility would be to compare dream reports to reports of waking mind wandering (Metzinger, 2013a; Fox et al., 2013; Wamsley, 2013).

Finally, at least two studies provide preliminary evidence supporting Rechtschaffen's observation on the comparative ease of recalling and reporting dreams. Stickgold et al. (2001) collected 1,748 reports from sixteen subjects over a period of fourteen days. During the daytime, subjects responded to a pager; at night, their sleep stages were monitored using a Nightcap sleep-monitoring system (Mamelak & Hobson 1989). Aside from between-subject variation, Stickgold and colleagues found that median report lengths varied more than twofold over the sleep–wake cycle, with REM reports being the longest, followed by reports from active wakefulness, quiet wakefulness, NREM sleep, and sleep onset. This finding has recently been replicated by Siclari and colleagues (2013), who used a serial awakenings paradigm to gather reports from different sleep stages, as well as a small number of reports from wakefulness. In addition, when they asked their subjects how far back in time they could recall the content of experience and how rich and complex it was (measured by how long it would take to recount it), they found the same pattern, with the highest rating occurring for REM sleep reports. This last finding is particularly interesting, because it suggests that dream reports may be better suited for capturing the temporal dynamics of experience than is the case for retrospective waking reports, thus overcoming one of the limitations, for instance, of descriptive experience sampling. A tempting conclusion is that if such findings were further substantiated in the future, this might suggest that dreaming, far from being a particularly clear target for skepticism about first-person reports, is in fact more readily recalled than waking experience. If so, then analyzing the trustworthiness of dream reports might not only be interesting because it offers a metatheoretical contribution to the debate on the trustworthiness of first-person reports and first-person methods. In addition, the comparative analysis of dream and waking report might be a fruitful new strategy suggesting new experimental paradigms for interdisciplinary consciousness research.

4.6 Conclusions

Over the course of the last four chapters, we have moved from naive, undifferentiated trust in dream reports, via philosophical skepticism about dream reporting, a detailed look at changing concepts of sleep and dreaming in the history of philosophy and of dream research, as well as an analysis of the methods of contemporary scientific dream research, to a defense of antiskepticism about dream reporting. This last chapter has rounded off my critique of skepticism about dream reporting by proposing inference to the best explanation as a theoretical justification of the transparency assumption and by articulating the main tenets of an antiskeptical approach not only to dream reports but perhaps even to first-person reports in general. This was important to show that trusting dream reports is neither a mere act of desperation resulting from methodological necessity nor a position of undifferentiated and dogmatic trust; rather, antiskepticism inherits important lessons from skepticism about dream reporting. In particular, by combining moderate distrust with a cautious and differentiated endorsement of the transparency assumption that regards transparency as a relative and dynamic property, varying with different reporting conditions and over different types of reports, I have proposed an account that affords a central place to dream reports in scientific dream research. Table 4.2 summarizes the main tenets of antiskepticism about dream reporting developed so far.[33]

This account has several advantages. It provides, I hope, an empirically informed defense of scientific dream research against the skeptic with the aim of putting persistent discomfort about the use of dream reports to rest. It also provides an affirmative answer to the problem of dream experience by suggesting why the experiential assumption is theoretically justified. Aside from aspiring to empirical plausibility, it also aims to be historically adequate by acknowledging how the view of dreaming as a genuine sleep phenomenon and its allocation to different sleep stages has changed along with changing conceptions of sleep and changing methodologies employed in sleep and dream research. This historical perspective highlights the contingency of whatever is the received view of dreaming of a given historical period. Likewise, what is considered the best explanation of dream reporting at a given time is not static but can change in response to scientific discoveries, and with it, the concept of dreaming will change, as well. Accordingly, my endorsement of the experiential and transparency assumption is open to enrichment from future scientific discoveries and theoretical developments.

Finally, antiskepticism provides a research strategy not only for further improving the use of dream reports in scientific dream research but also for answering more detailed questions about dream experience. So far, though antiskepticism presents a solution to the problem of dream reporting, it provides only a very basic defense of the claim that dreams are experiences. The second major goal of this book is to provide a conceptual framework for describing conscious experience in the dream state.

Table 4.2

Antiskepticism about dream reporting and the methodological background assumptions behind scientific dream research

Scientific dream research uses dream reports, *under the assumption of transparency*, to investigate *conscious experience during sleep*; in doing so, it is *methodologically constrained by the space of reportable dreams.*
Dream reports are the product of (verbal or nonverbal) behaviors conducted with the *sincere intent of conveying or recording certain relevant information about a specific dream.*
Experiential assumption: Dreams are conscious experiences in the sense that they are phenomenal states, or that it is *like* something to dream.
Transparency assumption: Dream reports are epistemically transparent in the sense that they are trustworthy sources of evidence about the occurrence and phenomenal character of experience during sleep *when gathered under ideal reporting conditions.*
Relative transparency: The *relative transparency* of a given (type of) dream report is determined by the extent to which the conditions under which it was gathered *approximate ideal reporting conditions.*
Reportability assumption: Dreams are reportable experiences in that the occurrence of experience during sleep can be reported *under ideal reporting conditions.*
Ideal reporting conditions: • *Temporal proximity* • Determining the appropriate level of *detail* and *specificity* required in a given study • Increasing the *expressive granularity* of dream reports (e.g., by developing experiential categories together with the participants in a given study, using questionnaires about specific dreams, supplementing verbal with nonverbal reports, etc.) • Using training and expertise to probe the depth of dream experience and investigate how it relates to the breadth of dreaming • Optimizing the sleeping environment, timing and method of awakening, interpersonal relationship between subjects and experimenters, minimizing censorship or incomplete reports, etc. • Using future studies to further investigate the factors underlying dream reporting and working toward further improvement
The **conservative approach** to dream reports: the transparency assumption is limited to those aspects whose presence (or absence) is *explicitly* described in dream reports, and only in the absence of evidence to the contrary (e.g., later recantations, uncertainty, etc.). Future research should aim to render unreported aspects of dream experience reportable; however, no inferences beyond what is explicitly described in dream reports are warranted.

Antiskepticism provides a theoretical foundation, a point of departure from which to pursue this project. Yet, as an answer to the problem of dream experience, it is still incomplete. All that can be said at this point is that dreams are phenomenal states occurring in sleep; as an answer to the problem of dream experience, this is as essential as it is uninformative and sketchy. In the following chapters, I put antiskepticism about dream reporting, combined with the conservative approach, to the test. The broader goal is to question a number of existing assumptions about dreaming, but also, more importantly and constructively, to clarify the concept of dreaming and say what type of experience, exactly, it refers to. What kinds of conscious experiences are dreams, and how does dreaming relate to imagination, hallucination, and perception? What is the role of self-consciousness in the dream state? Do dreams involve a thinking self, and perhaps even real beliefs and insight? Are dreams, as Descartes thought, deceptive experiences? How strongly, if at all, does the phenomenology of being or having a self overlap with standard wake states, and what is the relationship between the phenomenal dream self and the sleeping body? Finally, how can the analysis of self-experience in dreams contribute to our theoretical understanding of self-consciousness? How exactly can dreaming, in other words, be located on the conceptual map of concepts commonly used to describe waking experience?

While antiskepticism is, in an important sense, a refined version of the received view of dreaming and dream reporting, this view can no longer function as a guide to answering these more detailed questions about the phenomenology of dreaming. Over the next chapters, it will become clear that the intuitions of both philosophers and researchers discussing these questions are too different to claim that intuitive plausibility could be even a preliminary guide to their solution. Here, quite simply, there is no homogeneous received view of dreaming, even in contemporary philosophical and empirical theories. We are, in a sense, entering a conceptual no-man's-land, embarking to fill in the blank spots on the conceptual map of dream experience.

Much of what I have to say in the following chapters depends strongly on antiskepticism about dream reporting. Moreover, I will sometimes be fairly lenient—more lenient, in fact, than I would like and than I have recommended here—in using dream reports or results from report-based research to support my claims. Consequently much of what I have to say is highly preliminary and contingent on future findings. Yet, in the current situation, I think that relying on dream reports even from less-than-ideal reporting conditions is a practical necessity. I also think that this strategy enables genuine progress. The success of my approach depends, in part, on its productivity: if it is able to resolve existing conflicts and controversies, clarify open questions, and develop a coherent theory of dreaming—or if it at least suggests a strategy for how this might be done—then it counts as successful. At the very least, I hope that the following chapters might be considered as an extended (and theoretically and empirically informed) thought experiment on what can be said about dream experience given the status of scientific dream research today. Hopefully, this will serve as a fruitful basis for future theoretical work on dreaming in philosophy of mind but might also highlight interesting open questions for future research and suggest new experimental paradigms. If this approach enables me to at least make the broad outlines of an interdisciplinary theory of dream experience visible—to sketch, as it were, the rough outlines and position of dreaming on the map of concepts referring to conscious experience and the phenomenology of selfhood—this will be enough.

5 Dreaming as Quasi-Perceptual Experience: The Traditional View

5.0 Introduction

The *conceptualization problem*, or the question of how to conceptualize conscious experience in the dream state, dates back to antiquity and continues to be a matter of philosophical controversy. Unlike the philosophical debate on dreaming initiated by Malcolm (1956, 1959/1962), the question here is not whether dreams are experiences at all but rather, metaphorically speaking, where to locate dreaming on the map of concepts commonly used to refer to waking experience. The question of whether dreaming should be described as quasi-perceptual or imaginative experience, which I explore over the next two chapters, is a particularly good place to start. There are different ways of making this distinction, but as will become clear over the following two chapters, the philosophical debate typically hinges on the question of whether the phenomenal character of dreaming is best likened to that of standard perception or to that of imaginative states such as daydreaming.

Perception is commonly used as a success word. Visual experience, for instance, only counts as perceptual if it is sufficiently accurate with respect to the mind-independent objects it represents and appropriately sensitive to changes in the subject's actual environment. Because quasi-perceptual experience, as I use the term, only refers to the phenomenal character of an experience, it can be applied to illusions or hallucinations that from a third-person, epistemological perspective do not count as genuine instances of perception. To say that dreaming involves quasi-perceptual experience, then, is to say that dreaming is phenomenally indistinguishable from genuine perception. In this view, dreams and waking perception, but also illusions and hallucinations, form a common kind of mental state (cf. Macpherson, 2013): they are all quasi-perceptual experiences in a phenomenological sense.

This naturally raises the question of whether dreams are better described as hallucinatory or as illusory experiences, and this is also the most important distinction within the traditional view. While requiring the same phenomenological analysis, hallucinations and illusions are distinguished by their different degrees of external stimulus correlation. Proponents of the view that dreams arise independently of a contemporaneous and appropriate external stimulus take a position according to which dreaming fits the philosophical concept of hallucinating: dreaming seems to put us in contact with mind-independent objects where no such objects are actually being perceived (Crane, 2011). Others regard dream imagery as a distorted perception of an actual contemporaneous stimulus, such as residual peripheral or bodily sensations occurring during sleep. This fits the philosophical concept of illusion, defined as an inaccurate perception of an object as having different properties from the ones it actually has (Crane, 2011; Macpherson, 2013). For dreams, the distinction between hallucination and illusion thus typically hinges on the question of whether dream imagery has its source in external stimuli or bodily sensations (and hence is illusory) or arises only from internal signal generation in the brain (and hence is hallucinatory).

Aside from the debate on the sources of dreaming, two different strengths of the quasi-perceptual view can be distinguished. The stronger version takes dreaming to replicate waking perception by giving rise to exactly the same type of experience that characterizes standard wake states. The weaker version says only that dreaming feels like perceiving, but does not commit to the view that dreaming actually replicates all aspects of waking experience. Importantly, neither version requires that dreaming is deceptive: realizing that one is dreaming does not necessarily change its quasi-perceptual phenomenal character.

Because quasi-perceptual experience is a purely phenomenological description, saying that dreams are quasi-perceptual experiences is compatible with saying that they are nonetheless products of the imagination. As will become clear, the claim that dreams are imaginative experiences does not per se entail any phenomenological claims about their similarity with waking fantasy and daydreaming. Yet the ambiguity between phenomenological and nonphenomenological readings of the imagination view is a potential source of confusion. From now on, only those authors who claim that dreams are imaginative experiences in the phenomenological sense will be considered as espousing the imagination view of dreaming.

Framing the debate about dream experience as a debate about the quasi-perceptual versus imaginative phenomenal character of dreaming points to an inherent problem. Both views assume that the phenomenology of dreaming can be modeled on the phenomenology of waking consciousness, albeit in different ways. Indeed, this assumption often stands in the way of a detailed analysis of the phenomenology of dreaming in its own right. To the extent that it does, it potentially introduces oversimplification and what I call wake-state bias: the practice of projecting beliefs about waking experience onto dreaming without critical reflection or solid empirical evidence. An important claim defended over the following two chapters is that premature comparative statements about the phenomenology of dreaming and waking experience of the type built into the quasi-perceptual and the imagination views are best suspended.

Antiskepticism, combined with the conservative approach, makes much of the existing philosophical controversy surrounding both views appear misguided. What we need instead is a description of the phenomenology of dreaming that at least at the outset is neutral about its similarity with waking experience. This description can then function as a point of departure for developing a novel and more differentiated solution to the conceptualization problem. I will pursue this strategy over the subsequent chapters. For now, an important metatheoretical insight is that the conceptualization problem has to be solved *before* questions about the relationship between dreaming and quasi-perceptual and imaginative experience can be answered. Indeed, the conceptualization problem of dreaming can be regarded as a test case for theories describing this relationship (I return to this question in chap. 12). Beginning one's investigation of dreaming by modeling it on waking perception or imagination, by contrast, is putting the cart before the horse.

Finally, a note on my methodology in the following two chapters is in order. In my review of the literature, I contrast theoretical statements about dreaming from very different philosophical traditions. In doing so, I focus on the specific portrayal of dreaming by different philosophers rather than on its broader theoretical context. Importantly, many of the authors discussed here were not at all, or not primarily, interested in advancing a theory of dreaming in its own right, and it would be anachronistic to think they were. Rather, they pursued quite different theoretical aims. However, in discussing dreams, whatever their reasons for doing so, they often subscribed to background assumptions that are relevant for the conceptualization problem, and it is these implicit background assumptions that I am interested in here. By taking these arguments out of their original context and placing them in a new and slightly different one, I hope to illustrate developments in the history of ideas and interconnections between philosophy, psychology, and empirical research that would otherwise easily be overlooked. This comes at a price: I will not always be able to do justice to the individual theories I discuss, but rather aim to show, in fairly broad strokes, how the philosophical treatment of dreaming itself has developed and how it is related to scientific work on dreaming. In keeping with this aim, my choice of authors is selective rather than comprehensive,[1] with the aim of charting predominant historical changes and continuities, as well as introducing readers to different existing variants of the quasi-perceptual and the imagination views.

5.1 The Philosophical Literature

5.1.1 Antiquity

The traditional view of dreaming as quasi-perceptual experience can be traced back to Aristotle's (350 BCE/1997) *Parva Naturalia*, which presents the first detailed and systematic attempt to formulate a naturalistic theory of dream experience (Dreisbach, 2000;

Barbera, 2008; cf. sec. 2.1). Aristotle (350 BCE/1997, pp. 101–115) begins by discussing the relationship between waking and sleeping, which he describes as opposites defined, respectively, by the presence and absence of perception. He specifies that only the *actual* state of perception is absent during sleep—a point that later becomes decisive for his classification of dreaming. Importantly, the absence of perception in sleep is related not to any individual sensory organ but rather to the fact that the general perceptual faculty, which regulates the different sensory organs and is located in the area of the heart, has temporarily ceased to function. In this, sleep resembles other states of unconsciousness, though Aristotle does not specify exactly how these are distinguished from sleep.

In *On Dreams*, Aristotle asks whether dreams are caused by an affection of the faculty pertaining to intelligence or to sense perception. He quickly dismisses both possibilities: in dreams we can think about things that have nothing to do with the content of the ongoing dream, and if dreams were caused by the faculty of intelligence or mere opinion, we could not divert our attention from our dreams in this manner (Aristotle, 350 BCE/1994–2009a, Pt. 1). But because sleep is defined by the absence of actual perception, dreams also cannot be straightforwardly attributed to the faculty of perception: if so, we would be able to hear and see during sleep in the simple sense. Consequently the manner in which perception is involved in the generation of dreaming has to be one that does not put us in direct contact with mind-independent objects in our actual environment: this would be perceiving in the simple (or success) sense. Instead, "dreaming is an activity of the faculty of sense-perception, but belongs to this faculty qua presentative" (Aristotle, 350 BCE/1994–2009a, Pt. 1).

At this point, Aristotle could be read as saying that dreaming is a form of imagination or *phantasma* because it does not involve actual instances of perceiving. In his treatise *On the Soul*, however, Aristotle (350 BCE/1994–2009b) explains that though we can form separate ideas of the faculties of presentation and of sense perception, they are in fact identical. Apparently he takes this to be a fairly superficial conceptual distinction only, and one without any immediate consequences for the phenomenal character of dreaming. More generally, though the Greek *phantasma* (plural *phantasmata*) is often translated as "mental image," it seems to collapse the distinction between mental image and percept and perhaps should better be translated as "appearance" (Lycos, 1964) or "presentation" (Beare, 1906; Thomas, 2011). Several have argued that for Aristotle, *phantasmata*—which can refer not only to mental images but also to dreams or apparitions (i.e., the precursor to the newer concept of hallucination; Aleman & Larøi, 2008)—are involved in perception itself (for discussion and further references, see Thomas, 2011). This suggests that Aristotle's attribution of dreaming to the perceptual faculty qua presentative is not a statement about the phenomenology of dreaming.

Two further points seem particularly relevant. First, according to Aristotle, afterimages, hallucinations, and dreams arise when the movements of the sensory organs that are related to genuine perception persist after their external cause has disappeared: "The affection continues in the sensory organs, both in their deeper and in their more superficial parts, not merely while they are actually engaged in perceiving, but even after they have ceased to do so" (Aristotle, 350 BCE/1994–2009a, Pt. 2). The reason dreams or afterimages disqualify as instances of actual perception, then, is related not to a difference in the functioning of the sensory organs per se but rather to the way their movements are related to external stimuli in the subject's immediate environment. Dreams do not arise independently of such external stimuli but are their aftereffects. They are, metaphorically speaking, a kind of multisensory afterimage, a temporally delayed illusion.

Second, dreaming is phenomenally similar to waking perception and illusions precisely because it relies on the same kind of internal process: "So much at least is plain on all these points, viz. that the faculty by which, in waking hours, we are subject to illusion when affected by disease, is identical with that which produces illusory effects in sleep" (Aristotle, 350 BCE/1994–2009a, Pt. 1). Here, we see an early ancestor of the argument from brain activation (cf. sec. 3.1), or the assumption that wakelike brain activation leads to wakelike experience during sleep. Although Aristotle would not yet have put this in brain-related terms, the claim that the phenomenology of dreaming can be explained by appealing to internal processes that are deeply similar to those underlying actual instances of perception is implicit in his reasoning. For Aristotle, it is not just the phenomenology of dreaming that is quasi-perceptual.

Aristotle also thinks, however, that in sleep, the "simulatory movements" of the sensory organs can give rise to images of "even greater impressiveness" (Pt. 3) than in wakefulness, when they are obscured by competing movements caused by actual sensory stimulation. This also explains the deceptive character of dreams: a clear or healthy dream image "makes the dreamer think, owing to the effects borne in from the organ of sight, that he actually sees, and owing to those which come from the organ of hearing, that he really hears; and so on with those also which proceed from the other sensory organs" (Pt. 3). Aristotle explains that "as a rule, the governing sense affirms the report of each particular sense, unless another particular sense, more authoritative, makes a contradictory report" (Pt. 3). In sleep, when no such contradictory evidence comes from other sense organs, appearance is taken for reality.

Yet Aristotle does not take deception to be inevitable. In a passage often referred to as the first mention of lucid dreaming, he suggests that even if one realizes that one is dreaming and no longer mistakes the dream image for the real thing, the phenomenal character of the experience stays the same:

> If the sleeper perceives that he is asleep, and is conscious of the sleeping state during which the perception comes before his mind, it presents itself still, but something within him speaks to this effect: 'the image of Koriskos presents itself, but the real Koriskos is not present'; for often, when one is asleep, there is something in consciousness which declares that what then presents itself is

> but a dream. If, however, he is not aware of being asleep, there is nothing which will contradict the testimony of the bare presentation. (Aristotle, 350 BCE/1994–2009a, Pt. 3)

Note that in this passage, Aristotle does not specify what exactly causes lucid insight. In particular, he allows that dream imagery can fail to replicate its waking-life counterpart even without leading to lucidity, noting that "the dream presentation, though but little like it, appears as some actual given thing" (Aristotle, 350 BCE/1994–2009a, Pt. 3). In this view, dreams "possess verisimilitude after the manner of cloud-shapes, which in their rapid metamorphoses one compares now to human beings and a moment afterwards to centaurs" (Pt. 3). While dreaming subjectively feels like perceiving and continues to do so even when we realize that we are now dreaming, the things perceived in dreams do not necessarily resemble the perceptual objects encountered in wakefulness and may even be bizarre by waking standards. Here we have the beginnings of a highly differentiated description of the phenomenology of dreaming and of the relationship between quasi-perceptual dream imagery and waking perception, but also between dream deception and lucid insight. We also have a first indication that classifying dreams as quasi-perceptual experiences may not be the same as saying that dreams replicate waking experience in all of its detail. Indeed, according to Aristotle, the clarity of dream images and their verisimilitude vary with the strength of the movements of the sensory organs. Likening dream imagery to reflections in a pool of water, Aristotle notes that if these movements are sufficiently large, for instance, during fever or indigestion, no dream image whatsoever may appear, or only one that is distorted, whereas small movements, like a still pool of water, enable the formation of clear images.

Finally, it is interesting to note that Aristotle does not take just any presentation in sleep to be a dream. Instances of actual perception, even when they are feeble or remote, are not instances of dreaming, nor are the true thoughts that occasionally occur in lucid dreams or sleep behavior such as sleep talking: "The dream proper is a presentation based on the movement of sense impressions, when such presentation occurs during sleep, taking sleep in the strict sense of the term" (Aristotle, 350 BCE/1994–2009a, Pt. 3).

The view that dreaming shares the phenomenal character of waking perception was widespread in antiquity. The ancients, who still believed in the divine origin of dreaming and attributed dreaming to external causes, still spoke of "seeing" a dream (Barbera, 2008). A kind of objective reality was attributed to dreams, and the analogy between dreaming and perceiving was taken in a literal sense. Only around 500 BC did philosophers begin to consider dreams as subjective states arising from internal processes (Barbera, 2008). Democritus and Epicurus believed that dreams, like waking perception, are caused by external influences emanating from the surfaces of bodies and affecting the sensory organs. Others, like Plato and Aristotle, thought that dreams arise from the persistence of the internal movements caused by previous waking perception during sleep (for details, see Barbera, 2008). In the *Timaeus*, Plato writes:

> But when night comes on and the external and kindred fire departs, then the stream of vision is cut off; for going forth to an unlike element it is changed and extinguished, being no longer of one nature with the surrounding atmosphere which is now deprived of fire: and so the eye no longer sees, and we feel disposed to sleep. For when the eyelids, which the gods invented for the preservation of sight, are closed, they keep in the internal fire; and the power of the fire diffuses and equalizes the inward motions; when they are equalized, there is rest, and when the rest is profound, sleep comes over us scarce disturbed by dreams; but where the greater motions still remain, of whatever nature and in whatever locality, they engender corresponding visions in dreams, which are remembered by us when we are awake and in the external world. (Plato, 360 BCE/2009, 44d–45e)

Dreaming is also discussed in the *Theaetetus* as an example of sensory deception alongside illness and delusions, showing that perception cannot be equated with knowledge (Plato, 1996, 157d–158a). Anticipating Descartes's formulation in the *Meditations*, Theaetetus and Socrates agree that there are no certain marks to distinguish dreaming from wakefulness. Even their theoretical discussion of the possibility of dream deception could take place, in exactly the same manner, in a dream (Plato, 1996, 158b–158c). Essentially the same position is found in Cicero's (1994) *Lucullus*. Again, it is the subjective indistinguishability of dream and waking experience that is epistemologically worrisome. Hearing something in a dream, for instance, is no different from hearing something in wakefulness.

While this brief survey suggests a continuity between Greek and Roman philosophy regarding the phenomenal character of dreams, Roman philosophers generally placed a stronger emphasis on cognitive factors, which were thought to intermingle with residual perceptions during sleep, to explain the putative sources of dreaming. Dreams thus moved from being glimpses of a quasi-objective world of divine origin to being internal and subjective experiences, whose causes could nonetheless be linked to residual aftereffects of genuine perceptions or even to concurrent external or bodily influences during sleep, and finally also to the dreamer's anxieties and waking concerns (for discussion and further references, see Barbera, 2008). Dreams were seen either as the hallucinatory aftereffects of previous perceptions arising in the calm of sleep, as in Aristotle and Plato, or as illusory perceptions of contemporaneous stimuli during sleep.

5.1.2 Modern Philosophy

In modern philosophy, the quasi-perceptual view of dreaming was primarily linked to the epistemological problem of dream skepticism, as most prominently discussed by Descartes. In section 1.1, I pointed out that the crucial background assumption behind Descartes's discussion of dream skepticism in the *First Meditation* is that dreams not only are experiences but potentially replicate waking experience in all its detail. This is true both on the level of dream content (or what the dream is about, e.g., sitting at his desk by the fire) and on the phenomenological level of description. In particular,

Descartes emphasizes that his dream of sitting by the fire could have exactly the same distinct phenomenal quality as the corresponding waking experience:

> I am the same being who perceives, that is, who apprehends certain objects as by the organs of sense, since, in truth, I see light, hear a noise, and feel heat. But it will be said that these presentations are false, and that I am dreaming. Let it be so. At all events it is certain that I seem to see light, hear a noise, and feel heat; this cannot be false, and this is what in me is properly called perceiving (*sentire*), which is nothing else than thinking. (Descartes, 1641/1901, II.9)

Whether he is dreaming or not, he is having the same kind of conscious experience. Dreams cast doubt on his knowledge of the external world but also, to the same extent, show that he has indubitable knowledge of the kind of mental state that he is now undergoing.

Even Descartes's resolution of the problem of dream deception in the *Sixth Meditation* does not contradict the quasi-perceptual view of dreaming. Here he argues that because dreams are incoherent and lack connection to waking life, they can in fact be distinguished from waking experience. The difference between dreaming and waking is located on the levels of dream content and narrative coherence, but not on the phenomenological level of description.

Note also that the Aristotelian and the Cartesian characterizations of dreaming correspond to two different strengths of the quasi-perceptual view: whereas Aristotle only claimed that dreams involve quasi-perceptual experiences, Descartes (at least in the *First Meditation*) additionally claims that dreams typically or frequently replicate waking experience in all its detail. In the weaker Aristotelian view, the phenomenal character of dreaming qualifies as quasi-perceptual because its clarity *potentially* approaches that of waking experience and because it relies on the same proximal causes, namely, the internal movements of the sensory organs, as waking perception. Yet the verisimilitude of dream imagery is variable and can be quite different from waking experience. In the stronger Cartesian view, dreams confront us not only with a seemingly realistic though completely hallucinatory world but also with the hallucinatory experience of being a thinking and bodily self in it. The problem is not, as for Aristotle, that we fail to notice the discrepancy between dream imagery and its waking counterpart, but that dreams provide an exact replica of waking experience. For Descartes, dreams are deceptive precisely because they resist closer scrutiny; even when we inspect our current experience to determine whether we are dreaming or awake, the similarity between dreaming and wakefulness is such that we can find no certain marks to distinguish them (Descartes, 1641/1901, I.5). This assumption is both stronger and more specific than Aristotle's, and thus more vulnerable to empirical disconfirmation.

Another modern philosopher interested in the relationship between perception, dreaming, and imagination was Leibniz. Whereas Descartes was primarily troubled by the deceptive character of dreams, Leibniz offers a more positive treatment of the topic:

> There is one very remarkable thing in dreams, for which I believe no one can give a reason. It is the formation of visions by a spontaneous organization carried out in a moment—a formation more elegant than any which we can attain by much thought while awake. To the sleeper there often occur visions of great buildings which he has never seen, while it would be difficult for me, while awake, to form an idea of even the smallest house different from those I have seen, without a great amount of thought. . . . Even such unnatural things as flying men and innumerable other monstrosities can be pictured more skillfully than a waking person can do, except with much thought. They are sought by the waker; they offer themselves to the sleeper. There must therefore necessarily be some architectural and harmonious principle, I know not what, in our mind, which, when freed from separating ideas by judgment, turns to compounding them. (Leibniz, 1956, vol. 1, pp. 177–178)

While Leibniz appears to be suggesting here that dreaming is sharply distinguished from waking imagination both in its quasi-perceptual character and because it is beyond our control, he has Theophilus take a more continuous and moderate view in *New Essays on Human Understanding*. Theophilus allows that images in the different sensory modalities can also occur spontaneously during wakefulness, though we generally have much more control over them in wakefulness than during the dream state. We can at least stop their formation during wakefulness, though different people vary in their ability to do so. Unbidden imagery occurring during wakefulness is "like a magic lantern with which one can make figures appear on the wall by turning something on the inside" (Leibniz, 1704/2005, II.XXI.12). Because sleep generally obliterates our ability to control our thoughts and actions, Theophilus also takes up the question of whether we are morally responsible for our behavior during sleep:

> There is a great deal of difference between the actions of a drunk man and of a true and acknowledged sleepwalker. We punish drunkards because they could stay sober and may even retain some memory of the punishment while they are drunk. But a sleepwalker is less able to abstain from his nocturnal walk and from what he does during it. (Leibniz, 1704/2005, II.XXVII.22)

Apparently control offers not a sharp but only a gradual distinction between dreaming and imagining, but also between waking actions and sleep behavior. In the *New Essays*, Leibniz also explicitly endorses a gradual phenomenological distinction between imagining and perceiving, allowing that we can at least occasionally be deceived by waking imaginings, as well as by dreams: "although sensations are ordinarily livelier than imaginings, still we know that sometimes imaginative people are as much impressed by their imaginings as others are by the truth of things, and perhaps more so" (Leibniz, 1704/2005, IV.II.14). In refusing to sharply distinguish dreams and imaginings by their liveliness or vivacity, Leibniz is arguing not that dreams lack quasi-perceptual character but that both dreams and waking imaginings can occasionally be subjectively indistinguishable from genuine perception. For Leibniz, vivacity is only one criterion distinguishing imagining and perceiving, and it only works in standard cases. The same is true for control, which we can fail to have over waking imagery—and which we can occasionally have over dreams, as in lucid dreams, whose existence he explicitly acknowledges (cf. sec. 3.5). Leibniz's emphasis on the potential continuities between imagining and perceiving is important because, as we will see in chapter 6, the assumption of a sharp rather than a gradual distinction characterizes the rival imagination view.

Another important distinction, for Leibniz, consists in the complexity and internal coherence of waking experience (Leibniz, 1956, Vol. II, p. 603). Dreaming shows that "the truth about sensible things consists only in the linking together of phenomena, this linking (for which there must be a reason) being what distinguishes sensible things from dreams" (Leibniz, 1704/2005, IV.II.14). Because the vividness criterion fails so consistently for dreams, "we can distinguish dreams from sensations only because they aren't connected with sensations—they are like a separate world" (II.XIX.3). This remark commends him as a proponent of a strong version of the quasi-perceptual view: dreams create the experience of a separate world, and while they last, this world is subjectively indistinguishable from the real world. It is only from a third-person epistemological perspective (or rarely in lucid dreams) that we can realize the lack of connection and coherence of dreams and distinguish them from waking experience. Unlike Descartes, however, Leibniz was not troubled by the deceptive potential of dreams. As Hacking (2001) notes, Leibniz never entertained skepticism as a live option. We might even, according to Leibniz, be stuck in a coherent lifelong dream—but if we were, this would make no difference. Indeed, it seems that Leibniz was fascinated by the vivacity of dreams and our inability to control them. His remark that dream imagery exceeds our waking imaginative capabilities suggests that he also regarded dreams not so much as a state of uniform cognitive deficiency but as a state in which new cognitive capacities might become available. Unlike Aristotle or Descartes, Leibniz had a decidedly positive view of dreaming: to him, dreams were neither a topic for dullards (cf. sec. 2.1) nor a threat to our knowledge but spontaneously and even skillfully offered something actively sought during wakefulness.

A somewhat less enthusiastic position is suggested by Hobbes's remark, at the beginning of *Leviathan*, that "IMAGINATION is nothing but *decaying sense*; and is found in men, and many other living Creatures, as well sleeping as waking" (Hobbes, 1651/1985, p. 88). At first sight, this commends him as a proponent of the imagination view of dreaming. However, I argue that while Hobbes describes dreaming as a form of imagining, this does not commit him to the view that dreaming *feels* like waking imagining or daydreaming; rather, I suggest that he can be considered a proponent of the quasi-perceptual view.

To begin with, Hobbes uses the concepts of imagination and fancy quite liberally. While perception has external causes, the resulting experience of a sensation as being externally caused is a phenomenological characterization, and Hobbes (1651/1985, p. 86) remarks that "this *seeming*, or *fancy*, is that which men call *Sense*." Importantly, this phenomenological characterization is not unique to genuine cases of perception: "Their appearance to us is Fancy, the same waking, that dreaming" (p. 86). In its most general sense, fancy comprises both perceptual and imaginative states.

Concerning the causes of dreaming, Hobbes adopts the Aristotelian view that dreams arise from the "agitation of the inward parts of mans body; which inward parts, for the connexion they have with the Brayn, and other Organs, when they be distempered, do keep the same in motion" (Hobbes, 1651/1985, p. 90), even after the object that originally caused these motions has been removed. What, then, to make of his remark that dreams, "the imaginations of them that sleep" (p. 90), are a "decaying sense"? Elsewhere Hobbes suggests that this decay is mainly related to waking imagination and memory, not dreams:

> The decay of Sense in men waking, is not the decay of the motion made in sense; but an obscuring of it, in such manner, as the light of the Sun obscureth the light of the Starres; which starrs do no less exercise their vertue by which they are visible, in the day, than in the night. (Hobbes, 1651/1985, p. 88)

By contrast, in the dream state, this comparative obscuring is absent. Like stars, the full brightness of dreams is only perceptible in the calm of the night (Hobbes, 1651/1985, p. 88). Dreams rely on the same motions of the senses as waking imagination, but because they arise during the calm of sleep, the resulting experience is not the same as in wakefulness. Only in the dream state is the real brightness of the imagination fully visible. Dreams, to Hobbes, are an intensified form of waking imagination. He also emphasizes that the degree of decay is not uniform in wakefulness and is especially pronounced for memory. Though memory is an imaginative state, "when we would express the *decay*, and signifie that the Sense is fading, old, and past, it is called *Memory*" (p. 89). More vivid instances of imagination caused by strong sensory impressions are afterimages, as caused by gazing at the sun, or hypnagogic hallucinations experienced at sleep onset. For instance, "from being long and vehemently attent upon Geometricall Figures, a man shall in the dark, (though awake) have the Images of Lines, and Angles before his eyes" (p. 90).

Dream imagery is related to previous perception because like all imagination, dreams cannot arise unless they "have been before, either totally, or by parcells in the Sense" (Hobbes, 1651/1985, p. 90). In this respect, Hobbes's view is soundly empiricist. It is also naturalistic, in that he even refers to what is now called the sensory input blockade during REM sleep (Hobson et al., 2000): because the sensory organs "are so benummed in sleep, as not easily to be moved by the action of Externall Objects, there can happen in sleep, no Imagination . . . but what proceeds from the agitation of the inward parts of mans body" (Hobbes, 1651/1985, p. 90). Because dreams arise from the residues or aftereffects of waking perception, and "as there is no new object, which can master and obscure them with a more vigorous impression, a dream must needs be more cleare, in this silence of sense, than are our waking thoughts" (p. 90). Again, this phenomenological characterization is linked to the possibility of dream deception: in dreams, imaginations "appeare as if a man were waking" (p. 90). Yet Hobbes is not too worried about the deceptive quality of dreams and merely reiterates Descartes's remark (from the *Sixth Meditation*) that dreams are typically not about the same persons, places, objects, or actions experienced during wakefulness and rarely involve "so long a trayne of coherent thoughts" (p. 90).

> And because waking I often observe the absurdity of Dreames, but never dream of the absurdities of my waking Thoughts; I am well satisfied, that being awake, I know I dreame not; though when I dreame, I think my selfe awake. (Hobbes, 1651/1985, p. 90)

According to Hobbes, the phenomenological similarity between waking and dreaming only presents a problem "when by some accident we observe not that we slept" (Hobbes, 1651/1985, p. 91). This happens mainly to fearful and superstitious people, leading them to believe that they experienced visions or apparitions rather than mere dreams. Superstitious beliefs in "Satyres, Fawnes, nymphs and the like" (p. 92) can, he thinks, be traced to this type of retrospective dream deception. These somewhat deprecatory remarks add a new dimension to the epistemological problem of dreaming: dream deception, at least in certain people, can extend into wakefulness and can have practical rather than merely theoretical repercussions (cf. sec. 10.2.3).

While this suggests that dreams are the fullest expression, in a phenomenological sense, of the imagination, Hobbes's remarks on the causal pathways underlying dreaming and waking experience suggest otherwise. Hobbes notes that while sleeping in either a cold or a warm environment raises dreams of fear or anger, during wakefulness, the opposite occurs: anger causes heat in the body. On the one hand, this suggests that "our Dreams are the reverse of our waking Imaginations; The motion when we are awake, beginning at one end; and when we dream, at another" (Hobbes, 1651/1985, p. 91). On the other hand, this means that to Hobbes, dreams are not wholly dependent on the continued effects of previous sensations but can also be caused by contemporaneous ones.

Finally, the cognitive mechanisms involved in the production of dream narratives are also different from waking imagination: dreams are an example of an unguided and inconstant train of thought lacking design. Though this type of associative thinking can also occur in wakefulness in people who "are not onely without company, but also without care of any thing" (Hobbes, 1651/1985, p. 95), our waking train of thought is typically more constant and more strongly regulated by desire. Again, we find a tendency to associate dreams with the feebleminded or superstitious who have incomplete control over their thoughts.

Historically, the similarity between Hobbes's and Aristotle's views on dreaming is surprising given Hobbes's harsh critique of the Aristotelian doctrine of perception. A first difference is that Hobbes rejects the view that sensible qualities exist in the objects of perception. To Hobbes, sense is original fancy and hence distinguished from the qualities inherent in the objects of perception. Second and relatedly, Hobbes rejects the view that sensation is caused by a visible but immaterial species, "a *visible shew, apparition,* or *aspect,* or *a being seen*" (Hobbes, 1651/1985, p. 86) sent to the eye; Hobbes's theory of sensation is strictly mechanistic, involving immediate (as in touch) or mediate (as in seeing, hearing, and smelling) pressure sent via the nerves to the brain. Unlike late Aristotelian models of perception, the soul is not an independent causal factor in this process; sensation is explained by continued and counteracted movement, not by an actualization of the inner potentialities of the soul. Finally, Hobbes rejects the scholastic distinction between sense, imagination, and intellect. For him, these are not three distinct faculties: sensation, because it is not to be confused with the mind-independent properties of the objects of perception, is nothing but original fancy. Imagination does not mediate between sense and intellect, but rather fancy can refer to sensation as well as to the understanding, which is "the imagination that is raysed in a man . . . by words" (p. 93). Historically, it is interesting to note that these important differences between Hobbes's mechanistic account and the scholastic account of sensation and imagination (for details, see Gert, 2007; Leijenhorst, 2007) exist alongside palpable similarities between Aristotle's and Hobbes's accounts of dreaming.

Hobbes's discussion of dreaming is also notable because it foreshadowed important aspects of Berkeley's and Hume's views on the relationship between imagination and perception. Berkeley, though he had little to say about dreaming as such, perhaps took the idea of perception as fancy furthest, emphasizing that "our sensations, be they never so vivid and distinct, are nevertheless IDEAS, that is, they exist in the mind, or are perceived by it, as truly as the ideas of its own framing" (Berkeley, 1710, I.33). For Berkeley, the ideas of sense only have more reality because they are stronger, more orderly, and more coherent than the "creatures of the mind; but this is no argument they exist without the mind" (I.33). Dreams, "phrensies" (I.18), and the like show

> that IT IS POSSIBLE WE MIGHT BE AFFECTED WITH ALL THE IDEAS WE HAVE NOW, THOUGH THERE WERE NO BODIES EXISTING RESEMBLING THEM. Hence, it is evident the supposition of external bodies is not necessary for the producing our ideas; since it is granted they are produced sometimes, and might possibly be produced always in the same order, we see them in at present, without their concurrence. (Berkeley, 1710, I.18)

Within Berkeley's idealist framework, dreams are a paradigm example of our general predicament.

Hume's taxonomy of the mind distinguishes between impressions—including sensations, passions, and emotions, which enter into the mind "with most force and violence"—and ideas, or "the faint images of these in thinking and reasoning" (Hume,

1739/2000, 1.1.1.1). This dichotomous distinction between impressions and ideas at times colors Hume's view of the phenomenology of dreaming. In the first paragraph of *A Treatise of Human Nature*, he writes:

> In sleep, in a fever, in madness, or in any very violent emotions of the soul, our ideas may approach to our impressions: As on the other hand it sometimes happens, that our impressions are so faint and low, that we cannot distinguish them from ideas. (Hume, 1739/2000, 1.1.1.1)

This suggests an ambiguity in Hume's view. On the one hand, he seems to be saying that because ideas at best *approach* the vivacity of waking impressions, they also characteristically fall short. On the other hand, his subsequent remark on the occasional indistinguishability of impressions and ideas suggests that they are gradually rather than sharply distinguished by their vivacity. This ambiguity points to a deeper underlying problem: the concept of vivacity is as central to Hume's theory as it is problematic (for critical discussion, see Waxman, 1994; see also Ryle's famous and concise critique in Ryle, 1949/2000, chap. 8).

To begin with, not only impressions and ideas are distinguished by their vivacity, but also different types of ideas. For instance, memories and waking imagination "are only distinguished by the different *feeling* of the ideas they present" (Hume, 1739/2000, 1.1.3.5), with ideas of memory being stronger and livelier than those of fancy. While Hobbes took the opposite to be true, for Hume, the greater verisimilitude of memory apparently carried its own phenomenological signature. The same is true of beliefs, which for Hume are a form of sensation (1.4.1.5) and "consist in nothing, but the vivacity of an idea" (1.4.2.41). Writing about the vivacity of belief, Hume confesses that "I find a considerable difficulty in the case; and that even when I think I understand the subject perfectly, I am at a loss for terms to express my meaning" (1.3.7.7). He thus feels

> obliged to have recourse to every one's feeling, in order to give him a perfect notion of this operation of the mind. An idea assented to *feels* different from a fictitious idea, that the fancy alone presents to us: And this different feeling I endeavor to explain by calling it a superior *force*, or *vivacity*, or *solidity*, or *firmness*, or *steadiness*. This variety of terms, which may seem so unphilosophical, is intended only to express that act of the mind, which renders realities more present to us than fictions. (Hume, 1739/2000, 1.3.7.7)

Though the precise characterization of vivacity, according to Hume, is impossible, it clearly refers to a felt aspect of experience (Broughton, 2006, p. 45), and he trusts that "every one himself will readily perceive the difference between feeling and thinking" (Hume, 1739/2000, 1.1.1.1). Based on this reading, one might say that Hume's project is to distinguish mental states by their phenomenal character: conceptual distinctions, for instance, between impressions and ideas, but also between memory and imagination, or belief and conception, should map onto phenomenological distinctions. Yet several remarks suggest that he thought that vivacity, understood as a felt property, can vary *independently* of the dichotomous distinction between impressions and ideas. For instance, Hume writes:

> As an idea of the memory, by losing its force and vivacity, may degenerate to such a degree, as to be taken for an idea of the imagination; so on the other hand an idea of the imagination may acquire such a force and vivacity, as to pass for an idea of the memory, and counterfeit its effects on the belief and judgment. (Hume, 1739/2000, 1.3.5.6)

This suggests that the distinction between memory and imagination must rest on something other than vivacity: clearly, Hume is using these terms in a sense that is not only phenomenological. As Broughton (2006, p. 45) points out, the example of dreaming presents a problem for Hume precisely because it blurs the phenomenological distinction between impressions and ideas: if the two are really distinguished only by their vivacity as a felt aspect of experience, then there is no obvious reason why dreams should not count as impressions. Yet this is clearly not what Hume wants to say: his assertion that faint impressions may be phenomenologically indistinguishable from ideas and that memories and imaginings may be confused under certain conditions only makes sense if some further criterion exists for saying that they are nonetheless distinct. Similarly, Waxman notes that though the distinction between impressions and ideas by their vivacity is gradual, the instances in which we fail to distinguish them because of either the vividness of ideas or the faintness of impressions

> do not seem to be true exceptions. For, by presenting them as a failure or inability to *distinguish* idea from impression, Hume makes it seem as if they were merely cases of mistaken identification. . . . If so, the difference between impressions and ideas cannot be eliminated simply by an increase or diminution of vivacity; that is, never can *the idea itself* become an impression or *the impression itself* an idea. (Waxman, 1994, p. 30)

If this is correct, then Hume's taxonomy of the mind can only be understood by considering vivacity as referring to the phenomenal character of experience *independently* of the distinction between ideas and impressions—even though this contradicts Hume's own statement at the beginning of the *Treatise*. In this reading, Hume classifies dreams as ideas, alongside imaginings and memories, not because of their phenomenal character but because of their dependence on prior impressions. According to the famous copy principle (Hume, 1739/2000, 1.1.1.4–8; for discussion, see Broughton, 2006), simple impressions have temporal priority over simple ideas and simple ideas cannot be formed without the previous occurrence of corresponding simple impressions. This argument is closely related to Hume's rejection of innate ideas, the point being that the imagination cannot create wholly new ideas but rather depends on the materials provided by previous impressions (Hume, 1739/2000, 1.1.1.12; Waxman, 1994, p. 61). Thus, saying that dreams are ideas in this sense would not commit Hume to the position that there is a principled phenomenological distinction between ideas, including dreams, and impressions.

Yet the failure of vivacity to mark a clean distinction between impressions and ideas—to provide a clear-cut phenomenological signature for this conceptual distinction—presents a problem for Hume, because it threatens to leave him without *any* unambiguous criterion for a distinction he clearly wants to make. He is left with a dilemma. On the one hand, the distinction between impressions and ideas is motivated by common sense: "It is acknowledged that, in fact, many of these perceptions arise not from anything external, as in dreams, madness, and other diseases" (Hume, 1748/2004, p. 99). On the other hand, his adherence to empiricist principles and his reluctance to make any causal claims about the external sources of perception—which for him would be ad hoc and ungrounded—also potentially get him into trouble because the resulting position blocks him from saying that the commonsense distinction between impressions and ideas is theoretically justified. In the *Enquiry*, he writes:

> Here experience is, and must entirely be silent. The mind has never anything present to it but the perceptions, and cannot possibly reach any experience of their connexion with objects. The supposition of such a connexion is, therefore, without any foundation in reasoning. (Hume, 1748/2004, p. 99)

I propose that the uncomfortable classification of dreams as ideas epitomizes this tension between Hume's commonsense belief in an external world and his admission that this belief is essentially ungrounded. That impressions and ideas can typically be distinguished by their vivacity offers a welcome solution, but Hume is also honest enough to realize that this phenomenological distinction does not work all the way in, thus admitting that ideas (as in dreams) can approach the vivacity of impressions or indeed be indistinguishable from them. His insistence that, in such cases in which the phenomenological distinction is eroded, the conceptual distinction nonetheless holds is ultimately, I think, a concession to common sense. As such, it is intuitively appealing and theoretically unsatisfying in equal measures. This tension is also directly reflected in Hume's remarks on dream skepticism. Referring to Descartes, he writes:

> When he awakes from his dream, he will be the first to join in the laugh against himself, and to confess, that all his objections are mere amusement, and can have no other tendency than to show the whimsical condition of mankind, who must act and reason and believe; though they are not able, by their most diligent enquiry, to satisfy themselves concerning the foundation of these operations, or to remove the objections, which may be raised against them. (Hume, 1748/2004, p. 104)

While Hume's thinking was too deeply rooted in common sense for him to feel the pull of dream skepticism or even take it particularly seriously, he was also honest enough to admit that this instinctive dismissal lacked theoretical justification. What is fascinating about Hume's skepticism may well be this inherent ambiguity between epistemic modesty and common sense. "Sceptical doubt, both with respect to reason and the senses, is a malady, which can never be radically cur'd" (Hume, 1739/2000,

1.4.2.57); it is, in fact, the natural result of thinking about these problems in the first place. In this situation,

> carelessness and in-attention alone can afford us any remedy. For this reason, I rely entirely upon them; and take it for granted, whatever may be the reader's opinion at this present moment, that an hour hence he will be perswaded there is both an external and internal world. (Hume, 1739/2000, 1.4.2.57)

Hume, unlike Descartes, was resigned to, or perhaps even comfortable with, this residual uncertainty; and if I am correct, then this attitude is reflected not just in his discussion of skepticism but also in his taxonomy of the mind in general, and his treatment of dreaming in particular. At least the unruly phenomenal character of dreams did not prevent Hume from placing dreams alongside other ideas in another respect. In his discussion of the association of ideas, he notes that the train of ideas is never completely arbitrary but is governed by certain principles:

> And even in our wildest and most wandering reveries, nay in our very dreams, we shall find, if we reflect, that the imagination ran not altogether at adventures, but that there was still a connexion upheld among the different ideas, which succeeded each other. (Hume, 1748/2004, p. 13)

At least in this respect, then, order is imposed on dreaming.

In the eighteenth and nineteenth centuries, several philosophers discussed dreams in the context of visions, apparitional experiences, and superstition, thus continuing the tradition of Aristotle and Hobbes. Kant's remarks on dreams are sparse and mostly made in passing (for details, see Gehring, 2008, pp. 92–104), but nonetheless important because they foreshadow the contemporary comparison between dreaming and psychosis (cf. sec. 5.2.2). In *Essay on the Maladies of the Head*, Kant metaphorically notes that madmen are waking dreamers (Kant, 1988, p. 894). Similarly, In *Dreams of a Spirit-Seer*, he uses dreaming as a metaphor for those lost in the idiosyncratic worlds of their imagination (cf. Kant, 1766/2002, pp. 39–48). Here he also comments briefly on the phenomenology of genuine sleep dreams, which he takes to result from an intermingling of phantasy and external perceptions persisting during sleep (p. 34). In this view, dreams are part illusion, part hallucination, and likely they are too confused to allow for a clear identification of their external sources. While these dreams take place during half sleep, the dreams of deep sleep, according to Kant, are almost completely forgotten. He does, however, point out that the waking impression that such dreams involved only dark images does not license any inferences about their actual phenomenal character (p. 34). He even muses that experiences occurring during deep sleep may be clearer than waking perceptions; as in Aristotle's and Hobbes's theories of dreaming, the calm state of the soul during deep sleep and its isolation from external sensations may actually enhance the clarity of the experienced images.

In later writings, Kant suggests that dreaming has a function (Kant, 1833, §34). Dreams, for Kant, arise from a natural agitation of the inner organs during sleep. He even takes them to be a necessary accompaniment of sleep, quite dramatically suggesting that without dreams, sleep would be the same as death, and dreams have survival value. In this view, reports of dreamless sleep are ruled out; necessarily, we dream throughout sleep. For good measure, he even adds some supposedly autobiographical remarks, noting that often one dreams of getting stuck in the middle of a speech, or of wearing one's sleeping cap to a meeting instead of one's wig. Even these unpleasant side-effects of the sleep of reason are, it seems, worth their while: Kant is quick to point out that nightmares in particular ensure the circulation of blood throughout sleep.

Schopenhauer (1851/1988, pp. 227–310) follows Kant in linking dreams to phenomena such as apparitions, clairvoyance, second sight, prophetic dreams, and somnambulism, which he suggests all have their common origin in the "dream organ," but discusses dreams much more extensively. Though many of Schopenhauer's statements on apparitional experiences are obscure, his remarks on dreaming are nonetheless interesting because unlike Kant's early writings, they are not polemic but aspire to a seriously minded, brain-based theory of dreaming (cf. Gehring, 2008, pp. 137–142). Toward the beginning of his analysis of apparitional experiences, Schopenhauer rhetorically asks whether imagery generated by purely internal brain activity really is indistinguishable from sensory perception. Dreams, he argues, clearly show this to be the case: to consider dreams as mere phantasy or akin to waking imagination is "a lack of reflection, or of honesty: for it is evident that they are specifically different from these" (Schopenhauer, 1851/1988, p. 230; translation mine). Even the liveliest instances of waking imagination are, he claims, much more faint, incomplete, and elusive than the tangible reality presented by our dreams. Unlike waking imaginings, dreams effortlessly appear as something completely foreign, as an external reality, and often run counter to our expectations and wishes. He even claims that dreams have all the detail of waking perception, including the shadows cast by dream objects. Because of this, dream deception is so strong that when we wake, reality must fight to reassert itself (p. 230).

This suggests that Schopenhauer endorses a particularly strong and explicit version of the quasi-perceptual view: dreaming replicates waking experience in all its detail and subtlety. Accordingly, the only certain criterion distinguishing dreaming from wakefulness is the empirical criterion of awakening (Schopenhauer, 1859/1988, p. 48). Even this, however, is called into question by false awakenings, or realistic dreams that replicate our sleeping environment "as if our skull had been rendered transparent" (Schopenhauer, 1851/1988, p. 240; translation mine). Schopenhauer adds that in some false awakenings, the visual field is broadened to include experiences beyond one's bedroom and beyond physical obstacles such as walls, curtains, or blinds, allowing for a clear view of the courtyard, garden, or houses on the other side of the street. This is possible because the false awakening is not an instance of "physical seeing" (p. 241; translation mine); "yet it is a dream of what is really there . . . a form of perception through the dream organ" (p. 241; translation mine). Here, he seems to be saying that realistic false awakenings do not simply result from memories of one's current surroundings but involve a genuine (though unique) form of perception.

This assumption forms the connecting point between dreaming and clairvoyance. For Schopenhauer, all these phenomena are attributable to what he calls the dream organ. Dreams typically lack any direct relationship to the external world, but prophetic dreams or false awakenings, which he also calls "true dreams" (Schopenhauer, 1851/1988, p. 240; translation mine), are similar to somnambulist perception, clairvoyance, and sleepwalking in that they bear a direct and veridical relationship to the external world (p. 271). Schopenhauer's endorsement of the quasi-perceptual view, then, is strong not only because he claims that dreaming replicates the subtleties of waking perception on the phenomenological level of description, but also because he considers at least some dreams as genuine instances of perception in an epistemological sense.

These rather optimistic remarks do not prevent Schopenhauer from emphasizing the similarity between dreaming and madness. In both cases, he notes that the ability to coherently and reflectively connect ongoing experience to waking memories is compromised. Indeed, "dreaming can be considered as a short madness, and madness as a long dream" (Schopenhauer, 1851/1988, p. 232; translation mine). Yet, though dreams are often marked by poor judgments and a reduced power of reasoning (p. 238), all the other mental faculties are at least occasionally available during the dream state (pp. 32–33).

Regarding the sources of dreaming, Schopenhauer defends the naturalistic view that sleep-related changes in sensory processing and brain activity provide the conditions for dreaming to arise. In this view, dreams are caused neither by external sensory stimuli (the occasional integration of sounds or smells into dreams being exceptions occurring during light sleep) nor by the association of ideas; because deep sleep involves a quiescence of the brain, dreaming cannot, he thinks, be a product of the imagination (Schopenhauer, 1851/1988, pp. 232–233). Instead Schopenhauer adopts the Aristotelian view that dreams result from the echoes of organic impressions that are weak during wakefulness and during sleep once more excite the sensory activity of the brain. Importantly, though the source of excitation is different in dreaming and wakefulness, the brain "always speaks the same language: and it is in this language that it interprets these weak, internally generated impressions during sleep, just as it does the strong and determinate ones regularly coming from the outside during wakefulness" (p. 237; translation mine). The result is that dreaming involves the same type of experience as ordinary waking perception. The only difference is that the brain is now functioning in reverse:

> The dream organ is the same as the organ of waking consciousness and apprehension of the external world, with the difference that it is used, as it were, in reverse order and from the other

end; and the sensory nerves functioning in both can be activated either from the inside or from the outside. (Schopenhauer, 1851/1988, p. 251; translation mine)

Historically, we find Schopenhauer's analysis of dreaming poised between Aristotelian dream theory and contemporary brain-based accounts. On the one hand, Schopenhauer is clearly influenced by the Aristotelian view that dreaming results from the echoes of organic impressions that only in the quiet of sleep give rise to vivid and intense quasi-perceptual experience. The main difference is that Schopenhauer relocates the dream organ to the brain, whereas Aristotle located the central perceptual organ in the area of the heart. On the other hand, Schopenhauer's claim that dream imagery arises because during sleep, brain activity is interpreted as if it were caused by external sensations foreshadows a view held by contemporary dream researchers (e.g., Hobson, 1988; Hobson et al., 2000; cf. sec. 5.2.2) and thus lifts the argument from brain activation to the next level. Of course, Schopenhauer was not yet aware of brainstem-driven autoactivation of the sensory cortical areas during sleep (cf. sec. 2.2); but his remark that during dreams, the brain functions in reverse to waking perception—from the inside to the outside, rather than vice versa—comes close. It also echoes Hobbes's view of dreams as relying on causal connections that are reversed compared to waking imagination. Schopenhauer's discussion of dreaming thus nicely illustrates the metatheoretical claim that via some detours, and abstracting from occasionally quaint or outright odd and antiquated elaborations, deep continuities in philosophical theorizing, but also salient connections between the history of philosophical dream theories and contemporary dream research can be made visible.

5.1.3 Twentieth Century to Present

Moving into the twentieth century, we find, first, a number of authors interested in dreams from the perspective of philosophical phenomenology. In a series of lectures from 1904 and 1905, Husserl (2006) briefly discusses dreaming in the context of the imagination, noting that dreams, hallucinations, and illusions are often described as phantasy even though phenomenological analysis clearly suggests otherwise. To Husserl, phantasy and imagination are distinguished from waking perception because they do not involve the experience of presence and reality. In this view, there is something like a common core of phantasy, exhibited not just in its artistic expressions, and not only in deliberate cases, but also in spontaneous fantasy, memory, or expectation. This common core only becomes visible, as it were, once phantasy is distinguished from perceptual experience—including hallucinations, illusions, and, tellingly, dreams. While the latter are experienced *as* perceptions—they are, in the terminology introduced earlier, quasi-perceptual experiences—phantasy lacks this character: its objects are not experienced as present or as real in the same way. Dreams, alongside other quasi-perceptual experiences, are a limiting case for Husserl's discussion of phantasy, suggesting that the common, everyday use of the term is misleading to the extent that it fails to distinguish between phantasy and quasi-perceptual experience (Husserl, 2006, pp. 6–7).

This is not to say that to Husserl, the distinction between phantasy and quasi-perceptual experience is absolute. He later explains that we may become so absorbed in phantasies as to react to them behaviorally, for instance, by engaging in a conversation with an imaginary person, making a fist, and so on. It is noteworthy, however, that Husserl calls such cases of being completely immersed in our phantasies waking dreams. Here, as in visions or trance, the world of phantasy temporarily merges with the real world and may even give rise to corresponding beliefs and actions. Yet Husserl notes that this experiential blurring of phantasy and reality is temporary and highly unstable; waking dreams are quickly terminated by actual perception. Moreover, their blurring is not complete: "The world of reality *almost* disappears while we pursue these phantasies, but still subtly makes its presence felt, so that the phantasy is continuously colored by a slight consciousness of its illusory character" (Husserl, 2006, p. 44; translation mine). At the moment at which phantasy loses its imaginative experiential character altogether, it ceases to belong to the domain of phantasy proper, now requiring a different phenomenological description. Perhaps this is the reason why Husserl does not discuss dreams more extensively in his work on phantasy: phenomenological analysis shows that dreaming lies outside the target phenomenon he is interested in.

Other authors who analyze dreams from the perspective of Husserlian phenomenology follow suit. Globus notes that the experiential character of dreaming is such that "we perceive a life-world while dreaming and live a life in it," and "dreaming and waking life worlds are indiscernible as unreflectively lived" (Globus, 1987, p. 89). According to Uslar, the dream "*as* dream, while it is dreamt, is not dream but reality" (Uslar, 1964, p. 16; translation mine); it is an experienced reality involving a self—experienced, quite simply, as *me*—interacting with persons and objects in the dream (pp. 163–164). Similarly, Conrad (1968) views dreams as analogous to our waking experience of reality: in dreams, the self is the center of experience in much the same way as in wakefulness. Insofar as their experiential character is concerned, "the question of whether dream visions are perceptions or imaginations is not a problem at all. They are obviously perceptions" (Conrad, 1968, p. 62; translation mine). It is only the connection between conscious experience and the external world that is lacking (p. 57), and only in this broader, nonphenomenological sense is dreaming a form of imagining. To Conrad, as to Husserl, the transition between waking phantasy and genuine dreaming is gradual, with reverie (*Träumerei*, or dreamlike wake states) taking an intermediate position. Waking reverie is dreamlike, according to Conrad, in that it lacks the goal orientation of many phantasies and the overt split between the self controlling the phantasy and the imagined self. But reverie also lacks the strong sense of immersion that characterizes dreams; only in dreams are we entirely at the mercy of the phantasy world, seemingly (and dramatically) drowning in it (p. 59). Dreaming, according to Conrad, begins when the ability for control is lost. In this sense, dreams are a more perfect, more absolute form of reverie (p. 60). To these authors, then, a strong version of the quasi-perceptual view was so obvious that it did not require a separate argument. The consensus is that dreaming potentially replicates the waking experience of a self in a world; in addition, dreams are interesting for theories of the imagination precisely because they show where waking imagination falls short and thus help delineate the borders of what counts as waking imagination. Moreover, in the phenomenological tradition, we find that claims about the indistinguishability of dreaming and wakefulness, traditionally raised in an epistemological context, are now emphasized as claims about subjective experience in their own right. The motivation for philosophical interest in dreams has once more shifted.

Bergson (1914/2007) combines an analysis of dream experience with a theory of dream generation. His characterization of dream experience reiterates familiar points:

> A dream is this. I perceive objects and there is nothing there. I see men; I seem to speak to them and I hear what they answer; there is no one there and I have not spoken. It is all *as if* real things and real persons were there, then on waking all has disappeared, both persons and things. (Bergson, 1914/2007, p. 5)

Yet his argument adds something new, because he more thoroughly explores the illusory character of dreams than the other authors discussed so far. Dreams, according to Bergson, are linked both to external and internal sensations during sleep, predominately in the visual and auditory modalities, but also to real-body experiences. He likens dream imagery to the "ocular spectra," "colored spots," or "phosphenes" (Bergson, 1914/2007, p. 5) that one sometimes sees when one closes one's eyes during wakefulness. These, he writes, are "the principal material of which we shape our dreams, 'such stuff as dreams are made on'" (p. 5). In addition, there are external sensations such as sounds or the sensations of light that we can distinguish even when our eyes are closed. These external sensations are less precise in dreams than in wakefulness, but it would be false to say that we are completely cut off from them while dreaming. To Bergson, they are part of dreaming proper. Yet the internal and external sources of sensations in dreams do not explain why these vague images or sounds are integrated into a particular dream narrative the way that they are. Here memories are the additional ingredient, and

> among the phantom memories which aspire to fill themselves with color, with sonority, in short with materiality, the only ones that succeed are those which can assimilate themselves with the color-dust that we perceive, the external and internal sensations that we catch, etc., and which, besides, respond to the affective tone of our general sensibility. When this union is effected between the memory and the sensation, we have a dream. (Bergson, 1914/2007, p. 9)

Memory and affective tone function as a selection process, as it were, determining the precise way in which external stimuli shape our dreams. Moreover, Bergson notes that the union of sensation and memory is not unique to dreaming but "resembles the birth of all our perceptions. The mechanism of the dream is the same, in general, as that of normal perception" (Bergson, 1914/2007, p. 9). The phenomenological similarity between dreaming and perception thus is grounded not only in the fact that dreams integrate both internally and externally caused sensations but also in the fact that they both rely on the integration of sensory stimuli with memory and thus have a distinctly cognitive flavor:

> Thus, in the waking state and in the knowledge that we get of the real objects which surround us, an operation is continually going on which is of quite the same nature as that of the dream. We perceive merely a sketch of the object. This sketch appeals to the complete memory, and this complete memory, which by itself was either unconscious or simply in the thought state, profits by the occasion to come out. It is this kind of hallucination, inserted and fitted into a real frame, that we perceive. (Bergson, 1914/2007, p. 10)

In this view, dreams are interesting because, far from being completely isolated from the sleeping subject's environment, they shed light on the interplay between external and internal sources and thus on the internal, mnemonic constraints on sensory perception itself. Again, dreaming is presented as an argument for the view that perception can only be understood with reference to internal cognitive processes and indeed is a form of online hallucination, a virtual reality. Yet at the same time, dreams are not wholly internal but modulated by contemporaneous sensory inputs.

What, then, distinguishes dreaming from wakefulness? According to Bergson, it is not the abolition of the intellectual faculties. In the dream state, we become indifferent to but not incapable of logic, and though "the logic of a dreamer is feeble enough and often resembles a mere parody of logic" (Bergson, 1914/2007, p. 11), it is not completely or uniformly abolished. The same cognitive faculties exercised during waking are also exercised during dreaming, but they are in a state of relaxation during dreams, not in a state of tension, as in wakefulness.

> The dream consists of the entire mental life minus the tension, the effort and the bodily movement. We perceive still, we remember still, we reason still. All this can abound in the dream; for abundance, in the domain of the mind, does not mean effort. What requires an effort is the precision of adjustment. To connect the sound of a barking dog with the memory of a crowd that murmurs and shouts requires no effort. But in order that this sound should be perceived as the barking of a dog, a positive effort must be made. It is this force that the dreamer lacks. It is by that, and by that alone, that he is distinguished from the waking man. (Bergson, 1914/2007, p. 12)

This lack of tension explains the chief characteristics of dreaming, namely, incoherence, the distorted sense of duration, and the order in which memories present themselves to the dreamer. For Bergson, this effortless quality of dreaming, which for Leibniz was the source of the fascination and richness of dream imagery, is mainly a shortcoming. Yet, to understand the quasi-perceptual character of dreaming, one has to turn to the cognitive sources of dreaming.

In Russell's discussion of dreaming, though sparse and largely fragmented, we find a strong connection between the epistemological problem of dream deception and the consequences of dreams and hallucinations for theories of perception. Again, this is related to a strong phenomenological background assumption. Though we can never, according to Russell, conclusively rule out that we are now dreaming, the perhaps more important point is that in another way, dreams are not all deceptive. Echoing the familiar Cartesian point, Russell notes:

> In dreams I have all the experiences that I seem to have; it is only things outside my mind that are not as I believe them to be while I am dreaming. There are in fact no illusions of the senses, but only mistakes in interpreting sensational data as signs of things other than themselves. Or, to speak more exactly, there is no evidence that there are illusions of the senses. (Russell, 1948/2009, pp. 149–150)

More than thirty years earlier, in *The Problems of Philosophy*, Russell related dreams to sense data, noting that the sense data involved in dreaming and waking experience are the same, and it is of these—and only these—that we have primitive certainty (Russell, 1912/1999, p. 11). The attempt to secure the reality of physical objects over and above our sense data is futile: dreams show that sense data do not require the presence of mind-independent objects. Regarding the differences between dreaming and wakefulness, Russell reiterates the standard Cartesian argument, claiming that only dreams' inconsistency distinguishes them from waking life. At the same time, echoing Descartes's original worry, he allows that dreams at least potentially replicate waking experience: "This whole scene, with our inference, may occur in a dream" (Russell, 1914/2009, p. 77). For Russell, however, this means that dreams, like waking hallucinations and illusions, "are every bit as real as the objects of waking life" (p. 69); even in dreams, after all, the objects of sense have an "indubitable momentary reality" (p. 69). Dreams illustrate the indirectness of perception: even in wakefulness, our knowledge of the objects of perception—as well as of the minds of other people—is always inferential. Dreams and waking life "must be treated with equal respect; it is only by some reality not *merely* sensible that dreams can be condemned" (p. 69). In addition, Russell's neutral monism commits him to the metaphysical claim that "the ultimate constituents of matter are not atoms or electrons, but sensations" (Russell, 1921/2008, p. 83). Consequently no deep metaphysical difference distinguishes the perceptual objects encountered in dreams and in wakefulness. Yet this acknowledgment does not lead to skepticism: though "there is no logical impossibility in the supposition that the whole of life is a dream . . . there is no reason whatever to suppose that it is true" (Russell, 1912/1999, p. 13). On the one hand, common sense offers a much better, or lovelier, explanation than the dream hypothesis, enabling us to reject the view that we are caught in a lifelong dream (sec. 4.1). On the other hand, the consideration of quasi-perceptual experience, including dreams, leads to a profoundly altered conception of perception and indeed of reality itself: the immediate objects of perception are sense data, and physical objects are logical constructions out of a series of possible perceptions, no more and no less.

Though Malcolm's work on dreaming focused the philosophical discussion for most of the second half of the twentieth century on the question of whether dreams are experiences at all (cf. chap. 1), it is noteworthy that the rift between pre- and post-Malcolmian philosophical treatments of dreaming is nowhere near as large as one might think. In particular, some of the main themes of earlier philosophical theorizing have analogues in the contemporary discussion on dreaming. Fast-forward to Antti Revonsuo (1995, 2005, 2006), quite possibly the strongest contemporary proponent of the quasi-perceptual view and at the same time the philosopher to have discussed dreams most extensively from the perspective of philosophy of mind—not only in the twentieth century but likely in general. In his work, we find the trend of moving away from the epistemological problem of dream skepticism (Revonsuo, 2006, pp. 122–125) toward an experiential analysis of dreaming in its own right continued. Yet this shift in interest leads to a characterization of dreaming that is entirely compatible with the phenomenological background assumptions underlying the classical epistemological project. In particular, Revonsuo claims that "there is nothing in the experience itself, in the actual qualitative character of the experience, that necessarily distinguishes the dream experience from a corresponding perceptual experience in the waking state" (p. 82): even more so, "the qualities of dream experience are identical with the qualities of waking experience" (p. 84) in that dreams contain the full range of multimodal sensory qualities found in wakefulness. He acknowledges that the frequency with which these qualities occur in dreams differs from wakefulness, but emphasizes that their structural organization is the same: both dreaming and wakefulness involve the experience of a dream world, which is taken as the real world and has a representation of the dreamer at its center.

Importantly, Revonsuo derives a new strategy from this familiar phenomenological characterization: that of using dreaming as a theoretical and research model of consciousness. Because "dreaming is nothing but consciousness—or to be more precise, the macro-level of consciousness—at its barest, operating on its own" (Revonsuo, 2006, p. 45), dreaming reveals that consciousness has a basic and state-independent structure, namely, the experience of a self in a world. Consequently, because dreaming exemplifies pure consciousness, it can be used as a model of standard waking consciousness. Dreaming is thus treated not as a special type of conscious state but as an exemplar of conscious experience itself. Dreaming consciousness and waking consciousness differ with respect to the causal paths of production, but otherwise "the ontology of dreams is the ontology of consciousness" (p. 46). Moreover, because dreams, according to the virtual reality metaphor, can be seen as offline simulations of waking experience, consciousness itself is essentially simulational: "Not only are dreams experiences but, in a way, all experiences are dreams" (p. 55). Specifically, dreaming reveals presence and full immersion (p. 119), or the conscious experience of being a self in a world, to be universal characteristics of consciousness as such. This view not only echoes Bergson's remarks on the essentially hallucinatory character of waking perception and Hobbes's and Berkeley's description of perception as fancy but also is reminiscent of phenomenologists working on dreams in the Husserlian tradition who emphasized that dreaming is essentially world experience. Revonsuo also evokes the Cartesian vocabulary of clarity and distinctness:

> The dreaming brain brings out the phenomenal level of organization in a clear and distinct form. Dreaming is phenomenality pure and simple, untouched by external physical stimulation or behavioural activity. Furthermore, dreaming depicts consciousness first and foremost as a subjective world-*for-me*. (Revonsuo, 2006, p. 75)

Revonsuo's view of dreaming as pure consciousness thus involves the most explicit and far-reaching answer to the conceptualization problem so far, in that he not only defends a particularly strong version of the quasi-perceptual view but also discusses the consequences of this view for an interdisciplinary theory of consciousness more generally. In this view, the near equivalence of phenomenal experience in dreaming and wakefulness not only suggests a conceptual framework for describing dream experience—namely, in terms analogous to standard wakefulness and perception—but also supports a metatheoretical claim about the relevance of theoretical and empirical work on dreaming to consciousness research.

It is noteworthy, however, that the suggestion of using dreaming as a research model for consciousness research is not new. And it is even more noteworthy that a similar proposal was suggested in a context that is diametrically opposed to Revonsuo's position of biological realism, or the claim that consciousness is "a real biological phenomenon in the brain" (Revonsuo, 2006, p. xvii): namely, in Patricia Churchland's reductive or outright eliminative materialist account of conscious experience. In her 1988 book chapter "Reductionism and the Neurobiological Basis of Consciousness," Churchland leaves open the question of whether a reductionist account will ultimately turn out to be possible or whether an elimination of the folk-psychological concept of conscious experience will be necessary. Either way, she suggests that research on dreaming and REM sleep may provide a research model for making progress on consciousness: "If there is a domain relevant to consciousness which has sufficient supporting infrastructure and surrounding theory to enable experimental discovery, it is the sleep-dream-awake cycle" (Churchland, 1988, p. 290). She gives three reasons: one, the existence of robust introspective and phenomenological differences between wakefulness, dreaming, and dreamless sleep; two, the existence of behavioral criteria for the objective identification of the states of waking, dreaming, and sleep, enabling the use of animal models; three, progress on the identification of a neuronal generator producing shifts between these states (here, and throughout the chapter, she refers mainly to Hobson's work). Importantly, this suggestion is fueled not only by an optimistic assessment of the status of research on sleep–wake states but also by the view that dream experience is not just contrasted with dreamless sleep but also marked by characteristic differences compared to standard wakefulness. In particular, and contrary to Revonsuo, Churchland (1988, p. 286) suggests that dream bizarreness presents a powerful argument for claiming that consciousness, rather than constituting a single, unified, or natural kind, lacks a homogeneous underlying organizing principle. In fact, she suggests that a careful consideration of dreaming may lead to a diversification of the target phenomenon, casting doubt on "pre-scientific intuitions concerning what it is to be conscious" (p. 310). She briefly returns to this claim in more recent work, once more noting that changes in consciousness across the sleep–wake cycle are central to the "contemporary mind/brain problem" (Churchland, 2005, p. 285), in part because "being awake, paying attention, explicitly remembering, emerging from anesthesia, smelling mint, regaining awareness following an absence seizure, regaining awareness following a complex partial seizure, being aroused from sedation, dreaming and hallucinating, are, quite possibly, different ways of being conscious" (p. 287).

This is not to say that the different characterizations of the phenomenology of dreaming suggested by Revonsuo and Churchland are mutually exclusive, and both could admit that similarities and differences coexist alongside each other. Perhaps they are merely focusing on different aspects of dreaming: where Churchland is driven by the interest of diversifying consciousness as an explanatory target, Revonsuo's work is motivated by the search for a unifying principle. Yet a deeper problem is that dreaming lends itself to contradictory research projects such as Revonsuo's thesis that the phenomenal level of organization can actually be found in the brain and Churchland's eliminativism about the folk-psychological concept of consciousness. This points to a deeper metatheoretical problem: without a unified framework for describing conscious experience in dreams, dreaming can apparently be used as a kind of wild card, motivating very different research projects.

Karlsson's (2001) philosophical dissertation on dream experience suggests a strategy for adopting a quasi-perceptual view of dreaming while drawing a phenomenological distinction between dreaming and waking perception. Karlsson (2001, pp. 68–83) argues that the selective deactivation of the DLPFC during REM sleep and associated cognitive deficiencies (cf. sec. 3.2) make a phenomenological difference to quasi-perceptual experience in dreaming as compared to wakefulness. In particular, he notes that the wide content of visual awareness, which extends beyond the current visual field to previous visual experience and anticipation of future visual experience, is largely lacking in the dream state. Contrary to Bergson, Karlsson claims that while dreams involve quasi-perceptual visual experiences in the narrow sense, the cognitive contributions to perception, such as memory and expectations, are lacking.

> It is part of the traditional view that the dream is experienced as a *world*, a world that is taken in through perceptual acts. Now, without wide visual awareness, this world would shrink considerably. It would only consist of what is presently within the visual field of the dreamer. So, whereas the dream as remembered upon waking up may strike us as a coherent world in which I, the dreamer, was moving around, getting involved in various types of situations, it might thus be that the dream was not *experienced* as such a world, but rather as a series of more or less isolated scenes. This would explain why we, as dreamers, seldom react when discontinuities occur and the dream world becomes incoherent. (Karlsson, 2001, p. 80)

He concludes that "being-in-the-dream is not quite like being-in-the-world; there is at least one aspect of perceptual awareness missing in dream experience" (Karlsson, 2001, p. 84). This view is also interesting for another reason: it suggests that the more fine-grained and differentiated a view one takes on the phenomenology of dreaming, the more this view will be restricted to certain types of dreams. To the extent that lucid dreams differ from nonlucid ones in the cognitive capacities involved and their neural correlates (see sec. 3.5), they would necessitate a separate account; if Karlsson is right, the phenomenology of quasi-perceptual experience varies not only between dreaming and wakefulness but also across different types of dreams. Finally, where Bergson emphasizes the cognitive contribution to quasi-perceptual experience in dreaming and wakefulness and indeed takes waking memory to be the crucial ingredient for explaining the process of dream formation, Karlsson argues that this cognitive-mnemonic dimension of experience is what distinguishes waking perception from dreaming.

Metzinger's (2003, 2009, 2013c; see also Windt & Metzinger, 2007) work on dreaming, developed in the broader context of the self-model theory of subjectivity, is yet another example of a complex position on the phenomenal character of dreaming that allows for a differentiated analysis of different types of dreams. Unlike the philosophers discussed so far, Metzinger considers dreaming as a contrast condition or research model not for waking imagination or for consciousness in general but for theories of self-consciousness in particular. His key claim is that the phenomenology of selfhood, of being or having a self, is impoverished in dreams as compared to standard wake states. Metzinger echoes the familiar point that "dreams are conscious because they create the appearance of a world" (Metzinger, 2009, p. 135). However, he spells this out in technical terms, according to which dreaming involves the activation of a global world-model within a window of presence where this model is experienced not as a model but as an actuality (Metzinger, 2003, p. 251): the world-model experienced in dreams is phenomenally transparent.[2] Lucid dreams are distinguished from nonlucid ones, but also from standard wake states, because this quality of phenomenal transparency is almost completely lost: the world-model, including the self-model, is now experienced *as* a model. Importantly, in this view, the onset of lucidity is related not just to a purely intellectual insight but to a more thoroughgoing change in the phenomenal character of the overall experience. Indeed, according to Metzinger, lucid dreams can metaphorically be described as "one big *thought*" (p. 538).

This view has interesting consequences for questions regarding the quasi-perceptual versus imaginative character of dreaming. First, while the description of dreaming as involving the appearance of a world suggests a basic similarity between dreaming and waking experience, other constraints that Metzinger uses to characterize standard wake states are not or only partially satisfied in dreams. Others are not fully applicable to dreams at all. For instance, Metzinger (2003, p. 259) notes that questions concerning the fine-grained phenomenal character of nonlucid dreams, such as questions about the intensity or homogeneity of phenomenal content, may be unanswerable owing to the widespread lack of high-level attention in the dream state and dream amnesia. He also argues that dreams are less stable and more dynamic than waking experience because of their hyperassociative and bizarre character. For the same reason, the holistic and internally coherent character of the overall experience is expressed to a weaker degree in the dream state.

Second, the aspects of waking experience that are most severely compromised in the dream state are related to self-consciousness. The cognitive deficiencies that characterize nonlucid dreams and poor dream recall after awakening prevent the integration of dreams with the autobiographical self-model (Windt & Metzinger, 2007) but also contribute to the metacognitive deficit, or the failure of dreamers to realize that they are currently dreaming. Metzinger also argues that while dreamers can act on desires or impulses, the phenomenal-functional property of agency is not fully expressed in dreams (Metzinger, 2003, p. 254): dreamers are not agents in the stronger sense of being able to experience themselves as deliberately directing their attention, thoughts, or actions. Relatedly, the first-person perspective is only weakly instantiated and notoriously unstable: dreamers experience themselves as the center of the dream world but are unable to form a stable conscious model (or what Metzinger [2003, pp. 411–426] calls a "phenomenal model of the intentionality relation") of their epistemic or agentive relation to this world. If they could, they would realize that they were dreaming and would become lucid.

The emerging picture of dreams as phenomenal experiences distinguished from standard wakefulness through a number of multilevel constraints allows for a fine-grained and highly differentiated analysis. Moreover, different types of dreams (e.g., lucid and nonlucid dreams, but also prelucid dreams and false awakenings) can be distinguished by their different levels of constraint satisfaction (Windt & Metzinger, 2007). In Metzinger's view, dreams fail, in particular, to replicate one particular aspect of standard wakefulness: the cognitive deficiency of nonlucid dreams means that while dreams involve the experience of a world, they are only subjective states in a conceptually weak and uninteresting sense. Yet the degree to which a phenomenal self is present in dreams is variable (Windt & Metzinger, 2007), and lucid dreams once more involve a wakelike self. In understanding dream experience, the explanatory target has shifted, then, from the dream world to the dream self.

Finally, like Revonsuo's claim that waking experience itself is dreamlike or virtual, the self-model theory emphasizes the simulational character of conscious experience, including waking perception: conscious experience relies on internal representational mechanisms, and in this very general sense even standard wake states are appearance, fancy, or, as Metzinger puts it, "online hallucinations" (Metzinger, 2003, p. 51). The central claim of the self-model theory, after all, is that there is no such thing as a self in the sense of a substance or an entity that is continuous over time. The self is a simulation, a model that we mistake, in experience, for the real thing. Dreams merely illustrate the concept of a global simulational state, involving a simulation of a world-model and of a self in it, in a particularly clear manner. Yet Metzinger also provides the conceptual resources for describing the differences between dreaming, standard waking perception, hallucinations, and imaginative wake states. From a third-person epistemological perspective, dreams are better compared, in his view, to daydreams and waking hallucinations because they lack the close degree of external stimulus correlation characteristic of perception: dreams are offline simulations. Phenomenologically, this fact is not available during dreams or hallucinations: both feel like genuine perceptions, even though they fall short from a third-person epistemological perspective. This phenomenal character also distinguishes dreaming from waking imagination: both are offline simulations, but waking imagination is also experienced as such. Daydreams and deliberate mental imagery, but also thought, are "phenomenal simulations" (Metzinger, 2003, p. 58), not phenomenal realities. Again, as in the classical literature, we find that dreaming cuts across the common distinction between perception and imagination: it is both offline simulation and experienced reality.

In sum, in the contemporary philosophical literature, dreaming is increasingly regarded as a contrast condition, potentially enabling insights about imagination and perception, but also about consciousness and the self. At the same time, the metatheoretical assessment of dreaming has changed considerably. Far from only being a threat to knowledge, as in Cartesian epistemology, the theoretical discussion of dreaming is increasingly viewed as a potentially fruitful tool for generating new insights about the structure of consciousness, but also about the nature of the self. This use of dreaming as a test case for a taxonomy of the mind was already evident in modern philosophy, for instance, in Hume's distinction between impressions and ideas. But it only developed into a full-blown, interdisciplinary research strategy in the late twentieth and early twenty-first century. Yet we can also see continuities in the philosophical discussion of dreaming, reaching from Aristotelian dream theory via Descartes to contemporary theories. In particular, the kinship between dreams, hallucinations, and illusions, which lies at the heart of the quasi-perceptual view, is often taken to support the cognitive deficiency view of dreaming. Where dreams, for Descartes, posed a threat to knowledge, we find newer authors increasingly focusing on the absence of a thinking or even knowing self in dreams. With few exceptions (i.e., Leibniz and Bergson), the quasi-perceptual view of dreaming is complemented by the view that dreaming is a state of cognitive deficiency perhaps even analogous to pathological wake states. In this tradition, dreams are defined as much by their similarity to waking perception as by their failure to replicate waking thought.

5.2 The Empirical and Psychological Literature

5.2.1 The Phenomenology of Dreaming and Its Comparison with Waking Perception

In assessing the empirical plausibility of the quasi-perceptual view of dreaming, three types of evidence are particularly relevant. These are evidence from dream reports, findings on the putative neural correlates of dreaming, and external stimulus incorporation in dreams. Report-based studies investigating the phenomenology of dream imagery are a particularly obvious place to start.

To begin with, the formal analysis of dream reports promoted by Allan Hobson (1988; cf. sec. 2.2) shows that a majority of dream reports contain references to visual imagery, followed by auditory and movement sensations. By contrast, reports of touch, taste, smell, thermal, and pain sensations are exceedingly rare in dreams. For these rare types of imagery, it is initially unclear whether their low frequency in dream reports reflects their rare occurrence in dreams or rather is due to forgetting and underreporting. Antiskepticism, combined with the conservative approach (cf. sec. 4.4), suggests that this question can be answered, for instance, by systematically using affirmative probes: if the frequency, for instance, of touch sensations remains low when participants are explicitly asked to report such sensations immediately after awakening, this would strengthen the hypothesis that they occur only rarely in dreams. If, on the other hand, the frequency with which rare sensations are reported is considerably higher under such circumstances, this would indicate that they occur more frequently in dreams than suggested by the analysis of free dream reports.

Based on the existing evidence, a preliminary assumption is that the variability and different frequencies of modality-specific imagery in dream reports reflect robust differences in their occurrence in dreams. This is supported by the fact that the different frequencies with which modality-specific types of imagery occur in dream reports are confirmed by both contemporary (Strauch & Meier, 1996; McCarley & Hoffman, 1981) and historical (Calkins, 1893; Weed et al., 1896; Monroe, 1905) studies, suggesting that this is a general and highly stable feature of dream phenomenology (Schwartz, 2000).

While this tells us something about the distribution of multimodal imagery in dreams, two further questions are not addressed by these results. The first is whether this distribution is similar to or different from the frequency of modality-specific imagery in wakefulness. To support a strong version of the quasi-perceptual view according to which dreaming typically replicates waking experience, one would have to show that the same distribution of modality-specific sensory experiences is found in reports of standard wake states. The second question is how exactly modality-specific types of imagery are experienced in dreams.

Taking the second question first, a strategy for investigating the quality of modality-specific imagery in dreams is to adjust the detail, specificity, and expressive granularity of dream reports (cf. chaps. 3 and 4). Rechtschaffen's (1983) study, in which subjects were presented with photographs with different degrees of color saturation, brightness, and visual clarity and asked to choose the ones that best resembled their dreams, offers a particularly good example. In this study, visual dream imagery obtained higher scores of visual clarity following awakenings from phasic REM sleep than following awakenings from tonic REM sleep; other measures such as color saturation and brightness showed no changes. In a later study using a similar setup, Rechtschaffen and Buchignani (1992) found that 40.1 percent of dreams were judged to resemble normal photographs, 20.2 percent were judged to resemble achromatic photographs, and 11.2 percent were judged to resemble photographs with moderate color desaturation. They did not find a correlation between visual measures of dream imagery and self-judgments of how well the dream was recalled, suggesting that varying levels of recall were not related to variations in visual imagery in dreams. They concluded that "the single most frequently reported category of dream image was one that visually resembles external reality. . . . The most frequent departures from a realistic appearance were a loss of color saturation and a loss of background detail" (Rechtschaffen & Buchignani, 1992, p. 150). Similarly, Antrobus et al. (1987) found that the brightness of dream imagery approximated that of waking perception more strongly than the clarity of focus; rates for both measures were highest during wakefulness, slightly lower following REM sleep, and lowest following NREM sleep awakenings (see also Antrobus, 1990; Antrobus & Wamsley, 2009). Variations in the intensity of dreaming have also been found within REM dream reports, with different measures of intensity such as vividness of imagery and emotional intensity increasing throughout individual REM periods (Kramer et al., 1974, 1975; Kramer, 2007) and throughout the night (Foulkes, 1966; Verdone, 1965).

What about the comparative analysis of dreaming and waking perception? To begin with, note that to frame questions about the experiential character of dreaming by asking about its quasi-perceptual as opposed to imaginative character is to phrase the question in comparative terms from the beginning. A variety of background assumptions are at play here. In particular, this formulation presumes that dreams, waking perception, hallucinations and illusions (in the quasi-perceptual view), or imaginings and dreams (in the imagination view) are mental states of a common phenomenological kind, for instance, because they converge on some identifiable, homogeneous, and highly invariant experiential property rather than being umbrella terms for loosely related but diverse kinds of experiences. The attempt to apply a unified label such as quasi-perceptual experience to perception, hallucinations, illusions, and dreams, however, may be oversimplified: without a detailed account of what it is to be a quasi-perceptual experience, this label is almost completely empty. Given the apparent variability even of visual dream imagery (and possibly a detailed investigation of other modality-specific types of imagery would paint an even more differentiated picture), an alternative might be to substitute coarse-grained questions about the quasi-perceptual versus imaginative nature of dream imagery with a number of more fine-grained ones concerning color saturation, clarity, brightness, and so on. Perhaps dreams more strongly resemble visual perception in some dimensions, and waking imagination in others. Variations between REM and NREM dreams, as well as between phasic and tonic REM dreams and early versus late-night REM periods, should also be taken into account. And, as suggested by the debate about dream color (see chap. 4), even variations *within* these dimensions are possible: perhaps some dreams are colored, while others are black-and-white or mixed; and perhaps dream color even varies within REM periods, early night REM periods, and so on. If so, making progress on the question of dream experience may mean giving up on—or at least suspending—the philosophical question of whether dreams are quasi-perceptual experiences altogether. Though quasi-perceptual experience is a helpful label for structuring the philosophical debate about dreaming, it may be too coarse-grained to be useful for the project of developing an empirically informed conceptual framework of dream experience. A somewhat disturbing metatheoretical conclusion would be that this might be a case in which the conceptual granularity of (large parts of) the philosophical debate lags behind that of scientific dream research.

Moreover, even a detailed study comparing visual dream imagery to photographs can only be used to support comparative statements about dreaming and wakefulness if one assumes that visual experience in wakefulness resembles, for instance, a high-resolution color photograph. A more careful approach—especially given philosophical doubts about the appropriate characterization of visual experience in wakefulness (cf. Noë, 2002, and the other contributions in the same collection)—is to regard this as an open question requiring further research. Even if the phenomenology, say, of visual dreaming had been rigorously investigated, it would be premature, without an equally rigorous investigation of reports of waking visual experience gathered under the same conditions and using the same measures, to make any comparative statements about dreaming and waking perception. At least we can now say what would answer such questions: for instance, we might envision researchers asking subjects to compare their visual experiences, as well as their visual imaginings and their dreams, to photographs over the sleep–wake cycle. Similar comparative studies would then have to be conducted for the other sensory modalities as well, for instance, using audio recordings, and so on. At least theoretically, this could be done.

Indeed, this may be a promising (though demanding) research strategy; but as we have seen, this is not the way the quasi-perceptual view is commonly portrayed in the philosophical literature. Typically, the philosophical variant of the quasi-perceptual view has the status not of a research methodology or of an open question but of an intuitively obvious and commonly accepted fact. To be fair, it might be somewhat anachronistic to demand scientific tractability of historical theories of dreaming. Still, the standard philosophical way of thinking about dreams carries with it a problematic heritage. The attempt to describe the phenomenology of dreaming by analogizing it to that of waking perception without further investigation does little more than project uninvestigated background assumptions about waking perception onto dreaming. Given that questions can be raised about the phenomenology of both dreaming and waking perception, this is to put the cart before the horse. Consequently, the quasi-perceptual view, understood as a phenomenological claim, is too vague and unspecific to be of much use. To the extent that it is not coupled with a rigorous research strategy, it amounts to little more than an intuitive and unexamined characterization of dreaming. Because of the variability of dreaming (e.g., between phasic and tonic REM sleep), it is likely oversimplified and incapable of accounting for the spectrum of different types of experiences in dreams. It also glosses over a number of more interesting and fine-grained questions, for instance, regarding clarity, focus, and color saturation in dreams. Moreover, because this is nonetheless the standard way of thinking about dreaming, at least in the history of philosophy, it carries with it a potentially misleading air of authority and commonsense acceptance. We will have to be careful, then, to separate the useful elements of this philosophical view from excess baggage.

An alternative to contrasting modality-specific types of imagery in dreams and wakefulness would be to suspend, for the time being, premature comparative statements. One could ask, for instance, whether some general phenomenological description of dreaming is available, perhaps even one that picks out an invariant phenomenological property that underlies variation in the distribution and quality of modality-specific types of dream imagery. If so, the claim that dreams form a common phenomenological kind would have changed from an implicit background assumption to an empirically informed description of what it is actually like to dream. Moreover, such an account of dreaming would form a more secure basis for a contrastive analysis of dreaming and wakefulness (Windt & Noreika, 2011). Indeed, this is exactly the strategy I will pursue over the following chapters. It will require refocusing the debate from the description of dream imagery to the relation between the dream self and the experienced dream world.

For now, note that attempts to assess the empirical plausibility of the quasi-perceptual view are hampered by yet a further problem. As suggested in section 4.4, there may be something like a maximum resolution of dream phenomenology. Nielsen recently reviewed existing findings in support of the "reality simulation perspective" of dreaming, noting that "the task of differentiating dreaming from other forms of imaginative cognition . . . is largely incomplete" (Nielsen, 2010c, p. 595). While the core assumption of the reality simulation perspective is that "dreaming consists of a convincing simulation of waking reality experience" (p. 595), an inherent problem, as he points out, is that existing measures of dream content and bizarreness "only remotely address the possibility that dreaming might simulate details of reality apprehension that fall beyond the introspective and reporting capacities of experimental subjects" (p. 601). Nielsen explains:

> Numerous forms of subtle perceptual activity contribute to this spatially and temporally coherent here-and-now experience of reality while awake: all actions linked to perceptual search, such as looking, listening, touching, smelling, and so forth, are of this type. Less-obvious activities are the ubiquitous orienting reaction to novel stimuli, the appreciation of gravity and other orientational information, the background awareness of posture and kinesthesia, and the sense of proprioceptive feedback. These perceptual activities operate largely outside of focal awareness and are typically subjugated to the central contents of consciousness—usually what is perceived to be seen, heard, felt, smelled, or tasted in the outer world. The background perceptual activities are nonetheless a constant contributor to the complex contour of subjective awareness. (Nielsen, 2010c, p. 601)

In particular, he notes that dream content analysis is insufficient to determine the frequency of such perceptual activities in dreaming, suggesting the need for methodological refinement:

> Only probe questions about subtle perceptual activity that are directed at these specific junctures in the dream—and preferably administered to subjects while they are giving the report, not afterwards—could determine definitively if such activities were, in fact, represented in the background of dream experience. (Nielsen, 2010c, p. 601)

A further possibility for making progress might be to use "subjects who possess superior skills in self-reflection and communication," as they "may be likely to provide more accurate, more detailed reports of reality simulation information of this subtler kind" (Nielsen, 2010c, p. 602). Of course, even such probe-based measures would likely have a maximum resolution—and an interesting point for the comparative study of dreaming and wakefulness is that this maximum resolution may not be the same for dreaming and wakefulness. If subjects were able to describe subtle aspects of waking experience but systematically reported being unable to do so for dreaming because they simply could not remember, this might suggest that certain successors to the question of whether dreaming involves quasi-perceptual experience would have to remain unanswered.

5.2.2 The Neural Correlates of Dream Imagery: Imaging Studies, Lesion Studies, and the Analogy between Dreams and Hallucinations

Given these difficulties in investigating whether dreaming and waking perception form a common phenomenological kind, a better and more precise strategy might be to ask whether the quasi-perceptual character of dreaming can be assessed by investigating the neural correlates of dream imagery. As discussed in section 3.2, the visual areas are highly activated during REM sleep, fitting in well with descriptions of vivid, predominately visual imagery following REM sleep awakenings. Perhaps a fairly coarse-grained phenomenological description suffices to make underlying similarities in cortical activation patterns visible. This would also be in keeping with the argument from brain activation, which again, as we have seen, has a long philosophical history.

Probably the most elaborate statement of a functional equivalence between dreaming and wakefulness, and one that is closely related to Revonsuo's analysis of dreaming, is suggested by Llinás and colleagues (Llinás & Paré, 1991; Llinás & Ribary, 1994). Llinás and Paré (1991, pp. 526–527) regard reentrant thalamocortical activity in the range of 40 Hz as a probable functional basis of consciousness and argue that in REM sleep, this thalamocortical loop mimics the reality-emulating activity of waking perception, thereby generating the "world analogues" experienced in dreams (p. 114). Waking perception, in this view, is an "oneiric-like state modulated by the senses" (Llinás & Ribary, 1994), suggesting that conscious experience is essentially dreamlike, with external sensory input only exerting a modulatory influence on waking perception. They present two main arguments for this view. The first builds on the general similarity in neurophysiological processing in wakefulness and dreaming. The second links changes in the phenomenology of dreaming to pathological conditions such as unilateral neglect and prosopagnosia. Note that while neither condition is related to a loss of vision per se, both can be described as involving an inability to *see*, not just to imagine, certain parts of the visual field or faces. Because both conditions result in a partial loss of the phenomenology of seeing, and because they are complemented by a corresponding change in dreaming, this lends some support to the quasi-perceptual view.

The quasi-perceptual character of dreaming is also sometimes taken to be supported by the implication of the so-called PGO waves in REM sleep dreaming (cf. sec. 2.2), which result in heightened activity in the oculomotor, vestibular, and visual brain regions (Hobson et al., 2000). Hobson has argued that rather than being entirely random, the complex activation patterns associated with PGO waves have a high degree of spatial specificity, and there is a close correspondence between cellular activity at the level of the brain stem and the direction of rapid eye movements (Hobson, 1988, p. 207). If one further assumes that rapid eye movements are related to visual experience within the dream, this seems to suggest that dream imagery itself is determined by brain-stem-driven activation patterns. Indeed, Hobson has argued that the multimodal sensory quality of dreams arises because during REM sleep, internally generated signals are interpreted as if they were external stimuli affecting the sensory organs:

> If the higher-level neurons of the visual system are subjected to the same type of phasic excitatory signal that they "see" during the wake state, they will process that signal as if it came from the outside world. Our cortical neurons read the signals as visual inputs. . . . In REM sleep, the brain has no choice but to interpret its internally generated signals in terms of its previous experience with the outside world. (Hobson, 1988, pp. 210–212)

The structure of the argument is importantly similar to the one suggested by Schopenhauer (1851/1988, p. 237): the brain speaks the same language in dreaming and in wakefulness and therefore interprets internally and externally generated stimuli in the same manner.

Horikawa and colleagues' (2013a) suggestion of a "principle of perceptual equivalence," according to which perception and dreaming share a common neural substrate (cf. sec. 3.7), is another example of this view. In particular, it suggests that by investigating cortical activation patterns observed during visual stimulus presentation in wakefulness, reports of visual dream experiences can be predicted from neuroimaging data gathered during the preceding sleep stages. Yet because this study targeted not REM sleep but sleep onset, the generalizability to REM sleep dreams is unclear. Also, without a systematic comparison with waking imagination, it is not clear that this study offers clear-cut evidence against the imagination model of dreaming. If, for instance, researchers were able to use the analysis of brain activity during waking stimulus presentation to predict daydream reports with a similar accuracy as dream reports, the experiment would not be able to differentiate between the two rival hypotheses. Indeed, in one study, researchers were able to predict object perception from neuroimaging data gathered during object imagination, so this does not seem implausible (Cichy et al., 2012; for discussion, see Haynes, 2015). Hence a proponent of the imagination view could argue that the predictions were successful because there is a sufficient degree of overlap in the neural activity underlying imagination and perception to begin with (cf. sec. 6.1.3).

Further reason for caution stems from the fact that evidence of the correspondence between rapid eye movements and voluntary gaze shifts during the dream state continues to be controversial. Lucid dreamers can control the direction of real-eye movements by performing previously determined gaze shift patterns within their lucid dreams, thus lending preliminary support to the scanning hypothesis (LaBerge, 1988b, 1990; cf. sec. 3.5). Still, this finding may not be generalizable to nonlucid dreams, and it also does not exclude the possibility that REMs might nonetheless be quite different from eye movements involved in standard waking perception. Based on a comparison of eye movements in REM sleep and wakefulness, Aserinsky and colleagues (1985) present evidence suggesting that REMs are significantly slower than comparable waking saccades, and argue that these differences cannot wholly be explained by the effect of closed eyelids during REM sleep or their occurrence in darkness. And while Doricchi et al. (2007) found evidence for a correspondence between dreaming and waking eye movements in a patient with unilateral spatial neglect, they also found evidence for the existence of morphologically distinct and dissociable types of REMs, possibly requiring a more differentiated approach to the analysis of REMs in future studies. The need for a more differentiated analysis is also suggested by Herman (2000), who notes that to a considerable degree, REMs are the result of low-level orienting activity rather than of voluntary scanning. A strong version of the scanning hypothesis assuming an exact correspondence between REMs and visual dream imagery is likely oversimplified.

Despite these differences, one might still argue, as suggested by LaBerge and Zimbardo (2000), that the eye movements involved in scanning visual dream imagery are more similar to perception than imagination. However, others have shown that scanning activity involved in mental imagery is itself similar to waking perception, at least when the imagery is dynamic rather than static and experienced as being localized externally rather than internally (Antrobus & Singer, 1964; Marks, 1972; Zikmund, 1972; Borst & Kosslyn, 2008). Finally, one study (Sprenger et al., 2010) compared the velocity of eye movements during REM sleep to waking imagery tasks performed by subjects with eyes open and closed. They found that there is a general similarity between eye movements in REM sleep and waking imagination with eyes closed, and REMs are similar to eye movements during semantic imagination with eyes open, as well as to eye movements during the exploration of visible stimuli. A cautious conclusion would be to say that questions concerning the relationship between REMs and voluntary scanning, but also the relationship between eye movements in dreams, waking perception, and imagination, require more research, with all three types of eye movements overlapping in some respects. Yet the available evidence on REMs allows for no unequivocal argument in favor of the imagination or the quasi-perceptual view.

What about neuroimaging studies? First, note that the terminology used in these studies differs. While Braun et al. (1998) use the neutral term *imagery* to refer to dreaming, Dang-Vu et al. (2007, p. 103) relate the same imaging results to the occurrence of visual *percepts* and "hallucinoid imagery" (Dang-Vu et al., 2009, p. 34) in dreams, suggesting that they take dreaming to have a quasi-perceptual phenomenal character. Hobson also typically describes dreams as involving either internally caused *perception* (Hobson, 2009, p. 304) or primarily visual, motor, and auditory *hallucinations* (Hobson et al., 2000).

Second and more importantly, a more detailed look at regional activation patterns during REM sleep suggests that the evidence from neuroimaging is not as clear-cut as one might wish. In particular, it seems that similarity between dreaming and wakefulness only applies to activation patterns in cortical areas implicated in higher portions of the visual hierarchy, that is, the visual association areas in the extrastriate cortex. By contrast, the primary visual (striate) cortex is deactivated during REM sleep and functionally dissociated from higher visual areas (Braun et al., 1998). Because frontal regions are also selectively deactivated during REM sleep, Braun et al. suggest that "during REM sleep, the extrastriate cortices and paralimbic areas to which they project may be operating as a closed system, functionally disconnected from frontal regions in which the highest order integration of visual information takes place" (Braun et al., 1998, p. 94). Consequently, evidence suggests that during REM sleep, the closed system formed by the extrastriate cortices and paralimbic areas is dissociated from both environmental stimuli (mediated by visual areas V1 and V2, which are comparatively deactivated in REM sleep) and prefrontal areas implicated in highest-order integration:

> REM sleep may represent a state in which the brain engineers selective activation of an interoceptive network, which is dissociated from primary sensory and heteromodal association areas at either end of the visual hierarchy that mediate interactions with the external world. (Braun et al., 1998, p. 94)

Given that most neuroimaging studies have not, so far, attempted to relate measures of brain activity to dream reports gathered from the same sleep stage, a cautious interpretation is that they suggest that visual dream imagery, whatever its phenomenal character, is associated with *different* patterns of brain activation than waking perception. The lack of activation in the primary visual areas suggests a plausible explanation for the functional isolation of dreaming from external inputs (though we will see, in the next section, that this part of the story is also complex). The question for the quasi-perceptual view, then, is whether the comparative deactivation of these areas makes a phenomenal difference, as well. If so, then given that the primary visual areas are implicated in visual perception, this would seem to speak *against* the quasi-perceptual view of dreaming; if not, it would strengthen the view that dreams are quasi-perceptual experiences. Either way, because the phenomenal character of dreaming is precisely the matter that is open to debate, neuroimaging data from REM sleep alone cannot help decide the question.

The argumentation in the paper by Braun et al. (1998, p. 92) is interesting in this context. The authors argue that part of the reason for investigating neural activation patterns in the visual areas during REM sleep is precisely that researchers do not know whether imagery can be evoked at higher levels of the visual hierarchy or rather requires the contribution of early visual areas. They interpret their results as supporting the former view: neuroimaging shows that the primary visual areas are deactivated in REM sleep, and independent studies show that "intense visual imagery-laden dream activity is most prevalent" (p. 92) during REM sleep. They conclude that the primary visual areas do not appear to be necessary for vivid visual imagery. However, note that this conclusion assumes that the phenomenological similarity between dreaming and perceiving (rather than imagining) is an established fact.

The situation is further complicated by the continuing controversy surrounding the role of the primary visual areas (V1) in visual awareness (for a recent discussion, see Haynes, 2015). In 2000, Lamme and colleagues concluded their review of the relevant literature by claiming that neither cell activity in V1 nor V1 synchrony reflects visual perception and hence cannot be regarded as a correlate of visual awareness. However, the more interesting question is whether V1 activation is nonetheless necessary for quasi-perceptual visual experience. Arguably, this cannot be determined by experiments involving the presentation of visual stimuli to human or animal subjects, as this necessarily introduces a confound between external stimulus processing and internally generated, quasi-perceptual experience. For this reason, such studies can only investigate the role of V1 in standard waking perception but remain silent about dreaming and hallucinations. As Tong observes, "One possible interpretation is that V1 is necessary for normal aspects of visual awareness but not for atypical forms such as hallucinations, visual auras, dreams and visual imagery" (Tong, 2003, p. 226).

More recently, a number of studies using transcranial magnetic stimulation (TMS) have been taken to suggest that recurrent processing in early visual areas (V1 and V2) is necessary for conscious visual experience (Koivisto et al., 2010, 2011; Salminen-Vaparanta et al., 2012). However, contradicting these findings, ffytche and Zeki (2011) present evidence from three patients with lesions to the primary visual cortex who report preserved (though degraded) visual experience in their so-called blind fields. Aside from giving verbal reports describing their visual experiences (and ones that correlated significantly with the actual presence of stimuli), the patients also produced drawings. The authors conclude that "V1 is not necessary for a crude but conscious visual awareness of visual occurrences in a blind field, and hence neither are the return projections to it" (ffytche & Zeki, 2011, p. 248). So while it is possible that early visual areas contribute to the phenomenal quality of standard visual experience, it is not clear that they are strictly necessary even for the awareness of external stimuli. Methodologically, this means that dreaming itself may be an important contrast condition for determining the contributions of the primary visual areas to visual *experience* (as opposed to external stimulus processing); but as we have seen, rather than solving the question of whether dreams are quasi-perceptual experiences, this would presume that the quasi-perceptual character of dreaming had already been established.

Given the difficulties in comparing dreaming and waking perception, could evidence from waking hallucinations tip the balance in favor of the quasi-perceptual view? The analogy between dreaming and hallucinating gains initial support from Hobson's (1999, 2002; Hobson & Voss, 2011) theory that the formal features of dreaming are equivalent to psychotic wake states. In this view, the formal features of dreaming, such as hallucinations, strong emotions, bizarreness, and delusions, are comparable to organic psychosis or delirium. The phenomenological similarity between the "normal delirium" of dreaming and abnormal or pathological delirium is grounded in the fact that these states "always have . . . the same kind of cause. . . . The common cause of normal and abnormal delirium is a sudden shift in balance of brain chemicals" (Hobson, 1999, p. 62).

This hypothesized analogy between dreaming and psychotic wake states is supported by several publications by Claude Gottesmann (1999, 2002, 2006, 2007). In a review article, Gottesmann (2002) suggests that the disinhibition of dopaminergic neurons in REM sleep, in particular in the nucleus accumbens, may explain the hallucinatory and delusional quality of both dreaming and waking psychosis. At the same time, the near-complete absence of the negative symptoms of schizophrenia during dreaming may be related to the increase of dopamine influence in the prefrontal cortex. Hobson's and Gottesmann's hypothesis relies predominately on the neurophysiological and neurochemical similarities between REM sleep and schizophrenia. Importantly, they do not question the phenomenological similarity between dreaming and waking hallucinations but rather attempt to explain both dreams and waking hallucinations in common neurophysiological and neurochemical terms.

While this is a compelling strategy, the analogy between dreaming and psychosis should, again, be treated with caution (Windt & Noreika, 2011). One reason is that there are several obvious phenomenological differences between schizophrenia, which is characterized by a predominance of auditory verbal hallucinations (Aleman & Larøi, 2008, pp. 48–49), and dreaming, where visual imagery is the predominate modality. As Noreika (2011) has argued, assessing the continuity, but also the discontinuity, between dreaming and the waking experiences of schizophrenic patients may require a more differentiated, multidimensional approach and should remain open to the possibility that some aspects may vary independently of each other. And again, this raises the familiar problem that attempts to model dreaming on psychotic wake states—or vice versa—may be oversimplified and premature.

Rather than comparing dreaming to psychotic wake states and schizophrenic hallucinations, another interesting perspective is to compare dreaming to hallucinations more generally. Nielsen (2010b) notes not only that different medical conditions are accompanied by an increase in the subjective intensity of dream imagery and emotions, sometimes even described as surpassing their waking counterparts, but that changes in dreaming are also often associated with waking hallucinations. For instance, alterations in dreaming associated with RBD, including vivid dreams, nightmares, and dream-enactment behavior, are more prevalent in patients with Parkinson's disease who experience waking hallucinations than in those who do not, suggesting that in these patients, REM-sleep-related mechanisms may be associated both with sleep dreams and with waking hallucinations. Similarly, intensive care unit patients often experience not only vivid dreams but also delusions, disorientation, and hallucinations. Thus, comparing changes in dreaming with different types of hallucinations may make additional evidence on phenomenological and neurophysiological similarities available.

If one looks at the details, however, the situation quickly becomes much more complex. In particular, hallucinations themselves are highly heterogeneous, spanning different sensory modalities (for excellent introductions, see Slade & Bentall, 1988; Aleman & Larøi, 2008; Blom & Sommer, 2012; for an overview of functional neuroimaging of hallucinations, see Allen & Modinos, 2012; Aleman & Vercammen, 2012; for an overview of neurophysiological research, see van Lutterfeld & Ford, 2012). This is even true if one concentrates on modality-specific hallucinations. Visual hallucinations, for instance, have a highly heterogeneous phenomenology, ranging from simple (e.g., phosphenes) to complex and fully formed hallucinations of people, animals, or objects (with geometrical hallucinations of shapes and patterns being somewhere in between); and from partial (where one aspect, such as color, is hallucinatory) to more global hallucinations (such as panoramic hallucinations of entire sceneries; see ffytche et al., 2010; Collerton et al., 2012).

As is the case for dreaming, it is not clear that hallucinations (as treated in the clinical-psychological literature; for discussion, see sec. 8.4) fall safely on the perceptual rather than on the imaginative side of the spectrum. Indeed, a complex interaction between bottom-up and top-down mechanisms may contribute to the formation of hallucinations. While certain hallucinations, as in Charles Bonnet syndrome (ffytche et al., 2010), can be related to sensory deficits, others are perhaps better explained by top-down factors such as attention and perceptual expectations. Understanding the underlying mechanisms giving rise to waking hallucinations may require an integrative model taking different factors into account, ranging from lower-level, bottom-up processes (including release phenomena due to lesions in sensory pathways or irritative processes affecting the integration of perceptual information in cortical areas, as in epilepsy) to higher-level, top-down processes, such as attentional deficits or disturbed source/self-monitoring (for details and discussion, see Aleman & Larøi, 2008, chap. 7). Emotions, expectations, and memories of previous episodes may also contribute to hallucinations (Collerton et al., 2012). As ffytche (2013) suggests, a family of related theories may be required to do justice to the heterogeneity of hallucinations. And whatever the answer regarding the lower-level perceptual as opposed to higher-level cognitive sources of hallucinations, there would still be room for asking whether all these different types of hallucinations share a quasi-perceptual phenomenology, or whether some are better compared to imagination and daydreaming on the phenomenological level of description.

At least some of these uncertainties regarding hallucinations may be related to the fact that generally, integrative approaches spanning a wide range of visual disorders involving hallucinations rather than focusing on specific psychiatric or neurological disorders are lacking (ffytche et al., 2010). ffytche (2007) attempts to provide such a broad overview of visual disorders by suggesting a neurophenomenological classification system. Importantly, the assumption that a careful phenomenological description is a first step toward constructing an integrated model of hallucinations is implicit in his reasoning:

> Why add further complication to an already complex field? The answer lies in the possibility that the neurophenomenological approach, and syndromes derived from it, reveal features hidden in our traditional taxonomies by bringing us closer to the underlying pathophysiology. The emerging neurobiological insights may ultimately fulfil the early 20th-century aspiration that specific hallucination phenomenology points to a specific etiology. (ffytche, 2007, pp. 186–187)

According to his model, veridical perceptions, hallucinations, illusions, and dreams are similar on three dimensions, namely, in their experienced vividness, external localization, and the lack of agentive control, whereas imagery during wakefulness is typically described as less vivid, located internally or "in the mind's eye," and under the subject's control. In this view, dreams can be classified, tentatively, as forming a common neurophenomenological kind with both hallucinations and perception (cf. Macpherson, 2013). ffytche's model is an important first step toward capturing salient differences between visual hallucinations and perception on the one hand and imagination on the other. Yet it is important to note that it relies on a highly stereotyped view of dreaming. Indeed, ffytche does not provide a detailed phenomenological description of dreaming in its own right. Hence his proposal of grouping dreams alongside hallucinations and perception is better regarded as a promise, an invitation to systematically compare dreams and hallucinations associated with various visual disorders on the phenomenological and neurophysiological levels of description. Moreover, in a more recent publication, he devotes more attention to the specific question of whether dreams are hallucinations (ffytche, 2014). He concludes, quite cautiously, that at least visual dream imagery counts as visual hallucination—although he denies that the opposite is true, namely that all visual hallucinations are dreamlike. He also emphasizes that a subgroup of hallucinations may arise from misinterpreted imagery and that whether this is also true of dreaming is unclear. It thus seems that hallucinations across different psychiatric and neurological disorders are almost as poorly understood as dreams, suggesting that they cannot provide any quick-and-easy contrast conditions for equally heterogeneous dreams.

Finally, comparing dreams and waking hallucinations also involves a conceptual difficulty: saying that dreams and hallucinations are both quasi-perceptual experiences is not the same as saying that dreams *themselves* are hallucinations. Because the concept of hallucination is often used in a clinical context, some authors explicitly suggest that it should be confined to wakefulness, arguing that classifying dreams as hallucinations might ultimately be more misleading than helpful (Aleman & Larøi, 2008, p. 17). Certainly there are differences. For instance, subjects undergoing hallucinations still retain awareness of their environment and interact with others, while the same is not true for dreams, and hallucinations also have a much stronger impact on everyday functioning. Perhaps this is a reason for preferring the neutral concept of quasi-perceptual experience over that of hallucination for the discussion of dreaming. At the same time, applying the concept of hallucination to dreaming can help in specifying the stimulus source of dream imagery—that is, whether it depends wholly on internal stimuli, as suggested by the philosophical concept of hallucination, or on a distorted perception of external stimuli, as in illusions. At least, in describing dreams as hallucinatory, one should be careful to distinguish one's use of the term from its more narrow clinical use.

Given these uncertainties, the best strategy may be to leave comparative statements aside altogether and focus, for the time being, on the neural correlates of dream imagery itself. One type of study seems particularly relevant and targets, once more, visual dream imagery. Jakobson, Fitzgerald, and Conduit (2012a) recently aimed to investigate the neural correlates of dream imagery by applying transcranial direct current stimulation (tDCS) over the frontal and right posterior parietal cortex during stage 2 sleep. They contrasted different stimulation conditions (high and low intensity) with sleep stages without stimulation; subjects were then awakened and asked to report any images and thoughts they had experienced and to rate their vividness on a scale from one to ten. EEG was obtained to ascertain that sleep following stimulation was not disturbed. Results indicated that tDCS over the posterior parietal (anode) and frontal (cathode) cortex during stage 2 sleep led to a significant increase in the frequency of visual dream reports as judged by independent raters who were blind to the experimental conditions. However, a follow-up experiment found no differences in the frequency of imagery reports following two different stimulation conditions, suggesting that the effect of tDCS might be nonspecific and independent of electrode position (Jakobson, Fitzgerald, & Conduit, 2012b). tDCS during REM sleep also had no significant effect on reports of dream imagery (Jakobson, Conduit, & Fitzgerald, 2012). A possible interpretation, suggested by the authors, is that cortical activity and associated visual imagery are harder to manipulate during REM sleep than during stage 2 sleep.

Although the existing results are highly preliminary, this type of approach might be particularly helpful, because it could allow researchers to move beyond merely correlative statements toward an investigation of the causal contributions of specific brain regions to dream imagery (Karim, 2010; Noreika, Windt, Lenggenhager, & Karim, 2010). If so, this would be an important step toward developing a solid basis for the comparative study of dreaming, hallucinations, and waking perception. Yet it seems that researchers are only taking the very first steps toward achieving this goal.

5.2.3 External Stimulus Incorporation in Dreams

Perhaps the most direct way to investigate the quasi-perceptual view is to study the integration of external stimuli in dreams. A first type of evidence comes from experimentally induced sensory stimulation during sleep. This type of approach has a long history. Researchers have long suspected that dreams are caused by external stimuli. In 1830, Abercrombie reported the case of a man in whom dreams could be induced by whispering into his ear (cited in Woods & Greenhouse, 1974), and in 1892, Ladd suggested that visual dream imagery is caused by the retinal *Eigenlicht.* While early dream researchers mostly experimented with vestibular stimulation during sleep (e.g., by using a rotating chair) or by fixating the sleeper's body in a certain position (Hoff, 1929; Hoff & Plötzl, 1937; for further references and discussion, see Schönhammer, 2004, as well as sec. 8.2), later researchers focused on auditory, visual, and tactile stimulation during REM sleep. In one such study (Rechtschaffen & Foulkes, 1965; see also Kerr, 1993, p. 29), subjects slept with their eyes taped open and their pupils chemically dilated. During REM sleep, experimenters presented them with illuminated objects such as books and coffee pots. Ten seconds later, subjects were awakened and asked to report any previous mentation. When judges tried to match the presented objects to the dream reports, they did only slightly better than chance, and there was not a single unambiguous case of incorporation. While this dramatically illustrates the strength of the sensory input blockade, many other studies have found evidence for the integration of light flashes, sounds, sprays of water applied to the skin (Dement & Wolpert, 1958), and thermal (Baldridge, 1966), electrical (Koulack, 1969), and verbal stimuli (Berger, 1963; Breger et al., 1971, p. 27; Hoelscher et al., 1981) in dreams. Sensory incorporation has also been studied in children's dreams and during NREM sleep dreams (for details and further references, see Arkin & Antrobus, 1991).

Generally, incorporation can take different forms and, as suggested by Kramer et al. (1983; see also Kramer, 2007), may depend not only on stimulus type and strength but also on the personal meaningfulness of stimuli. They presented their subjects with the names of familiar versus unfamiliar persons during REM sleep. Incorporation of any kind was significantly more frequent for familiar than for unfamiliar names. The most frequent type of incorporation was representational: rather than hearing the names in their dream, most subjects dreamt about the corresponding person. This pattern suggests that processing during dreams is sensitive to meaning, but also that the connection between external stimuli and dream experience is less direct and more associative than in wakefulness. To the extent that dream imagery is responsive to external stimuli, it is apparently not as strongly constrained by such stimuli as standard waking perception. Conceptually, such instances of indirect or even cross-modal integration (as in seeing a person rather than hearing his or her name) are particularly interesting: they do not fit the philosophical concept of hallucination because an external stimulus is present; yet they also do not fit the philosophical concept of illusion, because they do not involve a mere distortion of the actual stimulus (e.g., hearing the name in a slightly different voice). Rather, such indirect incorporation seems to cut across different forms of quasi-perceptual experience.

Further evidence for visual stimulus incorporation comes from studies of lucid dreaming. The Lucidity Institute, founded by Stephen LaBerge, developed a specific device designed to submit light signals during periods of REM sleep. The so-called DreamLight (also NovaDreamer), a kind of sleep mask that detects rapid eye movements, submits flashes of light when the subject enters REM sleep. These light flashes are supposed to remind the dreamer to perform a reality test and thus help induce lucidity (LaBerge & Levitan, 1995). Initial studies show results to be ambiguous, and the effectiveness of this induction technique is unclear (Price & Cohen, 1988; Stumbrys et al., 2012). Some dreamers successfully use the cues as "reminders" to become lucid; others fail to become lucid because they concoct alternative explanations for the flashing lights (Keelin, 1995). For the quasi-perceptual view of dreaming, these examples are nonetheless interesting because the integration into the ongoing dream appears to be seamless. For instance, participants describe seeing flashes of red, sparkling chandeliers, or someone playing with a light switch (for further examples, see Keelin, 1995), indicating that the stimulus often merges with dream imagery. Importantly, subjects' frequent failure to detect the external cause of their visual experience suggests that the light flashes are not immediately recognized as qualitatively different from dream imagery. Hunt and Ogilvie (1988), however, note that lucidity cues often lead to awakening in experienced lucid dreamers, perhaps suggesting that these dreamers are more sensitive to the differences between external stimuli and dream imagery, after all. At least, sensory stimulus integration in lucid dreams would present a good opportunity to study the exact quality of such external stimuli compared to dream imagery in more detail.

Aside from visual stimuli, auditory stimuli have also been applied during lucid dreams. Strelen (2006) showed not only that auditory stimuli evoked event-related potentials during lucid dreams but also that subjects were able to distinguish between high and low tones they were listening for during their dreams and were able to communicate their judgment as to which type of tone they were hearing via eye movement signals. Even veridical perception, it would seem, is possible in dreams.

Or is it? What exactly do these types of studies reveal about the phenomenal character of dream imagery? In particular, does the seamless integration of external stimuli in dreams support the quasi-perceptual view? Two points seem particularly relevant. One, this type of approach has not just methodological, but also metatheoretical consequences for the study of dreaming and its analysis in philosophy. Recall that much of the philosophical literature on the quasi-perceptual view regards dreaming as arising independently of external stimuli, and some authors even deny, for conceptual reasons, that instances of sensory incorporation during sleep are genuine dreams. By contrast, it now seems that investigating external stimulation during sleep may be the most straightforward and systematic way of investigating dream generation. In particular, it provides a methodology for investigating how dreams put their gloss, as it were, on external stimuli, enabling researchers to chart the factors determining which stimuli are integrated in dreams, and how (i.e., by contrasting direct and indirect forms of incorporation, the meaningfulness and intensity of stimuli, etc.).

Two, because many instances of stimulus incorporation in dreams are indirect, it is unclear what exactly they say about the phenomenal character of dreaming. If names presented during sleep are transformed into dreams of the corresponding person, does this tell us anything about whether the person looked exactly like the waking counterpart, or about whether seeing this person in the dream felt like seeing someone in wakefulness? Relatedly, if light flashes submitted by a lucid dream mask are integrated into a dream, for instance, in the form of a glowing chandelier, does this show that the dream chandelier inherited its perceptual quality from the light flashes submitted by the dream mask? Is the association of dream imagery with an external source sufficient to prove its quasi-perceptual quality? Or is it possible that external stimuli, to be integrated into dreams rather than lead to awakening, lose their quasi-perceptual quality and are, as might be suggested by a proponent of the imagination view,[3] dampened so as to merge seamlessly with imaginative dream imagery? Ellis provides a characteristically poetic formulation of this idea:

> Whatever the stimuli from the physical world that may knock at the door of dreaming consciousness, that consciousness is apart from them, and stimuli can only reach it by undergoing transformation. They must put off the character which they wear as phenomena of the waking world; they must put on the character of phenomena of another world, the world of dreams. (Ellis, 1911/1931, pp. 18–19)

Of course, whether this transformation is strictly necessary is an empirical question: if external stimuli can even occasionally be integrated into dreams without being transformed, this would be enough to refute the necessity claim. Yet it is possible that Ellis is entirely right with respect to indirect stimulus incorporation. So a prediction might be that dreams typically put a dreamlike gloss on external stimuli, and dream experience is generally more loosely constrained by external stimuli than standard waking perception (cf. chap. 8). Apparent counterexamples, such as the ability of lucid dreamers to discriminate between high and low tones, would have to be accommodated by this account, for instance, by claiming that this does not strictly require that the tones were experienced in exactly the same way as during wakefulness. My interest here is not so much to endorse either one view or the other as to point out that finding satisfying answers to these phenomenological questions proves frustratingly elusive. Indeed, the more pressure one puts on the quasi-perceptual view, the less clear it becomes how exactly it translates into empirically investigable questions. If, however, the empirical evidence is systematically unable to address philosophical questions about the quasi-perceptual versus imaginative character of dreams—and if even external stimulus incorporation in dreams, as the arguably most direct candidate for answering them, raises more questions than it answers—then perhaps the philosophical controversy over these two views is best abandoned. Perhaps the dichotomous question of whether dreaming involves quasi-perceptual or imaginative experience, though intuitively compelling, is a nonstarter.

Aside from sensory incorporation in dreams, sensory stimulation can also be used to investigate the mechanisms underlying reduced responsiveness to environmental stimuli across different sleep stages. While the so-called sensory input blockade is often taken to be one of the main differences between REM sleep and wakefulness, but also between REM sleep and NREM sleep (cf. the AIM model proposed by Hobson et al., 2000; see sec. 2.2 for discussion), newer studies paint a more differentiated picture. Wehrle and colleagues (2007) used fMRI to investigate thalamocortical reactivity to auditory stimulation during REM sleep. They found two very different patterns of reactivity during tonic and phasic REM sleep: during tonic REM sleep, they found residual activation of the auditory cortex, suggesting that some processing of external stimuli was preserved. By contrast, during phasic REM periods, marked by bursts of REMs, they found evidence for heightened activity of a thalamocortical network, which they suggest might be functioning as an isolated and closed intrinsic loop. This suggests that to understand fluctuations in external stimulus processing during REM sleep, we might need a distinction between different functional microstates within REM sleep (see also Ermis et al., 2010). Results from sensory stimulation during NREM sleep suggest a similar need for a more refined theoretical account. Neuroimaging studies have shown that during NREM sleep, thalamocortical responsiveness to auditory stimulation is preserved, suggesting, once more, that the disconnection from the environment is not complete. As Dang-Vu (2012, p. 164) puts it, "Spontaneous neural events in sleep, such as spindles, determine the fate of incoming external stimuli during sleep and therefore might ultimately participate in the consolidation of previous experience." Far from being a state of uniform passivity, "brain processes during NREM sleep appear to be non-stationary and highly dependent upon spontaneous brain activity such as prominent sleep spindles or the phase of the slow oscillation" (Schabus et al., 2012, p. 10). This supports an argument already introduced in chapter 2: it is becoming increasingly clear that even in deep sleep, we are not "dead to the world," but rather that throughout sleep there are windows of responsiveness to external stimuli, carefully orchestrated by spontaneous fluctuations in brain activity.

An even more interesting question is whether this responsiveness is selective. Does any evidence connect these fluctuations to meaningful stimuli? And how, if at all, are these findings connected to dreaming? A series of related studies (Eichenlaub et al.,

2013; Ruby et al., 2013) might provide the sketches of a highly preliminary answer. Participants who had previously been classified as high and low dream recallers were presented with randomized auditory stimuli (frequent standard tones, rare deviant tones, and very rare first names) during wakefulness and sleep. Differences in dream recall frequency were confirmed in the laboratory, with high recallers reporting dreams following 94 percent and low recallers reporting dreams following 33 percent of morning awakenings. The event-related potentials (ERPs) elicited by first names differed between the two groups both in wakefulness and in sleep, suggesting that unexpected but meaningful stimuli induce more complex cognitive processing in subjects with high dream recall across the sleep–wake cycle (Eichenlaub et al., 2013).

An analysis of EEG alpha reactivity showed slightly different results: first names led to a decrease in alpha power in wakefulness and an increase during REM sleep (Ruby et al., 2013). In wakefulness, this decrease was more sustained in high recallers. No group difference between high and low recallers was found in REM sleep, a finding that the authors suggest might be related to more frequent awakenings in high recallers in response to first names. Overall, the authors interpret their findings as providing evidence for an increased reactivity to meaningful stimuli in high recallers over the sleep–wake cycle. While much research has focused on determining personality differences between high and low recallers, these studies suggest the possibility of neurophysiological trait differences between the two groups.

Importantly, if high dream recall is indeed associated with a deeper and more complex processing of meaningful stimuli, then dreaming may be associated not so much with a blocking of, as with responsiveness to, external stimuli. If so, the classical view, dating back to Aristotle, of dreams unfolding in the silence of sleep may have to be reconsidered: perhaps external stimulus processing is not so much the opposite of dreaming as an integral part of it.[4] Be that as it may, the studies discussed here suggest that the investigation of external stimulus processing may lead to a refined understanding of different sleep stages. In addition, as suggested by Ruby et al. (2013, p. 7), researchers may have to look not only beyond REM sleep but even beyond sleep itself to understand dreaming, relating differences in dream production and dream recall to interindividual differences in functional brain organization. This could lead to a profound rethinking of the methods involved in sleep and dream research, but also of the concepts of dreaming, sleep, and wakefulness itself.

5.3 Conclusions

Are dreams quasi-perceptual experiences? Throughout this chapter, it has become clear that quasi-perceptual experience is a helpful label for structuring variants of the traditional view in philosophy. Many philosophical discussions of dreaming, reaching from antiquity to the present, converge on the assumption that dreaming feels, in some relevant sense, like perceiving in wakefulness. This assumption lies at the heart of at least four important philosophical problems: the epistemological problem of dream deception, the distinction between imagination and perception (or ideas and impressions), the problem of perception, and the characterization of conscious experience as an essentially virtual or dreamlike kind of online hallucination. In all these cases, phenomenological assumptions about dreaming feed into specific theoretical claims. Yet, in this process, the percept-like quality of dream imagery itself is mostly taken as a given and not justified or even explicated in detail. The concept of quasi-perceptual experience captures this general and often unarticulated background assumption about the phenomenology of dreaming, and it is helpful because of its neutrality about the sources of dreaming and the question of whether dream imagery is hallucinatory or illusory. In structuring the philosophical debate, this is an advantage.

What, then, about empirical plausibility? My survey of the empirical evidence appears to have raised more questions than it has answered. A problem with the quasi-perceptual view is that with few exceptions, the philosophical discussion does not lend itself to precise empirical predictions. Saying that dreaming involves the appearance of a world—a characterization that can be found both in classical phenomenology and in empirically informed discussions of dreaming from the perspective of philosophy of mind—may capture a central feature of dream experience but hardly lends itself, without further qualification, to being operationalized in an empirical study. Indeed, it seems very hard to find points of contact between the philosophical literature and contemporary dream research. While the neutrality and all-encompassing nature of the quasi-perceptual view is helpful for structuring the philosophical literature, the same factors render it frustratingly elusive for assessing the empirical evidence.

At least, we can now discard two variants of the quasi-perceptual view as empirically implausible. The first is the claim that dreaming replicates waking experience in all its detail. There simply is no solid evidence for this claim, and given the distribution of different types of imagery in dreams, it is very likely false. This is not to deny that dreams potentially replicate waking experiences; it is only to say that they do not typically appear to do so. The second is the claim that dreams are wholly independent of contemporaneous external stimuli, either as a matter of empirical fact or as a conceptual claim to the effect that a distorted perception of stimuli in the sleeper's environment forms no part of dreaming proper. Given the fluid integration of external stimuli in dreams, and given evidence from sleep-only studies on the fluctuations in external-stimulus processing during NREM and REM sleep, this claim seems too strict. If one wanted to insist that sleep necessarily involved the absence of perception and that sleep and wakefulness are opposites, this could result in the utterly absurd view that sleep, as a matter of fact, does not exist. Rather, it seems plausible to say that sleep is characterized by varying degrees of responsiveness to external stimuli and that dreaming can, and perhaps even frequently does, fluidly integrate external sensory stimuli with internally generated ones. Methodologically, this suggests that a particularly valuable way to investigate dreaming in more detail might be to look beyond the dreaming brain to stimuli in the sleeping subject's environment, and perhaps even beyond sleep itself (i.e., for the investigation of trait differences between high and low recallers).

In addition, the analysis of the empirical literature suggests how progress could be made: namely, by studying the phenomenology of dreaming in more detail and systematically comparing dream and waking experience reports. After all, a central problem with the quasi-perceptual view is not only that detailed studies of the phenomenology of dreaming are lacking but that the same is true for the suggested contrast conditions of standard waking perception and hallucinations. Even if a careful phenomenological characterization based on such studies were available, a further theoretical problem would be that it would not be clear where exactly to draw the cutoff line: How similar would dreaming have to be to standard perception to count as quasi-perceptual? How similar would hallucinations have to be to both dreaming and perceiving, but also among each other? Do phosphenes count as quasi-perceptual, or is this only the case for fully formed hallucinations of objects or persons? What about bizarre objects, as in dreams? And so on and so forth.

An alternative possibility would be to reverse the strategy: given that detailed contrastive studies of dreaming and wakefulness are lacking, perhaps a better strategy might be to simply assume that dreaming itself plays a role in identifying the relevant phenomenological characteristics of quasi-perceptual experience, but also for identifying its neural correlates. This would have the advantage of leaving detailed questions about modality-specific types of imagery to the side; rather, the question would become whether dreams themselves form a common phenomenological kind, whether they converge on some common and maximally invariant phenomenal feature. If so, and if this feature could meaningfully be applied to perception, but also to hallucination and imagination, such a theory would count as successful. I will argue that this is indeed the case; progress can be made by shifting the focus of the debate from modality-specific types of dream imagery to a relational property: the experienced relation between the phenomenal dream self and the dream world. This requires suspending comparative statements about dreaming and waking experience and analyzing self-experience in dreams in its own terms. The next step, however, is to analyze the imagination view as a rival to the traditional view. This will also provide additional structure to the subsequent discussion.

6 Dreaming as Imaginative Experience: The Rival View

6.0 Introduction

The traditional view in philosophy is that dreaming, in some form or another, involves the appearance of a world: phenomenologically, dreaming falls on the side of perception, illusion, and hallucination rather than on the side of imagination. Yet we have seen that this view—perhaps precisely because it has long been the traditional view—is often poorly articulated, metaphorical, and has not, so far, been translated into clear-cut empirical questions. For an empirically informed conceptual framework of dreaming, saying that dreams are quasi-perceptual experiences is as all-encompassing as it is unsatisfying: the label of quasi-perceptual experience may well be too slippery and underdetermined to lead to any straightforward predictions. Moreover, we have seen that throughout the history of philosophical theorizing, the quasi perceptual view of dreaming has coexisted with the assumption that the closest wake-state relative of dreaming is imagining or fantasy. This is a core assumption in empiricist philosophy but can also be found in the contemporary internalist or representationalist claim that conscious experience is essentially virtual or dreamlike.

The problem of how to bridge the gap between the quasi-perceptual view as a phenomenological claim about dreaming to the relationship between dreaming and imagining is left unexplained in this process. For philosophy of mind, the central problem appears to be the question of how a conscious state that is in some sense imaginative can *feel* like perceiving. As is the case for the debate on dream experience (cf. chap. 1), this has seemed, to many, to be a contradiction. This chapter proposes a preliminary solution to this problem. I argue that the putative gap between dreaming as quasi-perception on the phenomenological level of analysis and neurocognitive accounts of dreaming as an imaginative process rests on an erroneous conception of imagining and perceiving as sharply rather than gradually distinguished. This dichotomous conception of imagining and perceiving is a philosopher's construction. It is empirically implausible, creates more problems than it solves, and is best rejected. Dreams, as it turns out, are a particularly clear case for rejecting a dichotomous view of imagining and perceiving altogether: dreaming blurs any sharp boundaries one may want to introduce between imagining and perceiving, and a coherent account of dreaming will have to explain how dreaming is both an imaginative and a quasi-perceptual process. This is not at all a conceptual contradiction but an empirically fruitful theoretical assumption. Moreover, this approach will bring us closer to operationalizing and rendering empirically tractable a successor to the philosophical versions of the quasi-perceptual view discussed in the previous chapter. To see why this is the case, it is necessary to first engage with the rival view: the claim that dreams are imaginative experiences in the phenomenological sense. Like the quasi-perceptual view, there are different variants of this view, depending on which aspects of dreaming are claimed to be imaginative. The strongest version claims that dreaming involves neither quasi-perceptual experiences nor instances of (false) beliefs (Ichikawa, 2008, 2009). A weaker version allows that dreaming involves false beliefs, but denies that it involves quasi-perceptual experiences (McGinn, 2004, 2005a, 2005b). And a third version allows that dreaming involves quasi-perceptual experiences, but denies that it involves beliefs (Sosa, 2007). In this chapter, I concentrate on arguments claiming that dream imagery is unlike quasi-perceptual experience (I discuss the question of whether dreaming involves actual or imaginative beliefs in chap. 9).

This proliferation of different versions of the imagination should not be surprising; indeed, it points to a problem inherent in the philosophical debate on imagination itself. The concept of imagination, both in folk psychology and in philosophy, is used too loosely to allow for the identification of sharp boundaries (for an overview, see Gendler, 2013; for a recent attempt to bring order into the heterogeneity of the imagination, see Dokic & Arcangeli, 2015; for discussion, see Brüggen, 2015). Ryle famously rejected the view that imagination denotes any single kind of mental state. Instead he argued that there "are hosts of widely divergent sorts of behaviour in the conduct of which we should ordinarily and correctly be described as imaginative" (Ryle, 1949/2000, p. 256). According to Ryle's deflationist account of the imagination:

> There is no special Faculty of Imagination, occupying itself single-mindedly in fancied viewing and hearing. On the contrary, "seeing" things is one exercise of imagination, growling somewhat like a bear is another; smelling things in the mind's nose an uncommon act of fancy, malingering is a very common one, and so forth. (Ryle, 1949/2000, p. 258)

All these different types of imaginative activities do not have at their core "one common nuclear operation" (Ryle, 1949/2000, p. 256), and there is nothing that binds them together in any deep sense. Similar claims can be found in approaches to the imagination maximally different from Ryle's. Thompson (2007), who discusses mental imagery from the perspective of classical phenomenology and enactivism, argues that prominent treatments of the topic in the imagery debate are plagued by oversimplified phenomenological descriptions. He notes than an unquestioned background assumption adopted by different proponents in this debate is that mental imagery in some relevant sense resembles picture viewing. Much of the debate on the allegedly pictorial nature of mental imagery thus results from confusing a metaphorical term—mental imagery—with a phenomenological description. As Thompson points out, "The only time visual experience is straightforwardly pictorial is when one has the visual experience of looking at a picture, or the experience of remembering or imagining looking at a picture" (Thompson, 2007, p. 297). While this might sound trivial, the consequences are important, as the actual differences between perceiving, visualizing, and remembering are obscured in this process.

Here I argue that a similar problem plagues phenomenological characterizations of dreaming. Indeed, we will see that the attractiveness of the imagination view relies, to a considerable extent, on the deceptive ease and apparent comfort of the armchair fallacy, or the tendency to confuse conceptual or even metaphorical descriptions with phenomenological observations.[1] Put in the simplest terms, the imagination view results from the assumption that if dreams are a form of imagination, then dreaming must also *feel* like imagining. Relatedly, there must be something it feels like to engage in different forms of imagination in the first place. This assumption, in turn, is importantly connected to the claim that perception and imagination are sharply distinguished: drawing a sharp phenomenological distinction between perception and imagination and showing that different kinds of imagination converge on a shared phenomenal core is part of what it takes to show that perceiving and imagining are different in kind. This reasoning, however, is fallacious. Even if sharp criteria distinguishing perceiving and imagining could be given, it would not follow that dreaming necessarily falls safely on one or the other side of this putative divide. Perhaps dreaming is a different kind of mental state from both. Moreover, even if dreaming is best conceived of as an imaginative state, this has no direct and necessary consequences for the phenomenology of dreaming. The imagination view, as will become clear, not only mistakes a conceptual claim about the imaginative nature of dreaming for a phenomenological one but also fails to account for the phenomenological variability underlying both dreaming and waking imagination.

Yet important lessons can be learned from the imagination view. In particular, the imagination view functions as a corrective for the traditionally more widespread quasi-perceptual view and puts pressure on it to develop a more differentiated account of the phenomenology of dreaming in its own right. An important advantage—and part of the reason why it is more easily refuted—is that, unlike the quasi-perceptual view, the imagination view makes more easily testable predictions. Also, many specific points discussed by its proponents lead to an enriched and more differentiated understanding of the phenomenology of dreaming and highlight the ways in which dreaming fails to replicate standard waking perception. Contrasting the imagination view with the traditional view shows the extent to which both rely on ungrounded and empirically implausible background assumptions about dreaming. Because both distort the phenomenology of dreaming by espousing premature comparative statements, a contrast-neutral description of dreaming is needed, which describes dreaming in its own right rather than modeling it on waking experience. Such analogies, if at all, should be an end point of an empirical investigation, but not the beginning of inquiry. While the main upshot of my discussion of the quasi-perceptual and imagination views is negative, it is still, I think, constructive. In particular, by engaging with both views and confronting them both with each other and with the empirical and psychological literature, we can salvage their respective strong points and slowly work toward a better, more empirically plausible view, as well as toward more productive and empirically tractable types of questions. I explore these questions over the following chapters.

6.1 The Philosophical Literature

6.1.1 The Imagination View as an Answer to Cartesian Dream Skepticism

As so often in the history of philosophical theorizing about dreaming, the imagination view departs from the epistemological problem of dream skepticism. This connection is particularly evident in John Locke's *Essay Concerning Human Understanding*. Here Locke comments on the problem of dream skepticism twice (Locke, 1689/1997, IV.II.14, IV.XI.8). Both times, his reaction is driven by an almost impatient incredulity: he takes the phenomenological background assumptions behind Descartes's scenario of dream deception to be so obviously false that they warrant no detailed refutation at all.

> But yet here, I think, we are provided with an evidence, that puts us past doubting: for I ask anyone, whether he be not invincibly conscious to himself of a different perception, when he looks on the Sun by day, and thinks on it by night. (Locke, 1689/1997, IV.II.14)

Conceptually, it is interesting to note that Locke very naturally equivocates between thinking and dreaming in this passage. To Locke, dreaming is clearly an instance of thinking—and as such, it ought to feel like thinking, as well. What troubles Locke is why anyone might deny this obvious point in the first place.

> If anyone say, a dream may do the same thing, and all these ideas may be produced in us without any external objects, he may please to dream that I make him this answer, 1. That 'tis no great matter, whether I remove his scruple, or no: where all is but dream, reasoning and arguments are of no use, truth and knowledge nothing. 2. That I believe he will allow a very manifest difference between dreaming of being in the fire, and being actually in it. (Locke, 1689/1997, IV.II.14)

To Locke, this difference is, at least in part, phenomenological: our "certainty is as great as our happiness, or misery, beyond which, we have no concernment to know" (Locke, 1689/1997, IV.II.14). There is clearly a difference between dreaming of being in the fire and really being in it, because happiness and misery, pleasure and pain, are simply not experienced in the same manner in dreams. Sarcastically, Locke invites those convinced by Descartes's scenario to put their conviction to the test:

> If our dreamer pleases to try, whether the glowing heat of a glass furnace, be barely a wandering imagination in a drowsy man's fancy, by putting his hand into it, he may perhaps be wakened into a certainty greater than he could wish, that it is something more than bare imagination. (Locke, 1689/1997, IV.XI.8)

Aside from these brief comments, Locke's remarks on dreaming are not much different from those of his fellow empiricists: dreaming is "the having of ideas (whilst the outward senses are stopped, so that they receive not outward objects with their usual quickness) in the mind, not suggested by any external objects, or known occasion" (Locke, 1689/1997, II.XIX.1); it is a "loose and incoherent manner of *thinking*" (II.XIX.4). Yet where Hume struggled to reconcile the vivacity of dreams with their conceptual status as ideas rather than impressions, Locke bears what he takes to be the phenomenological consequences of this conceptual point.

Here is the philosophical problem: can Locke's description of dreaming be discredited simply because it is nonstandard and deviates from what we have encountered as the traditional view? Clearly one would wish for a more detailed argument. Yet the fault is not only Locke's. In providing nothing more than a general and what he takes to be intuitive characterization of the phenomenology of dreaming, Locke is doing exactly the same as proponents of the traditional quasi-perceptual view. There is no prima facie reason—other than, perhaps, intuitive gut reaction—for preferring one view over the other. And given the diversity of intuitions about the phenomenology of dreaming, intuitiveness hardly seems a prudent way to resolve the question. Moreover, we have already seen that dreaming is unlike standard perception in the distribution of multimodal imagery, suggesting that a strong version of the quasi-perceptual view fails. Given these uncertainties, why not even say that dreaming is better described as imaginative experience and thus avoid the problems of the quasi-perceptual view altogether?

Note also that Locke's discussion of dreaming is empirically plausible in an important respect: pleasure and pain are rarely mentioned in dream reports (Hobson, 1988). Locke need not even deny that dream fires can potentially be painful: for him, it is enough that typically they are not. He also makes the commonsense observation that even if we are deceived by our dreams while they are occurring, it does not follow that we can doubt the evidence of sensory perception while we are awake. Both points might compel us, initially, to lean toward the imagination view.

A similar attack on Cartesian dream skepticism is mounted by Austin (1962). According to him, dreams are not illusory, nor do they typically involve the delusional belief in the reality of dream events (Austin, 1962, pp. 12, 27); "dreams are dreams" (p. 27), a category of their own. Like Locke, Austin takes this to be a commonly accepted part of folk psychology:

> In fact, it is just because we all know that dreams are *throughout un*like waking experiences that we can safely use ordinary expressions in the narration of them; the peculiarity of the dream-context is sufficiently well known for nobody to be misled by the fact that we speak in ordinary terms. (Austin, 1962, p. 42)

This is a clever reversal of how to interpret dream reports: according to Austin, the way we narrate dreams not only fails to provide evidence for the phenomenological similarity between dreaming and wakefulness but, to the contrary, shows how certain we are of the phenomenological differences between the two states (Austin, 1962, p. 49). Apparently it is only certain philosophers who are deceived about this fact. In this view, the Cartesian specter of dream deception is a myth, a theorist's illusion:

> Is it the case that "delusive and veridical experiences" are not "qualitatively different"? Well, at least it seems perfectly extraordinary to say so in this sweeping way. . . . I may have the experience (dubbed "delusive" presumably) of dreaming that I am being presented to the Pope. Could it be seriously suggested that having this dream is "qualitatively indistinguishable" from *actually being* presented to the Pope? Quite obviously not. (Austin, 1962, p. 48)

A more careful defense of the imagination view of dreaming is provided by Stone (1984). His central claim is that being wide awake is self-recognizing: the state of wakefulness is such that when we are in it, we have certain knowledge that we are awake. Yet "this is compatible with my mistakenly believing I am wide awake when I am not" (Stone, 1984, p. 353). The main challenge is to show that there is indeed an epistemological advantage to being awake. According to Stone, this is only the case if dreams are not hallucinations. "The obvious alternative," he notes, "is that dreams are self-generating fantasies during sleep very much like day-dreams. When we dream then we aren't presented with visual objects in perceptual space, we merely fantasize that we are" (p. 363).

Why endorse this view? Stone is not content to claim that it is phenomenologically obvious, but rather argues that it has explanatory virtues compared to the traditional view.

> This theory has obvious advantages: it provides a more unified and conservative account of mental life than its rival, for it assimilates dreaming to a familiar mental process, fantasizing, common in waking subjects, as opposed to a bizarre process, hallucinating, something most waking subjects never do. It provides a simpler account of the relation between dreaming and daydreaming: they are essentially the same phenomenon. (Stone, 1984, p. 364)

Stone allows that we can mistake dreams for vivid sense experiences while asleep; indeed, he thinks this is a fact that a theoretical account of dreaming must account for. Yet he does not think that this threatens his claim that being wide awake and enjoying vivid sense experience is self-recognizing: we may be deceived when absorbed in vivid fantasy, but when we are wide awake, we cannot help but recognize vivid sense experience for what it is. In this view, we cannot rule out that we are hallucinating, or that we are a brain in a vat or deceived by an evil demon. Even so,

> It is still indubitably certain that I am wide awake and not merely fantasizing. . . . If dreams are fantasies, sense-experience is a certain indication by which we can clearly distinguish the fact that we aren't dreaming. Descartes need only pinch himself. (Stone, 1984, 366)

Of course, it is not clear that this unidirectional distinguishability would satisfy the Cartesian skeptic: it would still be the case that in certain conditions, we would be unable to distinguish fantasy from real sense experience. Even if sense experience during wakefulness is self-recognizing, dreams might still leave us in doubt as to whether we are really in a state that is self-recognizing or only in one that seemingly is (cf. chap. 10; for a similar argument, see Ichikawa, 2008). Yet Stone's argument underscores an important motivation for endorsing the imagination view, namely, the hope that it provides a quick and easy escape from Cartesian dream skepticism. In doing so, it echoes important aspects of perhaps the best-known proponent of the imagination view: Jean-Paul Sartre.

In Sartre's analysis of dreaming, the imagination view is developed from a rebuttal of dream skepticism to a full-blown phenomenological account of dreaming. Sartre begins his discussion of dreaming with Descartes's familiar example, asking, like so many before him, whether dreams are the special case of an imagined world presenting itself as real and as actually perceived (Sartre, 1940/2005, p. 309). His answer, however, is negative. For Sartre, "It is possible to imagine, while dreaming, that I perceive; but it is certain that I cannot doubt, while I am awake, that I perceive" (p. 309; translation mine). The very attempt to feign such doubt in our waking perceptions, as implied by the Cartesian scenario of dream deception, is absurd: perception affords the same immediate and indubitable evidence as Descartes's *cogito*, whereas dreams lack this experiential quality and the certainty that goes with it. According to Sartre, perception has its own phenomenal signature: imagination does not mimic or reduplicate the world of perception; instead, imagining and perceiving are different in kind, and the difference between them is given in experience. In particular, imaginary objects are experienced as unreal or absent because they have their own space, separate from the space of perceptual objects (p. 287).

As Sartre notes, dreaming is important in this context because one might think, as did Descartes, that dreaming is a counterexample, a case in which an imagined world is experienced as real. Unlike other phenomenologists, who took dreaming to be a limiting case of accounts of the imagination (cf. sec. 5.1.3), Sartre insists that there is, properly speaking, no imaginary world, including the world of dreams. To him, deflating the assumption that dreaming involves world experience is important because it threatens to undermine his entire account of the imagination (Sartre, 1940/2005, p. 309).

Part of Sartre's strategy is to deny that reflexive consciousness is possible in dreams. In doing so, he endorses the view that dreaming is a state of cognitive deficiency, and in this respect, his position converges with that of many philosophers discussed in the previous chapter. All dreams, according to Sartre, have narrative structure, but cognitive abilities such as attention, reflection, and memory are absent and indeed would disrupt the dream state if they occurred (Sartre, 1940/2005, pp. 324–325). Hence we do not explicitly believe in the reality of our dreams, nor are we capable of correctly realizing that we are now dreaming. At best, imaginary reflexive acts are possible in dreams, but not actual ones (p. 336). Whenever reflexive consciousness asserts itself in a dream, this is a momentary awakening, and if the dream is not interrupted, any reflective awareness of the fact that one is now dreaming will immediately disappear (p. 311). Here Sartre effectively denies the possibility of prolonged lucid dreams.

The absence of reflection in dreams also explains why dreaming is different from waking imagination. In particular, to Sartre, dreaming is characterized by a form of epistemic closure that waking imagination lacks: dreams are experienced as fictions, but ones that have captured us so completely that the very concept of reality eludes us, and in which we are condemned to creating unreal, fictional worlds. Sartre likens dreaming to the experience of a naive reader who is completely immersed in a story: he experiences it as fiction, but not as reality, and is simultaneously unable to escape this fiction without terminating the experience (cf. my discussion of cognitive immersion in sec. 11.3.3). The resulting position on dreaming as imaginative experience is twofold: the phenomenology of dreaming excludes the phenomenology of perception, but also the occurrence of reflection. Dreaming is essentially a phenomenon of belief, but it lacks the reflective quality of waking consciousness. In particular, dreaming is not, as Descartes assumed, the appearance of a world: being in the world, understood as an experiential property, is necessarily lost in dreams (Sartre, 1940/2005, p. 339).

Yet Sartre somewhat confusingly allows that there is a weaker sense in which being in the world is crucial to dreaming after all. He notes that his own dreams frequently begin as stories that he is reading or being told; "All of a sudden, I identify with one of the characters in the story, which now becomes *my* story" (Sartre, 1940/2005, p. 320; translation mine). He now has a dream self in the dream world, where both are experienced as fictional, not as real. Indeed, Sartre notes that the only way in which the dreamer can penetrate into the dream world is by identifying with a character or object in the dream (p. 333). Even the experience of being in an unreal world, as in dreams, apparently requires the experience of a dream self in this world.

Does Sartre's insistence that dreaming is phenomenologically distinct from perceiving result from the armchair fallacy? That is, is Sartre mistaking a conceptual point about the distinction between imagining and perceiving for a phenomenological observation? Several critics have defended this view. Hering (1947, p. 204) notes that

Sartre should at least have formulated the inability of dreams to involve reflection as a hypothesis, rather than merely stipulating this to be the case. Referring to the folk-psychological assumption that dreams are experienced as real and share the phenomenological character of perception, Hering comments:

> We cannot lay them aside with Sartre's observation (not explicitly stated but perhaps implicitly presumed) that there are no real sensations which would "render possible" a perception. Expressed in this form the objection would be a quite unphenomenological, sensualistic prejudice, which we should not allow to obstruct our view of the phenomenal situation. (Hering, 1947, p. 198)

Similarly, Globus points out that even if Sartre's argument on the lack of reflection in dreams were correct, it would "not bear upon our perception of a dream world which is prereflective. . . . Instead, the point is that the life-world while dreaming 'looks like' an authentic world and is 'taken' to be an authentic world, no matter what reason later says about this perception" (Globus, 1987, p. 68).

At this point, it may seem that as so often in phenomenological debates, we have reached the level of foot stomping (Kriegel, 2007), where two authors defend putatively obvious yet vastly different and perhaps even contradictory phenomenological claims. Clearly such a confrontation is not helpful for settling the points of disagreement between different theories. Because neither the quasi-perceptual nor the imagination view, at least in large parts of the philosophical debate, is systematically informed by dream reports, their divergence suggests that intuitions about the phenomenology of dreaming are best left to the side. Insofar as the phenomenology of dreaming is concerned, one philosopher's intuition is, apparently, another's conceptual absurdity. Yet this confrontation holds the potential for progress. One reason is that the imagination view functions as a corrective of the quasi-perceptual view, and vice versa. Whichever position one is drawn to, it seems that both rest on untutored assumptions, and their confrontation can help illustrate this point. Another reason is that Sartre's somewhat strained remarks on the importance of self-identification in dreams suggest that to relate dreaming to imagination and perception, it may be best to focus not on the quality of imagery but on the relationship between the dream self and the dream world. This is the strategy that I will later pursue. But first, it is important to clarify the relationship between imagining and perceiving itself.

6.1.2 The Imagination View of Dreaming and the Dichotomous Distinction between Perceiving and Imagining

Several contemporary authors writing on dreams have continued the project, evident in Sartre's work, of using dreaming as a test case for upholding a sharp distinction between imagining and perceiving. Importantly, they assume that modeling the phenomenology of dreaming on that of waking imagination is part of what it takes to show that imagining and perceiving are sharply distinguished.[2] Analyzing their arguments will lend additional structure to the problem and show what exactly is at stake.

Ichikawa writes:

> Dreams needn't involve false beliefs and misleading sensory experiences. Dreams are very much like vivid daydreams. . . . Lose yourself enough in your daydream, and you will feel, in some sense, as if you are really there. . . . Our dreams in sleep are, on the imagination model, *like that.* (Ichikawa, 2009, p. 119)

Similarly, McGinn notes that dreaming should be distinguished from both perceiving and hallucinating because

> dreams consist of mental images, analogous to the mental images of daydreams, not to those of sensory percepts. Dreaming is a type of imagining, not a type of (objectless) perceiving. . . . I think it should be intuitively quite clear that visualizing my mother's face in my mind's eye is very different from having a sensory impression of my mother's face, that is, actually seeing her. And I also think that most people intuitively recognize that dream experiences are imagistic not perceptual in character. (McGinn, 2005a, p. 65)

As in Sartre's work, dreams are not characterized by the phenomenology of presence or the appearance of a world. Instead the analogy between sleep dreams and daydreams suggests that the sense in which we lose ourselves in our dreams should be understood not as involving a robust experiential quality but metaphorically as the cognitive experience of absorption or fictional immersion experienced in waking fantasy, but also in reading a novel or watching a movie. McGinn (2004, pp. 92–93) adds an interesting twist to this discussion, arguing that dreams give rise neither to the phenomenology of presence, as in perception, nor to that of absence, as in daydreams and fantasy. In this respect, it seems that dreaming is neither here nor there. At the same time, McGinn somewhat unsatisfyingly suggests that for reasons of simplicity, dreaming be conceptualized as imaginative experience and not as a kind of experience in its own right or as a sort of hybrid (e.g., as imaginative seeing) (pp. 94–95).

An important motivation for defending the imagination of dreaming, as becomes clear in McGinn's analysis, is the claim that perception is conceptually tied to waking consciousness. Dreams, in this view, cannot be percepts because while we are asleep and dreaming, we are not aware of our current surroundings. At the same time, McGinn allows that hallucinations are percepts, even though they only seemingly put the hallucinating subject in perceptual contact with objects in her environment. Why, then, accept that hallucinations, but not dreams, are percepts? According to McGinn, it is part of common sense that dreaming terminates both perception and hallucination: if a brain in a vat went to sleep and started to dream, its state of consciousness would change, both conceptually and phenomenologically (McGinn, 2004, p. 81). Without further justification, this appeal to common sense, however, is somewhat bewildering. Given the historical and contemporary popularity of the view that dreaming is akin to hallucinating, this would have the strange consequence that most philosophers who historically wrote on dreams were impervious to common sense. Of course, this is always a possibility. Yet it is not one that is prone to bolster confidence in yet another argument from alleged common sense. This is especially true because the same argument would seem to work against the imagination view: if hallucinating is conceptually tied to wakefulness in a manner that prevents its application to dreaming, then certainly, so are imagining and daydreaming. Why, then, think that either one or the other can more smoothly be applied to dreaming?

In O'Shaughnessy's analysis of dreaming, we again find that the primary motivation for classifying dreaming as an "imagining-of consciousness" (O'Shaughnessy, 2002, p. 430) is that consciousness is conceptually tied to wakefulness. According to O'Shaughnessy, consciousness requires knowledge of the external world, reactivity to external stimuli, and perceptual awareness. Like perception, consciousness is a success word: conscious beings are "epistemologically empowered, they are in a position to make discoveries about the world" (O'Shaughnessy, 2000, p. 12; for discussion, see Crowther, in press). By contrast, dreaming is "exhausted by its being a merely as-if consciousness" (O'Shaughnessy, 2002, p. 430); more generally, visual imagination, including hallucinating and dreaming, is

> (i) an *as if* seeing, (ii) it *is not* seeing, (iii) and is so *of necessity*, and (iv) *that is all* it is. And that is to say that this perfectly real phenomenon is *pure* "as if": its "as if"-ness exhausts its being. (It is a sort of ghost of its model.) . . . What these facts tell us is that the use of "see" in explaining the above senses is of necessity a secondary or *derivative* use. It informs us that the phenomenon in question is *essentially modelled upon* seeing. (O'Shaughnessy, 2002, pp. 427–428)

O'Shaughnessy, unlike Sartre and Austin, thinks that dreams can be genuinely deceptive, involving mistaken beliefs in the reality of dream objects as well as actual emotions, and "such thoughts and emotions and desires are real examples of their kind" (O'Shaughnessy, 2002, p. 418). Yet he assumes that the cognitive deficiency of dreaming explains the phenomenological differences between dreaming and being awake:

> Dreaming *is not* consciousness, and *of necessity* . . . whereas consciousness is a rational condition dreaming is irrational; and whereas the one is overall mentally active and in particular encompasses the steady activeness of thinking, the other is essentially inactive; and while consciousness involves awakeness of the perceptual attention, dreaming depends upon its closure or sleep; and whereas when conscious we see ahead to and engineer our immediate mental future (under a limited description), in dreaming the mental future simply comes at us out of the blue. (O'Shaughnessy, 2002, p. 429)

In this view, the phenomenology of dreaming is distinguished from wakefulness because the dreamer does not have the experience of being an epistemic agent; again, the relationship between the dream self and the dream world takes center stage for solving the conceptualization problem of dreaming. What is unclear, however, is why O'Shaughnessy apparently regards this as a matter of conceptual necessity rather than as an empirically testable prediction.

Generally, the issue of control is perhaps the most difficult and hotly debated point not only for the imagination view of dreaming but in the entire philosophical debate about the distinction between imagining and perceiving. Historically, it has been common to assume that imagination involves "a special effort of the mind" (Descartes, 1641/1901, VI.2). Similarly, Wittgenstein (1967, pp. 621, 633) notes that imagining is an activity and is unlike perception because we can banish mental images, but not percepts.

The obvious problem is that if control indeed characterizes imagination, then dreams obviously fall short. Most agree that dreams happen to us, rather than being under our control. Because the issue of control is the most obvious objection to the imagination view, this is where its contemporary proponents concentrate their efforts, claiming that dreams are brought about by a process of unconscious authorship and that dreams, though typically not under voluntary control, are nonetheless subject to the will (McGinn, 2004; Ichikawa, 2009). As Ichikawa puts it, "Dreamers create their dreams without recognizing their own agency" (Ichikawa, 2009, p. 116); we unconsciously author our dreams without ever realizing that we do so. Moreover, the type of control we have over our percepts is, according to the imagination theory, different from the direct type of control we typically have over mental images, including dreams. "Although we do have some control over our perceptual experiences, any such control is indirect" (p. 107). In addition, according to Ichikawa, lucid control dreams show that dreams are at least occasionally amenable to direct deliberate control.

This argument has several problems. First, dream control is one of the aspects that most clearly distinguish lucid from nonlucid dreams (Voss et al., 2013; sec. 3.5); hence it is not at all clear that the argument from lucid control dreams generalizes to nonlucid ones. The obvious way of understanding lucid dream control, namely, as the ability to actively and intentionally control the ongoing dream, is not available to nonlucid dreamers. Saying that nonlucid dreamers have this ability but do not realize it seems strained and glosses over a very real distinction, which can be investigated in the laboratory and gives rise to empirical predictions.

Moving to nonlucid dreams, a naturalist assumption will be that they are generated by subpersonal brain processes; metaphorically, one might even say that in this sense, they are created by the sleeping subject. However, the ascription of dream generation to the dreamer, that is, to the dream *self*, says something very different. Unless one wants to commit to overburdened metaphysical claims, claims about the dream self should be understood in a purely phenomenological sense. Speaking of the dream self is nothing other than convenient shorthand for the phenomenology of selfhood, the experience of being or having a self, in dreams. Saying that the dream self creates its dreams in this phenomenological sense, however, is clearly false: phenomenologically, only lucid dreamers experiencing lucid control dreams do, but nonlucid dreamers do not.

McGinn addresses this problem. He tries to account for what he calls the "illusion of psychological passivity" by saying that "the dream producer is unconscious relative to the consumer" (McGinn, 2004, p. 89), or the passive audience of the dream. But this raises the further problem of who the dream producer is supposed to be. It is not the phenomenal dream self, or the consumer of the dream, who, according to McGinn, is unconscious of the dream production process and enjoys the illusion of passivity. Hence McGinn's proposal would require positing an additional dream producer with intentions of her own; perhaps this producer is even a phenomenal self in its own right, but only one that is overlooked, as it were, by the consumer. Or perhaps it is a nonphenomenal self, whatever that may mean. Either way, the introduction of such a homuncular explanation does not appear justified in the absence of any evidence in its favor. This view is metaphysically overburdened, and phenomenologically apparently quite untestable. McGinn is right, of course, that the process of dream generation requires a theoretical explanation; but a parsimonious and empirically plausible one will not ascribe the process of dream generation to the self (I return to this question in chap. 12). Consequently, it is at best metaphorical and at worst misleading to say that *we* control or author our dreams.

Moreover, even the claim that lucid dreams are amenable to *direct* control in the same manner as waking imagination is questionable. A critical look at the literature suggests that lucid insight, as well as concomitant dream control, is in fact a fleeting and unstable phenomenon, often leading to other than the expected results and initiated not by means of direct "thought control" but by engaging with the objects and characters in the dream and momentarily suspending disbelief (LaBerge, 1985; LaBerge & Rheingold, 1990; Brooks & Vogelsong, 1999; see also sec. 9.3.3). For instance, lucid dreamers typically do not simply *will* certain changes in the dream environment to occur but employ specific techniques such as flying or passing through walls. Although dream lucidity is a learnable and thus to an extent controllable skill, it is not a reliable one: according to one study, lucid dreamers only remember waking intentions, for instance, to perform dream experiments, about half the time, and less than half of these remembered intentions are successfully carried out in dreams (Stumbrys et al., 2014). In addition, the occurrence and duration of lucid dreams are largely dictated by the duration of REM phases. Lucid dreamers can wake themselves up by fixing their gaze on something within the dream and thus disrupting the rapid eye movements that are a physiological feature of REM sleep, and they may even develop techniques to maintain lucidity, but they cannot prolong their lucid dreams at will. They also cannot voluntarily suppress visual imagery for prolonged periods, or at least not without waking themselves up.[3] Though they may be able to induce other types of imagery (e.g., Worsley, 1988), visual imagery may be so central to the process of dream generation that its suppression leads to a disruption of REM sleep. Thus, while lucidity may enable experienced lucid dreamers to influence the ongoing dream, the degree to which even lucid dreams are amenable to direct control is limited at best (Schatzman et al., 1988; Brooks & Vogelsong, 1999; for a more detailed discussion and further references, see Windt & Metzinger, 2007). To the contrary, lucid dreams are mostly controlled by engaging with the dream world, by taking it as real. If so, then lucid dreams, far from strengthening the imagination view, appear to strengthen the quasi-perceptual view.

Even this conclusion, however, may be oversimplified. This is so because the general distinction between imagining and perceiving as involving direct versus indirect forms of control is itself problematic. Waking imagination can have an involuntary quality; we can feel that a particular image, like a thought, occurs to us spontaneously and persists despite our wish to banish it. Ichikawa acknowledges these cases but argues that "the fact that sometimes we imagine things we'd rather not be imagining does not show that imagining is not an action and subject to the will" (Ichikawa, 2009, p. 107). This is the case not just for compulsive thoughts but also for the familiar experience of having a tune running through one's head though one neither likes it nor wishes to "hear" it. Ichikawa likens these cases to unwelcome habits or addictions: "Even when our imagery is unwelcome and we cannot banish it, we can *try* to banish it; we know what it is to banish it. We are failing to perform an act" (p. 107). This description is correct, I think, in that we do have some control even over involuntary imagery; but it is not clear that this control is thereby direct. For instance, when I find myself unable to stop an unwanted fantasy or annoying earworm by direct "thought control," I may nonetheless succeed by engaging in some other complex activity, for instance, concentrating on a novel or listening to music, thus diverting my attention from the unwanted image or tune by using some source of external sensory stimulation. This type of control is not, it seems, fundamentally different from the indirect type of control I can take over sensory perception by averting my eyes, using earplugs, or turning up my stereo to better enjoy a favorite song. Nor does this inability to terminate unwanted imagery at will seem to be caused by my own ineptness. Loss of control over imagery not only is a major complaint in psychiatric patients but also occurs in healthy subjects. Excessive daydreaming and compulsive fantasizing can be a very real source of distress: subjects report that it interferes with their relationships, and it may lead to behaviors (pacing, speaking out loud to oneself) that are embarrassing and strenuous to hide (Bigelsen & Schupak, 2011; see also Schupak & Rosenthal, 2009). Loss of control over fantasy can affect the quantity, but also the contents and the intensity, of mental images and may involve both the inability to form images one wishes to have or the appearance of unbidden and even disturbing imagery (Horowitz, 1972). All these examples question the assumption that we have direct control over mental imagery and that imagery can always be described as having an active character. The claim that we have direct control over fantasy, even as a purely theoretical point, simply does not help us to understand these cases of unwanted or even compulsive fantasy, and it stands in stark contrast to the very real distress experienced by subjects suffering from this condition.

It is also unclear that we never have direct control over perception, or that perception is inherently passive rather than active. The latter view is defended by proponents of sensorimotor or enactive theories of perception (O'Regan & Noë, 2001; Noë, 2004; Thompson, 2007). To the extent that knowledge of sensorimotor contingencies and expectations shape ongoing perception, perception is not just an action but an *inter*action with the external world. But one does not have to subscribe to such accounts to find less controversial examples of taking direct control over one's percepts. Bistable images, in which subjects can switch between two different percepts voluntarily (e.g., between seeing a duck and a rabbit, to use Wittgenstein's [1953] well-known example), show this clearly (for details, see Windmann et al., 2006). Here control over the visual percept is direct in that it requires only shifts in attention and fixation patterns. In other cases, such as the familiar example of "zooming" in on a particular conversation, previously barely distinguished from the background sound at a party, mere shifts in attention may be sufficient. If so, then the weakest interpretation is that in these rare cases, perception has an active quality. This is enough to refute the claim that perception, unlike imagination, is always passive. Finally, as Thomas (2010) points out, the distinction between altering one's percepts through cognitive processes such as attention and bodily movements such as turning one's head *itself* can be rejected by calling the dichotomy between cognitive and bodily action into question. A preliminary conclusion is that perception, imagination, and dreaming are only gradually distinguished by different types of control; no sharp distinction can be drawn between our ability to exert direct or indirect control over the contents and the occurrence of perception, imagination, and dreams.

Finally, why think that dreaming requires authorship and control at all? An underlying assumption is that the narrative structure of dreams is such that it gives "every sign of intelligent design" (McGinn, 2004, p. 90). Again, this claim involves two background assumptions. The first is that dreams really do have narrative, storylike structure. This is an empirical and, as will become clear, controversial issue. The second is that dreams have the kind of narrative structure that requires intelligent authorship. Clearly, all other things being equal, an account that does not require an unconscious dream producer would be preferable. For the time being, the most cautious approach is to formulate this as a question: given that there is no clear sense in which nonlucid dreamers author their dreams, and given that even in lucid dreams, control is fleeting and unstable, how can the process of dream generation be explained without making tedious assumptions about unconscious control and authorship?

Putting this question to the side, for now, note that the analysis of dream control and authorship has already enabled a number of important insights. First, we have seen that by focusing on this issue, we now have a better idea of what kinds of questions are involved in the contrast between the quasi-perceptual and the imagination view. In particular, by investigating the issue of control, it has become clear that this factor cannot even be used to distinguish between perceiving and imagining. But if this is correct, then asking on which side of the putative divide between perceiving and imagining dreaming falls is equally flawed: control, as one of the classical candidates for drawing this distinction, cannot clearly distinguish between imagining and perceiving. A much more fruitful project is to describe continuities in control across perception, imagination, and dreaming. Doing so, however, changes the nature of the debate: the contrast between the imagination and the quasi-perceptual view is no longer exclusive. Acknowledging this fact will require a messier and more complicated, but also more sophisticated account of dreaming. For the same reason, it will also have the potential to be more empirically plausible.

Second, the issue of dream control has allowed for the formulation of more specific questions. In particular, an important desideratum for an empirically informed theory of dream formation is to explain how dreams are generated without assuming them to be the product of a dream producer, a dream author cleverly concealed, as it were, behind the scenes.

Third, we can now proceed to ask a number of more informative questions that follow from the imagination view. How exactly does dreaming resemble the phenomenology of waking imagination, and how does it differ? And which processes underlying dream generation are related to waking imagination? Now that we have freed ourselves from the tedious assumption that dreaming must neatly be described either as quasi-perceptual or as imaginative experience, it will be easier to see what is good about the imagination view, and how it functions as a corrective of the equally tedious assumption that dreaming replicates waking experience in all its detail.

6.1.3 Lessons from the Imagination View of Dreaming

In this section, I focus on several aspects of the imagination view that I take to be the most valuable for the question of dream experience: indeterminacy, recognition by just knowing, attention dependence, observational attitude, and emotion. My main claim is that by focusing on features of dream experience commonly taken to support the imagination view, we in fact find further support for the continuity between perceiving and imagining. In this process, perception will occasionally appear more imagination-like—or vice versa—than commonly assumed. This shows that the investigation of dreaming is theoretically fruitful because it has the potential to alter our theoretical understanding of standard wake states. In some cases, these can even be assimilated to dreaming, rather than the other way around.

To begin with, a widely shared observation is that dream imagery is often indeterminate and that this property distinguishes dreams from waking perception (Stone, 1984, pp. 363–364; McGinn, 2004, pp. 85–86). In this view, dreams have no more detail than is experienced. Questions about detail in perceptual objects have an answer; questions about detail in dreamed and imagined objects need not.

The indeterminacy of mental imagery has long been a topic of philosophical discussion. Descartes alludes to it in the *Sixth Meditation*, noting that we cannot form a mental image of a figure with a large but determinate number of sides. An imagined chiliagon, for instance, is no different from a figure with ten thousand sides. The resolution of mental imagery is too poor, it seems, to sharply bring a determinate number of sides into view. More recently, indeterminacy has frequently been discussed in the imagery debate (see Block, 1981; Tye, 1991, 2009, 2010; Nanay, 2009b). As Dennett (1981a) points out, while it is possible to imagine a striped tiger, an imaginary tiger need not have a determinate number of stripes. According to Dennett, this means that mental images are better compared to descriptions than to pictorial representations: descriptions, but not pictures, could represent a tiger as having numerous stripes without specifying the exact number of (visible) stripes, and the same is true for imagined or, as Dennett adds, dreamed tigers.

It also seems plausible that visual dream imagery can be indeterminate in certain respects, and Ichikawa (2009, pp. 108–109) supports the imagination view by taking up Schwitzgebel's (2002c, 2011c) example of color indeterminacy in dreams. Indeed, we have seen that dreams are often marked by a loss of color saturation and of background detail and that they may appear unfocused (cf. sec. 5.2.1), and at least a small percentage of dreams might be indeterminate with respect to color (cf. sec. 4.4; Murzyn, 2008). More evidence comes from lucid dreams. Lucid dreams often do not end abruptly but fade into wakefulness, taking on a washed-out, muted quality. Brooks and Vogelsong (1999, pp. 144–145) write that they frequently experienced a fading of the visual scene or certain blank spots with an almost neutral color tone during their lucid dreams and suggest that such blank spots are also present in nonlucid dreams, but simply go unnoticed.

In addition, vagueness plays an important role in dreams, being one of the subtypes of bizarreness (together with incongruity and discontinuity; see Hobson, 1988; Revonsuo & Salmivalli, 1995) that have been targeted by a number of studies. Perhaps a subgroup of vague dream contents, such as vaguely seeing a landscape or another dream character, can be redescribed as phenomenally indeterminate.

Instances of character identification in dreams by just knowing can possibly also be interpreted in this manner (Kahn et al., 2000, 2002; Revonsuo & Tarkko, 2002). If a dreamer describes *just knowing*, in her dream, that she was seeing her mother, is unable to provide any specific information about the visual experience of her mother, and even denies having had such detailed visual imagery, then according to the conservative approach, the most cautious interpretation would be that she saw her mother indeterminately.[4]

What about the other sensory modalities? Voss et al. (2011) analyzed the dream reports of congenitally deaf-mute individuals and found that nearly 80 percent lacked any discernible reference to sensory loss. In fact, their participants frequently described understanding speech or speaking themselves in their dreams. Voss and colleagues note that one dreamer, upon questioning, explained that she had not actually heard anything in her dream, but communication had been of a blurred, telepathic nature. This is interesting for two reasons. The first is methodological: the occurrence of speech in dreams should not be conflated with the occurrence of auditory experience. This fact, however, may easily be overlooked by raters unless they explicitly ask about the quality of sensory experience in dreams. The second is the prediction that generally and even in healthy subjects, phenomenal indeterminacy in dreams may be much more widespread than assumed (in sec. 7.3.3, I develop this argument in more detail with respect to bodily experiences in dreams). Again, this possibility raises a number of empirical research questions. As discussed in section 5.2.1, it is unknown to what extent dreaming simulates the subtle aspects of perceptual experience (Nielsen, 2010c), and perhaps phenomenal indeterminacy provides an answer to why dreams seem realistic while nonetheless failing to replicate the detail of the waking world—and perhaps of our experience of it. How much of dream experience, exactly, requires detailed imagery, and how much can be explained by phenomenal indeterminacy and feelings of just knowing? Are healthy subjects able, for instance, to confidently describe hearing dream speech, or does dream speech generally have a thoughtlike or "telepathic" quality (for an argument claiming that dream speech involves the mechanisms of speech production but not of auditory perception and hence cannot properly be described as hearing, see Symons, 1993; see also sec. 8.2.2)?

For now, and for the purpose of argument, let us take it for granted that phenomenal indeterminacy is widespread in both dreams and waking imagination. Even so, note that similar claims about indeterminacy can be made about waking perception. In particular, indeterminacy has been used to question the snapshot view of visual perception (cf. Noë, 2002). If visual experience itself is gappy, and if this gappiness is unnoticed, then we might erroneously attribute more detail to visual experience than is actually present. In this view, while detail is indeed present in the objects of perception, it is not present, in the same manner, in our experience of them, and this is even true of standard waking perception. For now, we can word this as an open question: how much detail really is present in visual experience, and how good are we at identifying phenomenal indeterminacy in visual perception?

Dennett (1991, pp. 53–54) uses the example of a playing card, moved slowly from the periphery of the visual field toward its center while the eyes are focused straight ahead. As Dennett notes, one can detect the motion of the card without being able to tell what color it is. Closer to center of the visual field, one can see that the card represents a face before being able to tell whether it is a jack, queen, or king. The idea is that by carefully attending to one's experience of the playing card (or, in the terminology introduced in chap. 4, by systematically improving the reporting conditions), one can become aware of the relative lack of visual information pertaining to color, face type, and so on. Yet our experience of the card was phenomenally indeterminate all along; we just did not notice. According to Dennett (1991), such examples of phenomenal indeterminacy show that phenomenal experience in general, and visual experience in particular, is much less detailed than commonly assumed. Physiological facts about seeing—such as the existence of a blind spot, caused by a lack of photoreceptors on a circumscribed region of the retina, and the limited resolution of vision in the parafoveal and peripheral regions—lend additional support to this claim (see, e.g., Staznicker, 2011). Because detail is readily available in the world in response to behavioral exploration, we mistakenly infer that this detail is also present in experience. Yet, at any given moment, the amount of detail in visual experience is limited: visual experience is not snapshotlike and does not provide uniformly detailed information about the world at any given moment. Importantly, this is true not only for dreams and imaginings but also for visual perception.

The indeterminacy of visual perception was already discussed by Husserl (1966). He emphasizes that the degree of determinacy in perception is highly dynamic. In broad daylight, the color of a house is seen more determinately than at dusk or in the fog. Similarly, my initial view of a table will be indeterminate in certain respects (e.g., regarding the grain of the wood), and looking at the table more closely or from the other side will reveal new aspects of what was previously indeterminately perceived or not perceived at all. My initial view of a table, then, involves a determinable kind of indeterminacy. According to Husserl, whatever is seen involves an anticipatory intention of how its indeterminate or empty aspects could be filled or brought into better view (for discussion, see Madary, 2012b, 2013). Yet this second perception of the table, even as it becomes more determinate in certain respects, becomes less so in others: for Husserl, determinacy and indeterminacy are gradually distinguished, and every perception involves partially filled and partially empty intentions.

Related to indeterminacy is the alleged attention dependence of dreams (McGinn, 2004, p. 79). Indeed, given that there is nothing more to the dream world than what is experienced, it seems attractive to say that there is no more to dreaming than what is the focus of attention. A related claim is that it is impossible for one's mind to wander away from the topic of one's ongoing dream: once attention is distracted from a dream object, it disappears, and the same supposedly is true of imagined objects (McGinn, 2004, p. 79). According to McGinn, dreams are an instance of modal blindness: "The possible world I am conjuring in my dream is the only world that I can represent" (p. 80). As in Sartre's theory, the fictional world of dreaming is epistemically closed.

This claim, however, is empirically false. Lucid dreams not only show that one can divert one's attention away from the ongoing dream (i.e., by realizing that or wondering whether one is dreaming, as in prelucid dreams) but suggest that having divergent thoughts may well be necessary for successful attempts at dream control. Lucid dream control, after all, involves thinking about what one will do next and how one will change the ongoing dream, as well as remembering the lucid dream experiments one plans to conduct. This is especially true for signal-verified lucid dreams, which clearly involve directing one's thoughts away from the dream and to the waking world. While early authors often took dreams to be "single-minded" (Rechtschaffen, 1978), more recent work has shown that cognitive activities occur even in nonlucid dreams (see sec. 9.3 for detailed discussion). And though dream thinking often has an ad hoc and confabulatory quality, once thinking is allowed to occur in dreams at all, it is unclear why such thoughts should be constrained as to their content (i.e., being about the dream rather than about something else), especially as empirical evidence suggests otherwise.

It is also unclear whether we can regard even waking imagination as a creature of attention in any simple sense of the term. Though mind wandering is often described as occurring when attention is directed to internal thoughts and away from external stimuli, episodes of mind wandering are frequently unintentional and may even go unnoticed (for details, see Schooler et al., 2011). Indeed, an obstacle to the study of mind wandering is that people fail to notice—and hence to report—that they are doing it. This is why special probes are needed to determine the frequency of mind wandering (for a review of methods used for the study of mind wandering, see Gruberger et al., 2011). Recent research on mind wandering suggests that two core processes are involved, where the first involves a decoupling of attention from perception, and the second involves fluctuations in the ability to become aware of the contents of consciousness, including the ability to realize that one's mind has wandered (Schooler et al., 2011). The two apparently vary independently of each other, suggesting that no simple account of attention with respect to mind wandering is possible. Hence, saying that even waking imagination, which arguably overlaps with much (though possibly not all; see Gruberger et al., 2011; and sec. 6.2.3) of what we do during mind wandering, is attention dependent is oversimplified, and an account is needed of how mind wandering can involve a reorientation of attention while itself going unnoticed.

If dreams and waking imagination cannot be equated with attention in a simple sense, and if at least some cases exist in which attention is not wholly devoted to the ongoing dream or daydream, can anything more be said about the relationship between the dream self and dream imagery? A prediction of the imagination view is that dreamers do not take an observational attitude toward dream imagery. Because dream objects are merely imagined and not perceived, dreamers do not widen their eyes or turn their heads to see dream objects more clearly, as they would in waking perception or while hallucinating (McGinn, 2004, pp. 76–77). This claim conflates, however, the behavioral passivity of the sleeping subject with actions of the phenomenal dream self. Indeed, there is no obvious reason why dreamers should not take an observational attitude toward dream events; it is just that in standard cases, due to REM sleep paralysis, this is not observable from the third-person perspective. At least occasionally, dreamers do describe situations in which they behave in ways that improve their perception of an object in a dream. For instance, in Freud's (1899/2003) famous dream of Irma's injection, he takes his patient to the window to better be able peer into her throat. Ellis reports the following dream:

> Thus I dreamed that I was in a drawing-room and saw a beautiful and attractive woman with an unusually low evening dress entirely revealing the breasts; then, between the breasts, three additional nipples appeared, and I realised in my dream that here was a case of supernumerary breasts of sufficient scientific interest to be carefully examined later on; and then, as I gazed, I saw a number of little fleshy nipple-like protuberances on the body, and thereupon I realized that I was really looking at a case of the rare skin disease *molluscum fibrosum*. Thus the perpetually wavering and developing image is at the same time a succession of quite different images. (Ellis, 1911/1931, pp. 45–46)

These types of reports certainly support the view that the dream self can take an observational attitude toward dream imagery: dream objects are not wholly attention dependent but respond, though in sometimes unexpected ways, to perceptual search. As Nielsen (2010c) notes, evidence from dream reports suggests that "acts of touching and looking both occur as they might in reality and, moreover, that the acts lead to the apparent acquisition of new 'perceptual' information" (Nielsen, 2010c, p. 601). This supports the view that dreaming may not only involve taking an observational attitude toward dream imagery but also support different types of exploratory perceptual behavior.

Again, the same might be true for waking imagination, though possibly to a weaker degree. Thompson (2007, chap. 10) notes that in imagining an object, we also coimagine the visual experience of such an object. Imagination itself cannot simply, in this view, be conceived of as picturelike but always includes, at least implicitly, an imagined perceiver, a perceptual interaction with the imagined object. Though Thompson emphasizes that imagined objects cannot be explored in the same way as perceptual ones, imagining is not wholly independent of sensorimotor processes but is at least implicitly concerned with imaginatively simulating them (pp. 291–297).[5]

A further prediction of the imagination view is that dreaming fails to give rise to intense, wakelike emotions, suggesting that dream emotions are only "muted versions of themselves" and lack the "sting" of real emotions (McGinn, 2004, p. 111). Similarly, Ichikawa writes that "dreams don't involve emotions, except in the way that fictions do" (2009, p. 119). Here, he carefully sidesteps the question of whether fiction involves genuine emotions, making the question of dream emotion contingent on the question of emotions generated in response to fiction (for discussion, see Gendler, 2013). By contrast, McGinn more strongly states that "dream fear is not *quite* the same as real fear" (McGinn, 2004, p. 110).

This description of dream emotions, however, is contradicted by the well-documented association of nightmares, both in frequent nightmare sufferers and in subjects experiencing the horrific recurrent nightmares that are one of the major symptoms of post-traumatic stress disorder (PTSD), with daytime anxiety, nightmare distress, and low levels of well-being, as well as by findings on general carryover effects between dream and waking emotions (Nielsen, Kuiken, & McGregor, 1989; Punamäki, 1999; Nielsen, Laberge, et al., 2000; Blagrove et al., 2004; Valli et al., 2005). Aside from the fact that dream emotions are often described as being at least as intense as waking emotions and covering roughly the same range as waking emotions (Nielsen et al., 1991), the fact that dream emotions can shade into and even have a causal effect on waking emotions and daytime levels of well-being is a strong argument for taking dream emotions seriously and regarding them as equivalent to waking emotions. At least there seems to be no principled obstacle to saying that dreams, like waking experiences, can elicit strong emotional reactions. The extent to which the same is true for waking fantasy is a further question. At least, cases of compulsive fantasy suggest that waking daydreams are not emotionally neutral (Bigelsen & Schupak, 2011; see also Schupak & Rosenthal, 2009).

What we can see, then, is that a number of predictions of the imagination view of dreaming fail. Yet they do so in a highly informative manner: alongside the candidates proposed to distinguish dreaming from waking perception, the candidates proposed to distinguish imagining from waking perception have failed as well. Not only does dreaming fail to fall soundly on one side of the (imaginary) divide between imagining and perceiving, but it erodes the very distinction itself. Clearly there is a phenomenological difference between imagining and perceiving; mostly, we do know whether we are engaged in one or the other. Yet none of the factors commonly taken to account for this difference is entirely satisfying or perfectly sharp: perceiving, dreaming, and imagining cannot be distinguished cleanly by the factors of control, phenomenal indeterminacy, recognition by just knowing, attention dependence, observational attitude, or emotional reactivity.

This claim is further supported by a number of findings on the continuity between imagining, perceiving, and dreaming. Numerous experiments have shown that visual imagery behaves like visual perception, for instance, in the time required to perform similarity judgments or mental rotation tasks, and in the overflow of the visual field, such as in seeing or imagining an object from close up or zooming in on it (for an overview and further references, see Kosslyn, 1994; Currie & Ravenscroft, 2002, pp. 71–88; Grush, 2004). A similar case can also be made for other types of imagery, including motor imagery (Currie & Ravenscroft, 2002, 84–88).

The similarity between imagery and perception is not restricted to the phenomenological level of description. A number of studies have shown that dreaming can be influenced by waking imagery and that both waking imagery and dreaming have similar effects on learning. Waking imagery rehearsal has been shown to reduce bad dreams and dream-related daytime distress in frequent nightmare sufferers (Germain & Nielsen, 2003a; Lancee et al., 2011), and a similar effect has been suggested for lucid dreams (Gavie & Revonsuo, 2010). Erlacher (2005) investigated the effects of executing certain movement patterns (such as throwing a coin into a cup) during lucid control dreams on subsequent waking performance and compared them to the effects of mental rehearsal (Erlacher, 2010). Though the general effects of sleep on motor skills are unclear (Blischke & Erlacher, 2007; Blischke et al., 2008), there is some support for the view that executing actions in lucid dreams, but also imagining their execution in wakefulness, has a positive effect on their subsequent execution (Erlacher, 2005). Importantly, such findings suggest that the effects of mental and dream rehearsal are similar to actual rehearsal during wakefulness, thus again supporting a continuous account. Erlacher and Schredl write:

> The studies reviewed in the present article demonstrate that REM (lucid) dreamed actions are comparable—like imagination during mental practice—to actual actions and that it is plausible that the dreamed action shares to some extent the same neural substrate on a cognitive level as if the action is carried out in the waking state. (Erlacher & Schredl, 2008b, p. 12)

Desseilles et al.'s (2011) discussion of neuroimaging data from REM sleep further supports an overlap between dream imagery, waking mental imagery, and illusions, and compelling evidence, both from neuroimaging and from lesion studies, points to an overlap in the neural mechanisms involved in visual imagery, hallucinations, and visual perception (Pylyshyn, 2002; Grush, 2004, p. 387; Kosslyn et al., 2006; but see Howard et al., 1998). It may even be possible to use fMRI response patterns for imagined objects to predict patterns for seen objects, and a decoder trained on visual imagination can be successfully used for decoding neuroimaging data obtained during visual perception. This suggest similarities in neural activation patterns, at least in higher visual areas (Cichy et al., 2012; see also Haynes, 2015). At the same time, subtle differences, for instance in hippocampal activation, may distinguish different states traditionally classified, in the philosophical debate, as forming part of imagination, such as remembering the past and imagining the future (Kirwan et al., 2014).

Phenomenologically, the well-known Perky experiment (Perky, 1910) is often taken to demonstrate that under certain conditions, subjects can fail to distinguish between mental imagery and perception, and this effect may be especially pronounced in certain patient groups (Currie & Ravenscroft, 2002, pp. 72–78). In Perky's original experiment, subjects were asked to form a mental image, say of a banana, and unbeknownst to them were presented with a picture showing the object they had been asked to imagine. Though most subjects failed to detect the difference, the stimulus was found to influence the orientation and vividness of the imagined objects. Segal (1972) reports a number of further studies showing that visual imagery can be experimentally influenced by visual stimuli, suggesting that waking imagery is more closely related to external stimulus processing than is often assumed (for critical discussion, see Hopkins, 2012; Nanay, 2012). In certain conditions, even the borders between waking imagery and perception, then, may be blurred to the point of subjective indistinguishability.

To conclude, dreaming commends itself as a wonderful target phenomenon for investigating the continuity between perceiving and imagining. This metatheoretical claim is, of course, at odds with many existing philosophical attempts to conceptualize dream experience by modeling it on either waking perception or imagination and regarding dreaming as a test case for introducing a sharp distinction between perceptual and imaginative experience. To the contrary, dreaming is interesting, or so I would argue, not because it fortifies the distinction between perceiving and imagining but because it erodes it. Certainly the distinction between perceiving and imagining is a helpful conceptual tool, both in folk psychology and in research. Yet it is an idealization and as such does not work all the way in. In particular, it does not entail the existence of any sharp phenomenological distinction between the two types of states. Rather, an empirically informed account of dreaming and its relation to imagining and perceiving should aim to do justice to the folk-psychological distinction between these states, for instance, by explaining why they typically differ from each other, while resisting the temptation to draw any sharp distinction between them.

Perhaps one reason why the imagination and the quasi-perceptual view have seemed so attractive to different thinkers is that dream imagery occasionally conforms to both: some dreams indeed involve vivid, wakelike visual imagery, while others may take on a faded, washed-out quality. Phenomenal indeterminacy may not be evenly distributed in all dreams or even over all sensory modalities in dreams. And even when it occurs, it may easily be overlooked (I return to this point in sec. 7.3.3). For now, our next step is to examine the empirical and psychological literature in more detail. This will help clarify the role of cognitive processes in dream formation and will clear up some further problems in contemporary theorizing about dreaming.

6.2 The Empirical and Psychological Literature

6.2.1 Dreaming as Thinking: The Language of the Mind during Sleep?

The philosophical discussion of dreams as imaginative experiences is paralleled by a discussion of the cognitive sources of dreaming in the empirical and psychological literature. The question is framed slightly differently than in the philosophical literature, however, and the central—and, as I have been arguing, problematic—claim that dreams are imaginative experiences in that they are *phenomenologically* like waking imagination does not play an important role in the empirical and psychological literature and indeed is outright rejected by many authors. At the same time, in the psychological literature, the analogy between dreaming and thinking gives rise to a different set of problems. The most important and most famous of these, perhaps, is the Freudian claim that dreams are inherently meaningful and have quasi-linguistic structure.

This claim is perhaps most clearly articulated in the work of David Foulkes (1978, 1993a, 1999), who is one of the most important proponents of a cognitive theory of dreaming. His central claim is that "dreams are thoughts. Dreaming is thinking" (Foulkes, 1978, p. 3). Foulkes explicitly distinguishes our ability to "think in pictures" (p. 5), as in waking imagination, from the phenomenology of dreaming, noting that in dreams, "the pictures are sufficiently perceptlike generally to lead us to believe, until the moment of our awakening, that we actually are seeing real events" (p. 5). Yet, to understand the contents of dreaming, or so he claims, one should look not to low-level perceptual but to higher-level cognitive processes: "Dreaming is realized thinking . . . and its perceptual constituents can be made to reveal, as surely as the words of wakefulness, the content of the underlying thoughts seeking expression" (p. 10). Though dreaming feels unlike waking thought, it shares its quasi-linguistic, propositional character.

A similar position is endorsed by Fromm (1957), who links dream symbolism to the assumption that dreaming is a language with its own grammar and syntax, and by Calvin Hall, who views dreams as "the embodiment of thoughts. They are the medium by which a psychological process, cognition, is transformed into a form that can be perceived" (Hall, 1953, p. 274). Though the explicit ascription of a grammar and syntax to dreaming is comparatively recent, note that the view of dreaming as a form of symbolic and meaningful language is also closely related to the practice of dream interpretation in general, which throughout history and to this day has given rise to "dream dictionaries," such as Artemidorus's (1975) famous *Oneirocritica*, or *Interpretation of Dreams*.

Yet, it is important to look at the arguments for this view in more detail and ask whether they justify the description of dreaming as quasi-linguistic thinking in a more than purely metaphorical manner. First, alluding to Rechtschaffen (1975), Foulkes suggests that dreaming may be "even more prototypic of linear, narrative-style thinking than much of our ordinary waking thought" (Foulkes, 1978, p. 9). Second, Foulkes claims that "*all* symbolic activity, *all* representational experience, needs to be traced back to a motivational base," and "perception itself is a higher symbolic process" (p. 7). However, the motivational sources and symbolic character of dreaming are more pronounced than is the case for waking perception: "It is . . . not that dreams are unique in their motivational-symbolic interrelatedness. It is only that they display, in a relatively flamboyant way, a motive-thought problem which is a major concern of all cognitive psychology" (p. 7). This last claim is also related to Foulkes's view that we are responsible for our dreams (p. 4), as well as to the Freudian assumption that dreams are symbolic and inherently meaningful. Like Freud, Foulkes views dreaming as analogous to language, arguing that dreams express their meaning in virtue of the grammatical and quasi-linguistic structure of dream content. Indeed, according to Foulkes (1978, p. 17), Freud's is the only classical theory of dreaming to fully acknowledge the need for an explanation of the inner structure or grammar of dream experience.

Recall that Freud's dream theory (Freud, 1899/2003, 1940/2003; for general discussion, see Marinelli & Mayer, 2000, 2002; for a recent discussion of the contribution of psychoanalytic dream theory to neuroscience, see Ruby, 2011) relies on the now-familiar distinction between manifest and latent dream content. Manifest content is the content of the remembered and reportable dream, and its function is to protect the dreamer, via the mechanisms of disguise and censorship, from the truly meaningful, but hidden or latent, content of the dream. Dreaming is a form of wish fulfillment, but because these wishes are unconscious and often repressed—preventing them from becoming the object of conscious waking thought—they can only be expressed in an indirect and concealed manner in the dream state. The latent dream content is the true but disguised meaning of the dream, and it can only be accessed via interpretation and free association and under the guidance of an expert, typically a therapist. Dream interpretation is only interested in manifest dream content insofar as it helps discover the deeper but hidden meaning of the dream. But how is the hidden meaning of dreams transformed into manifest dream content in the first place? Freud describes this process as involving a translation from one language to another:

> Dream thoughts and dream content lie before us like two depictions of the same content in two different languages, or better, the dream content appears as a transformation of the dream thoughts into a different manner of expression, whose signs and compositional rules we can discover by comparing the original and the translation. The dream thoughts are immediately comprehensible, as soon as we have discovered them. The dream content is given pictographically and each of its characters has to be transformed into the language of dream thoughts. One would obviously be led astray if one tried to read these characters according to their pictorial content as opposed to their symbolic relation. (Freud, 1899/2003, p. 284; translation mine)

To discover the underlying and truly meaningful dream thoughts, then, one has to understand the rules of translation according to which they are transformed into the manifest dream content. Only then can one translate the manifest dream content back into the underlying dream thoughts, revealing the original meaning of the dream. For Freud, these rules of translation are condensation, displacement, pictorial representability, and secondary revision. Foulkes describes these as the "transformational and combinational rules of a dream grammar" (Foulkes, 1978, p. 17). But why assume these rules to be necessary in the first place? Foulkes's central argument is that dreaming is a creative and generative process. Dreams are structured rather than random, and their recurrent constituents can readily be identified. Like drama, dreams are not only sequentially organized but also thematically coherent (pp. 13–18). This is true not just for the dream as a whole but also for single dream images. Like sentences, dream images have propositional structure and can thus be seen as formally equivalent to sentences: "Each image is a pictorial sentence in which such verbs and nouns are combined according to rules much like those applying to verbal sentences" (p. 15). In this view, bizarre and discontinuous dreams are

> degenerative organizational phenomena [that] are so extraordinarily uncharacteristic of REM sleep that they clearly must be regarded as defective cases. The rule, both between and within images, is that dream expressions are linearly, propositionally structured and that constituent elements of these structures can be reliably identified and assigned meaningful structural labels. (Foulkes, 1978, p. 15)

There is "a systematic grammar guiding surface dream expressions," and the "infinitive creativity of dream imagery" is comparable to and no more mysterious "than the demonstrated infinite creativity of verbal expression" (Foulkes, 1978, p. 16).

Aside from generativity and syntactic and semantic structure, dreaming is similar to speech in that both "involve the 'externalization' of thought in terms of a sensory modality" (Foulkes, 1978, p. 17), such as audible speech or visual imagery. Foulkes also notes that the distinction between latent and manifest dream content proposed by Freud bears similarity to Chomsky's (1957) distinction between deep and surface grammar, and suggests that there are numerous possibilities for cross-fertilization between the two theories (Foulkes, 1978, p. 17).[6]

Is the assumption that dreaming is "a form of natural language" (Foulkes, 1978, p. 14) tenable? A first problem is that this view requires the existence of a system of combinatorial rules and the attribution of grammatical competence to the dreamer. However, as discussed in relation to McGinn's argument that dreams are the product of intelligent design, this competence cannot straightforwardly be attributed to the dream self: clearly the quasi-linguistic structure of dreaming does not aspire to phenomenological plausibility. We certainly do not *feel*, while dreaming, that we are engaging in quasi-linguistic thought. Moreover, if it is not the phenomenal dream self that has these linguistic competences, then one has to posit some other nonobservable and unconscious agent who does—and this homuncular theory is certainly not attractive. Freud's explanation, predictably, is that this difficulty is due to the mechanisms of disguise and censorship; however, this results in the epistemologically uncomfortable position that the very competence that lies at the heart of the issue is hidden.

A related problem is that we have no straightforward way to assess the accuracy of dream interpretation. There is simply no way to check whether the proposed rules of translation used to reconstruct the latent content of a given dream are appropriate to its actual meaning (assuming such meaning exists). Because we lack independent means of accessing latent dream content, its very existence is a purely theoretical construct. Freudian dream theory has to be believed despite the lack of obvious evidence in its favor—indeed, the lack of such evidence is inherent in Freudian dream theory. Denying the hidden meaning of a dream is, after all, simply a further result of censorship. Because Freudian dream interpretation is essentially unfalsifiable, it has even been compared to religion (e.g., Hobson, 1988, 2002). For language, the situation is importantly different. Though philosophers may squabble about the correct use of language, these squabbles occur before a background of generally accepted grammatical rules and ways of determining linguistic competence.

Note that the same point applies to dream production. Just as there is no way of checking the competence of a dream interpreter, there is also no way of assessing the competence of a dreamer in producing dreams in the first place. Imagine a dream interpreter informing a dreamer that given the latent content of the dream, the manifest dream content should have been different from the content described in the report: in the future, the dreamer should try to dream more skillfully! The dream he has reported is the result of a bad translation and distorts its real content. Accusing a dreamer of being an inept translator of latent dream thoughts has the air of absurdity. This somewhat bizarre scenario points to a more serious problem: if dreaming is quasi-linguistic, there should be agreed-upon norms for doing it correctly and incorrectly, as well as more or less eloquently. Simply assuming that bizarreness is a degenerate form of dreaming is not enough; especially as it could be, for instance, a particularly clever and creative way of concealing the dream's real meaning. The problem might be that of the translator, not of the dreamer, and norms should be available for assessing this difference. Without such norms, the claim that dreaming involves a quasi-linguistic form of competence is empty. But this also means that the alleged productivity and systematicity of the language of dreams—an important motivation for comparing dreams to language—cannot be ascertained. The ability of speakers to understand novel sentences and produce new ones based on a set of learned syntactical rules is both impressive and testable. But for dreams, not only is there no one to whom this competence could be attributed, but its very existence must be accepted by default.

Note that there is yet a further disanalogy between dreaming and language, which is inherent in Freudian dream theory itself. Where linguistic competence involves the ability to both produce and understand language, this competence, in Freudian dream theory, is unevenly distributed for the language of dreams. In deciphering the meaning of dreams, competence is attributed to the psychoanalyst, not the dreamer. Strangely, the dreamer, in this view, masters the alleged translation rules well enough to produce the dream, but not to understand it. And even worse, not understanding the dreams he produces is part and parcel of the wish-fulfilling function of dreaming. This is why the deeper meaning of dreams requires professional decoding in the first place. But this is to claim, essentially, that while subjects competently produce speech in a given language, they are systematically unable to understand what they are saying. I conclude that the metaphoric description of dreaming as quasi-linguistic thinking requiring professional interpretation is best rejected.

The analogy between language and dreaming and the assumption that dreams have hidden meaning—despite their ultimate failure—help illustrate a further problem for Freudian dream theory. Recall Putnam's (1986) claim that depth-grammar analyticity, or the assumption that there are analytic or purely conceptual truths of a hidden nature, does not exist (sec. 1.3.1). We can use Putnam's point to highlight a nice parallel between Malcolm's analysis of the concept of dreaming and Freudian psychoanalysis. In both cases, the underlying logic—in Malcolm's case, of dream-related sentences; in Freudian dream theory, the message underlying manifest dream content—is hidden to the layperson, to the ordinary speaker or dreamer; the expertise of the professional philosopher or psychoanalyst is required to make it visible. Putnam's rejection of depth-grammar analyticity can now nicely be applied to dream interpretation, as well: if dreams have meaning, they have surface meaning of the kind that is transparent in dream reports, and not just to experts, but to laypersons as well. No special diviners of the meaning of dreams are necessary. The alleged depth meaning of dreams does not exist.

This is not to deny that dreams can be subjectively meaningful or that dream interpretation can generate genuine insights. It is also not to deny that certain social contexts—such as dream groups or even just talking a dream over with a friend—can generate new and even surprising insights that one might not have had on one's own (Edwards et al., 2013). In keeping with antiskepticism, we might even want to say that just as special reporting conditions are needed to ensure the transparency of dream reports with respect to previous dream experience, so special interpretation conditions may facilitate the assignment of subjective meaning to dreams. Importantly, however, the conservative approach once more functions as a corrective: just as phenomenological claims about dreaming are constrained by the space of reportable dreams, so the transparency of dream meaning assumes that the dreamer, not the dream interpreter, is the ultimate arbiter of the "correct" interpretation of a given dream. Of course, a theory of dreaming will ultimately have to account for the fact that dreams often seem to have symbolic or metaphorical meaning. But for now, the challenge for a naturalistically minded theory of dreaming is how to explain the apparent meaningfulness and symbolic nature of dreaming *without* evoking an underlying agent, intention, or concealed message.

6.2.2 Continuity, Discontinuity, and the Problem of the Missing Contrast Condition

Even if dreaming is not a quasi-linguistic kind of thinking, this does not exclude the possibility that interesting continuities between waking thought and dreams exist. Large portions of dream research are devoted to investigating the cognitive sources of dreaming, and much of this research makes dreaming appear as an ordered process rather than as a product of random, brain-stem-driven signal generation (Hobson et al., 2000). There are several ways of spelling out this claim. A particularly prominent one, associated with the continuity hypothesis, is that dream content is continuous with waking experience, but also with waking thoughts and concerns (e.g., Domhoff, 1996, 2003, 2011a; Strauch & Meier, 1996; Schredl & Hofmann, 2003; Schredl, 2006, 2010a, 2014; Hobson & Schredl, 2011). A first observation is that dream content appears to be continuous with waking activity. For instance, the frequency of driving in wakefulness is directly correlated with the frequency of driving in dreams (Schredl & Hofmann, 2003), and more generally, the frequency of thinking and physical activity in dreams appears to be related to daytime levels of physical activity (Breger et al., 1971, pp. 26–27). Similarly, studies have shown that patients dream about hospitals and health before operations (Breger et al., 1971), suggesting that daytime stress has an effect on subsequent dream content. Experimental manipulation (e.g., studying the effects of emotionally engaging versus neutral films) also has an effect on subsequent dreams (Schredl, 2008b). There is even evidence for a two-way effect, such that dreams incorporating waking events also have a stronger effect on postsleep mood (Schredl & Reinhard, 2009–2010).

However, several problems arise for the continuity hypothesis. Without additional factors explaining the continuity, it is exceedingly vague. While Schredl (Hobson & Schredl, 2011) emphasizes the continuity of dream content with waking experience, thoughts, and feelings, Domhoff (2011a) defends a narrower version of continuity with waking conceptions and concerns. Another problem is that dreams are discontinuous with waking activities in striking and seemingly systematic ways. Certain activities such as reading, writing, and arithmetic appear rarely in dreams (Hartmann, 2000), though they occupy much of waking life, whereas others, such as conversations, movement, and being outside in nature, frequently occur in dreams (Hartmann, 1998; Schredl & Hofmann, 2003). Perhaps this suggests that dreaming is less concerned with serial activities than with a loose, autoassociative mode of cognition (Hartmann, 1998, 2000). Dreaming also has a distinctly social flavor, with the average dream involving approximately four dream characters (Kahn et al., 2002). This suggests that the continuity with waking experiences, such as activities and social contacts, explains only a small part of dream content, and that continuity does not extend to all types of dream content. Other dream themes, such as unaided flying, also depart quite drastically from standard waking activities (Hobson & Schredl, 2011).

Schredl (2003, 2006) suggests that specific factors determine the integration of waking experience into dreams, such as temporal distance (with more recent experiences having a better chance of appearing in dreams) and emotional involvement. It is well known that we often dream about stressful past or future events, such as the death of a loved one or upcoming examinations or journeys. Indeed, all of these are among the typical dream themes that most people report having dreamt about at some time in their lives (Nielsen, Zadra, et al., 2003; Yu, 2008), a finding that likely reflects waking concerns and emotional involvement. A number of studies confirm the general effect of stress on dreams (for discussion, see Breger et al., 1971), and traumatic events can be incorporated into dreams in a relatively realistic manner even years after their occurrence (Zadra, 2001). Other factors, such as competence, may also play a role. While professional musicians apparently dream about music twice as often as controls, this is not related to differences in the quantity of daily exercise but apparently reflects professional competence (Uga et al., 2006).

Evidence also points to long-term consistency of dream content. Lortie-Lussier (2000) found some evidence for the internal consistency of dream content in dream diaries recorded at intervals of 10, 15, and 17 years; however, due to methodological limitations, the long-term consistency of most features did not reach statistical significance in this study. Domhoff (1999, 2001, 2003; Domhoff & Schneider, 2008a) cites evidence from both cross-sectional and longitudinal studies that observed absolute thematic constancy of dream content in individuals over years. Such long-term consistency may coexist alongside and perhaps even be masked by day-to-day fluctuations in dream content (Domhoff & Schneider, 2008a). Domhoff (2003, pp. 27–28) suggests that this long-term consistency supports the view that dreams conform to a repetition principle—a possibility that becomes particularly obvious in repetitive or recurrent dream themes, which are experienced by 50 to 80 percent of people at some point in their lives.

Fluctuations in dreaming may also be driven by factors that are related neither to waking experiences nor to waking concerns. Nielsen (2004, 2010d) suggests that chronobiological factors influence both the intensity and dreamlike quality of dream reports and might account for mid- to long-term fluctuations in dream content. For instance, in women, hormonal changes throughout the menstrual cycle may mediate changes in emotional dream content. Generally, the length and quality of dream reports from both REM and NREM sleep change throughout the night. Some evidence also suggests that the incorporation of memory sources into dreams depends on a circaseptan rhythm, such that daytime events are most likely to be incorporated into dreams either immediately after the event (thus partly confirming the Freudian concept of day residue) or after a five- to seven-day interval, whereas experiences from two to four days before the dream are less likely to be dreamed about (Nielsen & Powell, 1992; Blagrove, Fouquet, et al., 2011; Blagrove, Henley-Einion, et al., 2011). There is also evidence that this effect is specific to REM sleep dreams (Blagrove, Fouquet, et al., 2011).

In sum, while we have ample evidence that dream content is influenced, among other things, by waking memories, experiences, concerns, emotions, personality traits, and chronobiological factors, the precise mechanisms driving incorporation into dreams are not well understood. Importantly, thematic continuity can exist alongside thematic discontinuity (see also Hobson & Schredl, 2011), suggesting that hypotheses about the continuity and discontinuity of dream content should be seen not as exclusive but as complementary.[7]

Continuity on the level of dream content also should not be confused with discontinuity between REM sleep/dreaming and standard wake states on the level for dream form (cf. sec. 2.2). Investigating the thematic content of dreams is one thing; investigating the way these contents are experienced in dreams is another. One recent study attempted to investigate continuity and discontinuity of dream content and form by comparing the dreams of congenitally deaf-mute and paraplegic subjects to those of nonhandicapped individuals (Voss et al., 2011). Raters (a behavior therapist, a psychoanalyst, a nonclinical psychologist, and a nonpsychologist) did particularly poorly at identifying the dreams of deaf-mute subjects, and interrater agreement was generally low. While this would seem to speak against the continuity hypothesis, the authors suggest that their results are insufficient to support any strong conclusions: elements of waking life often appear out of context in dreams, and discontinuity with waking activity may coexist alongside continuity with waking concerns.

Despite its inconclusiveness, this latter point is interesting from a metatheoretical perspective, because it suggests that the debate about continuity versus discontinuity is plagued by some of the same problems as the philosophical debate about the quasi-perceptual versus imagination view of dreaming. Both debates revolve around the question of whether an appropriate wake-state contrast condition for dreaming exists, and if so, what it is. As we have seen, the plausibility of the continuity hypothesis depends on what exactly dreaming is taken to be continuous with, and so far, there is no uniform answer to this question. In keeping with this claim, Montangero (2012a) argues that the apparent discontinuity between dreaming and waking thought promoted by Hobson would disappear if dreaming were viewed as a metaphoric and imaginative form of thinking, comparable to waking imagination and daydreaming. Moreover, by starting out from comparative assumptions about dreaming and its (dis) continuity with wakefulness, both the continuity and the discontinuity hypotheses risk importing wake-state assumptions into a theoretical characterization of dreaming, by either assimilating dreams to waking experience or overemphasizing the differences between dreaming and wakefulness. Generally, contrasting the thematic and formal continuities and discontinuities of dreams and standard wake states is a matter not of either/or but rather of more or less. This conclusion is shared by Hobson and Schredl (2011), two of the chief protagonists in this debate. Part of the seeming disagreement, consequently, may not be factual, but may arise from different perspectives on the same target phenomenon. As in the philosophical debate about the quasi-perceptual versus imaginative character of dreaming, the acknowledgment of a gradual distinction between continuity and discontinuity might go a long way toward resolving the apparent disagreement. Again, my preliminary suggestion is to bracket, for the time being, comparative statements about dreaming and wakefulness.

Yet (and this is an important point) the fact remains that the debate surrounding the continuity hypothesis has proved helpful, because the studies concerned with its investigation are slowly sketching the contours of a more differentiated picture of dreaming in its own right. Even if we leave the comparative assumptions implicit in this debate to the side, it seems clear that cognitive explanations are needed to explain why dreaming, to a considerable degree, seems to be an ordered rather than a completely random process. Emphasizing the cognitive sources of dreaming seems more apt for explaining these findings than regarding dreaming as a state of uniform cognitive deficiency. And because claims about the cognitive deficiency of dreaming, as we have seen, are often associated with the quasi-perceptual view, the imagination view with its emphasis on the cognitive sources of dreaming may provide the better explanatory resources for addressing the ordered nature of dreaming. Perhaps this is enough. And perhaps the inability of empirical studies to adjudicate between the continuity and discontinuity hypotheses will eventually lead to a more precise formulation and more workable successor-theories, or even to the disappearance of the apparent contrast between the two.

6.2.3 Dreaming and Waking Imagery

Perhaps the closest relative of the philosophical imagination view of dreaming is research on the relationship between dreaming and waking imagery. Three kinds of studies are relevant: studies investigating the development of dreaming and waking imagery in children; the comparative study of dream and waking fantasy reports; and the investigation of the neurophysiological processes involved in dreaming and imagining.

Drawing on extensive studies of children's dreaming, Foulkes (1993a, 1999) argues that dreaming is a cognitive achievement that is gradually developed throughout childhood and is "continuous with our waking reflective ability to think in images" (Foulkes, 1999, p. 15). In the late 1960s and early 1970s, Foulkes and his colleagues collected 2,711 dream reports during 1,347 nights in the sleep laboratory from two groups of children, ages three and nine at the beginning of the study and ages nine and fifteen at the end. The major findings from the longitudinal study were replicated in a smaller cross-sectional study in the mid-1980s, which focused on dream development between the ages of five and eight (for details and further references, see Foulkes, 1999). In addition to dream reports, researchers also collected a variety of data on the children's social and cognitive skills, with a focus on visuospatial skills and language proficiency (Foulkes, 1993a, 1999). The studies' results were surprising in several respects, suggesting a gradual development of dream imagery, narrative complexity, and reporting ability during childhood and early adolescence.

In Foulkes's study (for details, see Foulkes, 1999, pp. 56–115), preschoolers between the ages of three and five reported dreams following only 15 percent of REM awakenings. Their dream reports were brief and often described static imagery; only a quarter of the reports described any kind of movement. Thematically, preschoolers' dreams were dominated by references to animals and to their own body state, such as hunger or fatigue. Reports of five- to seven-year-olds remained rare but became longer and described social interactions and kinematic imagery. At ages seven to nine, reports became more frequent, following 43 percent of REM awakenings, and showed a considerable increase in narrative complexity. Only at this point did children start describing thoughts and emotions, as well as active self-participation in dream events. While the dream reports of younger children consisted in single-story units and unrelated actions, older children described sequences of interrelated dream events. The basic formal features of dreaming seemed developed in this age group, but dream recall frequency did not approach adult levels until nine to eleven years, and only then did individual recall rate and report length appear as "stable and interrelated personal traits" (Foulkes, 1999, p. 99). Finally, from eleven to fifteen, there was an increase in bizarreness. An increase in unfamiliar settings and characters had already been observed in the younger age groups, but the study found a significant increase in vague or settingless reports and distorted imagery in early adolescence.

The claim that young children's dreams lack complexity, bizarreness, and emotions was disputed by Hobson et al. (2000) and Resnick et al. (1994). However, the careful methodological safeguards employed in Foulkes's studies seem to allay the worries that the dream reports were skewed by the laboratory situation or that children were simply unable to articulate adequate dream reports (for discussion, see Domhoff, 2005).[8] For instance, Foulkes found no difference between laboratory reports and systematic home reports gathered by parents following timed awakenings. Also, no difference was found between the dream reports of the studied group of children and the reports of a new group of children who joined the longitudinal study in its fifth year, suggesting that long-term participation in the study did not alter children's individual reporting abilities (Foulkes, 1993a, pp. 123–124).

Concerning the developmental stages of dreaming, two points are of interest. First, if we take these findings at face value, a clear progression occurs in the development of the formal features of dreaming, from first static, then kinematic, dream imagery to active self-participation, dream thoughts, emotions, and narrative complexity, to an increase in bizarreness. Second, these qualitative changes in dream reports are not accompanied by a concomitant increase in recall frequency but rather seem to precede increases in dream recall. In support of his claim that dream reports following laboratory awakenings reliably reflect actual dreams and not merely individual differences in reporting ability, Foulkes (1999) emphasizes that the children's verbal skills, for example, their ability to describe images from memory, did not predict their dream reporting ability. Rather, he found a positive correlation between the children's performance in tests of visuospatial skills and their dream recall frequency, suggesting that the development of dreaming, as reflected by reports, is associated with waking visuospatial cognition. Children who had average memory and verbal skills but performed badly in tests of visuospatial skills also had lower dream recall than their peers and reported dreams similar to those reported by younger age groups. Foulkes argues that this difference "surely can't be just a matter of poor remembering or describing; it has to be a deficit in dream-*making*" (Foulkes, 1999, p. 121). Dreaming, then, appears to depend on the ability to mentally re-create or simulate the world, as in waking imagination (p. 65). The same is true for the self: using different tests to determine the children's development of self-understanding, Foulkes found that mature levels of waking self-understanding correlated positively with the active participation of the self in dreams and with waking visuospatial skills (Foulkes, 1993a, p. 129).

These results, according to Foulkes, have far-reaching implications not only for the study of dreaming but for consciousness research in general. Relating his findings on the development of dreaming to general theories of cognitive development in childhood—for example, the shift in cognitive maturation between the ages of five and seven and infantile amnesia (Foulkes, 1999, pp. 142–144)—Foulkes suggests that dreaming may serve as an index of general cognitive development (Foulkes, 1993a, p. 114): in reporting their dreams, "the children were unwittingly also describing the range and limits of their visual-spatial imagination" (Foulkes, 1999, p. 67). Recent research suggests that the ability to engage in mental time travel, that is, to mentally project oneself into the future and use information learned in the past for problem solving, may develop in children around four years (Atance, 2008; Redshaw & Suddendorf, 2013). This fits in well with the shift in dream complexity around that age.

In their recent review of dreaming and REM sleep, Nir and Tononi (2010), while accepting the vivid, hallucinatory quality of dream imagery, conclude that because of its relation to visuospatial skills and memory, "dreaming might turn out to be the purest form of imagination" (Nir & Tononi, 2010, p. 97).

> If the flow of brain activity during dreaming were shown to be largely backwards, as one would expect in imagery, rather than forwards, as in perception, many of the seemingly bizarre properties of dreams, such as blended characters and scene switches, would be easier to explain, as they are standard features of one's imagination. Such a top-down mode might disrupt the encoding of new memories, and thus underlie dream amnesia. In addition, top-down mental imagery could obstruct the processing of incoming stimuli and disconnect one from the environment. (Nir & Tononi, 2010, p. 97)

This suggests that the comparison between dreaming and waking imagination is a fruitful project. Indeed, Foulkes and Fleisher (1975) found evidence for the intrusion of dreamlike bizarreness into waking fantasy. However, recent studies comparing dream and waking fantasy reports from healthy and schizophrenic subjects suggest that the parallel may be overstated (Scarone et al., 2007; Limosani et al., 2011a, 2011b). Bizarreness levels in dreams of schizophrenic and healthy subjects are similar, but schizophrenic subjects also show a strong correlation between bizarreness levels in dreams and waking fantasy reports, which is not the case for healthy subjects. This finding suggests that dreams are more bizarre than waking fantasy in healthy subjects (Williams et al., 1992), and only in schizophrenic subjects do waking fantasies become more dreamlike and bizarre.

Perhaps the most all-encompassing theory investigating the cognitive sources of dreaming has been proposed by Domhoff (2001, 2003). His neurocognitive model of dreaming emphasizes the dependence of dreaming on visuospatial skills and on a specific network involving the limbic, paralimbic, and association areas of the forebrain. This model attempts to integrate three main sources of evidence: Foulkes's research on children's dreaming, Mark Solms's (1997, 2000) research on alteration or cessation of dreaming in lesion patients, and results from dream content analysis supporting the continuity of dreaming with waking concerns and memories. According to Domhoff, dreams draw from the dreamer's waking memories and general knowledge, thus expressing "several key aspects of people's conceptual systems, especially conceptions of the self, family, and friends" (Domhoff, 2003, p. 32). This view nicely incorporates both lower-level experiential aspects and higher-level cognitive influences. It also makes room for the apparent meaningfulness and expressive quality of dreams, suggesting that some dreams may even be the product of figurative thought. Aside from emphasizing the top-down (as opposed to bottom-up) processes underlying dream formation (Foulkes & Domhoff, 2014), this approach is closely related to the newest development in current theorizing about dreams from a cognitive science perspective: the analogy between dreaming and waking mind wandering.

6.2.4 Wandering Minds: From Waking Mind Wandering to Nocturnal Dreaming

Recent years have seen a flurry of publications on the topics of mind wandering, defined as task-independent thought, and the default-mode network, a network of brain regions related to off-task, baseline brain activity (Raichle et al., 2001; Buckner et al., 2008). As Fox and colleagues (2013) note, the default-mode network was discovered serendipitously when researchers noticed that certain brain regions are actually more highly activated during periods of rest than during task performance. This initially surprising discovery is reminiscent of the finding that brain activation reaches wakelike levels in REM sleep (cf. chap. 2). It is also, properly considered, not at all surprising: despite the outward appearance of rest, people are far from doing nothing when they are not focused on external stimuli and ongoing tasks (though mind blanking may be an important and underexplored exception; cf. Ward & Wegner, 2013): experience sampling suggests that up to 50 percent of waking thought is made up of mind wandering (Schooler et al., 2011).

Much research has focused on the negative effects of mind wandering on mood and task performance. Recent publications, however, recommend a more balanced perspective, taking the positive functions and pleasant aspects of mind wandering (such as constructive and volitional daydreaming) into account, but also differentiating different types of mind wandering, both thematically and in their relation to different contexts (Smallwood & Andrews-Hanna, 2013; McMillan et al., 2013). A significant proportion of mind wandering appears to be dedicated to future planning. Other proposed functions include creativity, attentional cycling between multiple goal states, and dishabituation, where brief episodes of mind wandering are thought to facilitate learning (Schooler et al., 2011). Many mind wandering episodes may also elude temporal characterization, being neither about the past nor about the future, and are perhaps best classified as atemporal (Jackson et al., 2013).

Different types of mind wandering may also have different phenomenological profiles (Diaz et al., 2013). Stawarczyk and colleagues (2013) found that most mind wandering episodes involved either visual imagery or inner speech, but typically not both. Moreover, future-oriented mind wandering episodes involved more speech and less visual imagery and were more self-relevant, realistic, concrete intentional, and structured than non-future-oriented mind wandering. The important role of self-reflection in future-oriented mind wandering is believed to support the link between autobiographical memory and the ability to mentally simulate the future (Smallwood et al., 2011).

Evidence also suggests that dissociable subsystems of the default-mode network exist and are associated with different kinds of conscious mental activity. The dorsal medial prefrontal cortex system appears to be involved in thinking about one's present situation and current mental state, as well as in social cognition, and the medial temporal lobe system is thought to be involved in autobiographical memory, but also future planning (Andrews-Hanna et al., 2010; Andrews-Hanna, 2012).

So far, mind wandering is largely a blind spot in the philosophical discussion of dreaming and imagination. Key encyclopedia entries on imagination do not mention mind wandering at all (Thomas, 2011; Gendler, 2013), and philosophers have only recently begun to pay attention to mind wandering and its relation to dreaming (Metzinger, 2013a, 2013c). In dream research, the situation is importantly different, and several authors have suggested that there is an overlap between the brain areas implicated in dreaming and the default-mode network (Pace-Schott, 2007, pp. 137–141; Domhoff, 2011b; Wamsley, 2013; Braun, 2014). Pace-Schott (2013) suggests that the disconnection between separate subsystems of the default-mode network during REM sleep may explain why self-referential and social reasoning are less constrained by autobiographical memory in dreams. The most thorough analysis, however, is provided by Fox and colleagues (2013), who present the results of a meta-analysis of results on brain activation during REM sleep and during mind wandering and link them to phenomenological similarities and differences between the two states. They found the strongest overlap between REM sleep and waking mind wandering in the medial prefrontal cortex and medial temporal lobes (for a neuroimaging study providing evidence for a breakdown of cortical connectivity in the default-mode network during NREM sleep, see Sämann et al., 2011). The overlap is not perfect, and the differences are equally interesting: several regions beyond the default-mode network are active in REM sleep and may, as Fox and colleagues suggest, be linked to the intensely visual and bizarre character of dreaming. By contrast, activation levels in prefrontal executive areas including the DLPFC, which are already decreased in spontaneous as compared to goal-directed thought in wakefulness, reach their lowest levels in REM sleep. Fox and colleagues conclude that "dreaming can be seen as an intensified version of waking spontaneous thought—or conversely, that MW [mind wandering] during wakefulness could be seen as an attenuated, waking form of dreaming (or, as its colloquial moniker suggests, 'daydreaming'" (Fox et al., 2013, p. 11). They suggest that dreaming and waking mind wandering can be placed on a continuum, with dreaming appearing as an intensified form of mind wandering (Fox et al., 2013) and as the quintessential form of cognitive simulation (Domhoff, 2011b).

Yet, as always, counterexamples exist. Chow et al. (2013) found evidence for a more widespread and dynamic interaction between unimodal sensorimotor areas and the association cortices, including the default-mode network, during REM sleep than during NREM sleep or wakefulness. While this generally supports the involvement of the default-mode network in dreaming, they also found that activity in these regions alternated reciprocally and in an anticorrelated manner, suggesting that REM sleep is characterized by a unique pattern of interconnectivity between regions belonging to the default-mode network and other brain areas. They suggest that these rhythmically alternating patterns of brain activity distinguish REM sleep from both NREM sleep and wakefulness, evoking the familiar conception of REM sleep as a "third state" of consciousness (Chow et al., 2013; cf. Jouvet, 1999; sec. 2.1).

Again, putting empirical issues to the side, this suggests that the familiar tug-of-war between similarity and difference exists in the debate about dreaming and mind wandering as well, with one side pleading for the uniqueness of REM sleep/dreaming, and the other advocating a more continuous view. As Wamsley (2013) points out, this debate is not innocent: the tendency to emphasize the differences in brain activation between REM sleep and wakefulness and to regard dreaming as a state of cognitive deficiency is the successor to the historical view that dreams are unknowable and mysterious occurrences. As a purely methodological point, such separatist assumptions about REM sleep/dreaming are not helpful and may even prolong the prejudiced view that dreaming is irrelevant to the study of consciousness. While modeling dreaming on wake states introduces the risk of oversimplification and wake-state bias (i.e., the naive and unreflected practice of projecting beliefs about waking experience onto dreaming), emphasizing discontinuity thus may introduce the opposite danger of marginalizing dreaming as a target for serious scientific research.

For present purposes, the main question is whether the investigation of dreaming, mind wandering, and the default-mode network is simply more of the same, presenting the familiar debate about dreaming and imagining in a slightly modernized guise, or whether it holds the promise of genuine epistemic progress. I want to tentatively suggest that recent research linking dreaming to waking mind wandering, conceived of in the proper manner, can be tremendously constructive and can possibly help bridge the gap between the largely inflated differences between the quasi-perceptual and the imagination view, both in their philosophical and their empirical and psychological guise. In particular, the conceptual neutrality of mind wandering is an important strength: so far, the term, as pointed out earlier, only describes stimulus- or task-independent thought, where thought is used in a wide, phenomenologically neutral sense. By contrast, imagination, with its connotations from the history of philosophy and its association with (pictorial) mental imagery, is much more loaded. The philosophical innocence of the concept of mind wandering may be a particularly important advantage: as noted earlier, for philosophy, mind wandering is still largely uncharted conceptual territory. This neutrality, I suggest, holds the promise of a fresh and unbiased approach. Moreover, empirical research on mind wandering is increasingly recognizing the heterogeneity of the cognitive and neural processes involved, as well as the phenomenological variability underlying this umbrella term. This suggests that there are no principled obstacles against grouping dreaming under the heading of mind wandering. From now on, my suggestion is to conceive of dreaming as a form of mind wandering in the sense that dreaming, like waking mind wandering, is a task-unrelated and largely stimulus-independent form of conscious cognitive processing. Perhaps regarding both dreaming and waking imagination as related but in some yet-to-be defined sense distinct forms of mind wandering may, for the time being, offer a kind of compromise, and a first step toward a solution.

So far, this is only a terminological suggestion. The relevant question is how exactly to spell out the ways in which imagination and dreaming are continuous with each other and with waking perception, as well as the ways in which they are not. At the end of section 6.1.3, I suggested that the solution may be to refocus the debate from dream imagery to the relation between the dream self and the dream world. In Fox and colleagues' (2013) claim that dreaming is an intensified form of waking wandering, we find a similar point. They emphasize that dreaming is distinguished from waking mind wandering not only by being more bizarre and by the greater intensity of the sensory aspects of dreaming, but also by being more immersive: "In dreaming . . . external stimuli are almost entirely blocked, and the audiovisual content can take on the aspect of an immersive, three-dimensional simulated reality" (Fox et al., 2013, p. 11). Again, as in the quasi-perceptual view, we find the familiar theme that experientially, dreaming involves the appearance of a world: whatever the exact experiential character of dreaming, dreams are experienced as realities. A related point is almost equally important: Fox and colleagues note that where dreams are "temporarily extended, fairly cohesive narratives spanning several minutes or longer . . . waking MW [mind wandering] thoughts typically only last for several seconds" (Fox et al., 2013, p. 11). In this view, dreaming generates not only an immersive spatial reference frame but also an extended temporal one: a chain of events with a roughly narrative structure. I return to this theme at the end of the book. For now, note that this transition from nonimmersive waking thought to immersive dreaming, both in a spatial and in a temporal sense, can only be understood by looking more closely at the phenomenal dream self, or the phenomenology of selfhood in dreams.

6.3 Conclusions

This chapter has shown that dreaming is a test case not only for general theories of consciousness (Revonsuo, 2006) but also for theories of the imagination. The claim that perceiving and imagining are sharply distinguished not just conceptually, but also phenomenologically, and that this distinction is reflected in the phenomenology of dreaming, has a long philosophical history. A preliminary result of this chapter is that while the imagination view functions as a valuable corrective for the quasi-perceptual view, it only translates into an empirically plausible research strategy once the underlying assumption of a sharp distinction between perceiving and imagining is rejected.

At present, we are left with a number of open questions, and perhaps with a better sense of how they might be approached experimentally. This in itself is progress. As Hunter (1983) notes, questions about dream phenomenology are too often resolved in the armchair. By now it should be clear, however, that our untutored beliefs about dreams are simply not reliable. Any attempt to conceptualize dreaming in cognitive terms that are commonly used to characterize wake states, such as thinking or imagining, is prone to invite problems—and while such a strategy can form the endpoint of an inquiry into dreaming, the applicability of such wake-state terms to dreaming should not be assumed from the outset (for a similar argument regarding the application of concepts such as problem solving and creativity to dreams, see Blagrove, 1996). Yet, we might still ask whether imagination could at least be a guide to the range of possible experience in dreams. Hunter convincingly argues that it is not: in trying to imagine dreams, we inadvertently imagine waking experience.

> Someone might reply to this that he could elaborate upon, and thereby show the imaginability, that is, the possibility, of a fully realistic handshake in a dream. . . . Does this not show that such a dream is possible?
>
> No. All that has been done here is that a handshake has been described. . . . This proves nothing so much as that human beings can describe handshakes and call them dream events. It would show what it was intended to show only if it were known that human beings were designed in such a way that no falsehood could cross our lips. (Hunter, 1983, p. 92)

Hunter concludes that "the question of the difference between dreaming and waking life is very much more complicated than it has mostly been taken to be" (Hunter, 1983, p. 92). On the positive side, this means, however, that investigating the phenomenology of dreaming, unencumbered by waking expectations, bears real promise. Because dreams are nothing but experience—and because there is no more detail to the dream world than what is experienced—dreams may even be a particularly apt target for detailed phenomenological investigation. Moreover, we can now see why our failure to notice that we are dreaming may not require that our dreams replicate waking experience in all its detail—and also why the assumption that they do has often, in the history of philosophy, seemed particularly intuitive or attractive. As Hunter writes:

> Are dream people and dogs perceived as through a haze? Does the fire in the study burn on and on without needing more logs? Do parts of the wallpaper disappear behind the lamp as we move and reappear again on the other side of it? An inept dream genie producing fake incidents on the screen of the mind might slip up on such details, but if our dreams tend to lack such attention to detail, we do not notice that they do; and even if it were common knowledge that they were inauthentic in such ways, we could still always imagine dream experiences being altogether authentic. (Hunter, 1983, p. 88)

Earlier, I suggested that the imagination view, while relying on equally misguided background assumptions as the quasi-perceptual view, functions as a helpful corrective. Now we can see why. It forces us to consider the predictions of the quasi-perceptual view in all its detail; and as we do so, the intuitive loveliness of the quasi-perceptual view largely evaporates. In describing waking experience, we often tend to conflate detail in the world with detail in experience. We then import putative detail in waking experience into our descriptions of dream experience. It is sometimes argued that detailed experience, even in wakefulness, is an illusion resulting from behavioral access to a detailed world (see sec. 6.1.3). Yet in dreams there is no world independent of dream experience; so why assume that dream experience simulates a level of detail that might not ever be there, even in waking experience? And why assume that dreaming simulates behavioral access to a detailed world, even though while we are dreaming, this type of access is in fact lacking? Surely the more parsimonious answer—and a prediction for future research—is that phenomenal indeterminacy pervades dream experience. While phenomenal indeterminacy does not sharply distinguish imagining from perceiving, it offers a particularly helpful conceptual tool, originating from the imagination view, for describing the experience of dreaming, and perhaps even for redescribing the phenomenology of waking perception in this process.

Another advantage of the imagination view is that it provides the resources for explaining the ordered character of dreaming. By contrast, thinking that dreams are wholly the product of bottom-up processes unfolding in a state of overall cognitive deficiency renders the ordered nature of dreaming more rather than less mysterious. Even Hobson's (1988; Hobson et al., 2000) model, which emphasizes the random and generic nature of dreaming, explains dream formation, in part, with respect to higher-level, synthetic cognitive processes.

The main problem of the imagination view, then, is not the attempt to conceptualize dreaming as imaginative but the attempt to model the phenomenology and neurophysiology of dreaming on waking imagination. It is this latter attempt, or so I claim, that fails. The assumption, unique to the philosophical discussion of dreaming, that the phenomenology of dreaming can be assimilated to that of waking imagination is the result of the armchair fallacy. Turning the imagination view into a viable research project requires abandoning these phenomenological claims, as well as the underlying and empirically implausible attempt to introduce a sharp distinction between perceiving and imagining.

While there is no exact equivalent to the imagination view in psychology and empirical dream research, we saw that the continuity hypothesis runs into similar problems. In particular, it is unclear which aspects of wakefulness—experience, concerns, thoughts—dreaming is supposed to be continuous with. A similar comparative statement is also seen in the view that dreaming is a form of quasi-linguistic thought. Leaving these (largely problematic) comparative statements to the side, the imagination view predicts that dreaming can only be understood by taking the higher-level, top-down contributions to dream formation into account. Psychological and empirical work on the cognitive and mnemonic sources of dreaming pays an important contribution to a multilevel framework of dreaming.

The neutral concept of mind wandering may be particularly helpful in this context. The growing field of research on mind wandering provides an interesting theoretical context for the investigation of dreaming and the attempt to relate different forms of task- and stimulus-independent thought to activity in the default-mode network. At the same time, the concept of mind wandering is, at least at this point, theoretically neutral and so does not carry with it the same philosophical baggage as do the concepts of daydreaming or imagining. There also seems to be common ground between those who endorse the view that dreaming, as an intensified form of mind wandering, involves world experience (Fox et al., 2013) and proponents of the quasi-perceptual view who nonetheless regard dreaming as an intensified or more perfect form of imagining. Recall that the latter view was, after all, implicit in Leibniz's and Hobbes's discussion of dreaming, but also in classical phenomenology (e.g., in Husserl and Conrad; see secs. 5.1.2 and 5.1.3 for discussion). Of course, this is not to say that the question of how the musings of a wandering mind are transformed into an experienced reality in dreams has been answered. I return to this question in chapter 12. The following chapters attempt to build up to an answer by providing a contrast-neutral framework for describing dream experience by focusing on the relationship between the dream self and the dream world.

7 Are Dreams Subjective Experiences (I)? Phenomenal Selfhood and Bodily Experiences in Dreams

7.0 Introduction

Are dreams subjective experiences? What does it mean to say that dreaming typically involves the experience of being or having a self? How exactly can phenomenal selfhood, or the phenomenology of self-experience, in the dream state be described conceptually? And what is the relationship between the dream self and the experienced dream world?

We have seen that the debate about the quasi-perceptual versus the imaginative character of dreaming does not provide the resources for answering these questions. What we need instead is a conceptual framework describing the phenomenology of self-experience in dreams and its relation to the experienced dream world in a manner that does not succumb to the dual temptations of oversimplification and wake-state bias. Rather than projecting largely unsystematic and unquestioned assumptions about waking experience onto dreaming, what we need, at least initially, is a contrast-neutral approach that regards dreaming as a valuable target phenomenon in its own right and can account for the inherent variability and multiplicity of dreaming.

Over the following chapters, I examine these questions in detail and formulate a number of more specific ones. In particular, I distinguish different dimensions of self-experience and propose a conceptual framework for describing their occurrence in dreams, as well as their relationship to each other. The most important of these are bodily experiences, or the phenomenology of embodied selfhood (discussed in this chapter); the phenomenology of cogitative selfhood (or the experience of being a thinking and perhaps even a rational or knowing self, discussed in chap. 9); and spatiotemporal self-location (discussed in chap. 11).

While my analysis of self-experience in dreams departs from the phenomenological level of description, the conceptual framework developed here suggests questions for the functional and neurophysiological levels of description, as well. For instance, to be truly informative, an analysis of bodily experience in dreams will also have to investigate its functional relationship to the sleeping body. Specifically, how do different degrees of phenomenal embodiment map onto functional embodiment (chap. 8)? Relatedly, investigating whether dreaming involves the phenomenology of being a thinking self and a cognitive agent is one thing; asking whether this is sufficient for saying that dreaming involves an epistemic self capable of genuine beliefs, knowledge, and insights is another, requiring an analysis not just from the phenomenological but also from the third-person epistemological perspective. Investigating this latter question in detail will lead to a reassessment of the epistemological problem of dream deception (chap. 10). Finally, in chapters 11 and 12, I use the main results from the previous chapters to propose a new answer to the conceptualization problem and suggest several important consequences beyond dreaming.

Because chapters 7 to 10 have a similar structure, some introductory remarks are in order. Each chapter begins by providing a brief overview of the main positions in the philosophical literature on dreaming. Because I am interested in evaluating not just their conceptual coherence but also their empirical plausibility, I present these as working hypotheses. I then review the empirical literature with the dual aims of proposing an empirically informed philosophical framework and providing a metatheoretical critique of existing studies. The final section of each chapter suggests a revised and more empirically plausible version of the original hypotheses and makes suggestions for future research.

In the present chapter, I focus on bodily experience and its relationship to phenomenal selfhood. Because this approach draws heavily from dream reports, I advance several arguments for their trustworthiness with respect to bodily experiences in dreams. These can be considered a specific application of the conservative approach (see sec. 4.4). Based on a review of the scientific literature, I then introduce a conceptual framework for describing what I take to be a wide spectrum of bodily experience in dreams, ranging from phenomenal indeterminacy and phenomenal disembodiment, or the experience of lacking a body, to weak and strong forms of phenomenal embodiment. An important conclusion is that phenomenal embodiment is not necessary for the phenomenology of selfhood in dreams, and self-experience is preserved even in dreams in which phenomenal embodiment is lost altogether. In the final section, I propose the *weak-phenomenal-embodiment hypothesis*, according to which bodily experience, at least in a majority of dreams, departs from waking bodily experience not only in the distribution of different types of bodily experience and in how much of the body is represented, but also in that the integrated character of bodily experience is typically lost in dreams. I propose that bodily experiences in dreams can be described as resulting from disturbed multisensory integration and a disturbed integration of body-part/whole-body representations. This suggests that the study of bodily experience in dreams can enrich existing theories about the relationship between bodily experience and self-consciousness.

7.1 The Bodily Duplicate Hypothesis and the No-Body Hypothesis

The first step toward answering the question of whether dreams are subjective experiences in the sense of involving the phenomenology of embodied selfhood is to take a systematic look at differing views on this matter from the philosophical literature. The most general way of categorizing traditional views on the dream self is to contrast maximally opposing cases. For present purposes, and in largely simplified terms, I will classify those who claim that dreaming replicates standard waking self-experience as adhering to the *phenomenal-duplicate hypothesis*. Those who deny that dreaming involves the phenomenology of selfhood altogether will be classified as adhering to the *no-self hypothesis*. These two general hypotheses can be subdivided into more specific ones, and as will become clear over the following chapters, different variants are defended by different authors. Consequently they are better regarded as umbrella terms and do not necessarily correspond to consistent and uniform theoretical positions. Nonetheless, formulating them in this general manner is helpful because it shows how different variants of the phenomenal-duplicate hypothesis and of the no-self hypothesis are connected, and because taken together, they chart the extreme ends of the space of possible positions on the phenomenology of selfhood in dreams. Typically, or so I will claim, the most empirically plausible answer will fall somewhere in between.

Because the phenomenal-duplicate hypothesis is both older and more elaborate, I will begin my discussion here. Again, and perhaps not surprisingly, Descartes is one of its most prominent proponents. When Descartes introspects his experience of sitting by the fire and wonders whether he is dreaming or awake, he is not just amazed, after all, by the warmth of the fire and the crispness of the piece of paper in his hands. What he finds truly disturbing is that dreams mimic the experience of embodiment and of agentive interaction with the dream world. "I certainly look upon this paper with eyes wide awake; the head which I now move is not asleep; I extend this hand consciously and with express purpose, and I perceive it" (Descartes, 1641/1901, I.5).

Descartes's endorsement of the phenomenal-duplicate hypothesis is particularly strong because he takes dreams to replicate waking self-experience in its broadest and most inclusive sense, including both the experience of being a phenomenally embodied self and the more cognitive dimensions of phenomenal selfhood, such as being a thinking self and a cognitive and epistemic agent. Even if he is dreaming, Descartes, after all, is still able to introspect his ongoing experience, reflect on its realistic, wakelike quality, and conclude that realistic appearances notwithstanding, all of this could be a dream. This yields a first very general formulation of the phenomenal-duplicate hypothesis:

Phenomenal-duplicate hypothesis: Dreaming involves the wakelike experience of being a fully embodied self and a cognitive agent.

Focusing on the implicit claims about bodily experience, we can say that Descartes thereby also endorses the bodily duplicate hypothesis.

Bodily duplicate hypothesis: Dreaming involves the wakelike experience of fully embodied selfhood.

Construed in this way, this formulation raises more questions than it answers. It buys into problematic comparative assumptions about dreaming and wakefulness of the type identified in chapters 5 and 6 while leaving completely open what "fully embodied selfhood" is supposed to mean. For the time being, this is no more than a helpful idealization, a first landmark on the map of possible positions on bodily experience and phenomenal selfhood in dreams. My strategy will be to translate it into an empirically investigable assumption and a number of better-formed successor hypotheses over the course of this chapter.

For now, note that the bodily duplicate hypothesis not only leaves the concept of "fully embodied selfhood" undefined but is also coarse-grained in another respect. Recall that Descartes's argument requires only that dreams potentially include a realistic phenomenal duplicate of his waking self, not that they typically do so. Even if dreams only rarely involve a phenomenal duplicate of his waking self, their ability to do so is sufficient to cast doubt on waking self-experience in general, as the possibility of dream deception can then no longer be ruled out. A stronger position would claim that dreaming not just potentially but typically replicates all aspects of waking bodily self-experience. An example of this stronger position is found in Revonsuo's work on dreams:[1]

> The self in the dream is the character who represents the dreamer. This character, the dream self, *usually* possesses a *body image much like the one experienced during wakefulness*. The dream self is *positioned in the center of the dream world*. The dream setting and events are seen and experienced from his or her point of view. The dreamer *feels as if he or she were embodied inside our bodies during wakefulness*. Thus, the dream self has a *bodily existence and location in the dream world*. In this respect the dream self is *not all that different from the waking self*. (Revonsuo, 2005, p. 207; italics mine)

This description of the dream self gives the bodily duplicate hypothesis conceptual traction and takes a first step toward rendering it more meaningful. According to Revonsuo, self-experience in dreams is not just characterized by the phenomenology of embodiment. Rather, the perspectival character of dream experience—the sense of experiencing dream events from a particular point of view—is grounded in the experience of being a phenomenally embodied self. In this view, the dream self is not only the locus of the visual first-person perspective; like an avatar in a virtual second life, the dream self has a distinctly bodily flavor and is anchored in a phenomenal dream body (cf. Revonsuo, 1995). The bodily duplicate hypothesis, then, includes three dimensions: self-location, perspectivalness, and phenomenal embodiment.

Finally, note that the bodily duplicate hypothesis can be endorsed even if one denies the generalized version of the phenomenal-duplicate hypothesis. Metzinger's claim that dreams are weakly subjective states (cf. sec. 5.1.3) characterized by cognitive deficiency and that nonlucid dreamers are not bodily agents because they do not reflect on or deliberately control their behavior within the dream is a good example (Metzinger, 2003, p. 253). In particular, he still endorses the bodily duplicate hypothesis, claiming that dreaming "not only contains the appearance of a world but also (in most cases) creates a fully embodied, spatially extended self moving around in a spatially extended environment" (Metzinger, 2009, p. 135), and "dreams are internal simulations of a complete behavioral space, including target objects; complex, ongoing behaviors; and other agents" (Metzinger, 2003, p. 253).

That authors as different as Descartes, Revonsuo, and Metzinger converge in their endorsement of the bodily duplicate hypothesis highlights the fact that for present purposes, this a phenomenological claim only. While Descartes's claims about the phenomenology of self-experience in dreams are related to the metaphysical position of substance dualism, this is by no means necessary, and the bodily duplicate hypothesis is best regarded as metaphysically neutral. Indeed, it is entirely compatible with the claim that there is no such thing as a self in the sense of an entity or a substance that is continuous over time and separable from the physical body (Metzinger, 2003, 2009).

The most radical way of opposing the bodily duplicate hypothesis is to deny that dreaming involves any bodily experience whatsoever, or indeed that it involves a phenomenal self. This position fits nicely with the imagination view: if dreaming is phenomenologically unlike perceiving and like daydreaming, this will also hold true for the phenomenology of self-experience in dreams. The intuition that bodily experience is lacking or somehow deficient in dreams sometimes also surfaces in response to dream skepticism. It not only finds expression in the proverbial pinching test (which, as will become clear, is a matter of philosophical debate in its own right) but is also echoed by philosophical treatments of dreaming. Recall, for instance, that for Locke (1689/1997), dream pain lacks the bite of real pain, and presumably the same would be true for any bodily experience in the dream state (cf. sec. 6.1.1). Relatedly, Sartre's description of dreaming as the "perfect realization of a closed imagination" (Sartre, 1940/2005, p. 319; translation mine) involves a claim about the dream self. According to Sartre, when the dreamer identifies with a dream character, for instance, a slave, this character "does not lose its constitutive character of unreality. To the contrary it is I, projected into the slave, who becomes an imaginary self" (p. 330; translation mine). What is it like, then, to be a merely imaginary self? Sartre likens this to the experience of identifying with a character in a book but also points out an important difference. In reading, one may have the experience of identifying with the protagonist while still being distinct from him. In dreams, this residual ability of critically distancing oneself from the fictional self is lost. Now the dream self is "lived, acted upon and suffered as unreal" (p. 331, translation mine). Sartre concludes that "the dream is a privileged experience that can help us understand what consciousness would be if it lost its 'being in the world' and at the same time were deprived of the category of the real" (p. 339; translation mine).

Again, this account raises a number of questions. For instance, what exactly could it mean to say that the dream self is *experienced* as *un*real? In reading and in daydreaming, this might be spelled out by saying that the imagined self is experienced as unreal in comparison to the more robust sense of identification with and ownership for one's actual body. Yet Sartre's claim that in dreams, the imaginary world has closed on itself seems to suggest that he does not think that the experienced unreality of the dream self is a comparative matter. Here, Sartre's general position on the self might be helpful. Elsewhere, he endorses the view that the phenomenal self, or "I," is always the object of a reflective act and so does not appear in unreflective consciousness.

> When I run after a streetcar, when I look at the time, when I am absorbed in contemplating a portrait, there is no I. There is consciousness *of the streetcar-having-to-be-overtaken*, etc., and non-positional consciousness of consciousness. In fact, I am then plunged into the world of objects; it is they which constitute the unity of my consciousness; . . . but I have disappeared; I have annihilated myself. There is no place for *me* on this level. And this is not a matter of chance, due to a momentary lapse of attention, but happens because of the very structure of consciousness. (Sartre, 1957, pp. 48–49)

The situation of the dream self, then, is not unlike the situation of the unreflected self: recall that for Sartre, dreaming is necessarily unreflective. For present purposes, we can describe Sartre as endorsing the claim that the self is a creature of reflection, and in unreflective wake states, as in dreams, there is no phenomenal self in a more robust sense. We can sharpen this into a generalized version of the no-self hypothesis:

No-self hypothesis: Dreaming does not involve the realistic, wakelike phenomenology of self-location, phenomenal embodiment, or cognitive agency.

To pick out the specifically bodily variant of the no-self hypothesis, allow me to introduce the *no-body hypothesis*:

No-body hypothesis: Dreaming does not involve the phenomenology of embodied selfhood.

In McGinn's version of the imagination view, the no-body hypothesis takes center stage. In particular, McGinn's claim that dreaming involves the phenomenology neither of presence nor of absence is linked to his claim that dreaming lacks phenomenal embodiment:

> Since I have no (conscious) body awareness during the dream—no proprioceptive percept that enters the dream's content—there is nothing to bring together with the material of the dream to generate an impression of presence. . . . Since there is no self-location, there is no location of objects relative to the self. (McGinn, 2004, p. 93)

Contrary to Revonsuo's claim that the dream body places the dream self at the center of the dream world, McGinn notes that "if there is no perception of self to anchor the representation of presence, there is also no perception of self to give substance to the representation of *absence*" (McGinn, 2004, p. 93). The alleged absence of self-experience is linked to an absence of distinctly bodily experience: "In the case of dreams, *there is no such proprioceptive percept*, and hence no contrast point; thus the object cannot be 'posited as absent'" (pp. 93–94). Because dreams lack phenomenally embodied selfhood, dreams also lack a sense of perspectivalness and self-location relative to other dream objects and characters.

Why does McGinn think that we have no proprioceptive percepts in dreams? I suspect that this has to do with an underlying equivocation of the dream body and the sleeping body. This becomes particularly obvious in McGinn's distinction between dreaming and hallucinating. Recall that for McGinn, dreams, unlike hallucinations, are not percepts (in the phenomenological sense of quasi-perceptual experience). If dreams were hallucinations, then "genuine presence would be written into the experience" (McGinn, 2004, p. 94). In particular, he claims that hallucinations produce the same behavioral and emotional reactions as genuine perception, while dreams fail to do so: if I took a dream tiger to be present, "I might leap from the bed and run for the door" (p. 94). Because no such behavior is observed in sleeping subjects, dreams are not percepts. Here, however, McGinn clearly fails to distinguish between the actions experienced within the dream and those observed in the sleeping subject. Just because dreams occur in the absence of any easily detected contemporaneous behavioral evidence, this does not mean that they are not percepts, just as it does not mean that they are not experiences. Saying anything else is just Malcolm's argument in a new guise.

Yet things may not be quite so simple. McGinn quite rightly allows, for instance, that "I can dream *that* I am facing a tiger, but this is not a relation between my body as it then is, lying asleep in bed, and the tiger; it is a relation between the tiger and my body as represented in the dream" (McGinn, 2004, p. 93). Does this mean that dreaming involves the phenomenology of embodiment, after all? But if so, then why is the body as represented in the dream insufficient to anchor the experience of presence? Claims about bodily experience and its relation to presence are phenomenological claims only; at least in theory, they do not require that these experiences be veridical in the sense of corresponding to the current state of the physical body. Perhaps McGinn means to say that imaginative bodily experience of the type occurring in dreams is sufficient only for imagined, but not for real presence, whatever that might mean. As he does not clarify this point, it is hard to see, however, how this maps onto a clear-cut phenomenological distinction between real and imagined bodily experience and its relation to real and imagined presence of the type that his argument requires.

Yet, his argument may have diagnostic value. The close conceptual kinship between the sleeping body and the dream body might make it particularly hard to imagine that vivid bodily experiences, such as movement sensations, can occur in dreams in the absence of their externally observable counterparts. The counterintuitive nature of an active dream self frolicking through the dream world, experienced by a peacefully sleeping, inert subject, might thus lead to a rejection of bodily experience in dreams. This view might then leak out into the dream world, leading to the view that dream tigers and their ilk, experientially speaking, are not quite the real thing (or, as was the case for Malcolm, are not experiences at all). The imagination view might be, then, the product of incredulity, coupled with the desire to derive phenomenological consequences from the conceptual distinction between dreaming and wakefulness, and from the lack of observable evidence of vivid bodily experience during sleep. By contrast, the quasi-perceptual view might be characterized as the product of overzealousness: once bodily experience is accepted to occur in dreams, why not, after all, go all the way and assume that it is no different from bodily experience in wakefulness? An underlying problem for both views, however, is that they proceed without a careful consideration of empirical evidence on bodily experience in dreams.

While Sartre's and McGinn's analysis of dreaming amounts to a sweeping rejection of wakelike self-experience in dreams, the philosophical debate has occasionally focused on more specific questions. In the 1960s, we find a number of authors disagreeing on the proverbial question of whether pinching oneself in a dream results in real pain. Incidentally, this debate offers another nice example of the natural affinity of the imagination view, applied to bodily experience in dreams, and the armchair fallacy, or the practice of mistaking purely conceptual claims for phenomenological observations.

This becomes evident in Nelson's (1966) article "Can One Tell That He Is Awake by Pinching Himself?" Here Nelson argues that the pinching test provides a refutation of the Cartesian dream argument. Contrary to Descartes's claim, there is a felt difference or a "sensuous mark" (Nelson, 1966, p. 82) by which dreaming can be distinguished from wakefulness. The problem is that Nelson finds himself utterly unable to conceive of dreaming a pain:

> When . . . I try to conceive the dreaming of a pain I find that I cannot. I find that I have conceived, not the dreaming of a pain, but simply the experiencing of a pain. But that is not dreaming a pain: that is being in pain. (Nelson, 1966, p. 82)

This is not, he continues, due to a psychological inability to conceive of dream pain. The inability to dream a pain is a logical matter. Dreams are necessarily deceptive: saying that someone dreamt of X, according to Nelson, means either that X does not exist or that the person was unable to perceive it. However, because one cannot be mistaken about presently being in pain, one cannot hallucinate or dream a pain:

> For to suppose that I might now be dreaming a pain as I pinch myself would be to suppose that the pain I felt did not really exist—I only thought it did; or that it did exist but I was not in a position to perceive it. (Nelson, 1966, p. 83)

As Hodges and Carter (1969) point out, Nelson's argument hinges on the claim that dreaming that one is in pain entails not only that one is not really in pain but also that one falsely believes that one is. But without further evidence in its support, this claim about the deceptive nature of dreaming is dubious: it simply presupposes that feeling and dreaming a pain are mutually exclusive and that dreams are, as a matter of logical necessity, belief inviting. The argument is circular: it merely presupposes what it is intended to show. Kantor (1970) nicely sums up the problem by pointing out that Nelson's view of sleep, like Malcolm's, essentially confuses sleep with death: if sleep as a matter of logical necessity rules out any ability to respond to external stimuli, and thus any experience, then either no one ever sleeps, or sleep is indistinguishable from coma or even death. In a more experimentally oriented manner, Kantor notes that pinching a person in sleep leads to wincing, movement and sounds, and sometimes even dreams of being pinched. He concludes that the pinching test fails. The occurrence of pain—or of other bodily experiences—in dreams cannot be excluded based on purely conceptual considerations.

More generally, one can say that because of their categorical denial of self-experience and phenomenal embodiment in dreams, the no-self and no-body hypotheses are natural candidates for committing the armchair fallacy, and the looming confound between purely conceptual and phenomenological claims, as well as between bodily experience and the physical body, should be a constant warning sign in evaluating these and related principled claims about dreaming. The absence of phenomenally embodied self-experience in dreams is at best a working hypothesis, and if true, it should be confirmed by empirical research. By contrast, simply proclaiming the absence of bodily experience in dreams for purely conceptual reasons is phenomenologically empty.

(Yet there is also reason for caution. Judging from Kantor's remarks on the behavioral effects of the pinching test, one might not want to leave this matter in the hands of experimentally minded philosophers. Indeed, it would be entirely understandable, psychologically speaking, if some shied away from the demand for the experimental investigation of dream pain altogether. Seen in this light, the armchair fallacy might certainly be the more comfortable option, after all.[2])

In sum, the quasi-perceptual and imagination views rely on diametrically opposed assumptions about the phenomenology of self-experience in dreams. The imagination view hinges on the claim that dreaming fails to involve the experience of embodied selfhood; and while this is sometimes portrayed as an a priori conceptual claim, it is in fact better regarded as a phenomenological claim that is open to empirical refutation. By contrast, the quasi-perceptual view includes, in its strongest reading, the view that dreaming replicates wakelike, phenomenally embodied selfhood. Despite their contradictory claims about bodily experience in dreams, there is another sense in which both views are deeply similar: they converge on the assumption that self-location, perspectivalness, and presence are intrinsically connected to phenomenal embodiment. While proponents of the bodily duplicate hypothesis assume that the dream self is present qua phenomenally embodied, proponents of the no-body hypothesis assume that the absence of a phenomenal self in dreams is connected to the absence of phenomenal embodiment.

Whether phenomenal embodiment and phenomenal selfhood can actually be identified in this manner, however, is an open question. Investigating this relationship in more detail is the task of the remainder of this chapter. The first problem for the analysis of bodily experience in dreams is to determine the exact relationship between phenomenal selfhood and phenomenal embodiment, and to do so in a way that does not simply assume them to be necessarily connected. I begin by summarizing findings on the dream self and asking whether and to what extent they are informative about phenomenal embodiment in dreams.

7.2 A Review of Bodily Experiences in the Dream State

7.2.1 Phenomenal Selfhood versus Phenomenal Embodiment

What, then, is the relationship between phenomenal selfhood and phenomenal embodiment? And how can the two concepts be defined in such a way as to ensure that the relevant experiential dimensions can be identified in dream reports? For now, I will only suggest a very rough first pass at a working definition. The idea is that this rough-and-ready definition will be enough to guide my review of the empirical literature, enabling me to give a more refined and empirically informed answer at the end of the chapter.

To begin with, in investigating phenomenal embodiment, the explanatory target is bodily experience in a purely phenomenological sense. The question is not what it is like to experience one's *actual* body, or how the *physical* body is given in experience. The question is also not how we experience what, *in fact*, are bodily states, or to investigate the *bodily basis* of experience. Trivially, as soon as one adopts a naturalistic approach to phenomenal experience (including phenomenal embodiment), every phenomenal experience is embodied in the sense that it has a bodily basis, for instance, a specific neural correlate (this is what Dainton [2011, p. 203] means by minimal embodiment). Within a naturalistic framework, the mind is not disembodied in the Cartesian sense of being an immaterial substance. My explanatory target here is narrower: bodily experiences are experiences that have a distinctly bodily flavor on the level of phenomenal experience itself.[3] Yet bodily experiences are not necessarily veridical, and I bracket, for now, any further questions as to whether their content is representational or misrepresentational from a third-person perspective. Let us say, then, that phenomenal embodiment refers to cases of *seemingly* experiencing one's body, or of distinctly experiencing oneself as a bodily self or embodied agent, where this is understood in a phenomenological sense only. In keeping with antiskepticism, the study of phenomenal embodiment in dreams is based on the study of descriptions of bodily experiences in dream reports.

Which types of experiences, then, count as bodily experiences? A classical and fairly uncontroversial list of bodily experiences (adapted from de Vignemont, 2011) includes *touch* (carrying information not just about the shape of objects and pressure on certain body parts but also about texture, temperature, solidity, humidity, contact, weight, pressure, force, and vibration); *proprioceptive* sensations (of the position and movement of the whole body and specific body parts); *vestibular sensations* (e.g. of balance); *nociception* and pain sensations; and *interoceptive sensations* (e.g., of breathing or heartbeat). Recent work on the effect of multisensory conflict on bodily experience (e.g., in the rubber-hand illusion; Botvinick & Cohen, 1998; see chap. 12 for discussion) suggests that this list should be expanded to include the visual experience of seeing one's body. Visual experience shapes our experience of the size and position of the whole body and its parts (for further references, see de Vignemont, 2011, in press); it is also linked to the vestibular system, and its loss has a distinct effect on the sense of balance. Trying to stand on one leg, first with eyes open and then with eyes closed, provides a simple and yet poignant demonstration. While most research so far has focused on the importance of vision for bodily experience, there are also other ways of sensing one's body, such as hearing one's footsteps or even one's breath, and so on. This suggests that the list of bodily experiences should be expanded to include instances of seeing, hearing, et cetera, one's body, and also that multimodal integration is crucial for understanding bodily experience.[4]

The next question, then, is how to define phenomenal selfhood. At least at the outset, pursuing a reductive strategy, for instance, by grounding phenomenal selfhood in a specific form of experience (e.g., bodily experience, emotions, agency, ownership; cf. Blanke & Metzinger, 2009; Seth, 2013), would be to beg the question. If something like the Humean bundle theory of the self turns out to be correct, there may be no distinct and unified impression of the self over and above the train of thoughts, emotions, and impressions. While such a reductive or even eliminative account might well be the outcome of a research program on phenomenal selfhood, we should not assume it from the outset. A maximally permissive and theoretically neutral working definition is that the target property is, simply, the experience of being or having a self. Again, this is a purely phenomenological characterization: it does not assume that something like a self in the sense of a stable entity existing over time actually exists. The phenomenology of self-experience does not carry any metaphysical commitments about the actual existence of a self. This working definition also does not assume that phenomenal selfhood consistently or continuously exists: during states of unconsciousness, there is no phenomenal self. For present purposes, then, phenomenal selfhood is simply whatever we experience as our self in standard situations; it is also, among other things, what people describe when they say that they were present in their dream, that in their dream, they identified with or had a dream self, and so on. The first question, then, is whether subjects consistently describe being present in their dreams (e.g., by using the first-person pronoun), and if so, how exactly the phenomenal self is experienced in dreams. We now have the conceptual tools for approaching the question of whether phenomenal selfhood in dreams is consistently tied to bodily experience.

A number of studies have used the content analysis of dream reports and questionnaires to investigate varying levels of self-representation in the dream state. A common finding is that a dream self is present in the overwhelming majority of dreams and mostly acts as an active participant, though it can also, more rarely, observe the dream from a third-person, bird's-eye perspective (Snyder, 1970; Foulkes & Kerr, 1994). For instance, Strauch and Meier (1992, p. 112) found that the self is present in approximately 90 percent of dreams and that only 10.6 percent of dreams are described as selfless. Moreover, when a self is present, it tends to be an active participant (70.7 percent), and in 67.1 percent of all cases, participation has a social component in that the dream self either interacts or acts together with other dream characters.[5] More rarely is the dream self only passively involved in dream events (11.1 percent) or not involved at all (7.6 percent).

Some evidence points to variation across sleep stages. Foulkes and Schmidt (1983) found that sleep onset dreams contained less self-representation than REM sleep dream reports. Occhionero et al. (2005) used a combined methodology of asking previously trained subjects to give a free dream report, followed by a description of their level of self-participation and of "how they felt inside the dream scene" (Occhionero et al., 2005, p. 79). They found that 83.33 percent of REM and 47.78 percent of SWS (slow-wave sleep) dream reports described an actively participating dream self. They also found that different categories of self-representation varied to a considerably greater degree in SWS sleep than in REM sleep, suggesting that self-representation may vary both across and within sleep stages.

Developmental factors may also play a role. In a longitudinal study of children's and adolescents' dreams (ages nine to fifteen), Strauch (2005) found a change toward more realistic dream content and a shift from passive toward active behavior, with a particularly strong increase in interactions with other dream characters, thus confirming the earlier findings of Foulkes's (1993a, 1999) studies on children's dreams.

A number of other studies confirm the central role of the self in dreams. For instance, dream threats are typically directed at the dream self, and in adult dreams, the dream self often participates actively in threatening events (40–60 percent), exceeding reports of self-participation in threatening events from wakefulness (for discussion and further references, see Valli, 2008, pp. 70–81). McNamara, McLaren, and Durso (2007) found that REM and NREM reports described different types of social interactions, with the self more frequently acting aggressively in REM dreams and more frequently initiating friendly interactions in NREM dreams (see also McNamara et al., 2005). They suggest that behavioral strategies for dealing with social situations differ systematically between sleep stages. Again, there might be developmental differences. Crugnola et al. (2008) found that preadolescents' dreams contained more references to physical aggression and fewer references to friendly interactions and, interestingly, to body parts than normative adult samples (cf. Domhoff, 1996). Differences in social interactions may also distinguish dream and waking fantasy reports. Strauch and Lederbogen (1999) found that in children's waking fantasy reports, the self took an active role in both aggressive and friendly interactions, whereas in dream reports, it tended to be the target of friendly and aggressive actions of other dream characters.

What, if anything, do these studies tell us about the relationship between phenomenal selfhood and bodily experience in dreams? Because most of these studies do not explicitly distinguish between cognitive and bodily dimensions of phenomenal selfhood, a cautious interpretation would be to say that while they superficially share the same explanatory target—the dream self—they investigate the narrative content of dreams rather than phenomenal selfhood. Yet, by showing that different types of self-representation (active versus passive) and social interactions vary both across and within sleep stages and that developmental factors should be taken into account, they give a first indication that a similar (and perhaps even systematically related) variability in the phenomenology of (embodied) selfhood might exist. Still, to investigate the phenomenology of selfhood and its relation to bodily experience, much more detailed information about the individual dreams would be necessary.

To see why, let us return to the study by Occhionero et al. (2005). They used an eight-point nominal scale of different categories of self-representation in dreams, ranging from selfless dreams and dreams in which the self was either a passive observer or an active participant to dreams in which there was a double representation of the self. They distinguished different categories of self-representation, allowing, for instance, for the awareness of the self as a noncorporeal, thinking agent, but also for varying levels of *bodily* self-representation, ranging from the static representation of body parts to the "plurisensorial hallucination of Self, like the Self-awareness one experiences upon waking" (Occhionero et al., 2005, p. 80). The finding that a majority of REM dreams and nearly half of all SWS dreams are characterized by a complex representation of the self as an active participant consequently seems to suggest that in these cases, phenomenal selfhood in the dream state, including bodily experience, mimics bodily self-experience in wakefulness. Indeed, in an earlier paper introducing the self-representation scale used by Occhionero et al. (2005), Cicogna and Bosinelli (2001, p. 31) claim that "in most dreams, the dreamer actively participates in the event, and the pluri-sensorial hallucination of oneself does not allow a clear distinction between dreaming and waking self-awareness (i.e., the precise awareness of oneself under every aspect)." Occhionero et al. (2005, p. 80) use a similar formulation, writing that active participation dreams involve "precise awareness of oneself, both mental and physical, analogous to wakefulness" (p. 80). If true, this would support the claim that the bodily duplicate hypothesis characterizes a majority of dreams.

But what exactly is the basis for this assumption? Occhionero and colleagues give the following dream report as an example of phenomenally embodied participation in dreams:

> I was in the country and talking with a friend; my girlfriend was there and we were talking about a building. . . . I felt ill at ease because I was involved in a discussion I didn't like. . . . I was feeling as if I were in the real life. (Occhionero et al., 2005, p. 80)

Note that no bodily experiences are mentioned in the report and that the form of participation—that is, conversing with other dream characters—is not obviously connected to specifically bodily experiences. Just as the occurrence of dream speech may not require the occurrence of auditory experience but in some cases may have a "telepathic" quality (cf. sec. 6.1.3), so dreaming of a conversation may not require any specific bodily experiences, for instance, of moving closer and turning one's head to hear better, gesturing, or moving one's lips. In keeping with the conservative approach, the general observation that the experience was similar to waking experience does not in itself warrant any specific inferences about the phenomenal character of the experience. Moreover, because Occhionero et al. (2005) take this dream report to be a typical example of fully embodied self-participation, it would seem that the bodily duplicate hypothesis is implicit in their reasoning rather than directly targeted by their study.

This methodological shortcoming is not unique to Occhionero and colleagues' study. It is only particularly obvious because while other studies focus more strongly on narrative dream content, their study explicitly attempted to characterize the phenomenology of self-experience in dreams. By essentially identifying, however, active self-participation in dreams with full, wakelike phenomenal embodiment, they introduce a potential confound between the two. This line of reasoning is deeply problematic. It may also be an artifact of a similar confound common in thinking about waking experience. Do we continuously experience ourselves as richly embodied in wakefulness? Surely, the physical body is always present, and at any given moment, it seems we can voluntarily *attend* to our bodies and bring bodily experience, as it were, into the foreground. This does not show, however, that we consistently enjoy rich bodily experience when attention is directed elsewhere, for instance, to a conversation with a friend or to an engaging novel or film. Indeed, attentional resources may place an upper limit on the resolution of bodily experience, even in wakefulness. It is an open question, then, what type of bodily experience accompanies different types of waking experience, and at least at the outset, we should not assume that phenomenal selfhood, even in wakefulness, consistently or necessarily has a richly embodied flavor.

The tendency to mistake the continuous presence of the physical body for continuous presence of bodily experience is a variant of the problem from stimulus error (Schwitzgebel, 2007). When asked to describe conscious experience, subjects often describe objects in the world instead, confusing physical objects with the way they are experienced. A related problem is refrigerator-light error, in which a child believes that even when the refrigerator door is closed, the light is always on, because it is always on when she opens the door to check. It is tempting to make this error regarding conscious experience in general (Jaynes, 1976; Schwitzgebel, 2007, 2011c). For instance, when asked whether one has constant tactile experience of a particular body part, such as of one's feet in one's shoes (Schwitzgebel, 2007), one may answer affirmatively because one automatically shifts attention to one's feet (and, indeed, because one knows them to be present all along). This automatically introduces a confound between attentional focus and the unattended aspects of experience. As Jaynes writes:

> It is like asking a flashlight in a dark room to search around for something that doesn't have any light shining on it. The flashlight, since there is light in whatever direction it turns, would have to conclude that there is light everywhere. And so consciousness can seem to pervade all mentality when actually it does not. (Jaynes, 1976, p. 23)

Projecting this error to the phenomenology of dreaming invites the familiar problem of wake-state bias: the assumption that similarity in narrative contents in dream and waking reports implies similarity in phenomenal experience. If we assume that conversing with a friend involves bodily experience in wakefulness, the same, according to this reasoning, must be true for dreaming. In this manner, stimulus error and refrigerator-light error are unwittingly imported into theoretical descriptions of dreaming.

Note that this problem is not restricted to bodily experience. Let's say you have just woken up from a vivid dream of your current home. You are struck by the realism of the dream—your home in your dream, in fact, looked exactly the way it really does. Let's examine, for a moment, how far this claim takes us. Does it mean, for instance, that all of the pictures were hanging on the walls in exactly the right places? Did the portrait of your grandmother, which you distinctly remember seeing at one point in your dream, hang there, in the hall, throughout the dream, even after you turned your back on it (for a similar example, cf. O'Shaughnessy, 2002, p. 404)? Did the dream books in your dream bookcase have the same order as they really do—alphabetized or perhaps sorted by color, including the one you carelessly stuck in the wrong place yesterday while cleaning up? Note how at some point (and it hardly matters when exactly), these questions should undermine any certainty you might have initially had about how to answer them. Dream objects have no mind-independent existence—they only exist insofar as they are experienced. The appearance–reality distinction does not exist for dreams: at each moment, dream objects have only those properties they are experienced as having and can only be said to be present in the dream insofar as they are experienced in it. According to the conservative approach, the same insight must be applied to dream reports: for methodological reasons, no more should be inferred about dream experience than what is explicitly described in dream reports. This is not to say that there *is* no more to a given dream than what is in fact reported—typically there will be. But because dream research is constrained by reportable dream experience, special steps need to be taken to ensure that actually reported dreams mirror the space of reportable dreams as closely as possible (cf. chap. 4). For all practical purposes, the resolution of dream phenomenology as a scientific endeavor is limited by the detail of dream reports. But this also means that the impression that your dream home looked exactly the way your real home does—or that you felt, in the dream, exactly as you do now—will not take us very far. Surely, it is interesting to note that dreams create this subjective impression of verisimilitude—but this impression does not, in itself, justify the conclusion that dreaming replicates waking experience in all of its detail. Given the combined threats of stimulus error, refrigerator-light error, and wake-state bias, detailed reports are an indispensable safeguard.

To this, one might object that to sustain the illusion that the dream world is real, a certain level of background detail may be necessary. Given that we typically fail to realize that we are dreaming while we are dreaming, it might be fruitful to assume that the background perceptual processes involved in dreaming mimic waking experience after all, at least to a certain extent. If so, one might want to be more lenient in attributing experiential richness to dream experience. The conservative approach may be unnecessarily strict.[6]

This is an important objection, and I will ultimately have to answer it. For now, I think it is best regarded as raising an open question, namely, the question of how much experiential detail is required in dreams to prevent us from realizing that we are now dreaming, or indeed to create the semblance of a continuous dream narrative. I would also like to suggest a principle of parsimony in answering it: what is the *least* degree of detail and continuity required for the formation of seemingly continuous and detailed dream narratives? Relatedly, how much bodily experience is necessary to sustain the experience of being present in the dream? Perhaps different kinds of self-representation in dreams—active participation versus passive observation—are related to different degrees of phenomenal embodiment, or perhaps they are not. Either way, one should refrain from overhasty generalizations and conclusions, and the refrigerator-light error, stimulus error, and problem of wake-state bias highlight this point. They also suggest, however, why such generalizations have an intuitive pull, and why the questions highlighted here have frequently been overlooked by philosophers and researchers thinking about dreams.

Of course, nothing I have said so far precludes the possibility that the bodily duplicate hypothesis could still turn out to be correct. But if it does, and this point is equally important, this would be more akin to a stroke of luck than thanks to careful, empirically grounded considerations. And even if strictly adhering to dream reports for the study of dream experience sounds preposterous to some, studying bodily experience as reported in dreams is clearly a better approach than relying on untutored intuitions. At least in the current situation, the choice, I think, is one between a data-driven and scientifically minded approach to bodily experience in dreams based on dream reports, and the lack of a scientific answer at all. Antiskepticism about dream reports coupled with the conservative approach seems to be the safer bet.

7.2.2 Toward a Graded View of Phenomenal Embodiment in Dreams

The clearest results concerning bodily experiences in dreams are related to the occurrence of movement. Following visual and auditory imagery, movement is the most frequent type of imagery experienced in dreams and is described in three-quarters of all dream reports (Hobson et al., 2000; Hobson, 1988). According to one study, dream movement typically involves either the dream self or another dream character (86 percent), and only rarely an object (14 percent) (usually a vehicle). Fifty-five percent of dream movements involve upper limbs, 37 percent involve lower limbs, and 8 percent involve the whole body (Porte & Hobson, 1986b; see also Hobson, 1988, pp. 247–256). When whole-body movements do occur in dreams, they often have a circular, curvilinear, back-and-forth quality (Hobson, 1988) and may be associated with emotionally intense dreams involving vestibular sensations such as flying, falling, or spinning (cf. Schönhammer, 2004, 2005). The proportion of movements involving single limbs compared to whole-body movements is similar for REM and NREM dreams, but movements in general are much more frequent in REM dreams, occurring in only 18 percent of NREM dreams (Hobson, 1988, p. 254). It is important to note, however, that these numbers are hard to interpret because Porte and Hobson did not differentiate between movements involving the dream self, which would be crucial for the study of bodily experiences, and movements involving other dream characters, which are likely associated with visual dream imagery. With due caution, we might nonetheless say that the difference between frequent limb movements and comparatively rare whole-body movements gives a first indication that the representation of single body parts may be more frequent than a global representation of the whole body in dreams.

Several studies suggest that the frequency of movement sensations varies for different subject groups. For instance, truck drivers more frequently dream of driving than do controls (Schredl, Funkhouser, & Arn, 2005–2006), and for athletes, the amount of time spent on sports activities in waking correlates with the frequency of sports activities in dreams (Erlacher & Schredl, 2010). Yet continuity with waking activities cannot explain all occurrences of movement sensations in dreams. Two recent studies investigating the frequency of movement sensations in paraplegics (Saurat et al., 2011; Voss et al., 2011; cf. Voss, 2014) found that a majority dreamt about voluntary leg movements such as walking, running, dancing, standing, or riding a bike but were rarely paraplegic in their dreams. This is interesting because whereas visual dreaming is eventually lost in subjects who lose their eyesight before the age of seven (Kerr & Domhoff, 2004), lower-limb movements are apparently maintained in dreams even in subjects who have never been able to move their legs during wakefulness. Reports of movement sensations in paraplegics' dreams are also interesting, however, for what they do not show, thus once more highlighting the importance of the conservative approach. Take the following two dream reports as an example:

Example 2: Subject with congenital paraplegia for 22 years: "I was not in a wheelchair but walking to a night club, to go dancing." (Saurat et al., 2011, p. 1427)

Example 5: Subject with spinal cord injury for 7 years: "I was rollerblading with my father and my mother, in a town that I did not recognize. We were on the road. After that, we rode mountain bikes." (Saurat et al., 2011, p. 1428)

What, if anything, do these reports show about the occurrence of bodily experience in the respective dreams? Aside from implicit references to whole-body movement, not much. Without additional information, neither report licenses any inferences as to which other types of bodily experiences were or were not involved in the respective dreams. Might the experiential difference between walking and rollerblading or bike riding in one's dreams be nothing other than a difference in the perceived speed of self-movement and narrative context? Or might it simply be related to the dynamic nature of visual imagery? After all, congenital paraplegics will have observed the types of movements they dream about in others and will have experienced whole-body movements, for instance, when sitting in a car or wheelchair. For this reason, such short reports are not particularly informative about the associated bodily experiences. Other reports, however, more persuasively suggest the involvement of specific movement sensations in the legs of congenital paraplegics:

Example 1: Subject with congenital paraplegia for 40 years: "I was able-bodied, learning ballet at the 'cours Florent' [a Parisian drama school]. I was dancing the Swan Lake. I wore a white tutu and pink ballet shoes. I was dancing with a light step and doing what the teacher told me to do." (Saurat et al., 2011, p. 1427)

This report contains more specific information about the body (i.e., the shoes, the tutu, and the lightness of the movements) than the other two examples. Nonetheless even here one may wonder how realistic and how detailed the phenomenology of dancing actually was. Did the dream really include the tactile and proprioceptive sensations, the subtle postural adjustments or gait control that would be involved in actually dancing a particular piece of choreography from *Swan Lake*? Did the dreamer see her pink ballet shoes and white tutu, or did she just know that she was wearing them? It would be astonishing, indeed, for a dream to replicate the exact experience in all of its detail, especially if the dreamer were not a dancer intimately familiar with the choreography and the experience of dancing it herself. Note that I am not here claiming that the authors themselves endorse such strong conclusions. I am only using these examples to highlight what I take to be a more general methodological problem in investigating and theorizing about bodily experience in dreams.

To continue, Voss et al. (2011) failed to find significant differences between the dreams of congenital paraplegics and healthy controls along a number of dimensions. As indicated by free dream reports and affirmative probes, the only significant difference in movement-related variables was that paraplegics reported moving with the help of others more frequently than controls. Yet the study found no statistically significant differences in the frequency of movements involving the self or in the frequency of specific movements such as falling, flying, walking, running, or tripping. One possible explanation is that in dreams, when they are freed from the constraints of their physical body, paraplegics experience the same type of vivid and rich bodily experience that healthy subjects enjoy in wakefulness. An alternative interpretation, however, is that what enables both paraplegics and controls to experience vivid movements in dreams is that the constraints on seemingly realistic simulations of bodily movement in dreams are lower, as it were, than in wakefulness. Rather than simulating the exact motor, tactile, and proprioceptive sensations, postural adjustments, and so on involved in different movement patterns, schematic and phenomenally indeterminate representations of the overall movement may be enough. In this view, even such dream movements as have never been experienced before (such as unaided flying) could feel realistic (or rather, fail to feel unrealistic) precisely because not every aspect is felt, and deviations from the phenomenological profile associated with these movements in wakefulness would not necessarily be noticed.[7]

Recall, in this context, that performing actions in a lucid dream has a similar effect on actual performance as imagining them (Erlacher, 2005, 2010; Erlacher & Schredl, 2008a, 2008b). This suggests that the positive effect on performance does not depend on simulating the actual movement and associated bodily experiences in all their detail; a schematic mental simulation of the type we are capable of in waking imagination suffices. This is not to say that a ballerina practicing choreography from *Swan Lake* in a lucid dream would have the same experience as the one described by the paraplegic subject. Surely the ballerina familiar with the exact choreography would have different motor imagery in her dream, and varying degrees of determinacy are possible. Yet this is not the same as saying that her dream would exactly replicate the experience of actually dancing in wakefulness. The computational challenge associated with simulating the experience of dancing choreography from *Swan Lake* in the absence of the customary muscular feedback, physical effort, and sense of gravity would lead us to expect phenomenological differences, as well (cf. Erlacher et al., 2014).

Indeed, Erlacher et al. (2014) found that performing bodily movements in a lucid dream—walking 10, 20, and 30 steps at a regular pace and performing a gymnastics routine—took significantly longer than in wakefulness. A possible explanation is that this is due to computational demands on simulating a familiar movement sequence in the absence of feedback from the physical body. Relative time was preserved, suggesting that the overall structure of the movements was nonetheless realistic. Strangely enough, however, the effect was more pronounced for walking (52.5 percent) than for the gymnastics routine (23.2 percent). Why would the easier task take more time in a lucid dream, suggesting higher computational load? A possible explanation might be that the gymnastics routine—involving jumping and somersaults—may have been better suited to the curvilinear, back-and-forth movement pattern associated with dream movement than was the case for ordinary walking. Moreover, as Carlos Schenck (2005) is fond of emphasizing, RBD patients don't walk, they run, and in fact one data set from Erlacher and colleagues' study had to be excluded because the participant ran, rather than walked, in the lucid dream. Perhaps slow walking in a straight line is just not tailored to dreaming; and when one does walk in a dream (by simulating the exact movement pattern involved in counting the steps, as required in Erlacher and colleagues' study), this might simply require so much effort that it takes longer than it would in wakefulness. Indeed, 16 out of 37 lucid dream reports had to be excluded from data analysis because for one reason or another, the task was not successfully completed. Simple tasks—counting, walking, and gymnastics—were not easily performed in lucid dreams. Here is an excerpt from a report, also excluded from data analysis, of the gymnastics routine:

> I went to a garden where it was bright and I thought, "Okay, I am doing the experiment now." I gave a LRLR [left-right-left-right eye signal] and I jumped and I felt immediately that jumping was very different compared to wakefulness. Just a different perception of the body, also slower. I continued and I did the forward roll—which lasted almost eternally. When I finished the task I gave a LRLR again. (Erlacher et al., 2014, p. 4)

The gymnastics routine described here took 163 percent longer than in wakefulness (14.8 versus 5.6 seconds). Note that this case was exceptional because the dreamer also noticed that the task was taking longer. In the other cases included in data analysis, the gymnastics routine also took longer than in wakefulness, but the dreamers did not notice this fact. Apparently it took a more drastic objective difference in the actual duration for this difference to become noticeable. And if this is true for lucid dreamers concentrating on a task, it seems plausible that nonlucid dreamers would be even less sensitive to differences between dream movements and their waking counterparts.

There are further reasons to think that bodily experience in dreams is often sparse as compared to wakefulness. Judging from dream reports, bodily experiences other than movement sensations are extremely rare in the dream state. Thermal sensations of heat or cold are mentioned in only 4 percent, and sensations of touch in only 1 percent of dream reports; pain, physical pleasure, and visceral sensations are also extremely rare in dreams (Hobson, 1988, pp. 244–245). Even if the frequency of tactile and somatosensory sensations in dreams should turn out to be higher, it still seems that dreams are marked by a systematic underrepresentation of tactile sensations as compared to visual, auditory, and movement sensations. This is supported by results from independent studies from different historical periods (Schwartz, 2000), suggesting that the underrepresentation of such bodily experiences is an invariant feature of the phenomenology of dreaming.

Again, there is a high degree of inter- and intraindividual variation. Using a general questionnaire, Okada et al. (2005) distinguished dreamers according to the frequency with which they mentioned experiencing different sensory modalities in their dreams. About 20 percent of participants reported experiencing all sensory modalities equally often in their dreams; 48 percent reported experiencing predominately vision, sound, and kinesthetic sensations, and the rest dreamt predominately without experiencing any of these sensations. The all-modalities group scored highest on tests determining their ability to generate and control vivid visual imagery during wakefulness, suggesting an indirect link between visual imagery in wakefulness and multimodal, but not specifically visual, imagery in dreams. King and DeCicco (2007) found a correlation between low levels of physical and mental health and more frequent dream reports of body parts, but also of movements such as walking and running, possibly reflecting increased waking concerns with such activities. Similarly, DeCicco et al. (2010) found that women with breast cancer have significantly more dreams than healthy women of disease and death-related themes, as well as more frequent pain sensations and references to the torso. Garfield (1988) reviews a number of examples from different studies showing how women's dreams change in response to menstrual cycle, menopause, and pregnancy, often specifically affecting changes in bodily imagery in dreams. A recent study by Lara-Carrasco et al. (2013), however, found that the dreams of pregnant women in their third trimester differed from those of childless women only in a few respects. Pregnant women did dream more frequently of motherhood, babies, and childbirth, as would be expected, but other differences—such as body-related themes—were not found between the two groups.[8]

Rare sensations may even exceed visual and auditory imagery in their intensity. For instance, Voss et al. (2011) found that touch, which is generally rare in dreams, was rated as more intense by paraplegics than by the control group. The following dream report of a paraplegic subject gives a nice example of thermal sensations occurring in a dream:

I walk along a beach, my naked feet are immersed in the water. I walk further and further into the ocean. The water is very cold but I don't shiver, just the opposite, it is a very nice feeling. I stride into the ocean until it is deep enough to start making first swimming movements. Now I completely immerse myself in the water. I dive with even stronger strokes, deeper and deeper and I see the sun shine through the water a last time. (Voss et al., 2011, p. 682)

The occasional intensity of rare sensations is also illustrated by Helen Keller's (1908/2003) dream reports.[9] Left both blind and deaf following a childhood illness at the age of nineteen months, she was restricted to the sensations of touch, smell, and taste for most of her life and reported experiencing vivid, storylike dreams involving these modalities. While vision is the primary modality experienced in dreams of sighted subjects, the loss of vision and increased attention to rare sensations, including touch, in wakefulness apparently changes the frequency with which these sensations occur in the dream state. Indeed, a recent study found that auditory sensations occurred in 80 percent of dreams reported by congenitally blind subjects. Yet somaesthetic sensations were described in only 32 percent, and taste (8 percent), smell, pain, and temperature sensations (4 percent) were relatively rare (Staunton & O'Rourke, 2012; for evidence that the loss of visual dreaming affects the frequency of other types of imagery in dreams, see also Meaidi et al., 2014).

What about the visual representation of the dream body? Burroughs reports the following dream:

I am in a dark room where I can see my face in a mirror. It seems to be a metal mirror on a cabinet of some sort. My face is quite young, about eighteen, with a curious dark red color like faces in a painting I did today . . . dusky red and also scarred or damaged in some way around the mouth and chin. It may be some sort of skin eruption. There are several views—some quite close. (Burroughs, 1995, pp. 23–24)

More systematic evidence for a visual representation of the body comes from lucid dreams, and several of the different techniques of reality testing discussed in the lucid dream literature involve the body. Many lucid dreamers look at their hands in lucid dreams to determine whether they are dreaming or not (LaBerge, 1985). Other reality tests involve looking in the mirror (Levitan & LaBerge, 1993), pinching oneself, and sticking one's finger through the palms of one's hands (LaBerge, 1985; LaBerge & Rheingold, 1990). All these examples suggest that when the visual dream body becomes the object of explicit attention in dreams, it tends to behave in unexpected ways. And even when the attempt, for instance, to stick one's finger through the palm of one's hand in a lucid dream results in an appropriate visual experience, it may fail to give rise to associated sensations of pain or solidity.[10] The following excerpt from a dream reported by Clare Johnson highlights the difficulty of simulating a realistic visual representation of one's own body, even during a lucid dream.

I remove my hands from the textured wall and raise them before my eyes, remembering how I'd performed that same gesture the day before, while giving a presentation at the university about lucid dreams. To my surprise, although I raise my hands with palms turned to my face, I only see the backs of them, and moreover they look smaller than normal and weird. I examine them for a moment, then walk to a mirror to my left and look at myself. I look normal (albeit a little pale, my eyes slightly distant, observing me observing myself), and I raise my hands again to bring them into the reflection. One is fine, but the other is clenched into a club-fist. With an effort of will, I remove it from the reflection, determined that when I bring it in again it will look normal. After several attempts, it is normal. I meet my distant gaze in the mirror and decide not to fiddle around with my hands anymore—I want to explore more of this dream. (Johnson, 2005, p. 6)

The frequent lack of touch and pain sensations in lucid dreams can even be exploited to perform a series of slightly macabre dream experiments, such as the following:

I pushed my hands into the sides of my head until my fingertips met in the middle. I checked that they had met and were aligned by clicking my fingernails together. I pushed my hands forwards, so that they would emerge through my eyeballs and be in front of my eyes, and would then block my view. It seemed possible that they might block my view before that point. However, when my hands reached a point that I judged or sensed to be at the back of my eyeballs, I felt as if my hands were pressing against the backs of my eyeballs and that, if I moved my hands any further, I would pop my eyeballs out. Even though I knew that it was a dream, at this point I was in fact inhibited and have not tried that particular experiment again. The nearest I have got to doing so was to make my fingertips meet in the middle of my head, but I was distracted by what seemed to be an impenetrable lump that I felt below the penetrable lobes of my cerebral cortex, as if I had a brain within a brain. (Worsley, 1988, p. 330)

Worsley also reports that the integration of visual and haptic imagery can be difficult to achieve in lucid dreams, even if explicitly desired. For instance, Worsley (1988, p. 331) reports easily being able to extend his arm to a physically impossible length to reach an object at the other end of the room in a lucid dream. However, he notes that the ability to simultaneously feel the extended arm, rather than just to see it, requires considerable effort. To the extent that this supports the claim that there is indeed a felt difference between dreaming and wakefulness, note that the argument for saying so is quite different from the philosophical debate about dream pinching discussed in section 7.1: here the difference is a matter of empirical fact, not of conceptual necessity. Importantly, the difference is also not absolute. For instance, trying to cut one's arm with a knife in a lucid dream can produce a "disconcertingly real sensation" (Green & McCreery, 1994, p. 28), suggesting that more realistic forms of bodily experience occur in lucid dreams, as well. Indeed, an online study asked experienced lucid dreamers to pinch, caress, and apply pressure to their forearm during wakefulness, in lucid dreams, and in their imagination and to rate the intensity of the resulting sensations (Giguère & LaBerge, 1995). Results indicated that the dream sensation most like its waking counterpart was caressing, whereas pressing one's arm in a dream had an intermediate intensity between waking and imagination, and pinching in a dream felt only slightly more intense than imagined pinching. The authors hypothesize that the lack of pain and discomfort caused by pinching may have been related to motivational bias—especially as dreamed pinches involved tactile sensations and only selectively lacked the pain. While this suggests that the pinching test may not deserve the folk-psychological and philosophical attention that it has received, the results indicate that touch sensations can be induced voluntarily during dreams to an extent that is not possible during daydreams, though the intensity of such dream sensations may still fall short of their waking counterparts (for discussion, see Windt et al., 2014).

To conclude, this brief review of reports of bodily experiences in dreams suggests, more than anything, that a more differentiated conceptual framework for describing different degrees of phenomenal embodiment in the dream state is needed, and one that can do justice to the underlying variability of the target phenomenon. Importantly, it is not sufficient to describe the positive occurrence of bodily experiences in dreams. Rather, it is equally important to understand the sense in which bodily experience is absent in dreams. This requires a change in perspective. So far, in reviewing existing studies on bodily experiences in dreams, I have shown that several general theoretical conclusions that these studies are sometimes taken to support are not borne out by the available evidence. I used the conservative approach as a normative tool, suggesting that if one goes beyond what is explicitly described in dream reports, one easily falls prey to a double error, ultimately describing dream experience in terms applicable to the description of mind-independent objects, including the physical body, in wakefulness. The next step in describing phenomenal embodiment in the dream state requires a positive and more systematic argument for why to trust reports of bodily experiences in dreams. Although I summarized the general reasons for trusting dream reports in chapter 4, several more specific arguments regarding the trustworthiness of reports of bodily experiences in dreams are available. I start out by reviewing them and then propose a framework for describing both the presence and the absence of bodily experiences in dreams.

7.3 A Conceptual Framework for Describing Phenomenal Embodiment in the Dream State

7.3.1 Are Reports of Bodily Experiences in Dreams Trustworthy? Applying the Conservative Approach to Reports of Bodily Experiences in Dreams

Even if dream reports, when given under ideal reporting conditions, are generally trustworthy with respect to previous dream experience, we might still have reasons for being skeptical about reports of bodily experiences in dreams. If the refrigerator-light and stimulus errors, coupled with wake-state bias, really help explain the intuitive appeal of the bodily duplicate hypothesis, then perhaps this is true not only for researchers scoring dream reports and philosophers theorizing about dreams but also for subjects reporting their dreams. In proposing adherence to dream reports, the conservative approach may be perpetuating a biased, theory-laden way of thinking about dreaming.

This type of objection is fueled by two independent types of concerns about the trustworthiness of dream reports on the one hand and the trustworthiness of reports of bodily experiences on the other. The intuitive pull of skepticism about dream reports has already been discussed sufficiently. Similarly, bodily experiences have seemed, to many, to be particularly elusive. In the literature, we see substantial disagreement as to whether consciousness involves the ongoing experience of unattended objects, background feelings, and sensations (e.g., James, 1890/1950a,b; Searle, 1992) or is limited to objects and sensory modalities that are at the center of attention at any given time (e.g., Jaynes, 1976; Dennett, 1991; for discussion and further references, see Schwitzgebel, 2007). Here is an example of the rich or abundant view of bodily experience during wakefulness:

> The next thing to be noticed is this, that *every one of the bodily changes, whatsoever it be, is FELT, acutely or obscurely, the moment it occurs.* . . . Our whole cubic capacity is sensibly alive; and each morsel of it contributes its pulsations of feeling, dim or sharp, pleasant, painful, or dubious, to that sense of personality that every one of us unfailingly carries with him. (James, 1890/1950b, pp. 450–451)

According to James, it is not just that bodily changes are felt but that bodily experience is continuous: "Our own bodily position, attitude, condition, is one of the things of which *some* awareness, however inattentive, invariably accompanies the knowledge of whatever else we know" (James, 1890/1950a, p. 241). In this view, bodily experience is the nucleus of the self (p. 400); the continuous feeling of "the same old body always there" (p. 242) anchors, as it were, the stream of thought and suffuses it "with that peculiar warmth and intimacy that make it come as ours" (p. 242). Contrast this with an example of the thin or sparse view:

> We are constantly reacting to things without being conscious of them at the time. Sitting against a tree, I am always reacting to the tree and to the ground and to my own posture, since if I wish to walk, I will quite unconsciously stand up from the ground to do so. Immersed in the ideas of this first chapter, I am rarely conscious even of where I am. (Jaynes, 1976, p. 22)

Rather than adjudicating between these two views about waking experience—a project that would be well beyond the scope of this book—I want to point out that as they stand, and in the absence of further empirical evidence, both views are articulations of opinions about waking bodily experience that require investigation in proper experience reports. As Schwitzgebel remarks, a "war of philosophical intuition thus threatens" (Schwitzgebel, 2007, p. 10). Given these uncertainties about the adequate characterization of bodily experience in wakefulness, the project of developing a conceptual framework describing bodily experience in dreams may seem particularly unattractive because it combines two notoriously elusive kinds of reports: those of dreams and bodily experiences. It will not surprise readers that I take these concerns, though important, to be ultimately ungrounded. Yet because they threaten my account of bodily experience in dreams, and also because they raise important questions for future research, I think they deserve careful consideration.

A first way of allaying these worries is to point out that subjects are not *generally* unable to report different types of sensory experiences in dreams. If subjects were always committing the refrigerator-light and stimulus errors or were just generally bad at reporting sensory experiences in dreams, the different frequencies with which the sensory modalities are reported should not be so robust over different studies (cf. Schwartz, 2000). For instance, hearing should be as frequent as seeing in dreams, as there is no intuitively plausible reason to assume that a sizable proportion of dreams unfold in environments that would be devoid of sound were they experienced in wakefulness.

A related worry may be that while generally able to report sensory experiences such as seeing and hearing, we systematically fail when trying to report sensations of smell, taste, touch, or pain. In this view, our failure to report bodily experiences in dreams would not reflect their rare occurrence in dreams; rather, reports of bodily experiences would generally fail to reflect the occurrence of such experiences, both in dreaming and in wakefulness. A prediction would be that reports of bodily experiences should hover around the same low levels for dreams and waking experiences.

Note, however, that reports of bodily experiences in dreams sometimes differ in marked ways from bodily experience in wakefulness. If subjects were inadvertently reporting on their waking bodies rather than on bodily experiences in dreams, this would be surprising; for instance, paraplegics should not be able to report walking in their dreams, and lucid dreamers should not find their dream bodies behaving in bizarre and unexpected ways. If implicit bias were driving reports of bodily experiences in dreams, these types of deviant reports, but also the general underrepresentation of bodily experience in dream reports, would be left unexplained. While subjects may well commit stimulus error when asked to answer general questions about the way they typically dream, asking them specific questions about specific dreams and collecting free dream reports immediately after awakening likely comes as close to avoiding implicit bias as possible. Training subjects to report on their dream experiences rather than on the objects experienced in their dream, while bracketing any assumptions about their waking counterparts, might also help. One could also investigate whether subjects' theoretical beliefs about dreaming color the way they report their dreams.

A study conducted by Schwitzgebel (2007; see also Schwitzgebel, 2011c) attempted to assess the frequency of bodily experiences in wakefulness and thus to shed light on the question of whether conscious experience is rich or thin (Schwitzgebel, 2007) or abundant or sparse (Schwitzgebel, 2011c).[11] He used Hurlburt's beeper method (see Hurlburt & Schwitzgebel, 2007; cf. sec. 4.5) to determine whether subjects would be able to report experiencing their feet in their shoes in different conditions. These conditions included focusing their reports on experience in general and on visual experience, tactile experience, tactile experience in the left foot, and visual experience in the far right visual field. Averaged over all conditions, 100 percent of participants reported having visual experience immediately before the beep, 63 percent far-right visual-field experience, 76.5 percent tactile experience, and 49 percent tactile experience in their left foot. Schwitzgebel concludes that "experience is *at least* as rich as my participants report" (Schwitzgebel, 2007, p. 33), ultimately supporting a moderate view between the rich and thin views of experience.

Does this have any consequences for the comparison between dreaming and wakefulness? Because gathering retrospective reports from the moment exactly before the beep—followed by a detailed interview—is comparable to the type of reports gathered in laboratory studies of dreaming, at least an indirect comparison with dream studies is possible. Indeed, it is noteworthy that the frequency with which visual experience was reported in Schwitzgebel's study is similar to that reported in dream studies; at the same time, however, there is a large discrepancy for tactile sensations, which are mentioned in only 1 percent of dream reports. Consequently it would seem that based on the available evidence, waking bodily experience is richer in tactile sensations than bodily experience in dreams. In addition, Schwitzgebel's findings help allay the worry that we are *generally* unable to report tactile experiences, even when we are not attending to them. This, after all, is clearly something many of his subjects were able to do, even when they were asked to report only on tactile experiences in the left foot.

We have now narrowed the scope of the skeptical objection considerably. If one wants to attribute the rare occurrence of bodily experiences in dream reports to forgetting and insist that such experiences are consistently had in dreams but forgotten upon awakening, one will have to do so by appealing to a form of memory impairment that occurs *specifically* with respect to bodily experiences and only during the dream state. Without further evidence in favor of such a view, however, this objection seems unduly contrived. Hence the view that rare reports of bodily experiences in dreams reflect their rare occurrence better explains existing findings. At the least, the remainder of this chapter can be seen as the outcome of an experiment in applying the conservative approach to reports of bodily experiences in dreams, aimed at developing a conceptual framework that takes available empirical evidence from the analysis of dream reports maximally seriously. So even though the conservative approach may appear to place an inordinate weight on the trustworthiness of dream reports, it is actually the most cautious approach when the alternatives are considered.

There is, however, another worry. In saying that the frequency of reports of bodily experiences indicates the frequency of their occurrence in dreams, I am assuming not just that positive reports of their occurrence are trustworthy but also that subjects can reliably report the absence of certain types of experiences. I am assuming, in other words, that at least under ideal reporting conditions, bodily experiences in dreams are reportable experiences. If one wants to derive any positive conclusions from experience reports, this assumption is necessary; the problem is analogous to the problem of reports of nondreaming discussed in chapter 4. Consequently, to describe phenomenal embodiment in dreams, it will be necessary to account for the ways in which phenomenal embodiment can be lost in dreams, as well.

7.3.2 Phenomenal Disembodiment

Does it make sense to say that subjects can report the absence of certain types of bodily experiences after awakening from a dream? And in doing so, are they giving a positive report of an experienced absence or merely reporting the absence of something that was not experienced as present? Does asking these types of questions even make sense?

To begin with, as Dennett famously noted, "the absence of representation is not the same as the representation of absence" (Dennett, 1991, p. 359; cf. Brentano, 1874, p. 217). Likewise, the absence of experience is not the same as the experience of absence. If the experience of absence exists, then it requires the presence of phenomenal experience in the first place. It also does not suffice to say that a certain experiential property is missing from experience; rather, one will have to account for how it is possible to palpably experience the absence of this specific property. I will use the concept of phenomenal disembodiment to describe cases in which the property of being a bodily self is not just absent in experience but experienced as absent. The question, then, is whether phenomenal disembodiment is more than a theoretical possibility and is actually reported to occur in dreams.

Cicogna and Bosinelli give the following example from SWS: "I was inside a gigantic photocopying machine. I knew I was inside this machine, not as a physical human being but as *an abstract entity*, as a mind, so I couldn't see myself" (Cigogna & Bosinelli, 2001, p. 32). If we take this report at face value, then it describes an experience in which a *phenomenally disembodied* self was present: self-experience was present, but associated with the experience of *lacking* a body, or of being a disembodied self. In this dream, a visual representation of the body was clearly lacking; in addition, while the experience of self-location was preserved (the self is described as present, after all, in a photocopying machine), the quasi-Cartesian description of being an abstract mind indicates that the experience was characterized by the palpable absence of a dream body. Similar examples are found in lucid dreamers' occasional reports of being a disembodied point or freely moving center of awareness while nonetheless experiencing the dream from an internal visuospatial perspective (LaBerge & DeGracia, 2000; I discuss further examples in sec. 7.4. and in chap. 11).

Phenomenal disembodiment is interesting because it shows that though we always experience ourselves as *bodily* selves in standard situations, we can also experience ourselves as lacking a body. In these cases, there is still a basic sense of self-identification, for instance, with a disembodied point of awareness. There is a sense of occupying a spatiotemporal location in the dream, which is the locus of the visuospatial perspective from which the dream is experienced. This basic sense of self-identification and self-location is sufficient for saying that in such dreams, there is still a phenomenal self; at the least, it is sufficient for retrospectively describing the experience by saying that a phenomenal self was present (e.g., for saying, "I was in the dream, but I had no body"). Phenomenal disembodiment offers a particularly good example of how the study of dreaming can enrich consciousness research and shed light on the contingencies of waking consciousness. The structure of standard waking consciousness may be a poor indicator of the space of phenomenal possibility.

At the same time, to those who take reports of phenomenal disembodiment to be too outlandish and too far removed from standard waking experienced to be credible, it should be pointed out that phenomenal disembodiment is not without precedent. Indeed, numerous examples of a mismatch between the physical body and the experienced body can be found in the recent literature. "Sensations of absence" or "illusory amputation" (Dieguez & Blanke, 2011) occur in neurological patients, as well. In what appears to be a reversal of phantom limb experience, where an amputated limb is experienced as present, they may have the feeling that a body part that is actually present is missing. Likewise, for amputee patients, a "true sensation of amputation" seems to be the complementary phenomenon to phantom limb experiences.

Sensations of absence for body parts or a hemibody occurring independently of amputation are called hemiasomatognosia; for some patients, this robust sensation of absence is nonetheless accompanied by insight into the fact that this absence is merely illusory, that is, that the limb or hemibody experienced as absent is in fact still present. This shows that hemiasomatognosia is an experiential rather than a higher-level cognitive phenomenon: if it were merely delusional, then insight into the fact that the limb was still present should dissipate the illusion. By contrast, instances of delusional rejection of ownership for a body part or a hemibody (typically the left side of the body) are called somatoparaphrenia; here patients will additionally deny that the body part or hemibody is their own or will even attribute it to someone else (for a review of such cases, see Brugger, 2006; Dieguez et al., 2007, 2103; Feinberg et al., 2009; Vallar & Ronchi, 2009; Dieguez & Blanke, 2011).

Arzy et al. (2006) point out that somatosensory loss for a limb may also be accompanied by a loss of visual input from the limb, creating the experience of seeing through the limb that is also experienced as absent (asomatoscopy). Asomatoscopy, also called negative autoscopy and described as a negative hallucination, can also occur in isolation. Typically it describes a transient inability to see one's own image in a reflective surface such as a mirror (Blom, 2010). This suggests that body parts and limbs that are still physically present can disappear from experience by degrees, depending on whether they are only felt or also seen to be absent. Recall that the same may be true for the dream body, which may be represented only visually or only tactually.

The experienced absence of a limb is not restricted to neuropsychological patients. In the recent literature, xenomelia (formerly called body integrity identity disorder, or BIID), or the desire to amputate, paralyze, or disable a healthy limb, has been much discussed. While McGeoch et al. (2011) suggest that this condition arises from unrecognized right parietal dysfunction and thus is of neurological rather than psychological origin (Sedda, 2011), some debate surrounds the matter, and Giummarra et al. (2011) describe it as an identity disorder comparable to gender identity disorder. The feeling of estrangement from an otherwise healthy limb is often present since early childhood, limited to a precisely demarcated part of the limb, and patients perform normally in neurological, neuropsychological, and psychiatric testing (Hilti et al., 2013). Importantly, they are not psychotic or delusional. They desire amputation because it will make them "feel whole, complete, set right again" (Blanke et al., 2009, p. 184). Indeed, the subjective mismatch between the experienced body and the physical body is resolved following amputation, which paradoxically enables patients to feel complete and often leads to an increase in subjective well-being (First, 2005).

Hilti and Brugger (2010) discuss two examples of so-called integrity disorders that they suggest are mirror images of each other: one, the desire for amputating a healthy limb in BIID or xenomelia; and two, phantom sensations in dysmelia, or the congenital absence of limbs coupled with the experience that the absent limb is still present. In both cases, there is a mismatch between the physical body and the body as experienced; as Hilti and Brugger (2010, p. 315) write, where phantom limbs are an example of "animation without incarnation," the negative phantoms of patients with xenomelia are cases of "incarnation without animation." Alluding to the German term *Beseelung*, they note that in xenomelia, the disowned limb is not "ensouled"; it is experienced not as a live limb but as something bothersome and alien. This is interesting, first, because it shows that functionality does not map cleanly onto experienced ownership for limbs: the disowned limbs of xenomelia patients are fully functional, and amputees or subjects with congenital dysmelia may experience fully functional limbs (see the example of Aiha Zemp discussed later in the chapter) although no limb is in fact present. Second, because of the close similarity between "positive" and "negative" phantoms, it has been suggested that both might be caused by the disturbed integration of multisensory information from the affected limb (Blanke et al., 2009). Indeed, around 65 percent of subjects with xenomelia describe having strange sensations localized to the affected limb, suggesting that the limb is experienced as qualitatively different from the rest of the body: 46 percent describe feeling that the limb belongs to another (somatoparaphrenia); 23 percent feel that the limb is absent or has already been amputated (asomatognosia); and 38 percent describe sensations of tingling, tickling, prickling, pricking, burning (paresthesia), or reduced sensation (Blanke et al., 2009). Paresthesia is also often described as affecting phantom limbs (Brugger, 2006).

In the terminology just introduced, at least a subgroup of asomatognosia and delusional cases of somatoparaphrenia for a limb or a hemibody would be cases of partial phenomenal disembodiment. By contrast, phenomenal disembodiment in dreams can also occur with respect to the entire body. An interesting question would be how often experiences of partial disembodiment occur in the dream state and, in keeping with Schwartz and Maquet's (2002) proposal of a neuropsychological assessment of dreams, whether these cases are related to changed activation patterns similar those observed in patients. A preliminary hypothesis is that phenomenal disembodiment, both in dreams and in wakefulness, is related to disturbed multisensory integration. This hypothesis, suggested by Blanke et al. (2009) with respect to BIID, also seems plausible given evidence from dream reports especially from lucid dreams (sec. 7.2.2), as well as given that the processing of external stimuli and bodily sensations, while not completely absent, is altered during sleep (see chap. 8 for details).

Of course, there are differences. For instance, paresthetic sensations, which characterize both positive and negative phantoms in wakefulness, are systematically underrepresented in dreams. Moreover, while "negative phantoms" in wakefulness are typically limited to a circumscribed part of a limb, in dreams, phenomenal disembodiment may affect the entire body. Phenomenal disembodiment in dreams may thus be both stronger (in that different bodily sensations that are preserved in wakefulness—such as seeing the disowned limb, but also paresthesia—may be lacking in dreams) and more global (in that in dreams, the whole body, not just a body part, may be affected). Here a prediction would be that varying forms of *partial* phenomenal embodiment for single limbs in dreams can be explained by a dysfunction in multisensory integration, whereas *global* phenomenal disembodiment for the whole body may be the special case in which the processing of own-body-related stimuli has broken down almost completely.

Do whole-body variants of phenomenal disembodiment occur in wakefulness, as well? Cotard's syndrome, in which patients sometimes report feeling dead, unreal, or disembodied, is a possible candidate (for a review, see Debruyne et al., 2009). However, Ratcliffe (2008, p. 178) argues that the feeling of absence, nothingness, or nonexistence involved in Cotard's delusion and depersonalization should be understood in a more metaphorical sense. While these disorders may involve a reduction of bodily feelings, he claims that such feelings do not disappear completely (Ratcliffe, 2008, pp. 169, 182). These patients are *emotionally* disembodied (Metzinger, 2006) but do not necessarily lack bodily experience altogether (for a similar account, see Gerrans, 2015). This suggests that Cotard's delusion and depersonalization do not have the strength of global phenomenal disembodiment in dreams.

Finally, there are more pedestrian, nonpathological examples of experiencing the absence of a limb, for instance, if one's arm has fallen asleep. I once awoke to the bizarre and uncomfortable sensation of not being able to feel where my right arm was—in a moment of disoriented, half-awake panic I used my left arm to "find" it folded above my head. It had gone to sleep so completely that I had no sense of its position. Sacks describes the loss of bodily experience in the lower half of the body following spinal anesthesia as follows:

> If one is given a spinal anaesthetic that brings to a halt neural traffic in the lower half of the body, one cannot feel merely that this is paralysed and senseless; one feels that it is wholly, impossibly, "non-existent," that one has been cut in half, and that the lower half is absolutely missing—not in the familiar sense of being somewhere, elsewhere, but in the uncanny sense of *not-being*, or being nowhere. The terms that patients use communicate something of this incommunicable nothing. They may say that part of them is "missing," "evacuated," "gone": that it seems like dead flesh, or sand, or paste; devoid of life, of activity, of "will"; devoid of the organic, of structure, of coherence—without materiality or imaginable reality; cut off or alienated from the living flesh (with which, none the less, it forms an impossible continuity). One such patient, trying to formulate the unformulable, finally said that his lost limbs were "nowhere to be found," and that they were "like nothing on earth." (Sacks, 2004, p. 674)

Deafferentation, as in the famous case of IW (see Gallagher & Cole, 1995) or Sacks's (1985) "disembodied lady," may seem to be a further example; however, these cases are more correctly described as involving a loss of proprioception and touch below the neck, because patients continue to experience thermal and pain sensations, muscle fatigue, and vestibular sensations.

In sum, while a complete loss of bodily experience may well be possible in wakefulness, as well, it seems that most of the obvious candidates should rather be seen either as the loss or attenuation of particular kinds of bodily experience or as the experienced absence of a particular body part or a hemibody (partial phenomenal disembodiment). By contrast, global phenomenal disembodiment with respect to the entire body may be most complete in the dream state. Future research should focus on generating more detailed phenomenological descriptions, as well as attempting neuropsychological assessments that compare phenomenal disembodiment in dreams and wakefulness.

7.3.3 Phenomenal Indeterminacy

Phenomenal disembodiment, as noted in the previous section, is just one of two ways in which the body can be absent from experience. Indeed, it is the extremely special case in which the body is represented as absent on the level of phenomenal experience itself. But given that the experience of absence is not the same as the absence of experience, one cannot generalize from these cases of experienced absence to cases in which bodily experience was absent but not experienced as such. While phenomenal disembodiment appears to be relatively rare in dreams, it is thus possible that there is a different and perhaps even larger group of dreams that lack the phenomenology of embodiment without this fact being additionally experienced.

What would it feel like to have this type of experience? Note that if the lack of phenomenal embodiment were not itself experienced, the experience would be one *neither* of phenomenal embodiment *nor* of phenomenal disembodiment. The best way to describe such an experience would be to say that the body simply did not enter into experience at all. There would be no *experienced* fact of the matter as to whether one had a body or not; the experience itself would be entirely open in this respect. Importantly, if the body (or certain body parts) were experienced neither as present nor as absent, it seems plausible that they would be *described* neither as being present nor as being absent in retrospective reports. Consequently the simplest explanation for the widespread but implicit omission of bodily experiences in dream reports might be their equally inconspicuous absence in dream experience. These cases of the absence of bodily experience without the experience of absence, or of disembodiment, are instances of *phenomenal indeterminacy* with respect to the body.

Again, there may be precedents in waking experience. Nonconscious hemiasomatognosia (Dieguez & Blanke, 2011) refers to cases in which the disappearance of body parts from awareness is itself unnoticed. For instance, patients with personal neglect will forget to shave, comb their hair, or apply makeup to the left side of their face; patients with motor neglect underuse or completely fail to use the neglected limbs (despite there being no motor impairment); and patients with anosognosia for hemiplegia will behave as if they were not paralyzed and even deny that there is anything wrong with them. One way of explaining this neglect is to say that they have ceased to experience their hemibody but are unaware of this fact: the neglected body parts are experienced neither as present nor as absent but have dropped out of conscious awareness altogether. If so, then it might be possible to characterize the experience of at least a subgroup of such patients as phenomenally indeterminate with respect to the neglected body parts.

Note that phenomenal indeterminacy of the type described here is a type of indeterminacy in experience, in the sense that experience remains open with respect to the presence or absence of a body part by way of an *unnoticed* absence of the relevant kind of bodily experience. This is not a matter of an in-between or almost-but-not-quite case of bodily experience. It is not the same, for instance, as a borderline case of baldness, in which a person has somewhat too many hairs for being determinately bald, but not quite enough for determinately falling outside the category of baldness or counting as having a hairdo (for an introduction to paradoxes related to this problem, see Sainsbury, 1995). Phenomenal indeterminacy is not the same as semantic indeterminacy or conceptual vagueness. The problem is not indeterminacy as to whether the *concept* of bodily experience applies. Rather, phenomenal indeterminacy with respect to bodily experience is compatible with saying that the concepts of phenomenal embodiment and disembodiment *both* fail to apply: phenomenal indeterminacy does not occupy an intermediate region *between* phenomenal embodiment and disembodiment but refers to experiences best described as *a*bodily.

Phenomenal indeterminacy is often compared to the indeterminacy of descriptions (Dennett, 1981a). For instance, a literary description of a heroine in a novel may provide details about her looks and facial features without mentioning any details about her hair, and it can do so without alerting the reader to this omission. The description would then be indeterminate with respect to the type of hairdo—for instance, whether her hair was long or short, or which color it was—or even with respect to whether she had any hair at all. Not only could no specific claims about her hairdo (or lack thereof) be made on the basis of this description, but it would also be incorrect to say that she sort-of-but-not-quite had hair (as one might say for borderline cases of baldness). The matter of her having or not having hair would simply not be in the description. Yet this indeterminacy about her hairdo would itself be a determinate fact: the problem is not that the extent to which the description goes into detail concerning the heroine's hairdo itself is indeterminate.[12] In this respect, indeterminacy in descriptions is analogous to phenomenal indeterminacy. This analogy also, however, alerts us to a difference. If, while reading a novel, you suddenly notice that it contains no description of the heroine's hair, you do not suddenly imagine her as not having any. For phenomenal indeterminacy, or so I would claim, the situation is importantly different: once the absence of bodily experience becomes available to attentional processing, there is a change in experience itself. The experience slips from phenomenal indeterminacy into phenomenal disembodiment.

Let us return once more to Dennett's playing-card example, where moving the card toward the center of the visual field gradually makes information about color and shape available (cf. sec. 6.1.3). The trick of attending to the periphery while focusing your eyes straight ahead makes you aware that your visual experience in the periphery lacked color all along, but it does not make color experience suddenly disappear. Attention merely turns the previously unnoticed absence of color into an experienced absence. We can now say that realizing that color experience is absent in the periphery of the visual field is analogous to the special case of phenomenal disembodiment, in which the body is experienced as absent. As the playing-card example shows, experienced absence requires a higher degree of cognitive and attentional sophistication than the unnoticed absence related to phenomenal indeterminacy.[13]

Next imagine the type of visual experience you have in the periphery of your visual field spreading to the center. Imagine, further, that you were not specifically attending to color or focusing your eyes on any specific object. Imagine being in a state of unfocused distraction, your attention being captured by constantly changing, unstable persons and objects. Your visual experience would now be characterized by widespread indeterminacy, but you would also fail to notice this fact. This, essentially, might be the situation in dreams. It is nicely illustrated by Huxley:

> In dreams, one sees houses, trees and other objects, which are perfectly recognizable as such, but which remind one of the actual objects as seen "out of the corner of the eye," or of the pictures thrown by a badly focused magic lantern. A man addresses us who is like a figure seen in twilight; or we travel through countries where every feature of the scenery is vague; the outlines of the hills are ill-marked, and the rivers have no defined banks. They are, in short, generic ideas of many past impressions of men, hills, and rivers. (Huxley, quoted in James, 1890/1950b, p. 47)

The same thing, I claim, may be going on for the body in dreams: large portions of the phenomenal body are systematically missing in dreams; they are absent, but their absence is not experienced. Note that phenomenal indeterminacy is closely related to the epistemic notion of indeterminacy blindness:[14] typically, unless special steps are taken, we fail to detect the indeterminacy that is present in experience all along. Dennett's playing-card example suggests that even in wakefulness, phenomenal indeterminacy is both pervasive and typically unnoticed. If phenomenal indeterminacy is indeed more widespread in dreams than in wakefulness, the same might well be true for indeterminacy blindness.

The concept of indeterminacy blindness is helpful because it can be applied on different levels of description. Its most direct application is to refer to our habitual failure to notice the lack, for instance, of color experience in the periphery of the visual field. Indeterminacy blindness of this type alleviates computational load: if attention is systematically directed away from indeterminacy in phenomenal experience, then there is no reason to fill in the gaps, to create an illusion, as it were, of greater determinacy. Dennett explains that in dreams and hallucinations, "the only work that the brain must do is whatever it takes to assuage epistemic hunger—to satisfy 'curiosity' in all its forms. . . . If our brains can satisfy all our particular epistemic hungers as they arise, we will never find grounds for complaint" (Dennett, 1991, p. 16).

If I am correct in claiming that phenomenal indeterminacy is more pervasive in dreams than in wakefulness, then dreams are states in which our epistemic appetites, in Dennett's terminology, are even more easily satisfied than in wakefulness. Consequently dreams can get away with leaving out quite a lot of the type of detailed sensory information that accompanies waking experience while still seeming realistic. Because our epistemic appetites are small in dreams, phenomenal indeterminacy is simply overlooked. A prediction is that in lucid dreams, the situation could be interestingly different: perhaps, indeterminacy is more readily noticed in dreams. Moreover, indeterminacy blindness not only characterizes dream experience itself but may persist into dream reporting and even scientific theorizing about dreams. Even if dreaming was, phenomenologically speaking, quite different from standard wakefulness, this fact would never be reported unless the dreamer had first noticed that something was amiss.

Indeterminacy blindness can thus also be helpfully applied to theorizing about the phenomenology of dreaming: it is a theoretical oversight driven, via the descriptional indeterminacy of dream reports, by indeterminacy in dream experience itself.

The assumption that dreams are characterized by phenomenal indeterminacy with respect to the body fits in well with findings on the bizarreness of the dream self. In a study relating different forms of dream bizarreness (incongruity, discontinuity, and vagueness) to dream content categories, Revonsuo and Salmivalli (1995) found, first, that the self, though present in most dreams, accounts for only 3 percent of the elements described in dream reports (see also Revonsuo, 2006, p. 110). Compared to other dream content categories, the dream self was also only rarely bizarre. According to the bodily duplicate hypothesis, these findings suggest that the dream self is particularly immune to the bizarre distortions that affect as many as 53 percent of nonself dream characters (Revonsuo & Tarkko, 2002). However, an alternative interpretation suggests itself: the reason why the dream self is rarely bizarre (and why it rarely shows up in dream reports at all) is not that it is a boringly realistic look-alike of the dreamer's waking self, but that in many or perhaps even most dreams, the body image (the matter of what the dream self looks like, the size and proportions of the limbs, the facial features, etc.) does not enter into experience at all. The self is present, but indeterminately so.

Siri Hustvedt provides a literary formulation of this idea:

> In my waking life I'm a woman, but sometimes in my dreams I'm a man. My masculinity is rarely a question of simple anatomy. I don't discover that I've sprouted a penis and am growing a beard, but rather I realize that I'm a man in the same moment I am troubled by the vague memory that I was once a woman. My sex becomes important in the dream only when it's called into doubt. It is doubt, not certainty, that produces first the question of my sexual identity and second the need to be one thing or the other, man or woman. (Hustvedt, 2006, p. 95)

Unless the question of her sexuality is raised in dreams, or so she seems to be saying, her sexual identity is indeterminate. She participates in the dream, but her dream self is neither male nor female; and perhaps, if Hustvedt were a less careful observer, she would simply assume her dream self to be female (and retrospectively describe it as such) by default. Epistemic hunger, first expressed as uncertainty, is the condition for dreaming of herself *as* a man or woman in the first place. Moreover, there is no need to imagine that Hustvedt's male dream self is anatomically complete: a detailed, full-body representation of this sort is not needed. Perhaps dreams of being someone other than oneself—of occupying a body other than one's own, of being an animal, a different person, or even just a younger version of oneself—arise from a vague sense of *not* being quite one's waking self, of some passing awareness of a discrepancy from the way one habitually experiences one's body. If so, then what requires explanation are not so much cases in which the dream self takes on an appearance that is different from its waking appearance—or indeed in which phenomenal embodiment is missing altogether—but rather cases in which phenomenal embodiment is exactly the same as in wakefulness. In this view, dreams that approximate the requirements of the bodily duplicate hypothesis would seem to be the exception rather than the rule. We can also see how phenomenal indeterminacy might contribute to an account of the narrative construction of dreams. If Hustvedt's description of her dreams is accurate, it would seem that higher-level conceptual and lower-level sensory representations are closely connected in dreams, and dream narratives result from a subtle and highly dynamic interplay between imagination-like and quasi-perceptual cognitive activity. Dream imagery is produced on demand, tailored to satisfy perceptual exploration and thought-like musings alike.

Phenomenal indeterminacy is a conceptual tool for understanding one of the ways in which bodily experience is missing in many dreams. I suggest that the body can be absent from experience in two ways: one, the unnoticed absence of bodily experience, associated with phenomenal indeterminacy with respect to the body; and two, phenomenal disembodiment, which refers to the experienced absence of the body and that requires a higher degree of cognitive and attentional sophistication than phenomenal indeterminacy. Experiencing the lack of a body (or a body part) is both more complex and more clearly bodily than having no experience of the body or of its absence whatsoever. But this also means that phenomenal indeterminacy is both the simplest and the most complete way in which bodily experience can be absent: phenomenal indeterminacy with respect to the body arises when the body is absent in such a way as to not even draw attention to its absence. This means that the simplest form of self-experience in dreams is one that cannot be categorized as phenomenally embodied or disembodied at all.

The conceptual distinction between phenomenal disembodiment and phenomenal indeterminacy has practical consequences for how to deal with dream reports: just as one should not infer the presence of bodily experiences in dreams independently of what is mentioned in dream reports, so one should avoid making the same mistake with regard to phenomenal disembodiment. Rather, the best explanation for the widespread absence of the body in dream reports may be its inconspicuous absence in dream experience itself. This also means that the conservative approach should be applied not just to the analysis of dream reports but also to the practice of dream reporting: special reporting conditions, created by carefully worded instructions and questions, may be necessary to render phenomenal indeterminacy visible. For instance, asking explicitly about bodily experiences and including a "neither/nor" or an uncertainty option in questionnaires might be a step in this direction.[15] Asking only about the presence of a dream body might introduce an unnecessary confound, and if my account is correct, then dreamers might "just know" that they have a body without experiencing it in any distinctly bodily way. This possibility should be taken into account in future studies.

A preliminary conclusion is that at least a subgroup of dream reports characterized by an absence of references to the body can possibly be described as referring to an experience that was phenomenally indeterminate with respect to the body. Assuming that such reports were given under ideal reporting conditions, this could be true for a subgroup of active participation dreams, for instance, those in which the dreamer participates only by conversing with other dream characters: this type of participation is not obviously linked to bodily experience, but only to thoughts and maybe feelings (cf. sec. 7.2.1). The same type of analysis can also be applied to dream reports describing so-called passive observer dreams.

Take the following example: "I was at a gasoline station and I was observing this scene: a child was mounting an inflatable horse which had a motor inside. When the signal rang the horse bumped against a pole. The child was my nephew" (Cicogna & Bosinelli, 2001, p. 31). Occhionero et al. (2005) give this as an example of a passive observer dream, which they found to be the second most frequent type of self-representation in both NREM (24.4 percent) and REM (6.67 percent) dreams. Note that the report explicitly locates the dream self within the dream, as observing the events from a particular point of view. Cicogna and Bosinelli note that in such observer dreams, the dreamer "is placed in a more or less defined space. Even in the case in which that space is not defined at all, the dreamer's experience cannot exclude the awareness of 'being there' and observing" (Cicogna & Bosinelli, 2001, p. 31). They also suggest that observer dreams involve "the implicit or explicit awareness of the physical presence of oneself" (p. 31). We have seen, however, that the feeling of self-localization does not necessarily require the experience of oneself as a *physical* presence in the sense of being a phenomenally embodied self. Certainly, observer dreams are compatible with phenomenal embodiment or even phenomenal disembodiment; without a more detailed report, there is no way to decide the question. Yet phenomenal indeterminacy offers a parsimonious explanation for cases in which dreamers fail to describe any bodily experiences even upon further questioning.

The current prediction is that if phenomenal indeterminacy with respect to the body exists in the dream state, then a substantial percentage of dream reports gathered immediately after awakening will involve references neither to the body as present nor to its experienced absence. This will be true even if questionnaires and affirmative probes are used to determine the frequency of bodily experiences in the dream state. Descriptions of self-experience in dreams lacking references to the body will not disappear, even under ideal reporting conditions ensuring that the reports have a sufficient level of detail and specificity, and even if their expressive granularity is increased. If this is correct—and based on the available evidence, this does not seem wildly implausible—then phenomenal indeterminacy with respect to the body might be the best way to explain the inconspicuous absence of the body in many dream reports. And importantly, phenomenal indeterminacy might cut across a number of different types of self-representation in dreams. Passive observer dreams may simply be a particularly good place to look.

7.3.4 Weak Phenomenal Embodiment

While phenomenal disembodiment and phenomenal indeterminacy with respect to the body characterize different ways in which bodily experience is missing in dreams, *weak phenomenal embodiment* is one way of describing its presence. Once more, it cuts across a number of different categories of self-representation, such as active participation dreams, but also dreams in which there is only a static representation of the self involving a complete or partial body representation (Cicogna & Bosinelli, 2001, p. 31; Occhionero et al., 2005, p. 79).

Because of the underrepresentation of proprioceptive, tactile, and pain sensations in dreams (cf. sec. 7.2.2), weak phenomenal embodiment is most likely to be related to movement sensations and the phenomenology of spatial extension or of occupying a more or less determinate volume in space. Here is an example of the former type:

> After lying on my back for a long while, still seemingly awake, I suddenly feel as if I've turned into a bluish gas: actually a cloud of coarse blue spheres in the general form of my body that floats above the bed. (LaBerge & DeGracia, 2000)

This is an example of a false awakening giving rise to the experience of leaving the physical body and being embodied in a second body, as in out-of-body-experiences (see sec. 12.1 for discussion). Importantly, the experience is also closely related to instances of phenomenal disembodiment described earlier in that the body is experienced in a markedly nonstandard way. Dreams in which the dream self identifies with an extensionless, disembodied point may be only gradually distinguished from dreams of occupying a voluminous but diffusely shaped volume (such as a gaseous form), suggesting that phenomenal disembodiment and weak phenomenal embodiment should be placed on a continuum.[16]

Weak phenomenal embodiment also characterizes dreams in which only certain body parts are experienced. In addition to movement sensations and the feeling of occupying a voluminous space, there may be visual, or more rarely proprioceptive or tactile, sensations of specific body parts, and these may make such a strong impression as to be a primary motivator of the dream narrative:

> When living in Spain, I awoke early one morning, and heard a mosquito buzzing. I fell asleep again and dreamed that a huge insect—as large as a lobster, but flat like a cockroach, and scarlet in colour—had alighted on my hand. The creature had two long horns, and from each of these

proceeded numerous very long and delicate filaments which were inserted into my hand to a considerable depth. I had to cut the creature in half, and draw away the forepart, which was attached to my hand, with great care lest I should leave portions of the filaments in the flesh. This animal seemed all the more unpleasant because it was noiseless, and its attacks, I thought, imperceptible. I appeared to be attacked by a succession of them. (Ellis, 1911/1931, p. 75)

In dreams, however, a hand seen does not imply a hand felt, or even a hand connected, by way of an arm, to a torso.[17] Dream limbs, even when they appear, are not necessarily multimodal or integrated into a detailed whole-body representation; the experience may be weakly phenomenally embodied in both respects. Weak phenomenal embodiment can refer to the experience of isolated body parts or to the degree of multimodal sensory integration; it has a quantitative (or how much of the body is experienced) and a qualitative (or how the body is experienced) dimension.

Experiencing particular body parts, even vividly, is compatible with saying that the rest of the body was not in the dream at all. For this reason, phenomenal indeterminacy naturally complements weak phenomenal embodiment: while the latter positively characterizes the extent to which a phenomenal body is present, the former targets the ways in which it is incomplete or absent. Phenomenal indeterminacy suggests how to make sense of the fact that in dreams, we may experience only a body part, without experiencing either the rest of the body or its absence.[18] Indeed, note that weak phenomenal embodiment necessarily involves phenomenal indeterminacy: if no portions of the phenomenal body were indeterminate, then this would no longer be a case of *weak* phenomenal embodiment at all. At the same time, the converse is not true, and it is possible that a dream is phenomenally indeterminate with respect to bodily experience as such. Phenomenal indeterminacy can occur without weak phenomenal embodiment. And of course, phenomenal indeterminacy and weak phenomenal embodiment may not only coexist simultaneously and in a complementary fashion but can also gradually shade into each other. The following dream reported by Stephen LaBerge nicely describes a transition from phenomenal indeterminacy and perhaps phenomenal disembodiment to weak and possibly even strong phenomenal embodiment:

I had been dreaming—but then I suddenly realized that I must be asleep because I couldn't see, feel, or hear anything. I recalled with delight that I was sleeping in the laboratory. The image of what seemed to be the instruction booklet for a vacuum cleaner or some appliance floated by. It struck me as mere flotsam on the stream of consciousness, but I was focused on it and tried to read the writing, the image gradually stabilized and I had the sensation of opening my (dream) eyes. Then my hands appeared, with the rest of my dream body, and I was looking at the booklet in bed. My dream room was a reasonably good copy of the room in which I was actually asleep. Since I now had a dream body I decided to do the eye movements that we had agreed upon as a signal. I moved my finger in a vertical line in front of me, following it with my eyes. (quoted in Warren, 2009, p. 122)

7.3.5 Strong Phenomenal Embodiment

Can strong phenomenal embodiment occur in the dream state? And what would it mean to say that it did? Treating strong phenomenal embodiment as a hypothetical maximum, let us say that it involves a detailed (though possibly distorted) and multimodal representation of the whole body, including body parts. Strong phenomenal embodiment once more has a quantitative and a qualitative dimension, in that it is differentiated from weak phenomenal embodiment both by *how much* of the body is experienced and by *how* it is experienced. In strong phenomenal embodiment, the body would be, simultaneously, represented not only visually but in all the modalities (including touch, proprioception, vestibular sensations, nociception and pain sensations, and interoception) commonly used to characterize bodily experience. Because I am only interested in introducing the concept of strong phenomenal embodiment to characterize the upper end of the spectrum of bodily experience in dreams, this list need neither be complete nor define a sharp cutoff line. Importantly, the strongly phenomenally embodied dream self also need not be a look-alike of the waking self; let us just assume that it would be experienced in a richly and distinctly bodily manner.

Does this type of experience occur in dreams, and if so, how often? The existing evidence suggests that a majority of dreams do not involve a strongly phenomenally embodied self of this type. Yet this is a straightforwardly empirical question. One might wonder, however, whether the concept of strong phenomenal embodiment, sketchy as it is, demands too much. Perhaps the bodily duplicate hypothesis does not depend on the claim that even standard wakefulness involves strong phenomenal embodiment of this sort. Indeed, if standard wake states fail to give rise to strong phenomenal embodiment of the type described here—or if phenomenal embodiment in wakefulness were found to vary along similar dimensions as phenomenal embodiment in dreams—then the bodily duplicate hypothesis might be wholly compatible with weak phenomenal embodiment. Without a careful study of bodily experience in standard wake states that is differentiated enough to take variations into account—for instance, by charting variations in bodily experience across quiet wake states, such as lying on the couch while fully immersed in a novel, to active wake states, such as dancing or doing sports—we do not know how to assess the comparative statement implicit in the bodily duplicate hypothesis. Perhaps only dancers, athletes, or experienced practitioners of mindfulness body-scan meditation ever enjoy rich bodily experience of this sort, and perhaps the same is true in their dreams. Whatever the answer turns out to be, questions about the bodily duplicate hypothesis and about strong phenomenal embodiment should, at present, be regarded as separate.

Given these uncertainties, one might think that the best strategy for placing the concept of strong phenomenal embodiment on more solid empirical ground is to use reports from standard wake states as an upper measure of phenomenal embodiment. In this case, strong phenomenal embodiment would refer to the richest examples of bodily experience in wakefulness—either from certain subject groups or from the general population. But there is another possibility. Rather than looking to waking experience to define the upper limits of phenomenal embodiment, it might make sense to turn once more to altered states of consciousness such as dreams. If phenomenally disembodied and phenomenally indeterminate dreams help chart the simplest forms of bodily experience, might not dreams supply examples of the upper end of the spectrum, as well? Dreams in which there is a sense of embodiment in more than one body, that is, by switching either between two dream bodies or between the physical body and the dream body, may be an example.

In January 1898 I was able to repeat the observation. In the night of January 19–20, I dreamt that I was lying in the garden before the windows of my study, and saw the eyes of my dog through the glass pane. I was lying on my chest and observing the dog very keenly. At the same time, however, I knew with perfect certainty that I was dreaming and lying on my back in my bed. And then I resolved to wake up slowly and carefully and observe how my sensation of lying on my chest would change into the sensation of lying on my back. And so I did, slowly and deliberately, and the transition—which I have since undergone many times—is most wonderful. It is like the feeling of slipping from one body into another, and there is distinctly a double recollection of the two bodies. I remembered what I felt in my dream, lying on my chest; but returning into the day-life, I remembered also that my physical body had been quietly lying on its back all the while. This observation of a double memory I have had many times since. It is so indubitable that it leads almost unavoidably to the conception of a dream-body. (van Eeden, 1913)

Is this really an example of strong phenomenal embodiment in the sense suggested earlier? Note that the dream report does not mention specific bodily sensations, and the only body parts mentioned explicitly are the chest and back. Perhaps, then, this is a case of two weakly instantiated representations of the torso alternating or even superimposed on each other.[19] Moreover, while the report suggests that the dream switched between the experience of a dream body and of the physical body, the same experience could, presumably, also occur in a false awakening; the dreamer would then be rapidly alternating between two different dream bodies. Occhionero et al. (2005) note that dreams can involve a double representation of a bodily self, and the following report suggests that a dreamer may even *simultaneously* identify with two different dream characters: "I was in a South-American country, I was riding a horse, other people were with me. . . . We were pursuing a man, who was also myself, because he (I) had some money" (Occhionero et al., 2005, p. 80). Again, descriptions of specifically bodily sensations are lacking; yet it seems possible that dreams could give rise to the experience of a double or supernumerary body, or of supernumerary limbs. Waking subjects sometimes describe the experience of a supernumerary phantom limb, or the experience of an additional body part, such as a third arm, occupying a separate space from the actual limbs (Brugger, 2006; Dieguez & Blanke, 2011). And while ordinary phantoms are rarely visual, supernumerary phantoms may be visual as well as postural-kinesthetic. Perhaps, if similar experiences occur in dreams, the hypothetical maximum of phenomenal embodiment is better defined as involving the experience of a supernumerary limb or even a whole body. Examples from dream reports could then be used to inform our theoretical concept of strong phenomenal embodiment and render it more empirically plausible and precise. And researchers, using such dream reports as their guide, might then try to experimentally induce cases of dual or supernumerary embodiment in wakefulness (for a recent example, see Heydrich et al., 2013). For now, take the following example, once more reported by Clare Johnson, as a possible candidate for strong phenomenal embodiment in two bodies at the same time.

I am at University, working on an essay which is giving me that stone-in-the-stomach feeling of boredom. My head grows heavy and I rest it on my arms, which are folded on my desk. Waves of sleep roll over me and I realize I absolutely must have a nap. Eyes still closed, I force myself to stand up, and slump onto my bed where I fall instantly asleep. After minutes or hours, a curious feeling steals over me: I have two bodies. I am simultaneously lying stretched out on the bed, and sitting at my desk.

I can feel the weight of my head on my arms, the coolness of the floor under my feet. And I can feel the bed under the length of me, the pillow under my head. This is most odd. I wonder whether I did in fact move to the bed, or whether I am still sitting at my desk. For what seems like many minutes, I experience this dual consciousness. There is some imagery; flashes of my bedroom with its colourful wall hangings and my desk strewn with papers. The imagery is as confusingly simultaneous as the physical sensation, as it is viewed from two different perspectives. I reflect that this must be a bizarre type of lucid dream.

I try to tip the balance between the two bodies and succeed in becoming mostly the body at the desk for a moment or two, while the body on the bed loses reality slightly. Since I am no longer sure whether I lay down on the bed or not, I cannot decide where my real body actually is. In the end, I drag myself from this pleasant but perplexing double state to see what's happening. It is hard to wake up as I am experiencing sleep paralysis. In order to get some mobile energy moving around my body, I imagine I am wriggling my toes, and eventually manage to open my eyes to discover that my real body is . . . on my bed. (Johnson, 2005, pp. 3–4)

Whether or not this example is truly representative of the upper end of phenomenal embodiment, in dreams or in general, and whether it can lend empirical grounding to the concept of strong phenomenal embodiment, is one question. For now, it is equally interesting to note that it gives rise to a number of other more tractable questions. First, it would seem that dual embodiment in this case was associated with bilocation (in the sense of apparently simultaneous self-location in two different places), a reduplication of the origin of the visual first-person perspective, and simultaneous self-identification with two body models. Consequently this may be comparable to (though possibly more stable than) heautoscopy, where patients not only describe seeing a double of themselves but report that self-location rapidly shifts between, or is simultaneously located in, both the real body and the visual double (Blanke & Mohr, 2005; Furlanetto et al., 2013; Heydrich et al., 2013). Whether the dream described here also involved a double visual representation of both bodies is, however, unclear, and investigating such cases would be an interesting project for future research.[20] Second, one might raise a number of questions about the relationship between cognitive insight and the experiential quality of the episode. To begin with, note that Johnson's remark that she was able to tip the balance toward the body sitting at the desk, thereby rendering the body lying in bed less real, suggests a certain level of cognitive penetrability and control. It would be interesting to know, then, what degree of control over dual embodiment is possible and whether the experienced position of both bodies could be altered. Relatedly, one might ask about the relationship between Johnson's cognitive uncertainty about where her real body was during the episode and the felt quality of the experience. Was her inability to decide where she was really sleeping matched, for instance, by a felt indeterminacy in the sense of self-location? Might more detailed questioning reveal that the actual location and position of the physical body lying in bed left its phenomenal mark, as it were, on the phenomenal dream body lying in the same position? Or were the two bodies, at least temporarily, really indistinguishable on the level of phenomenal experience in the degree of experienced realness? Finally, do cases of supernumerary embodiment exist in which *both* body models are illusory, or is it always the case that one of the two is, in fact, at the location and in the position of the actual body? And is the experience of embodiment in more than two bodies possible? Perhaps leaving the concept of strong phenomenal embodiment, for the moment, fairly undefined is an advantage, precisely because reflecting on possible readings helps identify a number of follow-up questions for future research.

7.4 Taking Stock: The Weak-Phenomenal-Embodiment Hypothesis

We have seen that bodily experience is variable in the dream state and that a conceptual framework distinguishing different degrees of embodiment is needed to do justice to this fact. The conceptual tools suggested here—phenomenal disembodiment, phenomenal indeterminacy, weak and strong phenomenal embodiment—are a first pass at imposing order on this variability. The idea is that they can then be sharpened to enable more fine-grained conceptual distinctions as additional empirical evidence becomes available. As they stand, however, they have already helped raise a number of questions about bodily experience in dreams. This is an important step toward an empirically plausible framework describing bodily experience in dreams.

Perhaps the most surprising aspect of the conceptual framework introduced here is that not phenomenal disembodiment but phenomenal indeterminacy occupies the bottom end of the spectrum. The most complete way in which the body can disappear from consciousness is not to experience its absence but to have it drop out of experience altogether. Phenomenal indeterminacy as an experiential property is complemented by indeterminacy blindness, or the fact that phenomenal indeterminacy is typically overlooked unless special steps are taken to make it, as it were, visible. A second important discovery is that phenomenal indeterminacy and phenomenal disembodiment are compatible with phenomenal selfhood. This is of immense metatheoretical importance, because it suggests that the conceptual distinction between phenomenal selfhood and phenomenal embodiment maps onto a real-world phenomenological distinction. Phenomenal selfhood is not necessarily bodily, not even in the sense of giving rise to the experience of *lacking* a body. The concept of *a*bodily or nonbodily self-experience is not phenomenologically empty.

The upper end of the spectrum is marked by strong phenomenal embodiment, referring to either the multimodal experience of a whole body including body parts or the sense of embodiment in more than one body at the same time. The middle end of the spectrum, occupied by weak phenomenal embodiment, likely characterizes a majority of dreams, thus giving rise to the weak-phenomenal-embodiment hypothesis:

Weak-phenomenal-embodiment hypothesis: The dream self, in a majority of dreams and for a majority of dreamers, is phenomenally embodied only in a weak sense that is predominately associated with movement sensations, specifically as are associated with individual body parts rather than the whole body. Moreover, bodily experience in dreams is marked by disturbances in multisensory integration. For instance, body parts may be seen but not felt or vice versa, and so on.

Given the status of research today, the weak-phenomenal-embodiment hypothesis is best understood as a prediction. Yet it is a prediction that is informed by a careful reading of the existing literature. Even if it ultimately requires adjustments or is outright falsified by future studies, I would argue that at present, it provides a more empirically plausible account of bodily experience in dreams than either of the hypotheses discussed at the beginning of this chapter. There appears to be somewhat more to the dream body, understood in a strictly phenomenological sense, than would be predicted by the no-body hypothesis, and somewhat less than predicted by the bodily duplicate hypothesis. To the extent that both are intended to capture what it is typically like to dream, both fail. They do so not only by attributing too little or too much bodily experience, respectively, to dreams, but also by implicitly assuming that a uniform characterization can do justice to bodily experience in dreams.

To be informative, the weak-phenomenal-embodiment hypothesis requires further qualification. Indeed, it is weak in a number of related respects that I have only hinted at. First, its claims on offering a stereotyped characterization of what it is like to dream are weak; indeed, if it did not allow for a high degree of variation, it would certainly be false. At best, the hypothesis aims to characterize a majority of dreams but does not pretend to generalize to all dreams. A second and related point is that as it stands, it is extremely coarse-grained. This is in part due to the paucity of systematic research on bodily experience in dreams, and in part necessary to account for the variability of the target phenomenon. I take this form of weakness to be a strength, however, in that the hypothesis can easily be enriched by future findings and rendered more precise. Third, the weak-phenomenal-embodiment hypothesis claims that only a weak link exists between phenomenal selfhood and phenomenal embodiment: the relationship between phenomenal selfhood and phenomenal embodiment is probably more variable than in standard wake states, and in some dreams, phenomenal selfhood can be dissociated from phenomenal embodiment altogether. Fourth and in its primary sense, the weak-phenomenal-embodiment hypothesis translates into two different but related claims about how we experience ourselves as *bodily* subjects in a majority of dreams. Applied to *how much* of the body is experienced, it says that an integrated whole-body representation, including body parts, is often lacking in dreams; body parts may be experienced in isolation or only intermittently rather than throughout a given dream. Applied to *how* the body is experienced, it says that bodily experience in dreams typically lacks the multisensory quality of bodily experience in wakefulness.

We can use this last point to hazard a preliminary prediction of how bodily self-experience in dreams differs from waking self-experience: the ways in which bodily experience in dreams departs from standard wakefulness should typically be related to disturbances in multisensory integration and the integration of body-part/whole-body representation. If there are systematic differences in bodily experience between dreaming and wakefulness, then at least this is a good place to look. A brief detour into the literature on bodily experience in wakefulness will help illustrate this claim.

In recent years, a plethora of empirical studies and theoretical articles has been devoted to the multisensory character of bodily experience and its relation to self-consciousness. Multisensory conflict—for instance, between visual and tactile stimulation or vestibular cues (for details and further references, see Aspell et al., 2011), but also between visual input and interoception (e.g., heartbeat; see Aspell et al., 2013)—leads to experimentally investigable changes in self-identification and experienced self-location. Neuroimaging data and neurological evidence suggest that self-identification and self-location rely on distinct neural mechanisms (Blanke, 2012).

A careful consideration of the empirical literature has led Frédérique de Vignemont (in press) to propose the "multimodality thesis," according to which bodily experience, and specifically the spatial character of bodily experience, cannot fully be accounted for by the body senses, and vision additionally plays a crucial role in shaping standard waking consciousness. In the weak version of the multimodality thesis favored by de Vignemont, bodily experience is constitutively multimodal in the sense that a loss of multimodality leads to a profound change in (though not obliteration of) bodily experience. As de Vignemont notes, the spatial resolution of the body senses (e.g., touch, proprioception) is too poor to accurately represent the location of tactile stimuli on the body and the size and shape of body parts. Consequently, vision is crucial to ensure the veridicality of bodily experience, and the bodily experience of congenitally blind subjects is, she claims, different in kind from that of sighted subjects.

Another way of saying this is that in standard wake states, bodily experience is overdetermined: it is the product of the integration of different sources of visual, tactile, vestibular, and proprioceptive information about the body. By contrast, in most dreams, bodily experience—to the extent that it is present at all—is underdetermined: the body is often experienced only in one modality at a time (e.g., seen, felt, etc.), and even when different types of bodily experience occur simultaneously, they are poorly integrated with each other (recall the difficulty that lucid dreamers have when attempting to realistically telescope their arms not only visually but also in experiencing corresponding tactile sensations; see sec. 7.2.2). While this description captures a majority of dreams, departures from weak phenomenal embodiment can be described in a similar manner. Phenomenal indeterminacy arises when self-location occurs in isolation due to a near-complete breakdown in multisensory and body-part/whole-body integration. In phenomenal disembodiment, the absence of a multisensory, integrated whole-body representation is itself reflected in experience. Finally, in strong phenomenal embodiment, multisensory and/or body-part/whole-body integration is reestablished, or, as in cases of supernumerary embodiment, conflicting whole-body representations either alternate with each other or are experienced simultaneously.

Even if we assume, for the time being, that this account adequately captures the difference between bodily experience in dreams and in standard wake states, one might object that we should look elsewhere for a wake-state model of bodily experience in dreams. Indeed, it seems that there is a much better contrast condition than standard waking bodily experience: phantom limb experiences. Like bodily experience in dreams, phantom limbs involve a partial breakdown of the multimodality of standard waking bodily experience: phantom limbs are typically felt but not seen. Perhaps dreams can be described as a whole-body variant of phantom limbs in wakefulness. This would also nicely complement my claim (in sec. 7.3.2) that phenomenal disembodiment in dreams is comparable to "negative" phantoms, for instance, in xenomelia. It might also translate into a fruitful research strategy, as existing findings from the robust—and, contrary to bodily experience in dreams, well researched—occurrence of phantom limbs could be generalized to dreams. Though initially attractive, however, the analogy between bodily experiences in dreams and phantom limbs in wakefulness is potentially misleading. To see why, a more detailed discussion of phantom limbs is necessary.

One way to assess the analogy between bodily experiences in dreams and phantom limbs would be to compare the general phenomenological characteristics of phantom limbs and bodily experiences in dreams. As is the case for bodily experience in dreams, however, phantom limbs in wakefulness are variable: they occur not only following amputation but also following spinal cord injury and in subjects with congenitally absent limbs; supernumerary phantoms may also occur after brain damage (see Brugger, 2006, 2012). Following amputation, phantoms may gradually change in size, be associated with painful or nonpainful sensations, be frozen in a certain position or be movable, and so on. Therefore it is unlikely that phantom limbs provide a uniform model of dream experience. A more direct way of assessing this claim, therefore, is to ask whether subjects experiencing phantom limbs in wakefulness also experience phantoms in their dreams, and whether they experience them in the same way as in wakefulness.

Indeed, a number of studies have investigated this question (for a review, see Brugger, 2008). Mulder et al. (2008) found that 21 percent of amputees dreamt about themselves as after amputation, 31 percent as before amputation, and 37 percent in a mixed manner. This variability has precedent. Augustine (1997) already used the example of phantom limbs to illustrate the unpredictable nature of the dream body, remarking that people who have lost limbs sometimes dream of their missing limbs and sometimes do not (for details, see Sirridge, 2005). Despite this variability, Mulder et al. (2008) interpret their results as providing "modest evidence for the existence of a basic neural representation of the body that is, at least, partly genetically determined and by this relatively insensitive for changes in the sensory input" (p. 1266), suggesting that dreaming is marked by a greater "representational stability" (p. 1269) than wakefulness:

> In spite of the fact that the primary and secondary motor cortices have been "re-mapped" after amputation, leading to the plausible expectation that this would lead to a permanent alteration of the body scheme, other brain systems, especially those involved in dreaming, seem to re-create the absent limb over and over again, even years after amputation. In only 30 participants the system adapted fully to the actual body status, in the remaining 108 participants there was still a revival of the absent limb during dream activities, even when the amputation took place more than 12 years ago. (Mulder et al., 2008, p. 1269)

What exactly does it mean that the absent limb seemed to be created over and over again in the patients' dreams? How closely did the re-created phantom limb in the dream mimic the wakelike experience of its counterpart? And more importantly, was the reappearance of the missing limb in dreams typically associated with phantom limb experience during wakefulness? A study conducted by Frank and Lorenzoni (1989) provides evidence for a strong association between phantom limb experience in dreaming and in wakefulness: 92 percent of their subjects experienced phantom sensations in wakefulness, and 83 percent had a complete body in their dreams. However, the phantom experienced in dreams only corresponded to the waking phantom for half of the subjects. For instance, patients experiencing a reduced (or shrunken) phantom in wakefulness sometimes reported having a complete and normally sized limb in their dreams, and vice versa. Also, the missing limbs were mostly represented visually in dreams and more rarely involved tactile and never painful sensations. Occasionally the opposite could also be the case. For instance, one subject reported seeing that his left arm was missing in a dream but feeling touch sensations and being able to use the arm for manual activities. Despite such exceptions, the authors conclude that the representation of phantom limbs is importantly different in dreams and in wakefulness, with waking phantoms being represented predominately tactually and dream phantoms predominately visually.

This conclusion is supported by Vetrugno and colleagues' (2008) case study of a phantom limb patient with RBD. In wakefulness, he was able to imagine moving his phantom hand but experienced unpleasant and even painful tingling and burning sensations in it. In dream reports gathered immediately after awakening, he reported being able to move his missing hand normally during his dreams. During REM sleep, he manifested continuous movements of the left limb stump associated with seemingly purposeful movements in the other limbs: apparently his stump was following the movements of his dream arm. Yet the patient noted that the unpleasant tingling and burning sensations he experienced in his phantom in wakefulness disappeared in his dreams. As the authors suggest, this might be explained by the absence of conflicting afferent information about the amputated limb. In a later study investigating the dream reports of amputees following REM sleep awakenings, three (out of six) patients reported dreams in which the amputated limbs were intact and functional, though lacking the phantom limb sensations experienced in wakefulness (Alessandria et al., 2011). This disappearance of phantom sensations in dreams fits in well with the familiar pattern of sparse body representation and disturbed multisensory integration in dreams.[21]

These findings on phantom limbs in dreams nicely illustrate my claim that bodily experience in dreams cannot simply be described as a whole-body variant of phantom limbs in wakefulness. Rather, bodily experiences in dreams adhere to different representational principles. Dream research extends findings from existing work on phantom limbs, but we cannot simply generalize from waking phantom limb experience to dreams. Rather, understanding the subtle differences between bodily experience in dreams and waking phantoms, as well as the variability of both, is an important target for future research.

Before closing, I want to lend further support to this claim by presenting a brief case study of the dreams of a single subject. Aiha Zemp was born without legs and forearms but experienced vivid phantoms in her missing limbs since early childhood. She participated in a number of studies, ranging from clinical observation and behavioral testing to functional neuroimaging (for details and further references, see Brugger, 2012). These confirmed that her phantom sensations "were more than just wishful thinking, and reflected a cortical representation of body parts that had never physically developed" (p. 206; see also Brugger et al., 2000). Here is a description of one of her first memories of her phantoms from when she was three years old:

> I know exactly what it's like to kneel and feel my heels on my buttocks. Later, I experience phantoms in my hands, and I soon use my fingers to do things people didn't think I was capable of, like tying a bow. This enables me to develop a certain kind of empathy, because when I watch others, I know exactly what it feels like to make certain movements. As a child, I think this is funny. I talk about this access and about the use of my fingers and feet, and my parents, uncles and grandparents laugh and say, "That's wishful thinking, you're imagining things." I am offended, because these are specific experiences from my everyday life. At age five, I decide not to talk about my hands and feet anymore. My body image changes considerably with the orthopedic attempt to give me, at age seven, arm prostheses. Through these, I lose much of my access to my hands, which leads me to develop an enormous resistance against the prostheses. (unpublished material, provided courtesy of Peter Widmer; translation mine)

If objects were placed in the space where she felt her phantoms, she would move them; for instance, if a bag was placed in a car in the area where she felt her legs to be, she would move them, even if this meant sitting in an uncomfortable position (on "obstacle shunning" in phantom limbs patients, see Brugger, 2006). She also reported using her phantom hands to gesture while speaking and being able to improve the accuracy of movements in her stumps—for instance, when using an iPad—by executing the movement at the position of the stump with her phantom fingers.

In the last months of her life, Aiha Zemp was kind enough to take the time to answer a general questionnaire about her dreams developed by Bigna Lenggenhager and me. We were extremely interested in her dreams not just because of her phantoms. Aiha Zemp was also a gifted lucid dreamer, recalling her dreams three to four times per week and having lucid control dreams two to three times per week. She had practiced Zen meditation for more than thirty years and had learned to induce lucid dreams from Peter Widmer, her Zen master. She also reported that meditation had generally sharpened her awareness of her phantoms. She used her lucid dreams—especially those in which she herself dissolved—to prepare for her death. She died in December 2011. The following information is taken from her answers to our questionnaire (unpublished material), as well as from a number of unpublished dream reports that Peter Widmer generously shared with me.

Asked how she dreamt about her body, Aiha Zemp answered that in her dreams, her limbs were sometimes present and sometimes absent (i.e., stumps), as were her phantoms. Other dream figures had normal bodies and she did not dream of others having stumps. She experienced many different types of movement in her lucid dreams, such as flying, dancing, jumping, and kneeling. In dreams in which she had complete limbs, she would use her hands to execute movements; in dreams in which she had her stumps, she would use those. Asked how she typically experienced her body in her dreams, she wrote that she could both see and feel it. She also had lucid dreams, however, in which she was present only with her consciousness, without a body, or in which her body dissolved. In these dreams, she still felt herself to be located at a specific position; in other dreams, she would additionally see her dead body from the outside. In a particularly poignant dream, she describes that the visual body seen from the outside was made up from different periods of her life: it had the face of a child, the torso of a young woman, the stumps from her current life. At other times, she had vivid bodily experiences in her dreams. In one dream, she practiced walking meditation, carefully holding her arms and legs in a fixed position and moving as slowly as possible and with great concentration. Asked to rate the vividness of her body, including her phantoms, on a scale from 0 to 7, she noted that both in wakefulness and in lucid dreams, she experienced all her body parts with equal intensity (7).[22] In nonlucid dreams, however, this was different. Here her lower arms, hands, calves, and feet—her phantoms—were experienced as less intense (5) than the other parts of her body, which she experienced with the same intensity (7) as in wakefulness. After she awoke from a lucid dream, her phantoms were even more vivid than they normally were.

There are many reasons not to generalize from Aiha Zemp's dreams. Her phantom limbs are one reason; her experience with meditation, the associated heightened sense of bodily awareness, and the ability to induce lucid dreams are another. Clearly her dreams, especially toward the end of her life, when she was sick and preparing for death, also had a highly personal, even spiritual, meaning. Her bodiless dreams are certainly to be seen in this context, and for her, they were a source of joy and strength. Her descriptions of her dreams are informative for the depth of dreaming (Solomonova et al., 2014), for the analysis of dreaming in one particular individual and under very special conditions. Yet her descriptions of her dream body emphasize the familiar pattern of variability seen throughout this chapter. Perhaps due to her added sensitivity toward her phantoms, her bodily experiences in dreams were often particularly vivid, perhaps even involving strong phenomenal embodiment, as may have been the case in the dream in which she engaged in walking meditation. Interestingly enough, however, this intensity was not equally distributed, being most pronounced in lucid dreams and slightly weaker in nonlucid ones. It is tempting to suppose that this fits in with the representational sparseness of the dream body (or weak phenomenal embodiment coupled with phenomenal indeterminacy) in a majority of dreams, but such a conclusion would be purely speculative. Her descriptions of her bodiless dreams, in which she was nonetheless positioned at a specific location within the dream, once more show that phenomenal disembodiment is compatible with a robust sense of phenomenal selfhood. What is striking here is the way she very naturally described this type of experience as involving the presence of a self. Apparently there is a basic sense of selfhood that persists even when bodily experience is lost. And apparently this basic sense of self is sufficient to allow the dreamer to identify with it—to later report, after awakening, that "I" was present in the dream. At least, it is noteworthy that subjects do not, in reporting such dreams, linguistically distance themselves from their dream selves.

This suggests that while (bodily) self-experience is typically unified in that we experience ourselves as a single self, embodied in a single body (for details, see Bayne, 2012), neither self-experience nor bodily experience is in fact simple: self-experience is dissociable from bodily experience, and bodily experience itself is shaped by the interplay of potentially dissociable dimensions. This insight does not just stem from dreaming but is further supported by recent experimental approaches to bodily experience in wakefulness. Combining experimental manipulations such as the rubber-hand illusion with a psychometric approach to questionnaire data, Longo and colleagues (2008; Longo & Haggard, 2012) argue that different components of bodily experience can be distinguished. These are embodiment in the rubber hand, loss of one's own hand (perhaps similar to what I call phenomenal disembodiment), movement, and affect. They also argue that embodiment itself comprises ownership, location, and agency. Where in waking subjects, experimental manipulation is necessary to chart the potentially dissociable components of bodily experience, in dreams, such dissociations occur both frequently and spontaneously. This is why dream research can provide a valuable extension of existing research on bodily experience and its relationship to phenomenal selfhood.

For now, a preliminary conclusion is that the subjective unity of the bodily self should not be confused with structural simplicity: while bodily experience may seem to form a stable and constant backdrop of waking experience, it is both highly plastic, in that it changes over time, and structured, in that at any given moment, what is experienced as "the" self or "the" body is in fact made up of potentially dissociable and varying dimensions of experience. Of course, this illusion of stability and simplicity of the (bodily) self is itself an interesting explanandum; yet speaking of "the" phenomenal self or "the" phenomenal body as a single, simple, and stable type of experiential content is misleading to the extent that it glosses over the underlying variability of the target phenomenon.

Note how the example of dual embodiment discussed at the end of the last section emphasizes both points: here the dual sense of embodiment and of self-location does not lead to a breakdown in the experienced unity of the self. The dream report describes the superimposition of conflicting visual imagery as if seen from different perspectives as disturbing; yet the dreamer simultaneously identifies with two bodies, each of which are experienced as unified. Indeed, the experience of having two bodies logically requires that each body representation itself have a certain level of unification: if they did not, they would not be experienced as two *distinct* bodies in the first place. This highlights the extent to which the experienced unity of the self is both an illusory and a highly stable property of conscious experience, maintained even in the face of large-scale disruptions in sensory integration. The persistence of this sense of unity throughout the spectrum of experiences ranging from phenomenal indeterminacy to strong or even dual phenomenal embodiment suggests that phenomenal selfhood is a maximally invariant property that stays stable despite large-scale variation. Importantly, we are now beginning to see how the investigation of dreaming can contribute to the theoretical understanding of both the experienced unity and simplicity of the (bodily) self and its variability and dissociability.

7.5 Conclusions

Throughout the chapter, we have moved from a stereotyped and in many ways oversimplified characterization of bodily experience and phenomenal selfhood in dreams to a more refined and empirically plausible conceptual framework. In this process, the bodily duplicate hypothesis and the no-body hypothesis have been replaced by a successor hypothesis, the weak-phenomenal-embodiment hypothesis. This does not mean, however, that these initial hypotheses were simply discarded as unhelpful; rather, using them to sketch the space of possible positions on bodily experiences in dreams has helped structure the discussion of the empirical evidence and led to the identification of new types of questions.

While the weak-phenomenal-embodiment hypothesis introduced in the preceding section is located, in a sense, between the two original hypotheses outlined at the outset, it also involves a departure from an assumption on which both implicitly converge: the view that phenomenal embodiment and phenomenal selfhood are inextricably linked. The weak-phenomenal-embodiment hypothesis aims to capture the variability of bodily experiences in dreams and weaken the link between phenomenal embodiment and phenomenal selfhood in equal measures.

Whereas the proposed conceptual framework, ranging from phenomenal indeterminacy to strong phenomenal embodiment, provides conceptual tools for describing bodily experiences in dreams, the weak-phenomenal-embodiment hypothesis is best regarded as an empirical prediction about the type of bodily experience that likely characterizes a majority of dreams. Even if this prediction turns out to be false, it would not show the underlying conceptual distinctions to be inadequate. For present purposes, its main aim is to offer a first and fairly coarse-grained classification of different forms of bodily experience in dreams and their relation to the phenomenal self, not to be exhaustive. If it helps advance our understanding of the relationship between phenomenal embodiment and phenomenal selfhood, if it can be used to develop empirical successor questions, and if it is open to refinement in response to new findings, then it will have fulfilled its goal.

Importantly, we are also beginning to see how a theoretical understanding of bodily experience in dreams might contribute to research on bodily experience and phenomenal selfhood more generally. In particular, the proposed framework is not restricted to dreaming but can be used to contrast dreaming with bodily experience in standard and altered, pathological and nonpathological wake states (e.g., asomatognosia, somatoparaphrenia, xenomelia, phantom limbs, the rubber-hand illusion, etc.). An important desideratum is the comparative study of bodily experiences across the sleep–wake cycle, using the same methodology and the same conceptual framework. Such a research program would take an important step toward identifying the state-independent conditions for the phenomenology of selfhood and of phenomenal embodiment. Aside from affording an important role to scientific dream research, this type of research program would also, however, profit from an interdisciplinary perspective. This development is already becoming obvious in existing work on bodily experience in wakefulness, and the same is true for dreaming. We have already seen numerous examples for saying that philosophical work on dreaming, unless it is informed by empirical research, is at best speculative and at worst empty; and empirical research easily turns trivial unless it is informed by careful conceptual groundwork. The present chapter is intended, in part, as a case study of how an integrated and interdisciplinary approach might contribute to a deeper theoretical understanding of both dreaming and wakefulness.

8 Are Dreams Disembodied Experiences? The Role of the Body and of the Brain in Shaping Bodily Experience in Dreams

8.0 Introduction

Now that we have the conceptual tools for examining the phenomenology of embodiment and its relation to phenomenal selfhood both in dreams and across the sleep–wake cycle, the next important question is how to describe the relationship between the phenomenology of (dis)embodied selfhood in dreams on the one hand and the sleeping, physical body and the brain on the other hand. Again, the obvious place to start is the widespread assumption that dreaming is a state of functional disembodiment, in which phenomenal experience arises independently of external stimuli and real-body inputs. A first thing to note is that this shared assumption is paralleled by maximally opposing views on the phenomenology of dreaming. Here the no-body hypothesis predicts that when the functional connection to the sleeping body is lost, the phenomenology of embodied selfhood goes with it. By contrast, the bodily duplicate hypothesis claims that dreaming replicates waking bodily experience while being functionally decoupled from the sleeping body.

In chapter 7, we saw that both hypotheses misconstrue the phenomenology of embodiment and its relationship to phenomenal selfhood in dreams. I introduced the weak-phenomenal-embodiment hypothesis as a first step toward a more empirically plausible view. This suggests that the relationship between phenomenal embodiment and the sleeping body also has to be reconsidered. The all-or-nothing view of bodily experience in dreams—the view that dreaming either replicates waking bodily self-experience or lacks it altogether—sits nicely with the view that dreams are functionally disembodied states. The same is not obviously true, however, for the weak-phenomenal-embodiment hypothesis. At least at the outset, it is not clear why a state of complete functional disembodiment would result in the variability and widespread phenomenal indeterminacy of bodily experience predicted by the weak-phenomenal-embodiment hypothesis. If the phenomenology of embodied selfhood in dreams resists a uniform characterization, why not think that the same variability might reappear on the functional level of description, as well? In other words, why not at least consider the possibility that dreams are not only weakly phenomenally but also weakly functionally embodied states?

In this chapter, I address a number of questions about the bodily and brain basis of bodily experience in dreams. First, how exactly is the relationship between the phenomenal body and the physical body described in the literature? Second, which view of the functional relationship between bodily experience and the sleeping body is best supported by empirical findings? Third, what are the conceptual and metatheoretical consequences if dreams really turn out, as I will argue, to be best described as weakly functionally embodied states? And how might this lead to an integrated research program on phenomenal-functional embodiment across the sleep–wake cycle?

After reconstructing two standard versions of the *functional-disembodiment hypothesis* from the philosophical and empirical literature, I investigate the empirical evidence on the connection between real-body inputs and bodily experiences in dreams, as well as existing findings regarding the neural correlates of bodily experience in dreams. I then put pressure on the functional-disembodiment hypothesis by arguing that it is not just empirically implausible but also theoretically unsound and methodologically intractable. Contrary to the popular view, the phenomenology of embodiment in dreams is best explained by appealing to the continued processing of real-body inputs, rather than by trying to conceive of it as entirely independent of the physical body.

In the final section, I argue that the *weak-functional-embodiment hypothesis* better fits the empirical evidence not only on the phenomenology of dreaming but also on its bodily and brain basis. This view has important consequences for the concept of dreaming, suggesting that bodily experiences arising in dreams are better conceived of as illusory rather than as hallucinatory. This view also sheds light on a particular variant of the mind–body problem: the *body-brain-body problem*, which addresses the functional relationship between bodily experience, brain activity, and the physical body. The investigation of bodily experience in dreams is, in this view, a research strategy for investigating the relationship between bodily experience and its real-body basis. Theoretically, this point is of considerable importance because it suggests that the widely assumed dichotomy between internalist and externalist approaches to conscious experience is unhelpful and even misleading when translated into an explanatory framework and research strategy for the scientific investigation of conscious experience. In rethinking theoretical conceptions about dreaming, I suggest, we are rethinking commonly accepted tenets about the mind–brain-body relationship more generally.

8.1 The Functional-Disembodiment Hypothesis

The assumption that dreaming occurs in a state of near-complete functional disembodiment is widespread. Functional disembodiment is just a special case of the sensory input blockade commonly taken to characterize REM sleep (Hobson et al., 2000); it refers to the functional disconnection between dream experience and distinctly *bodily* inputs (e.g., tactile, thermal, pain, movement and vestibular sensations, proprioception, etc.). Functional disembodiment draws attention to the fact that the sleeping, dreaming subject is temporarily decoupled not only from sensory inputs from the actual environment but also from own-body perception. The second aspect of functional disembodiment is REM sleep paralysis, which prevents the external enactment of internally experienced dream behavior. The functional-disembodiment hypothesis can be formulated as follows:

Functional-disembodiment hypothesis: Dreaming is characterized by a functional disconnection from the sleeping body such that phenomenal experience is decoupled from real-body input and motor output.

In this view, dreaming is comparable to the predicament of a brain-in-a-vat, the only difference being that dreaming is a spontaneously and regularly occurring instance of *cranial envatment,* in which the brain is temporarily cut off from the body. By contrast, the brain-in-a-vat as traditionally conceived (Putnam, 1981) owes its envatment to the doings of evil scientists, and its predicament is more permanent than that of the dreaming brain.

The functional-disembodiment hypothesis raises important questions about the relationship between phenomenal embodiment in dreams and the sleeping, physical body. While proponents of the bodily duplicate hypothesis and the no-body hypothesis embrace the functional-disembodiment hypothesis, they profoundly disagree about the consequences for phenomenal embodiment. This suggests that the functional-disembodiment hypothesis is compatible with two diametrically opposed positions on the relationship between phenomenal and functional embodiment. The first is to conceive of phenomenal embodiment as independent of functional embodiment. Call this the *independence hypothesis* of phenomenal embodiment in dreams:

Independence hypothesis: Phenomenal embodiment is independent of functional embodiment in that (wakelike) bodily experience arises in dreams in the absence of sensory input and motor output.

This is the position implicitly endorsed by proponents of the bodily duplicate hypothesis.[1] Descartes's assumption, in the *First Meditation,* that dreaming casts doubt on his own existence as a bodily agent commits him to the independence hypothesis.[2] A similar position is seen in Revonsuo's (2006) and Metzinger's (2003) description of dreaming as a purely internal phenomenon, involving the experience of a virtual or hallucinatory self in a virtual world. Metzinger writes that dreamers are "not functionally embodied agents," and dreams demonstrate "how full-blown, complex reality-models can evolve from an exclusively *internal* stimulus source" (Metzinger, 2003, pp. 256, 259). Aside from dream-enactment behavior in RBD patients, the only exception is the correspondence between dream-eye movements and real-eye movements; this is the one sense in which "the phenomenal dream self is not completely disembodied" (p. 261; see also Revonsuo, 2006, p. 78). The connection between the functional-disembodiment hypothesis and the phenomenological level of description becomes even more obvious in Metzinger's claim that a majority of dreams involve

> a fully embodied, spatially extended self moving around in a spatially extended environment. The virtual self thus born is an exclusively internal phenomenon in an even stronger sense than that of the waking self: It is immersed in a dense mesh of causal relations, all of which are internal to the brain. Dreamers are self-aware, but functionally they are not *situated*. (Metzinger, 2009, p. 135; but see Metzinger, 2014; and sec. 8.4)

Similarly, Revonsuo states that "dreaming occurs behind a sensory barricade, isolated from both the external world and the physical body. The story concerning motor output is no less dramatic" (Revonsuo, 2006, p. 90), for the physical body is fully paralyzed during REM sleep. "Subjectively, however, we do not have the slightest idea that our physical body is in a state of complete paralysis while we dream: our dreams are full of vivid bodily movement" (p. 90).[3]

It is not surprising that the independence hypothesis is endorsed by the philosophers who have, so far, been most concerned with advancing an empirically informed account of dreaming. This is so because the independence hypothesis is popular in scientific dream research, as well. Hobson writes that dreams

> include a strong sense of self, of self-as-agent, and movement of that self-agent through a perceptual space, all integrated with emotional salience. Dreaming is thus a virtual reality experience with a remarkably predictive simulation of external reality. . . . A question of direct relevance to the virtual reality hypothesis is whether the enactment of dream behaviors utilizes the same brain circuits as those which mediate those very behaviors in waking. A review of the lucid dreaming literature supports the identity hypothesis. (Hobson, 2009, p. 43)

In this view, the bodily dream self is part of the virtual reality of dreams, and the finding that the same brain circuits are involved in dream and waking movement is taken to support the view that a simulation of wakelike behavior can be brought about internally during REM sleep. By contrast, the sleeping body is not taken to be relevant to a theoretical explanation of bodily experience in dreams.

Finally, something similar to the independence hypothesis of bodily experience in dreams finds its expression in the brain-in-a-vat thought experiment itself. Importantly, the envatted brain, like the Cartesian dreamer, continues to enjoy wakelike though fully hallucinatory bodily experience:

> The computer is so clever that if the person [the subject whose brain has been placed in a vat and is being appropriately stimulated by an evil scientist] tries to raise his hand, the feedback from the computer will cause him to "see" and "feel" the hand being raised. . . . It can even seem to the victim that he is sitting and reading these very words about the amusing but quite absurd

supposition that there is an evil scientist who removes people's brains from their bodies and places them in a vat of nutrients which keep the brain alive. (Putnam, 1981, p. 6)

It would seem, then, that it is only a small step from cranial envatment in dreams to literal envatment, as in the science-fictional thought experiment. Dreaming is simply a real-world example of the brain's general predicament. Where Cartesian dualism has the mind disconnected from the brain and the body, contemporary proponents of internalist or representationalist positions emphasize the disconnection of the brain from the rest of the body. Importantly, dreaming is simply a particularly clear example of the generally virtual character of experience, including the phenomenology of embodied selfhood.

The second way of unpacking the functional-disembodiment hypothesis is to combine it with the no-body hypothesis. In this view, wakelike bodily experiences, or indeed any bodily experiences, are just the type of thing that the cranially envatted, dreaming brain fails to generate internally. Perhaps functional embodiment is even a necessary condition for having bodily experiences at all. Call this the *dependence hypothesis* of phenomenal embodiment:

Dependence hypothesis: Phenomenal embodiment depends on functional embodiment in that (wakelike) bodily experience does not (or perhaps cannot) arise in dreams in the absence of sensory input and motor output.

The dependence hypothesis appears to underlie McGinn's equivocation between the sleeping body and the phenomenal dream body (see sec. 7.1). McGinn explains that if I dream of a tiger standing next to me,

It is not literally true that I, the dreamer, take the tiger to be right next to me; if I did I might leap from the bed and run for the door! It is perhaps *as if* I have the experience of seeing a tiger, but it doesn't follow that I really have such an experience. Hallucinating a tiger would produce the affective and motor reactions characteristic of veridically seeing a tiger, since genuine presence would be written into the experience; but dreaming of a tiger doesn't produce these reactions—so there is a difference between the two cases. (McGinn, 2004, p. 94)

Cranial envatment in dreams is not, in this view, analogous to actual envatment: according to McGinn (2005a, p. 64), the brain-in-a-vat experiences robust hallucinations, whereas dreaming only involves mental images. Moreover, dreaming fails to involve the experience of seeing a tiger because dreaming lacks the appropriate functional connection to the sleeping body.

Can the dependence hypothesis be adjusted to allow that when bodily experiences occur in dreams, they do so in a state of incomplete functional disembodiment? To the extent that the dependence hypothesis is motivated by the view that dreaming, qua imaginative experience, necessarily fails to give rise to "real" bodily experience, such a weakened version would be self-defeating. Yet I am interested here not so much in the actual arguments for endorsing the dependence hypothesis (indeed, we have already seen that the phenomenological claims made by the no-body hypothesis are empirically false) as in using the dependence hypothesis to map the space of possible positions on the functional relationship between bodily experience in dreams and the sleeping body. For this reason, asking about the compatibility of the dependence hypothesis with varying degrees of phenomenal embodiment in dreams is potentially informative. The answer will depend on the strength of the dependence claim. On the one hand, it would seem that bodily experiences in dreams can be accommodated by the dependence hypothesis as long as they occur in a state of incomplete functional disembodiment. On the other hand, there are many instances of bodily experiences in dreams that cannot plausibly be said to depend in any *direct* manner on real-body sensations, such as flying, falling, or whole-body movement sensations—which arguably do not require that the sleeping subject fall out of bed, let alone fly or levitate above it. Consequently any direct form of dependence quite obviously has to be rejected from the outset.[4]

In sum, in evaluating the functional-disembodiment hypothesis, we will also have to adjudicate between the independence and the dependence hypotheses. My survey of the empirical literature in the next section suggests that while phenomenal embodiment in dreams is indeed less tightly constrained by real-body inputs than is the case in wakefulness, it is not thereby completely independent. Rather, there are many instances in which a causal relationship exists between the continued processing of real-body inputs and bodily experience in dreams. Moreover, the investigation of the bodily (and not just brain) basis of bodily experiences in dreams is explanatorily relevant and methodologically fruitful. This in turn suggests that the independence and the dependence hypotheses require such substantial revision that the motivation for endorsing the functional-disembodiment hypothesis itself is obliterated. While the independence hypothesis is both empirically implausible and conceptually flawed, the way in which real-body inputs modulate bodily experience in dreams is also less direct than assumed by the dependence hypothesis. My interim conclusion in the final section of this chapter is that bodily experiences occurring in dreams are best conceptualized as illusory perceptions of the sleeping body. Dreams cannot uniformly be considered as instances of cranial envatment but rather are weakly *functionally* embodied states. A significant portion of dreaming cannot, in this view, be regarded as hallucinatory in the sense of being an entirely virtual and exclusively internal product of brain activity. Spelling out this claim and the motivations for endorsing it requires a profound departure from some of the most popular and widespread theoretical views of dreaming in the contemporary literature. We will also see, however, that in appealing to the bodily (and not just brain) basis of bodily experience in dreams, we are returning to a rich historical tradition; and indeed I will claim that much can be learned from taking these historical sources seriously.

8.2 The Bodily Sources of Dreaming

8.2.1 Historical Predecessors

The idea that dreams are shaped by real-body inputs dates back to antiquity. Dreaming was sometimes even thought to be marked by a heightened sensitivity to the sleeping body—an assumption that is the opposite of the functional-disembodiment hypothesis. Because changes in dreaming were thought to indicate ailments that remained below the threshold of experience in wakefulness, dream incubation and dream interpretation were even used to diagnose illnesses, most prominently in Epidaurus and at other Asclepian shrines (Van de Castle, 1994, pp. 62–63). Even Aristotle (350 BCE/1997, pp. 131–132), who was otherwise critical of the practice of dream prophecy, explained diagnostic and rare cases of prophetic dreams by saying that small affections, which are drowned out in wakefulness, have a stronger effect during the quiet of sleep. For example, soft noises might lead to dreams of thunder, or warmth might lead to dreams of walking through fire.

In *On Diagnosis in Dreams*, Galen (1996) writes that "the vision-in-sleep [*enhypnion*], in my opinion, indicates a disposition of the body." He gives the dramatic example of a man who

> dreamed that one of his legs had turned to stone, and many of those clever about such matters judged that the dream pertained to his slaves, but the man was paralyzed in that leg, although none of us expected that. (Galen, 1996)

The report bears a striking similarity to Sacks's report of a patient's dream that she had become "a living sentient statue of stone" (Sacks, 2001, p. 214). Upon awakening, she was parkinsonian and catatonic and was finally diagnosed with encephalitis lethargica (see also Sacks, 1999a).

Further examples of such diagnostic or even prodromal dreams are reported in the literature (see Haskell, 1985a, 1985b). Migraine auras may first manifest in dreams, preceding migraine attacks in wakefulness (Sacks, 1999b, 2001; Nielsen, 2010c). Carlos Schenck (2005) notes that a rise in aggressive dream content and the outward enactment of dream behavior, which is the major symptom of RBD, often precede neurodegenerative conditions such as Parkinson's disease. He speculates that the characteristic increase in violent dreams in RBD patients may actually be considered as

> the first sign of a neurologic disorder whose other ("classic") manifestations may not emerge until several years or decades after the onset of RBD. This is fascinating, and also ominous, as a distinct change in a person's dreams (a form of consciousness), and then the acting-out of those dreams, can be the harbinger of a full-fledged brain disorder that is decades from manifesting itself. (Schenck, 2005, p. 31)

Note that most of these examples, however, concern the manifestation of neurological symptoms in dreams—they may have an effect on the body, but they are not straightforward instances of a heightened sensitivity to the body during the dream state, as believed by the ancients. There is also a long tradition, however, linking more pedestrian bodily ailments such as indigestion to dreaming. Aristotle (350 BCE/1997) thought that heavy meals, wine intoxication, or fever increase the movements of the sensory organs during sleep, which at times could become so violent as to cause strange or confused dreams or even obliterate dream imagery altogether. Hobbes assumed that "dreames are caused by the distemper of some inward parts of the Body," with different distempers causing different types of dreams (Hobbes, 1651/1985, p. 91). For instance, "lying cold breedeth Dreams of Feare, and raiseth the thought and Image of some fearfull object" (p. 91).

The idea that dreams result from external sensory inputs or bodily sensations during sleep was especially popular in the nineteenth and early twentieth centuries. Robert Macnish (1838), author of a monograph titled *The Philosophy of Sleep*, thought that dreams are caused by an excitement of the inner organs, for instance, through fever or indigestion, and thus are useful in diagnosing illnesses. Like Hobbes, Macnish promoted a predominately negative view of dreaming, and one that is at odds with Aristotle's view that the clearest dream images arise in the calm of sleep:

> In health, when the mind is at ease, we seldom dream; and when we do so, our visions are generally of a pleasing character. In disease, especially of the brain, liver and stomach, dreams are both common and of a very distressing kind. (Macnish, 1838, pp. 47–48)

Macnish also suggested that dreams can be controlled by changing one's sleeping position. He relates a curious experiment in which a dreamer left his knees uncovered and consequently "dreamed that he travelled during night in a diligence: travellers, he observes, being aware that in a coach it is the knees that get cold during the night" (Macnish, 1838, pp. 55–56). Likewise, he observed that letting the feet hang over the side of the bed might lead to dreams of "standing upon the brink of a fearful precipice" (p. 57).

This view is echoed by Bergson, who explains that dreams of flying or floating occur when we become aware, in a dream, that our feet are not touching the ground. But according to Bergson, it is not just the soles of our feet that can make themselves felt in our dreams:

> Often in the midst of the night the contact of our body with its light clothing makes itself felt all at once and reminds us that we are lightly clothed. Then, if our dream is at the moment taking us through the street, it is in this simple attire that we present ourselves to the gaze of the passers-by, without their appearing to be astonished by it. We are ourselves astonished in the dream, but that never appears to astonish other people. I cite this dream because it is frequent. (Bergson, 1914/2007, p. 6)

Aside from appealing to the soles of the feet, as did Bergson, other authors thought dream flying to mirror the breathing pattern and movement of the lungs (in German aptly referred to as *Lungenflügel*, or lung-wings; see Scherner, 1861). Vold went to great lengths to conscientiously map his sleeping posture to his flying position in his own dream reports:

> Posture: Left lower extremities slightly retracted; the torso lying horizontally, turned slightly leftwards towards the light, one hand covering the eyes. . . . Dream: I am floating above the ground, the body in a horizontal or almost horizontal (slightly bent backwards) position, the lower extremities slightly retracted, so that the knees point upwards, the limbs dangling loosely from the knees, leaving only a small space between the soles of the feet and the ground. (Vold, 1910/1912, p. 797; quoted in Schönhammer, 2004, p. 58; translation mine)

Beyond the body, external sensations were also more generally appealed to in theoretical accounts of dreaming. At the end of the nineteenth century, Wilhelm Wundt wrote:

> The ideas which arise in dreams come, at least to a great extent, from sensations, especially from those of the general sense, and are therefore mostly illusions of fancy, probably only seldom pure memory ideas which hence become hallucinations. (quoted in Woods & Greenhouse, 1974, p. 179)

Havelock Ellis also believed most intense sensory images in dreams, as well as many dream emotions, to be related to an external stimulation source or to somatosensory sensations:

> There is something unusual or uncomfortable in the sleeper's position, or he has been subjected to some slight unusual strain which has modified his nervous condition, or there has been some deviation from his usual diet, or a physiological stress of some kind is making itself felt within him—careful self-questioning constantly reveals the actual or probable existence of some external or certainly peripheral stimulus of this kind. (Ellis, 1911/1931, p. 72)

At the same time, Ellis acknowledged that dreams often add their own creative element and thus are not mere "presentations" of external stimuli. Similarly, Silberer (1919, p. 28) somewhat poetically noted that while somatic sources may provide the starting point and the theme of a given dream, they do not, in themselves, explain the narrative spun around it by the dreaming mind. In this view, the process of dream formation may be triggered by, but not reduced to, somatic sources.

Perhaps the most famous example linking a dream to bodily sensations is Maury's (1861) guillotine dream. After a long and involved dream of the French Revolution that finally culminated in his execution, Maury awakened to find that the headboard of his bed had fallen on his neck. Because the dream seemed to build up toward the final dramatic event of the execution, this dream is sometimes used to illustrate the idea that dreams are instantaneous memory insertions or even have a precognitive element, anticipating future events (e.g., Goblot, 1896; Dennett, 1976; cf. chap. 1).

Some researchers went beyond such incidental observations and designed specific experiments to investigate bodily influences on dreaming. Under Wundt's supervision,

Weygandt (1893) studied the influence of breathing, blood circulation, temperature changes, urge to urinate, uncomfortable sleeping position, and visual or auditory stimulation on dream content. He did this by observing his own dreams, some of which had been experimentally manipulated by his assistant (for details, see Schredl, 2010b). Other researchers conducted informal experiments, with the dreamer acting as both experimenter and experimental subject.

Given their largely anecdotal nature, why are these historical sources interesting? *Leibreiztheorie* (somatic-stimulus theory), after all, had its heyday in the nineteenth and early twentieth centuries and was largely eclipsed by Freudian dream theory (for an excellent introduction with a wealth of references to the historical literature, see Schönhammer, 2004). Freud himself (1899/2003, pp. 38–56) devoted a long chapter to the putative bodily sources of dreaming, only to ultimately reject *Leibreiztheorie* in favor of the psychic sources of dreams. Moreover, even those emphasizing the somatic or bodily sources of dreaming were quick to point out that a bodily stimulus alone is not sufficient to explain the creative and narrative element of dreaming. What exactly, then, is the added value of reviving these largely unsystematic and prescientific observations on the bodily sources of dreaming?

Note that I did not simply relate them because they are quaint and enjoyable, but because they illustrate the contingency of currently popular theoretical assumptions about dreaming. Descartes no more *knew* that dreams arise independently of real-body inputs than the philosophers and pioneers of scientific dream research cited in this section *knew* the opposite to be the case. Proponents of both views were acting on hunches, intuitions, and largely incidental observations. At least initially, both views have equal prima facie plausibility. By briefly reviewing the history of a largely forgotten and marginalized way of thinking about the sources of dreaming, I wanted to lay the groundwork for a different, and in a certain sense conservative, alternative to the exclusively brain-based approaches that dominate today's discussion. Moreover, the following sections will show, I hope, that in trying to understand the phenomenology of bodily experience in dreams, there is genuine explanatory value in looking beyond the brain to the sleeping body. This is not to say that one should look at the body *rather* than at the brain, or *only* at the body, but to suggest that theoretical accounts of dreaming, as well as the methods used by scientific dream research, should be more inclusive than suggested by the functional-disembodiment hypothesis. Even this brief prelude suggests that the historical literature contains a wealth of research to be tapped.

8.2.2 Contemporary Studies

Empirical evidence on the bodily basis of dreaming in the contemporary literature falls into two groups: one, evidence for the incorporation of bodily stimuli in dreams, which in turn can be separated into direct evidence from stimulation studies and clinical patients, and indirect evidence from sleep paralysis, lucid dreams, nightmares, typical dream themes, and perhaps even emotions; and two, evidence for the external enactment of dream behavior. I will discuss these in turn.

8.2.2.1 Evidence for the Incorporation of Bodily Stimuli in Stimulation Studies and in Clinical Patients

As noted in section 5.2.3, several studies have found evidence for the incorporation of visual and auditory stimuli in dreams, though the frequency of incorporation and the effect on dream content are highly variable and hard to predict. But stimulation studies have also been used to investigate the correlation between the sleeping body and the dream body. In a study conducted by Dement and Wolpert (1958), sprays of water on the skin were incorporated in dreams in 42 percent of cases; by contrast, sounds and light flashes were incorporated in only 9 percent and 23 percent of all cases. Another study found that electric pulses to the hand were incorporated in 45 percent of cases (Koulack, 1969). Several studies investigated the effect of blood-pressure-cuff stimulation on the legs, which caused a mildly painful sensation during wakefulness, during sleep (Nielsen, 1993; Nielsen, Ouellet, & Zadra, 1995; Sauvageau et al., 1993, 1998). Blood-pressure-cuff-related themes were found in 40 to 80 percent of dream reports, depending on the length and intensity of stimulation. Such themes included references to the blood-pressure cuff, references to the legs (such as wearing strange shoes or experiencing trouble walking), and more intense, sometimes even painful sensations in the legs (importantly, there was no difference in the intensity of arm sensations), as well as a general increase in the bizarreness of body representation. Occasionally the stimulus was not only distorted but displaced, either to the nonstimulated leg or even to another dream character. For instance, one participant dreamt of trying to rescue a farmer stuck beneath a horse (Nielsen, 1993). Typically, however, blood-pressure-cuff stimulation led to painful sensations in the stimulated leg, and the pain seemed to be a motivating factor driving the dream narrative.[5] Limited evidence also indicated that stimulation caused an upset in virtual orientation in dreams and that the effect was less pronounced for gymnasts than for a control group, possibly owing to the extended training of the vestibular system in gymnasts (Sauvageau et al., 1998).

Early dream researchers also experimented with vestibular stimulation during sleep, for instance, by asking subjects to sleep in a rotating chair or by fixating the sleeper's body in a certain position (Hoff, 1929; Hoff & Plötzl, 1937; for further references and discussion, see Schönhammer, 2004), and found some evidence that vestibular stimulation leads to reports of flying in dreams. Leslie and Ogilvie (1996) investigated the effects of rocking on experienced lucid dreamers sleeping in hammocks, finding that vestibular stimulation led to an increase in lucidity and vestibular imagery in dreams. One study also investigated the efficiency of tactile stimulation (vibration) at the wrist, ankle, or index finger for the induction of lucid dreams (Paul & Erlacher, 2011). While stimulation was incorporated in more than 40 percent of all cases, it also frequently led to awakening and only successfully induced lucidity in 2 out of 30 reported dreams.

Further evidence for the incorporation of painful stimuli comes from studies with hospitalized burn victims. Raymond et al. (2002) found that approximately one-third of hospitalized burn patients reported painful dreams, usually relating to the burned body parts, but occasionally also to other body parts. The occurrence of dream pain was independent of levels of pain during the night or upon awakening. Although the assumption that dream pain resulted from an incorporation of real pain suggests itself, it is also possible that dream pain was related to the patients' waking preoccupation with pain. In stimulation studies, this worry, however, can be countered by alternating real and sham stimulation.

Finally, in his review of disturbed dreaming, Nielsen (2010b) notes that headache sufferers, especially migraine patients, experience characteristic changes in dreaming. They report more repetitive dreams, but also horrific nightmares, brilliant colors in dreams, and the carryover of emotions and sometimes hallucinations into wakefulness. In particular, the dreams preceding migraine attacks contain more anger, misfortune, and aggression than those that do not precede migraine attacks.

The interim conclusion is that we have much evidence for the incorporation of bodily stimuli in dreams. However, the precise form of incorporation and the intensity of the resulting dream experience are variable and sometimes only loosely constrained by the original stimulus. This is particularly true for cases in which stimuli are displaced to nonstimulated limbs or other dream characters.

8.2.2.2 Evidence for the Incorporation of Bodily Stimuli in Sleep Paralysis, Lucid Dreams, Nightmares, and Typical Dreams

Beyond stimulation studies, the most impressive evidence for saying that dream-related bodily experiences are modulated by REM-sleep-related changes in the physical body comes from sleep paralysis (SP) nightmares (Cheyne et al., 1999; Cheyne, 2003, 2005). At sleep onset or upon awakening, the experience of paralysis during subjective wakefulness may combine with highly stereotyped forms of nightmares whose content seems to be closely related to the physiological features of REM sleep (Girard & Cheyne, 2006). Here are two examples, which nicely illustrate both the subjectively realistic quality and the distinctly bodily flavor of these experiences:

Male, Aged 24. As I drifted off to sleep I felt someone climb up on the bed and lie spoon fashion against my back. I could feel this "Being" against the full length of my body and with its hand resting on my right shoulder. I could also feel its breath on my neck as well as hearing it breathe. I had a large pimple on the back of my neck and I felt the "Being" press its lips against this and begin to suck. As it did so my whole body especially the muscles in my upper back began to spasm and convulse. I was unable to move and found it difficult to breathe. The sucking would stop for a moment as the "Being" took a breath. The spasms also stopped although I was paralysed, but would recommence along with the sucking. This lasted for what seemed like an eternity, during which time I tried to call for help but couldn't make a sound. I fought to break the paralysis by concentrating my will, after some time I snapped out of it. I felt my self almost immediately fall back to sleep even though my heart was still racing from terror. (Parker & Blackmore, 2002, p. 50)

Female, Aged 24. I was woken as I heard my bedroom door open and someone/thing walked to my bed, it stepped over my clothes and leaned over me. I thought I could feel it breathing on me. All this time I couldn't move, or [s]cream, or see and I was very scared. The presence said to me "Don't worry, Clare (a friend in the same room) will be OK." Then the presence turned, stepped back over my clothes and walked out without shutting the door. I didn't hear it shut anyway. As soon as it left I was able to [move], see, scream, when I turned the light on I was able to see that the door was shut! I went straight back to sleep. (Parker & Blackmore, 2002, p. 50)

Note how both reports, aside from describing feelings of paralysis, also have a distinctly bodily flavor and refer to breathing, ranging from the sound of sucking and feeling the intruder's breath on one's skin to being unable to breathe or scream oneself. Moreover, both experiences are described as occurring in wakefulness, followed by a quick transition back to sleep—a point that seems surprising given the terror described by both subjects.

Intruder experiences and old-hag attacks of the type just described have a long and colorful history. In her engaging book, Adler (2011) traces intercultural consistencies and historical continuities reaching back to Assyrian and Babylonian sources, linking a wealth of folklore and superstition to scientific research on SP. She argues that beneath existing cultural variation there is a uniform experiential core, with the ingredients of the classical night-mare (spelled with a hyphen to distinguish it from comparatively more pedestrian sleep nightmares and bad dreams) being not just the experienced inability to move and the extreme realism of the episodes, but most prominently the sense of a presence, breathing difficulties, and vestibular-motor experiences. She notes that these core ingredients can be identified in accounts of nocturnal attacks from different cultures; the examples discussed by Adler include the Canadian Inuit, historical and present-day populations of Newfoundland (see Hufford, 1982) and Japan, and most prominently the Laotian Hmong. She also provides impressive evidence showing that many different languages refer to nightmares by using terms describing sensations of pressure, being choked, pressed, squeezed, bound, or ridden. The Spanish *pesadilla* and Portuguese *pesadelo*, for instance, are derived from the term for weight; the German terms *Albdruck* and *Hexendrücken* describe pressure exerted by elves or witches; and in Newfoundland, nightmares are called *ag rog*, or "hag ridden" (Adler, 2011, pp. 14–16). As Adler points out, the English term *haggard* may be derived from *ag rog,* and indeed, worried and exhausted is exactly what one would expect hag-ridden persons to look like.

The night-mare has a natural affiliation to supernatural narratives, and the experience has also frequently lent itself to religious or spiritual interpretations, as well as to paranormal ones (as in alien abductions). Occasionally, nocturnal, presumably night-mare-related encounters have left the realm of the supernatural to intrude on reality, with old hags, witches, and incubi often playing a prominent role in witch trials (Adler, 2011, p. 47). Nocturnal intruders may also appear in a more contemporary guise. For instance, in the 1980s, Palestinians experiencing night-mares reported nightly intrusions by Israelis, and vice versa, suggesting that the experience latched onto current fears (p. 26). Being informed about their true, sleep-related origin does not always dissipate the nightly visitations. In Japan, where the phenomenon of SP and its REM-related origin are fairly well known, the night-mare has been experienced by one-third of the population and is familiar from computer games, television, and books; there are even techniques for inducing the episodes, whose horrific nature is deemed desirable and even enjoyable by some (pp. 20–22). At other times, the experience may occur outside any narrative context, being described simply as awakening with the experience of pressure on the chest and inability to breathe (p. 11).

Night-mares have been the target of a number of highly interesting studies. Cheyne and colleagues (Cheyne et al., 1999; Cheyne, 2003, 2005) have argued that experiences arising during SP can be grouped into three clusters: one, intruder experiences, characterized by a sensed presence, as well as often by auditory, visual, or tactile sensations and the feeling that something is pulling at the bedcovers; two, incubus or old-hag attacks, characterized by the experience of being choked or strangled by an old hag or evil incubus that is felt or even seen to be sitting on one's chest, often associated with sensations of pressure, trouble breathing, the feeling of being smothered, pain, and death thoughts; three, vestibular-motor experiences, often described as pleasant and involving sensations of floating, flying, falling, movement, having left the body, or even seeing oneself (Cheyne, 2005). Cheyne et al. (1999) suggest that the stereotyped features of the old-hag attack have their source in the shallow and rapid breathing pattern characteristic of REM sleep and note that SP episodes generally occur in the supine position (Cheyne, 2001).

Cheyne and Girard (2007a, 2007b) propose a general model of the night-mare, according to which intruder and old-hag experiences can be placed on a continuum. Indeed, the details of SP night-mares are subject to some variation. For instance, the first report cited earlier differs from the classical old-hag attack in that the being is lying behind, not on top of, the subject; the second report is more akin to intruder experiences, though perhaps, due to its breathing theme, on the verge of turning into a full-fledged old-hag attack. This type of continuity is accounted for by Cheyne and Girard's model, according to which the combination of REM sleep and subjective wakefulness with feelings of paralysis and lying, typically, in the dark in a supine position leads to an activation of the threat-activated vigilance system, or increased and biased attention to possible threats. At low levels of activation, the threat-activated vigilance system may only give rise to the vague feeling of a presence; at increased levels, auditory, visual, and tactile cues may become incorporated and interpreted as the approach of an intruder. The experience of restricted breathing may then finally culminate in a full-fledged old-hag attack. By contrast, Nielsen (2007) has pointed out that the feeling of a presence may also occur in nonthreatening contexts and is perhaps better understood as a form of hallucinatory social imagery. He argues that rather than itself being triggered by a perceived threat, felt presence may lead to fear only in susceptible individuals and only given this reaction may give way to night-mare experiences. Despite making different claims about the relationship between felt presence and negative emotions, both models afford a prominent role to own-body perception. In the full-fledged old-hag attack, the sensation of choking may be nothing other than an awareness of the rapid and shallow breathing pattern characteristic of REM sleep, coupled with an SP-related inability to control one's breathing pattern, that is, to take a deep breath. Both Cheyne and Girard (2007a, 2007b) and Nielsen (2007) also propose that their divergent models can be generalized to explain the thematic construction of dreams and sleep nightmares occurring independently of SP experiences.

For present purposes, it suffices to note that intruder experiences and old-hag attacks can not only be explained in straightforwardly naturalistic and parsimonious terms but also offer an example in which REM-sleep-related inputs from the physical body are associated with highly specific, stereotyped, and often disturbing hallucinatory episodes. At least in these cases, the close association between subjective experience and the physical body—for instance between the experience of choking and sleep-related breathing patterns, or between the subjective experience of paralysis and actual body position and muscular atonia—suggests that these night-mares are not just occasioned by changes in the physical body, but count as involving distorted or illusory perceptions of the physical body.[6] This is further supported by the stereotyped content of SP night-mares. For instance, Parker and Blackmore (2002) found SP experiences to be more uniform than dreams without SP, with character interactions being more aggressive and the dreamer typically being the victim of attacks. SP experiences also contained more cognitive and auditory experiences, as well as four times as many references to body parts. Their findings were replicated by Maggiolini et al. (2007), who also found that flying and staircase dreams were associated with sexual content, as well as with more frequent bodily experiences. For now, an interim conclusion is that SP experiences are indicative of a systematic link between the physical body, the phenomenally experienced body, and possibly the overall emotional intensity of the episodes.

What is unclear, however, is whether these episodes count as instances of real-body perception during *sleep*, or whether they are instances of REM-related processes intruding upon wakefulness. In the latter case, the existence of a systematic link between the physical body and bodily experience would not be surprising. We expect such a link to exist during wakefulness; what is contentious is whether it exists in dreams. Indeed, SP is often attributed to a carryover of REM-sleep-related paralysis into wakefulness, and Mahowald and Schenck (1999) have suggested that SP is a dissociative state combining features of wakefulness and REM sleep. However, as Powell and Nielsen (1998) point out, the subjective impression of being awake during these experiences may be misleading, and at least some SP experiences may be instances of false awakenings, in which REM-sleep-related paralysis of the sleeping body is experienced *within the dream* rather than in wakefulness (see also Parker & Blackmore, 2002). In fact, the subjective impression of wakefulness is exactly what one would expect from reports of false awakenings, or realistic dreams of waking up. Moreover, it is possible that the overlap between false awakenings and SP is even greater. Spanos et al. (1995) point out that unless subjects try to move during a false awakening, the experience will not be reported as an instance of SP, suggesting that both false awakenings and SP might occasionally go unrecognized and hence unreported.

I cannot conclusively resolve the question of whether SP experiences or a subgroup thereof are unrecognized dreams and genuine sleep-phenomena here. For now, it suffices to note that the relation between night-mares and REM-sleep physiology as well as their similarity with certain dreams such as false awakenings support the hypothesis that a close relationship exists between REM-sleep-related distorted bodily perception on the one hand and hallucinatory, dreamlike imagery on the other. The question, however, is whether this can be generalized into a statement about the relationship between bodily experience and the physical body in dreams.

Allow me to point out an interesting fact about dreaming that is often overlooked and to my knowledge has not been studied systematically. Most dreams are decidedly *not* affected by the position and orientation of the sleeping body; typically the dream world, it would seem, is experienced not from a horizontal but from an upright position. Dream posture thus typically does not seem affected by the actual position of the sleeping body. False awakenings, erotic dreams, and certain flying dreams are counterexamples. In SP night-mares and OBEs, the self typically also has a horizontal position. However, while in OBEs, the phenomenal self is usually located at an elevated position, looking down at (what is perceived as) the real body (Blanke & Mohr, 2005), in SP night-mares, it is typically lying in bed, looking upward at an intruder or attacker experienced as being distinct from the self. A prediction would be that in dreams, variations in the degree of phenomenal embodiment should be describable as variations in the degree of functional embodiment, as well. Given that the supine position seems particularly closely associated with SP night-mares, investigating the relationship between sleeping posture and dream content would be quite interesting. For instance, perhaps strong phenomenal embodiment is more likely to occur in dreams in which the position and posture of the dream self correspond to that of the physical body: the more closely the dream body is brought into alignment with the physical body—for instance, by taking a horizontal rather than a vertical, upright position—the more intensely bodily the overall experience becomes. False awakenings and dreams of flying would be a good first place to look. So far, only a few studies have investigated the relationship between sleeping position and dream content. A recent study found that sleeping in the prone position may be associated with sexual themes and dreams of persecution, but also with feelings of being unable to breathe or move (Yu, 2012). Another study suggested that nightmares may be more prevalent in left-sided than in right-sided sleepers (Agargun et al., 2004).

The link between SP-related experiences and dreaming is strengthened by their similarity not only with false awakenings but also more generally with lucid dreams and OBEs. The voluntary induction of SP is sometimes suggested as a method of inducing lucid dreams (Sevilla, 2004), and LaBerge and DeGracia (2000) note that wake-induced lucid dreams are often accompanied by vibrations, tingles, waves of warmth, a sense of melting, floating, peeling, flipping over, flopping, or slipping. These sensations are similar to those occurring during sleep onset more generally, which may also involve tingling sensations, numbness, and body distortions and are likely indicative of the process of gradually losing touch with the physical body (Mavromatis, 1987, pp. 35–36). Both sleep onset and lucidity also appear to be associated with vestibular-motor experiences and sensations of flying. Barrett (1991) found that flying dreams were more likely to be reported by subjects who reported frequent lucid dreams, prelucid dreams, dreams of sleep, or false awakenings. Like lucid dreams, the induction of flying dreams is a learnable skill. The best and certainly most charming example is Mary Arnold-Forster (1921). She acquired the power "of making a favourite dream recur more or less at will, and of being able to greatly increase its features" (Arnold-Forster, 1921, p. 75), just by thinking about flying in wakefulness. Several authors have investigated the relationship between lucid dreams and OBEs (for details, see Irwin, 1988; Blackmore, 1988; Levitan & LaBerge, 1991; Green & McCreery, 1994), suggesting that OBEs might be either misinterpreted lucid dreams or inferior types of lucid dreaming (LaBerge, 1985, p. 236; LaBerge & DeGracia, 2000; see also Metzinger, 2009). This suggests that OBE-like experiences may cut across sleep onset and lucid dreams.

We have already seen that SP-related night-mares often have a pronounced bodily flavor, and the same is true of genuine sleep-nightmares, which have been suggested to be associated with high sympathetic activity, such as accelerated heart and respiration rate, but also periodic leg movements and increased blood pressure. Nielsen and colleagues suggest that "the hypothesized increase in sympathetic activity in NM [nightmare] etiology is likely an anomaly of normal REM sleep autonomic activity, which is itself quite volatile in nature" (Nielsen, Paquette, et al., 2010, p. 113). In their study, the association between heart rate variability and nightmares was especially pronounced following REM sleep deprivation. Nightmares are also, according to a recent study (Robert & Zadra, 2014), characterized by distinctly bodily experiences: the factors that most clearly distinguish nightmares from bad dreams are emotional intensity and threats to the dreamer's physical integrity, both of which are more pronounced in nightmares.

Further evidence comes from the study of sleep disorders. MacFarlane and Wilson (2006) found that almost 85 percent of subjects suffering from sleep apnea, narcolepsy (which is often associated with sleep paralysis; see Mignot & Lin, 2009), or periodic leg movements reported having nightmares and bad dreams; however, there seemed to be no correlation with specific forms of dream content, as patients with apnea did not, for instance, report dreams of choking or suffocation more frequently than the other subject groups (see also Schredl, Schmitt, et al., 2006; Schredl & Schmitt, 2009). Nonetheless MacFarlane and Wilson (2006) conclude that sleep disorders generally increase the frequency of nightmares. There is also some evidence that sleep behavior itself may be triggered by tactile stimuli. One study using video-polysomnography to monitor a patient with frequent sleepwalking episodes while she slept at home found that roughly one fifth of the episodes were triggered by physical contact and movement, either by the patient's husband or cat (Mwenge et al. 2013). Some of the episodes were also associated with sleep mentation.

As is the case for SP night-mares, bodily experiences associated with sleep-nightmares and lucid dreams may have a tendency towards stereotypicality. This is supported by studies of typical dream themes, which show a surprising degree of consistency not only across different cultures but for different historical periods (Nielsen, Zadra, et al., 2003, p. 226; Schönhammer, 2004, pp. 289–326; Yu, 2008). The single most prevalent dream theme, reported by 81.5 percent of participants, is the dream of being chased or pursued. It is closely followed by dreams of sexual experiences, falling dreams, and themes related to school or studying, being late, and trying again and again to do something. Just over 48 percent of subjects report dreams of flying or soaring through the air (Nielsen, Zadra, et al., 2003, p. 217). The most prevalent dream themes are also the ones that subjects are most likely to dream about frequently; again, the chase dream ranks highest. Nielsen and colleagues remark that the general

> evidence of consistency over ages, genders, regions, historical cohorts and cultures supports the notion that some typical themes are near-universal in occurrence. It also supports the claim that such themes are linked to biological, psychological and/or sociocultural events common to human beings regardless of culture, gender and other mediating circumstances. (Nielsen, Zadra, et al., 2003, p. 227)

Part of what makes these dreams typical is their relation to the current concerns of the participants; as most of these studies were conducted with university students, the prevalence of dreams of school and studying is not surprising. Other dream themes, such as dreams of pursuit and of falling or flying, bear no obvious connection to current concerns. Rather, they seem to be characterized by vivid motor imagery and bodily sensations. An analogous interpretation to SP experiences suggests itself. Indeed,

Schönhammer (2004, p. 292; see also Schönhammer, 2005) argues that many typical dream themes arise from a distorted or illusory form of own-body perception. He suggests that dreams of being unable to move, to escape from a pursuer, or dreams of engaging in persevering, but futile actions might result from a fleeting awareness of the sleeping body, lying motionless in bed. Dreams of disorientation, of sleeping, awakening, or lying in bed, as in false awakenings, might indicate that dreamers have reached a certain degree of situational awareness and are close to realizing that they are asleep. And dreams of being naked, erotic dreams, tooth dreams, and dreams of eating or searching for a toilet might be caused by corresponding physical sensations like hunger, cold, and so forth (Schönhammer, 2004, pp. 302–315).

Just as SP experiences are often extremely frightening, so typical dreams of falling or being pursued are mostly nightmarish and unpleasant. They also appear particularly prone to exact repetition: in a study focusing on recurrent dreams, Zadra (2001) found that as many as 70 percent of recurrent dreams involved typical dream themes. It is important to note that in this study, dreams were considered as recurrent only if they always or almost always occurred in exactly the same way over a period of six months. Once more, the most frequent recurrent dream theme was the dream of being chased, accounting for 23.3 percent of recurrent dreams (Zadra, 2001, p. 236). Recurrent dreams were predominately nightmarish, although recurrent dreams of flying, which are often experienced as extremely pleasant, presented a counterexample (Zadra & Donderi, 2000, pp. 1017–1018). These findings on recurrent dreams suggest that typical dream themes are more liable to exact repetition than other types of dreams. In this case, intense movement sensations and previous experience might work together to increase the likelihood of exact repetition. Once this pattern has been established, the recurrent dream might even be triggered by less-intense stimuli.

All of this suggests that there continues to be a systematic relationship between at least some bodily experiences in dreams, including vestibular-motor experiences, but also feelings of paralysis, and real-body inputs during sleep. A tempting conclusion is that such cases can be conceptualized as involving an illusory perception of the sleeping body, resulting either from an awareness of the loss of sensory inputs during sleep or from an increased awareness of real-body inputs and sleeping position, as in REM-related paralysis coupled with an awareness of breathing pattern, heart rate, and so on. The connection between vestibular sensations such as flying and lucid dreams may indicate that lucid insight into the fact that one is dreaming is related to a heightened awareness of the sleeping body, as well. Moreover, the process of losing touch with the sleeping body may actually *enhance* the phenomenology of (bodily) presence in dreams. Many of the examples discussed so far, such as flying, falling, tingling sensations, or paralysis, are ways in which the functional disconnection from the physical body appears to shape and even intensify the phenomenology of embodiment during the dream state. Such dreams may involve a realistic perception of the sleeping body or altered bodily experiences reminiscent of the subjective distortions of size, position, and texture of body parts that are sometimes experienced during anesthesia, but also following spinal cord injury, after amputation (Paqueron et al., 2003), or at sleep onset (Mavromatis, 1987). In dreams, such nonstandard bodily experiences can also fluidly integrate ongoing concerns of the dreamer. The following dream report is a nice example:

> At a time when we had to work hard to prepare for geography class, he dreamt that his body, which had grown to an endless, elongated size and was pulsing with life, was Europe; he felt his bones and ribs to be the Alps and Pyrenees, the streams flowed through his veins, he could follow the course of the Rhine, but especially the Danube ran, large and powerfully, through the whole length of his body. (Kurz, 1919; quoted in Jezower, 1928/1985, pp. 351–352; translation mine)

There may even be a systematic association between the subjective meaningfulness of dreams and bodily experience. Kuiken and colleagues (1996) and Kuiken and Nielsen (1996) propose that the subjective meaningfulness of so-called existential dreams—intense dreams associated with frustration, sadness, separation, and loss, which may prompt dreamers to rethink certain aspects of their lives upon awakening—may be related to neuromuscular inhibition associated with the orienting response. Kuiken and Nielsen (1996) suggest that awareness of movement inhibition may be reflected as a pervading sense of ineffectuality on the level of dream content, resulting in dreams of weakness, fatigue, being unable to move, or engaging in futile actions. Repeated experiences of incompetence and inadequacy may in turn accentuate negative emotions such as frustration, guilt, sadness, or hopelessness within the dream. As movement inhibition is attenuated upon awakening, subjects may then find themselves outwardly enacting the emotions experienced in the dream. This carryover of emotions into wakefulness, such as the experience of waking up crying, may endow the dream with a lingering sense of reality and meaningfulness. Thus the dream may prompt a shift in self-perception and sensitize dreamers to certain aspects of their waking lives. Though the impact of existential dreams can further be enhanced by reflecting on the dream after awakening, Kuiken and Nielsen (1996, p. 202) remark that "the self-perceptual depth prompted by existential dreams seems to be an effect of dreaming per se." If they are correct, the subjective meaningfulness of dreams is, at least in some cases, an artifact of distorted own-body perception imposing itself on the dream and promoting the carryover of dream-related emotions into wakefulness.

Nielsen and colleagues' (2003) survey of typical dream themes also found a close association between movement sensations and subjectively perceived meaningfulness. When asked which of the fifty-five different dream themes they personally considered to be the most important, participants answered with a high degree of variance, but a relatively large percentage (9.1 percent) chose the chase dream. A similar percentage of participants rated the themes of seeing a person now dead as alive (7.6 percent) and seeing a person now alive as dead as most important (9.7 percent). The finding that dreams whose themes are of immediate personal relevance are considered as personally important is not surprising. But the finding that a sizable proportion of participants rated the chase dream as their most important dream theme highlights the connection between movement sensations, repetition, and subjective impact (see also Nielsen, 1991).

The connection between flying dreams and perceived meaningfulness was already noted by Havelock Ellis (1911/1931, p. 130), who remarked that "no dreams . . . are so vivid and so convincing as dreams of flying," and described them as leaving behind "a vague yet profound sense of belief in their reality and reasonableness" (p. 132). He remarked that people "who dabble in the occult" may easily interpret these "aviation dreams" as "a real excursion of the 'astral body'" (p. 131), suggesting that dreams of this kind may even be the root of spiritual beliefs: "To float, to rise into the air, to fly up to heaven, has always seemed to man to be the final climax of spiritual activity. The angel is the most ethereal creature the human imagination can conceive" (p. 146). A similar connection between supernatural beliefs and dream-related experiences also exists, as we have seen, for SP night-mares, as well as possibly, as Metzinger (2005) has argued, for OBEs. Perhaps dreams involving intense bodily experiences are particularly belief inviting, and perhaps they induce a bias towards supernatural beliefs even upon awakening (cf. chap. 10).

Finally, while impressive evidence suggests that bodily experiences in dreams often arise from external stimulus-incorporation and can be described as involving own-body perception, there are also important objections to the claim that dreams are weakly functionally embodied states. Even if a body-based explanation can be extended from SP night-mares to lucid dreams, nightmares, typical dream themes, and recurrent dreams, it would be a mistake to assume that these examples are representative of the breadth of dreaming; compared to the majority of more mundane dreams, typical dream themes are actually atypical. In their analysis of a large sample of dreams, Domhoff and Schneider (2008b) found that only 4 percent of dream reports described flying, gliding, or floating, and only in a small percentage of these dreams did dreamers fly or glide without the aid of airplanes, boats, and so on. The authors estimated that unsupported flying occurs in less than 1 percent of dreams. As Schönhammer puts it, "'Typical dreams' would seem to be everybody's extraordinary dreams" (2005, p. 19)—and the strong sense of bodily presence in these dreams might also make them particularly memorable.

Moreover, bodily stimuli often fail to enter into dreams. An impressive example is penile erection, which standardly accompanies REM sleep in males. While the exact relation with sexual dream content is not clear, sex dreams in general appear to be relatively rare (Hobson, 1988, p. 245), and sexual interactions occur in only 12 percent of the dreams of male subjects (Hall & Van de Castle, 1966). Consequently, there appears to be at best a loose connection between physical arousal and dream content (Fisher, 1966). While REM sleep regularly supplies the physiological conditions, it would seem that dream content is less willing to follow suit than a Freudian might have hoped for.[7] However, saying that bodily stimuli often fail to enter into dreams is not the same as saying that they never do, or that they are irrelevant to the theoretical explanation and empirical investigation of such bodily experiences as do occur in dreams. From the available evidence, it would seem that dreams are neither functionally disembodied nor functionally embodied states in any strong sense of the word. The most empirically plausible position will fall somewhere in between, and for present purposes, this is enough.

There may also be alternatives to explaining bodily sensations in dreams in terms of illusory own-body perception. For instance, Schredl (2011a, 2011b; Knoth & Schredl, 2011) argues that flying, toilet, and pain dreams are better explained by appealing to the continuity with waking concerns and memories than to sensory incorporation during sleep. Schredl (2008a) found flying dreams, for instance, to be related to personality factors and dream recall frequency. Some evidence also points to gender differences, with women recalling more dreams than men and reporting more nightmares, nightmarish themes, and negatively toned dreams, as well as more frequently experiencing episodes of sleep paralysis than men (Nielsen, Zadra, et al., 2003). However, factors such as personality traits and waking concerns could also contribute to the incorporation of real-body stimuli, for instance, by enhancing susceptibility to bodily changes during sleep, and it is not intuitively clear that these explanations contradict each other.

A further important objection would be to say that my discussion of bodily experiences in dreams and their real-body basis has, so far, overlooked an important dimension of experience that is both characteristic of dreaming and distinctly bodily: emotions. Because this is indeed an important omission, I will take a brief detour through the theoretical and methodological challenges raised by this objection.[8]

8.2.2.3 Challenges for Investigating the Bodily Basis of Dream Emotions

Dreams are sometimes described as hyperemotional, with emotions being nearly ubiquitous in dreams and being seen as candidates for driving the construction of dream narratives (Merritt et al., 1994; Hobson et al., 2000). The full spectrum of emotions occurs in dreams, and much evidence points to carryover effects between dream and waking emotions (Nielsen, Kuiken, & McGregor, 1989; Nielsen, Deslaurier, & Baylor, 1999; cf. sec. 6.1.3). Recent years have seen a proliferation of research on the contribution of sleep in general, and REM sleep in particular, to emotional processing (cf. sec. 12.4). Walker (2009a; Walker & van der Helm, 2009) has proposed a "sleep to forget and sleep to remember" model, according to which REM sleep/dreaming decouples memories from their emotional tone, enabling us to retain memories without reliving the emotions associated with the original experiences. REM sleep, according to this view, provides an optimal state for "affective therapy" (Walker, 2009b). Some evidence also suggests that dream emotions are associated with overnight emotional adaption (Lara-Carrasco et al., 2009). Perogamvros and colleagues (2013; Perogamvros & Schwartz, 2012) have proposed a reward activation model of sleep and dreaming, suggesting how different sleep stages contribute to the consolidation of emotional and motivational memories. The model predicts a tendency toward motivationally relevant dream contents, but also toward the types of instinctual and motivation behaviors often seen in parasomnias, as well as a close relationship between mood disorders, psychiatric disorders, and disturbed sleep.

Although we have ample evidence for saying that emotions play a central role in dreams and possibly even that sleep and dreaming contribute to emotional processing, it is less clear how this relates to bodily experience in dreams. To begin with, in the philosophical literature, the relationship between emotions and bodily experience itself continues to be a matter of controversy. Whereas embodied, somatic, or perceptual theories describe emotions as the perception of bodily changes, cognitivist theories liken emotions to propositional attitudes (for details and further references, see de Sousa, 2013). Both types of theories have problems. While embodied, somatic, or perceptual theories have trouble accounting for the fact that emotions typically seem to have intentional content and meaning, cognitivist theories struggle to explain why emotions can be independent of, and even run counter to, judgments.

Let us assume, for the sake of argument, that a successful account of emotions can resolve these problems by acknowledging both the meaning of emotions and their bodily basis. Several such accounts exist. For instance, Damasio's (1994, 1999, 2010) somatic marker hypothesis proposes that emotions guide decision making by creating a bias toward certain behaviors or value judgments on the basis of learned associations between reinforcing stimuli and physiological affective states. This model gives a prominent role to emotions in reasoning, thus challenging traditional views of rationality and everyday reasoning abilities as being independent of and perhaps superior to emotions (Damasio, 1994, 1999). Prinz's (2004, 2012) somatic appraisal theory suggests that the perception of a bodily state is semantically akin to appraisal, thus allowing emotions to have both somatic and semantic content. And recently, Seth (2013) has proposed that emotion can be described as interoceptive inference, arguing that emotions arise from probabilistic, predictive models of the causes of interoceptive signals and are intimately linked to embodied selfhood and the experience of body ownership. Importantly, this approach rejects the apparent opposition between lower-level bodily or physiological and higher-level appraisal or cognitive theories of emotions. Rather, emotions are explained as resulting from multi-layered interoceptive predictions occurring on different levels of abstraction (Seth, 2014).

Such accounts, according to which the apparent meaningfulness of emotions can be reconciled with their bodily basis, lend initial support to the claim that dreams are weakly functionally embodied states. If these accounts are on the right track, then the intensely emotional character of dreaming might be explained by describing dream emotions as arising from a (distorted) perception (or prediction) of changes in the sleeping body. This also fits in well with the apparent association between intense emotions and bodily experiences in dreams, of which nightmares are perhaps the clearest example: after all, if dreams are weakly functionally embodied states, we might expect the distorted perception of unaccustomed bodily changes during sleep to have a disturbing and unsettling effect. If so, one might even want to say that the functional link between dream emotions and the sleeping body is even tighter than is the case for sensorimotor experiences, and thus that dream emotions are a primary example of the distinctly bodily (and not just brain) basis of dreaming. While this is indeed an intriguing possibility and an exciting topic for future research, there are a number of important challenges to such a view, at least given the current stage of research.

Even if emotions are perceptions of bodily changes, this is not quite the same as saying that on the phenomenological level of description, emotions—either generally or in dreams—have a distinctly bodily flavor. There could, for instance, be emotions that lack any obviously bodily flavor, such as the quiet joy and intellectual satisfaction experienced while reading a philosophical classic—and this could be true even if the respective emotion could ultimately be redescribed as involving a perception of one's current bodily state. Yet, for many emotions, it is clearly true that they do have a bodily flavor; there is clearly a close association, for instance, between fear and, say, trembling, elevated heartbeat, and sweaty palms. But note that the same association with bodily changes, at least on this coarse-grained level of description, holds for nervousness (for instance, before giving an important talk) or more positive forms of excitement (such as the joyful anticipation and giddiness before a first date). We should leave room for the possibility that emotions are underdetermined by perceptions of bodily changes. And even if bodily profiles can be used to systematically distinguish different emotions, this may still not be sufficient to distinguish their more fine-grained nuances. For instance, as de Sousa (2013) points out, physiology and phenomenology may not be fine grained enough to distinguish guilt from embarrassment or shame.

Prinz's somatic appraisal theory explicitly addresses this problem. According to Prinz (2012), emotions are individuated by mapping characteristic bodily patterns, such as changes in posture, respiration, facial expressions, or action tendencies, to associated triggers. Importantly, the same pattern of bodily changes can arise in response to different triggers, and it is only by taking the specific trigger into account that we can explain the meaning of a given emotion. Moreover, one should expect a high degree of variation in such associations between bodily changes and their triggers, not just intra- and interindividually but also interculturally.

We might ask, then, whether dream emotions are appropriately associated with their customary triggers. Indeed, it seems that while a correlation exists between bizarreness and shifts in emotion, dream emotions themselves are only rarely bizarre or inappropriate to dream content, suggesting that emotions drive dream plot construction (Merritt et al., 1994). If so, then the causal relationship may be inverted in the case of dreams: the point is not so much that dream emotions are appropriate to their customary triggers but that dream content is responsive to dream emotions.[9] This does not, however, answer the question of whether dream emotions are perceptions of real-body changes. To answer this further question, one would first have to answer the more basic question of whether the customary bodily profiles of specific waking emotions characterize dream emotions, as well. Without more detailed phenomenological studies, it is unclear, for instance, whether dream fear is also associated with sweaty palms, trembling, and elevated heartbeat. To my knowledge, this type of question about the distinctly bodily flavor of dream emotions has not been investigated in detail. Yet note that the weak-phenomenal-embodiment hypothesis predicts that the bodily profiles of specific dream emotions should be, to a considerable degree, phenomenally indeterminate. If the weak-phenomenal-embodiment hypothesis is on the right track, most dream emotions should not be expected to have a richly bodily flavor.

To this, one might object that even if dream emotions turned out not to have a distinctly bodily flavor, this might not per se count against an embodied approach to dream emotion. One might surpass the phenomenological differences altogether by looking to changes in the sleeping body. Indeed, the association between nightmares, heartbeat, and leg movements would seem to point in this direction. Perhaps it is not necessary that the negative emotions associated with nightmares are associated with the experience of elevated heartbeat and leg movements; perhaps the association between dream emotions and actual changes in heartbeat and leg movements is sufficient. For now, this is an open question: while some evidence points to increased physiological arousal accompanying dream emotions, results have been inhomogeneous (for details and further references, see Schredl & Erlacher, 2011a). Even so, note that the attempt to bypass a detailed phenomenological investigation of dream emotions changes the potential scope of claims about the bodily basis of dream emotions: unless one is able to detect a systematic association between the bodily flavor of dream emotions and their real-body basis, one could at best say that real-body changes cause dream emotions, but not that dream emotions themselves count as a perception of such bodily changes. For this latter claim, a stronger correspondence between emotional experience and real-body changes would be needed.

In thinking about the bodily basis of dream emotions, it is also important to point out that proponents of somatic theories of emotion need not claim that all emotions are necessarily associated with bodily changes; rather, it might be sufficient that the brain enter into the same type of state as it would have in response to actual bodily changes (Damasio, 1999). Damasio argues that the ability to enter into an "as-if body loop" and to bypass the body, for instance, in imagined emotions, is an important condition for explaining the appraisal function of emotions. The ability to anticipate bodily changes is also crucial to ensure fast responses to potentially dangerous stimuli. Moreover, because emotions can arise in response to changes in cognitive states, even this "as-if body loop" is not necessary for emotion (Damasio, 1999). Finally, note that in Damasio's view, emotions need not even be conscious at all; rather, they can involve unconscious neural responses to changes in bodily states. Such feelings are more basic and primordial than dreams (Damasio, 2014). But if this is correct, then somatic theories of emotions do not necessarily predict that dream emotions themselves have a distinctly bodily flavor: even if emotions are typically associated with the perception of bodily changes, the same might not be true for the special case of dream emotions. By contrast, to support the claim that dream emotions themselves arise from residual perceptions of changes in the physical body, something much stronger is needed. To support this stronger claim, one would first have to narrow the explanatory target to consciously experienced dream emotions that have a distinctly bodily flavor on the phenomenological level of description. Second, one would have to investigate whether this bodily flavor can be mapped onto changes in the sleeping body. Only then could one defend the claim that dream emotions themselves are perceptions of real-body changes.

Finally, even if the theoretical and conceptual issues pertaining to emotions and their relation to real-body changes had been resolved, and even if the target had been restricted to a subgroup of dream emotions, another challenge would be to surpass the methodological problems associated with mapping real-body changes to dream emotions. This problem is most pronounced for models linking emotions to interoception (Seth, 2013). Even if dream reports described the characteristic phenomenological profile associated with particular emotions in wakefulness, it would be nearly impossible to decide whether they referred to (illusory or even veridical) perceptions of real-body changes or rather to, say, interoceptive hallucinations of muscular or visceral sensations. As Blom and Sommer note:

> The conceptual distinction between sense perceptions and hallucinated percepts allows for a practicable operationalization in clinical and neuroscientific terms as long as the "distance senses" are involved but fails to do so in the case of many types of bodily sensation. The notion that any type of percept can be either 'real' or 'imagined' may well be true, but it quickly loses its meaning when our judgment must be made without the possibility of the external verification of sensory input signals. (Blom & Sommer, 2012, p. 166)

This problem would appear to be particularly pronounced for interoceptive or visceral sensations. Stimulation studies, which are arguably the backbone of research programs for systematically investigating the real-body basis of dream experience, would likewise be hard, if not impossible, to conduct. So even if one assumes that there is indeed compelling evidence for saying that dreams are hyperemotional states and that dream emotions are very likely closely associated with real-body changes, it is hard to see how this would translate into the kind of rigorous research program needed to make progress on the real-body sources of dream emotions.

In sum, while I think that theoretically, investigating questions about the real-body basis of dream emotions is an extremely promising field, and while prominent philosophical and neuroscientific theories of emotions do sit nicely with the view that dreams are weakly phenomenally-functionally embodied states, I think that this type of approach is also subject to important limitations. While thinking about the bodily basis of dream emotions can be used to motivate a number of fascinating questions for future research, the theoretical-conceptual and methodological difficulties discussed here prevent me from outright endorsing the claim that emotions are a good, or perhaps even a paradigmatic, example of dreams as weakly phenomenally-functionally embodied states.

8.2.2.4 Evidence for the External Enactment of Dream Behavior

We have seen numerous examples, both from sensory stimulation during sleep and from intense forms of dreaming, for saying that bodily experience in dreams is closely associated with, and perhaps even modulated by, sensory inputs from the sleeping body. Another line of evidence for the close connection between the dream body and the physical body comes from the correlation between real-body and dream-body movements.

The closest (though perhaps least dramatic) such correlation is that of the eye movement signals conducted by lucid dreamers (see sec. 3.5), which can be read off the EOG and are retrospectively confirmed by dreamers. The signaling method has also been used to investigate the effects of other types of dream movement on the sleeping body. A study conducted by LaBerge et al. (1981) found a correlation between fist clenches in lucid dreams and EMG measures of muscle twitches in the forearms. The sequence of right-left fist clenches performed in the dream corresponded to the sequence of right-left forearm muscle twitches as measured by EMG. Fenwick et al. (1984) showed that finger and hand movements carried out during lucid dreams had a stronger effect as measured by EMG than shoulder movements, suggesting that the real-body effect of dream movements may be strongest for distal muscles. LaBerge et al. (1983) found a correlation between sexual activity during lucid dreams and respiration rate, skin conductivity, vaginal EMG, and vaginal pulse amplitude. Other findings suggest that voluntarily controlling one's respiration rate in a lucid dream has a measurable effect on actual respiration rate (LaBerge & Dement, 1982). Similarly, Erlacher and Schredl (2008a,b) found that both heart rate and respiration rate increased while lucid dreamers performed squats in a lucid dream, as compared to pre- and postexercise periods of counting. Taken together, these studies suggest that movements conducted during lucid dreams produce voluntary and measurable changes in the sleeping body. Although the enactment of these movements is prevented by REM sleep paralysis, it would be false to assume that dream movements were purely internal simulations in the sense of being completely decoupled from motor output.

While such subtle correlations are almost impossible to investigate during nonlucid dreams, this is a methodological problem, and it is tempting to assume that bodily movements during nonlucid dreams have the same real-body effects. Baldridge et al.'s (1965) finding that fine muscular activity in the throat, hand, and foot concurred with periods of rapid eye movements and was absent during periods of ocular quiescence would seem to support this view, though the authors unfortunately did not awaken their subjects to collect dream reports. Cherpillod and colleagues (1965), by contrast, did awaken their subjects to ask them about their dreams. They not only found that REM sleep was accompanied by an increase in fine body movements, but also that the density of these movements, together with the density of eye movements, was significantly associated with their participants' ability to recall dreams. Other studies have confirmed the existence of sleep-stage-dependent shifts in movement during sleep, with movements being most frequent during sleep onset, but more frequent during REM sleep than during slow-wave sleep, and between 200 and 300 small movements occurring throughout sleep (Wilde-Frenz & Schulz, 1983). Dement and Kleitman (1957a) already pointed out that REM periods are both initiated and terminated with postural shifts, and noted that REM sleep itself, though marked by a decrease in gross body and limb movements, is accompanied by numerous small distal limb and digital movements. These were so small, however, that they were not picked up by the EEG and were only noted by an experimenter who went to the trouble of watching subjects throughout the night. Sleep in general, and REM sleep in particular, is not a state of uniform bodily immotility but rather marked by the occurrence of numerous micromovements. Some evidence also points to a link between dream movement and elevated EMG activity in the arms and legs, as well as elevated muscular activity in the chin and lips associated with dream talking (for details and further references, see Schredl & Erlacher, 2011a).

In RBD enactment dreams, the correlation between dream and real-body movements is more dramatic and more easily observable. As discussed in section 3.4, though the specificity and apparently goal-directed nature of the movements is compelling, the correlation with dream content rests mainly on anecdotal evidence. That said, it is interesting to note that RBD dreams may be similar to SP experiences in that there is both a close relationship between the physical body and the dream body and a rise in aggressive and threatening dream content.

In addition, several other parasomnias may be related to dream-enactment behavior, such as sleepwalking, sleep terrors, or sleep-related eating disorder. As they mainly occur in NREM sleep, however, their correlation with dreaming should be treated with caution: it is possible that they are only occasionally associated with dreaming, and again, the evidence for their correlation with dreaming is mainly anecdotal (for details, see Schenck, 2005; and sec. 3.4).

As noted earlier, a general correlation exists between nightmares and body movements. Both idiopathic and posttraumatic nightmares are associated with gross body movements such as periodic leg movements (Germain & Nielsen, 2003b). Unlike in RBD, these movements do not seem to be directly associated with the movements performed within the dream. Here, there may be only a general and indirect association between bodily experiences in nightmares and real-body movements.

Finally, describing dream-associated movements as instances of dream *enactment* (or dream-*enacting* behaviors), as typically done in the scientific literature, runs the risk of blurring subtle theoretical distinctions. Are such behaviors really instances of *enacting* internally experienced dream behavior? Or should we rather say that the dream is driven by a (distorted) perception of real-body movements? What exactly would it mean to say that dream-associated movements are cases of dream enactment in any interesting sense? The following dream reported by Ernst Meumann nicely blurs the distinction between real-body perception and the external enactment of dream behavior:

> In my dream, I was sitting on the pinnacle of a high tower. I was beginning to feel dizzy and tried to climb down, but couldn't find any secure point to place my foot. "I will have to jump," I thought, and attempted a jumping movement towards the interior of the tower. But I sat as if spellbound, unable to execute the leg movement. I then dreamt that a figure (whose looks I was vaguely conscious of) was threatening to pull me from behind into the abyss beneath the tower. The sense of exertion was escalating to the point of desperation, and suddenly I managed to jump. At that moment, I awoke, because I had hurled both legs so violently as to kick away my blanket. (Jezower, 1928/1985, pp. 306–307; translation mine)

Did Meumann's successful dream-jump cause his real legs to move? Or did he rather "succeed" in jumping because he sensed his real legs moving? A first problem, here, is that the causal relationship between real-body and dream-body movements is unclear. The problem is in part empirical, and stimulation studies—for instance, investigating the effect of externally induced body-part and whole-body movement on dreams—might contribute to their elucidation.[10]

Progress on questions about the causal relationship between dream and real-body movements may come from an unexpected area of research: research on muscle twitches, which are most prominent in infancy, but continue to permeate adult REM sleep in humans and other mammals. Blumberg (2010) argues that both folk-psychology and common scientific theorizing misdescribe the relevance of muscle twitches in two important respects: first, because twitching is preserved even when the cortex has been disconnected from the brain stem, twitches cannot, he claims, be seen as mere by-products of internally experienced dream movements. To the contrary, Blumberg argues that because of their purely brain-stem driven nature, muscle twitches are more basic than dream experience and arise independently of it. Second, while the received view says that the sensory consequences of muscle twitches are blocked during REM sleep, Blumberg argues that the opposite is true: the sensory feedback generated by muscle twitches is an indirect source of stimulation of the forebrain. This suggestion not only contradicts evidence that the processing of external sensory stimuli is reduced during REM sleep, but also goes against the general principle of sensory attenuation for the consequences of self-generated movements. Blumberg claims that for sleep-related muscle twitches, neither is true: not only do the sensory consequences of muscle twitches continue to be processed in REM sleep, but they are also processed differently from self-generated movements in wakefulness, namely as if they were unexpected and externally generated (see also Tiriac et al., 2014). This suggests that muscle twitches, far from being an irrelevant by-product of REM sleep-related brain activation, may be central to brain development. In a process akin to "motor babbling," self-generated twitches may play a central role in sensorimotor integration: "By generating a movement and observing its proprioceptive consequences—including sensory consequences arising from passively moving joints—the animal can learn kinematic and biomechanical relationships among muscles, joints, and effectors and their perceptual correlates" (Blumberg et al., 2013, p. 2108). As Blumberg notes, this account of muscle twitches dramatically changes the direction of explanation: "with the addition of sensory feedback, twitching is potentially transformed from a functionless by-product of dreaming to an active contributor to sensorimotor integration within the spinal cord and brain, including the development, refinement, and maintenance of somatotopic maps" (Blumberg, 2010, p. 5). If this theory is correct, then it may well be that at least in nonlucid dreams, bodily experiences and movement sensations are a mere reflection of automatic, brain-stem driven muscle twitches, rather than vice versa. Attempts to conceive of sleep-related behavior as dream enactment behavior would then be profoundly misguided. Indeed, Blumberg (2010) suggests a new interpretation of RBD, noting that even the specific behaviors observed in RBD patients may be the product of a damaged brainstem generating particularly robust motor activity. Sensory feedback might then lead to the incorporation of these movements in violent nightmares, thus creating, essentially, the illusion of dream enactment.

Can this type of approach be reconciled with evidence from lucid control dreams? Note that even if dream movements can be described as resulting from sensory feedback from muscle-twitches—as a mere reflection, as it were, of bodily movements during sleep—this would not preclude the opposite from being true, at least in the special case of lucid control dreams. There could still exist counterexamples, in which real-body and real-eye movements and muscle twitches are transiently driven by intentional dream actions. Yet, these conflicting interpretations point to a deeper problem. Asking about the causal relationship between real-body movements on the one hand and experienced movement in dreams on the other presupposes that the two are strictly separate. The more tightly one conceives of the connection between the dream body and the sleeping body, the less sense such a distinction will make. The weak-functional-embodiment view I am working toward questions this strictly dichotomous distinction, though importantly without collapsing it altogether.

There are also other reasons to be cautious about the concept of dream enactment. In philosophy, it is common to distinguish action from behavior by saying that action, but not behavior, requires intentionality.[11] Even if there are cases in which we want to say that dream movements themselves are intentional, the same is not necessarily true of their outwardly observable counterparts. In RBD in particular, dream-related real-body movements, even when associated with intentional dream actions, are not themselves intentional and thus are better described, more neutrally, as dream-related behaviors. By contrast, the eye movement signals of lucid dreamers appear to be a case where not just the execution of a movement within the dream but its outward execution itself is intentional. These may be an example of dream enactment in a more interesting sense. Finally, other micromovements or limb movements during sleep may be too loosely and unsystematically associated with dream movements to be described as dream related at all. Again, the exact association between movements, including muscle twitches, and dream content should be investigated in more detail, possibly even giving way to a gradual distinction ranging from movements accompanying (but not related to) dream content, to dream-related behaviors (as in RBD), to genuine dream enactment (as in some lucid control dreams), where not just the dream action itself, but its outward execution is intentional.

8.3 The Neuronal Basis of Bodily Experience in Dreams

Where the previous section concentrated on the correlation between the sleeping body and bodily experiences in dreams, this section briefly summarizes findings on the neural correlates of bodily experiences in dreams and raises some methodological issues related to their investigation.

PET imaging shows that regional cerebral blood flow (rCBF) in the primary sensorimotor cortex significantly increases during REM sleep as compared to quiet wakefulness (or lying still in the scanner; see Maquet et al., 2000; Dang-Vu et al., 2007, 2009). Although the increase was highest in a group of subjects who performed a presleep task that activates the sensorimotor cortex, the same effect was also observed in a control group that had not received any specific training before falling asleep (Maquet et al., 2000). Activity in the somatosensory and somatomotor regions also shows a higher degree of variability than is the case for other sensory areas during REM sleep (Braun et al., 1997). Intuitively, this finding supports both occasional reports of intense bodily experiences in dreams and their variability across dreams.

The continued activation of the primary sensorimotor areas also suggests that real-body inputs continue to be processed, to a certain extent, during REM sleep dreaming. By contrast, the primary visual areas, thought to be implicated in external stimulus processing, are deactivated during REM sleep (Braun et al., 1998). Also, as one would expect based on findings that peripheral stimulation leads to the incorporation of stimuli in dreams, some researchers have reported that peripheral somatosensory stimulation of different body parts can alter the functioning of the sleeping brain. For instance, stimulating the fingers of human newborns results in characteristic patterns of evoked potentials during different sleep stages (Desmedt & Manil, 1970).

The neural basis of dream movements can also be investigated with the help of experienced lucid dreamers. In a case study with a single lucid dreamer, Erlacher et al. (2003) found that hand clenching during a lucid dream (the occurrence of which was signaled by the dreamer via predetermined eye movements) correlated with decreased EEG alpha power over the lateral motor areas. Because a decrease in alpha power is associated with cortical activation and increased excitability of neurons (Pfurtscheller, 1992), this suggests that the activation of the motor cortex is correlated with hand clenching in lucid dreams. Though the study would have to be replicated with larger samples, it presents preliminary evidence for saying that dreamed and real movements rely on the same neural mechanisms.

This hypothesis is further supported by an fMRI study conducted by Dresler et al. (2011). They contrasted lucid and nonlucid REM sleep dreaming to investigate the neural correlates not only of dream lucidity but of specific contents of lucid dreams. Participants were asked to signal the onset of lucidity by performing gaze shifts (left-right-left-right). They were to continue by clenching their left and then their right fist for ten seconds, performing a gaze shift before switching from left to right, and so on, as long as possible. When eye movement signals were followed by a period of forty seconds without signaling, participants were awakened and asked for a report. Results from lucid dreaming were compared with actual and imagined hand clenching during wakefulness. Dresler and colleagues found that the activation patterns in the sensorimotor cortex observed during dreamed hand clenching overlapped with those during actual and imagined hand clenching. They take this to support the view that REM sleep paralysis is generated downstream, at the level of the brain stem. During dreaming, cortical activation was more localized than in wakefulness, "representing either generally weaker activation or focal activation of hand areas only" (Dresler et al., 2011, p. 1834). This latter finding seems to fit in well with my claim in chapter 7 that isolated body-part representations can occur in dreams, and neuroimaging studies with lucid dreamers could be used to investigate this in more detail.

A number of studies have shown that the vestibular system is involved in REM sleep. The activation of pontine vestibular neurons in REM sleep is associated with REM on–off centers (Pompeiano & Morrison, 1965, 1966; Morrison & Pompeiano, 1970; Hobson et al., 2000), and Braun et al. (1997) hypothesize that the activation of the cerebellar vermis in REM sleep may play a role in generating rapid eye movements and motor inhibition. Morrison and Pompeiano (1970) demonstrated that bilateral lesions in the medial and descending vestibular nuclei in cats abolished REM bursts and changes in heart rate associated with phasic REM sleep, while tonic REM sleep was preserved.

While these studies suggest that the vestibular system is involved in REM sleep generation, its association with vestibular sensations in dreams has not been systematically investigated. Though the curved, circular, back-and-forth or reversed quality of movement trajectories in dreams strongly suggests an involvement of the vestibular system (Hobson, 1988, pp. 247–256), this evidence is indirect at best (Hobson, 2005, p. 113). Mild support for such an association, however, comes from patients with labyrinth lesions, who report frequent vestibular sensations and dizziness during dreams (Eisinger & Schilder, 1929).

Finally, what is known about the sensory input and motor output blockade? In their review of neuroimaging on REM sleep, Maquet et al. write that "as far as a precise anatomical localization is allowed by PET, the topography of the relatively quiescent areas during REM sleep is similar to [the] ventral attentional network" (Maquet et al., 2005, p. 224), which functions as an alerting mechanism when salient stimuli arise outside the focus of top-down attention. They explain:

> These considerations suggest that the ventral attentional network is relatively quiescent during REM sleep. If this hypothesis were true, one predicts that the focus of attention during REM sleep should be less sensitive to salient, behaviorally relevant, external stimuli than during wakefulness. The focus of attention is difficult to assess experimentally during REM sleep. However, it may transpire in dream reports. *One would expect that the dream narrative (obtained after awakening from REM sleep) would not be easily modified by external stimulation*, even if behaviorally relevant. Some observations support this hypothesis. *During REM sleep, external stimuli instead of interrupting the flow of the dream storyline, are incorporated into it* (Foulkes, 1966). Similarly, *responsiveness to external auditory stimulation is reduced during sleep*, at least in part, *due to incorporation of the external information into ongoing cognitive activity*. (Maquet et al., 2005, p. 225; italics mine)

I quote this passage at length because the line of argumentation is interesting. First, note that the evidence for the relative quiescence of the ventral attention network in REM sleep is indirect and presented with due caution. Second and more importantly, the rationale for assuming its implication in the sensory input and motor output blockade is that dreaming is not *easily* modified by external stimuli, not that it fails to be modified by them altogether. This assumption is further supported by the finding that dreaming incorporates such stimuli into the unfolding narrative: the sensory input blockade (Hobson et al., 2000) or, as Dang-Vu et al. (2000, p. 1000) more correctly put it, the "reduced sensitivity to external salient stimuli" functions, at least in part, by integrating external stimuli into the ongoing dream and thereby preventing awakening. This integration, in turn, helps explain the lack of responsiveness during REM sleep.

This claim gains further support from a recent study. Bastuji and colleagues (2012) used intracortical electrodes implanted in the brains of epilepsy patients to record cortical activation in response to painful laser pulses throughout the sleep–wake cycle. While evoked potentials were attenuated in all three areas studied during stage 2 sleep, during REM sleep, the lateral operculo-insular system associated with the analysis of somatic stimuli remained active, and only the mid-anterior cingulate system associated with orienting and avoidance behavior was comparatively quiescent. The authors propose that this dissociation between sensory and orienting networks might explain why nociceptive stimuli can be incorporated in dreams without leading to awakening.

We can now translate these empirical insights into an important methodological principle: stimulus incorporation in dreams is crucial for investigating the exact mechanisms underlying (partially) preserved sensitivity and reduced responsiveness to external stimuli during sleep. This type of research, importantly, does not show that the processing of external stimuli is obliterated. Rather, reduced responsiveness to external stimulation during sleep is explained, in part, by appealing to sensory incorporation in dreams. Moreover, this naturally follows from the available means of studying the sensory input blockade: any empirically grounded hypothesis on the functional (dis)connection between the sleeping body and the dream body will have to take instances of stimulus incorporation into account and thus will fall short of supporting any strong version of the functional-disembodiment hypothesis. If this analysis is correct, then the functional-disembodiment hypothesis is prevented from being true by the very methodology involved in its investigation.[12]

Based on the evidence so far, it would seem that the input–output blockade (cf. Hobson et al., 2000, p. 824) is a misnomer. And it is one with dangerous consequences, because it leads to the assumption, popular in philosophy (cf. sec. 8.1), that dreams occur in a state of complete functional disembodiment, as well as to the view that the dream body is a purely and exclusively internal product of the brain. What started out as a mere terminological convention thus has far-reaching theoretical consequences.

8.4 Taking Stock: The Weak-Functional-Embodiment Hypothesis and the Body-Brain-Body Problem

Throughout the chapter, we have been chipping away at the functional-disembodiment hypothesis and the view that dreaming is a naturally occurring state of cranial envatment. We have seen that a more empirically plausible view is to acknowledge that in many cases, there is indeed an interesting and informative relationship between bodily experiences in dreams and the physical body. In this last section, I spell out this claim and some of its theoretical implications in more detail. To begin with, we can now introduce the weak-functional-embodiment hypothesis as a complement to the weak-phenomenal-embodiment hypothesis:

Weak-functional-embodiment hypothesis: Dreams are weakly functionally embodied states in that the processing of real-body inputs and the production of motor outputs are attenuated, but not completely absent, in the dream state, and in that a majority of bodily experiences occurring in dreams are best understood not as arising independently of the physical body but as illusory perceptions of the sleeping body.

This hypothesis is weak in a number of ways. First, the strategy of appealing to external stimulation to explain the occurrence of bodily dream imagery likely cannot be generalized to other types of imagery. Visual dream imagery is an obvious example. Though visual stimuli (e.g., light flashes) are occasionally incorporated in dreams, experiments suggest that the extent to which visual stimuli are occluded even when the eyes of sleeping subjects are taped open is impressive (Rechtschaffen & Foulkes, 1965; cf. sec. 5.2.3). And while some precedent exists for linking visual dream imagery to ocular phenomena and phosphenes (e.g., Ladd, 1892; Bergson, 1914/2007), it is not obvious how this would suffice to explain complex visual dream imagery. The vast majority of objects and characters populating our dreams simply have no obvious counterparts in our sleeping environment. For bodily experiences, the situation is importantly different. Not only are they generally more rare than visual dream imagery (as predicted by the weak-phenomenal-embodiment hypothesis), but the rate of stimulus incorporation is also higher for somatosensory/vestibular stimulation than for visual, auditory, or olfactory stimulation (see also Schredl, Atanasova, et al., 2009), suggesting a tighter link between bodily experiences and an external stimulation source.

Second, the weak-functional-embodiment hypothesis is weak in that it only claims that, based on the available evidence, it is plausible that a large part—perhaps even a majority—of bodily experiences in dreams are systematically related to real-body inputs. This is different from the claim that there is any kind of *direct* correspondence or even dependence between bodily experiences in dreams and real-body inputs, and it is also not to deny that bodily experiences in dreams could arise, in principle, independently of appropriate real-body inputs. In particular, the weak-functional-embodiment hypothesis does not claim that bodily experiences in dreams can be reduced to real-body inputs, or that there is any simple causal relationship between real-body stimulation and dreams. Though clearly a descendant of the *Leibreiztheorie*, the weak-functional-embodiment hypothesis is not committed to the existence of any direct and straightforward link between dream imagery in general, and bodily experiences in dreams in particular, and external sensory stimulation, thereby avoiding some of the pitfalls that the *Leibreiztheorie* was traditionally accused of.

Third and relatedly, the weak-functional-embodiment hypothesis should be understood not as a metaphysical claim about the supervenience base of bodily experience, either in general or in dreams in particular, but only as an explanatory claim and methodological observation.[13] This comparatively modest aim does not, however, make the hypothesis trivial. Showing why requires a short detour into the contemporary debate on dreaming.

As indicated earlier, dreaming is sometimes regarded as a real-world example of conscious experience in general, and bodily experience in particular, occurring independently of any external stimulus source, including the physical body. Revonsuo (2006, pp. 142–144) uses the example of dreaming to argue that "radical embodiment" approaches to consciousness are empirically implausible because they mistakenly locate phenomenal consciousness outside the brain. In his own preferred view, biological realism, this is metaphysically confused:

> As the data from dream experiences show, we now know for a fact that *some* system strictly confined to the brain *can* present experiences where perceptual objects are related to oneself in exactly the same manner as in conscious visual perception. (Revonsuo, 2006, p. 143)[14]

By locating perception and bodily experience outside the brain, proponents of radically embodied views conflate, or so he argues, the phenomenal body image (or virtual self) with the physical organism. However, according to Revonsuo:

> The lived body, or the phenomenal body image, is not *identical* with the organism's physical body of flesh and bones. . . . The contents of both perceptual and bodily awareness are, during REM dreaming, *totally dissociated* from the corresponding states of the physical body. (Revonsuo, 2006, p. 92; italics mine)

Note that the weak-functional-embodiment hypothesis agrees partially with Revonsuo in that it accepts the distinction between the phenomenal body and the sleeping body. From this, however, it does not follow that there is a *complete* dissociation between the two, and indeed, the latter claim should be rejected. Moreover, in this respect, the weak-functional-embodiment hypothesis departs not just from internalist views but also from externalist or embodied views, whose proponents can accept Revonsuo's claim that dreams are a case of functional disembodiment. Noë, for instance, accepts that dreams "*are* produced by neural activity alone" (Noë, 2004, p. 214). He argues, however, that this is exactly what distinguishes dreaming, phenomenologically speaking, from perceptual experience, which he claims is shaped by the interaction with the environment and the skillful exercise of sensorimotor knowledge. He writes:

> Let us take it as settled that when we dream there is no dynamic exchange with the environment (although this might turn out not to be true) and let us accept that, therefore, neural states alone are sufficient for dreaming. . . . Given these assumptions, what the fact of dreaming shows is *not* that neural activity alone is sufficient for perceptual experience, but only that it is sufficient for *dreaming* experience. (Noë, 2004, p. 213)

Both sides, then, agree in their characterization of dreams as functionally disembodied states of cranial envatment; they disagree, however, on the appropriate phenomenological characterization of dreaming, as well as on the question of whether insights from dreaming can be generalized to perceptual experience in standard wake states. The most important point, for present purposes, is that both deny that an interesting explanatory relationship holds between the physical body and dream experience.[15] By casting doubt on this basic claim, the weak-functional-embodiment hypothesis aims to show that the dichotomy between internalist and externalist views, insofar as they are applied to the theoretical explanation and experimental investigation of dreaming, stands on shaky ground. In particular, the weak-functional-embodiment hypothesis says that both sides fail to acknowledge that, as a matter of fact, an interesting causal relationship holds between the physical body and bodily experiences in dreams. In the contemporary debate, this possibility has typically been overlooked. While leaving contentious metaphysical issues to the side, the weak-functional-embodiment hypothesis provides a theoretical framework for reducing (though not obliterating) the chasm between bodily experience in dreams and the physical body. It also predicts that this framework translates into a fruitful research program for investigating changes in phenomenal-functional embodiment across the sleep–wake cycle. This, I claim, is an important advantage over both internalist and externalist theories, or indeed over any theory that endorses the functional-disembodiment hypothesis.

For the same reason, the weak-functional-embodiment hypothesis is an entry point to much larger and less tractable questions about the relationship between dreaming, the brain, and the body, and between mind, brain, and body more generally. The idea is that by focusing on a tiny part of this vast theoretical problem—namely, on the relationship between bodily experiences in dreams, the brain, and the physical body—and by showing how this specific problem can be rendered scientifically tractable, a first step might be taken toward resolving these larger issues, as well. The weak-functional-embodiment hypothesis is thus importantly related to a specific variant of the mind–body problem: the *body-brain-body problem*, or the problem of how bodily experience in dreams, but also in wakefulness, can be understood as arising from dynamic causal interactions between the brain and the (rest of the) body.

Note that this formulation of the problem makes sense only if one rejects the cranial envatment view of dreaming. This is why the way in which the weak-functional-embodiment hypothesis departs from the functional-disembodiment hypothesis, though theoretically modest, is not trivial. Not only is the acceptance of the weak-functional-embodiment hypothesis the condition for an integrated research program on bodily experience and its real-body basis across the sleep–wake cycle, but it also deprives internalist conceptions of conscious experience of their central real-world example. Importantly, in doing so, the weak-functional-embodiment hypothesis remains entirely neutral on the question of whether brain activity alone suffices for conscious experience in general or bodily experience in particular, either in dreams or in wakefulness. Nothing that I have said about bodily experiences in dreams disproves the possibility of a completely virtual or hallucinatory bodily self-model. Indeed, I find myself drawn to the idea that brain activity alone suffices for bringing about conscious experience, including strong phenomenal embodiment. It is at least conceivable that in dreams, bodily experience could arise independently of real-body stimuli. For present purposes, I am interested only in defending the much more pedestrian claim that we have no good evidence for saying that, as a matter of fact, bodily experience in dreams typically unfolds independently of the physical body.

This leads to an important conceptual point. According to the weak-functional-embodiment hypothesis, bodily experiences in dreams are not hallucinatory but illusory or distorted perceptions of the sleeping body. To this, the internalist might object that given the alleged metaphysical neutrality of the weak-functional-embodiment hypothesis, there is still room for asking whether full-body hallucinations exist, either in dreams or more generally. The answer will depend, in part, on what exactly one means by the concepts of hallucination and illusion. To determine how exactly the weak-functional-embodiment hypothesis departs from existing claims about the hallucinatory or virtual character of the bodily dream self, some conceptual groundwork is necessary.

To begin with, the concept of hallucination in its *philosophical* sense refers to cases of seemingly perceiving an object where no mind-independent object of the relevant kind is in fact being perceived; and illusions refer to the perception of an object as having properties or qualities that it does not actually have (Crane, 2011). Illusions, in other words, are cases of misperceiving an object, whereas hallucinations are not instances of perception (in the success sense) at all. Clearly it is not inconceivable that full-body hallucinations in the philosophical sense could exist; a brain-in-a-vat enjoying strong phenomenal embodiment would be an example. The worry here is mainly of a methodological nature. Even if dreams at least on occasion involve full-body hallucinations in the philosophical sense, it is not clear how this fact could ever be the object of scientific research. This is enough, I think, to shift the burden of proof.

Take the example of a false awakening involving completely realistic visual imagery of one's current surroundings. Here one might say that an appropriate, mind-independent object really does exist. The reason for saying that the dream-bedroom is nonetheless hallucinatory, however, is that by ascertaining that the subject's eyes remained closed throughout the episode, we can rule out that visual imagery during the false awakening involved genuine visual perception. An appropriate object was indeed present, but not perceived.

For the body, the situation is importantly different. Because the sleeping body is the only object that is continuously and reliably present during the dream state, dreams involving the presence of a bodily self are analogous to false awakenings in that the seeming object of perception is actually present. Consequently, to defend the claim that the dream body is a hallucination rather than an illusion, one would have to argue that the relevant causal pathways for the transmission of real-body inputs are blocked. But this is exactly what one cannot do: there is no single organ of body perception, the functioning of which could be determined by checking whether it is open, as is the case for the eyes. To the extent that bodily experiences are present, there is no obvious way to deny that, in keeping with the philosophical terminology, they involve an illusory but distorted perception of the physical body. Even cases of nonstandard embodiment—dreams of having a body that is very different from one's actual body, or perhaps even cases of dual embodiment or of lacking a body—might count as illusory misperceptions of the sleeping body. Because in the real world, an appropriate object (the physical body) is guaranteed to be present, there is no obvious way to deny that it is being perceived, albeit in a distorted manner. By contrast, the proponent of the hallucination view would have to show why the physical body does not count as an appropriate object or why bodily experience in a given dream does not count as its perception. But as long as the concept of illusion is lenient enough to accommodate large-scale distortions (e.g., feeling that one is a gaseous volume, etc.; cf. chap. 7), and as long as empirical evidence suggests that the processing of bodily stimuli is not wholly obliterated in dreams,[16] it is hard to see how this could be done.[17]

In the *psychological* literature, the difference between hallucination and illusion is sometimes regarded as a matter of degree (Aleman & Larøi, 2008). For instance, auditory hallucinations of voices may be triggered by sounds that are qualitatively dissimilar to voices; they have a trigger but are considered hallucinatory because the stimulus is not appropriate in the relevant sense. The concept of hallucination is also used to refer to idiosyncratic experiences, describing instances of misperceiving an object in a way that would likely not be shared by other subjects under the same conditions. The sound of a vacuum cleaner triggering voices, for instance, might count as a hallucination, whereas mistaking the sound of a drill for a scream might under certain conditions be considered as illusory, especially if other subjects are prone to similar misperceptions. Importantly, stimulus dependence as such does not count against saying that a given experience was hallucinatory in the psychological sense.

If the distinction between hallucination and illusion is regarded as gradual, dependent on the degree of external-stimulus correlation and the idiosyncratic nature of the experience, the question of whether the dream body is hallucinatory or illusory may not have a uniform answer. Aside from depending on what counts as an appropriate stimulus, one could investigate, for instance, whether bodily stimulation in dreams leads to predictable experiences in dreams (thereby strengthening the illusion view) or rather to highly idiosyncratic and utterly unexpected ones (thereby strengthening the hallucination view). The answer will likely lie somewhere in between.

While the philosophical concept of hallucination sets the bar for body hallucinations occurring in dreams somewhat too high, psychological usage threatens to undermine the possibility of a clean and uniform answer. Of course, saying that some dreams involve body hallucinations while others involve illusions is not necessarily a disadvantage. Indeed, using two different concepts to describe bodily experience in dreams may increase the precision of our conceptual tool kit. Yet I would argue that the use of two different concepts should be well motivated by explanatory considerations. A conceptual distinction only makes sense if it helps pick out a real-world distinction. If we had evidence of two systematically different types of bodily experiences in dreams, where one was highly idiosyncratic and unrelated to an appropriate stimulus and the other more straightforwardly and predictably related to experimental stimulation, and if grouping them under different concepts made sense within a broader explanatory framework, I think this approach would be justified. The weak-functional-embodiment hypothesis would have to be adjusted accordingly. Yet, given the available evidence, I am not convinced that bodily experiences in dreams fall neatly into one category or another in this manner. Rather, the weak-functional-embodiment hypothesis is weak, among other things, because it posits only a weak link between bodily stimulation and bodily experience in dreams while allowing for a large degree of variation in the degree of stimulus correlation. Given the present stage of research, I think this permissive approach is more cautious than assuming, from the outset, that two different and competing accounts of bodily experience in dreams exist. Working out the details is a project for future research, and if it turns out that my initial strategy is false, this would be genuine progress. Yet even in this case, starting out with a maximally inclusive and unitary concept of bodily experience in dreams as illusory may be the most promising choice, if only for strategic reasons.

For now, I would like to introduce a reading of the concepts of hallucination and illusion that is slightly different from both standard philosophical and psychological usage in that it is motivated by methodological and explanatory considerations. Because both empirical plausibility and methodological adequacy are important goals of the present theory and indeed the condition for enabling an integrated research program in the future, I think this departure is warranted. From now on, I will say, then, that bodily experiences in dreams should be considered illusory (rather than hallucinatory) to the extent that they cannot be satisfactorily explained or fruitfully investigated by appealing to brain activity alone; rather, to wholly understand the unique phenomenology of bodily experience in dreams, one must understand how the brain processes real-body inputs to produce the phenomenology of embodied selfhood in dreams, and why the brain responds the way it does to external stimuli during sleep. By saying that bodily experience in dreams is illusory, I want to draw attention to the fact that any satisfying explanation of its occurrence will have to look beyond the brain.[18] Note how this will be true even if bodily experiences in dreams, or a subgroup thereof, ultimately turn out to be genuinely independent of the physical body and to rely wholly on brain activity: establishing this fact in itself would require an investigation of the breakdown of the causal loop between the brain and the physical body. Even if this ultimately leads to the reintroduction of the concept of bodily hallucinations in dreams, the point would still hold that initially, in understanding bodily experience in dreams, the states of the sleeping body are not explanatorily inert: rather, the processing of real-body inputs figures into an explanation of bodily experiences in dreams. Conceptualizing bodily experience in dreams as illusory would be a viable research strategy even if this conceptual convention ultimately turned out to be inappropriate in light of further evidence.

But things might also turn out differently. Perhaps one might ultimately even defend the slightly stronger claim that real-body stimulation not only causally enables bodily experience in dreams, but that bodily experience is just the kind of experience that the dreaming brain does not generate on its own and in the absence of appropriate real-body (or functionally equivalent) prompting. Moreover, if this were true for dreaming, it might well be true more generally. Perhaps bodily experience in general is just the kind of experience that the brain never generates entirely on its own, but only in response to external stimulation that is functionally equivalent to real-body stimulation (e.g., direct brain stimulation). Given that dreaming is often regarded, as we have seen, as the most plausible real-world example of cranial envatment, it would seem that if this assumption is false for the case of dreaming, there may well be no other plausible real-world candidates. I think there is much in favor of this stronger claim, but for now, it is only a prediction, an empirically informed hunch.

Either way, this brings us back to my point that investigating bodily experience in dreams is a new—albeit a practical and highly restricted—version of the mind–body problem: the body-brain-body problem, or the problem of how to describe the functional relationship between the brain and the physical body in bringing about the phenomenology of embodiment. This may appear to be a relatively innocent and superficial adjustment of the functional-disembodiment hypothesis, but it is not. First, note that when we conceptualize dreams as weakly functionally embodied states, we reduce the chasm between dreams, typically conceived of as offline states, and standard wake states. This is important, because it suggests that changes across the sleep–wake cycle might be described as involving variations along a continuum of strong to weak functional embodiment, perhaps with bodily illusions like the rubber-hand illusion or full-body illusions taking an intermediate position between standard and dreamlike functional embodiment. The important new insight is that even in REM sleep, the functional loop between the physical body and the brain is not obliterated; rather, even in REM sleep, it is still plausible to speak of degrees of functional embodiment or functional situatedness.[19] For the same reason, the weak-functional-embodiment hypothesis shifts the position of dreams on the map of mental concepts away from offline states and toward weakly functionally embodied or functionally situated states, leaving the position occupied by global offline states, at least at present, empty.

How might one investigate whether bodily experiences in dreams are really, as a matter of fact, causally dependent on real-body stimulation? Given that there is no obvious way to test whether all bodily experiences arising in dreams are occasioned by external stimuli without applying such stimuli and thus tampering with the unfolding dream, this claim would seem to evade empirical confirmation. Even assuming that typical dreams, nightmares, and lucid dreams involve illusory own-body perception, these are extraordinary dreams, both in that they are particularly memorable and in that they are relatively rare. So it could still be true that some or even a majority of bodily experiences in dreams are genuinely hallucinatory in the sense of arising independently of the physical body, as predicted by the functional-disembodiment hypothesis. Note, however, that the independence variant of the functional-disembodiment hypothesis is in exactly the same situation: it is equally impossible to prove, experimentally, that spontaneously occurring bodily experiences in dreams *lack* an external source and hence are purely hallucinatory or virtual. Thus the question will have to be decided indirectly, by inference to the best explanation. My point is simply that given the available evidence on bodily experience in dreams, the weak-functional-embodiment hypothesis seems preferable, thus shifting the burden of proof to the proponent of the functional-disembodiment hypothesis.

The weak-functional-embodiment hypothesis is not only preferable because it offers a unitary account of bodily experience in dreams and resists the temptation to pry phenomenal embodiment and functional embodiment apart. It is also preferable because it suggests a research program for its own investigation. The basic idea was formulated by Edgar A. Singer back in 1924. He argued that dreams present a formidable challenge to the attempt to "treat sensation as an observable object" (Singer, 1924, p. 596), identified and measured by its association with a stimulus. The main tenets of psychophysics, or so Singer argued, are threatened by dreams, because if dreams are sensations rather than imaginings, the concept of sensation would become decoupled from its correlation with stimulus intensity. Moreover, "No definition . . . is in the full spirit of the empirical method if it suggest no experimental way by which a question of fact may be answered within any required degree of accuracy" (pp. 595–596). Singer suggests that the question can be investigated as follows. If dreams are sensations, then a stimulus—he uses the example of a horn blast—will occasion more frequent dreams in nearby sleepers than would have occurred during a period of silence. Second, a larger proportion of these dreams will involve a sound episode than dreams occasioned by inaudible events. Third, the range of quality and intensity of such stimulated dreams will be smaller than those occasioned by different types of stimuli: if dreams are sensations, then stimulation should influence not only the occurrence of dreams and auditory sensations in dreams but also their intensity. And while Singer concedes that the proposed experiment is ideal, he thinks this is only due to practical limitations; the underlying problem of dream sensation, to his mind, is empirical and hinges on exactly such experimental issues. If his prediction were confirmed, then "the occasion of the dream-sound fulfills the conditions we have laid upon a *sensory stimulus*" (p. 598).

Of course, Singer's predictions would have to be adjusted to the findings of contemporary dream research. Specifically, findings on REM sleep and dreaming make the view that dreaming itself is occasioned by external stimulation seem outdated. Also, using a horn blast as a stimulus may not be the most fortunate choice, as this is more likely to obliterate both dreams and sleep altogether. Nonetheless Singer's second and third predictions are applicable to bodily stimulation during sleep. For instance, if vestibular (or tactile) stimulation during sleep increases the frequency of vestibular (or tactile) sensations in dreams, this result would confirm his second prediction. And if the range of quality and intensity of bodily experiences in dreams really is smaller than is the case for dreams in which external stimulation is lacking, his third prediction would be fulfilled. Given that instances of stimulus incorporation in dreams can be recognized by independent judges (cf. sec. 8.2.2.1), there is reason for optimism, at least from the perspective of the weak-functional-embodiment hypothesis. But more importantly, Singer's challenge, albeit in a slightly modernized guise, suggests a research strategy for resolving the relationship between bodily experiences in dreams and their real-body basis.

There is, however, another problem for the weak-functional-embodiment hypothesis. An important objection would be to accept that rare bodily experiences in dreams (e.g., tactile or interoceptive sensations) are related to real-body inputs, but to deny that the same point is true of all bodily experiences in dreams. Instances of seeing the dream body, for instance, would seem to lack any obvious relationship to bodily stimuli. In addition, it would seem that certain types of bodily inputs are more likely to be incorporated in dreams than others. For instance, though tactile stimuli can be recognizably incorporated in dreams, this is not true for all forms of tactile input during sleep (e.g., the feeling of the bedsheets or the pillow). It would also seem that body position and orientation have a negligible effect on a majority of dreams (for discussion, see sec. 8.2.2.2). Perhaps the weak-functional-embodiment hypothesis should be restricted to specific types of tactile sensations and specific sleeping positions, with spatial orientation and movement in dreams occurring mostly independently of the sleeping body.

I do not think, however, that this follows. Research on the role of the vestibular system in OBEs suggests an account that is compatible with the weak-functional-embodiment hypothesis and can be applied to dreams, as well. Schwabe and Blanke (2008) appeal to an inherent ambiguity in vestibular signals to explain vestibular sensations of having left the body during OBEs. The vestibular otoliths in the inner ear are sensitive to horizontal, vertical, and semicircular movement. Importantly, however, they cannot distinguish between gravity and acceleration and thus are inherently ambiguous between lying down—for instance, in a supine, stationary position—and accelerating forward in an upright position. As Schwabe and Blanke suggest, this ambiguity may result in illusory changes in self-location and the feeling of moving out of one's body, though the body is in fact lying stationary in bed. In greatly simplified terms, the idea is that we have a prior, top-down expectation for being in an upright position, such that in OBEs, ambiguous signals from the otoliths are interpreted in a Bayes-optimal way as indicating upright whole-body movement. Computational modeling can then be used to investigate how "the brain computes a phenomenon like self-location by the integration of basic bodily sensory signals" (Schwabe & Blanke, 2008, p. 6). The plot thickens further with the finding that otolithic processing is disturbed in neurological patients reporting OBEs (Blanke & Mohr, 2005).

A similar account can be used to explain movement sensations and upright orientation in dreams. Aside from ambiguous vestibular stimuli from the otoliths, we can add an attenuation of tactile sensations and sensations of pressure, for instance, from the bedsheets, as well as dynamic visual imagery to the mix. Coupled with an expectation of moving through the world in an upright position, this would create a natural bias for experiencing upright movement sensations in dreams. Alternatively, dynamic visual imagery may even result from sensations of movement; and indeed, we will see in a moment that evidence points in this direction. For now, it is mainly important to note that dream movement does not necessarily present a counterexample to the weak-functional-embodiment hypothesis but can be nicely accommodated by it.

Moreover, the comparison between dreams, OBEs, and full-body illusions may be a way of broadening the weak-functional-embodiment hypothesis beyond dreams. Hohwy, who discusses Schwabe and Blanke's (2008) Bayesian approach to OBEs from a predictive processing perspective, argues that rubber-hand and full-body illusions are paradigmatic examples of the fragility or flimsiness of bodily experience: "To experience the world—to engage in perceptual inference—we do not need a firm grasp of the material body" (Hohwy, 2013a, p. 139). This flimsiness may be the mark of weak-functional embodiment; and if Hohwy is correct, then it may not be limited to dreams and altered states of waking consciousness, but may characterize even standard wake states. Simple experimental manipulations, after all, are sufficient to induce a dramatic departure of the phenomenal body from the physical one, suggesting that bodily experience is generally only weakly constrained by inputs from the physical body.

This is further supported by recent work on body representation in wakefulness. In a series of publications, Longo and colleagues (2010; Longo & Haggard, 2010) have proposed that two mechanisms of higher-order body representation beyond the primary sensorimotor cortex can be distinguished. The first, which they call somatoperception, includes what I have been treating here as bodily experience proper. The second, somatorepresentation, refers to cognitive and semantic knowledge about bodies in general and one's own body in particular, but also emotions and attitudes about one's body. Longo and Haggard (2010) argue that semantic knowledge about one's body, including its location, is separable from the implicit body model. While explicit judgments about the size of body parts are highly veridical, representations subserving position sense, at least in the example of the hand, are massively distorted.

> These distortions, furthermore, are not idiosyncratic or random, but are highly stereotyped across individuals and appear to retain vestigial traces of the primary somatosensory homunculus of Penfield. . . . Although the homunculus is defined as the cortical map of tactile stimuli on the skin, our results suggest it may also provide the mental map of the body itself. (Longo & Haggard, 2010, p. 11729)

The distortions were also specific to stimulated skin regions, with perceived tactile distances being related to tactile sensitivity. This line of research is directly relevant for the body-brain-body problem. Indeed, as Longo et al. (2010) suggest:

> Localising a touch on the body surface is a two-stage process. First, the stimulus must be localised within a somatotopic map. This involves identifying its relative position with respect to other locations in the map. We suggest this relative position computation is a purely somatosensory process, making no reference to the physical body as an object in the world, potentially being computed entirely within a single map in primary somatosensory cortex or even subcortically. . . . Second, this somatotopic location must be mapped onto a corresponding bodily location; this is an essentially somatoperceptual process, referenced to a representation of the body (the superficial schema). (Longo et al., 2010, p. 658)

If one applies this framework to dreaming, one would expect widespread distortions in bodily experience to occur. But more specifically, one could use Longo and colleagues' model to more precisely locate where and how exactly bodily experience in dreams departs from waking bodily experience. The lack of integrated whole-body/body-part representations and widespread phenomenal indeterminacy in dreams might indicate that the corresponding somatotopic maps are not activated during the dream state. By contrast, in dreams in which tactile stimuli are referred to another body part or even another dream character, as was sometimes the case in Nielsen's (1993) blood-pressure-cuff study, the second stage of processing might be disturbed. Future stimulation studies could be used to investigate more systematically at which level of processing such distortions are most likely to occur in dreams. In this way, the analysis of bodily experience in dreams might be a wonderful opportunity to tease apart these two stages of processing and investigate the contribution of somatosensory processes under conditions in which they are only weakly constrained by the physical body.

Here is another prediction from this line of research. Longo et al. predict that "functional interactions between body representations are more likely to occur between two somatoperceptual representations or two somatorepresentational one[s] than between one of each class" (Longo et al., 2010, p. 665). Again, this translates into a prediction about dreaming, suggesting, first, that semantic knowledge about the body will be dissociable from bodily experiences and, second, that the presence of one type of bodily experience may increase the likelihood of other types of bodily experiences, as well. Preliminary evidence for the former prediction is supplied by cases of *just knowing* one had a certain type of body in a dream, even though this was not accompanied by specific tactile or movement sensations, or by the experience of seeing the body. Such purely semantic beliefs about the dream body also fit in well with indeterminacy blindness (see chap. 7), or the tendency to attribute more detail to the body than was present in dream experience. Here phenomenal indeterminacy on the level of somatoperception would easily be overlooked because of the comparatively greater richness of somatorepresentation. Dreams may also provide examples of functional associations between different somatoperceptual processes, thus supporting the latter prediction. Recall that in intense forms of dreaming, such as nightmares, lucid dreams, or typical dream themes, emotional intensity correlates with intense vestibular sensations, but also frequently with tactile sensations (cf. sec. 8.2.2.2), suggesting that having one type of bodily experience creates a bias toward having others, as well.

It might also be fruitful to combine this work on distorted bodily experience in dreams with the tentative hypothesis that (ambiguous) vestibular signals are a driving force of dream imagery. This seems promising, as recent research on the interaction between the vestibular system and multisensory cortical networks suggests that vestibular signals modulate somatosensory cortical processing, perhaps even increasing sensitivity to tactile input and thus promoting somatoperceptual processing (Ferrè et al., 2012; Ferrè, Vagnoni, & Haggard, 2013). As Ferrè and colleagues tentatively suggest:

> The vestibular system participates in a form of sensory signal management, changing the balance between the various sensory systems as the relation between the body and the external environment changes. This sensory rebalancing may be a crucial element in the brain's capacity to reorient towards novel or salient features in the environment. (Ferrè, Bottini, Iannetti, & Haggard, 2013, p. 748)

For instance, vestibular activation following head rotation may lead to reorienting toward changed sensory signals, thus facilitating the processing of information for new or altered perceptual environments. The story, roughly, would be that in dreams, vestibular processing—prompted by ambiguous signals from the otoliths—drives the integration of visual imagery and at the same time heightens sensitivity to real-body input. Vestibular processing, alongside emotion, may thus be a primary factor of dream production. Similarly, Cheyne and Girard (2009) suggest that OBEs may result when vestibular-motor experiences first lead to feelings of having left the physical body, and second to out-of-body autoscopy, or visual imagery of the body as seen from the outside.

Emphasizing the role of the vestibular system in dreams also fits in well with recent claims according to which the vestibular system is central not only for the processing of self-related movement, but more generally for phenomenal selfhood. In order to accurately represent movement through the world, it is necessary to bring self-movement into alignment with other bodily experiences, but also with visual input from the environment, as well as with movements and actions observed in others. Because of its multimodal character, the vestibular system is optimally poised to contribute not just to self-location and the spatial aspects of the bodily self (Pfeiffer, 2015), but also, as Lenggenhager and Lopez (2015) note, to more fine-grained aspects of experience such as the sense of owning and controlling a body and the self-other distinction. This line of research might be another way in which the weak-functional-embodiment hypothesis can be rendered more precise by applying insights from bodily experiences in wakefulness to dreaming.

Finally, I want to point out that the weak-functional-embodiment hypothesis is good at explaining not only the occurrence of bodily experiences in dreams but also why they are frequently absent. The British idealist philosopher F. H. Bradley foresaw this point in 1894. He was driven by the question "why, when we strive to move in dream, do we not always move?" (Bradley, 1894, p. 373). The problem is not, he writes, that our motor ideas, or our intentions to move, are generally weak in dreams. Instead, "We may reply that in the position of our limbs there is physical inertia, and, so far as that position implies feelings, there is psychical inertia too" (p. 374). He explains:

> If no sensation of any kind tells me that my arm is raised and bent, rather than hanging by my side, I cannot, I presume, go on to strike and to strike in a particular manner. But in dream this defect is normal. The sensations required to carry out the series do not occur, and the idea remains in consequence general and suspended. It is opposed by the body because, so to speak, it is out of relation with the detail of the bodily machinery. (Bradley, 1894, p. 376)

In 1918 we find Gregory applying a similar explanation to dreams of frustrated effort more generally, or to dreams in which the dreamer's intentions are thwarted by the reluctance—or inability, as he suggests—of dreams to provide the requisite imagery.

> The professor dreams that he wanders through the university passages, unable to find his laboratory. The visual representation of the scene is probably vague and usually rather kaleidoscopic, but . . . sufficient to convince the savant that he is in his familiar haunts. (Gregory, 1918, p. 127)

What is missing, however, are the background sensations of touch and movement that would be required for a truly realistic facsimile of the actual experience. Hence "the situation will *seem* right to the dreamer but it will *feel* wrong" (Gregory, 1918, p. 127). It is the imperfect reproduction of the background sensations—by which he predominately refers to bodily experiences—in dreams, or so he claims, that "lead to a feeling of confusion or wrongness that finds expression in the failure of our attempts" (p. 128). He is quick to add that this is not necessary; "sometimes the dream so suits the bodily posture or is so adapted to such sensations as it can receive, that the tendency hardly exists at all" (p. 128). It is only when the dream is out of harmony with the bodily posture that dreams fail to provide the background sensations Gregory takes to be necessary for successful dream action. Dream flying, he notes, is just such an example, for it is often done in a prone position, corresponding to actual sleeping posture.

According to both Bradley and Gregory, there is a link between the functional disconnection between the sleeping body, the phenomenal dream body, and the ineffectuality and indeterminacy of dream movements. This basic idea is inherited by the present theory, according to which changes in phenomenal and functional embodiment can be meaningfully described as occurring in concert. It also nicely anticipates my suggestion that dreaming is often phenomenally indeterminate with respect to bodily experiences and links this indeterminacy, or the widespread absence of bodily experience in dreams, to the sleeping body. Whereas a century ago these ideas remained largely anecdotal, both the theoretical environment and the scientific infrastructure have changed sufficiently to lift this to the level of a rigorous and systematic research program. Indeed, note how these two historical examples form a natural complement to Nielsen, Kuiken, and McGregor's (1989; Kuiken et al., 1996) suggestion that ineffectuality in dreams is linked to neuromuscular inhibition, which in turn facilitates a carryover of dream emotion into wakefulness and increases the subjective meaningfulness of the associated dreams. By joining scientific dream research to recent work on full-body and body-part illusions, and by adding the theoretical background and conceptual tools provided by philosophy of mind, this type of integrated research program could help make genuine progress on the body-brain-body problem.

8.5 Conclusions

Is dreaming a real-world case of cranial envatment? And can bodily experience in dreams be understood as arising independently of the sleeping body? At the beginning of the chapter, I introduced two variants of the functional-disembodiment hypothesis and showed how they are connected to two diametrically opposing accounts of the phenomenology of embodied selfhood in dreams: the bodily duplicate hypothesis, associated with the independence hypothesis, and the no-body hypothesis, associated with the dependence hypothesis. I then used a critical survey of the historical and contemporary literature to work towards a rival theory, the weak-functional-embodiment hypothesis, whose basic claim is that bodily experiences in dreams can meaningfully be described as illusory own-body perception. An early ancestor of this view is the *Leibreiztheorie*, or the view that dreaming in general is caused by bodily sensations. The largely anecdotal historical examples discussed at the beginning of section 8.2 helped illustrate the contingency of the currently popular adherence to the functional-disembodiment hypothesis and take a first step toward undermining its plausibility.

Next I summarized empirical findings on the bodily sources of dreaming and specifically of bodily experiences in dreams. Stimulation studies have a measurable effect on dreaming, though the precise manner of incorporation is variable. At the same time, stereotyped bodily experiences occurring in the hypnagogic state, during lucid dreams, typical dreams, and nightmares—such as SP, sensations of flying or falling, or dreams of being chased—may be explained in terms of illusory own-body perception. There is some suggestion in the literature that the occurrence of one type of bodily experience, such as SP or flying, generally leads to more frequent body references and higher levels of aggression or emotional intensity. Such dreams may also be experienced as subjectively more meaningful than others. Moreover, sleep is replete with micromovements, some of which may be related to dreaming or even described as instances of dream enactment, as in certain lucid control dreams. A tentative conclusion from these findings is that phenomenal and functional embodiment in dreams are intimately linked. Contrary to currently popular views, dreams are weakly phenomenally-functionally embodied states, and the unique pattern of bodily experience in dreams is best explained by appealing to partly preserved but altered processing of real-body inputs.

Preliminary findings on the neural correlates of bodily experiences in dreams support a number of these phenomenal-functional considerations. The variability of activation in the primary sensorimotor regions fits in well with the variability of bodily experiences in dreams but also suggests a stronger role of external stimulus processing for bodily than for visual stimuli. This is also borne out by the higher incorporation rate for bodily than for visual stimuli in dreams. Moreover, the empirical rationale for understanding the mechanism of reduced sensitivity to external stimuli itself depends to an important extent on studies of stimulus incorporation. The most important conclusion is that the input–output blockade and the functional-disembodiment hypothesis are misnomers: even in sleep, there continues to be a systematic and theoretically relevant relationship between bodily experience and the sleeping body, and in order to make progress on the experimental investigation of dreaming, this relationship cannot be ignored.

I closed the chapter by introducing the weak-functional-embodiment hypothesis, placing it in a broader theoretical context and defending it against several objections. Though the hypothesis is weak in a number of respects, it marks a substantial departure from contemporary theories of dreaming; it has its roots in historical theories but differs from them in being grounded in systematic conceptual considerations and by taking empirical findings into account. In particular, by conceptualizing bodily experiences as illusory rather than as hallucinatory, the present theory presents a departure from both internalist and externalist views of dreaming. It is also the first step toward locating dream research in the context of an interdisciplinary research program on a particular variant of the mind–body problem: the body-brain-body problem, or the problem of how bodily experiences (in dreams, but also across the sleep–wake cycle) can be mapped to changes in causal coupling between the brain and the sleeping body. The key idea is that the weak-functional-embodiment hypothesis not only presents an empirically plausible theoretical account of bodily experiences in dreams but also translates into a fruitful and innovative research strategy on the mind–body relationship.

9 Are Dreams Subjective Experiences (II)? The Phenomenology of Thinking and the Problem of Dream Belief

9.0 Introduction

So far, my investigation of the subjective character of dreaming has focused on the relationship of phenomenal selfhood and bodily experiences in dreams, as well as on their grounding in the sleeping, physical body. A second and in some ways more classical way of understanding questions about the subjective character of dreaming is to say that they concern the extent to which the dream self is a thinking self and a bearer of beliefs. The problem of dream deception naturally follows suit.

As in the foregoing chapters, I start out by sketching the landscape of philosophical positions by reconstructing maximally opposing hypotheses about dream thinking and dream belief from the philosophical literature. In its strongest reading, the phenomenal-duplicate hypothesis claims that dreaming replicates, among other things, standard waking experience of being a cognitive agent and thinker of rational thoughts. Dreams, in this view, additionally equip the dream self with an array of beliefs about the dream world, some or even most of which are false and contradict long-standing beliefs held in wakefulness. By contrast, the *no-self hypothesis* denies that dreaming involves the experience of being a thinking self and cognitive agent or a possessor of beliefs in any interesting sense.

Even this superficial characterization of the two opposing views gives rise to a number of questions. Some of these are of a conceptual nature, asking whether instances of dream thinking and dream belief can meaningfully be described as genuine instances of thinking and believing. The answer will vary with different accounts of what it is to think or believe, suggesting that dreaming is an interesting test case for theoretical accounts of thought, belief, and different attitude types more generally. Other questions are of a more empirical nature, asking whether, given a certain account of what it is to think or believe, dreaming actually (ever or even typically) involves instances of thinking or believing.

In sketching these two pairs of hypotheses about dream thinking and dream belief and discussing their theoretical and conceptual implications, I will again be particularly concerned with framing these positions in such a way as to make them amenable to empirical investigation. Moreover, tracing the connections between these pairs of hypotheses itself will lead to important insights. In particular, while the philosophical discussion has focused on dream belief, I argue that the problem of dream thinking is actually the more basic and theoretically more interesting phenomenon. Given an account of dream thinking, an answer to the problem of dream belief naturally suggests itself.

In section 9.3, I introduce an empirically informed conceptual framework for describing thinking, self-reflection, and metacognition in dreams and distinguishing them from cognitive processing and behavioral responsiveness during sleep. I also suggest how this framework can be applied to lucid dreams. Section 9.4 discusses evidence on the neuronal basis of dream thinking. I close by introducing an intermediate account of thinking in dreams, called the *weak-cogitation hypothesis* (sec. 9.5), and sketching the conditions under which we can meaningfully speak of dream belief (sec. 9.6).

9.1 The Doxastic-Duplicate Hypothesis and the No-Belief Hypothesis

Questions concerning dream belief play a prominent role in the philosophical discussion because of their close connection to dream skepticism. The reason why dream skepticism is troubling, after all, is not just that dreams supposedly endow us with a double body or place us in a seemingly realistic environment; the real problem is that in doing so, they also endow us with a complete set of beliefs about the dream world and the dream self in it, and they do so in the same way as waking experiences do. As Descartes tells us, his realization that he might be dreaming alerts him to the possibility that "all those opinions are entirely false and imaginary" (Descartes, 1641/1901, I.11).

Crucially, the dream hypothesis threatens to undermine the epistemic status of his beliefs and his status as a knowing self, *without* thereby infecting his status as a thinking and doxastic self, or a self capable of *having* thoughts and entertaining beliefs. Dream-Descartes can have all the same beliefs as his waking counterpart; he could, in other words, be a doxastic duplicate of waking Descartes. Moreover, the problem is not just that dream-Descartes could have a complete set of beliefs (in the sense of having a network of interconnected beliefs, just as he would in wakefulness)[1] but that he could form new ones: it is their belief-inviting character that makes dreams truly troubling. Call this the *doxastic-duplicate hypothesis*:

Doxastic-duplicate hypothesis: Dreams invite beliefs in the same sense as waking experiences; moreover, in dreams, we have a complete set of beliefs concerning the dream world and the dream self in the same sense as we do in wakefulness.

A tempting way to evade the skeptical threat is to deny that dreaming involves real beliefs. If the dream self is not capable of acquiring and possessing beliefs, then it cannot entertain or acquire any false beliefs while dreaming. In particular, it cannot believe, while dreaming, that it is awake. Consequently there is no threat of dream deception. Call this the *no-belief hypothesis*:

No-belief hypothesis: Dreams do not invite beliefs, and in dreams, we do not form any new beliefs or indeed believe anything in the full sense of the word.

The no-belief hypothesis has been a popular way of denying the threat of dream skepticism. It is, for example, the key to Malcolm's (1956, 1959/1962) rebuttal of dream skepticism. However, the no-belief hypothesis does not require an outright rejection of the experiential hypothesis of dreaming. A subtler and for present purposes more interesting challenge comes from the claim that dream beliefs are imaginative and that believing in a dream does not entail (and even precludes) actually believing (Sosa, 2007; Ichikawa, 2008, 2009). Sosa, for instance, allows that "unquestionably, I do believe and intend things *in my dream*" (Sosa, 2007, p. 3), but argues that "even if *in my dream* I intend to keep running, *in actuality* I have no such belief or intention" (p. 4).

Why exactly might dream belief fall short of real belief? There is, of course, a diverse and highly technical discussion of what it means to have a belief (for an introduction, see Schwitzgebel, 2011a), and the answer will vary with different accounts. One might think, then, that the status of dream belief hinges on the question of whether dream belief fits the concept of belief in a special technical sense. A more sweeping denial of dream belief, however, is to suggest that the notion of dream beliefs contradicts commonsense assumptions about what it means to have a belief. This account, if successful, would be much more damaging because it would suggest that dream beliefs fail to qualify as real beliefs as a matter of principle.

A first argument for this view is that dream beliefs are often inconsistent with longstanding waking beliefs. Roughly, the argument says that there is a sense in which I retain beliefs about myself, but also about my current sleeping environment, my plans for the next day, and so on, during sleep. Yet in my dreams I can have completely different beliefs, seemingly without engaging in a process of belief revision, either upon acquiring the aberrant dream belief in my dream or upon rejecting it after awakening (Ichikawa, 2009, pp. 112–113). As Sosa puts it, if it is correct to say that I retain the belief that I am in my bedroom during sleep, "it is hard to see how one could then concurrently believe that one is being chased by a lion, rather than lying in bed" (Sosa, 2007, p. 5). This is absurd not because contradictory beliefs are impossible per se but because it would mean that "the whole person might believe that p and concurrently believe also the very negation of that first belief, i.e. that not-p" (p. 5).

A first worry about this position is that there is widespread inconsistency even among standard waking beliefs. Hence the attempt to make beliefhood contingent on consistency and the claim that it is impossible to simultaneously hold contradictory beliefs may be too strong or at least require additional motivation. Bortolotti (2005, 2009) discusses a number of cases showing that different forms of irrationality characterize not just delusions but even ordinary beliefs. This suggests that rationality and consistency do not place any hard constraints on the conditions for belief possession and ascription (see also Schwitzgebel, 2011b). At least, the proponent of the view that they do owes an account of why dream beliefs are inconsistent in a way that sets them apart from more acceptable forms of belief inconsistency.

For now, a more important point is that two different conceptions of belief are at play here. The beliefs I retain during sleep, for instance, my belief about my bedroom, my job, or my last vacation, are latent; I have them in the dispositional sense that I would affirm them or act on them if certain conditions of manifestation were fulfilled. For instance, if I woke up, my behavior would (or so we might assume, at least in standard situations) fit the dispositional profile of believing that I am in my bedroom rather than being chased by a lion. Consequently my behavior upon awakening counts as evidence for saying that I maintained the belief of being in my bedroom latently or dispositionally even while asleep. By contrast, the dream belief that I am being chased by a lion is (if anything) an occurrent belief; it is a belief fervently believed and acted on, or so we may assume, within the dream.[2] The claim, then, is that having a dispositional or latent belief during sleep rules out having contradictory occurrent beliefs while dreaming.

Now that the challenge to dream belief has been sharpened, we can begin to see why it is problematic. In particular, note that the logic of ascribing dispositional beliefs depends, in part, on their ability to match occurrent beliefs, as well as appropriate behavior. The ascription of dispositional beliefs is contingent on their occurrences, not vice versa; indeed, assessing their occurrences is the only way we have of ascertaining the presence of dispositional or latent beliefs in the first place. But if this is correct, then occurrently believing, in a dream, that I am being chased by a lion and manifesting appropriate flight behavior would seem to count *against* my having the dispositional belief that I am asleep in my bedroom.

The no-belief hypothesis, however, approaches the problem of dream belief from the opposite direction. It assumes that we already know what the dispositional belief is, and that it is, indeed, a belief maintained during sleep; moreover, it additionally assumes that the dispositional belief has priority over the occurrent dream belief, to the point of disqualifying it from being such an occurrent belief at all. It also assumes that occurrent beliefs manifested upon awakening (i.e., resetting the alarm clock to stay in bed for just a few more minutes after awakening from a dream in which one was chased by a lion) have priority, in assessing the presence of dispositional beliefs throughout sleep, over those manifested during the dream state. Unless one makes these additional assumptions, one is left with two inconsistent sets of belief manifestations, but no account of which of them (if any) falls short of beliefhood. Inconsistency with dispositional beliefs cannot, in other words, be used to argue that dream beliefs are not instances of occurrent beliefs. Without further argument, this strategy backfires too easily, leading occurrent dream beliefs to count against the assumed dispositional ones. The point for now is that there is no easy, non-question-begging way to resolve the question. Unless one wants to disqualify dream beliefs from counting as real beliefs simply because they occur in a dream, an independent argument for doing so is needed.

Surely the type of inconsistency manifested by dream beliefs presents a problem. One might want to argue, for instance, that occurrent dream beliefs are too short-lived to count against long-standing dispositional beliefs. If in a dream I believe for a duration of perhaps ten minutes that I am being chased by a lion, this does not suffice to count against a long-standing belief that, when sleeping in my bed, I am at little to no risk of being assaulted by wild animals. One would then need an account of the relationship between short-termed occurrent attitudes and long-standing beliefs (for such an account, see Schwitzgebel, 2013a). Yet note that this type of strategy would already change the nature of the debate. Dream beliefs would no longer be disqualified from beliefhood as a matter of principle. Rather, we might now say that it is, in part, an empirical question whether dream beliefs are indeed too short-lived to count against long-standing dispositional beliefs; if at least a subgroup of dream beliefs turned out to persist beyond the borders of sleep, for instance, by modifying long-standing waking beliefs, we might want to allow that this is sufficient for genuine beliefhood.

Aside from appealing to common sense, one might also deny that dream beliefs fit the concept of belief in a technical sense. Ichikawa (2009), for instance, defends the no-belief hypothesis by claiming that dream belief is incompatible with standard philosophical theories of belief. He notes that dream beliefs do not have the same functional role as real beliefs because they lack connection with perceptual experience and fail to motivate actions: "Beliefhood would have to be consistent with a disconnect from both action and perception, in order to let dream beliefs count as beliefs; but then on what basis would prototypical imaginings be excluded as beliefs?" (Ichikawa, 2009, p. 115). Similarly, he argues that interpretationist or dispositionalist accounts contradict the view that dream beliefs are real beliefs:

> If we observe Laura, who is dreaming that she is in England, on what basis can we ascribe to her the belief that she is in England? She is not behaving as if she is in England—she is not looking to the right before crossing streets. Neither does she give us utterances that are best interpreted as expressions of the belief that she is in England. Indeed, she is exhibiting very little behavior at all, since she is asleep. What is it about Laura in virtue of which she believes she is in England? No answer favoring *beliefs* over *imaginings* seems here to be forthcoming. (Ichikawa, 2009, p. 115)

While the claim that dream belief violates the conditions of belief in a special technical sense might look, at first sight, like a principled way of ruling out the occurrence of dream belief, I think this would be premature. To avoid the familiar equivocation between the dream body and the sleeping body, we should allow that the reason why Laura does not exhibit behavior indicating that she believes herself to be in England does not have to do with a disconnection between the belief and her current behavior—in her dream, she may well be getting on the bus to Big Ben—but has to do, quite simply, with REM sleep paralysis. Moreover, if Laura had RBD, she might well exhibit behavior appropriate to her belief of being in London (which would, incidentally, be wildly inappropriate regarding her dispositional or latent belief that she was sleeping in her bedroom in, say, Germany). It seems quite strange, however, to say that Laura's ability to have a belief depends on the contingent fact of whether she exhibits dream-related behavior during sleep, as she might if she were afflicted with RBD. Indeed, if the weak-functional-embodiment hypothesis is correct, then we can further generalize this point, claiming that the disconnection between Laura's observable behavior during sleep and her dream beliefs is not absolute, but a matter of degrees. This is enough, I think, to argue that the failure to manifest observable behavior is not sufficient to disqualify dream belief. Rather, investigating whether dream belief is connected in an appropriate way to dream behavior—including dream-related behavior of the outwardly observable type as in RBD, but also microbehavior in the form of muscle twitches—would be a scientific way of making headway on the problem of dream belief.[3] We would now have translated the question of dream belief into an empirical one and, with it, diminished the likelihood of a straightforward yes/no answer applicable to all instances of dream belief.

Now, one may balk at the claim that dream action counts as a manifestation of belief, insisting that beliefs must be tied to actual, publically observable instances of behavior. This is the position commonly endorsed by interpretationist positions or, for instance, by Dennett's intentional stance (e.g., Dennett, 1987, 1991; cf. Schwitzgebel, 2011a). By contrast, liberal dispositionalists may include private mental episodes, such as the feeling of surprise, in the set of dispositions relevant to belief possession (Schwitzgebel, 2002a, 2011a). In such an account, dream belief might more easily fit the profile of beliefhood. This is especially true as according to the weak-functional-embodiment hypothesis, the privacy of dream behavior is incidental, not necessary. It would seem arbitrary, however, to say that this incidental privacy of dream behavior—contingent on the strength of motor paralysis—should count against its being considered as behavior. If one relaxes the requirement that outwardly observable behavior is necessary for beliefhood, then dream belief can be accommodated by interpretationist accounts. The real question is whether dream belief is connected to dream behavior in the right manner, not whether this behavior is outwardly observable. This is a straightforwardly empirical question, and I return to it in section 9.6.

More importantly, however, note that according to both the commonsense and the technical version of the no-belief hypothesis, dreaming fails to involve beliefs through no fault of its own. The argument comes dangerously close to the principled claim that dream beliefs fall short of beliefhood no matter what the connection between dream belief, waking belief, and dream behavior, just by occurring in a dream. One can, of course, make the purely conceptual claim that dream beliefs principally fall short of beliefhood; my point here is merely that this is not particularly enlightening.

Nor would such principled objections to considering dream beliefs as real beliefs be particularly troubling for Descartes. The relevant and truly troubling point for Descartes is that he *takes* himself to have certain beliefs, whether dreaming or not. Whether these qualify as beliefs in some technical sense of the word is only of secondary importance. As Lewis notes, a person might "in fact believe or realize in the course of a dream that he was dreaming, and even if we said that, in such case, he only dreamt that he was dreaming, this still leaves it possible for someone who is asleep to entertain at the time the thought that he is asleep" (Lewis, 1969, p. 133).

If dream-Descartes erroneously took himself to have beliefs, this would simply add to the list of ways in which dreams are deceptive (for a similar argument, see Ichikawa, 2008). So instead of asking whether dream belief fits the concept of belief in some technical sense, one must now ask whether dreaming fulfills the conditions for saying that dream-Descartes *takes* himself to have beliefs. Moreover, this is a question with which proponents of the no-belief hypothesis can agree. Both Sosa and Ichikawa assume, after all, that dream beliefs exist, but only question their status as real beliefs.

9.2 The Cogitative-Duplicate Hypothesis and the No-Cogitation Hypothesis

Descartes's worry, then, is not so much whether he has beliefs in a technical sense while dreaming as whether he takes himself, while dreaming, to have certain beliefs. Following Lewis, we might frame the problem in the general sense as asking whether it is possible to think in dreams, including having the thought that one is now awake. I suggest that we can further unpack this claim by asking whether Descartes enjoys, while dreaming, the phenomenology of engaging in different types of thoughts, including the phenomenology of occurrently believing that *p*.

Recall that dream-Descartes, sitting at his desk in his dressing gown, does not just have the phenomenology of being a bodily self but also experiences the rich phenomenology of being a thinking self. From the beginning, dreaming endows him with a specific belief, namely, that he is awake. This belief is not abstract but has a distinct phenomenal flavor:

> I certainly look upon this paper with eyes wide awake; the head which I now move is not asleep; I extend this hand consciously and with express purpose, and I perceive it; the occurrences in sleep are not so distinct as all this. (Descartes, 1641/1901, I.5)

In purposively looking at the piece of paper in his hands, he shows himself to be a cognitive and behavioral agent, able to form and act on intentions. His initial certainty of being awake is then dispelled by his realization that he has dreamt such situations before, leaving him in a state of uncertainty about his current situation. He thinks about the matter and then *perceives* "so clearly that there exist no certain marks by which the state of waking can ever be distinguished from sleep, that I *feel* greatly astonished; and *in amazement I almost persuade myself* that I am now dreaming" (Descartes, 1641/1901, I.5; italics mine). He is left with nagging doubt about the truth of all his beliefs about himself and the world.

This emphasis on the phenomenology of engaging in different acts of cogitation continues in the *Second Meditation*. Much has been written about the *cogito* and about Descartes's claim that he is a thinking thing (see, e.g., Frankfurt, 1970/2008, chaps. 10–11; Williams, 1978, chap. 3; Broughton, 2002, chap. 7), but here several brief remarks should suffice. First, being a thinking thing not only refers to thinking in the narrow sense of rational thought or intellection but is broadened to refer to "a thing that doubts, understands, [conceives], affirms, denies, wills, refuses; that imagines also, and perceives" (Descartes, 1641/1901, II.8). These various forms of thinking in the wide sense render Descartes certain of his own existence as a thinker: "For it is of itself so evident that it is I who doubt, I who understand, and I who desire, that it is here unnecessary to add anything by way of rendering it more clear" (II.9). Whether he is dreaming or not, the phenomenology of thinking in this wide sense, or of engaging in cogitations, renders him certain of his existence. Finally, it does so only "as often as I think; for perhaps it would even happen, if I should wholly cease to think, that I should at the same time altogether cease to be" (II.6). Of course, in this case, he would not be deceived about anything at all—in the very act of doubting or wondering, he is rendered certain of his own existence. In an important sense, his very existence hinges on the phenomenology of being a cogitative self: being a thinking thing is the very essence of his being.

The claim that dreaming involves the phenomenology of being a thinking self is intimately related to the rationalist assumption that thinking is the mental process we are most certain of, the one thing that is inseparable from our existence. Moreover, being a doxastic self is crucially related to being a cogitative self: to attain beliefs, including the *cogito*, one must first be aware of one's thoughts. The insight implicit in the *cogito* is special in that it is not only acquired but also epistemically justified through acts of thinking. In engaging in any type of cogitation, Descartes gains not just a belief but knowledge of his own existence.

This phenomenological reading is essential to the *cogito*, which de Sousa claims "is the first explicit use of something like a criterial feeling of rightness to justify a knowledge claim" (de Sousa, 2009, p. 144). Similarly, Loeb (2005) has suggested that the relevant sense in which the *cogito* is indubitable is psychological: it is a belief whose truth

Descartes cannot bring himself to doubt, even if he is now dreaming (but see Broughton, 2002). So even if there is much more to thinking, believing, and doubting, their phenomenology is the aspect that Descartes is interested in. This means that for Descartes, the doxastic-duplicate hypothesis is really only a special—albeit a very special—case of what I call the *cogitative-duplicate hypothesis*:

Cogitative-duplicate hypothesis: Dreaming involves a phenomenal self capable of engaging in cogitations, such as thinking, wondering, doubting, and so on, in the phenomenological sense; the dream self is a cogitative self to the same extent as during standard wakefulness.

Having cogitations, for Descartes, is tied to the phenomenology of engaging in acts of thinking. It involves an agentive component, that is, the experience of purposively thinking or being able to direct one's thoughts at will, and an ownership component: to ensure himself of his own existence, Descartes must experience his cogitations as his own.[4] Note that Descartes goes much further—he additionally claims, for instance, that he can recognize each cogitation for what it is. But a proponent of the cogitative-duplicate hypothesis is not committed to this additional claim.[5]

The cogitative-duplicate hypothesis also need not claim that the phenomenology of thinking (e.g., of engaging in silent soliloquy while deciding to do something) is all there is to thinking or that it necessarily accompanies all kinds of thinking. Thinking can be done in various ways, for instance, in conversation, writing, or drawing and does not strictly require a *separate* inner process. Many have argued that thinking has much to do with intelligent performance (see esp. Ryle, 1949/2000, chap. 2; Bennett & Hacker, 2003, chap. 6), and perhaps cognitive processes can even extend beyond the skull or integrate sociocultural practices (cf. Menary, 2007, 2010, 2015). Clearly this would not be Descartes's favored approach, but the cogitative-duplicate hypothesis as such remains neutral on these questions. For present purposes, the main point is that thinking at least on occasion occurs in one's head, metaphorically speaking, and it is *like* something to think in this manner. This subgroup of cogitations in the phenomenological sense is what Descartes is interested in, and for present purposes, it is enough to note that theorists of different stripes can accept their existence.[6]

The upshot is that to deny that dreaming involves beliefs in the relevant sense, one will have to deny that dreaming involves the phenomenology of believing, but also more generally of engaging in cogitations. As long as dreams involve the phenomenology of believing and of cogitating more generally, they induce epistemic discomfort. Moreover, in this view, believing that p is just a special form of thinking that p. Occurrently believing involves processes such as thinking, affirming, or judging that p, and it is often preceded by doubt or uncertainty. The problem thus extends far beyond the technical sense of belief possession to the question of whether dreaming involves a cogitative self with the ability to engage in different forms of cogitations in the phenomenological sense.

We can now use this sharpened understanding of the cogitative-duplicate hypothesis to introduce the maximally opposed view, the *no-cogitation hypothesis*.

No-cogitation hypothesis: Dreaming does not involve cogitations in the phenomenological sense.

While there are different ways of denying the claim that dreaming involves cogitations, this phenomenological reading narrows the range of possible positions considerably. For instance, one might claim that acts of judging, thinking, or reasoning are incompatible with sleep and hence that instances of dream judging, asserting, or affirming do not count as real instances of judging, asserting, or affirming for conceptual reasons (as argued by Malcolm, 1959/1962, chaps. 9–10; cf. sec. 1.2). But this type of objection is not directly relevant to Descartes's worry about dreaming: without further argument, claiming that a purely conceptual argument against the existence of cogitations in dreams tells us anything about the phenomenology of dreaming would be to commit the armchair fallacy. Consequently a proponent of the no-cogitation hypothesis will have to say that as a matter of fact, dreaming does not involve cogitations in the phenomenological sense.

Here, one might look to the imagination view of dreaming for a more nuanced position. According to Sosa, imagining that one thinks, or dreaming that one thinks, is "subjectively much like the state of thinking that one thinks" (Sosa, 2007, p. 12). This renders the two states hard to distinguish, but still importantly different: "One is in an apparent state of thinking that one thinks, doing so (thinking one thinks) *only in a dream*, so that it is really only a state of *dreaming* that one thinks one thinks" (p. 12). The problem is not, then, that dream thinking necessarily lags behind real thinking phenomenologically, but that it does so conceptually, at least in a certain reading. Sosa allows that there is a thin sense of thinking according to which thinking that one is thinking is self-verifying. Even if dreams are imaginings, they count as thoughts in this sense. This does not, however, suffice for knowledge. Affirming that one is affirming, for instance, "*could* be mere appearance" (p. 11); in a dream, it would not count as real affirming. "So things might in a way seem subjectively just as they do now, although I would just be dreaming: thoughts would be crossing my mind, without my really affirming anything" (p. 11).

Ichikawa defends the similar position that "from an internal point of view, for the dreamer *quasi*-affirmation is importantly like affirmation and indistinguishable from it. One cannot, then, rationally affirm that one is awake, ignoring the possibility that one is *quasi*-affirming something false" (Ichikawa, 2008, p. 522). The problem, in this view, is how to describe the relationship between the phenomenology of thinking and the concept of thinking. *That* dreaming involves the phenomenology of thinking is granted.

If the no-cogitation hypothesis is not, at least in the context of the imagination view of dreaming, to be understood as a phenomenological claim, then how might one go about driving a wedge between dream thinking, dream affirming, and their ilk and real instances of their kind? The idea is that the relationship between thinking and dream thinking (and affirming and dream affirming, and so on) is similar to that between believing and dream believing, but also between perceiving and quasi-perceiving: real instances of thinking, believing, and perceiving depend on criteria such as the overall functional profile in the belief case and the presence of a mind-independent object of perception in the percept case, whereas dream thinking, believing, and perceiving depend on their phenomenal character alone. Phenomenology alone, however, does not suffice: thinking, affirming, and believing are, like perceiving, success words.

Does it really make sense, however, to say that whether one has a thought is contingent on its connection to action or its functional profile? As Williams notes, "I am thinking" is self-verifying, whereas "I am not thinking" is pragmatically self-defeating. "I am thinking" is true whether asserted, doubted, considered, or entertained, "for all these are modes of thought, so the fact that I doubt or consider anything, and in particular, doubt or consider the proposition that I am thinking, will make it true that I am thinking" (Williams, 1978, p. 74); and this includes, according to Williams, dream thinking. Thinking, at least as commonly understood, is an umbrella term for various modes of cogitative activity, and the *cogito* is even wider (p. 76). So whatever form of thinking I engage in, "I am thinking" will be correct. I might still be mistaken about the type of thinking I am engaging in, but I will, nonetheless, be thinking. There is no distinction between quasi thinking and real thinking analogous to the distinction between quasi believing (or occurrently believing) and belief possession in the fuller, technical sense.[7] Quasi thinking in the phenomenological sense *is* thinking.

One might want to accept this analysis of thinking but reject it for affirming, assenting, judging, or avowing, and perhaps even for other cogitations such as doubting, wondering, and so on. One might want to say, for instance, that some forms of cogitating (e.g., affirming, assenting, judging, or avowing) count as expressions of beliefs (Sosa, 2007, p. 4) and so only count as real instances of their kind when they express real beliefs, not just dream beliefs. The conditions for counting as genuine cogitations are not equal, in this view, for all kinds of cogitating. At this point, however, the difficulty falls back to the one already encountered in the debate about dream belief: unless one wants to deny that dream beliefs are real beliefs, because, after all, they are dream beliefs—which is not very informative—one should make room for considerations of empirical plausibility.

Another worry is that it does not really seem to make sense to drive a wedge between thinking and other cogitations such as affirming, assenting, judging, or avowing. The problem is not that it is impossible to define the concepts in such a manner as to make them contingent on the conditions of beliefhood—clearly this can be done—but rather that this runs strongly counter to common sense. This becomes clear when we consider further cases. In particular, note that in the imagination view, dreaming is itself a form of cogitating—hence the imagination view presupposes that there is a cogitative self in the dream state. This is, however, a very special kind of cogitative self. On the one hand, because dreaming, according to the imagination view, is a form of make-believe (Sosa, 2007, p. 8), the dream self can engage in propositional imaginings, and in doing so it really, and not just seemingly, has these propositional imaginings. It may also be able to engage in real instances of thinking. On the other hand, however, the dream self merely imagines engaging in other forms of cogitations, such as affirming, judging, asserting, or avowing, and in doing so consistently falls short of enjoying actual instances of their kind.

Things become especially difficult when we are forced to decide where to draw the line between cogitations that really can occur in dreams and those that do not (or even cannot). Note, for instance, that propositional imagining is importantly related to (though some would claim not identical with) supposing and conceiving (for discussion, see White, 1990). So one might wonder whether, in the imagination view, one can suppose that *p* in dreams, or whether one also only quasi-supposes that *p* while dreaming. And if one allows that supposing that *p* can occur in dreams, why not say that the same is true for assuming or judging that *p*? At least it is not obviously clear why one would dig one's heels in at some examples but not at others; here, the imagination view runs the risk of seeming unduly contrived. Moreover, the preoccupation with belief (and expressions of belief) leaves the status of other dream cogitations underdefined. For instance, how can wondering whether *p* or doubting that *p* in dreams be accounted for? What would count for or against dream wondering and dream doubting as being real instances of their kind?

These considerations are enough, I think, to shift the burden of proof. Importantly, nothing I have said so far amounts to a straightforward case for dream belief or various other types of dream cogitation. All I hope to have shown, so far, is that there is no obvious conceptual argument for denying the occurrence of thoughts and related kinds of cogitations in dreams. The answer to questions about dream cogitations will depend on whether dreaming involves the phenomenology of engaging in different cogitations, and if so, whether the overall functional profile of these cogitations is roughly comparable to their waking counterparts. To meaningfully adjudicate whether dream cogitations are real instances of their kind, considerations of empirical plausibility are relevant. What we have gained is that there is no principled, quick-and-easy way to reject the cogitative-duplicate hypothesis and defend the no-cogitation hypothesis. In its strongest form, the no-cogitation hypothesis is so strong that it is incompatible even with the imagination view of dreaming.

An alternative way to reject the cogitative-duplicate hypothesis and perhaps even to defend the no-cogitation hypothesis is to deny that as a matter of fact, the experience of being a cogitative self occurs in the dream state. This type of argument has a long history, and as we will see, it is also popular in the scientific literature. An early version is found in the *Essay on Consciousness*, anonymously published in 1728, allegedly as the first detailed account of consciousness (Pseudo-Mayne, 1728/1983, p. 2). Here, while allowing that it is like something to dream, the author argues that dreams lack the "least Resemblance or Similitude of a Conscious, or Self-knowing Perception" (p. 52). For this reason, "a Man can never know or be certain that he dreams. . . . Whereas, on the contrary, when awake, a Man cannot but assuredly know he is so" (p. 50). In dreams, we do not even experience ourselves as thinkers.

We find a similar claim in the more empirically oriented philosophical literature on dreaming. According to Metzinger, the nonlucid dreamer

> is severely disoriented about places, times, and persons, and constantly produces ad hoc explanations for the events he encounters. The phenomenal dream-self is almost incapable of conscious self-reflection and metacognition that could make the lack of intellectual consistency become a recognizable *fact* for herself. Additionally, short-term memory is greatly impaired and generally unreliable. Therefore, dreamers are only cognitive subjects in a weak and, arguably, philosophically uninteresting sense. (Metzinger, 2003, p. 252)

In this view, not only are nonlucid dreams characterized by a long list of cognitive deficiencies, but these also prevent the dreamer from realizing this fact. And even when dreams do involve certain cogitations, they lack an important part of the phenomenology of being a full-fledged cogitative self:

> The representational content missing in ordinary dreams is one of the *self-in-the-act-of-deciding* to take a specific course of action (the volitional subject), the *self-in-the-act-of-deliberately-attending* to certain perceptual or cognitive states (the attentional subject), and the self as *rationally* thinking about events currently taking place in the dream (the cognitive subject). (Metzinger, 2003, p. 254)

According to Metzinger, the crucial difference is the possession of a stable first-person perspective, which in turn enables

> the possession of a concept of oneself as the *thinker* of these thoughts, as the *owner* of a subjective point of view. In short, what is needed is not only reference from the first-person point of view but the capacity to mentally "ascribe" this act of reference *to* oneself while it is taking place. (Metzinger, 2003, p. 396)

Being a cogitative self in an interesting sense, then, depends on the ability to have particular kinds of self-referential thoughts. This ability, moreover, is closely linked to the phenomenology of being a cogitative self. In standard wakefulness, we both have this ability and experience ourselves as thinkers of thoughts, as having doubts, entertaining suppositions, or engaging in wonderings. Importantly, the same is true in lucid dreams. According to Metzinger, the onset of lucidity is marked by a change in the content of the phenomenal self-model. Together with the reappearance of autobiographical memory, cognitive self-reference once more becomes possible, leading to the reestablishment of the phenomenology of agency and of executive control (Metzinger, 2003, p. 540). This view also presupposes a close relationship between the phenomenology of being a cogitative subject and the epistemic status of the dream self: with the reappearance of a full-fledged cogitative self, as in lucid dreams, the dream self ceases to be deluded and comes to enjoy genuine insight. Even in dreams, the phenomenology of being a thinking self thus is closely tied to one's status as a knowing self.

For now, we might regard this as an empirical hypothesis: having a stable first-person perspective is a condition not just for having certain cognitive abilities and attaining genuine insight but also for enjoying the relevant kind of cognitive phenomenology. This view predicts that certain kinds of cognitive phenomenology in dreams have a high degree of epistemic reliability: we do not, in this view, experience ourselves as cogitative selves unless we have the relevant cognitive capacities that enable genuine insight.[8] This is an intuitively attractive view, because if it were correct, it would lend strong support to the reliability of our conscious thought processes.

9.3 A Conceptual Framework for Describing the Cogitative and Doxastic Dream Self

To decide whether dreaming involves the phenomenology of cogitative and doxastic selfhood, one has to investigate the occurrence of different forms of thinking in dreams. Do people have the phenomenology of thinking and believing in dreams? Do they thereby fulfill the criteria for belief possession? And what is the relationship between the cogitative dream self and dream lucidity? Answering these questions requires a number of conceptual distinctions, as well as a careful reading of the scientific literature.

9.3.1 Dream Cogitation, Dream Cognition, and Nonconscious Cognition and Behavioral Responsiveness during Sleep

Perhaps the most widely cited characterization of dreaming in the scientific literature is Rechtschaffen's (1978) now classical dictum that dreaming is marked by "single-mindedness" and a lack of reflective awareness. This view has its roots in early scientific theorizing about the sleep-stage correlates of dreaming. Early dream research was marked not only by the identification of dreaming and REM sleep but also by the assumption that NREM sleep mentation is more thoughtlike and REM sleep mentation more hallucinatory and dreamlike (e.g., Hobson et al., 2000; for discussion, see Nielsen, 2000a; cf. chap. 2). This distinction between NREM sleep thinking and REM sleep hallucinatory dreaming was paralleled by the assumption that thinking occurs less frequently in REM sleep dreams than in NREM sleep.

In his influential 1978 paper, Rechtschaffen defined single-mindedness as "the strong tendency for a single train of related thoughts and images to persist over extended periods without disruption or competition from other simultaneous thoughts and images" (Rechtschaffen, 1978, p. 97). While he allowed that cognitive activities such as judging, comparing, and evaluating occur in dreams (p. 99), he argued that these generally take place within the context of the dream. By contrast, he noted that instances of reflective awareness arising independently of the dream's content rarely occur in dreams. In this view, dream thinking is doubly constrained by the ongoing dream, one, in its content, and two, in the types of thinking that occur in dreams. An instance of the first constraint is the thematic consistency of dream thinking with the dream narrative, and an instance of the second is the lack of reflective awareness in dreams. This is similar to McGinn's claim that because dreaming is a form of imagining, and because imagining is a form of (attention-dependent) thinking, one's mind cannot wander away from the dream: as in imagining, one cannot think about other things within the dream or have concurrent imagery while dreaming without ceasing to dream (McGinn, 2004, pp. 32, 77–80).

For now, I want to step back from these theoretical claims and note that on a purely conceptual level, there is a difference between considering dreaming itself as a cognitive process (or as a form of thinking; cf. sec. 6.2.1) and the occurrence of conscious thoughts or cogitations within the dream.[9] Cognitive processing in the dream state (or dream cognition) refers to the cognitive processes involved in the construction of the dream narrative, the generation and bizarreness of dream imagery, the integration of memory sources, the incorporation of external stimuli, and so on. Thinking during the dream state (or dream cogitation) is something done by the dream self, or possibly by other dream characters (though in the latter case, it would not count as thinking in the phenomenological sense). Even if it should ultimately turn out that dream cogitation is closely constrained by dream cognition, this is an empirical question, and its investigation requires distinguishing the two, at least at the outset. A possible outcome could be that the distinction between dream cognition and dream cogitation ultimately turns out to be unnecessary because they are so closely linked that they can, once more, be collapsed into each other. This outcome, however, is not guaranteed.

To begin with, there is certainly a phenomenological difference between the cognitive processes underlying dream formation and dream cogitation proper. Even assuming, for the moment, that a strong form of the imagination view were correct and dreaming could properly be conceptualized as involving sensory and propositional imagination (rather than quasi-perceptual experiences and beliefs; cf. Ichikawa, 2009), it would still be important to investigate the relationship between the two—especially because imagination theorists disagree about whether dreaming involves sensory imagination, propositional imagination, or both (cf. McGinn, 2004; Sosa, 2007).

Hobson et al. (2000, p. 799) list nine characteristics of dreaming, five of which relate to what can broadly be considered the cognitive aspects of dreaming. Aside from being hallucinatory, bizarre, emotional, and often involving content reminiscent of instinctive fight-or-flight behavior, dreams are delusional; self-reflection is generally absent and when present is characterized by post hoc, illogical, and erratic reasoning; dreams lack orientational stability; dream events are integrated into a confabulatory narrative; and volitional control is attenuated. Note that some of these, such as narrative integration and orientational instability (as in the bizarre fusing of persons, times, and places) are instances of dream cognition and as such should not be confused with the claim that the dreamer *feels* disoriented within the dream or thinks about this fact. Other aspects refer to dream cogitation, such as the mistaken belief that one is awake, diminished self-reflection, and erratic reasoning. Importantly, Hobson et al. (2000) note that the attenuation of volitional control refers to the type of control seen in lucid dreams, but allow that "more mundane self-control of thoughts, feelings and behavior may be fairly common in dreams" (Hobson et al., 2000, p. 799).[10]

There is also a phenomenological and conceptual difference between cogitations performed by the dream self and the thoughts of other dream characters. Stumbrys et al. (2011) recently conducted a study investigating the arithmetic abilities of non-self dream characters in lucid dreams. Experienced lucid dreamers were instructed to ask other dream characters to solve simple mathematical problems. All things considered, the dream characters' arithmetic abilities were no better than those of primary school students. What is more interesting, however, is how the dream characters reacted to these questions. One participant reported that when he asked a dream character to calculate 18 – 6, she refused, complaining that the question was too private a matter. The participant found this answer both baffling and inspiring. Two other participants reported that in their dreams, dream characters, when prompted to solve mathematical problems, ran away. One even started to cry. At other times, the answers provided by dream characters were actually better than those anticipated by the participants themselves. One dreamer, who asked a dream character to calculate 15 / 3, reported that the dream character "gets back quite quickly with the answer '5.' It is funny as I [the dreamer] was thinking that the correct answer is '3'" (Stumbrys et al., 2011, p. 38). If we take these results at face value, then dream characters can seemingly have a mind of their own, reacting in unexpected ways and on occasion supplying (correct) answers, even when these answers momentarily elude the dreamer. This suggests that the phenomenological distinction between the thoughts of the dream self and those of other dream characters may be associated with underlying differences in cognitive processing, for instance, in the availability of different types of information to conscious cognitive activity.

Finally, both dream cogitation and dream cognition should also be distinguished from non-dream-related and perhaps even nonconscious cognitive processes during sleep. There is a wealth of research on sleep-related memory consolidation, and it has even been suggested that sleep be regarded as a state of cognitive enhancement (Diekelmann, 2014). There is also evidence that cognitive processing during sleep is sensitive to meaningful stimuli presented during sleep in a highly selective manner. For instance, participants are able to respond during all sleep stages to sounds by taking a deep breath; when told, before going to sleep, that this response will terminate the sound, response rates reach 90 to 100 percent (Harsh & Badia, 1990). In another study (Evans, 1990), participants reacted to verbal stimuli (such as instructions to scratch their noses or readjust their pillows) in 21 to 48 percent of all cases. According to EEG measures, they did not awaken, and their sleep was undisturbed. Neither study asked for dream reports, and subjects typically did not remember hearing the stimuli or responding to them in the morning, suggesting that at least some of these behavioral responses may have occurred independently of any ongoing phenomenal experience. Arzi et al. (2012) found that participants were able to learn novel associations between pleasant/unpleasant odors and sounds presented during sleep, as measured by the sniffing response. Following initially paired presentation of the stimuli, participants later also sniffed in response to the tones only. This behavior persisted into wakefulness and, once more, was not accompanied by awakenings or by any awareness that sounds or odors had been presented during sleep. There is also evidence that task-relevant information learned before sleep continues to be processed during NREM stages 1 and 2 and leads to covert preparation of motor responses even in the absence of overt behavioral responsiveness (Kouider et al., 2014). These studies demonstrate that not just selective and task-sensitive cognitive processing, but also intelligent behavioral responsiveness is possible during sleep. Yet, at least some (or perhaps even most) of this probably involves nonconscious cognitive processing that should not be confounded with dream cognition or dream cogitation.

As Evans (1990, pp. 77–78) points out, the extent to which complex cognitive processes including learned behavior occur during sleep is often underestimated. For instance, despite a considerable amount of body movement during sleep, one rarely falls out of bed, and habitual bed partners typically respect each other's territorial boundaries. This shows, once more, that the view of both sleep and dreaming as completely decoupled from external-stimulus input and behavioral output is overly simplified, and in some cases just plain false. But it also shows that cognitive processing in sleep, as measured by behavioral responses, is not to be confounded with dreaming or with thinking in dreams. To understand the subjective character of dream experience, it is of primary importance, first, to distinguish dream cogitation from dream cognition, and both of these from nonconscious cognitive processing and behavioral responsiveness during sleep, as well as, ultimately, from conscious but non-dream-related cognition during sleep (e.g., sleep thinking or stimulus perception). Second, one can then begin to systematically investigate how these different forms of cognitive processing are related to each other. Dream cogitation is, of course, a special case of dream cognition, and as such, it would be particularly interesting to know how conscious thinking in dreams contributes to the process of dream formation or, to the contrary, how it is itself driven by the cognitive processes underlying dream formation. Moreover, both dream cogitation and dream cognition are special instances of conscious cognitive processing during sleep, and both are distinct from nonconscious cognitive processing and behavioral responsiveness during sleep. These are simple conceptual distinctions, but they will prove to be enormously helpful in disentangling some common misconceptions about dreams in the literature and formulating new questions for future research.

While the distinction between dream cogitation, dream cognition, and non-dream-related or even nonconscious cognitive processing during sleep is primarily motivated by conceptual and theoretical considerations, evidence suggests that it is also empirically plausible. Fosse et al. (2001) investigated the prevalence of thoughts and hallucinations across the sleep–wake cycle. Daytime reports were gathered using a pager, and dream reports were elicited following timed awakenings. The prevalence of thinking was highest in quiet (90 percent) and active (79 percent) wake states, slightly reduced at sleep onset (64 percent) and during NREM sleep (38 percent), and reached a low point during REM sleep (22 percent). By contrast, hallucinations occurred only rarely during wakefulness and peaked during REM sleep. Fosse et al. (2004) found that the frequency of thought and hallucinations also changed throughout the night, with thinking being similarly infrequent in late-night NREM and REM sleep and hallucinations increasing in both sleep stages throughout the night.

On its face, this suggests that in REM sleep, cognition simply becomes less thought-like and more hallucinatory or imagistic: dream cogitation disappears with the overall change in cognitive processing associated with REM sleep. However, things are not that straightforward. In Fosse et al.'s (2001) study, thinking still occurred in one-fifth of dream reports; moreover, the occurrence of multiple thoughts was roughly similar across all states, suggesting that while the dream self is less likely to think than the waking self, when it does so, it is able to entertain multiple thoughts in much the same manner as during wakefulness. This means that the switch in cognitive processing related to the onset of REM sleep dreaming does not completely obliterate the occurrence of dream cogitation. Rather, in dreams, thoughtlike activity can coexist alongside dream-related imagery. In keeping with this hypothesis, the authors conclude that "variations in mentation seen across the wake–sleep states must be controlled not by one but by multiple underlying physiological factors" (Fosse et al., 2001, p. 35).

9.3.2 Thinking, Self-Reflection, and Metacognition in Dreams

Cogitation, as used so far, refers to the capacity to engage in a wide range of thought processes, such as planning, deciding, wondering, doubting, deliberating, supposing, and so on. An important question for the analysis of both lucid and nonlucid dreams is how thinking in dreams is related to self-reflection and metacognition. Being a cogitative self in the fullest sense involves not only mental reference but also, as Metzinger (2003, p. 396) notes, "the capacity to mentally 'ascribe' this act of reference to oneself while it is taking place." As Baker (1998, 2013) argues, this requires the ability to conceive of oneself *as* oneself, but also to conceptualize the distinction between oneself and others. Being a cogitative self cannot be divorced from the ability to engage in a particular kind of reference, namely, self-reference or self-reflection. This should not be taken to mean, however, that all forms of cogitating involve an *additional* act of self-reflection: thinking that *p* does not require additionally thinking that one is currently thinking that *p*. Clearly, a tight link exists between the phenomenology of thinking, which involves the feeling of ownership for one's thought processes, and the ability to think of oneself *as* the thinker of one's thoughts. Yet these are not the same, and not every act of thinking involves a further act of self-reflection. Whereas the sense of ownership for one's thoughts is nonconceptual (Metzinger, 2003, p. 399), self-reflection refers to explicit and conceptual forms of self-reference. The sense of ownership for one's thoughts is a feeling, whereas self-reflection is a form of thinking.

A first step toward investigating the cogitative dream self, then, is to distinguish different types of thinking, including self-reflection and metacognition. In particular, only some forms of cogitating are instances of self-reflection, namely, those that are directed at oneself, that is, at one's body, thoughts, emotions, wishes, and sensations, but also at one's personal history or plans for the future. Other cogitations, such as thoughts about persons, objects, and events, are not instances of self-reflection; they involve an awareness of oneself as a cogitative agent and a sense of ownership for one's thoughts but are directed at something other than the self. From now on, I reserve the term *thinking* for these types of cases. Finally, a special subgroup of self-reflection involves metacognition, or the ability to think about one's thoughts, wishes, plans, intentions, beliefs, and doubts. Metacognitive self-reflection is the special case in which self-reflection is directed at one's own cogitations or at one's intentional behavior. If thinking involves taking a step back to reflect on the world, metacognition involves taking a further step back to reflect on one's thoughts. As Flavell (1976) uses the term:

> Metacognition refers to one's knowledge concerning one's own cognitive processes or anything related to them, e.g., the learning-relevant properties of information or data. For example, I am engaging in metacognition if I notice that I am having more trouble learning A than B; if it strikes me that I should double check C before accepting it as fact. (Flavell, 1976, p. 232)

Metacognition has a monitoring component and an executive or agentive component, enabling one to take control of and optimize one's learning processes and bring one's actions into alignment with one's goals. Being a cogitative self in the fullest sense, then, involves the ability not just to self-reflect but to engage in a particular kind of metacognition, namely, thinking that one is currently thinking, and using this ability to monitor and optimize one's own actions. By contrast, thinking about one's body, about one's appearance in a photograph, and so on, are instances of self-reflection, but not of the metacognitive kind.

These distinctions invite problems when applied to dreaming, because in theorizing about dreaming, there is a tendency to equivocate between the cognitive processes involved in dreaming (or dream cognition) and explicit acts of thinking, self-reflection, and metacognition. To give an example, if dreaming is regarded as a form of thinking, then thinking in dreams (or dream cogitation) appears as an act of metacognitive self-reflection, because it is a case of thinking about one's own thoughts. For the same reason, the failure to distinguish dream cognition from dream cogitation threatens to obscure the distinction between self-reflection and metacognitive self-reflection in dreams: if dreaming is thinking, then when we engage in self-reflection (or indeed in thinking about anything at all in dreams), we automatically engage in metacognitive self-reflection.

There are, however, good reasons for upholding these conceptual distinctions, even in the case of dream cogitation. Although dreams, including all the characters, objects, and events experienced in them, are the product of cognitive processing, they are not part of dream cogitation proper; consequently, thinking about them is not an instance of self-reflection or metacognition in the sense introduced earlier. It may seem that all instances of dream thinking are instances of self-reflection because dreams depend, to a considerable extent, on internally generated stimuli. But unless one identifies the self with the brain, dreams are not generated by the self. The self, as I use the term, is always (and exclusively) a phenomenal self; the phenomenology of self-experience is the product of information processing in the brain and so should not be confused with the brain. But if this is correct, then there is no clear sense in which thinking about a dream character, for instance, involves thinking about oneself or one's own thoughts. During the dream state, the phenomenal self is the dream self, and in thinking about another dream character, one is not engaging in an act of self-reflection.

For the same reason, thinking in dreams is not always an instance of metacognitive self-reflection. In thinking about my appearance in a mirror in a dream, I am not engaging in metacognition. Certainly the mirror and my appearance in it are in fact the product of cognitive processing. But this does not turn instances of thinking about perceptual objects in dreams into instances of thinking about one's thoughts.

This point is related to the transparency of experience (see Crane, 2011), or the fact that in standard cases, we attend to the objects of perception rather than the process or act of perception itself. Only rarely, as in seeing double, do we become aware of the process of perception, and only then do we experience the perceptual world as constructed rather than as given. Because in these latter cases, one attends to and thinks about the process of perception itself, rather than about the objects of perception, they can be described as instances of metacognition. They involve thinking about one's ongoing perceptual experience *as* a cognitive process. But this is substantially different from thinking about the objects of perception, and the same is true for thinking about the objects or characters encountered in dreams. In order to be meaningful, the concept of metacognition should be reserved for cases in which we are not just thinking about what is in fact the product of a cognitive process (such as my visual experience of the coffee cup in front of me) but thinking about what we experience *as* our own cognitive processes.[11] Metacognition, as I use it, cannot be divorced from self-reflection, or the experience of thinking about oneself.[12] Failing to make this distinction leads to the assumption that all instances of thinking, either in dreams or in wakefulness, are in fact of the self-reflective, metacognitive kind.

These distinctions are important because they can help describe different kinds of dream cogitation and their variation across nonlucid and lucid dreams. It is often assumed, for instance, that self-reflection and metacognition do not occur in dreams at all, or that they mark the difference between nonlucid and lucid dreaming. Kahan (2001) gives a number of examples of researchers who argue that dreaming is marked by a loss of self-reflection and metacognition, and carefully argues against this view. Here I only discuss some particularly interesting examples.

Rechtschaffen (1978), as mentioned earlier, assumes that dreaming is inherently nonreflective. What he means is not that thinking does not occur in dreams, but rather that dreamers are typically unaware of their current sleeping environment and of the fact that they are dreaming. Only lucid dreaming, according to Rechtschaffen, involves instances of reflective awareness. By contrast, he likens nonlucid dreaming to psychosis because in dreams, we "lose so completely the road map of our own consciousness" (Rechtschaffen, 1978, p. 100). This view, however, rests on an equivocation between first- and third-person readings of reflective awareness. Rechtschaffen is correct to point out that by definition, nonlucid dreamers lack reflective awareness of the fact that they are asleep and dreaming. This means that they lack reflective awareness in a third-person epistemological or success sense. They are not aware of and hence unable to reflect on the fact that they are now dreaming. This conceptual point is not, however, particularly informative about the occurrence of self-reflection in dreams. In particular, it does not prevent nonlucid dreamers from enjoying the phenomenology of reflecting about their current situation even without becoming lucid. Reflective awareness, or self-reflection in a first-person, phenomenological sense, has no particular epistemic status, and dreamers can enjoy the phenomenology of reflective awareness while being profoundly deceived about their current situation, for instance, while thinking that they are awake. This type of thinking is an interesting target phenomenon, and the concept of self-reflection is helpful in picking it out.

Similarly, Weinstein et al. (1988, p. 445) argue that "the suspension of reflective self-representation is a specific correlate of REM phasic activity." They argue that the loss of self-reflection in nonlucid dreams causally contributes to the sense of presence in the dream and to the subjectively realistic, immediate quality of dream imagery. Their study, however, did not explicitly ask about instances of self-reflection in dreams; rather, it asked about the vividness of imagery and emotions and the overall realistic quality of the dream. Self-reflection in the dream was assessed only indirectly, for instance, if subjects used expressions such as "I was thinking that . . ." or "I was dreaming that . . ." in their dream reports. If they did not, this was rated as a loss of self-reflection.

This methodology has several problems. First, as Kahan (2001, p. 352) points out, subjects may underreport instances of thinking in dreams unless explicitly asked to describe them, so that self-ratings and affirmative probes may be necessary to more accurately assess the occurrence of self-reflection in dreams. Second, instances of thinking in dreams are not necessarily instances of self-reflection, and the study conducted by Weinstein and colleagues did not distinguish between the two. Third, the expression "I was dreaming" is a retrospective qualification of the dream and may not refer to cogitation during the dream at all. Fourth, seeing a loss of self-reflection as a corollary of self-participation in the dream, as Weinstein and colleagues propose, overlooks the fact that thinking can play a relevant role in, and even be a form of, self-participation in dreams. Conversing with other dream characters, for instance, is a form of self-participation that arguably involves at least some thinking, and the same is true for the typical dream of taking an examination. Thinking about one's actions can also play an important role in more active forms of self-participation in dreams, as in deciding whether to make a run for the train one is about to miss. Meumann reports a dream in which, while chasing after his train, he reflected on how often he had dreamt about the exact same situation, but reassured himself that this time he was surely awake (Jezower, 1928/1985, p. 306). Here we have an instance of metacognitive self-reflection coupled with active participation in a dream. By contrast, if one describes self-reflection and self-participation or presence in dreams as occupying opposite points along a continuum, one runs the risk of obscuring what is in fact an interesting target phenomenon, namely the interaction of these different cognitive processes in dreams. Indeed, the phenomenology of lucid dreaming suggests that the sense of presence in a dream can even be enhanced by the onset of lucidity (LaBerge, 1985; LaBerge & Rheingold, 1990), suggesting that even when taken in a third-person epistemological or success sense, self-reflection is not inimical to the sense of presence in dreams.

Purcell et al. (1986, 1993) commit the opposite mistake in their investigation of self-reflective awareness in dreams. They found self-reflection to be more prevalent in REM dreams than in NREM dreams and noted that it could take various forms, such as thinking about an idea or communicating with another dream character, noticing oddities while dreaming, being able to control the flow of dream events or even consciously reflecting on the fact that one is now dreaming. While these findings are certainly informative about thinking in dreams, they are not all instances of self-reflection. More importantly, several categories in their self-reflectiveness scale are not obviously related to thinking at all. For instance, observer dreams are not obviously more reflective than participation dreams but rank higher on Purcell and colleagues' scale of self-reflectiveness. Similarly, undergoing a transformation of body, role, emotion, or age need not involve more self-reflection than thinking about an idea or communicating with another dream character. According to their scale, the dreamer's presence in the dream already counted as involving some self-reflection, and the highest score of self-reflectiveness was reached in lucid dreaming. So while Weinstein et al. (1988) consider self-reflection to be the opposite of participation, Purcell et al. (1986, 1993) treat self-participation as a form of self-reflection independently of whether any thinking is involved. This view then easily leads to an equivocation between self-reflection and nonreflective forms of self-experience. Finally, because lucid insight into the fact that one is dreaming is seen as a form of self-reflection, it also leads to an equivocation between lucidity and self-reflection (e.g., Purcell et al., 1986; Darling et al., 1993). As we will see in the next section, there are, however, different degrees of lucidity, and not all are necessarily self-reflective.

What, then, is known about the incidence of different forms of cogitation in dreams? Kahn and Hobson (2005) asked subjects to rate the quality of their thinking in dreams and indicate whether it would have been similar in wakefulness, one, if the dream event had occurred in wakefulness (thinking *within* the dream), and two, concerning the occurrence of such an event (thinking *about* the dream). They conclude:

> Within the dream event cognition is often not qualitatively different from waking cognition no matter how outlandish the hallucinatory dream event may be. On the other hand, we are unable to detect how outlandish or bizarre the dream event may really be because our ability to access knowledge about how the world works is impaired. (Kahn & Hobson, 2005, p. 435)

This is an important step toward a more careful and differentiated assessment of thinking in dreams. In this view, dreamers are not generally irrational but systematically blinded to the fact that the events they are thinking about are only dream events. Dreamers are able to think rationally about events but are largely unable to notice the irrationality involved in the occurrence of such events (Kahn, 2007, p. 255). Kahn and Hobson explain this by saying that metacognition is lacking in dreams:

> Here, metacognition is both akin to its usual definition of "knowing that we are thinking," but takes on the special meaning of "knowing that we are hallucinating." In the dream state we are not aware that we are hallucinating. Our ability to have metacognition is state dependent. Only

when awake (or if we successfully transfer into a lucid dreaming state), are we able to distinguish between hallucinatory and non-hallucinatory thinking. We are able to step back and notice that indeed we have been hallucinating after daydreaming or after engaging in some fantasy. But during dreaming we are unaware that we are hallucinating, we cannot step back and reflect on the hallucination. (Kahn & Hobson, 2005, pp. 435–436)

The claim, then, is that metacognitive self-reflection is associated with dream lucidity. Note that this leaves room for saying, however, that metacognitive self-reflection occurs in nonlucid dreams, as well; it is only the special form of metacognition that involves thinking about hallucinations *as* hallucinations (or of dream imagery *as* dream imagery) that according to Kahn and Hobson's study does not occur in nonlucid dreams. One might also wonder whether dreaming really is all that different from wakefulness in our ability to realize that we are hallucinating. Surely the occurrence of hallucinations is much less frequent in standard wake states than in dreams, but when they occur, it is not clear that they are typically recognized *as* hallucinations.[13]

In a further study, Hobson and colleagues (2011) used an inferential reasoning paradigm to compare thinking in dreams to planned, focused waking thought. Three independent raters scored dream reports selected from Hobson's dream diary. Thoughts were rated as nonlogical if they made no reference to the preceding perceptual phenomena (e.g., if, in dreaming that one has taken on a tree pruning job, one does not wonder why one has done so, given that this does not at all fit one's professional profile). While their main finding was that dream thought is mostly impoverished and nonlogical (in the sense that unexpected percepts did not provoke rational thinking or logical analysis in the dream state), they also found some inferential reasoning to be present in dreams. They concluded that roughly one-third of dream thinking fit their definition of logical thought. Moreover, as Kahn clarifies, the dreamlike

combination of using and not using logic while dreaming also happens in the wake state. Nonlogic may occur when our minds wander, when we daydream, imagine, fantasize, become anxious, panic, or take mind altering drugs. In this sense, there is a continuum between wake and sleep states; daydreaming thoughts can be illogical just as dreaming thoughts can be logical. (Kahn, 2012, p. 79)

The evidence from these studies does not, then, merely highlight differences between dreaming and waking cogitation. Instead, it is best regarded, first, as pointing to differences between dream cogitation and a particular kind of waking thought, and second, as a first step toward investigating similarities and differences in thinking across sleep–wake states.

Indeed, several studies have found thinking to be relatively frequent in dreams. Snyder (1970) found that as many as 50 percent of dreams involve instances of decision making, and as many as 75 percent involve musings or silent observations about dream events. Kozmová and Wolman (2006) and Wolman and Kozmová (2007) found dreaming to support a wide range of cogitative activities and, together with several other authors, claim thinking to be similar in dreams and wakefulness (Fitch & Armitage, 1989; Moffitt, 1995; for further references, see Kahan, 2001). Dreamers can even carry on conversations with other dream characters in more than one language: Meier (1993) investigated the prevalence of speech in the dreams of bilingual subjects and found that in their dreams, they tended to show no clear-cut preference for a specific language, and in a majority of cases, the language used in the dream was appropriate to the context of the dream plot.

In a series of studies, Kahan and colleagues (Kahan & LaBerge, 1996, 2011; Kahan et al., 1997; Kahan & Sullivan, 2012) investigated the frequency of different forms of thinking across dreaming and wakefulness. The methodology was similar to the one employed in Fosse et al.'s (2001) study, using a beeper during the daytime and timed awakenings from different sleep stages. However, the results were markedly different, suggesting a considerably higher incidence of thinking in dreams than in Fosse's study. Generally, different kinds of cogitative activity were least frequent in NREM sleep, intermediate in REM sleep, and highest in wakefulness. This was the case for choosing; internal commentary; sudden as well as focused attention; and self-reflection on one's own thoughts (i.e., metacognition), feelings, and behavior. Thinking about external events was even more frequent in REM sleep (52 percent) than in wakefulness (42 percent), as was public self-consciousness (38 percent in REM versus 27 percent in wakefulness; see Kahan, 2001). As Kahan points out, a possible limitation is that the participants were experienced and novice lucid dreamers. Although the results do not come from lucid dreams, it is therefore possible that for those who do not typically experience lucid dreams, the incidence of thinking would be somewhat lower. Yet the overall trend of a high incidence of thinking in nonlucid dreams has been confirmed by more recent studies. Kahan and LaBerge conclude that "none of the high-order cognitive skills that we assessed is absent in dreaming and individuals reported the same range of cognitive activities in their dreaming and waking experiences, including the high-order cognitive activities" (Kahan & LaBerge, 2011, p. 508). According to the authors, this supports the view

> that there are stable similarities in the cognitive and sensory-perceptual processes across dreaming–waking. At the same time, our results show variation in the content and structure of reported dreaming and waking experiences. The latter findings suggest that episodic recollections of dreaming and waking experience are more similar in their process qualities (e.g., particular cognitive and sensory processes rated as having occurred during the experience) than in their structural or content qualities (e.g., locations, actions, reality orientation). (Kahan & LaBerge, 2011, p. 508)

The suggestion is that dreams—including dream thoughts—diverge from wakefulness in their contents more strongly than in the range of cognitive and cogitative processes.

Methodologically, the development of a questionnaire dedicated to investigating metacognitive, affective, and cognitive experience across the sleep–wake cycle is important progress (Kahan & Sullivan, 2012).

How do these studies reflect back on the hypotheses outlined at the beginning of the chapter? The no-cogitation hypothesis as a characterization of the phenomenology of engaging in cogitation in dreams is clearly false. What, then, about the cogitative-duplicate hypothesis? While Kahan and colleagues' findings support the view that no principled discontinuity exists between dream cogitation and waking thought, this is not to say that all varieties of thinking are continuously present in dreams: they likely fluctuate and, given the numbers cited earlier, can even be absent in some dreams. In particular, more conservative estimates suggest that thinking makes up only a small part of dream activities. In an analysis of five hundred REM dreams, Meier (1993) found that cognitive activities were mentioned in 20 percent of dream reports but only made up 5 percent of dream activities. This suggests that the dream self is not always a cogitative self, and different forms of thinking can be lacking in dreams. Of course, if the same were true of wakefulness, the cogitative-duplicate hypothesis could still be correct. For now, note that the plausibility of the comparative statement implicit in the cogitative-duplicate hypothesis depends both on the contrast condition (e.g., wakefulness in general, or only focused wake states) and on the range of dreams investigated.

Might phenomenal selfhood in dreams arise without cogitative selfhood? In light of the empirical evidence discussed here, this may well be possible. Havelock Ellis's dream of inspecting a gigantic tooth might be an example:

> I imagined that I was with a dentist who was about to extract a tooth from a patient. Before applying the forceps he remarked to me (at the same time setting fire to a perfumed cloth at the end of something like a broomstick, in order to dissipate the unpleasant odour) that it was the largest tooth he had ever seen. When extracted I found that it was indeed enormous, in the shape of a cauldron, with walls an inch thick. Taking from my pocket a tape measure (such as I carry in waking life), I found the diameter to be not less than twenty-five inches; the interior was like roughly-hewn rock, and there were sea-weeds and lichen-like growths within. The size of the tooth seemed to me large, but not extraordinarily so. (Ellis, 1911/1931, p. 40)

Note that the dreamer is here participating in the dream and even engaging in seemingly purposeful behavior (e.g., measuring the tooth). However, he makes no mention of surprise or wonder; the only explicit cogitative activity mentioned is the remark, at the end, that the tooth did not, after all, seem extraordinarily large. Affirmative probes would be needed to decide whether this is an instance of underreporting cogitative activity, and in their absence, the conservative approach recommends caution. Yet, I find this report interesting for another reason. In particular, note how the dream imagery creatively and dynamically seems to be doing Ellis's thinking for him. The dream-tooth responds to the dentist's remark about its unusual size and unpleasant odor by appearing, to Ellis, as a gigantic, cauldron-shaped thing, reminiscent perhaps of a cave or an underwater rock. Because there is little evidence of overt cogitation throughout the episode, it is almost as if the process of dream formation was overriding or even preempting any explicit and deliberate acts of thinking on the part of the dream self. As evidenced by Ellis's closing remark that the tooth's size did not seem particularly large to him, dream cogitation seems to lag behind dream cognition.

This observation is further supported by the characteristic instability of dream thinking. Dreamers are easily distracted by dream events and often fail to notice discontinuous jumps in dream narration. This is illustrated by the following report:

> I am in a plane; my wife is also in the plane, seated farther back. The plane lands at Logan Airport. Then the plane continues to go on before I can get off. But now it is a train. I wonder how far the next stop is from Logan. The train, which now is more like a subway, stops, but I miss getting off. I try going from car to car but find that I am in a sort of large empty supply-like room on the train. I wonder if my wife is worried. I finally get off and I am about to call her on my cell phone when I see that there is a call coming in that requires me to hit the answer button. (Kahn & Gover, 2010, pp. 187–188)

Throughout the dream, the dreamer maintains his preoccupation with getting off, getting back to Logan, and contacting his wife. Yet he is surprisingly unconcerned by the transformation from plane to train to subway. Indeed, his increasing preoccupation with getting to Logan and contacting his wife is progressively thwarted by the dream, as if it was confronting him with an externalized, visuospatial version of his ongoing worries. At the same time, the dream is not devoid of thinking and remembering, and the dreamer's goals remain consistent throughout the dream. This is sufficient for saying that a cogitative—though not necessarily a rational—self was present. As Kahn and Gover (2010) note, the dreamer, while remaining oblivious to the overall bizarreness of the situation, is still able to respond to it with at least a modicum of logical thinking (e.g., wanting to use a cell phone to contact his wife). Perhaps the progressive transformations of the dream preempt the dreamer's ability to respond in a more calm and reflective manner and realize what is really happening (i.e., a dream).

The dream report is also interesting because it hints at the deceptive quality of dream cogitation: in waking life, the dreamer is not married at all (Kahn & Gover, 2010, p. 188), so the dream self not only forgot details about his waking life but also enjoyed a somewhat deviant autobiographical memory. Yet note that even here, the way in which the dreamer's goals are thwarted is not wholly inconsistent: after all, the dream prevents him from contacting a wife that he does not in fact have, just as it prevents him from reaching a destination—Logan Airport—which in fact is likely not the place where he is now sleeping. Dream cogitation is slightly off-key, and perhaps dream cognition is subtly acting as a corrective. Perhaps dream cognition and dream cogitation, then, can be described as accessing different sets of autobiographical and situational information.

The dreamer's concern with his dream-wife brings us to another important point, already hinted at earlier. This is that dreamers are not only capable of thinking about the dream world and about themselves but are also able to engage in conversation with others. Thinking about and conversing with others, but also attributing beliefs, desires, hopes, and intentions to them, are important forms of social interaction. McNamara, Harris, and Kookoolis (2007; see also McNamara, McLaren, & Durso, 2007) obtained samples from wakefulness, REM sleep, and NREM sleep over a two-week period. They found that theory-of-mind attributions were even more frequent in REM dreams than in wakefulness, probably reflecting the social character of dreaming. In REM sleep, such attributions were mostly associated with aggressive interactions initiated by the dreamer, whereas in NREM dreams, theory-of-mind attributions were generally less frequent, but associated with dreamer-initiated friendly interactions. It seems not only that the dream self thinks about other dream characters, but that social cognition in dreams is also related to the initiation of different forms of social interaction. Social cognition and thinking about the minds of others thus play an important and unique role in dream cogitation.

9.3.3 Lucid Insight, Dream Control, Just Knowing, and the Strange Familiarity of Dreams: Toward a Graded View of Nonlucid and Lucid Dreams

Lucid dreams are often described as the opposite of nonlucid dreams in all matters related to thinking and the phenomenology of cogitative selfhood. Lucidity, in this view, involves wakelike cognition and is distinguished from nonlucid dreaming along a number of dimensions. According to Tholey and Utecht (1995/2000, p. 61), lucidity is defined not just as *knowing* that one is currently dreaming but also as having insight into the ability to control one's own actions and the events and characters in the ongoing dream; wakelike functioning of all the senses (including rare sensory experiences such as smell and taste); and memory of waking life and previous lucid dreams. In this view, lucid insight is not just an altered but a higher state of consciousness. The same type of strong definition is given by van Eeden (1913), who originally coined the term *lucid dreaming* (cf. sec. 3.5).

As Metzinger (2003, p. 532) points out, realizing that one is dreaming can be understood in two different ways. One, it can be an intellectual achievement; that is, one can reflect on the improbable nature of dream events, conclude that one is dreaming, and subsequently use this knowledge in thinking about and even actively controlling the dream. This type of lucidity is clearly related to dream cogitation. The following dream report offers an example of lucidity as a cognitive achievement, or C-lucidity (Windt & Metzinger, 2007; see also Noreika, Windt, Lenggenhager, et al., 2010):

> From the top of a fairly flat unknown mountain I look out over a wide plain towards the horizon. The thought occurs to me that I do not know at all what time of day it is. I examine the position of the sun. It is almost vertically overhead in the sky and its usual brightness. This surprises me, as it occurs to me that it is already autumn and a short time ago it was much lower in the sky. I think: "The sun is now standing vertically at the equator, so here it must be at an angle of about 45 degrees. Therefore if my shadow is not equal to my own height, I must be dreaming." I look at it: it is about 30 cm long. It costs me a fair effort to regard the whole almost dazzlingly bright landscape with all its villages as an illusion. (Green & McCreery, 1994, pp. 17–18)

But lucidity does not always involve explicit reasoning of this type. Dreamers often report becoming lucid because the dream took on an unrealistic, dreamlike quality. In these cases, the realization that one is dreaming may better be described as a *feeling* than as a cognitive achievement; lucidity in this sense is an experiential quality. According to Metzinger (2003, p. 532), this second type of lucidity results when the fact that one is dreaming becomes available to attention; in keeping with this view, Windt and Metzinger (2007) call this A-lucidity. As a result, the dream is no longer experienced as mind independent but as constructed.[14]

> I was standing in a field in an open area when my wife pointed in the direction of the sunset. I looked at it and thought, "How odd, I've never seen colors like that before." Then it dawned on me: "I must be dreaming!" Never had I experienced such clarity of perception—the colors were so beautiful and the sense of freedom so exhilarating that I started racing through this beautiful golden wheat field waving my hands in the air and yelling at the top of my voice, "I'm dreaming! I'm dreaming!" (LaBerge & Rheingold, 1990, pp. 1–2)

Note that in this dream report, a certain level of cognitive insight is present, as well: if it were not, the dreamer could not conceptualize the experience as a dream, and it could not, retrospectively, be scored as a lucid dream. Yet cognitive insight seems to be driven by an experiential change, a feeling that this is a dream that does not require a separate reasoning process. So while full lucidity will involve both A- and C-lucidity, an important point is that cognitive insight into the fact that one is dreaming can be reached without any elaborate cognitive reflection or reasoning process. The situation may then better be described as *just knowing* that one is dreaming, where knowing refers to an epistemic feeling rather than to a cognitive or intellectual achievement.

A-lucidity as a precursor to full lucidity (including C-lucidity) is not, then, itself an instance of dream cogitation. In many cases, it is a causally enabling condition for C-lucidity, and thus for a particular kind of cogitative process involving the ability to conceptualize the ongoing experience as a dream, as in the dream report cited earlier. Yet the simplest forms of lucidity may not form part of dream cogitation proper at all.

According to a study conducted by Gackenbach (1988, p. 193), the onset of lucidity is associated with a "dreamlike sense" in 48 to 67 percent of lucid dreams. By contrast, the recognition of an incongruent element triggers lucidity in only 11 to 19 percent of all cases, and 18 percent of lucid dreams arise out of nightmares. Several authors note that especially for trained lucid dreamers, the difference between "tacit" and full lucidity may become increasingly blurred. Experienced lucid dreamers may simply recognize the dream state by sheer familiarity, and lucidity may even be initiated without the explicit realization that they are dreaming; they feel that they are dreaming without conceptualizing this fact. In such cases, deciding whether or not a dream was actually lucid on the basis of a dream report may be ambiguous, and subjects may even be uncertain as to whether or not a given dream of their own was lucid (Worsley, 1988, p. 341; Brooks & Vogelsong, 1999, pp. 26–31; LaBerge & DeGracia, 2000). A preliminary conclusion is that the differences between lucid and nonlucid dreams, but also between cogitative and noncogitative forms of lucidity, are gradual rather than absolute.

If the simplest forms of lucidity are not yet sufficient for the emergence of a cogitative self, how can we make sense of the claim that dream lucidity involves metacognitive insight? After all, even the simplest form of A-lucidity involves awareness that one is dreaming, even if this insight cannot yet be conceptualized. One way of making sense of the claim that lucidity can be a form of metacognitive insight even if it does not involve cogitation is to say that not all forms of metacognition involve metacognitive self-reflection. Clearly there is a type of metacognition that involves conscious and explicit acts of thinking, and metacognitive self-reflection of this kind forms part of cogitation proper. However, the same need not be true for all instances of metacognition, and one might argue that metacognitive (or epistemic or noetic, as they are also sometimes called; cf. Koriat, 2000; Dokic, 2012) feelings exist that are of a nonconceptual, subsymbolic nature and arise from unconscious forms of metacognitive monitoring. These include, for instance, feelings of knowing and of certainty, feelings of doubt, tip-of-the-tongue phenomena, feelings of familiarity, and déjà vu. Dokic (2012, p. 303) argues that metacognitive feelings are "'seeds' of self-knowledge, i.e., can provide knowledge or justified beliefs about one's own mental and epistemic life." They are metacognitive, according to Dokic (p. 310), first, by yielding "information (or misinformation) concerning one's own epistemic states, processes, and abilities," and second, by involving "implicit *monitoring* mechanisms that are sensitive to non-intentional properties of first-order cognitive processes." They can be exploited in reasoning, but importantly, metacognitive feelings are not necessarily metarepresentational in that they do not require the ability to form a concept of or self-ascribe mental representations. Such metacognitive feelings arise, in other words, independently of conceptual capacities and cogitation proper.

An advantage of this distinction between metacognition and metarepresentation—or between metacognitive feelings and metacognitive self-reflection—is that it suggests an account of how self-representational capacities can gradually be acquired, for instance, in animals and small children (Proust, 2007, 2013). According to this view, metarepresentation is just the tip of the iceberg, a small and conceptually particularly obvious part of much broader metacognitive capacities. For the same reason, however, the distinction can also fruitfully be applied to lucid dreaming. In particular, we can now describe A-lucidity as a metacognitive feeling arising independently of the ability to conceptualize the experience as constructed, as a dream. Given the empirical evidence, it even seems plausible to say that in many (or even most) cases, metacognitive feelings of this type are precursors to and causally enable the emergence of conceptually mediated forms of lucid insight. These metacognitive feelings are themselves, however, noncogitative or, perhaps better, precogitative. In investigating how dream cognition gives rise to dream cogitation and full-blown lucidity, focusing on such precursors to dream cogitation, including metacognitive self-reflection, might be particularly useful.

Moreover, this type of analysis is not restricted to dream lucidity but can be applied to nonlucid dreams, as well. In particular, cases in which lucid dreamers just know that they are dreaming resemble other forms of recognition in nonlucid dreams, such as character recognition. Kahn et al. (2000) note that dream characters are often recognized by just knowing, rather than by their visual appearance or behavior. This can also lead to instances of misidentification, as in dreams of seemingly recognizing a close friend or family member despite the lack of any resemblance. In a follow-up study, Kahn et al. (2002) suggest that character recognition in dreams is often feeling-based, meaning that the identity of dream characters is felt independently of appearance or behavior. It is a form of identification arising independently of recognizing facial features, which can possibly be explained through disturbed connectivity in the cortical areas dedicated to face processing (Schwartz & Maquet, 2002). This explanation has also been suggested for misidentification in neuropsychological syndromes such as Fregoli or Capgras syndrome (for discussion, see Ratcliffe, 2008; for a case study of the Fregoli delusion, see Ellis & Szulecka, 1996). Gerrans (2014) has recently argued that hyperfamiliarity for persons and places in dreams is importantly related to delusions of misidentification and déjà vu (see also Schwartz & Maquet, 2002).

For now, I want to tentatively suggest that these instances of just knowing or hyperfamiliarity in dreams might be importantly related to the kind of metacognitive feeling that characterizes A-lucidity. The idea is that the feeling of just knowing who a dream character is—or just knowing that this is your apartment though the dream version looks nothing like its waking counterpart—results when the self-generated or constructed character of the ongoing experience becomes available to subsymbolic dream cognition, but not in a way that is sufficient to generate lucidity. More precisely, the feeling of just knowing results when ongoing unconscious monitoring of the self-generated nature of experience is misinterpreted as familiarity not for the process of imagery construction but for the dream's content, for the persons, places, or objects it represents. The feeling of just knowing is metacognitive to the extent that it arises from monitoring of self-generated information and concerns one's epistemic states (cf. Dokic, 2012): the feeling of recognizing, for instance, a person or place, even in the absence of any obvious resemblance. Feelings of familiarity and misidentification in dreams can then be redescribed as metacognitive feelings that on the level of overt dream cogitation are misinterpreted as just knowing the identity of a dream character, recognizing a place, or knowing that one has to reach certain goals. Dream cognition, in this view, has much to do with integrating the (misinterpreted) deliverances of metacognitive monitoring of self-generated imagery into a narrative context.[15]

This account is, of course, highly speculative, and many details remain to be filled in. For instance, how can the specificity of just knowing in character recognition in dreams be explained? Why do we often not just feel, in dreams, that certain dream characters are familiar, but additionally have the feeling of knowing who they are? And why is just knowing seemingly more likely to be associated with some aspects of dreaming (such as character recognition) than with others? Revonsuo and Tarkko give a possible explanation:

> We may interpret the abundance of pseudo-familiar persons in dreams as an overactivation of face recognition units: they fire even when the face percept does not in fact correspond to any of the descriptions in the store. This creates the inappropriate feeling of familiarity. If the face recognition units fail to activate any semantic knowledge connected to the visual representation of the person (since none in fact exists), then the dream person remains an unspecified "friend" or "acquaintance"; experienced as very familiar, but still nobody in particular. (Revonsuo & Tarkko, 2002, p. 17)

This account sits well with the occasional phenomenal indeterminacy of dream imagery (cf. chaps. 6, 7); in a sense, a feeling of familiarity for a generic dream character involves similar indeterminacy on the semantic level. How, then, to explain the occasional specificity of semantic knowledge activated in response to generic or indeterminate dream characters? Revonsuo and Tarkko continue:

> If some representations at the semantic level are nevertheless activated, the pseudo-familiar person will be falsely identified as a particular person whose identity may be either completely confabulated or partly borrowed from some real-world person(s) whose identity nodes have become inappropriately activated. (Revonsuo & Tarkko, 2002, pp. 17–18)

They suggest that the hyperassociative and emotional nature of dream cognition might explain this overactivation of face recognition units. To this, the present account adds that misinterpreted metacognitive monitoring might provide a mechanism that drives this process. In particular, the best explanation for a feeling of familiarity is that it is related to a specific person—one is not, after all, typically acquainted with generic persons. The activation of semantic knowledge may then be driven by the attempt to make sense of the initial feeling of familiarity, where this attempt may be largely unconscious but occasionally might also be driven by consciously wondering who the seemingly familiar person actually is. More detailed phenomenological accounts might help fill in the details. For now, the idea is that the specificity of such instances of just knowing rides on, and perhaps is prompted by, processes of largely unconscious metacognitive monitoring.

If anything like this account is correct, then it leads to a surprising and unexpected insight: aspects of dreaming that are often described as involving a breakdown in metacognitive monitoring and reality testing (for critical discussion, see Gerrans, 2014) such as feelings of familiarity and just knowing can actually be placed on a continuum with lucid insight. These superficially very different examples can be redescribed as involving different degrees of metacognitive access to the self-generated nature of dream imagery.[16] A weak and highly modularized form of this access can be found in feelings of familiarity in nonlucid dreams; a stronger and more global one in A-lucidity; and, finally, an even stronger form in conceptually mediated forms of full lucidity. Lucidity may just be the special case in which the feeling of familiarity is restricted not to a particular type of dream content—such as persons or places—but spreads to the process of dreaming itself, thus enabling the dreamer to hit on the right explanation of this strange feeling, namely, that all of this is a dream. This is a new way of making sense of the claim that dream cogitation lags behind dream cognition: metacognitive monitoring shapes the construction of dream imagery long before it becomes available on the level of dream cogitation or overt, metacognitive self-reflection.

Gregory's (1923) discussion of memory in dreams lends historical context to this idea. Dreams, as he notes, are often examples of cryptamnesia: they are fed by memories that in the dream are experienced as reality. Indeed, the mnemonic sources of dreams are typically, if at all, recognized only upon awakening. This phenomenon of unrecognized memories experienced as reality in dreams can be described, in the current model, as resulting when metacognitive monitoring of the sources of self-generated dream cognition is absent or greatly diminished. In these cases, access to the mnemonic sources of dream imagery is minimal: what is in fact a memory will be experienced as real. The hypermnestic character of dreams often manifests in this manner, namely, when memories, long forgotten in wakefulness, are accessed in the dream state and reflected on the level of dream content. Freud relates several such cases in which he would dream of a building that only later and by chance he discovered to have been an unidentified memory from long ago (e.g., the memory of a particular church tower seen years earlier on a train ride; see Freud, 2003, p. 31). Gregory even goes so far as to say that dreaming "is obviously, constantly and vividly a periodic cryptamnesic visitation: a familiar recurrence of recollection posing as perception" (Gregory, 1923, p. 29). Dreams also, however, occasionally involve instances of paramnesia, "a confusion of mind, a doubtfulness whether he were remembering the past or perceiving the present" (p. 30). He adds that "in many such dreams the paramnesic moment is clearly distinguishable before the dreamer definitely realizes that his dream is a dream" (p. 30). Again, we have a metacognitive feeling that this might be a memory rather than a perception acting as a precursor to lucid insight, once more suggesting that a low-level change in metacognitive monitoring may be a precursor to conceptual, metacognitive self-reflection. *Déjà rêvé*, or the impression of having dreamt this before (which may or may not be veridical), might be yet another example (cf. Funkhouser & Schredl, 2010). At earlier stages, there might simply be the feeling that the current events are strangely familiar, or a passing awareness, for instance in a horrific nightmare, that this is not really happening, that everything will be OK in the end.

One consequence of this analysis is that it makes the interpretation of even strong lucidity as an intellectual *achievement* in any strong sense appear questionable. Rather, C-lucidity increasingly looks like an occasional side effect of more frequent instances of subsymbolic metacognitive monitoring and metacognitive feelings in dreams. If many instances of lucidity can really be described as just knowing that one is dreaming, and if lucidity is frequently brought on by a change in the experiential quality of the dream rather than by a conscious reasoning process, then even C-lucidity does not necessarily involve the presence of a cogitative subject in any strong sense. Knowing that you are dreaming requires the ability to think that you are dreaming, and this involves some of cogitative abilities, but this is not enough to describe lucidity as a sophisticated intellectual achievement.

If the realization that one is dreaming can be relatively isolated and does not require a profound change in dream cogitation as compared to nonlucid dreaming, this would explain why thinking in lucid dreams often remains substantially dreamlike and is marked by frequent cognitive lapses (or "lucidity lapses"). The most common ones, even for experienced lucid dreamers, arise because dreamers do not grasp the full consequences of the fact that they are dreaming. They continue to believe that the dream is somehow related to waking life, that their dream actions have an effect on reality, or that dream characters have some kind of objective existence—for example, that they are sharing a dream with other lucid dreamers (Levitan, 1994). They might feel embarrassed by their actions or be afraid of engaging in "dangerous" actions, such as cutting themselves with a knife (Green & McCreery, 1994, pp. 28–29). A particularly charming example of such a lucidity lapse is provided by Mary Arnold-Forster. A gentlewoman even in her dreams, she noted that in her flying or gliding dreams,

> I find myself always now in my "flying dress"—it is a dress of straight close folds which fall three or four inches below my feet. The reason for this is that, once or twice when I have been moving among crowds in busy streets, gliding just clear of the ground, I thought that people must notice that my feet never move like theirs. In Oxford Street one day, when the pavement was very crowded, I feared that it would attract attention disagreeably if people noticed this fact, and that the curiosity of it might lead to an inconvenient amount of notice. I left the thronged thoroughfare and went quickly down Duke Street to avoid observation. Once in the quieter street, I flew along comfortably, but the next time I began to dream I found that I wore a long dress which hid my feet entirely, and no one can now see that I am not walking just as they are. (Arnold-Forster, 1921, pp. 74–75)

Such lucidity lapses are interesting because they reveal some of our most deeply ingrained intuitions and concerns; furthermore, it appears that some of these reactions are almost impossible to overcome. Aside from dangerous actions, social situations appear particularly prone to invite lucidity lapses:

> The next thing I know, Steve and I are looking up at the sky and there's a white parachute coming down—as it gets closer, I can see two people on the chute—one has skis on and is doing flips. I'm wondering aloud to Steve how this is possible and explicitly say, "This must be a dream—we're dreaming—this is a lucid dream! We're both in the same lucid dream." I waited for Steve to come to the realization that he was dreaming (i.e., the logic was that we're in the same dream because we each put ourselves there, not because I, the dreamer, had constructed this experience). I wanted Steve to write down that this was a dream so we'd remember. (Levitan, 1994)

Note that these lucidity lapses do not necessarily set dream thinking apart from waking thought—to the contrary, they involve exactly the same type of thinking that would occur were such events encountered in wakefulness. In wakefulness, one might well be excited about sharing an unusual experience with one's friend and ask the friend to take notes on it; and one might also well be concerned about bystanders looking up one's skirt while floating through the streets of London. Recall, in this context, Voss and colleagues' (2013; cf. sec. 3.5) finding that realism did not significantly distinguish lucid and nonlucid dreams. In our study, participants noted that emotions and thoughts in their dreams were much the same as they would be in wakefulness, and this was largely the same for lucid and nonlucid dreams. It is only in the context of a lucid dream that these count as cases of irrationality or lucidity lapses at all.

Memory, even for the most basic facts, does not appear to be reliable in lucid dreams and in the study by Voss and colleagues (2013) did not distinguish strongly between lucid and nonlucid dreams. Some lucid dreamers report being able to remember, within the dream, the activities of the previous day or where they are sleeping, but have trouble recalling other facts about their waking lives (Brooks & Vogelsong, 1999, pp. 124–126). There are counterexamples. For instance, Erlacher (2009) found that lucid dreamers were able to recall world lists learned during wakefulness in their dreams. Generally, however, when waking memories are retrieved within lucid dreams, they are often related to the context of the dream plot and occasionally turn out to be false (Gillespie, 1991; Brooks & Vogelsong, 1999, pp. 121–123). For instance, a lucid dreamer may "remember" that he is sitting at his desk while he is in fact asleep in bed (consequently reliving Descartes's nightmare of dream deception; Green & McCreery, 1994, p. 107). Other lucid dreamers report encountering difficulties when trying to perform cognitive tasks, such as solving relatively simple mathematical problems or continuously counting into the hundreds (Gillespie, 1988b), suggesting that the required attention span exceeds their abilities. This supports the view that complex lines of reasoning and logical thinking are difficult in dreams, whether lucid or not (see also

Hartmann, 1998, 2000). Similarly, Stumbrys and Daniels (2010) found that lucid dreams support creative rather than logical forms of problem solving.

One might want to say that dreams involving lucidity lapses, false memories, or faulty reasoning count against saying that the dream was fully lucid and that the term *lucidity* is best reserved for cases in which the dreamer not only realizes she is dreaming but realizes the full consequences of this fact. Metzinger (2003) and Windt and Metzinger (2007) promote such a strong concept of lucidity. If, however, the present account is correct, then this requirement may be too strong: thinking that one is dreaming, aside from its content, may not be all that different from other thoughts occurring in dreams, and it may be continuous with metacognitive feelings that do not count as instances of dream cogitation at all. In particular, as frequent examples of lucidity lapses show, thinking that you are now dreaming does not require the ability to further think about this fact. This would mean that dream lucidity would be related to having a thought with a specific content—namely, the thought that one is currently dreaming—but would not necessarily require a change in the overall type of cognitive processing.

Moreover, these kinds of lucidity lapses are compatible not only with conceptualizing one's ongoing experience as a dream but also with using lucidity to carry out prearranged experiments, as in the following dream reported by Worsley:

After a few moments, perhaps half a minute, I signalled lucidity onset with 5 eye movements. I searched for something to draw triangles on. There were several pieces of paper but most of them looked too small and had things written on them. I shouldn't use them, I thought, in case they were somebody's valuable notes. . . . I walked through two . . . rooms and found a blackboard with writing on it and some red chalk. I hesitated to use the blackboard, again in case somebody had left the writing there for further use. Nevertheless, I started to draw the triangles on the blackboard. The chalk seemed greasy and didn't write very well, which momentarily concerned me. . . . I tried to mark each attempt to draw a triangle with eye flicks: one flick for the first attempt, two flicks for the second, and so on. (Schatzman et al., 1988, p. 157)

Other times, a lucid dreamer may deliberately suspend disbelief in the reality of dream events to facilitate dream control:

I forgot how the dream started, but by lucid-dream standards it went on for a long time—maybe as long as 5 minutes. . . . While I was getting organized, Morty [Schatzman] wandered off. He went to look at a poster on which someone had made little drawings of a rather relaxed uninhibited woman. It seemed as if he was looking for something to occupy his mind while waiting. "Come on, Morty," I said, "you've got to watch me to see that I do these numbers right." I thought that request was a good thing to do to fill in the situation, that is, to build up the richness of the dream, as well as to make sure, under the eyes of the "master," of doing it right. "Come on," I said. "Pay attention. You're not the real Morty, so don't think there's any point in doing otherwise." He grinned. I drew the first series, I think, on top of a pile of new paving stones at the end of this cul-de-sac we were in. (Schatzman et al., 1988, pp. 171–172)

Brooks and Vogelsong (1999, pp. 92–95) note that the suspension of disbelief is both dangerous and helpful in exerting dream control. Going with the flow of a dream—for instance, checking for a bicycle in the garage rather than simply willing it to appear—can make dream control more effective. By contrast, changes evoked by mere willpower not only are more difficult to generate but may also be experienced as literally thought produced and less real. Generally, lucidity depends on the active maintenance of the dream plot. Uncontrolled dreams may simply fizzle or turn into nonlucid dreams (Brooks and Vogelsong, 1999, pp. 106–107). Lucid dreaming thus appears as a balancing act between active participation and critical, reflective distance (LaBerge, 1985, pp. 104–108). These examples show that the relationship between dream control and lucid insight is complex and reciprocal. One the one hand, lucid control dreams are an example of dream cogitation in the form of overt insight and deliberate control partly overriding and shaping lower-level dream construction; on the other hand, lucidity lapses and the deliberate suspension of disbelief may facilitate the maintenance of lucidity and of dream control.

Another reason for considering dreams involving lucidity lapses as lucid is that they likely represent the bulk of lucid dreaming, whereas full lucidity, in which lucid awareness combines with full intellectual clarity, constitutes a merely "hypothetical maximum" (Malamud, 1988, p. 309; Brooks & Vogelsong, 1999, pp. 29–30). Moreover, because lucidity lapses not only are compatible with, but sometimes may even facilitate, dream control, it is likely that most lucid dreams in laboratory studies of lucid dreaming are less than fully lucid.

This is supported by recent findings on the continuity between nonlucid and lucid dreams. Recall that Voss et al. (2013; cf. sec. 3.5) found that thinking and memory, but also certain aspects of insight, occurred in nonlucid dreams, as well. In another recent study, Dresler et al. (2014) investigated one specific aspect commonly thought to characterize lucid dreaming and wakefulness but to be absent in nonlucid dreams: volition. Ten frequent lucid dreamers were asked to complete a questionnaire assessing different components of volition in lucid and nonlucid dreams and in wakefulness. The results were as interesting for what they found as for what they did not. Self-determination, or the ability to act freely according to one's will, was higher in lucid dreams and wakefulness than in nonlucid dreams, confirming the commonsense expectation that volition, as typically understood, is less pronounced in nonlucid dreams. Planning ability, or the ability to act in accordance with plans and intentions, was high for wakefulness but low for lucid and nonlucid dreams; however, intention enactment, assessing how promptly intentions are executed, was highest in lucid and comparable in wakefulness and nonlucid dreaming. In keeping with existing findings on dream control (Stumbrys et al., 2014), Dresler and colleagues suggest that this might indicate that lucid dreamers are aware, while dreaming, of being able to enact their plans and intentions independently of real-world constraints, but mostly enact spontaneous intentions. Indeed, anecdotally, the onset of lucidity is sometimes described as being accompanied by a feeling of freedom; at the same time, lucid dreamers often forget their waking intentions or fail to execute those that they do recall. Strangely, neither powers of concentration (how easily subjects get distracted from a line of thought) nor self-access (to intentions and feelings in stressful situations) nor integration (the occurrence of seemingly contradictory thoughts and emotions) differentiated between wakefulness and nonlucid and lucid dreaming. These results are surprising in view of the common assumption that cognitive abilities are generally impaired in nonlucid dreams (e.g., Hobson et al., 2000). Again, the important message is that the differences between sleep–wake states, but also between lucid and nonlucid dreams, may be less pronounced than commonly assumed. Clearly the results are preliminary, and more detailed research is needed. Still, Dresler and colleagues' study takes an important methodological step toward a report-based contrastive analysis of changes in experience across wakefulness and lucid and nonlucid dreaming.

All of this suggests that lucid dreaming, understood only as realizing that one is dreaming, cannot be sharply distinguished from nonlucid dreaming in terms of the types of cogitation involved. Nonlucid dreams involve a wide range of cogitative activities, including self-reflection and metacognition, and lucid insight into the fact that one is dreaming does not, in itself, guarantee the reliability of cogitation and memory more generally. The balance of lucid insight, dream control, and cognitive clarity may determine the overall degree of lucidity, and accordingly one might distinguish different degrees of lucidity. In a fully lucid dream, all three of these factors may combine; however, in the vast majority of cases, lucid insight into the dream state seems to exist alongside naive realistic beliefs about the events and persons in the dream. Thus the only feature that reliably distinguishes lucid and nonlucid dreams is subjective insight into the dream state. While there is no question that lucid dreams are fascinating phenomena, it may be more helpful, in trying to understand the various forms of thinking in dreams, to highlight their continuity with nonlucid dreams. In particular, it would be a mistake to restrict the study of dream thinking (or dream cogitation), or even of metacognition and metacognitive self-reflection in dreams, to the study of lucid dreaming.

Why, then, is lucidity nonetheless interesting? It is interesting, or so I would claim, because it is the rare case in which largely automatic metacognitive monitoring gives way to explicit, conceptually mediated insight into the fact that one is dreaming, as well as, often, the ability to control the dream. C-lucidity, however, can coexist alongside widespread lucidity lapses, and it is only a small—indeed, a very small—part of cogitative activity in dreams. Yet, by investigating the relationship between different degrees of lucidity, one can begin to disentangle the relationship between dream cogitation and dream cognition. The same is also true for dream cogitation more generally: thinking, both in nonlucid and in lucid dreams, is only the tip of the iceberg, and beginning to relate instances of thinking, self-reflection, and metacognition in dreams to lower-level, experiential aspects of dream formation is an important step toward understanding not just dreams but the regulation of conscious thought across the sleep–wake cycle.

A particularly promising project might be to compare lucidity lapses and their relation to lucid insight to the relationship between mind wandering and meta-awareness of mind wandering. In this view, mind wandering without meta-awareness would be roughly analogous to nonlucid dreaming, and lucidity lapses might be comparable to mind wandering lapses, or instances of lapsing back into a mind wandering episode. This leads to some interesting considerations about lucid control dreams. As Schooler et al. (2011) note, there are at least three current hypotheses as to how meta-awareness, or noticing that one's mind has wandered, might be involved in the regulation of conscious thought. First, intermittent meta-awareness of mind wandering may contribute to the regulation of mind wandering, for instance, by aiding in the identification of mind wandering lapses and refocusing attention on tasks. Second, not only may meta-awareness minimize the negative effects of mind wandering, but the capacity to notice that one's mind has wandered might also be related to the ability to control conscious thought more generally, for instance, by initiating voluntary changes in one's thoughts. And third, mind wandering may produce the illusion of control over one's conscious thoughts and of metacognitive insight. Schooler et al. (2011) write:

> Mind wandering can terminate for reasons unrelated to meta-awareness (such as when an external event disrupts the internal train of thought or the action of an unconscious monitoring process). Once the episode has ended, individuals might reconstruct their recent conscious experience and in so doing realize that mind wandering had taken place. In this case the strong sense that we have "caught" our minds wandering could be an illusion of control. (Schooler et al., 2011, p. 323)

If meta-awareness of mind wandering can be described analogously to lucidity, this might once more strengthen the view that conceptually mediated forms of lucidity arise as side effects of largely subpersonal, automatic metacognitive monitoring. The impression that lucidity is an intellectual achievement, the outcome of a reasoning process, or that lucidity involves the ability to control the outcome of one's dream might then be largely illusory. Again, this fits the finding that dream control, though one of the factors most clearly distinguishing lucid from nonlucid dreams (Voss et al., 2013), is rarely successful. Stumbrys and colleagues (2014) found that the overall success rate for carrying out waking intentions in lucid dreams hovers around 20 percent, with waking intentions being forgotten about half the time and only carried out successfully in fewer than half of the instances in which they are remembered. Yet lucid dreamers, according to Dresler et al.'s (2014) study, feel empowered to act freely. If Schooler et al. (2011) are right, this might, at least in many cases, be no more than an illusion of control.

Indeed, Metzinger (2013a) has recently used findings on mind wandering, but also on dreaming, to argue that during two-thirds of their conscious lives, humans suffer a cyclically recurring loss in "M-autonomy," or the ability to engage in rational mental self-control and terminate and direct their thoughts at will. He uses this insight to attack the "myth of cognitive agency," or the claim that humans, in standard wake states, are cognitive and attentional agents in the sense of being able to control goal- or task-related deliberate thought and the focus of attention. The majority of conscious thought, he claims, arises from subpersonal processes. Again, the conclusion is that conscious thought is driven and prompted by largely unconscious cognitive processing, and if Metzinger is correct, this is true, in almost equal measures, for dreaming and wakefulness. The study of dream thinking may then be regarded as a natural extension of existing work on conscious thought and mind wandering and as a particularly obvious (though by no means unique) opportunity to investigate the interplay between conscious thought and subpersonal cognitive processing.

9.4 The Neuronal Basis of the Cogitative Dream Self

A relatively consistent finding regarding the neural correlates of the cogitative dream self is that the dorsolateral prefrontal cortex (DLPFC) is selectively deactivated in REM sleep (Nofzinger et al., 1997; Maquet et al., 2005). In wakefulness, activity in the DLPFC is associated with executive abilities, such as self-observation, planning, and decision making, as well as with more basic cognitive functions such as attention, working memory, and information retrieval (Muzur et al., 2002). Consequently it seems plausible to assume that its relative deactivation in REM sleep is related to the uncritical acceptance of dream events, poor self-reflection, and poor control over dream events (Muzur et al., 2002; Dang-Vu et al., 2005; Pace-Schott, 2007; Desseilles et al., 2011). At the same time, the persistent activation of the medial prefrontal cortex (mPFC) in REM sleep may underlie social cognition in dreams and the ability of the dream self to engage in theory-of-mind attributions (McNamara, McLaren, Kowalczyk, & Pace-Schott, 2007; Desseilles et al., 2011). Dream lucidity, as suggested by a number of recent studies, may accordingly be related to a reactivation of the DLPFC (see also Voss et al. 2014; Voss & Hobson, 2015; cf. sec. 3.5). As Voss et al. write:

> Since the DLPFC is thought to be the site of executive ego function, it has been suggested that the loss of volition, self-reflective awareness, and insight that is typical of normal dreaming, may be related to DLPFC inactivation. In lucid dreaming, these psychological functions return leading to the prediction that lucid dreaming will involve reactivation of the DLPFC. (Voss et al., 2009, p. 1198)

By contrast, Neider et al. (2011) suggest an involvement of the ventromedial prefrontal cortex in lucid dreaming and contrast this with the DLPFC hypothesis.

While these observations refer mainly to dream cogitation, the relative quiescence of the frontal regions in REM sleep is also thought to be related to the cognitive processes underlying dream formation. For instance, Pace-Schott writes that dreaming involves "a particularly potent form of spontaneous confabulation in which imaginary events are not only created and accepted as truth but are vividly experienced as organized, multimodal hallucinations" (Pace-Schott, 2007, p. 127). Because the DLPFC is linked to bottom-up attentional control in wakefulness, its quiescence in REM sleep might explain not only departures from wakelike cogitation but also "why external stimuli delivered during REM sleep are either ignored or automatically incorporated into the dream narrative, instead of interrupting the dream storyline" (Desseilles et al., 2011, p. 1004). Similarly, Dang-Vu et al. (2007, p. 104) relate frontal deactivation to personal and subpersonal changes in cognitive processing during REM sleep dreaming, such as the uncritical acceptance of bizarre dream content, the delusional belief of being awake, the lack of control over the unfolding dream, discontinuity and incongruity of dream content, and reduced sensitivity to external stimuli.

All of this suggests that the unique pattern of DLPFC activation during REM sleep is closely related to both dream production and dream cogitation. This does not mean, however, that this conceptual distinction is useless. Rather, I would like to suggest that the distinction between dream cogitation and dream cognition may help disentangle the relationship between lower-level dream formation and higher-level conceptual thought in dreams and identify their respective neural correlates. For now, the important point is that to truly understand these processes and their mutual interaction, it is crucial to first keep them apart. Ultimately we will need a much more differentiated and more fine-grained account of how fluctuations in cognitive and cogitative activity in dreams can be mapped onto changing cortical activation patterns.

Aside from selective deactivation of frontal areas during REM sleep, activity in the default-mode network may help us to understand similarities between dreaming and waking mind wandering, but also delusional wake states (Gerrans, 2014). The default-mode network may be particularly important for understanding the process of narrative dream formation (Pace-Schott, 2013). As discussed in section 6.2.3, the default-mode network is implicated both in waking mind wandering and in REM sleep dreaming (Pace-Schott, 2007, pp. 137–141; Domhoff, 2011; Wamsley, 2013), and this association has given rise to attempts to account for both dream cognition and dream cogitation. According to Gerrans (2014), dream narratives are driven by the activation of a frontally disinhibited default-mode network and may even, as Pace-Schott (2013) suggests, provide evidence of a hardwired storytelling instinct. The underlying idea is that in dreams, much as in waking mind wandering, the default-mode network contributes to the process of situating diverse elements in a personally relevant narrative context and thus is directly relevant for dream cognition. On the level of cogitation or conscious thought, frontally disinhibited default-mode network activity of the type seen in REM sleep may lead to the inability to distance oneself from the unfolding dream narrative and realize that one is only dreaming. This suggests that the largely automatic, lower-level, and imagistic production of dream narratives and the comparative inhibition of dream cogitation go hand in hand.

How can this account be applied to dream lucidity? Note that in the accounts discussed here, the ability to evaluate events in a decontextualized manner is thought to be associated with prefrontal activation and diminished default-mode activity (for discussion and further references, see Gerrans, 2014). Lucidity, however, would seem to require both of these processes: C-lucidity requires the ability to evaluate the ongoing events in a decontextualized manner and to conceptualize them, despite their vivid experiential character, as a dream, as mediated by frontal activation; at the same time, lucidity occurs before a backdrop of otherwise vivid dream imagery and narrative integration, which is likely associated with continued default-mode activity. What is surprising, from this perspective, is not that we rarely notice that we are dreaming but, to the contrary, that in lucidity, default-mode-driven story production coexists alongside the ability to distance oneself from the ongoing dream. This description sits nicely with the characterization of lucidity as a balancing act between participation and cognitively distancing oneself from the dream, as well as with the general fragility of lucidity, including lucidity lapses. It also fits in well with research on mind wandering, which is thought to involve both the production of autobiographically relevant stories, attributed to the default-mode network, and access to background knowledge, commonly attributed to the frontal-parietal network (cf. Gerrans, 2014; Smallwood et al., 2012). The current hypothesis would then be that degrees of connectivity between default-mode network and frontal-parietal regions might be involved in the transition from nonlucid to lucid dreaming.

Given the graded view of dream lucidity introduced earlier, it seems plausible that a more complex understanding of the relationship between activity in the default-mode network and the DLPFC will ultimately be required. Because cogitation is not uniformly absent in nonlucid dreams and because the types of cogitation occurring in lucid dreams are continuous with those occurring in nonlucid dreams, one would also expect variations in prefrontal activation to occur in nonlucid dreams. This might also explain why the experimental induction of 40 Hz activity over frontal areas leads not only to an increase in lucidity, but more generally to an increase in self-reflection in dreams (Voss et al., 2014; cf. sec. 3.5). One might also expect different types of lucid dreams to be associated with different neural activation patterns. Mota-Rolim et al. (2010) measured different EEG patterns during lucid dreams, depending on whether lucidity was reached spontaneously, induced by visual stimulation (acting as a "reminder" to become lucid), or resulted from presleep suggestion.

Finally, note that an inherent methodological problem is involved in studying the neural correlates of dream lucidity. This is that any study investigating signal-verified lucid dreams potentially confounds a strong type of lucidity involving dream control with a weaker definition of lucidity as realizing that one is dreaming. Studies investigating signal-verified lucid dreams—that is, all truly vigorous investigations of lucidity in the laboratory—count as investigating the neural correlates of dream lucidity only if one understands dream lucidity in a strong sense involving at least some dream control. By contrast, studying the neural correlates of insight into the fact that one is dreaming independently of dream control may be impossible for methodological reasons: without relying on the externally observable signals of lucidity, it would be impossible to determine the duration of lucid episodes with the required precision. This might mean that weaker degrees of lucidity not yet involving at least a sufficient degree of dream control to perform eye movement signals may systematically elude precise scientific investigation.

This methodological limitation is deplorable, because a better understanding of the continuities between lucid and nonlucid dreaming on the levels of dream cognition and dream cogitation could help clarify the relevance of lucid dream research for dream research in general (Desseilles et al., 2011). Here there are once more differing opinions regarding the generalizability of results from lucid dreams to nonlucid ones. For instance, Dresler et al. (2008) suggest that neuroimaging during lucid dreams might be used to identify the neural correlates of specific dream contents, the suggestion being that lucid dreaming might present a window on the sleeping brain more generally. By contrast, Dresler et al. (2009) and Spoormaker et al. (2010) assume that lucid dreaming is differentiated from nonlucid dreaming by involving higher order consciousness, including aspects like meta-awareness. "In this line of thought, the REM sleep-internal contrast between lucid and non-lucid dreaming can be seen as a paradigmatical contrast between primary and higher-order consciousness" (Spoormaker et al., 2010, p. 50). The latter assumption would question the generalizability of findings from lucid to nonlucid dreams (cf. sec. 3.5).

This question is in part empirical, insofar as the answer will depend on the degree of continuity between nonlucid and lucid dreams. It is also, in part, conceptual, depending on where to locate lucidity relative to the concepts of dreaming, sleep, and wakefulness. Whether Voss et al.'s (2009) suggestion of viewing lucidity as a hybrid state between dreaming and wakefulness or a description of lucidity as continuous with nonlucid dreaming will ultimately turn out to be the most useful remains to be determined. For now, note that these options are not necessarily mutually exclusive. Indeed, to the extent that cogitation likely plays a more important role in nonlucid dreams than commonly assumed, the gap between dream and waking thought itself may be reduced, because nonlucid dreaming could then no longer be characterized as a state of uniform cognitive deficiency. At the same time, if dreaming is merely the fullest expression of mind wandering (Fox et al. 2013) and waking mind wandering itself is essentially dreamlike, then the study of dream cogitation on the side of sleep and the study of mind wandering on the side of wakefulness might pave the way toward a continuous view of dreaming and waking thought by providing intermediate examples.

9.5 Taking Stock (I): The Weak-Cogitation Hypothesis

The foregoing analysis suggests that the question of whether dreaming involves a cogitative self requires a more differentiated answer than suggested either by the cogitative-duplicate hypothesis or by the no-cogitation hypothesis. In keeping with previous chapters, I want to introduce an alternative view, the *weak-cogitation hypothesis*.

Weak-cogitation hypothesis: Dreams are weakly cogitative states in that the frequency and reliability of different forms of cogitating (e.g., thinking, self-reflection, and metacognition) are only weakly pronounced and highly unstable in a majority of dreams; in particular, dream cogitation is typically only weakly differentiated from unintentional, automatic, and largely imagistic dream cognition.

The hypothesis is weak in a number of ways that run parallel to the weak-phenomenal-embodiment hypothesis (cf. sec. 7.4). First, it does not aspire to characterize dreams across the board but rather explicitly acknowledges a high degree of variability, both in lucid and in nonlucid dreams. Second, as it stands, it is extremely coarse-grained and to be truly informative will have to be enriched by theoretical considerations and further empirical findings. Third, it complements the weak-phenomenal-embodiment hypothesis by claiming that there is only a weak link between phenomenal selfhood and cogitation: exactly because of the instability and variability of cogitative activity in dreams, it would seem that phenomenal selfhood in dreams does not require the presence of a particular type of cogitative activity. Fourth, one way of spelling out the claim that the frequency and reliability of dream cogitation are unstable in dreams is to say that dream cogitation is incompletely differentiated from involuntary dream cognition. This last point is worth elaborating in some detail, both because it potentially offers an explanation of why cogitative activity is only weakly expressed in dreams and because it suggests the outlines of an account of narrative dream production.

What, then, does it mean to say that dream cognition and dream cogitation are only weakly differentiated? Dreaming, in the view defended here, is a form of conscious mental activity that is only weakly coupled to external and bodily stimuli and yet not experienced as thinking. With the exception of lucid control dreams, dreams are also not experienced as self-generated; the phenomenal property of agency for the process of dream generation is missing. Dreaming, like waking mind wandering, is a form of unintentional mental behavior (cf. Metzinger, 2013a): phenomenologically speaking, it is not something that we do, but something that happens to us. This loss of the phenomenal-functional property of mental agency, or the ability to control one's thought processes, is not, however, all that sets dreaming apart from intentional waking thought. In addition, the objects, characters, and places encountered in dreams are experienced not as the products of our own cognitive processing but as mind-independent entities. This unique phenomenal-functional profile is why it is not quite correct to say that dreaming is a form of thinking. Indeed, recall that an important desideratum for an empirically informed theory of dream formation is to explain how dreams are generated without assuming them to be the product of a dream producer, a dream author cleverly concealed, as it were, behind the scenes.

Yet we have also seen that there are numerous reasons for thinking that a cognitive perspective is essential for making sense of the process of dream generation (cf. chap. 6). This is true for waking concerns manifesting in dreams, but also for cryptamnesic dreams, or unrecognized memories experienced, while dreaming, as real. Much has been made of the associative nature of dreams, of their symbolic nature, and of the suggestibility of the dreaming mind. Recall how Havelock Ellis's dream of examining a tooth seemed to do his thinking for him, by naturally giving way to visual imagery of an ever-larger tooth that dream-Ellis himself did not take to be extraordinarily large. To Ellis, this "perpetual and unceasing change" (Ellis, 1911/1931, p. 20) of visual imagery was one of the most impressive aspects of dreaming. This is especially true, as rather than involving a succession of unconnected images, dreaming involves a train of fluidly and dynamically interconnected images. Ellis himself compares to dreaming to a

> constantly revolving kaleidoscope in which every slightest turn produces a new pattern, somewhat resembling that which immediately preceded it—so that, if the kaleidoscope were conscious we should say that each picture has been suggested by the preceding pattern—but yet definitely novel. (Ellis, 1911/1931, p. 21)

Dreaming, then, can be described as a type of unintentional, associative, imagistic, fluid (or highly dynamic), and often intensely emotional type of mental behavior. Moreover, dream cognition dominates and drives cogitative activity in dreams, with conscious thought often being overridden by automatic and dynamic dream imagery. At other times, as in lucid dreams, it would seem that metacognitive feelings are precursors to conceptually meditated insight and metacognitive forms of self-reflection. Occasionally this relationship may even be reversed, with dream imagery fluidly responding to dream cogitation. Finally, dream cogitation itself often takes on an associative quality that is markedly different from focused and rational waking thought. Lucidity lapses are a further example of how dream cogitation is quickly captured by lower-level cognitive activity.

How can the distinction between dream cognition and dream cogitation be integrated into the broader discussion on different styles of reasoning? In recent years, theories of higher cognition and reasoning processes have devoted much attention to whether two different systems of reasoning exist. System 1, which is said to exist in both animals and humans, is characterized by rapid, parallel, and automatic processes;

it involves associative, intuitive, and experience-based forms of reasoning. System 2 is taken to characterize evolutionarily more recent and distinctly human types of conscious, rule-based, and abstract thinking driven by slow, sequential, and error-prone processes. Dual-process accounts may differ as to the exact characterization of these two systems, and in particular, there is some debate as to whether there really exist two distinct systems of reasoning or whether these should be seen as continuous processing styles (for discussion and further references, see Evans, 2003; Evans & Stanovich, 2013). For now, even this superficial description of dual-process accounts is instructive because it seems clear that dream cognition is a paradigmatic case of the associative and fluid processing characteristic of System 1–style reasoning, whereas System 2–style reasoning, more akin to dream cogitation, appears to be less prominent and highly unstable in dreams. Further investigating the application of dual-process accounts to dreaming might be a way of placing dream research in a broader theoretical context. A slightly stronger claim would be that investigating the relationship between dream cognition and dream cogitation, as well as the transition from nonlucid to lucid dreaming, may extend existing work on System 1 and System 2-style reasoning and their mutual interaction. In particular, much of what I have said so far leans toward a continuous view of the relationship between these two systems. A highly speculative prediction resulting from the weak-cogitation hypothesis would be that dreaming might be a paradigm case in which these two styles of reasoning, while often still distinguishable, are blurred considerably and marked by intimate mutual interactions. Perhaps, these two systems of reasoning are best regarded as helpful idealizations and the distinction between the two collapses as we move to a more fine-grained level of analysis or to certain examples from dreaming and mind wandering.

Let us briefly consider another example that supports such a continuous view. Jackendoff (2012) has recently suggested that conscious thought can be separated into three components, where the first two, namely, verbal imagery and the feeling of meaningfulness, are conscious, and the third, which consists of the meaning of one's thoughts, is unconscious. The key idea of his "unconscious meaning hypothesis" is that the heavy cognitive work—inference and reference—is done unconsciously (Jackendoff, 2012, chaps. 15–16). In this view, "even when we're aware of our *thoughts*, we're not aware of the process of *thinking*" (p. 212); what we experience as rational thought is in fact grounded in intuitive judgment and largely unconscious processing. An important consequence is that the distinction between intuitive and rational thought collapses; intuition is the foundation of rational thought rather than an alternative form of thinking. What we experience as thinking—or what here I call cogitation—according to Jackendoff is only the linguistic expression linked to our thoughts:

> The thoughts themselves aren't conscious. Rather, what's conscious is the "handles" of pronunciation that are linked to the thoughts, plus some character tags that lend the pronunciation a sense of meaningfulness and conviction. And the conscious sense that one sentence logically

follows from another—that your reasoning is rational—is itself an intuitive judgment. (Jackendoff, 2012, p. 243)

This sits well with my claim that dream cogitation is only the tip of the iceberg in understanding dream cognition; perhaps dream cogitation is simply the rare case in which largely unconscious cognitive activity in dreams is associated with propositional thought and verbal imagery. By far the larger portion of dreaming, however, is associated with nonlinguistic types of visual, movement, and auditory imagery. For this reason, it does not properly fit Jackendoff's characterization of conscious thinking (or of cogitation, as used here); yet it rides on the same underlying, automatic, associative or intuitive processes often related to System 1–style processing. Dream imagery, to use Jackendoff's terminology, is the conscious handle on the underlying unconscious cognitive activity during sleep.

It is not surprising, then, that occasional instances of linguistic thinking in dreams, or cogitation proper, will also be of a largely associative or intuitive nature. If the basic and perhaps more fundamental cognitive work is being done unconsciously—as suggested by dual-process theories, but also by recent work on mind wandering (Metzinger, 2013a) and meaning (Jackendoff, 2012)—then it would seem that the most natural account of dream cogitation is to say that it is the conscious and verbal expression of underlying, largely automatic cognitive processes. As we saw earlier, Rechtschaffen was not quite right to describe dreaming as single-minded: indeed, if it were, one would expect dreaming to be characterized by an absence of cogitative activity altogether. What the weak-cogitation hypothesis claims is rather that dream cogitation, while failing to collapse entirely into dream cognition, is still constrained by unintentional cognitive activity in a meaningful way. This is also why the investigation of dream thinking is a fruitful entry point for investigating the process of dream formation in general.

If my account of A-lucidity, but also of feelings of familiarity and of just knowing in dreams (cf. sec. 9.3.3), is on the right track, then another way in which the type of imagistic cognitive activity that characterizes dreaming is linked to cogitation is by being a precursor to linguistically mediated forms of metacognitive insight. Indeed, as Proust (2013) suggests, the distinction between System 1– and System 2–style reasoning may also apply to metacognition, thus further supporting my claim that in a subgroup of lucid dreams, metacognitive feelings (on the level of dream cognition) give way to metacognitive self-reflection (on the level of dream cogitation). Whereas the standard interpretation of feelings of familiarity in dreams, but also of the general gullibility of the dreamer, is that they are due to the cognitive deficiency characteristic of the dream state, we can now present a more neutral reading, according to which the acceptance of bizarre dream events and seemingly inappropriate feelings of familiarity result from misinterpreted metacognitive monitoring. Dreaming is, of course, ultimately self-generated—but it is typically not experienced as such. The occasional feeling that the persons and places experienced in one's dreams are strangely familiar is the trace left by low-level, System 1-style metacognitive monitoring, where information about the self-generated nature of dreaming is erroneously attributed to the content of the dream rather than to the process of dream formation. What requires explanation is then not so much why dreamers typically fail to be surprised by the events unfolding in their dreams but rather why sometimes, as in prelucid or lucid dreams, they are. It is these occasional departures of dream cogitation from dream cognition—wondering why, for instance, one suddenly finds George Clooney sitting in one's kitchen—that require explanation. We might even want to say that these apparent instances of a reappearance of critical thinking in dreams involve a more complete breakdown of lower-level, System 1-style metacognitive monitoring, such that the self-generated nature of dream imagery is no longer available on the level of experience.[17] For now, a working hypothesis is that when dream cogitation separates from dream cognition, this is due to a transient functional dissociation in the types of information that are accessible to these two kinds of cognitive processing. This is most obvious, perhaps, for cases of cryptamnesia, in which dream cognition draws from memory sources that are not accessible on the level of dream cogitation and thus fail to be recognized as memories. C-lucidity (in which the dreamer can think about the fact that everything she is currently experiencing is a dream) in the absence of A-lucidity (or a dreamlike experiential quality) is the opposite case: here, information about the dreamlike nature of the experience can be accessed on the level of dream cogitation but not on the level dream cognition.

Methodologically, this is important because it suggests that regarding dreaming as a state of cognitive deficiency, as is common in the literature, is neither empirically plausible nor particularly fruitful. Focusing on the aspects that dream cogitation lacks as compared to waking thought may distract attention from the unique type of cognitive processing underlying dream generation and the dynamic interaction between dream cogitation and dream cognition. Kuiken (2010) notes that while conscious forms of self-regulation involving self-reflection and decision making are often only weakly expressed or absent in nonlucid dreams, this does not exclude the possibility that dreams are driven by "an unconscious but fluid form of self-regulation" (Kuiken, 2010, p. 22), even in the absence of DLPFC activation. Based on the observation, among other things, of a carryover effect following REM sleep awakenings, he notes that "it is possible, for example, that non-lucid dreaming entails self-regulated but fluid openness to 'what comes,' rather than the direct self-monitoring and response inhibition that enable 'rational' planning and decision making" (p. 22). In this respect, dreaming might be similar to musical improvisation or involuntary poetry (Kuiken, 2014). This is reminiscent of an idea that Rignano formulated in 1920, that "dreams . . . are *an ideative anarchy consequent on the cessation of all affective control*" (Rignano, 1920, p. 322). Rignano even suggests that "the most incoherent dream is exactly the one which most nearly approaches a purely intellective process, that is to say, a process of pure mechanical association of ideas, not influenced by the introduction of an affective element" (p. 319). If this is correct, dreaming might be a paradigmatic case and research model for investigating fluid, associative cognitive processing.

A final idea is that in keeping with my remarks on metacognitive feelings in dreams, a subgroup of dreams can possibly be described as involving metacognitive imagery representing the dreamer's current state of consciousness. Again, there is historical precedent for this idea. Silberer (1919, p. 14) distinguished three different types of symbolic phenomena in dreams and in sleep-onset imagery: First, imagery during sleep or half sleep may symbolize the contents of underlying thoughts; a familiar example would be dreaming of a trip that one is about to take, where the dream might be related to worries about packing, missing one's flight, and so on. The second class is made up, as discussed in section 8.2.1, of somatic phenomena, in which dreams are driven by real-body inputs (e.g., a dream of walking on Mount Etna while sleeping with a hot-water bottle under one's feet; cf. Silberer, 1919, p. 27). The third category comprises what Silberer calls functional phenomena. They are not, he writes, related to the contents of one's thoughts but rather reflect the functionality of ongoing cognitive processes. For instance, the inability to remember details about one's current research, caused by fatigue, might lead to a dream of asking a cranky secretary for some relevant piece of information, whereupon the secretary refuses, claiming to have lost the documents in question (p. 17). Another example would be a sleep-onset dream of trying to climb up a hill, but slipping back down after every step, symbolizing, as Silberer suggests, the futile attempt to develop an idea in his current state of fatigue (p. 13). False awakenings, but perhaps also the famous examination dream, in which one fails to remember the relevant answers, might be a further example. Dreams of trying to find one's dream diary or of thinking one is on the verge of an important discovery may be another.

This also suggests an alternative interpretation of dreams of frustrated effort (cf. chap. 8). In particular, not all dreams of trying to do something and failing need have a distinctly bodily flavor. While some of them—such as trying to flee from a pursuer—may be related to REM sleep paralysis becoming incorporated in the dream, others may be the result of a fleeting awareness of one's current conscious state. These two factors may also interact. John Mourly Vold gives a poignant example:

> I am attending some kind of lecture or drama; because I am unsatisfied with the lecture, I explain to those present that I want to present a dramatic work myself. I stand on the stage, on which there is a bed; I am only wearing a nightshirt, and apparently am not embarrassed in front of the ladies present; I lie down in bed, the book in my hand; a relative brings me a fresh shirt, which I put on over my dirty one. I start to read from the book, giving explanations as I go on; I experience difficulties, however, search for the right words and even find the process of reading itself difficult; I pause for a long time before an Assyrian picture. During my lecture, I fall asleep several times; the audience is restless . . . I go off stage . . . return; and this whole time it seems to me

that I am the author of the dramatic work, that I have authored it, respectively, in dreamtime. Upon awakening, I can only remember hints of the work. (quoted in Jezower, 1928/1985, p. 317; translation mine)

The possibility that dreams occasionally reflect not just an awareness of the physical body, but also metacognitive awareness of the sleep state is particularly fruitful because it suggests a way of bringing together different strands of ideas discussed over the last chapters. Dokic, after considering different accounts of metacognitive or, as he calls them, noetic feelings, argues that they

> are neither higher-order beliefs or memories . . . nor introspective experiences about first-order epistemic states. . . . Rather, they are first-order bodily experiences, namely non-sensory affective experiences about bodily states, which given our brain architecture co-vary with first-order epistemic states, in such a way that they can be recruited, through some kind of learning or association process, to represent conditions hinging on relevant epistemic properties of one's own mind. (Dokic, 2012, p. 317)

He continues:

> Within this model, noetic feelings can be seen to be associated with two kinds of metacognitive abilities, which I called "procedural" and "deliberate." At the procedural level, our brain realizes mechanisms whose function is to monitor the quality of our cognitive processes in order to produce spontaneous mental and/or physical behaviour (such as attempting to remember a name, reading more slowly, or moving one's head from side to side to resolve visual ambiguity). At the deliberate level, the same mechanisms can generate conscious noetic feelings, which can be further exploited in controlled reasoning to produce more context-sensitive behaviour (such as going through the alphabet to provoke remembering, pointing to difficult sentences, or using a magnifying glass). (Dokic, 2012, pp. 317–318)

Importantly, both procedural and deliberate metacognition can involve conscious feelings; the difference is that in procedural metacognition, these feelings remain largely inefficacious, whereas in deliberate metacognition, they can rationally be exploited, for instance, in drawing further inferences and in action. As Dokic (2012, p. 312) notes, "Deliberate metacognition is something that the subject herself does, rather than a mechanism inside her." Most dreams, perhaps, can be regarded as involving procedural metacognition; by contrast, only rarely, as in lucid control dreams, can metacognitive feelings also be exploited for reasoning and action within the dream.

Finally, if it is correct to say that dreams are weakly phenomenally-functionally embodied states, and assuming that metacognitive feelings in dreams are ultimately of a bodily nature, then the rarity of these latter examples of deliberate metacognition is not surprising. Indeed, the reason why dream cogitation departs from waking thought might be, once more, linked to the absence both of strong-phenomenal embodiment and to the weak, nonstandard grounding of bodily experience in real-body inputs. Perhaps Locke was on the right track when, in a sarcastic jape against Descartes, he wondered "whether the soul, when it thinks thus apart, and as it were separate from the body, acts less rationally than when conjointly with it, or no," and mused that the frivolity and irrationality of dreams might make it seem plausible that "the soul owes the perfection of rational thinking to the body" (Locke, 1689/1997, II.I.16; cf. sec. 4.5).

While Locke's remark clearly subscribes to the cognitive deficiency view of dreaming, a more neutral view might be to say that in dreams, weak phenomenal-functional embodiment and the instability of conscious self-regulation on the level of dream cogitation are linked. The intriguing and decidedly anti-Cartesian conclusion would be that experiencing oneself as a cogitative self, as a thinker of thoughts, in the fullest sense, might be importantly related to the phenomenal-functional property of embodiment after all.

9.6 Taking Stock (II): The Problem of Dream Belief Revisited

What, if any, are the consequences of the weak-cogitation hypothesis for the problem often thought to matter most in Cartesian dream skepticism, namely, the ability to entertain beliefs in dreams? As argued in sections 9.1 and 9.2, the problem is not so much whether dreaming involves beliefs in some technical sense as whether dreaming involves the phenomenology of occurrently believing that *p*. Being a cogitative self, or being able to engage in certain cogitations, is a necessary condition for occurrently believing that *p*. But though occurrently believing that *p* is a variant of thinking or asserting that *p*, not every cogitation will qualify as a belief in the phenomenological sense. Departing from a Humean account of belief, this section investigates the conditions under which it makes sense to say that beliefs occurred within in a given dream and tentatively suggests how a technical account of belief might be linked to the phenomenology of believing.

Hume famously argued that belief is a feeling. Belief, in this view, is distinguished from fiction, or propositional imagining, not in its content but in the manner of conception:

> It is something felt by the mind, which distinguishes the ideas of the judgment from the fictions of the imaginations. It gives them more force and influence; makes them appear of greater importance; infixes them in the mind; and renders them the governing principles of all our actions. (Hume, 1739/2000, 1.3.7.7)

While imagination and free association have a quality of looseness about them, the mind, in causal reasoning, is determined to pass from an impression to the associated idea, and both the looseness of imagining and free association and the determination of belief are felt by the mind (Hume, 1739/2000, 1.3.9.6). In this view, there is no sense in saying that one mistakenly takes oneself to occurrently believe that *p* while in fact merely imagining that *p*. In acquiring the greater force that makes a particular thought appear as a belief rather than as a propositional imagining, the thought also turns into a belief in the occurrent sense. Consequently, enjoying the phenomenology of occurrent belief requires having a specific type of phenomenal self, namely, a cogitative self that is able to engage in a number of different cogitations such as doubting, wondering, thinking, judging, asserting, and so on.

According to Hume, the transition from conception to belief occurs when a thought acquires the greater force and vivacity that makes it appear, subjectively, as a belief. This link between having a belief in the occurrent sense and enjoying the phenomenology of believing ensures that thinking that you occurrently believe that *p* is self-verifying in a similar way as thinking that you are currently thinking that *p*. Importantly, the phenomenology of occurrent belief is not causally inert but has a certain functional profile. Belief acquisition is often the outcome of a period of deliberation and terminates inquiry. The feeling of certainty and of determination involved in belief signals that the belief can now be used in action control, including further reasoning processes. Arguably, the occurrent belief could not fulfill this function if it were not clearly identifiable as a belief, and the easiest way for this to happen is to tag it as a belief on the level of experience itself. The view that belief is importantly related to epistemic feelings of certainty is also endorsed by de Sousa:

> There are feelings of certainty or rightness, which can also bear on specific beliefs or on the validity of inferences. The feeling of certainty freezes inquiry, to the extent that we may now feel we have the answer we were looking for. But it frees us for action based on the factual or normative propositions to which we now assent. (de Sousa, 2009, p. 146)

Similarly, Jackendoff notes:

> From the cognitive perspective, assenting to a sentence—judging it's true—amounts to assigning it a character tag that we might call "commitment" or "conviction." The conscious feeling correlated with this feature is that the sentence out there in the world depicts reality accurately—the sentence is "objective." (Jackendoff, 2012, pp. 197–198)

Objectivity, in this context, refers only to the experience of objectivity in a phenomenological sense and as such has no direct or necessary relationship to actual truth or objectivity. The question, then, is first, whether dreamers have occurrent beliefs in this phenomenological sense; and second, whether the phenomenological profile of dream beliefs is linked to their ability to fulfill the characteristic functional profile of belief, for instance, by terminating inquiry, influencing further reasoning, or being used for action control within the dream.

Take the following two dreams reported by Havelock Ellis and Ernst Meumann as an example:

> On another occasion, I dreamed that I was in the large *plaza* of a Spanish city (Pamplona possibly furnishing the elements of the picture), and that the governor emerged from his residence facing the square and began talking in English to the subordinate officials who were waiting to

receive him. The real reason why he talked English was, of course, the simple one that he spoke the language native to the dreamer. But in my dream I was extremely puzzled why he should speak English. I looked carefully into his face to assure myself that he was not really English, and I finally concluded that he was speaking English in order not to be understood by the bystanders. (Ellis, 1911/1931, p. 62)

I was on a quickly moving train. As I saw houses, trees and people flying past the window, the train suddenly entered a tunnel. The tunnel was apparently meant to be a helical tunnel, as I immediately sensed that the train was passing through a sharp curve whilst tilting in a worrisome manner. For a moment, I thought about this strange lateral position of the train, which I attempted to compensate by leaning forward: "It must correspond to the radius of the curve," I told myself. Suddenly, I saw (despite the tunnel) people and trees standing above me at a right angle, so that I had to strain my eyes upwards in order to discern them. "Right," I thought—"this projection has to occur because of our horizontal position." The circular motion of the train had now increased to such an extent that I found the dizziness (which had been present the whole time) unbearable and awoke. (Jezower, 1928/1985, pp. 305–306; translation mine)

In both dream reports, a belief is acquired following a period of uncertainty and wondering; it makes sense of a puzzling event and resolves the dreamer's previous uncertainty. The transition from puzzlement and feelings of wonder, via inquiry (i.e., scrutinizing the governor's face), to the resolution of doubt is particularly evident in Ellis's dream. In Meumann's dream, there is an explicit chain of reasoning and of interconnected beliefs. His surprise at the strange position of the train is related to his attempt to compensate for it by adjusting his body posture. His conclusion that the train's horizontal tilt corresponds to the radius of the curve is connected to his belief that he must see people and trees as standing perpendicularly above his head, and fascinatingly enough, the dream imagery follows suit. While Meumann is not at all bothered by the disappearance of the tunnel, there is still a certain coherence to his beliefs concerning the position of the train and his altered visual perspective.

If belief is taken in a first-person-phenomenological sense of occurrent belief, then these instances clearly qualify as beliefs. Nothing in the reports suggests that they felt somehow more like propositional imaginings than like real beliefs, and conservatism suggests that in the absence of first-person reports to the contrary, they should be seen as no different from standard occurrent beliefs in wakefulness with respect to their phenomenology. Moreover, the connection of these dream beliefs to previous uncertainty or puzzlement and subsequent reasoning suggests that dream belief can have a similar functional profile as waking belief. If this suffices for dream beliefs to qualify as beliefs in a technical sense, then there is no principled reason why dream beliefs should fall short of beliefhood. Instead their status as real beliefs will have to be decided on a case-by-case basis.

There is also good reason to think that this approach is methodologically tractable for the scientific investigation of dream belief. Kahan (1994) found that self-ratings of different types of thinking and control in dreams were generally similar to the ratings of independent judges. However, the frequency of dream thinking was much more frequent in self-ratings than when judges rated narrative reports. This suggests that as is the case for dream emotions (Merritt et al., 1994), the best methodology may be to ask participants to explicitly report instances of thinking and believing in dreams. And though *reports* of thinking and believing in dreams are, of course, not the same as thinking and believing in dreams, they are the arbiter of what can be said about thinking and believing in dreams, and can be considered, at least when given under ideal reporting conditions, as transparent.

Note that the emphasis on the phenomenology of occurrent belief also nicely solves the problem of how many beliefs to ascribe to a given dream. A qualm one may have about dream belief is that as soon as one takes a given dream to involve any beliefs at all, this might seem to entail the ascription of a host of latent beliefs independently of whether they were ever actually entertained during the respective dream and independently of any evidence that they were. This line of thinking may be what underlies the doxastic-duplicate hypothesis. For instance, one might think that the belief that the governor of a Spanish city is speaking English because he does not want to be understood by bystanders cannot occur in isolation but rather unfolds before the backdrop of a whole network of surrounding beliefs, such as the belief that this is really happening and that one is awake rather than dreaming.[18]

In the view defended here, however, this network requirement and the commitment to the doxastic-duplicate hypothesis it ultimately implies is too strong. Just as the dream world and the objects and characters in it have no existence beyond a given dream, so it only makes sense to speak of dream beliefs to the extent that a given dream involves the phenomenology of believing, where this phenomenology is embedded in a suitable functional context. Indeed, we have no reason to think that individual dreams, though having a limited duration of only a few minutes and often involving narrative episodes that are not directly continuous with waking life or with other dreams (recurrent dreams and false awakenings being exceptions), involve a host of interrelated beliefs. There would be no way to determine the content of these latent beliefs, as a large number of them, ex hypothesi, would not be reportable. Did dream-Ellis also have particular beliefs about the governor himself, for instance, about whether he had seen him before? Or did he have specific beliefs about the city he was in, what he himself was doing on the square, and what he had eaten for breakfast that morning? If having any beliefs at all involves having a network of beliefs, one would expect there to be an answer to these questions. Yet, in the absence of sufficiently detailed descriptions in dream reports, ascribing any specific, unreported beliefs would amount to divination. Instead, allowing for the occurrence of isolated dream beliefs naturally follows from the weak-cogitation hypothesis. Of course, this is not to rule out that individual dreams can involve a network of interrelated beliefs; rather, I am here only defending the weaker claim that the ascription of a network of beliefs is not entailed by the ascription of individual beliefs, but rather depends on appropriate empirical evidence.

What about alternative methods of belief ascription? In dispositionalist accounts of belief, belief possession is contingent on behavioral dispositions. If one believes that *p*, and given certain conditions of manifestation, one will manifest the appropriate type of behavior. For instance, if I am thirsty, and if I believe that a bottle of water is in the kitchen (and if additional conditions obtain, for instance, if there is nothing to drink in my office and I think that water is an optimal thirst quencher), then I will go into the kitchen to get the water bottle. Dispositionalists may hold that phenomenal dispositions, such as experiencing feelings of surprise if a belief turns out to be false, as well as cognitive dispositions, or using the belief in subsequent reasoning, also form part of the dispositional profile of belief (Schwitzgebel, 2002a). Even Ryle took this view, writing that belief "is a propensity not only to make certain theoretical moves but also to make certain executive and imaginative moves, as well as to have certain feelings" (Ryle, 1949/2000, p. 135). For now, this raises the question of whether dream beliefs might be ascribed solely on the basis of dream behavior, even in the absence of the phenomenology of belief. Similarly, one might think that there is a rationality assumption involved in belief ascription, such that one will ascribe the beliefs and desires to a system that, given its current situation, previous experiences, and observed behavior, it would be rational for the system to have (Dennett, 1981b). If so, then the phenomenology of belief might even be wholly irrelevant to belief ascription.

Both possibilities would suggest the possibility of circumventing the phenomenology of belief, as described in dream reports. This is not, however, necessarily advantageous. Earlier I noted that experienced lucid dreamers often become lucid without explicitly acknowledging this fact. Consequently it seems possible that a dreamer might engage in lucid dream behavior without explicitly forming the belief that she is dreaming. Does this type of lucid dream behavior in itself suffice for saying that this belief was nonetheless present and that the dream was, as a matter of fact, lucid? Does it make sense to say that the fact that one is dreaming can be selectively available to behavioral control but not to conscious thought, thus introducing the concept of B-lucidity (Windt & Metzinger, 2007; Noreika, Windt, Lenggenhager, et al., 2010)? And is this type of behavior sufficient to ascribe the *belief* that one is now dreaming, as well? Take, once more, the case of Mary Arnold-Forster's use of lucidity to induce flying dreams at will. She even developed a specific flying technique:

> The actual process by which I fly in my dreams has always been the same since the earliest days when I first fluttered down the nursery staircase. . . . By giving a slight push or spring with my feet I leave the ground and fly without further effort, by a simple act of volition. A slight paddling motion by my hands increases the pace of the flight, and is used either to enable me to reach a greater height, or else for the purpose of steering, especially through any narrow place, such as

through a doorway or window. If I am at all fatigued by a long flight, this motion of the hands is of great assistance and gives confidence and increased power. (Arnold-Forster, 1921, pp. 65–66)

Dream flying, for her, was an acquired skill, and it was clearly one activated only in the context of dreaming. Yet let us assume that at least occasionally, she flew in her dreams without first realizing that she was dreaming. Again, one might appeal to unconscious metacognitive monitoring to explain how even in the absence of any conscious beliefs about her current state, the fact that she was now dreaming was available on the level of behavioral control, thus enabling her to exercise her flying skills. Indeed, why not say that this is true for all flying dreams, as well as for all dreams including characteristically dreamlike behavior such as passing through walls, communing with the dead, and so on? Why not say that such types of dream behavior are sufficient for ascribing the *belief* that one is now dreaming, as well? Indeed, given that one is flying, it would be rational to assume that one was now dreaming.

While the concept of B-lucidity may be helpful for describing a precursor to full lucidity, it would be a mistake to think that it was thereby sufficient for belief ascription. As argued earlier, not all forms of metacognition involve cogitation or belief. Consequently, to ascribe dream belief on the sole basis of dream behavior is to obliterate the distinction between dream cogitation and dream cognition. In particular, it would seem that phenomenal beliefs can even run contrary to dream behavior. Over and above terminological quibbles over what counts as belief, this type of dissociation requires explanation; this is enough, I think, to insist that beliefs in the phenomenal sense are explanatorily relevant and require an account of their own. Consider the following example:

Perhaps this is a place that doesn't enter the race. A neutral timeless space-less place of shadows. I can levitate because there is no gravity here. This is then a place at the other end of the spectrum from a black hole where gravity holds even light in its immeasurably impacted weight. I find that I can levitate right off the sidewalk in front of the hotel *and I know this is not a dream*, that I am really floating up off the ground, and gaining speed now above the hotel, which is five stories tall—I'm five hundred feet above the sidewalk. I come back down in front of the hotel, and the streets are still empty and nobody sees me. (Burroughs, 1995, p. 4)

Importantly and perhaps somewhat disturbingly, this type of report is not an exception (cf. Bossard, 1987, p. 67; and sec. 10.3.2). Indeed, flying dreams are occasionally described as involving a particularly strong "sense of the reality of the experience" (Ellis, 1911/1931, p. 130; see sec. 8.2.2.2). This suggests that experiences that can, in fact, only occur in dreams (or at least in altered states of consciousness) not only can occur independently of the belief that one is now dreaming, but are even compatible with believing that one is awake. The apparent absurdity of the assumption that one is levitating off the ground while awake does not prevent the belief from *feeling* wholly objective, and this feeling can even be used in further reasoning and action control. If in a lucid dream I erroneously believe, for instance, that I am sharing the dream with someone else, then it makes sense to ask this other person to help remember the dream after awakening (see my discussion of lucidity lapses in sec. 9.3.3). Clearly the coexistence of this belief with the belief that I am dreaming is puzzling and requires explanation. Yet, to make progress toward understanding it, the most careful approach, for now, is to say that dream behavior is not sufficient and perhaps not even necessary for the ascription of dream belief. To count as a belief in an interesting sense, a dream belief will not, of course, be causally inert; it will, for instance, typically prompt other thoughts and actions within the dream. However, the status of dream beliefs *as* beliefs is not contingent on their functional link to any *particular* thoughts or actions. Rather, it is an empirical question how exactly different types of thoughts and actions are typically linked to dream beliefs.

All of this suggests that the phenomenology of believing cannot be circumvented in the ascription of dream belief. What is crucial for the ascription of dream belief is the phenomenology of believing, where this is reportable after awakening. A consequence of this view is that dream beliefs must be ascribed individually and possibly even in isolation; whether a given dream belief occurs before the background of a network of beliefs remains, for now, an open question. Moreover, while one might claim that to count as a belief in a technical sense, a phenomenal belief must have some degree of causal embedding or interconnectedness with other thoughts and actions, there is no principled restriction on the types of thoughts and actions that either induce or are induced by particular dream beliefs. Levitating off the sidewalk, apparently, can induce the belief that one is not dreaming. By contrast, alternative methods of belief ascription on the sole basis of behavior or by regarding the system in question as a maximally rational system do not seem adequate for capturing the variable nature of dream belief.

The real question is whether this phenomenologically based method of belief ascription is sufficient to place dream beliefs anywhere in the vicinity of waking beliefs. The answer will depend on whether their functional interconnectedness with other thoughts and actions within the dream is anything like that of waking beliefs, and also on whether any functional connections exist between individual dream and waking beliefs. At least there are no principled objections to saying that this might be the case, and occasionally dream beliefs appear to fulfill this requirement. The next chapter further discusses this question using the classical example of dream deception.

9.7 Conclusions

At the beginning of the chapter, I introduced two pairs of hypotheses about dream belief and dream thinking: the doxastic-duplicate hypothesis and the no-belief hypothesis, and the cogitative-duplicate hypothesis and the no-cogitation hypothesis. I argued that the crucial question, at the outset, is not whether dream belief fulfills the conditions of belief in any particular technical sense but whether dreams involve the phenomenology of occurrently entertaining beliefs. Consequently the question of dream beliefs collapses, for the time being, into the empirical question of whether dreaming involves the phenomenology of engaging in cogitations in general and the phenomenology of believing in particular.

In section 9.3, I introduced a conceptual framework for describing different forms of cogitation in the dream state, such as thinking, self-reflection, and metacognition, but also lucid insight and dream control, and illustrated these distinctions using examples from dream research. In particular, I argued that dream cogitation, or conscious thinking in dreams, should be distinguished from dream cognition, or the largely imagistic type of cognitive processing that makes up much of the phenomenology of dreaming. Both should also be distinguished from the nonconscious types of cognitive processing underlying dream formation, but also memory consolidation in sleep. An important conclusion was that despite a lack of conceptual clarity in the empirical literature, we do have empirical evidence for saying that different forms of thinking occur frequently both in lucid and in nonlucid dreams, and these two types of dreaming should be gradually rather than sharply distinguished in this respect.

I closed by reassessing the evidence on the cogitative and the doxastic dream self. Section 9.5 introduced the weak-cogitation hypothesis and suggests a novel way of accounting for the relationship between dream cognition and dream cogitation. The key idea is that dreaming is best viewed not as a state of cognitive deficiency but as a paradigmatic kind of associative and imagistic cognitive processing. A particularly surprising insight was that certain dreams, such as dreams of frustrated effort or false awakenings, but also feelings of familiarity and of just knowing in dreams, can be understood as arising from misinterpreted metacognitive monitoring. At least certain dreams can be described as involving metacognitive imagery reflecting the dreamer's current bodily and cognitive state. These types of imagery, as well as related metacognitive feelings, may be precursors to cogitative, conceptually mediated thought and even lucid insight.

In section 9.6, this account was translated into an account of dream belief, claiming that dream belief is only present to the extent that the dream self enjoys the phenomenology of occurrent belief. The ascription of beliefs to the dream self is thus constrained by the actual cogitative activity occurring in a given dream. Alternative methods of belief ascription—such as the rationality assumption, dispositionalist accounts focused on behavior, and the assumption that dream beliefs occur before the background of a network of beliefs—were, at least for the time being, rejected as premature. While functional integration with reasoning and behavior is crucial for assessing the status of dream beliefs as real beliefs, I argued that no strong rationality constraints can be placed on dream beliefs. For now, a preliminary conclusion is that while no principled objections prevent us from saying that dreams involve beliefs, the relationship between dream beliefs and behavior, as well as the rationality of dream belief, remains an open question. The same is true for the question of whether dreams are deceptive in the classical sense of involving false beliefs about one's current state of consciousness. I discuss the consequences of this position for Cartesian dream skepticism and for understanding the relationship between dreams and delusions in the next chapter.

10 Are Dreams Deceptive Experiences? Deception, Delusion, and Insight

10.0 Introduction

The problem of dream belief is inseparable from the problem of dream deception. Accepting that dreams, at least occasionally, involve beliefs opens the door to the possibility of radically false beliefs in dreams, and hence to dream deception. Under which conditions, then, do dreams count as properly—and threateningly—deceptive? In this chapter, I argue that a majority of dreams provide no foothold for dream skepticism in any interesting sense. Yet the instances of dream deception that do exist are more various and in a sense more worrisome than those considered by Descartes. Where Cartesian dream deception has us deceived primarily about the nature of the dream world, I argue that there are cases in which we are deceived, while dreaming, about our own status as rational subjects. This is worrisome not just because it brings the threat of dream deception closer to home, but also because my discussion of dream deception is informed by real-world examples from the literature. Dream deception is not just a theoretical possibility but an actuality. In some cases, it is even a cause of clinical concern. This chapter is an exercise in empirical epistemology, investigating how philosophy of mind and psychiatry, epistemology, and scientific dream research can mutually inform one another.

My strategy is the same as in the preceding chapters. Again, I begin by introducing two pairs of hypotheses about deception, delusion, and insight in dreams from the literature. Because this chapter is not just about dream deception as an epistemological problem but introduces an empirically informed and interdisciplinary perspective, I focus on versions of dream skepticism that place dream deception in the vicinity of wake-state pathologies. I then introduce a conceptual framework for describing different forms of concurrent and retrospective dream deception, as well as for assessing the strength of dream deception.

I close by reassessing the skeptical threat posed by dreaming and the prospects for a nosological analysis of dream belief. While the literature, so far, has focused on the comparison between dream deception and wake-state delusions, I suggest that a nosological analysis of insight in lucid dreams is particularly fruitful. This idea has its roots in antiquity and opens exciting new perspectives for future research.

10.1 The Deception Hypothesis and the No-Deception Hypothesis

The *deception hypothesis* lies at the heart of the philosophical discussion of dreaming. Its intuitive appeal is such that it does not seem to require much in the way of theoretical justification. As Descartes notes in the *First Meditation*, even in a best-case scenario of sensory perception, such as sitting by the fire and looking at a piece of paper in his hands, his initial certainty regarding the truth of his beliefs is shattered by his realization that he might now be dreaming (Descartes, 1641/1901, I.5).[1] All of this is familiar. Yet much hinges on the exact way in which dreams are claimed to be deceptive. Already it would seem that the deception hypothesis involves two separate claims, where the first concerns the phenomenal character of dreams, and the second the involvement of false beliefs.

Deception hypothesis: Dreams are deceptive experiences because they are subjectively indistinguishable from wakefulness and because they involve beliefs, some of which might be false.

Dream deception, for Descartes, is tied to the potential, not the actual, falsity of dream beliefs. The problem is not that all dream beliefs are in fact false but that the subjective indistinguishability of dreaming and wakefulness renders it impossible to identify those that are (Descartes, 1641/1901, I.2). Aside from requiring that dreaming involves beliefs, this assumes that dreams are deceptive because they at least potentially mimic standard wakefulness. This, of course, is the key intuition behind the phenomenal-duplicate hypothesis and its variants. To make matters worse, Descartes believes that dreams retain their deceptive potential even in the face of closer scrutiny. Hunter illustrates the key intuition with a particularly graphic example:

> In a dream we may visit a grandmother who is in fact long dead, but if we help her up out of her chair, the task will call for some exertion on our part, and her hand will prove warm, solid and perhaps a little sweaty, just as if we were not dreaming. (Hunter, 1983, p. 80)

The eerie conclusion is that even when confronting a dead grandmother, we are at the mercy of dream deception.

The *no-deception hypothesis* denies this claim, typically by arguing that dream beliefs fall short of real beliefs.

No-deception hypothesis: Dreams are not deceptive because they fail to involve false beliefs.

The most radical way of defending this claim is, of course, to deny that dreams are experiences at all. If dreams are not experiences, then they cannot be deceptive experiences, so there is nothing much to worry about on their account. In particular, if dreams are not experiences, then they cannot include instances of wondering, doubting, or believing anything whatsoever. This antiskeptical position was Norman Malcolm's (1956) original motivation for denying the experiential hypothesis of dreaming (cf. chap. 1). A similar idea is found at the end of *On Certainty*, where Wittgenstein writes that there are certain beliefs about which one cannot be deceived. For instance, my belief that I am sitting at my desk and writing leaves no room for error:

> But even if in such cases I can't be mistaken, isn't it possible that I am drugged? If I am and if the drug has taken away my consciousness, then I am not now really talking and thinking. I cannot seriously suppose that I am at this moment dreaming. Someone who, dreaming, says "I am dreaming," even if he speaks audibly in doing so, is no more right than if he said in his dream "it is raining," while it was in fact raining. Even if his dream were actually connected with the noise of the rain. (Wittgenstein, 1977, §676)

As is the case for Malcolm, Wittgenstein seems to be suggesting that dreaming is no different from a drugged state of unconsciousness. In particular, dreaming does not support genuine thoughts and beliefs.

Sosa mounts a more subtle attack on the deception hypothesis. His first counterargument is the familiar claim that dreams are not deceptive because they do not involve real beliefs (cf. sec. 9.1). Dreams are also prevented from being deceptive, however, for another reason:

> When we sleep and dream . . . our situation is inappropriate for the manifestation of perceptual competence. Hence, even assuming that we do have perceptual beliefs in our dreams, these are not apt beliefs, since even if and when they accidentally hit the mark of truth, they fail to do so in a way creditable to the believer's competence. But this does not affect the aptness of our perceptual beliefs in waking life. (Sosa, 2007, p. 42)

Sosa's argument is that dream beliefs pose no threat to the safety of our general doxastic competence because, even when true, they cannot be considered as the outcome of apt performances. Dream beliefs do not manifest the skill or competence involved in real belief, and so it is neither surprising nor particularly worrisome that they are unreliable. We would not, for instance, blame a skilled archer for the fact that his ability to hit the mark would be threatened by temporarily reduced competence, for instance, due to drug ingestion, or by an unexpected gust of wind (Sosa, 2007, chap. 2). Dreams show that belief formation is fragile, but they do not undermine our competence to form apt beliefs in wakefulness.

Ichikawa (2008), though he follows Sosa's rejection of dream belief as real belief, argues that this line of argument is insufficient to refute the deception hypothesis. He points out that dreams may well be deceptive even if they involve not real but only imaginative beliefs (Ichikawa, 2009). In this case, the problem would no longer be that we entertain false beliefs while dreaming, but that while dreaming, we mistake mere dream beliefs for real beliefs. This argument has us deceived not so much about the dream world as about the status of our own mental attitudes; it undermines our ability to distinguish real beliefs from mere imaginings.

Aside from the conceptually motivated claim that dreams are not deceptive because they are not experiences or do not involve beliefs, one might also defend the factual claim that dreams are not deceptive. Perhaps dream beliefs don't fall short of beliefhood for principled reasons; rather, as a matter of fact, we typically don't entertain any beliefs (of the relevant kind) in our dreams. Perhaps, to the extent that we do entertain beliefs in dreams, these beliefs are either true or, if they are false, they don't have the type of content required for dream skepticism. I argue later that this variant of the no-deception hypothesis adequately describes a majority of dreams. Yet this is not enough to defuse the threat of the dream argument: there is still room, as we will see, for diverse and worrisome forms of dream deception.

10.2 The Delusion Hypothesis and the Insight Hypothesis

This section introduces two variants of the deception and no-deception hypotheses that are of particular interest from an interdisciplinary perspective. While the deception hypothesis appeals to the phenomenal character of dreaming, an alternative view traces the subjective indistinguishability of dreams and waking experiences to the inability of the dreamer to tell the difference. If our discriminative abilities are, while dreaming, crippled by the instability of dream imagery, then we might fail to realize that we are dreaming even if dreams have different contents and a different phenomenological profile from waking experiences (cf. Hill, 2004a). This type of view immediately invites a nosological analysis, placing dreams in the vicinity of madness and delusions. Call this the *delusion hypothesis*:

Delusion hypothesis: Dreams are deceptive because they involve delusional beliefs.

Recall that Descartes presents the problem as follows:

> But how could I deny that I possess these hands and this body, and withal escape being classed with persons in a state of insanity, whose brains are so disordered and clouded by dark bilious vapors as to cause them pertinaciously to assert that they are monarchs when they are in the greatest poverty; or clothed [in gold] and purple when destitute of any covering; or that their head is made of clay, their body of glass, or that they are gourds? I should certainly be not less insane than they, were I to regulate my procedure according to examples so extravagant. (Descartes, 1641/1901, I.4)

In dreams, he realizes, he is duped into believing those very "same things, or even sometimes others less probable, which the insane think are presented to them in their waking moments" (Descartes, 1641/1901, I.5).

There is much controversy as to whether Descartes took the analogy between the delusions of insanity and dream deception literally or not. Broughton writes that "Descartes himself sees the occurrence of dreams as having the same basic kind of explanation as the occurrence of deranged waking sensation" (Broughton, 2002, p. 65). She argues that Descartes gives equal credit to the lunacy argument and the dream argument, suggesting that lunacy and dreams pose equal threats to sensory based beliefs (Broughton, 2005; for a similar account, see Radden, 2011, pp. 2–6). By contrast, Frankfurt (1970/2008, chap. 9) argues that Descartes only briefly considers the lunacy argument to ultimately dismiss it as flawed. The first-person narrator of the *Meditations*, Frankfurt claims, is an everyman; his epistemic situation is no better or no worse than that of others. The defects considered by Descartes—that he might be dreaming, or he might be the victim of an evil, deceiving demon—are representative of the typical defects of any human mind; they do not disqualify the cognitive capacities of the meditator in any particular, idiosyncratic way, as would be the case for insanity. In this view, dreaming is relevant to Descartes's epistemological argument in a way that idiosyncratic delusions are not. Frankfurt (1970/2008, pp. 114–115) also argues that the lunacy argument would undermine the very project of rational inquiry: if human rationality were inherently defective in the manner of rendering us all comparable to madmen, the project of rationalism would be reduced to absurdity. According to Frankfurt, for Descartes's project to make sense, he must assume both that we are capable of rational inquiry and that rational inquiry can produce true belief and real knowledge. If dreams are to be genuinely and dangerously deceptive, then they must be unlike waking delusions.

Independently of whether or not Descartes himself endorsed the delusion hypothesis, many others have. Kant (1766/2002) metaphorically likened the madman to a waking dreamer, and Schopenhauer (1847/1988) thought dreams to be a brief madness. Hughlings Jackson even suggested that the study of dreams might help understand insanity (Gottesmann, 2006). More recently, Hobson suggested that "dreaming . . . is not like delirium. It is delirium. Dreaming is not a model of a psychosis. It is a psychosis. It's just a healthy one" (Hobson, 1999, p. 44). Similarly, Gottesmann (2006) suggests that dreaming may be a new neurobiological model of schizophrenia. In the dream research community, the possibility of using dreaming as a model of psychosis has recently received much attention (D'Agostino & Scarone, 2011; see also D'Agostino et al., 2013). This suggests that the relevance of the delusion hypothesis extends beyond the epistemological problem of dream skepticism to scientific dream research and the philosophy of psychiatry. A tempting idea is that for this reason, the delusion hypothesis might offer a neurobiologically and phenomenologically plausible account of dream deception.

The possibility of using dreaming as a model system, either of standard wake states or of psychosis, is, however, problematic for a number of reasons (Windt & Noreika,

2011). An implicit danger, for instance, is that the analogy between dreaming and pathological or nonpathological wake states might bias theoretical treatments of dreaming toward an oversimplified view of the target phenomenon. For now, the most important point is that to count as a variant of the deception hypothesis, the delusion hypothesis will have to assume that not just any dream belief, but the specific belief that one is currently awake is delusional. This relocates the problem of dream belief, itself fraught with controversy, to the equally controversial problem of delusional belief.

I cannot offer an extended discussion of delusions here, nor will I defend a particular technical concept of delusion. I only want to point out that important parallels exist between the debate about dreaming and the debate about delusions. These parallels will provide the theoretical basis for deciding whether the analogy between dream belief and delusional belief is fruitful.

To begin with, the concept of delusion is used in many different theoretical contexts and lacks a widely accepted definition (Bortolotti, 2009, 2010). It can be understood in a folk-psychological, everyday sense, as in having the delusional belief that one could lose fifty pounds in a single week on a miracle diet, or in a clinical sense referring to delusions associated with psychiatric disorders. Even if the concept of delusion is reserved for pathological cases, however, it is unclear whether everyday reasoning errors are continuous with delusions seen in clinical contexts (Bortolotti, 2009; Radden, 2011).

As is the case for dreaming, the theoretical discussion is hampered by the considerable heterogeneity of the target phenomenon. Delusions are frequently characterized by their bizarreness and by the fact that they are false, as in the Capgras delusion, where patients believe a relative has been replaced by an impostor. But neither bizarreness nor falsity is necessary for delusions. For instance, a patient with persecutory delusions may think that her jealous spouse is following her around, and this belief may count as delusional even if, incidentally, it is true (Bortolotti, 2010). Generally, delusions are rigid and maintained despite evidence to the contrary, and they are typically reported sincerely and with conviction. They are also often a cause of considerable distress. Their effect on behavior is, however, inconsistent, and delusions may coexist alongside apparently contradictory behavior, as in patients with the delusion that they are being poisoned who nonetheless continue eating their food (Bortolotti, 2010). Delusions can be monothematic and relatively circumscribed, as in Capgras delusion, or polythematic and elaborated, for instance, in patients developing further paranoid thoughts surrounding the original delusion (Bortolotti, 2010; Radden, 2011).

Both the origins of clinical delusions and their relation to rationality, belief, and imagination are controversial. First, it is unclear how to assess the impairment of reasoning in the formation of delusions, or indeed whether delusions involve an impairment in reasoning abilities at all. One-factor accounts claim that delusions result from unusual experiences without requiring an additional deficit in reasoning (e.g., Gold & Hohwy, 2000). In this view, delusions can be likened to illusions arising from faulty perception (Hohwy & Rajan, 2012; Hohwy, 2013a). By contrast, two-factor accounts claim that an additional reasoning bias is implicated in the formation of delusions (for further references, see Bortolotti, 2010).

Second, the status of delusions as beliefs has been debated. The reasons why delusions are sometimes claimed to fall short of beliefhood are similar to those found in the debate about dream belief. Arguments often focus on their lack of integration with other beliefs, unresponsiveness to evidence, and failure to guide action (for discussion, see Bortolotti, 2009). As Bortolotti (2010) points out, however, many delusions do have some degree of inferential connection with other beliefs, are sensitive to arguments and evidence to the contrary, and influence behavior to some degree. Also, subjects can self-ascribe delusions and are often able to offer at least some arguments in their defense. Bortolotti herself describes delusions as "mental states grounded on experience and formed via an inferential process that involves beliefs accepted as true by the subjects" (Bortolotti, 2005, p. 192). For instance, in Capgras delusion, anomalous experience may be taken as evidence that a relative has been replaced by an impostor (p. 194), and the delusional subject will try to persuade others that the perceived change in appearance of the relative is evidence for this fact (p. 195). Like actual beliefs, such delusions can be remarkably persistent (p. 197), and patients' behavior can be successfully predicted by ascribing the respective delusional belief to them:

> They behave as believers when they act on their delusion, defend the content of their delusion with simple arguments and relate the content of their delusion to other beliefs they hold. Their behaviour is "anomalous" in that they resist counterevidence and their delusional states are partially compartmentalised in their belief systems. But resistance to change and compartmentalisation are features often shared by non-delusional beliefs. (Bortolotti, 2005, p. 206)

Similarly, Bayne and Pacherie note:

> The first lesson to be drawn from this picture of human "minimal" rationality is that the existence of some degree of compartmentalization or encapsulation—and the concomitant failure to draw certain inferences or notice and resolve certain inconsistencies—need not count as evidence that an agent is irrational or that his purported beliefs do not really qualify as beliefs. Rather, given our finitary predicament, some local irrationality may be the price to pay for global rationality. (Bayne & Pacherie, 2005, p. 180)

They conclude that "given our finitary predicament—the computational, memory, and time limitations we are subject to—it is actually irrational for us to aspire to ideal rationality" (Bayne & Pacherie, 2005, p. 180).

Third and relatedly, and much as is the case for dreaming, delusions are sometimes described as misidentified imaginings (Currie, 2000; Currie & Ravenscroft, 2002). Currie writes:

> What we normally describe as the delusional belief that P ought sometimes to be redescribed as the delusional belief that I believe that P, attendant on the cognitive hallucination that this is a belief that P, when in fact it is an imagining that P. (Currie, 2000, p. 176)

As is sometimes claimed for dream belief, delusions are, in this view, a disordered form of imagining masquerading, as it were, as belief (for discussion, see Gendler, 2013).

To make matters even more complicated, one can allow that delusions originate in a disordered imagination, but claim that they arise when imagining is transformed into belief. As Bayne and Pacherie explain:

> The crucial issue then is not whether some delusions might start their career as imaginings, whether perceptual or propositional, active or passive, but whether an impairment in identification-mechanisms would result in these imaginings being mistaken for beliefs (second-order mistaken belief) or in their being transformed into, or giving rise to, first-order mistaken beliefs. (Bayne & Pacherie, 2005, pp. 177–178)

McGinn endorses a similar view. He claims that delusions originate in a disturbed or overactive imagination giving rise to genuine false beliefs. He also claims that this is precisely what is going on in dreams, thus embracing the view that dreaming involves beliefs akin to wake-state delusions. This echoes the classical claim that "dreaming is sleeping insanity; insanity is a waking dream—roughly speaking" (McGinn, 2004, p. 113).

What about the phenomenology of delusion? Given the account of dream belief defended in chapter 9, this question is particularly important in the present context. According to one study, delusion is

> associated with high levels of conviction; it forms rapidly in the nature of an inspiration with little or no focused thought but, reliant upon perception, is invested with great importance and regularly considered. Emotionally charged delusion has a great capacity to influence thought and action, by the subject's own assessment. (Jones, 1999, p. 12)

Indeed, delusions are a topic not just of philosophical but more importantly of clinical concern precisely because of their disturbing nature. At least in clinical contexts, there is no question as to whether delusions *feel* like beliefs. In fact, they often feel like particularly important beliefs, and subjects often form a strong attachment to delusions despite their idiosyncratic and frequently distressing nature (Radden, 2011). This suggests that the account of dream belief suggested in chapter 9 can be smoothly applied to delusional belief, as well. For now, we can say that delusions, insofar as they fulfill the phenomenal-functional profile of standard beliefs, can be considered as beliefs (Schwitzgebel, 2011b). This does not place any strong demands on consistency with background knowledge, behavior, or rationality but only assumes that delusional beliefs cannot, as a matter of principle, be sharply distinguished from nonpathological beliefs. As is the case for dream belief, their status as beliefs may be decided on a case-by-case basis. Moreover, at least some, if not a majority, of delusions seem to fit the general requirements for beliefhood by manifesting some degree of integration with other beliefs and by being connected to emotion and behavior. If they were causally inert, they would not, after all, be identifiable as delusions in the first place.

This is not yet to say that dream beliefs can be likened to delusional beliefs; so far, I have only argued that the status of both *as* beliefs depends on similar criteria and that the two debates can inform each other. Given the clinical context of delusions, the key question for assessing the delusion hypothesis of dream belief is not just whether the everyday concept of delusion can be applied to dreams, but whether dream beliefs resemble delusions seen in psychiatric disorders. Here proponents of the delusion hypothesis of dreaming should tread carefully, because if clinical relevance and individual suffering and anxiety are indeed the hallmarks of delusions, then it is not clear that this concept can meaningfully be applied to dreaming at all. To be of explanatory value, the analogy between dream belief and delusion will have to be shown to rely on more than a loose, metaphorical, and extended usage of the concept of delusion. If this is not the case, then one might accept that dreams are deceptive but deny that they can be likened to wake-state delusions.

If the delusion hypothesis pathologizes dream deception, it is worth noting that a variant of the no-deception hypothesis does precisely the opposite. Indeed, if there is a view that is even older and more widespread, at least in the popular literature, than the claim that dreams are deceptive, it is the claim that dreaming involves a deep form of insight and provides access to new forms of knowledge. Call this the *insight hypothesis*:

Insight hypothesis: Dreaming is a unique source of genuine insight and knowledge.

This claim has a rich and colorful history; here a few brief remarks will have to suffice. In the *Republic*, Plato made some interesting remarks on the relationship between rational thought and the outrageous, often immoral character of dreams:

> When all the rest of the mind—the rational, regulated, controlling part—is asleep, then if the wild, unruly part is glutted with food or drink, it springs up and longs to banish sleep and go and satisfy its own instincts. . . . On the other hand, I imagine, when someone who's self-disciplined and who keeps himself healthy goes to sleep, he's made sure that his rational mind is awake and has eaten its fill of fine ideas and arguments, which he's brought to an agreed conclusion with himself. (Plato, 1994, 571c–571d)

When such self-disciplined persons go to sleep, the desirous part can be put to rest, and the best part, namely, the rational part, "is free to get on with the enquiries it carries out all by itself . . . and to try to fulfill its impulse for perceiving something in the past or present or future that it doesn't know. . . . In this state he can maximize his contact with the truth and minimize the lawlessness of the visions he sees in his dreams" (Plato, 1994, 571c–572a).[2]

Plato did not explicitly connect this view to the discussion of dream deception and remained silent on the matter of whether one might also engage in the type of metacognitive reflection that would allow one to correctly realize that one is currently dreaming. Nonetheless this passage suggests that sleep can function as a vehicle of genuine insight, at least under certain conditions and in certain subjects. It also suggests that the epistemic potential of dreams is closely linked to their deceptive character and that there might be a thin line between the two.

The close conceptual relationship between madness, deeper forms of insight, and prophecies of divine origin is highlighted in the *Phaedrus*. Here Socrates comments on the common conceptual roots of prophecy (*mantike*) and madness (*manike*), stating that "the two words, mantike and manike, are really the same, and the letter t is only a modern and tasteless insertion" (Plato, 360 BCE/1994–2009). Though prophecy is a form of madness, "there is also a madness which is a divine gift, and the source of the chiefest blessings granted to men." This is a "madness superior to a sane mind (sophrosune) for the one is only of human, but the other of divine origin." Another type of madness is related to the arts, and it arises in "those who are possessed by the Muses; which taking hold of a delicate and virgin soul, and there inspiring frenzy, awakens lyrical and all other numbers" (Plato, 360 BCE/1994–2009).

Much has been written about the creative potential of dreams and about dream insights and inventions (for discussion and further references, see Edwards et al., 2013). For now, I only want to note that an interesting connection exists between the view that dreams are deceptive and the idea that they are a source of genuine insight. How this relates to the quasi-pathological, delusional nature of dream belief is a further question. But an intriguing possibility is that perhaps both the deceptive character of dreams and their potential to convey genuine insight and knowledge are inseparable, and perhaps both are rooted in the kinship between dreaming and madness. If so, then the delusion and the insight hypothesis might not be opposites, as indicated here, but might actually complement each other.[3]

10.3 A Conceptual Framework for Describing Different Scenarios of Actual Dream Deception

10.3.1 Dream Deception, Dream Belief, and Doxastic Situatedness

The classical view, as we have seen, is that dream belief opens the door to dream deception. Yet we also saw, in chapter 9, that belief is part of a larger subgroup of cogitations. Beliefs are cogitations characterized by the phenomenology of certainty or rightness and standing in a functional relationship to prior and subsequent reasoning and actions. This phenomenal-functional profile is also shared, for instance, by affirmation, assurance, and conviction. Focusing on dream belief may be oversimplified to the extent that it detracts attention from the fact that the terms used to describe the target phenomenon are variable. Moreover, to understand the phenomenon of dream deception, we have to understand in more detail how belief is altered by other cogitations and by experience generally. This section introduces a new technical term, *doxastic situatedness*, as a conceptual tool for describing the target phenomenon underlying the presumable vulnerability to dream deception.

Doxastic situatedness refers to a subgroup of intentional attitudes. Beliefs are a paradigmatic case, but other formulations are available. Doxastic situatedness refers to exactly those attitude types that carry with them particular epistemic commitments. If I believe or affirm that *p*, am convinced or firmly assume that *p*, and so on, then I am thereby committed to the truth of *p*. This commitment influences how I subsequently reason about or act on *p*. If I am strongly convinced of a particular theoretical claim, I will defend it against objections; and if I clearly remember having left my keys in my purse, I will search for them there when needed, which deplorably is not always the most direct path to success.

Doxastic situatedness also has a phenomenological profile involving what are often called epistemic feelings.[4] Epistemic feelings flag different attitude types, and they do so directly and quickly. A long-standing conviction *feels* firm and unchangeable, indicating that I am unwilling to budge or change my position with respect to a certain claim. Surprise signals that a new stance toward the content of my prior belief is necessary: I was convinced that the keys were in my purse, but now, having emptied out and thoroughly searched its contents, my former conviction has lost its strength. If this happens for long-standing convictions, we may feel disoriented; our former position is no longer tenable, but we cannot yet see which new one to take. The spatial metaphor of doxastic situatedness expresses the basic intuition that a subgroup of intentional attitudes involves positioning or, occasionally, repositioning ourselves with respect to the content of certain propositions.

Note that this is not the case for all types of cogitations. We can mentally play with certain ideas, try them on, as it were, without really endorsing them or changing our original position. This makes up a large part of philosophizing. I can consider zombie arguments or even substance dualism without in the least feeling the need to endorse them. Imagining and conceiving are two examples of cogitations that carry with them no particular epistemic commitments as to the truth of the things conceived.

Doubt and uncertainty are particularly interesting in this context, because they affect our prior beliefs without themselves providing new ones. They show that our prior position is untenable, but fail to point toward a new one or may even induce us to feel torn between different alternatives. In doing so, they affect our doxastic position by disrupting existing beliefs. Doubt and uncertainty may, for instance, lead us to suspend or deliberately withhold belief or to forestall the formation of new belief. They may also lead us to explicitly disbelieve what previously we thought was true.

A new way of saying that dreams are deceptive, then, is to say that dreaming is a state of doxastic disorientation: dreams mislocalize us in our doxastic framework by preventing access to long-standing beliefs and memories, but also by temporarily enticing us to endorse new ones. Dreams are, literally, misleading experiences. In the Cartesian scenario of dream deception, the dreamer believes or affirms propositions such as "I am awake," "I am *really* holding this piece of paper in my hands and not just dreaming that I am," and so on. The dream self is now *doxastically situated* toward the content of these propositions in the manner of believing or affirming their truth. Again, not all dream cogitations involve doxastic situatedness, and only a subgroup that do will affect how one is situated toward the fact that one is now dreaming. If dream beliefs can occur in isolation, as argued in section 9.6, then believing, in a dream, that my mother has grown a beard does not entail that I thereby take myself to be awake. Other dream beliefs will undermine one's prior doxastic position without substituting a new one. One can be doxastically situated toward the proposition "I am awake" by simply doubting its truth. In this case, one will no longer believe that one is awake, but one will also not believe that one is dreaming. What is essential for dream deception, then, is being doxastically situated toward a certain type of content, expressed in propositions about one's current state of consciousness and the reality of ongoing experience, in a manner that is at odds with the actual truth or falsity of the respective propositions.

All of this suggests that dream deception is not merely a matter of believing, in a dream, the proposition "I am awake." Rather, we must inquire into the different forms of being doxastically situated toward the proposition "I am awake." We must also investigate the different causes of dream deception and ask whether deception is really, as Cartesian dream skepticism would have it, grounded in the phenomenal similarity between dreaming and wakefulness. But perhaps the most pressing question concerns the epistemic consequences of those cases in which doxastic situatedness is lost altogether. What is special about doxastic situatedness, after all, is that it is not just a phenomenal-functional property but an epistemic one. Being doxastically situated holds the promise of genuine insight, but also the threat of deception.

10.3.2 Doxastic Indeterminacy and the Absence of Dream Deception

To fall victim to deception, one has to experience oneself as a knowing self, and one's doxastic situation has to systematically misrepresent what is actually the case. Deception in an interesting sense requires not just that one's beliefs are false but also a modicum of systematicity. The falsity of one's beliefs should be more than a matter of superficial oversight, carelessness, or clumsiness. All of these situations are too easily avoided to count as interesting cases of deception.

My assessment of dream deception starts with cases in which the condition of doxastic situatedness is violated. This can be the case for dreams lacking any cogitative activity whatsoever, but also for those lacking beliefs or belief-like states in particular. In addition, because the occurrence of any particular belief does not in itself entail the occurrence of others, there might also be cases in which the dream self has beliefs, but fails to be doxastically situated with respect to the nature of her ongoing state of consciousness. From now on, I will use the concept of *doxastic indeterminacy* to characterize cases in which doxastic situatedness of the relevant type is lost.

A number of interesting parallels exist between doxastic indeterminacy and the concept of phenomenal indeterminacy as introduced in section 7.3.3. In a sense, doxastic indeterminacy is a kind of phenomenal indeterminacy: it is an indeterminacy or openness in experience. What is missing from experience is, again, a particular phenomenal property: that of believing, affirming, doubting, and so on, that *p*. Again, doxastic indeterminacy is the most complete form in which the phenomenal property of doxastic situatedness can be lost. As is the case for phenomenal indeterminacy with respect to bodily experience (see sec. 7.3.3), doxastic indeterminacy is not a matter of almost-but-not-quite belief but rather points to the inapplicability of the *concept* of belief. Doxastically indeterminate states are, strictly speaking, *a*doxastic states, just as states that are phenomenally indeterminate with respect to the body are not disembodied but *a*bodily states. Doxastically indeterminate states are not perched on the divide between conception and belief, and they are not cases of uncertainty or indecisiveness about which of two things to believe. They are also not cases of deliberately suspending belief or bracketing assumptions about the reality of ongoing experience; all of these, instead, are different ways or even techniques of doxastically repositioning oneself toward previously held beliefs.

In particular, note that just as phenomenal disembodiment still has a distinctly bodily flavor, so disbelief always involves belief. As James notes, "We never disbelieve anything except for the reason that we believe something else which contradicts the first thing" (James, 1890/1950b, p. 284). To James, doubt, not disbelief, is the opposite of belief. And while this is true in a certain sense, even doubt is directly relevant to belief and logically dependent on it; doubts are directed at particular beliefs and affect our ability to endorse them. This is why disbelief and doubt, though they are ways of characterizing the absence of belief, involve different forms of doxastic situatedness. Doxastic indeterminacy, by contrast, is something even less, referring to cases that have no doxastic or epistemic flavor whatsoever, that are utterly bland in this respect.

For this reason, doxastic indeterminacy will typically involve an unnoticed absence of doxastic situatedness and of specific epistemic feelings; noticing the absence of belief, after all, itself involves an epistemic feeling and is likely to prompt subsequent reflection and belief formation. Returning to Dennett's (1991, p. 16) metaphor of epistemic hunger, we might again say that our epistemic appetites, in dreams, are minimal. Dreamers fail to engage in belief formation even when confronted with bizarre occurrences, and when they do form beliefs, they are often of a perfunctory nature. Indeed, indeterminacy blindness on the doxastic level nicely explains why indeterminacy blindness is generally so pervasive in dreams. If I do not form any beliefs, for instance, about my body or about the color of a dream object, then I may not notice the inconspicuous absence of a dream body, or of color, and so on. In this way, doxastic indeterminacy may explain why other kinds of phenomenal indeterminacy often go unnoticed in dreams. And again, indeterminacy blindness may well persist after awakening, predicting that doxastic indeterminacy, like phenomenal indeterminacy in general, may be underreported unless subjects are prompted to describe both the occurrence and absence of dream belief and other belieflike states in dreams. Indeed, if I am correct that doxastic indeterminacy is pervasive in dreams, both in general and with respect to the question of whether one is dreaming or awake, then much of the theoretical debate about dream skepticism suffers from blindness denial, or a particular form of theoretical anosognosia, by persistently overlooking the fact that a majority of dreams do not involve any beliefs of the relevant kind at all.

Another important point is that doxastic indeterminacy describes not only cases in which doxastic situatedness is lost altogether but also cases in which individual beliefs occur in relative isolation. Recall that in chapter 7, I noted that phenomenal indeterminacy is required to make sense of the concept of weak phenomenal embodiment, explaining how isolated bodily experiences can occur without thereby being experienced as occurring in isolation. Indeed, unless one takes experience to be maximally and uniformly detailed, some degree of phenomenal indeterminacy will be required as a backdrop of ongoing experience. The same is true for doxastic situatedness. If it is correct to say that beliefs can occur in isolation and that occurrent beliefs do not necessarily entail the existence of a densely structured network of interrelated beliefs, then there will always be regions in our doxastic landscape that are indeterminate or uncharted and with respect to which we fail to be doxastically situated. Perhaps one can even say that dreams are *weakly doxastically situated* states, in which our doxastic landscape is marked by more blank spots than is the case for standard wake states, and in which the relations between different beliefs, convictions, and doubts are poorly developed and unstable. Given the putative connection between dream cogitation and epistemic feelings on the one hand and phenomenal-functional embodiment on the other, this would be predicted by the present theory.

Now that the concept of doxastic indeterminacy is in place, we can ask whether we have reason for saying that dreams are doxastically indeterminate with respect to the fact that one is now dreaming. Because doxastic situatedness is inseparable from the phenomenology of occurrently believing, this will be the case for dreams in which the question of whether one is dreaming did not enter the dreamer's mind at all. If this is correct, there will be many cases in which there is no fact of the matter as to whether dreamers believe they are dreaming or awake. As discussed at the end of chapter 9, dreamers can hold a number of beliefs without thereby forming any beliefs about their current conscious state. We also have a clear criterion for scoring the presence of dream deception in dream reports: instances of thinking or believing that or wondering whether one is now dreaming are reportable, and if, upon adequate probing and in sufficiently ideal conditions, no such instances are reported, then we have no evidence that the respective dream was deceptive in any interesting sense. To be maximally strict, instances of wondering whether or doubting that one is awake will count not as cases of doxastic indeterminacy but as cases of doxastic situatedness holding at least the potential for deception.

Doxastic indeterminacy offers a simple solution for cases of belief ascription that on their face allow for competing interpretations. As discussed at the end of chapter 9, dreams involving magical behavior such as flying or passing through walls, or B-lucidity, are an example. Using the conceptual tools of doxastic situatedness and doxastic indeterminacy, we can now offer a more precise analysis distinguishing different types of B-lucid dreams. In some cases, the dreamer realizes that this is a dream and consequently is able to deliberately engage in behavior that would be impossible in wakefulness but is appropriate to the context of dreaming. This would be the transition from B-lucidity to a stronger form of C-lucidity, or doxastic situatedness with respect to the fact that this is a dream. In other cases, the dreamer does not realize that activities such as flying or passing through walls are indicative of dreaming and engages in such behaviors *without* realizing that this is a dream or indeed without reflecting on these strange, newly acquired skills at all. This would be a case of B-lucidity coupled with doxastic indeterminacy regarding the dreamer's current conscious state. In still other cases, the dreamer erroneously believes that this is really happening, that this, perhaps surprisingly, is a real case of flying. The dreamer would then be doubly deceived, not just about the nature of the current conscious state but also about the range of possible behaviors in wakefulness. We would now have B-lucidity in the presence of doxastic situatedness and deception in an interesting sense.

Doxastic indeterminacy potentially narrows the scope of dream deception to dreams in which positive evidence indicates that the dreamer was doxastically situated toward the nature of the ongoing conscious state, and where the content of the respective beliefs was false (e.g., "I am awake," "This is really happening," etc.). Doxastic indeterminacy of this sort likely characterizes the majority of dreams and is a way of conceptualizing what Rechtschaffen (1978) calls the isolation of dreams, or their discontinuity with waking life. Recall that for Rechtschaffen, dreaming is marked by an absence of reflection because dreamers typically do not reflect on the fact that they are dreaming. As discussed in chapter 9, much existing research on self-reflection and metacognition in dreams has focused on the particular question of whether dreamers typically realize that they are dreaming, and the popularity of the cognitive deficiency view owes itself to a considerable extent to the fact that in most dreams, they do not. I have argued that interesting forms of thinking, self-reflection, and metacognition can occur in dreams independently of lucid insight. Here, I argue that they can also occur independently of deceptive beliefs that one is actually awake.

Globus (1987) expresses this point by saying that dream reflection typically takes place within the horizon of dreaming. He gives the following example:

> I was at an academic meeting where I comported myself academically. My dream past was by and large confined to the meeting, with an ill-defined sense of myself as having an academic history. I was single-mindedly caught up with the events of this meeting throughout the dream, and this dream life had its own academic past, quite isolated from my immediate waking past but quite consonant with my general past. (Globus, 1987, p. 81)

The lack of connection to his waking life is underscored by the fact that the dream occurred while he was staying in a cabin in the woods with his wife and children. These facts about his current location and recent activities do not enter into the dream; dream-Globus is not, for instance, surprised about his sudden presence at the conference. The dream nonetheless equips him with an alternative, though somewhat sketchy and undefined, personal history. Of course, this is not so much a dream report as, very probably, a carefully selected and reworded description of a dream, perhaps loosely based on an actual report. Yet while I think Globus's dream nicely captures the indeterminacy and sketchiness of his doxastic situation in the dream, I do not agree with his statement that "the world that I encounter within this horizon is full, complete" (Globus, 1987, p. 81), or that "my dream Being constitutes its own past, present and future, just as does my waking Being" (p. 81). Dream-Globus has a past, present, and future only insofar as he thinks of them in his dream; and this past is complete only in the sense that he fails to notice its sketchiness. The situation is analogous to the unnoticed absence of the dream body in phenomenally indeterminate dreams: the absence of a fully defined past is silently overlooked, as is the dreamer's inability to remember facts about his current waking life. We might still want to agree, with Globus, that dreams take place in a different horizon from wakefulness in the sense that waking memories and long-standing beliefs are not accessed, on the level of dream cogitation, in the dream. We should not take this to mean, however, that a new, determinate horizon, complete with an alternative autobiographical history, is produced every time we dream. Instead, we now can offer a parsimonious explanation of why it seems that our dreams equip us with a new past, present, and future: weak doxastic situatedness, coupled with widespread doxastic indeterminacy and indeterminacy blindness, gives rise to an illusion of greater completeness and glosses over all those details that are missing.

What about intermediate states between doxastic indeterminacy and doxastic situatedness? A possible objection to my account is that dreams involve unquestioning assent rather than full-fledged belief and that this is enough for deception. For instance, Reed, while granting that "ordinary dreams involve no considered opinion or strongly appraisive belief" (Reed, 1979, p. 45), argues that the concept of *taking for granted* should be applied to dreaming. Here "the dimension of appraisal is minimal . . . and the possibility of deception is assured" (pp. 43–44), since one can take something for granted that is false. According to Reed, this is enough to vindicate the traditional view: "Ordinary dreaming can involve belief, in the perhaps extended or uncommon sense of 'taking for granted'" (p. 44).

To this, I would respond that taking for granted, or unreflected assent, may well be a precursor to more sophisticated forms of belief. The distinction between doxastic indeterminacy and doxastic situatedness is intended to be not sharp but gradual. Taking something for granted in a dream—for instance, that this is not a dream, but is really happening—is a simple and unarticulated form of doxastically situating oneself toward the fact that one is now dreaming. In doing so, one fulfills the condition for the simplest form of deception. What I would question, however, is that we should ascribe even this simple form of belief to dreams in the absence of any positive evidence. Rather, it might be useful to distinguish such cases of taking for granted from doxastically indeterminate dreams. The concept of doxastic indeterminacy makes room for the possibility that experience does not automatically invite beliefs and does not per se involve even a primitive form of belief such as taking for granted.

In defending this claim, I disagree with James's assertion that we have a "primitive impulse to affirm immediately the reality of all that is conceived" (James, 1890/1950b, p. 319). Russell, discussing James's view that experience automatically invites belief, notes that were this the case, the need for any particular belief-feeling would be obliterated. He concedes that unreflecting assent without specific belief-feelings might be involved in dreams and hallucinations. He also allows that "a mere image, without the addition of any positive feeling that could be called 'belief,' is apt to have a certain dynamic power, and in this sense an uncombated image has the force of a belief" (Russell, 1921/2008, p. 166). Russell argues, however, that this does not suffice to explain all phenomena of memory and belief, but only their simplest precursors, as for instance in dreams. To explain fully formed beliefs, "there must be belief-feelings of the same order as those of doubt or disbelief" (pp. 166–167). These belief-feelings, according to Russell, are at least of three kinds, namely, memory, expectation, and bare assent, where each is "constituted by a certain feeling or complex of sensations, attached to the content believed" (p. 167), and their "difference lies in the nature of the belief-feeling" (p. 167).

The idea is that for explaining the transition from doxastic indeterminacy to doxastic situatedness, the phenomenal-functional profile of the associated doxastic (or pre-doxastic) states is crucial. The current theory predicts that even for simple precursors to belief, such as unreflected assent or taking for granted, there will be an associated phenomenology. My discussion of feelings of familiarity and of just knowing in dreams in chapter 9 suggests that this is indeed the case. Of course, the applicability of a fine-grained conceptual framework describing this transition to dreams will ultimately depend on the expressive granularity of dream reports. For now, however, we have no reason to assume that all dreams involve even a primitive form of either just knowing that one is dreaming or of unreflected assent to the seeming reality of the dream. If they did, this would not explain why this primitive form of belief is described in only some but apparently not in all dream reports. In cases where such belief-feelings are missing altogether, belief attribution should be avoided. A graded account of dream deception is the natural consequence. Without the phenomenology of doxastic situatedness, there are simply no grounds for saying that a given dream was deceptive in any interesting sense.

10.3.3 Concurrent Dream Deception and Apparent Insight

From an epistemological perspective, lucid dreams are interesting because they show that genuine insight and knowledge are possible in dreams. For the same reason, they also show that dreams at least occasionally fulfill the conditions for dream deception. Prelucid dreams are particularly relevant in this context. Just as lucid dreams are marked by the realization that one is currently dreaming, dreams in which one wonders whether or suspects that one might be dreaming are called prelucid dreams (Green, 1968; Green & McCreery, 1994). Prelucidity can refer to the phase of dreaming preceding full lucidity. Some, though not all, lucid dreams arise following a prelucid period of doubting or wondering whether one is dreaming. In other prelucid dreams, the dreamer may be unable to resolve the question of whether this is a dream, for instance, by becoming distracted and returning to a nonlucid dream or waking up. Prelucidity is also used to refer to dreams in which dreamers erroneously conclude that they are in fact awake (Green & McCreery, 1994; Brooks & Vogelsong, 1999).

Prelucid dreams in which the dreamer explicitly believes that she is awake are of particular interest in the context of dream deception because they are, in a sense, the mirror image of dream lucidity. Both involve doxastically situating oneself toward the fact that one is now dreaming, and both give rise to the phenomenology of knowing. Yet they require a diametrically opposed analysis from the third-person epistemological perspective. As pointed out earlier, a precondition for dream deception is that one be doxastically situated toward the propositions "I am dreaming" or "I am awake" in a particular manner. Doubting that one is awake or wondering whether one is dreaming are relevant to cognitive situatedness but do not, in themselves, involve false belief. Transient periods of prelucidly wondering whether one is dreaming that fail to shape subsequent beliefs place us in the vicinity of dream deception but do not yet involve deception in an interesting sense. Rather, genuine cases of dream deception require, for instance, being doxastically situated toward the proposition "I am awake" in the manner of believing it, or toward the proposition "I am dreaming" in the manner of disbelieving it. It is this subgroup of prelucid dreams that is most interesting from the perspective of Cartesian dream skepticism.

As is the case for the (lucid) belief that one is dreaming, the erroneous (prelucid) belief that one is awake can be reached in different ways. First, just as lucidity can sometimes take the form of "just knowing" that one is dreaming, so the (prelucid) belief that one is awake can sometimes be driven by the realistic quality of the dream. Experiential character can enable or prevent true insight. In the following example, the dreamer is already thinking about the topic of dreaming and wakefulness but cannot bring himself to accept that he is now dreaming:

> All these claims about the completely realistic perception of the dream world are untenable! How could it be possible to perceive this cold, clear morning and this colorful autumn forest in all their detail in a dream? How could one be able to feel this cold and clear sensation in one's lungs after taking a deep breath? This would be impossible! (Tholey & Utecht, 1995/2000, pp. 87–88; translation mine)

Let's call this *deception from perceptual realism.*[5] It describes cases in which the conclusion that this cannot be a dream follows directly from the realistic experiential character of the dream. No period of uncertainty necessarily precedes this conclusion; rather, the perceptual realism of the dream convinces the dreamer of being awake from the beginning.

Deception from perceptual realism can also succeed a period of doubt or of wondering whether one is currently dreaming. Many authors have remarked that so-called reality tests are unreliable because of the vivid and highly realistic quality of dreams (LaBerge, 1985, p. 123; Green & McCreery, 1994, p. 15). Even if dreamers already know that they are dreaming, they may find it difficult to detect the ways in which the dream world falls short of replicating the waking world:

> On Sept. 9, 1904, I dreamt that I stood at a table before a window. On the table were different objects. I was perfectly well aware that I was dreaming and I considered what sorts of experiments I could make. I began by trying to break glass, by beating it with a stone. I put a small tablet of glass on two stones and struck it with another stone. Yet it would not break. Then I took a fine claret-glass from the table and struck it with my fist, with all my might, at the same time reflecting how dangerous it would be to do this in waking life; yet the glass remained whole. But lo! when I looked at it again after some time, it was broken.
>
> It broke all right, but a little too late, like an actor who misses his cue. This gave me a very curious impression of being in a fake-world, cleverly imitated, but with small failures. I took the broken glass and threw it out of the window, in order to observe whether I could hear the tinkling. I heard the noise all right and I even saw two dogs run away from it quite naturally. I thought what a good imitation this comedy-world was. Then I saw a decanter with claret and tasted it, and noted with perfect clearness of mind: "Well, we can also have voluntary impressions of taste in this dream-world; this has quite the taste of wine." (van Eeden, 1913)

Indeed, this realistic quality may underlie many of the lucidity lapses discussed in chapter 9. In other cases, dreamers, though wondering whether they are dreaming, may be deceived despite their best efforts to decide the question:

> "But am I dreaming this?" I asked myself in a frightened state of excitement related to my attempts to move forward and being unable to do so. Then I noticed that in the bright southern sunlight, which was shining on the marble in the open hall, the columns were throwing dark shadows. "No, this is no dream," I told myself, "in dreams you cannot see light and shade, but only a diffuse, evenly distributed light." (Kurz, 1919, pp. 79–80; translation mine)

In some cases, the ongoing dream seems to respond to the dreamer's momentary doubt, almost as if dream cognition were tailored to lay skeptical worries arising on the level of dream cogitation to rest:

> I noticed that immediately in front of me, a house appeared to be standing on its head, and concluded that I must be dreaming. But then I noticed that I was wearing glasses, and immediately thought that the reversed visual effect must be caused by the glasses. In order to ascertain whether this was the case, I removed the glasses, whereupon I saw the house in its correct position. This led me to falsely conclude that I was awake after all. (Tholey & Utecht, 1995/2000, p. 88; translation mine)

Deception from perceptual realism can involve varying degrees of reflection, and occasionally dreamers are deceived despite their best efforts to the contrary. Some dreams almost seem driven by deceptive intent. This illusion is explained through the high cognitive penetrability of dream formation: dream cognition and dream cogitation are fed by the same sources, so the occasional responsiveness of dream imagery to consciously experienced thoughts is not surprising and is indeed predicted by the weak-cogitation hypothesis (sec. 9.5). This also suggests that in at least some of these cases, deception is systematic rather than incidental. This is why deception from perceptual realism has all the ingredients of Cartesian dream deception: it involves a dreamer who is capable of engaging in rational reflection and is deceived despite best efforts to the contrary. The seeming inescapability of deception is, of course, somewhat eerie, but while threatening dreamers' knowledge, it does not threaten their status as rational agents. Consequently such instances of deception could also be called *deception despite rational thinking*.

A slightly more uncomfortable form of dream deception involves cases in which the dreamer's failure to realize that this is a dream results from bad reasoning that is nonetheless experienced as completely rational. Unlike the cases discussed so far, this type of deception does not require a realistic mimicry of waking reality. As indicated in section 9.6, flying dreams appear to be particularly vulnerable to such faulty reasoning. This is illustrated by a dream reported by the dream researcher and philosopher John Mourly Vold, in which he was floating horizontally above the ground and thought, "How interesting that I can observe this floating now, while I am awake; it is strange, but it must be true" (Jezower, 1928/1985, p. 316; translation mine). In another example, John Mourly Vold dreamt he was attending a party and suddenly noticed that he was wearing nothing but his underpants. Embarrassed, he went into another room and asked himself whether he might be dreaming. "To decide the question, I take a close look at myself, find that I am also wearing a wool jacket, which is not even buttoned, and conclude that I must be awake" (p. 317; translation mine).

Let's call this *deception from cognitive corruption*. It arises when dream cogitation is corrupted, quite literally, by dream cognition; it is the mirror image of deception despite rational thinking, where dream cognition smoothly glosses over conscious worries and wonderings. What is deceptive here is not the dream world but the impression of being a rational subject. Note that this does not affect the dreamer's ability to engage in cogitations and form beliefs. The dream does not involve an illusion of doxastic situatedness; it involves an illusion of rationality taking place before a backdrop of real cogitations and beliefs. The real problem is that exercising cogitative abilities in dreams not only offers no safeguard against deception but also does not guarantee one's status as a rational or epistemic subject. Deception, in such cases, is located on the level of self-reflective metacognitive monitoring, leading dreamers to drastically misjudge the rationality and reliability of their ongoing reasoning processes and resulting beliefs. It involves a sophisticated type of metacognitive illusion on the level of consciously experienced thought.

These instances of deception from cognitive corruption are disturbing from an epistemological point of view because they show that the mere availability of cognitive capacities says nothing about their reliability. In a dream, one can have the impression of engaging in rational thought or remembering something about one's waking life and be completely wrong. The phenomenology of knowing, thinking, and remembering seems particularly vulnerable to this type of corruption in the dream state. This type of dream deception, then, is not so much deception about the nature of the dream world as deception about the reliability of one's current cognitive abilities. It is epistemologically troubling because it brings the threat of deception even closer to home: whereas Cartesian dream deception has us deceived about the perceptual world and our bodies, deception from cognitive corruption has us deceived about our minds. Consequently we can never be sure of being truly rational at any given moment.

Does lucidity offer protection against deception from cognitive corruption? Lucidity lapses suggest that the answer is no. Take the following example, in which the dreamer is lucid but deceived by a false memory within the lucid dream:

> At a desk by a window I think: "I am asleep." I hold up my hand. "If I wake myself now, will I find that I am sitting at the desk with my hand really in some other position?" I prefer to go on dreaming. "I am near waking, here at this desk," I think, "but if I am careful I may avoid full awakening." I look out of the window and picture myself standing outside on the veranda roof; *but this piece of dream-daydreaming does not become a dream reality*. I am still at the desk. I have a

> book with an index. . . . I walk about the room. . . . I continue the dream but without lucidity, waking at 7:00 a.m., surprised that I was really sleeping not in a desk chair but in bed. (Green & McCreery, 1994, p. 107)

The very possibility of this type of deception from cognitive corruption and deception from false memory is overlooked by many authors discussing dream skepticism. In his own attempt to defuse the skeptical dream argument, Descartes, in the *Discourse*, argues that the evidence of reason is a reliable source of knowledge whether one is dreaming or awake. In this view, dreaming does not threaten the reliability of clear and distinct ideas: the problem is not that reasoning is corrupted in dreams but that it is typically absent. Indeed, Descartes's reliance on clear and distinct ideas as certain indicators of knowledge commits him to the position that not all dreams are deceptive, but dreams can occasionally be the source of genuine knowledge. If the link between clear and distinct ideas and knowledge is to be reliable, it must prove immune to the objection that one could now be dreaming. Accordingly, Descartes argues that "although I might suppose that I was dreaming, and that all which I saw or imagined was false, I could not, nevertheless, deny that the ideas were in reality in my thoughts" (Descartes, 1637/2008, IV.4). This corresponds to his view that dreaming threatens only sensory-based beliefs but does not threaten the certainty of sciences like geometry and arithmetic, which are independent of experience. Even previously unknown truths, like a new proof in geometry, can be discovered in a dream:

> For if it happened that an individual, even when asleep, had some very distinct idea, as, for example, if a geometer should discover some new demonstration, the circumstance of his being asleep would not militate against its truth. . . . For, in fine, whether awake or asleep, we ought never to allow ourselves to be persuaded of the truth of anything unless on the evidence of our reason. (Descartes, 1637/2008, IV.8)

Elsewhere Descartes departs from his official position to respond to objections raised by his critics. In his replies to the *Meditations*, Hobbes asks whether it is not possible that a dreamer, in doubt whether he was dreaming or not, might dream that his dream was part of a long sequence of past events. Here the coherence test would fail: the dreamer would merely have the illusion of performing the test. This would render the coherence test worthless: we would never know whether we were really or seemingly performing it. Responding to Hobbes, Descartes grants:

> [Dreamers] cannot *really* connect the contents of their dream with the ideas of past events, although they can *dream* that they are making the connection. Does anybody deny that people can make mistakes in their sleep? But later, on waking up, they will readily see that they had been wrong. (Hobbes & Descartes, 1641/1975–1999, *Last Reply*)

Here Descartes has retreated to the view that although dreams are deceptive, this does not threaten our ability to know that we are waking. Deception is now unidirectional and restricted to dreams.

At the same time, Descartes's concession to Hobbes glosses over a problem. This is that Descartes has to assume that clear and distinct ideas are reliable, even in dreams, to escape from his skeptical doubts in the first place. Bourdin elaborates this point in detail:

> It is not certain that what appears as distinct and certain to one who is in doubt whether he is awake or asleep is clear and certain. . . . Or has it never happened to you, as it has to many others, that in your dream something appeared as certain and clear that you later realized was dubitable and false? It is a sign of wisdom never to trust those by whom one has been deceived even once; but, you say that this is different for that which has the highest degree of certainty. Are you serious when you say that that which is absolutely certain cannot even seem dubitable to a dreamer or madman? . . . Why shouldn't that which is false and dubitable appear to them as certain, even as absolutely certain? (Descartes, 1972, p. 393; translation mine)

Bourdin adds, for good measure, a description of a person who in sleep counted as he heard the clock chiming one, one, one, one. This led the sleeper to conclude that the clock was broken, having chimed one o'clock four times in rapid succession. There is no necessary connection, as Bourdin notes, between the appearance of clarity and distinctness and actual truth. Even 2 + 3 = 5—a truth that Descartes in the *First Meditation* had claimed was not threatened by the dream example—is, in this view, dubitable.

Descartes sarcastically responds that he had never dreamed of asserting such a thing: "But because only intelligent people correctly distinguish what is conceived [clearly and distinctly] and what only appears to be, I am not surprised that this felicitous man takes the one for the other" (Descartes, 1972, p. 398; translation mine). The impatience manifest in his reaction obscures the point that "clarity" and "distinctness" are success words: essentially, Descartes is admitting that a thought only counts as clear and distinct if it tracks the truth. This is important, because if true, it makes the criterion of clarity and distinctness unusable because from the first-person perspective, we can never distinguish between correctly and incorrectly applying it. What my remarks add to this problem is that concrete evidence indicates that deception from cognitive corruption is not a mere theoretical possibility but actually occurs in dreams.

This is not an arcane historical discussion. Grundmann (2002) focuses on exactly this problem and uses results from dream research to argue that there is an epistemic criterion for wakefulness. Based on Rechtschaffen's (1978) claims about the isolation and single-mindedness of dreams, as well as evidence from lucid dreaming, he argues that empirical evidence suggests that introspective thought rarely occurs in dreams. The problem, as he sees it, is not that the deliverances of introspection are themselves untrustworthy in dreams, but that we rarely form introspective opinions while dreaming. As soon as the ability to engage in critical reflection is reinstated, dreamers can use this ability to correctly realize that they are now dreaming and become lucid. Grundmann suggests the following criterion:

> When we (introspectively) notice that we have the ability to engage in critical reflection and can deliberately direct our attention, and we additionally (introspectively) notice that we can remember the history of our experiences from the first-person perspective, then we must check whether our ongoing experience is internally coherent and seamlessly integrated with our remembered experiences. If so, we have good reason to think that we are not now dreaming, but awake. (Grundmann, 2002, p. 73; translation mine)

Importantly, Grundmann's introspective criterion is a considerable improvement over Descartes's original coherence test. It is based, for instance, not on anecdotal evidence but on a careful reading of the empirical literature. He also notes that the question is not whether we could, in theory, dream that we are successfully conducting the test but whether there is in fact any evidence that we do. Because his is an empirically informed criterion, he allows that it is fallible; yet, he argues that its fallibility does not prevent it from being sufficiently reliable to justify our belief that we are currently awake.

There is much I agree with in this analysis. I agree, for instance, that it is in part an empirical question whether or how frequently we are deceived by our dreams. I also agree that given widespread doxastic indeterminacy, a majority of dreams likely provide no foothold for skepticism. I disagree, however, that there is no evidence suggesting that the formation of introspective beliefs is corrupted in the dream state. Certainly, in assuming this view, the fault is not Grundmann's; his argument reflects, I think, the widely accepted dichotomy between lucid and nonlucid dreams. Because of this way of thinking about dream lucidity, it is easy to overlook the fact that prelucid dreams of the type discussed here are the mirror image of fully lucid dreams: subjectively, they can involve the same types of cogitative activity and the same kinds of doxastic situatedness but require a diametrically opposed analysis from the third-person epistemological perspective. A subgroup of such prelucid dreams involving cognitive corruption can be regarded as real-world examples of using Grundmann's introspective criterion, and failing from a third-person, epistemological perspective. This is why, I think, such prelucid dreams are of immediate relevance for the epistemological problem of dream deception.

In sum, while lucid dreams show that Descartes was correct to claim that sleep and dreaming do not *necessarily* prevent knowledge claims in dreams from being true, the deeper problem is that, just as genuine instances of reasoning and remembering occur in dreams, so do instances of mock reasoning and mock memories, in which the dreamer merely has the impression of being rational. In many dreams, the "evidence of reason" is mere phenomenal evidence, without epistemic value. This, in turn, invites a deeper epistemological problem: although rational thought is possible in dreams, it may not be recognizable. Introspective evidence about the quality and reliability of one's own thought processes does not furnish epistemic justification.

Given that the issue of dream deception has already been transformed into an empirical question and that the fallibility of Grundmann's introspective criterion has already been accepted, how worried should we be? Perhaps my largely anecdotal examples of deception from cognitive corruption only point to rare freak examples and so at best constitute a slight and largely irrelevant amendment to Grundmann's claim that introspection, when present in dreams, is largely reliable. Fair enough; the frequency and relevance of deception from cognitive corruption are open to empirical investigation. But at the least, I think my examples and theoretical discussion so far show that this is a matter worth investigating in the future. A next step would be to investigate how likely dreamers who are already thinking about whether they are dreaming are to correctly conclude that they are, and how frequently dreamers actually mistake corrupted reasoning for rational thought. If it turns out, for instance, that wondering whether you are dreaming is similarly likely to lead to cognitive corruption and successfully attained lucidity, this would, I think, be a more substantial threat to Grundmann's introspective criterion. At this point, this is highly speculative, but here are three reasons for thinking that this might be the case. The first is that lucidity, as we saw in chapter 9, offers no safeguard against lucidity lapses or against cognitive corruption arising alongside lucid insight. The second is that in many cases, lucidity cannot properly be described as an intellectual achievement, suggesting that reasoning is at best a causally enabling, but not a necessary or sufficient condition for lucidity. The third is that based on the weak-cogitation hypothesis and its claim that dream cogitation is only weakly distinguished from dream cognition, one would predict that cognitive corruption in dreams would be not the exception but the rule. For now, then, we have reason for concern that Grundmann's introspective criterion does not offer a satisfying escape from the dream problem.

10.3.4 Retrospective Dream Deception and Dream-Invited Belief

All the examples discussed so far have involved concurrent types of deception, that is, deception occurring while dreaming. But dream-induced false beliefs or feelings of doubt and uncertainty can also persist into wakefulness, extending the scenario of dream deception beyond the borders of sleep. This lifts dream skepticism onto an entirely new level. While the philosophical literature so far has focused on whether we can be certain that we are now awake rather than dreaming and has assumed that dreams are readily identifiable as such after awakening, retrospective dream deception may present an entirely new type of skeptical threat. Importantly, to some, the question of how they can be certain whether a given experience occurred in wakefulness rather than in a dream is a cause not just of theoretical but of clinical concern and personal distress.

The clearest and perhaps best-known examples of dream deception reaching beyond the borders of sleep are feelings of doubt and uncertainty occasioned by false awakenings. False awakenings are examples of ultrarealistic Cartesian dream deception in that, at least occasionally, they re-create a near-perfect facsimile of dreamers' actual sleeping environments and the events that would likely occur if they woke up (Green & McCreery, 1994). According to an online study (Buzzi, 2011), a majority of false awakenings, at least for subjects sleeping in their habitual bedrooms, involve realistic imagery of their actual sleeping environment. The most frequent scenarios are "start-of-the-day" routines, but false awakenings may also involve the experience of waking up in the middle of the night. Subjects habitually experiencing lucid dreams also frequently experience false awakenings, often following lucid episodes.

Much like prelucid dreams, false awakenings often involve the belief that one has just awakened from sleep. Moreover, the realistic content of the episodes suggests that there is a strong integration of autobiographic memory sources. Both factors explain why false awakenings have the potential to develop the scenario of dream deception to its fullest by continuing to deceive the waking subject as to the nature of the experience. False awakenings are theoretically interesting because they involve a highly specific manner of being doxastically situated with respect to one's current conscious state: the dreamer is aware of just having awakened and often of just having dreamt, but not that the ongoing experience is still a dream. Brooks and Vogelsong (1999, pp. 27–28) call the awareness that one was just dreaming "postlucidity." But even following actual awakening, false awakenings can be profoundly disturbing, occasionally even leading subjects to doubt whether they are experiencing yet another false awakening. Such false awakenings can even occasion genuine philosophical doubt. Bertrand Russell used the example of false awakenings to argue that we can never prove that we are not dreaming at this very moment:

> It may be said that, though when dreaming I may *think* that I am awake, when I wake up I *know* that I am awake. But I do not see how we are to have any such certainty; I have frequently dreamt that I woke up; in fact once, after ether, I dreamt it about a hundred times in the course of one dream. . . . I do not believe that I am now dreaming, but I cannot prove that I am not. (Russell, 1948/2009, p. 153)

Note that none of the philosophical objections to the possibility of concurrent dream beliefs (cf. sec. 9.1) work for such retrospective types of dream deception. Retrospective dream deception involves the belief that a given experience occurred in wakefulness when in fact it occurred in a dream; this belief itself, however, is held in wakefulness, and so can't easily be disqualified as a mere dream belief. Moreover, there is a perfectly observable marker of the strongest form of retrospective dream deception: the confusion one feels after undergoing one or even a series of false awakenings and its effect on waking behavior. The philosophical discussion of whether dreams invite beliefs thus misses an important point: dreams have doxastic effects beyond the borders of sleep, and these effects are not limited to the induction of belief but include feelings of doubt and uncertainty. Dream-induced uncertainty as to whether you are now, finally, really awake or whether you are still dreaming is not, or at least is not exclusively, a theoretical problem but can be something felt. And it can have real consequences: aside from making you prone to engage in philosophical meditations, it can influence personal convictions and actions beyond the proverbial armchair.

If realistic false awakenings can lead us to question whether we have now really awoken from sleep or are just dreaming that we have, could there be cases of unnoticed false awakenings? Only then would we have an actual case of retrospective dream deception. Consider first a variant of deception from cognitive corruption involving instances in which dreamers correctly realize that they are undergoing an altered state of consciousness but fail to classify the experience as a dream. For instance, dreamers might think they are experiencing an OBE and have actually left their body, though the experience in fact occurred in a type of prelucid dream. LaBerge has argued that many OBEs are in fact misinterpreted dream experiences:

> In about one percent of the lucid dreams in my record, I felt I was in some sense out of my body. In every case, when examining the experience after awakening, I noted some deficiency in memory or my critical thinking during the experience. In one such situation, I tried to memorize the serial number of a dollar bill to verify later whether I really had been out of my body or not. When I awoke, I couldn't recall the number, but it hardly mattered. I remembered that I hadn't lived in the house I thought I was asleep in for several years. (LaBerge, 1985, pp. 236–237)

Metzinger (2009, pp. 133–134) reports a similar example of an OBE that he experienced as crystal clear and cognitively lucid. Despite failing to make any verifiable observations in the out-of-body state, he was exhilarated upon awakening and excitedly told his sister, who was sleeping in the same room in his parents' house, about the experience. She did not share his enthusiasm about his OBE and, while fumbling with the alarm clock, accidently set it off.

> At that moment, I woke up. I was not upstairs in my parents' house in Frankfurt but in my basement room, in the house I shared with four friends about thirty-five kilometers away. It was not quarter to three at night; the sun was shining and I had obviously been taking a short afternoon nap. For more than five minutes, I sat on the edge of my bed almost frozen, not daring to move. I was unsure how real *this* situation was. I did not understand what had just happened to me. I didn't dare move, because I was afraid I might wake up again, into yet another ultrarealistic environment. (Metzinger, 2009, p. 134)

Apparently one can realize that one is undergoing an altered state of consciousness while mischaracterizing the type of experience, for instance, by thinking one has just left one's body, while in fact one only dreamed that one did. Similarly, at least a subgroup of near-death experiences, shared dreams, or soul travel may be instances of misinterpreted dreams. For now, the more relevant question is whether one can fail to characterize a false awakening, OBE, or other altered state as a dream even after its occurrence. Consider the following example:

> I was lying on a sofa in the drawing-room; at about 3 p.m. I was reading a book, when the light seemed to be slightly darkened, and looking up I saw, leaning in at the window farthest from me, about three feet from the ground, and beckoning, a gentleman whom I had only seen once, about a fortnight or three weeks previously. Supposing that my father wanted me to sign my name (as a witness to a lease, or something of that kind) I got up, went out of the window (which led down into the garden), and passed along in front of the house, and up six steps into my father's study, which was empty. I then went into the yard and garden, but found nobody; so I returned to my sofa and my books. (Green & McCreery, 1975, p. 9)

Had the report ended here, there would be no reason to identify it as a dream report, and in fact the subject herself, even after realizing that the events in question had not really occurred, apparently believed this to have been an apparitional experience or waking hallucination rather than a dream. The report continues:

> When father came in, two hours afterwards, I said, "Why did you send Mr. H. to call me, and then go away?" My father replied, "What are you talking about? H. is down in Wales." Nothing more was said. I did not like to dwell on the subject to either of my parents, and I did not mention the occurrence to any one for several years. (Green & McCreery, 1975, p. 9)

My point here is not that this clearly was a dream; without further information, there is no way to be certain. Yet, assuming that the episode involved either an intrusion of REM-related imagery into wakefulness or an unnoticed episode of falling asleep on the sofa would elegantly explain the source of the experience.

Indeed, an impressive body of evidence suggests that dream imagery can merge with waking perception during SP night-mares (Hishikawa & Shimizu, 1995), where hallucinatory visual imagery, presumably related to REM sleep mechanisms, can take on a particularly realistic quality (Cheyne, 2003, 2005; Cheyne et al., 1999; see sec. 8.2.2.2). This carryover of REM-sleep-related hallucinations and paralysis into subjective wakefulness suggests that dream imagery can be integrated with, and be subjectively indistinguishable from, waking perception even when they occur simultaneously. An alternative interpretation avoids the problem of merging hallucinations and genuine perceptions by suggesting that in at least some of these cases, the entire experience is hallucinatory (Green & McCreery, 1975; Powell & Nielsen, 1998). Rather than occurring in the vicinity of sleep, these may be genuine sleep experiences—mere dreams of wakefulness rather than dreamlike intrusions on the actual wake state. Powell and Nielsen (1998) discuss the famous case of Anna O's snake hallucination, described by Breuer and Freud (1893/2003). They suggest that the case was not, as suggested by Breuer and Freud, a waking hallucination but was in fact an unidentified SP night-mare. Similar examples for the retrospective indistinguishability of dreaming and wakefulness can also be found in narcolepsy, where subjects often report suffering from sudden sleep attacks during the day, accompanied by vivid dream imagery that subjectively is continuous with waking experience (Mignot & Lin, 2009; Nishino, 2009;

Wamsley et al., 2014; I return to this point later). Subjects may only discover by chance that they dreamt at all, so real are these daytime dreams (Fosse, 2000; Nielsen, 2000b; Szücs et al., 2003).

This suggests that a subgroup of experiences classified as waking hallucinations, apparitional experiences, or intrusions of REM-related imagery into wakefulness may in fact involve genuine sleep dreams retrospectively not recognized as such. Note that if this is true, it has interesting theoretical consequences. If subjects occasionally mischaracterize their own experiences even retrospectively, then it at least seems possible that this type of mistake might sneak into the scientific literature, as well. This is exactly what Powell and Nielsen (1998) suggest may have been the case for Anna O's snake hallucination. Consequently it at least seems conceivable that retrospective dream deception could open the door to a more general kind of scientific dream deception. If dream deception is capable of carrying over into wakefulness, then it could presumably lead to the erroneous categorization of a whole *class* of dreams as waking experiences, namely, of exactly those experiences that are in fact dreams but that even after awakening are not categorized as dreams by the subjects experiencing them or by researchers analyzing their reports. Certain dreams may not only deceive the dreamer but be deceptive on the level of scientific inquiry, as well.

So far, I have been treating retrospective dream deception mainly as a theoretical possibility. It is important to note, however, that we have evidence not only for its actual occurrence but also for saying that it is a genuine cause of clinical concern. In the literature, the confusion between dream and reality is described as a unique disorder of dreaming. Nielsen (2010b) notes that dream–reality confusion, associated with an increased intensity and vivacity of dream imagery, can arise during bereavement, but also in mothers experiencing postpartum dreams in which their infants are in peril (which may also lead to dream-enactment behavior; cf. sec. 3.4); in intensive care unit patients, who may subsequently entertain delusional beliefs that the medical staff is trying to harm or even kill them; and in psychotic or borderline psychotic patients, in whom realistic dreams may precede violent and aggressive behavior. Solms (1997, pp. 176–193) discusses further cases, nothing that dream–reality confusion is marked by a radical change in the patients' relationship to their dreams, as well as by an increased frequency and vivacity of dreaming. In a majority of the cases he discusses, dream-reality confusion was associated with lesions in the anterior limbic regions and a significantly raised incidence of epilepsy (pp. 229–230). The dreams reported by his patients were predominately nightmarish, and Solms suggests that the disorder overlaps with recurrent nightmare syndrome. Tellingly, one patient reported that his dreams would continue after nighttime awakenings, thus leading him to mistake them for real experiences. Importantly, this patient could not determine whether he was actually awake or dreaming during the episodes (pp. 180–181).

Narcolepsy is also associated with deceptive daytime dreams. Narcolepsy is a neurological disease characterized by daytime sleepiness, cataplexy (or the sudden and transient loss of muscle tone, often in response to emotional excitement), sleep paralysis, and sleep-related hallucinations (Szücs et al., 2003; Fosse, 2000; Nielsen, 2000b). Daytime bouts of sleeping may be accompanied by particularly vivid dreams that are seamlessly integrated with waking experience. Narcoleptics also report longer and more complex dreams than healthy subjects when awakened from the first period of nighttime REM sleep (Mazzetti et al., 2009). This suggests that narcoleptics' daytime dreams, as already noted by Curley (1977), are candidates for a philosophically interesting form of dream deception. In a dream content analysis study, Schredl (1998) found the dreams of narcoleptic patients to be more bizarre and negatively toned than those of healthy controls. Evidence also indicated that patients may be particularly prone to realizing that they are dreaming while dreaming, suggesting that learning lucid dream induction techniques may be of therapeutic value. However, other patients, especially when they are unaware of their disorder, may remain oblivious that they are having daytime sleep and dream attacks, even after their occurrence. Szücs et al. (2003) report two case studies involving misleading and unrecognized dreams during undetected daytime sleep attacks. One patient, unaware of her condition, told the police she had been assaulted by a bus driver. Because her report was contradicted by eye witnesses and a gynecological examination found no evidence of sexual abuse, she later, for fear of appearing insane, retracted her report, claiming it to be a practical joke, and was accused of misleading the authorities. Another patient suffered extreme consequences from his daytime sleep attacks, including a car crash, financial and social losses, and marital conflicts:

> He frequently had curious experiences during the day—the neighbor throwing litter into the patient's bin; his wife throwing precious objects away. Sometimes he saw himself trying to clean dirt on the side of a ditch. These memories and experiences were confusing. They gave rise to a surprised and suspicious state of mind. Improbable and incomprehensible things happened, leaving him in doubt. Sometimes he gave sensitive-paranoid interpretations to the events, he also denounced the neighbour for filling his bin. His paranoidity drove his psychiatrist to the diagnostic conclusion of a delusional psychosis. Recently he had a severe conflict with his chief on account of a vivid experience of having had sexual intercourse with the chief's wife, which he mentioned to colleagues. Remembering every detail, he was convinced that his story was true, but the reactions of those around him gradually convinced him that this experience could be a hallucination. (Szücs et al., 2003, p. 315)

Luckily, upon being diagnosed with narcolepsy, the patient responded to treatment. Yet both examples show that there is a very real risk that narcoleptic patients, unaware of their disorder, may not only engage in detrimental behavior but also doubt their own sanity. The danger of misdiagnosing their condition is not restricted to self-diagnosis. Szücs et al. note that even though narcoleptic patients are not psychotic,

> In unrecognized narcolepsy, daytime hypnagogic/hypnapompic hallucinations may lead to a false diagnosis of psychiatric disease including schizophrenia or pseudologia phantastica. . . . The pathological experiences may confound juridical evaluation on account of false accusations and misinterpretation of reality. (Szücs et al., 2003, p. 316)[6]

And finally, to make matters even worse, several authors (Nightingale et al., 2005; Knudsen et al., 2010) have noted an association between narcolepsy and RBD. Attarian et al. (2000) report a case of RBD arising out of cataplexy and wakeful dreaming in a narcoleptic patient.

> In December of 1998 she returned to our center with a new complaint: for several months her cataplectic attacks had become more frequent and were not triggered by any identifiable stimulus. These attacks remained, however, identical in character to the ones she had experienced for the past 20 years. As she recovered from her cataplexy, she experienced vivid violent dreamlike imagery and acted out in response to these dreams, resulting in her being physically and verbally abusive to people around her, who had been incorporated into her waking dreams. Although she remained in a semiconscious state and was somewhat aware of what was going on around her, she was unable to control or alter these behaviors. These distressing events occurred daily lasting for a few minutes following a cataplectic attack. (Attarian et al., 2000, p. 132)

Despite realizing, on some level, that she was dreaming, the patient was unable to control her behavioral responses to her dream imagery. Note how aside from completely blurring the distinction between dreaming, sleeping, and wakefulness and raising related conceptual issues (cf. chap. 2), this type of dream enactment behavior during daytime sleep attacks, if not correctly identified by others as sleep-related, might have social and legal repercussions (the same may be true for confusional arousals; for a discussion, see Schenck, 2007, pp. 85–99). Again, this is much more than a purely theoretical problem.

Importantly, not all cases of retrospective dream deception are as spectacular as the ones discussed here. Rassin et al. (2001) found that 68 to 78 percent of healthy subjects report having had realistic dreams that at least for a short time after awakening they thought were real. Such dreams occurred on average twice per month. Six to 10 percent reported acting against others on the basis of what turned out to be a dream, and 7 to 16 percent had memories about whose origin in a dream or wakefulness they felt unsure.[7]

Aside from such instances of unrecognized dreams, dreams may also invite waking beliefs when they are correctly classified as dreams, and these cases may be regarded as more pedestrian examples of retrospective dream deception. It is surprising that this fact has been almost completely overlooked in the recent philosophical discussion of dream belief.[8] If it is possible to say anything indisputably obvious about dreaming at all, then it is, it would seem, that at all times and in all places, people have had a tendency to feel strongly and entertain specific beliefs about their dreams, and sometimes with strange consequences. In many cultures, attempts to incubate prophetic dreams have played an important role in pivotal points in life and such practices are often regarded as a coming-of-age ritual (Van de Castle, 1994; Garfield, 1995). There is and always has been a whole industry of books and experts on dream interpretation, dream symbolism, and dream prophecy; and people's willingness to believe in the meaningfulness of dreams has often been capitalized on by booksellers, authors, and so-called dream experts. This was true for the ancients, who would sleep at the temples with the hope that priests might later use their dreams to diagnose and heal their bodily ailments (Van de Castle, 1994). It is also true for patients in psychotherapy who collect their dreams at home so that their therapists might use them to diagnose and heal their psychological and emotional ailments. This is not to say that dream-induced beliefs or beliefs in the meaningfulness of dreams are necessarily or consistently false or deceptive; dream interpretation can yield valuable insights and may play a helpful role in therapy (for a recent discussion and further references, see Edwards et al., 2013). It is only to say that dreams commonly invite beliefs even beyond the borders of sleep. At the same time, the tendency of dreams to invite potentially misleading beliefs has been known since antiquity. In the *Lucullus*, Cicero describes the dream of Iliona, in which she "so firmly believed that her son had spoken, that she believed it even after waking up" (Cicero, 1994, 2.88). Similarly, recall that Hobbes (1651/1985), as well as Kant (1766/2002) and Schopenhauer (1847/1988), thought dreams to be the source of superstitious beliefs, maintained even after awakening.

For now, it is enough to note that at least some of these dream-invited beliefs are almost certainly deceptive. Beyond dream interpretation, people have long believed in shared dreaming, or in meeting another person in a dream dreamt by both at the same time. They have believed in precognitive dreams that could reveal the future, in dream telepathy, and in soul travel in dreams. Sometimes, as in dreaming of the recently deceased, believing that you really saw and spoke to your loved one in your dream can be a source of real emotional comfort (Garfield, 2001). Such dreams may be related to the psychological process of adapting to one's loss, and believing they were somehow more than just a dream might be helpful in some cases and for some people. Even so, this does not render such beliefs true from an epistemological perspective. It only suggests that some cases of dream deception might be adaptive.

Aside from such obviously dream-related beliefs, dreams may have had a deeper influence on culture and folklore than often realized. I have already discussed the example of old-hag attacks during sleep paralysis (cf. sec. 8.2.2.2), where the structure of the experience and superstitious beliefs are closely related (Hufford, 1982; Cheyne, 2003). As Adler (2011, chap. 1) notes, SP night-mares lend themselves to religious and spiritual, but also paranormal interpretations. But the same may be true for many more beliefs, some of them with religious or even philosophical import. Jones summarizes his survey of dream-related beliefs as follows:

> Dreams have played an important part in the genesis of the beliefs in a soul that can live and move apart from the body, in fabulous and supernatural beings, in the continued existence of the soul after death—with its power of returning from the grave and visiting the living, especially by night—in the connection between this idea and the spirits of departed ancestors (leading to the worship of these), in the possibility of human beings being transformed into other persons or into animals, in the identity of the spirits of animals with those of ancestors, and in night flights through the air. (Jones, 1951, p. 65)

Alongside other altered states such as OBEs or NDEs (Metzinger, 2005), the notion of the separability of mind and body may also have been fed by nocturnal dreams (Windt, 2011). Because most people are at least occasionally able to recall dreams, the folk-psychological concept of the soul can truly appeal to everyone's first-person experience of having glimpsed a world beyond the real world and having left their physical body during sleep. In this way, dreaming may support a form of "common-sense dualism" (Cheyne & Girard, 2009). Compared to more impressive and more rare experiences such as OBEs and NDEs, this more pedestrian and commonplace character commends dreams as the origin of the widespread belief in the soul. Indeed, in many cultures, sleep and dreaming are regarded as states in which the soul routinely leaves the physical body behind (for discussion, see Blackmore, 1983, chap. 8). This key intuition is also reflected in Descartes's endorsement of the retirement view, or the claim that when the soul separates from the physical body during sleep, we are unable to recall any dreams upon awakening (Hill, 2004b; sec. 4.5). In *Human, All Too Human*, Nietzsche traces both the belief in the soul and the origin of metaphysics to the belief that in dreams, we experience a second world. Dreams, in this view, are a return to the more primitive, primeval levels of humanity (for discussion and further references, see Wolfson, 2011). Dreams may thus explain the intuitive appeal of such beliefs and maybe even our proneness to couch them in corresponding philosophical theories. Of course, even if one could trace beliefs in the soul or in the afterlife to dreams, this would not show them to be false. From a strictly logical point of view, the origin of a belief says nothing about its epistemic status (see Metzinger, 2005). But tracing such beliefs to dreams also says nothing in support of their truth; receiving a message from a dead person in a dream, no matter how impressive, does not make the message any more trustworthy, just as experiencing SP night-mares does not license one to infer that one was really attacked by an old hag or vicious incubus. Dream-induced beliefs, even when they are true, bring us dangerously close to deception to the extent that their dream-related origin makes subjects more susceptible to unquestioningly accepting their truth. If it should turn out that our standards for accepting dream-invited beliefs are lower than our standards for accepting other beliefs, and if the gullibility that often characterizes dreaming extends to beliefs about dreaming even after awakening, this in itself would be a cause for epistemological concern.

We even have a technical term for dream-invited beliefs: Nielsen calls them "oneirogenic effects" and notes that what sets dreams apart from other types of imagery and enhances their effects on waking beliefs is their particular experiential character, which he calls "the reality sense." He proposes that "the reality sense is a primary oneirogenic factor that influences waking spiritual beliefs" (Nielsen, 1991, p. 233). Specifically, dreams, like standard wake states but unlike waking imagery, can give rise to subjective, wakelike impressions of reality, involving a sense of being oriented in the dream environment and participating in the ongoing flux of events, sometimes even to the point of making the dream world seem more real than the waking world (p. 234). Importantly, and in keeping with my focus on the bodily dream self (cf. chaps. 7–8), Nielsen notes that the reality sense is derived from somatosensory sources. He links this to the concept of belief defended by James (1890/1950b), according to which belief involves a sense of reality and is ultimately grounded in emotional and bodily states. Nielsen argues that in dreams, bodily sensations and the resulting sense of reality may give rise to animist belief, or beliefs in souls, spirits, and gods, but also in old hags, demons, and witches, as in nightmares and SP experiences. He also links the ancient Greek belief in healing dreams and visionary dreams to the reality sense of dreams and related oneirogenic effects.

The belief-inviting effect of dreams may have been even more pronounced in preindustrial times. Ekirch (2001) notes that in early modern societies, before electric lighting became widely available, sleep was commonly divided into two roughly equal periods, called first and second sleep, interrupted by a prolonged period of wakefulness. The intermittent period of wakefulness was used for praying or conversing with bed partners, for sexual intimacy, but also, importantly, for contemplation. Often this was accompanied by a state of confused thoughts and mind wandering rather than alertness and perhaps even, as Ekirch suggests, involved an altered state of consciousness (Ekirch, 2001, pp. 361–362). Importantly, this fragmented sleep pattern meant that dreams were more likely to be remembered and provided a natural opportunity to reflect on one's dreams, possibly with a profound effect on an individual's social relationships. As Ekirch quite dramatically puts it:

> Had pre-industrial families not stirred until dawn, remaining instead asleep in their beds, many of these visions of self-revelation, solace, and spirituality would have perished by the bedside—some lost in the throes of sleep, others dissipated by the distractions of a new day. (Ekirch, 2001, p. 367)

Dreams were also an important source of religious inspiration. However, the effects of dreams were not always positive, and perhaps the darkest chapter in dream deception is the close connection between SP night-mares and witchcraft. In his review of the historical records from witchcraft trials, Davies (2003) demonstrates a close similarity between the description of nocturnal assaults by witches and other supernatural beings and SP night-mares. At least for a certain historical period, dream-induced beliefs may have had repercussions far beyond folklore and myth, and if these beliefs were indeed one of the factors contributing to witch trials, then they had very real consequences for those accused.

Again, there may be scientific variants of this form of dream deception. Scientific efforts to prove the parapsychological potential of dreaming are still under way, and they continue to remain inconclusive. Ullman et al. write that "the psyche of man possesses a latent ESP [extrasensory perception] capacity that is most likely to be deployed during sleep, in the dreaming phase" (Ullman et al., 1973, p. 227), and end their book by hinting at the "exciting and limitless possibilities in the application of an interdisciplinary approach to the linking of parapsychological research to the basic, applied, and social sciences" (p. 228). Similarly, Krippner (2007), while noting that laboratory studies of anomalous (e.g., telepathic) dreams have remained inconclusive, emphasizes the value of parapsychological approaches to dreaming. This optimism has not been rewarded by decades of parapsychological research on dreaming but has nonetheless persisted. In his assessment of dreams and parapsychology, Domhoff notes that parapsychological research is special in this respect:

> There already has been an enormous expenditure of energy in the past 75–100 years to demonstrate the existence of paranormal processes in a variety of fields, but with very few positive results. . . . On most research questions, such a small pay-off would lead to the abandonment of the effort because science is characterized by a continuing critique of traditional ideas and the search for new ideas. Other new hypotheses initially rejected by the current scientific consensus are either absorbed into the mainstream within 10–20 years or they fade away, so parapsychology is clearly unique in this regard. . . . In conclusion, then, there is no good reason to believe that dreams have any relation to any type of paranormal phenomenon. (Domhoff, 2000)

The persistent effort to find a scientific proof for the paranormal potential of dreaming might thus be a testament not to the truth of this assumption but to the potentially (and persistently) deceptive character of dreaming (for a skeptical perspective on parapsychology, including dreams, see Blackmore, 1996).

In sum, independently of whether we entertain beliefs while dreaming, and independently of whether our dreams are deceptive in the specific sense of leading us to erroneously believe ourselves to be awake while dreaming, dreams have given rise to beliefs entertained beyond the borders of sleep from time immemorial. Many of these dream-induced beliefs, though difficult if not impossible to disprove, are likely false. Even to the extent that they are not, the tendency to regard dreaming as a source of insight may create a positive bias toward dream-invited beliefs, making subjects particularly prone to accept them as true. The philosophical tradition of regarding dreaming as a state of cognitive deficiency should not lead one to forget that there is a much broader and much older tradition of valuing dreams as a source of insight. At least occasionally, this tendency may even enable dreams to extend their deceptive reach beyond the borders of sleep. The beliefs occasioned by dreams are so heterogeneous and so plentiful that they have tinged the entire topic of dreaming with a strange aftertaste: dream research is often considered to be somewhat fringy and not quite to be taken seriously. Indeed, much of what one sees and reads about dreams supports this preconception. Books on dream interpretation, dream symbols, dream telepathy, and so on vastly outnumber scientifically or philosophically minded books and research articles on dreaming. The philosophical discussion has so far mostly neglected to acknowledge this large and murky area of potential cases of dream deception.

I want to close by briefly relating the most drastic example of dream-invited beliefs having real-world consequences that I have come across in my reading of the literature so far. In her discussion of SP-related night-mares in different cultures, Adler (2011) offers the intriguing suggestion that under extremely specific conditions, superstitious beliefs surrounding SP-related nocturnal attacks can endanger the lives of those that hold them. She argues that the high incidence of sudden unexpected nocturnal death syndrome in Southeast Asian male immigrants to the United States might be associated with SP-related beliefs. Having conducted extensive interviews investigating the specific cultural context of the *dab tsog*, the Hmong variant of the old-hag attack, Adler explains the nocturnal deaths by appealing to a nocebo effect, in which actual harm arises from an expected negative outcome. This biocultural and biomedical perspective, she argues, explains why the victims were apparently healthy, mostly male, with an average age of thirty-three, all of whom had recently emigrated to the United States. The particular combination of the stress of emigration, the felt responsibility for their families, the specific cultural context of the *dab tsog* attacks, and isolation from the strategies that in traditional Hmong society are used for dealing with *dab tsog*, together with a possible genetic cardiac disorder, might explain, she argues, the mysterious pattern of nocturnal deaths. She concludes that "sleep paralysis is certainly not fatal; however, under certain conditions, *dab tsog* can kill" (Adler, 2011, p. 133). If this is correct, then what starts out as an instance of concurrent and retrospective dream deception, leading subjects experiencing SP to endorse superstitious beliefs about nocturnal attackers, might turn into a self-fulfilling prophecy.

This points to an underlying conceptual problem: are the most complete cases of dream deception those that can no longer be properly described as deceptive at all? And given this possibility, is my critical discussion of dream-invited belief as involving a subtle form of retrospective dream deception not itself overly simplified and culturally chauvinistic? More generally, to what extent is the discussion of dreams as deceptive experiences presented here biased by the specific philosophical tradition of placing dreams in the context of philosophical skepticism? I cannot begin to answer these questions here, as they fall far outside the scope of the perspective taken throughout this book. I do want to point out, however, that they raise the possibility that considerable metatheoretical bias plagues any discussion of dream deception. Which types of dream-invited beliefs count as instances of dream deception may be subject to intercultural variation, and if Adler is correct, then the same might be true for the real-world effects of dream-invited beliefs as well. This point may be systematically missed, however, by the type of perspective taken here, which is concerned with connecting the epistemological problem of dream deception to findings from sleep and dream research.

I also want to emphasize that not all instances of dream-invited belief are deceptive. Dreams have long been described as a vehicle of personal insight (for discussion and further references, see Edwards et al., 2013), and recent research has begun to investigate this claim in more detail. Edwards et al. (2013) recently investigated the effect of dream group discussions on self-ratings of deepened insight and self-perception. Participants gave high ratings to items on the exploration-insight subscale, which assessed whether they felt they had explored their dream thoroughly during the sessions, whether they had learned about issues from waking life through working with their dream, and more generally whether they gained any insights during the sessions that they might not have had on their own. Participants' answers were significantly correlated with positive attitudes toward dreams. However, the ratings for the action gains subscale, asking whether they felt that they would be able to change bad dreams in the future or use insights from the session to change their waking lives, were low. Apparently, dream-related insight is not quite the same as acting on it, at least according to participants' own self-ratings. Moreover, Edwards and colleagues suggest that their study supports a distinction between two kinds of dream-related insight, where the first concerns the waking-life sources of dream content, and the second concerns insight about waking life resulting from the dream. Their results are highly preliminary; however, they suggest that under special circumstances—namely, discussing selected dreams in the context of a dream group—dreams may be a source of insight without thereby necessarily guiding actions and without requiring insight about the sources of the dream's content. This type of investigation of dream-invited belief and perhaps even genuine insight might usefully complement the investigation of dream deception. A particularly intriguing question concerns the effects of dream-invited beliefs—including genuine insight, as well as deceptive beliefs—on behavior. If Edwards and colleagues' results point in the right direction, then perhaps insights arising from dreams are mostly isolated from subsequent behavior, and perhaps it is only dramatic cases, as in superstition, but also dream-reality confusion or narcolepsy, in which dream-invited beliefs guide waking behavior. If so, the problem of dream skepticism may not so much be a problem of false beliefs arising from dreams but rather concerns their ability to influence subsequent action.

10.3.5 Assessing the Strength of Dream Deception

Now that a conceptual framework distinguishing different forms of dream deception is in place, is there any way to conceptualize the strength and severity of dream deception? If the degree of epistemic discomfort experienced upon awakening from a series of false awakenings is any indication, then it would seem that the psychological impact of dream deception has much to do with the degree to which dream cognition integrates waking-memory sources in a realistic manner. Realistic false awakenings are disturbing because they so exactly anticipate actual awakenings; they are close to waking experience in a way that more bizarre or fantastic dreams are not. Moreover, the dream self behaves and reasons much as it would following actual awakening, making it natural to identify with the dream self. This is, of course, exactly what Cartesian dream skepticism would predict: the psychological impact of deception is strongest when the dream, including the dream self, most strongly resembles our waking self and draws from the same memory sources.

At the same time, this description stands in marked contrast to the way dreams are commonly described in the philosophical literature. Many philosophers working on dreams have argued that a chasm separates dreams and waking life, as well as the dream and the waking self. For instance, Valberg argues that the subject of a dream, or what I have been calling the dream self, is not the same as the dreamer of the dream, or the sleeping person (Valberg, 2007, pp. 64–70):

> There is something ungraspable about emerging from a dream. First I am in the world of the dream, then I am in the world of reality (the world). Yet the two worlds are discrete, with discrete spaces and times. Thus the times at which I am "first" in one and "then" in the other are not related as earlier to later; nor is the "I" at these times a single individual who crosses from one world to the other. (Valberg, 2007, p. 69)

In this view, the transition from dreaming to wakefulness and from dream to waking self involves a break in consciousness. Moreover, though we use the first-person pronoun to refer to both the dream and the waking self, there is, in fact, no single individual that crosses the threshold from dreaming to wakefulness. Similarly, Rosen and Sutton argue:

> In vicarious dreams, the perspective of the dream is that of one of the dream characters, the one we call the protagonist, but this protagonist does not identify himself as the dreamer, and correlatively, the dreamer does not treat or identify herself as the protagonist. (Rosen & Sutton, 2013, p. 1047)

They add that vicarious dreams are rare, explaining that "in most remembered dreams the protagonist is, or is in some sense continuous with, the dreamer," leaving "no doubt that the observed protagonist is still *me*. Vicarious dreams . . . are unlike standard dreams in that the protagonist from whose perspective the dream occurs is someone other than the dreamer" (Rosen & Sutton, 2013, p. 1047).

According to Rosen and Sutton, the philosophical puzzle is what to say about the identity of the dream self. One might want to say, for instance, that "the occupant of the first-person perspective in vicarious dreams really is, despite all appearances, the dreamer" (Rosen & Sutton, 2013, p. 1048). Or one might want to "posit temporary and highly fleeting subjects of experience in vicarious dreams" (p. 1048), saying that vicarious dreams involve a "temporary separate entity, a fleeting self" (p. 1047). Rosen and Sutton explain:

> Given the fluidity and discontinuity of dreams as shown by our examples above, a dream protagonist may exist only for a short part of a dream. . . . Each protagonist, both within and across dreams, may have varying levels and kinds of connectedness with other dream protagonists and with the dreamer, and many such protagonists may have no such connectedness at all. (Rosen & Sutton, 2013, p. 1048)

I want to suggest, however, that there is an even more basic problem. Framing the question as one about the identity of the dream and waking self or about the existence of a separate entity during the dream state assumes that it makes sense to speak of the existence of selves at all and that there is an underlying fact of the matter on the basis of which questions about the identity of the dream and waking self could be decided. It also assumes that in at least some or even, as Rosen and Sutton suggest, a majority of dreams, the dream and waking self are in fact identical.[9]

Throughout this book, in talking about the dream self (or indeed the waking self), I have been using "self" in a strictly metaphorical sense referring only to the phenomenology of self-experience at any given moment. Different layers of phenomenal selfhood have been distinguished, ranging from bodily experience to the experience of being a cognitive agent and perhaps even a knowing self, and a recurring theme has been that these fluctuate both within and across dreams and across the sleep–wake cycle. Talking of the dream self is convenient shorthand for referring to how the experience of being or having a self arises from the fluctuating and highly dynamic interplay of different layers of phenomenal experience. It is not, however, to assume that any underlying entity that could be called the self actually exists; rather, the self is just the outcome of this pattern of fluctuating and dynamic information processing.

Once we accept this point, it is no longer necessary to assume that in a majority of dreams, the dream and waking self are indeed identical. Rather, questions concerning the identity of the dream and waking self can be suspended in favor of questions concerning the phenomenal-functional profile of self-experience in dreams and its variation across the sleep–wake cycle. While I have been more concerned with describing dream experience in its own right than with engaging in comparative statements, it would seem that phenomenal indeterminacy is more pervasive in dreams than in wakefulness. This appears to be true for phenomenal embodiment, but also for the phenomenology of being a thinking self and a possessor of beliefs. Doxastic indeterminacy, as we have seen, complements indeterminacy with respect to bodily experience. But if this is correct, then it may well turn out that in a majority of dreams, the phenomenology of self-experience differs from standard wakefulness.

One might still think, of course, that one could meaningfully ask whether in one's dreams, one has the same history as in waking life, whether one shares the same beliefs and desires, or whether one really looks exactly as one does in wakefulness. There do seem to be cases, after all, in which in one's dreams, one is a slightly different version of one's waking self or a different person entirely, as in Rosen and Sutton's vicarious dreams. Here, I want to draw attention to three points. One, note that some level of determinacy is required to experience and retrospectively describe oneself as having been another person, for instance, of another gender or with a different history, in one's dream, or even of having been a slightly different version of oneself. By contrast, more generic and indeterminate forms of self-experience may more easily be assimilated, in retrospect, to the waking self. It would seem to be a natural default assumption, after all, that in one's dreams, one is roughly the same person as in waking life; this is implicit in the customary use of the first-person pronoun to narrate dreams. Given indeterminacy blindness, there may not be much to contradict such an assumption in dreams involving indeterminate self-experience. Hence a sufficient level of determinacy may be required to realize the conflict between the dream self and the waking self and to say that in one's dream, one was not oneself but another person. This would also offer a parsimonious account of why we typically identify, to some extent, with the protagonist of our dreams: there is nothing to contradict such an identification. We would now have replaced the metaphysical question concerning the identity of the dream and waking self with a tentative hypothesis about the psychological processes underlying this identification. Two, vicarious dreams of being another person (or even an inanimate object) will not be sharply distinguished, in this view, from ones in which the dream self more strongly resembles the waking self, but can be placed on a continuum describing different degrees of similarity. Such vicarious dreams may even resemble waking self-experience more strongly than a majority of dreams in the greater determinacy required to create the impression, after awakening, that in a given dream, one was not oneself but another person. Three, there will not always be a determinate answer to questions concerning the beliefs, desires, or looks of the dream self; while it does make sense to ask about the phenomenology of self-experience, it does not always make sense to say that in a given dream, I had certain beliefs and desires, a particular hairdo, and so on. These facts may not have been part of the overall experience at all. Indeterminacy, coupled with indeterminacy blindness, explains why these questions nonetheless seem to make sense: they create the illusion of dreams being more determinate than they actually are.

Aside from reframing questions about the identity of the dream and waking self as phenomenological questions, this last point may also help us make progress on understanding the psychological dimension of identification, asking to what degree and under which conditions the waking self identifies with the dream self. A measure, as suggested at the beginning of this section, might be the degree of confusion experienced after awakening from a false awakening and the occurrence of dream-invited beliefs. One might also ask, departing from Valberg's description of dreaming, under which conditions dreamers experience their dream and waking selves as discontinuous and distinct, and under which conditions they do not. These are investigable phenomenological questions. For instance, we have seen that in many cases, the transition between dreaming and wakefulness is gradual, with dreams fading into wakefulness or hypnagogic imagery blossoming into a full-fledged dream. Kuiken and Nielsen (1996; Kuiken et al., 1996) have suggested that the perceived meaningfulness of dreams is related to the carryover of the sleep-paralysis-related orienting response into wakefulness. Again, the idea is that continuity in bodily experience—riding on the integration of real-body inputs in the dream—might be the vehicle transporting doxastic continuity and a causally enabling condition for dream-invited belief, as well.

An interesting project in empirical epistemology would be to investigate how these different dimensions interact in generating the experience of falling victim to dream deception. A possible explanation of why false awakenings are often described as a paradigmatic case of dream deception is that they are experienced as continuous with waking life and are fairly close, in terms of integrating waking-memory sources, but also in the form of self-representation, to waking self-experience. In particular, self-experience in false awakenings is likely characterized by a stronger form of phenomenal-functional embodiment and of doxastic situatedness than is the case for a majority of dreams. Again, the continuity of bodily self-experience may play a particularly important role in this process. One reason why false awakenings are so disturbing, after all, is that information about one's mental state and bodily position is reflected on the level of dream content while being unavailable on the level of dream cogitation. False awakenings involve metacognitive imagery, but this information is unavailable to conscious thinking, thus leading subjects to erroneously believe that they have really woken up.

The idea, then, is that the psychological strength of dream deception can be assessed by using different measures such as the degree of confusion experienced upon awakening, the degree of experienced continuity, detailed phenomenological descriptions of self-experience in a given dream as well as the willingness of subjects to identify with the dream self and say that in their dream, they were themselves rather than another person. More objective measures might be the degree of correspondence between the dream body as experienced and the physical body (including the sleeping position) and the degree to which the false awakening was actually realistic (and not just experienced as such), for instance by accurately mimicking the sleeping subject's actual environment. To assess the strength of retrospective dream deception, one could additionally correlate these measures with the dreamer's willingness to adopt dream-invited beliefs. According to the view defended here, the real problem, after all, is not whether the dream self is identical with the waking self, or even whether we entertain false beliefs during sleep, but whether dreams are belief inviting even beyond the borders of sleep.

The ability to integrate dreams into our waking narratives about ourselves may be an important factor. According to Schechtman's narrative self-constitution view,

> We constitute ourselves as selves by understanding our lives as narrative in form and living accordingly. This view does not demand that we explicitly formulate our narratives (although we should be able, for the most part, to articulate them locally when appropriate) but rather that we experience and interpret our present experiences not as isolated moments but as part of an ongoing story. (Schechtman, 2011, p. 398)[10]

Dreams can form part of this story, and to the extent that they do, they can come to be regarded as being on a par with, or even as more significant than, waking experiences. As Knudson et al. write, "Significant dreams motivate or animate personal repositioning" (Knudson et al., 2006, p. 221; see also Knudson, 2001): they may not only become part of our waking self-narrative but also lead to its reshaping. Many different factors—such as the cultural significance of dreams, their phenomenology, and so on—may play a role here. Cultural narratives and superstitions—recall the old hag—may offer a framework for integrating dream experience into waking narratives about ourselves. Note that the question here is not whether dreams themselves are stories and have narrative structure, or whether dreams are the product of a storytelling instinct (see Pace-Schott, 2013; sec. 12.3). Here the point is only that dreams clearly lend themselves to narrative formulation after awakening, in the form of verbal or written dream reports and dream diaries. Because of the linguistic form of dream reports and their formulation in the first-person, past tense, we inadvertently import wake-state bias into theoretical accounts of dreaming, thus laying the groundwork for regarding dreams as meaningful and personally significant experiences alongside waking experiences. The sense of identification with the dream self and the appropriation of dream beliefs and emotions naturally follow from this practice. It is easy to see how the format in which dreams are commonly reported might not only contribute to dream-invited belief but also feed theoretical misconceptions about dreams, for instance, by biasing researchers toward the phenomenal-duplicate hypothesis and its variants. The facility with which dreams can be formed into stories and can be accommodated in our waking narratives glosses over the fact that much of what we believe about dreams is a theorists' fiction. Understanding the mechanisms underlying dream deception may help us understand the psychological appeal of theoretical accounts of dreaming, as well.

For the same reason, a theoretical account of dream deception may be closely related to theoretical accounts of phenomenal selfhood. Gerrans (2015) uses evidence from psychiatric disorders and wake-state delusions to argue that the experience of being a continuous self over time arises when metarepresentations of body states are integrated with the representation of a temporally extended autobiography. The illusion of being a continuously existing and stable entity is, he thinks, an almost inevitable consequence of these integrative processes. If I am correct, then these factors also contribute to phenomenal selfhood in dreams, as well as to the experienced continuity between dream and waking self, which in turn is an important factor for assessing the strength of dream deception.

Finally, in assessing the factors contributing to retrospective dream deception, it will also be necessary to investigate cases in which retrospective dream deception and dream-invited belief do not take hold. The following dream report nicely illustrates that the vivid phenomenology of insight encountered in mystical lucid dreams is not, in itself, sufficient to guarantee the carryover of dream belief into wakefulness:

> Then I was flying again. I remembered to close my eyes and eliminate the visual environment. I did not remember to do any more. I remained floating with body awareness. I saw a bright light to my left. I remembered that a bright light does not need to mean that I am waking up. I was then surrounded by light. I seemed to float in the light and began to contemplate prayerfully what I was doing and might see. I called "Father" spontaneously, meaning God. I remained some time in this attitude and then woke up. While dreaming, I accepted what was happening unquestioningly as an experience of God. Upon waking reflection, although I recognized the event as having the characteristics of a mystical experience, it did not occur to me at all that it had been an actual experience of God. In fact, in my journal, I recognized in this dream two breakthroughs—that I could fall in a dream without crashing or waking up and that I could become aware of a light in a dream without figuring that I must be seeing the light of my room and waking up, as had been my usual response. (Gillespie, 1988b, p. 349)

Gillespie explains that "as a dreamer, I cannot but uncritically accept what happens. Upon awakening, I cannot but be the critic of both the dream and the dreamer" (Gillespie, 1988b, p. 350). Phenomenologically, this dream has the ingredients of a deep religious experience. Nonetheless the significance attributed to it upon awakening was completely different and apparently unrelated to the mystical quality of the dream.

There are interesting parallels here to phenomena seen in focal epileptic seizures. Picard reports two cases of ecstatic seizures associated with a heightened sense of self-awareness, well-being, and bliss, likely originating in the insular cortex. For one patient, these were associated with a deepening awareness, a sudden sense of clarity and of grasping the overall situation, of things seeming self-evident. Perhaps this can be called an epistemic hallucination, a hallucinated feeling of certainty and insight without epistemic justification or warrant. This latter point becomes especially clear in Picard's account of the second patient. This patient, a thirty-nine-year-old male philosopher, describes experiencing paranoid, repetitive, and even suicidal thoughts during his seizures that were of a frightening specificity and smoothly integrated any ongoing thoughts. He writes that "strangely, these thoughts, although totally convincing, always faded into obscurity when the spell was over. After a few minutes had passed, I knew quite clearly that these beliefs were totally false" (Picard, 2013, p. 2496). He likens these experiences to "aha!" moments:

> Although nothing around me seemed to have changed in any concrete way, every observation of my surrounding environment seemed to "make sense" in this way. . . . The great consilience, coherence, and vividness of everything in the world seemed to demand an explanation of its organization. And the explanation was immediate and completely convincing—some mind or purposive agency was at the root of the world's organization. This was a conviction that was both incredibly vague and totally compelling. Speaking as someone who is not normally given over to such—I am an atheist who happens also to be a professor specializing in logic—these beliefs are wildly out of character for me. But, looking back on these experiences, I am struck by two features of these beliefs, which I now believe may be related. First, they were absolutely immune to any rational doubt; indeed, they seemed not even to be possible candidates for any such doubt. Second, they seemed—perhaps ironically, given my earlier statement—to have many of the same features as the most justified, and rationally derived beliefs that a person could possibly hold. For instead of merely being justified by one or several other considerations or observations, these seemed to be irrefutably supported by literally everything in the world. Literally everything that I could experience seemed to cohere with, and lend support to, the belief that there was some sort of agency behind it all. The level of apparent organization possessed by everything in the world simply demanded such an explanation. (Picard, 2013, p. 2496)

The parallels between this description and the lucid dream reported by Gillespie are striking. They suggest that an interesting new way of thinking about dreams is to compare the phenomenology of belief, even of quasi-religious belief and mystic experience, in dreams to that seen in seizures. Understanding the conditions under which such experiences shape subsequent beliefs even after the episodes have passed—as well as the conditions under which belief is restricted to the individual dream or epileptic seizure—might be a step toward understanding the development of full-fledged delusions. An important question for future research, then, is whether investigating retrospective dream deception and contrasting it with cases of dream belief that do not extend into wakefulness might lead to a deeper understanding of wake-state delusions, as well.

10.4 Taking Stock (I): Reassessing the Skeptical Threat

Are dreams deceptive experiences? The analysis presented in this chapter suggests that there is no simple answer to this question. In doing so, however, it does not so much refute the skeptical threat posed by dreaming as relocate and broaden it. I have argued that a majority of dreams are not deceptive in any interesting sense: dreams characterized by doxastic indeterminacy with respect to one's ongoing conscious state provide no foothold for skepticism. At the same time, my tour of different scenarios of dream deception shows that dreams are deceptive in more ways than envisioned by Descartes. Deception from perceptual realism, which comes closest to Descartes's scenario, is also less worrisome, in some ways, than the other forms of dream deception sketched here. While deception from perceptual realism casts the dreamer as a hapless victim of dream experience, questioning the reliability of our judgments about the external sources of ongoing perception, deception from cognitive corruption threatens the reliability of judgments about our own reasoning abilities. Here it is not knowledge of the external world, but a particular kind of knowledge of our ongoing mental states that is threatened. Finally, retrospective dream deception, in which dream-invited beliefs carry over into wakefulness or dreams continue to be mistaken for waking experience even after awakening, at least in its fullest form is a source not just of theoretical, but of practical and clinical concern.

My strategy of using examples from dream research to analyze the empirical plausibility of different scenarios of dream deception has important consequences for assessing the scope of the epistemological threat posed by dreaming. In particular, this strategy has involved recasting dream deception as a practical rather than as a merely theoretical problem. Whereas global skepticism questions the possibility of knowledge per se, the scenarios of dream deception discussed here at most pose a local, highly circumscribed threat, for instance, to our ability to assess the rationality of ongoing thoughts, but also, in retrospective dream deception, to our ability to accurately assess the sources of beliefs and memories. While this might sound like a modest threat from the perspective of epistemology, my analysis reveals that certain scenarios of dream deception are not only *logically* but *nomologically* possible and actually occur.

The attempt to reformulate the classical argument of dream skepticism in such a way as to render it empirically plausible invites a basic objection. If the possibility of dream deception poses a global threat to knowledge, then it has to function on its own, independently of empirical knowledge. An argument that purports to undermine the certainty of empirical knowledge per se cannot itself be strengthened or weakened by empirical facts (Stroud, 1984; Grundmann, 2002). This is an important objection. Note, however, that it is not limited to my reformulated version of dream skepticism but rather applies to the classical Cartesian version of the argument, as well. To the extent that Descartes's skeptical argument makes certain assumptions about what it is like to dream, it has an inherently empirical dimension to begin with (see sec. 1.1). Therefore the attempt to assess this argument in the light of actual examples from dream research does not further narrow its scope but rather investigates the plausibility of what was an argument about the *nomological* possibility of dream deception from the beginning. By contrast, the analysis presented here would indeed be largely irrelevant to skeptical arguments functioning independently of such empirical background assumptions. Note, however, that for a version of dream skepticism to present a global threat to knowledge, such an argument would necessarily have to be based on an extremely general understanding of the concept of *dream* deception (Clark, 2005). Given that dreaming is a real-world phenomenon, however, this would mean that any such argument would no longer be about *dreaming* in the narrow sense of conscious experiences occurring in sleep at all: ex hypothesi, it could not rely on any facts or assumptions about what it is really like to dream. If this is correct, then any version of dream skepticism that functions independently of empirically tractable assumptions about dreaming would thereby cease to be a version of *dream* skepticism; at best, it could count as a case of dream skepticism in a metaphorical sense.

Other skeptical arguments avoid this problem and function independently of specific empirical background assumptions—and for this reason remain restricted to the realm of logical rather than nomological possibility. Such global or industrial-strength scenarios of deception (Clark, 2005) include, for instance, the brain-in-a-vat thought experiment and arguments claiming that deception is the work of an evil demon, scientist, or computer programmer (as in *The Matrix*). While this type of argument may fall under the label of dream skepticism in its broadest sense, this involves an extended usage of the term and certainly cannot be identified with an empirically based conception of consciousness in the dream state. Assuming that the psychological pull of such arguments is nonetheless derived, to a considerable degree, from the appeal to empirical plausibility implicit even in metaphorical comparisons to real-world dreaming (Clark, 2005), a real-world plausibility check may be of interest even in these cases.

Nomologically possible scenarios of dream deception are epistemologically threatening despite their dependence on empirical claims about dreaming. If there are indeed dreams in which I erroneously take myself to be awake because of the realistic phenomenal character of dreaming, then I can never be sure that I am now awake. Even if such dreams are rare, I can never exclude that I am not subject to this type of deception at this very moment. Likewise, if sometimes while dreaming I come to believe that I am awake based on a faulty line of reasoning that I nonetheless take to be rational and hence reliable, then I can never tell whether my beliefs are founded on rational reflection rather than on confabulatory reasoning. Finally, if some memories that I take to be of waking experiences are actually memories of unidentified dreams, as is sometimes the case in retrospective dream deception, then I can never tell with absolute certainty whether a given experience occurred in a dream or in wakefulness. This is enough to profoundly threaten, or so it would seem, the security of many of my waking beliefs.

Have I really shown that no ways exist to rule out the different scenarios of dream deception presented here? In section 4.1, I discussed inference to the best explanation (IBE) as a response to Cartesian dream skepticism, which claims that the threat of dream deception, though it cannot be disproved, can be discarded because the hypothesis that I am now awake presents a better, and lovelier, explanation of my current sensory experiences than the dream hypothesis. One might conclude that the same is true for most of the cases presented here, and that they can be set aside as largely irrelevant, at least for all practical purposes. Note, however, that formulating empirically plausible scenarios of actual dream deception enhances the overall explanatory goodness of the skeptical dream hypothesis. Instead of diminishing the threat of dream deception, the dependence of the dream argument on real-world scenarios of dream deception increases its bite. And because explanatory goodness is dynamic and responsive to further empirical evidence, some skeptical scenarios may be more plausible, and thus more worrisome, than others.

A related objection might be to say that the proposed variants of dream deception are threatening only in the context of rationalist Cartesian epistemology. Perhaps it is only Descartes's demand for absolute certainty that make dreams seem so threatening; as soon as one adopts a more relaxed attitude toward the possibility of deception, one might even admit that we are occasionally deceived by our dreams, but deny that this undermines the reliability of our knowledge claims or is truly worrisome.

Botterill uses the notion of epistemic risk, or "the relative degree of importance, in relation to potential consequences, to be assigned to errors resulting from fallible methods of acquiring beliefs" (Botterill, 2008, p. 150), to deflate the problem of dream deception. The key idea is that

> for fallibilist accounts of justification, introduction of some form of weighting for relative degrees of epistemic risk seems indispensable. For any human method of belief acquisition, being fallible, is liable to result in some errors, and there is a strong intuition that the strength of evidential justification required will depend upon how damaging to our interests those errors may be. (Botterill, 2008, p. 150)

How risky, then, is it to think that one is awake while in fact one is dreaming? Botterill thinks not very. He allows that in dreams,

> We may sometimes be deceived, but if we are, the mistakes only occur when the epistemic risk is low, since we are incapable of acting. So while we may mistake dreams for waking experience, such mistakes can be corrected with hindsight. The epistemic risk is only heightened when we are awake, so it is important from that perspective that we should not mistake something we merely dreamed for something which actually happened. (Botterill, 2008, p. 156)

He concludes that "you do not know that you are not dreaming. But you have all the justification you need for believing that you are awake" (Botterill, 2008, p. 159). The point, then, is, again, that dream deception is largely unidirectional, leading us to believe, while dreaming, that we are awake without thereby affecting our ability to realize that we are awake when in fact we are. Moreover, the unidirectional indistinguishability of dreaming and wakefulness is not epistemically risky; it is a form of deception that has no serious consequences.

Let us briefly consider how this sits with the analysis defended here. First, I fully agree that in a majority of cases, concurrent dream deception is not of a particularly worrisome type. Indeed, the situation is even less worrisome than in Botterill's account, because a majority of dreams do not even involve the low-risk type of deception he considers.

Second and more importantly, Botterill's analysis paves the way for a case-by-case analysis for assessing the level of risk introduced by different variants of dream deception. This fits in well with the strategy pursued in this chapter of assessing the nomological plausibility of different forms of dream deception and possible ways of measuring the strength of deception and susceptibility to dream-invited beliefs. Deception from perceptual realism, again, would not seem to be particularly risky; there is no harm in mistaking a highly realistic experience for the real thing (subjects prone to dream-enactment behavior being a possible exception). Dream deception of this kind is largely adaptive: mistakenly believing oneself to be awake while dreaming, as opposed to mistakenly believing oneself to be dreaming while awake (leading one, perhaps, to attempt to fly out the window), may be the lesser evil.

Similarly, deception from cognitive corruption is not really risky as long as it is restricted to the dream state. A reason for more serious concern is that it might not be; the ease with which we fall prey to cognitive corruption in dreams suggests that similar forms of deception might occur in wakefulness, as well. If so, then deception from cognitive corruption may be relatively benign when it occurs in dreams (leading, mainly, to slightly amusing dream reports of the type "I am standing here in my underpants, so I must certainly be awake"), but not when it occurs in wakefulness. Moreover, it might turn out that certain types of judgments—for instance, judgments about one's current state of consciousness and about the reliability of one's current reasoning processes—are particularly prone to cognitive corruption. If it turns out that beliefs about one's current conscious state tend to occur in the vicinity of sleep and dreaming (for instance, upon awakening from a series of false awakenings) or in the vicinity of other altered states of consciousness (such as out-of-body experiences), and if it turns out that the very situations that invite reflection on our current conscious state undermine the reliability of the resulting judgments, there would be somewhat more cause for concern. Wondering whether you are now dreaming might in itself (at least when it does not arise in the context of a discussion of philosophical skepticism) be an indication that you are. Perhaps our judgments about our current state of consciousness are only reliable when they occur in specific contexts, safely removed from the conditions that undermine their reliability. If so, then philosophical reflection in standard wake states might be the special case in which reflections of this type are not susceptible to cognitive corruption (though the continued pull of philosophical skepticism in general and of dream deception in particular does not, it would seem, invite confidence). We might even want to say that thinking about these issues is itself inappropriate, because the very act of doing so undermines our ability to decide the question. In this view, though deception from cognitive corruption is fairly benign as long as it occurs in dreams, it might alert us to the widespread occurrence of such cognitive corruption across sleep–wake states. This, in turn, would be genuinely troubling, at least from a theoretical perspective. Maybe Hume (1739/2000) was right after all: maybe skeptical worries have a nasty habit of seeming insurmountable as long as one thinks about them, but have a way of resolving themselves once we leave them alone and do something else.

Third and finally, as Botterill points out, retrospective dream deception has the potential to be genuinely worrisome. As we have seen, at least for certain subjects, this is not just a matter of theoretical concern but may have real-world consequences. Moreover, the incidence of deceptive dreams may vary considerably for different individuals; for certain subjects, dream deception may be not an occasional and rare occurrence but a regularly recurring problem.

A recent study used structured telephone interviews to investigate the occurrence of dream–reality confusion in narcoleptic patients. Such confusions were found to be extremely frequent, with 83 percent of patients (as opposed to 15 percent of healthy controls) reporting such occurrences. The authors summarize their results as follows:

> The severity of these delusions was striking. One man, after dreaming that a young girl had drowned in a nearby lake, asked his wife to turn on the local news in full expectation that the event would be covered. Another patient experienced sexual dreams of being unfaithful to her husband. She believed this had actually happened and felt guilty about it until she chanced to meet the "lover" from her dreams and realized they had not seen each other in years, and had not been romantically involved. Several patients dreamed that their parents, children, or pets had died, believing that this was true (one patient even made a phone call about funeral arrangements) until shocked with evidence to the contrary, when the presumed deceased suddenly reappeared. Although not all examples were this dramatic, such extreme scenarios were not uncommon.
>
> All patients with narcolepsy reporting dream delusions provided multiple examples of such occurrences. Two thirds of patients (65%) who completed the follow-up interview reported experiencing dream delusions at least once a week, and all but two (95%) had the experience at least once a month. In contrast, of the six control participants who reported delusions, only two (5% of all control subjects) had experienced this more than once in their lives. (Wamsley et al., 2014, p. 420)

The instances of dream–reality confusion were also longer lasting than those reported by controls, often persisting for hours, days, or even weeks; narcoleptics also rated their dreams as more vivid and emotional than controls and reported higher dream recall. For these patients, the problem is not that their ability to distinguish dreaming and wakefulness is fallible; it is that they often, and sometimes dramatically, fail to draw the distinction.[11] These clearly seem to be cases in which the objection that dream deception is threatening only in the context of Cartesian epistemology loses its force. Even if we do not demand infallible and absolute certainty, the dream argument is a genuine cause of concern. For most of us, this will only be true rarely and in special circumstances, but for some of us, the threat of dream deception may leak into our everyday lives.[12]

That said, I do not claim to have offered a final assessment of the risk, both theoretical and practical, of dream deception. Rather, I hope to have suggested some new ways of thinking about the problem and of investigating it in the future. For now, my main claim is that the conceptual tools offered in this chapter suggest that an assessment of dream deception from the perspective of empirical epistemology can meaningfully be combined with scientific dream research. That this changes the scope and theoretical relevance of the epistemological problem of dream skepticism goes without saying. How worried we should really be—but also who should be worried, and in which situations—is to a considerable degree an empirical question. Investigating it may shed light on the relationship between dreams and wake-state delusions, though perhaps in an unexpected manner.

10.5 Taking Stock (II): Dreams, Delusions, and Insight

If dreams are deceptive, are they deceptive in the manner of delusions? We can compare dream beliefs to delusions in at least two ways. The first is to compare their content to the content of wake-state delusions; the second is to compare their respective origins, that is, the underlying psychological mechanisms and neurofunctional processes.

To begin with, take the prelucid dream-belief that one is now awake rather than dreaming. Can this properly be considered as a delusional belief in the clinical sense? At least, it is difficult to find a comparable wake-state delusion. In wakefulness, even if I am hallucinating, the belief that I am awake is never false. In mistaking a hallucination for an actual perception, I am not in error about the nature of my current conscious state but only mistaken about the origin of a particular aspect of ongoing experience. It stands to reason that if dreams involve beliefs that are comparable to wake-state delusions, we should look to more circumscribed examples of beliefs about particular dream contents rather than to the belief that one is now awake rather than dreaming. Here, it is interesting to note that waking hallucinations differ strongly from each other in that not all of them invite belief. Patients with Charles Bonnet syndrome (ffytche et al., 1998) report vivid and seemingly realistic hallucinations but fail to believe in their reality. A study of more than four hundred schizophrenic patients found that just under 40 percent regarded their hallucinations as real, though 60 percent displayed varying degrees of unawareness about having a mental disorder (Amador et al., 1991, 1994). By contrast, only 1 to 3 percent of laboratory dreams are lucid (Cicogna & Bosinelli, 2001). This would seem to suggest that waking hallucinations are more likely to be correctly identified as such than is the case for dreams. An interesting question for future research would be to investigate whether the concept of doxastic indeterminacy applies to hallucinations in the same ways as I have suggested is the case for many dreams.

Lucidity lapses offer a particularly interesting contrast condition for waking delusions. For instance, it seems that lucid dreamers, though realizing they are dreaming, are particularly prone to believe that other dream characters are real (Levitan, 1994; cf. sec. 9.3.3). If this type of lucidity lapse is indeed more frequent than others, it would be interesting to know whether visual hallucinations of people (or voices) are more likely to be believed than other types of hallucinations. This seems plausible: virtual reality research shows that interacting with other characters increases the subjective experience of presence in a virtual environment (Sanchez-Vives & Slater, 2005; Slater, 2009; cf. sec. 12.2 for details). In this way, a comparative study of delusional beliefs and their relationship to different types of dream beliefs could be theoretically fruitful.

What about dream–reality confusions of the type discussed in the previous section? Are they really, as suggested by Wamsley et al. (2014), properly described as delusional? The question here is not so much whether corresponding wake-state delusions exist but whether this qualifies as a unique kind of delusional belief. Given the psychiatric connotations of the concept of delusion, applying it to narcoleptic patients may be unfortunate because it recasts what is in fact a sleep disorder in psychiatric terms. Moreover, one might raise questions about the similarity between dream–reality confusion and delusions proper. Do instances of dream–reality confusion, for instance, share the phenomenal-functional profile of paradigmatic wake-state delusions? To what extent do they involve biased reasoning, confabulation, and resistance to countervailing evidence? And can they be traced to similar sources as waking delusions? The answers to these questions might help determine the adequacy of the concept of delusion in these cases.

Aside from comparing the content of wake-state delusions and deceptive dream beliefs, one can also inquire into their origins. Theories of delusion formation can be distinguished into one- and two-factor accounts. Two-factor accounts assume that perceptual distortion is a necessary first factor contributing to delusion formation. They assume that additionally, cognitive biases play a role in the formation of delusions (Coltheart, 2007; see also Coltheart et al., 2011). The idea is that while delusions are originally formed in response to anomalous experience, patients may be additionally unable to reject this belief, despite its implausibility and inconsistency with their existing beliefs (Davies et al., 2001).

Others have argued that delusions arise from a single factor, namely, anomalous experience (e.g., Gerrans, 2002). Gerrans (2012) recently used the example of misidentification in dreams, such as Fregoli-like instances of character misidentification, to argue that the phenomenological similarities between delusions and misidentification phenomena in dreams support the view that anomalous experience plays a role in delusion formation. He suggests that dreams offer a particularly clear example of the experiential basis of delusions because dreamers "do not really form any beliefs about their experiences. . . . As a consequence in dreams we are hostage to experience" (Gerrans, 2012, p. 223). Similarly, Gold and Hohwy (2000) argue that some schizophrenic delusions can be seen as a sensible response to anomalous experience. They argue that these delusions violate none of the traditional departments of rationality, such as procedural rationality, but rather belong to a special type of experiential rationality, where the delusion is directly related to the anomalous experience rather than to a faulty reasoning procedure. More recently Hohwy (2013a) has argued that delusions are best explained as arising from inferences under uncertainty, suggesting that a similar account can be applied to delusions and perceptual illusions. This idea gains preliminary support from Hohwy and Paton's (2010) finding that a variant of the rubber-hand illusion (Botvinick & Cohen, 1998; Tsakiris & Haggard, 2005) can be used to induce the experience of supernatural touch in healthy subjects. They found that once subjects had experienced the rubber-hand illusion, two variants could be introduced. In the first, touch sensations were caused by a finger elevated above the rubber hand; in the second, touch was felt on a cardboard box. The authors suggest that this finding may be interesting for theories of delusion formation because participants spontaneously provided supernatural explanations in response to open-ended questions, describing the experience as telekinetic, magic, or outright witchy. While participants clearly did not subscribe to corresponding delusional beliefs, Hohwy and Paton argue that delusional beliefs might arise in the face of more persistent anomalous experience and disturbed reality testing.

How does dream deception fit into the discussion on delusion formation? One way, as suggested by Gerrans (2013a, 2014), is to compare the activity of the default-mode network in dreams, delusions, and standard wake states. Here it turns out that important differences distinguish the three states. Gerrans argues that while delusions differ from standard waking cognition in that the interplay between the default-mode network and dorsolateral systems responsible for decontextualized thought is disturbed, in dreams, this dissociation is even stronger, as in dreams, the default-mode network is not just hyperactive, but almost wholly unsupervised by decontextualized processes and largely decoupled from external sensory inputs. Leaving details to the side, this suggests that on the neurophysiological level of analysis, relevant dissimilarities between dreams and wake-state delusions exist.

How, then (if at all), do one- and two-factor accounts of delusions apply to dreaming? Dream beliefs provide examples of both types of delusions. First, deceptive dream beliefs are often occasioned by anomalous experience and do not seem to require deficient reasoning. Deception from perceptual realism and deception despite rational thinking are likely driven by a single factor, suggesting that here the erroneous belief does not require an additional reasoning bias. Other instances of dream belief may be better explained by appealing to two separate factors. An example would be inferring from the fact that one can fly that one is certainly awake; such cases can best be explained by anomalous experience coupled with (and perhaps creating a bias towards) cognitive corruption.

Dreams have also been compared to delusions in other respects. Researchers have noted that several types of delusions characteristic of schizophrenia often take place in dreams, such as disorientation, the paranoid experience of threats, and delusional misidentification resembling the Capgras and Fregoli syndromes (Kelly, 1998; Schwartz & Maquet, 2002). Yu (2009a, 2009b, 2010) notes that themes of persecution, being attacked, chased, or spied on, and behaving as if one had magical powers or superior mental or physical abilities occur frequently in both dreams and delusional wake states and has suggested a delusional model of such dream themes, comparing them to psychotic wake states. This thematic similarity between dreaming and waking delusions is certainly noteworthy; it is importantly different, however, from saying that delusions occur in dreams: delusions are beliefs (or, depending on one's account of delusions, belief-like imaginings) directed at actual, imaginary, or hallucinatory experiences. But dreaming of persecution is not the same as believing that one is being persecuted in one's dream. Only a subgroup of persecution dreams, namely, those involving beliefs related to the events experienced in the dream, can be compared to persecutory delusions; in other cases, dream persecution may not be associated with belieflike states at all. Persecution dreams can also turn into lucid dreams and thus need not be associated with false belief at all. In investigating the delusional character of dreams, it does not suffice to investigate the general thematic similarity between dream content and wake-state delusions; one should concentrate, instead, on the relationship between doxastic situatedness and the relevant dream themes.[13]

Finally, I want to suggest that dreaming may provide a contrast condition for a further dimension of psychiatric illness: insight. So far I have mostly treated insight, as in lucid dreams, as the opposite of deception. Nonetheless it should be noted that insight is increasingly being studied in the context of psychiatric disorder, as well (Amador & David, 2004). A nosological analysis of varying levels of insight in lucid dreams consequently suggests a promising new perspective. Historically, this relates to the close connection between deception, madness, and insight that was already noted by Plato (360 BCE/1994–2009) in the *Phaedrus* (cf. sec. 10.2). The underlying idea is that we can make progress on understanding the delusional character of dreams by investigating insight in lucid dreams.[14]

Early authors working on psychiatric disease saw insight as opposed to delusion, to the point of considering insight and delusion as conceptual contradictions and claiming that insightlessness or unawareness was part of the definition of delusion and insanity (Berrios & Marková, 2004; Fulford, 2004). Surely this is reminiscent of the view that dreams are necessarily deceptive. But just as it has gradually been accepted that dreams can involve lucid insight into the fact that one is dreaming, there is an increasing awareness in the literature that insight is not opposed to, but rather part of, the psychopathology of psychiatric illness (Amador & Kronengold, 2004). Around the beginning of the nineteenth century, the notion of partial insanity surfaced, which could refer to intermittent or incomplete insanity. This was also called *folie lucide*, or lucid madness (Berrios & Marková, 2004, p. 32). Lucid madness could refer to intermittent periods of insight or even to concomitant insight, in which lucidity and delirium coexisted (p. 40).

Though the scientific exploration of insight has met with some resistance (David, 2004, p. 368), research on insight has proliferated in recent years. Contemporary authors argue that many different kinds of insight exist and that they can coexist with delusion (see Amador & David, 2004; esp. Amador & Kronengold, 2004; David, 2004). As is the case for dreaming, insight into psychosis can take different forms, ranging from concurrent to retrospective and partial to complete insight, where partial insight involves acknowledging the presence of some symptoms while remaining unaware of others. In dreaming, the parallel to concurrent insight would be lucid dreams, whereas retrospective insight exists for all dreams that are recognized as dreams after awakening. Full insight would be involved in fully lucid dreams, whereas it is noteworthy that insight in many lucid dreams appears to be partial, as indicated by the frequency of lucidity lapses. Also, while some patients can recognize that their symptoms are pathological, they may fail to attribute them to mental dysfunction (Amador & Kronengold, 2004, p. 5). Jaspers (1948) already distinguished between awareness of illness or of having undergone a change and insight proper, or the ability to estimate the type and severity of the illness (Berrios & Marková, 2004; Fulford, 2004). Similarly, dreamers may realize that some of the events they are experiencing are bizarre, or even that they are undergoing an altered state, but may fail to classify this state as a dream (instead of, for instance, as an OBE; see sec. 10.3.3). In keeping with this hypothesis, Dresler and colleagues (2014) argue that dream lucidity and insight into psychotic states involve similar patterns of frontal and parietal cortical activation. They suggest that this finding offers an important perspective for regarding dreams as a theoretical and research model of psychosis.

The nosological analysis of insight in lucid dreams is strengthened by the close analogy between deception from perceptual realism and lucid dreams in which lucidity is reached as a result of spontaneous and unreflected "just knowing" rather than as the result of a period of reflection. In such cases, both the insight that one is dreaming and the false belief that one is awake rely on the same type of purely phenomenological evidence, namely, the dreamlike quality of imagery in the lucidity case and the realistic quality of imagery in the deceptive case. The respective beliefs are distinguished by their truth or falsity, but not by the process of belief formation or by the phenomenology of insight and certainty. In particular, one cannot say that the false belief of being awake was reached by a worse or somehow faultier reasoning process than the correct belief of dreaming, as no process of reasoning was involved in either case.

Consequently it seems to be a more or less incidental matter whether the dreamer hits upon the truth (concluding that this is a dream) or stumbles on a falsity (concluding that she is awake or has awakened from a dream). This hypothesis is strengthened by Voss and Hobson's (2015) observation that the neural activation patterns associated with lucidity and false awakenings are, at least based on current measures, indistinguishable. Perhaps, in the cases discussed here, the difference between insight and deception is driven by low-level experience alone, as in one-factor accounts of delusions, whereas other cases in which lucidity is a result of reflection may involve a second factor.[15]

A final difficulty in the nosological analysis of dreaming concerns the applicability of neurological models to insight and insightlessness. Note that dreaming has both been compared to neurological disorders and been proposed as a model of psychosis. This in turn assumes that a uniform explanation for psychosis and neurological disorders exists. Some evidence points in this direction. For instance, Amador and Kronengold (2004, p. 21) argue that anosognosia, which refers to the unawareness of neurological disorders or brain lesions, is similar to poor insight in psychosis and that both may be explained by a common model. Others doubt that neuropsychological models explaining insightlessness in neurological disorders, for instance, by appealing to frontal lobe dysfunction, can be applied to the lack of insight, for instance, in schizophrenia patients (Marková & Berrios, 1994; for details and discussion, see Larøi et al., 2004). I cannot go into the details here, but I do want to point out that for a nosological model of insight in dreams to be successful, this issue would have to be resolved.

In sum, the question of whether dreams are delusional, much like the question of whether they are deceptive, requires a highly differentiated answer. Still, comparing specific dream beliefs to wake-state delusions on a case-by-case basis may be theoretically fruitful, and I have suggested a number of ways in which it might be interesting to do so. Perhaps surprisingly, a nosological analysis can also be applied to dream lucidity and insight, which plays an important role in the psychopathology of psychiatric disease, such as psychosis. It is a question for future research whether insight and its loss in psychosis can be likened to anosognostic states seen in neurological disorders; and it is a further question how these, in turn, compare to dreaming. Yet, for now, a preliminary conclusion is that a nosological analysis of insight and lucidity lapses in lucid dreams might raise novel and fruitful questions for future research.

10.6 Conclusions

Throughout the chapter, I have developed a new and differentiated answer to the question of whether dreams are deceptive experiences, and if so, whether they are deceptive in the manner of wake-state delusions. My strategy has been to confront the epistemological problem of dream skepticism with theoretical-conceptual considerations on the nature of belief and deception, but also with findings from scientific dream research and psychiatry. This strategy is importantly different from the skeptical arguments traditionally found in the philosophical literature: where for Descartes, the dream argument stands at the beginning of his inquiry, I argue that it can only be meaningfully evaluated before the background of an empirically informed theoretical account of dream experience in general and of dream belief in particular. Like the question of whether dreams are functionally disembodied experiences (cf. chap. 8), questions concerning the deceptive character of dreaming point beyond dreaming itself. Where the former question concerns the functional relationship between the dream body and the sleeping body, the latter concerns the analysis of dream beliefs from a third-person epistemological perspective. Assessing both questions only makes sense, however, once a theory of dream experience in its own right is in place.

After discussing standard philosophical accounts of dream deception and of the delusional character of dreams, the main body of the chapter was devoted to the analysis of different variants of dream deception. The conceptual tools of doxastic indeterminacy and doxastic situatedness helped delineate the conditions under which dreams can be considered as deceptive in an interesting sense. I then introduced different forms of concurrent deception (including deception from perceptual realism, deception despite rational cognition, and deception from cognitive corruption) and retrospective dream deception, including dream-invited beliefs. I also tentatively suggested ways of measuring the strength of dream deception.

I closed by reassessing the skeptical threat. While, contra Cartesian dream skepticism, I deny that dreams should generally be regarded as deceptive, my approach should be regarded not as a refutation but as an extension of Cartesian dream skepticism in that it both narrows the scope of the skeptical threat posed by dreaming and applies it to several real-world examples. Though dream deception is probably only applicable to a minority of dreams, my account exacerbates the epistemological problem in that it emphasizes the fact that dream deception is much more than a theoretical possibility. Retrospective forms of dream deception such as dream–reality confusion are particularly worrisome, because they question our ability to identify the sources of waking memories. The same is true for dream-invited beliefs and cognitive corruption extending into wakefulness, as both threaten the reliability of our doxastic states and our ability to assess the rationality of ongoing thought processes. The outcome is that we should regard discussions of dream skepticism not as a purely theoretical exercise but as an interdisciplinary research project in empirical epistemology. In the final section, I discussed perspectives for a nosological analysis of dream deception, highlighted possible points of contact between the debates on dream belief and delusional wake states, and suggested that somewhat surprisingly, a nosological analysis of insight in lucid dreams might be particularly fruitful.

11 From Oneiragogia to Full-Fledged Dreaming: The Immersive-Spatiotemporal-Hallucination Model of Dreaming

11.0 Introduction

Over the last chapters, we have come a long way toward solving the conceptualization problem of dreaming. Having left the contrast between the quasi-perceptual view and the imagination view behind, the bulk of the second part of this book has been devoted to the investigation of phenomenal selfhood in dreams. In the process of first examining and then reformulating common theoretical assumptions about dreaming, a new characterization of dreaming, centered on the dream self, has emerged. We now have a conceptual framework in place according to which a majority of dreams are weakly phenomenally-functionally embodied states, as well as states in which cogitative activity is only weakly expressed and incompletely differentiated from dream cognition. This view has important consequences. For instance, it casts doubt on the widely accepted claim that dreaming arises completely independently of external and bodily inputs. It also suggests a new and more differentiated reading of the claim that dreams are deceptive experiences characterized by cognitive deficiency. Importantly, we now have the conceptual tools in hand to accommodate the considerable variability of the target phenomenon. As my critique of existing theoretical accounts of dreaming has often been that they present an overly simplified and stereotyped view of dreaming, this is an important step toward an empirically plausible view.

The next natural question is whether there is anything like a single, unifying model of dreaming that can accommodate both its typical features and its considerable variability on the phenomenal-functional levels of description. What are the necessary and sufficient conditions for dreaming to arise? How exactly can dreams be distinguished from standard and altered wake states, and what, if anything, sets dreaming apart from sleep-related experiences that do not yet count as full-fledged dreams?

The aim of the next two chapters is to develop a conceptual model of dreaming that answers these questions. I introduce *immersive spatiotemporal hallucination* (*ISTH*) as a conceptual tool for capturing the phenomenal core and the distinctive phenomenal-functional properties of dreaming.[1] I propose that ISTH, while accommodating the high degree of variability of dreaming, is itself invariant and stable across different types of dreams. Moreover, it captures exactly the phenomenal-functional properties that distinguish dreaming from standard and altered wake states.

The ISTH model also offers a new way of thinking about the transition from wakefulness, via sleep onset imagery, to full-fledged dreams. To fully understand what it means to say that dreams are *immersive* spatiotemporal hallucinations, it is not sufficient to define contrast conditions and say how exactly dreaming is distinguished from standard and altered wake states. Rather, a comprehensive theory of dreaming will also have to account for the continuity between dreaming and experiences occurring during sleep onset and sleep–wake transitions. The second important conceptual tool introduced in this chapter is that of *oneiragogic experience*, or of experiences leading into dreams. I argue that these experiences are particularly relevant for understanding the central claim of the ISTH model, according to which dreaming involves a shift in spatiotemporal self-location from a largely veridical, perceptual reference frame to one that is largely hallucinatory. With both of these tools in hand, we can formulate a new answer to the conceptualization problem, which is, in turn, the precondition for turning dreaming into a well-behaved, respectable, and promising target for interdisciplinary consciousness research.

In the chapter's final section, I defend the ISTH model against what I take to be the most important objections against this view. The next step, taken in the book's final chapter, is to develop several consequences from the ISTH model for the description of hallucinatory and imaginative wake states and their relationship to dreaming. This also leads to a reassessment of theories on the function of dreaming and a new reading of the question of why we dream.

11.1 From the Phenomenal Core of Dreaming to the Immersive-Spatiotemporal-Hallucination Model

How might one go about identifying the phenomenal core of dreaming? Given its heterogeneity, it is important to realize, at the outset, that there may turn out to be no invariant and distinctive phenomenal core of dreaming. Dreaming might not turn out to be a natural kind. On pains of committing the armchair fallacy, one should not simply assume that the concept of dreaming picks out a class of experiences characterized by any single, highly invariant, and distinctive phenomenal property. The best we can do might be to define dreaming by pointing to a set of jointly sufficient and partially overlapping phenomenal properties, none of which are strictly necessary for dreaming to occur. Perhaps only a loose family resemblance exists between the different types of experiences that pass under the name of dreaming. Clearly, having a unitary account of dreaming would be preferable, in view of its greater simplicity and parsimony, to such a cluster account. Such metatheoretical considerations do not, however, justify pressing the target phenomenon of dreaming into a conceptual straitjacket. While definitions rarely proceed without at least a modicum of conceptual legislation, any workable account of dreaming should be closely constrained by empirical research findings, or else it will be set adrift from the phenomenon it was originally meant to capture. For this reason, one should not assume that all forms of dreaming center on a common core from the outset. Fortunately, as I will argue, it is possible to introduce a unified and distinctive account of dreaming that is maximally empirically plausible and minimally legislative, suggesting only slight adjustments in the use of the concept of dreaming.

Note also that even if it is possible to pick out a single set of highly invariant phenomenal properties underlying many different types of dreams, this may not suffice to distinguish dreaming from standard and altered wake states. The invariant phenomenal core of dreaming might be too general to be distinctive of dreaming. I will argue that this is the case: the distinctive features of dreaming only become visible by looking beyond the phenomenological level of description to its functional characteristics. This should not be surprising: the absence of a fail-safe phenomenal marker to distinguish dreaming from wakefulness has been a cause of philosophical concern for centuries.

My analysis will proceed in two steps. I begin by isolating the invariant phenomenal core of dreaming. I then investigate how this purely phenomenological description can be turned into a distinctive account of dreaming by taking the functional characteristics of dreaming into account. Starting with the first question, and based on the foregoing discussion, it should be clear that the phenomenal core of dreaming cannot be identified with any of the features that characterize a majority of dreams. Rather, looking at examples that deviate from a majority of dreams is a more promising strategy, because these examples will allow us to whittle away all those aspects that are not strictly necessary for dreaming. Indeed, the conceptual framework developed over the last chapters has already given a good indication of where *not* to look for the necessary conditions for dreaming to arise. Given this extensive background, a cursory discussion here will suffice.

To begin with, while dreams often integrate multimodal imagery, neither multimodality nor specific types of imagery occur in all dreams. A majority of dreams involve visual imagery (e.g., Hobson et al., 2000), but even visual imagery is not necessary for dreaming to occur. Charcot first described the loss of visual dreaming, typically associated with lesions in the occipital-temporal area, in 1883 (for discussion and further case studies, see Solms, 1997, 2000; cf. sec. 3.6). Though adventitiously blind subjects occasionally experience visual dreams even years after they have lost their eyesight, both adventitiously and congenitally blind subjects also report vivid nonvisual dreams (Kerr, 1993; Staunton & O'Rourke, 2012; Meaidi et al., 2014). This finding suggests that although visual imagery is the most frequent type of imagery to occur in dreams, it is not a defining characteristic of dreaming. Consequently there is no single type of modality-specific imagery that necessarily occurs in dreams.

Similarly, multimodality does not appear to be necessary for dreaming. If no specific sensory modality necessarily characterizes dreaming, then the same will be true for any combination of modality-specific types of imagery. Preliminary support for the claim that multimodality per se is not necessary comes from studies suggesting that a subgroup of participants characterize their dreams as involving no sensory experiences whatsoever (Okada et al., 2005). This claim is also supported by the general observation that multisensory integration is frequently disturbed in dreams (sec. 7.4). Consequently dreams do not necessarily involve multimodal imagery, and it even seems possible that dreams can fail to involve any sensory experiences whatsoever. I return to such imageless dreams later in this section.

What about other typical dream characteristics, such as cognitive and mnemonic deficiencies, bizarreness, and strong emotions (Hobson et al., 2000)? Again there are obvious counterexamples to the claim that they are necessary conditions for dreaming to arise. Fully lucid dreams show that dreams are not necessarily characterized by cognitive and mnemonic deficiencies, and realistic false awakenings show that dreams are not necessarily bizarre. There are no a priori conceptual reasons for claiming that dreaming is necessarily a state of cognitive deficiency or that dreams are necessarily bizarre, and indeed, there are empirical grounds for doubting that this is the case (cf. secs. 9.3 and 10.3).

Emotions are a more convincing candidate for a stable characteristic of dreaming. Studies using self-ratings and affirmative probes suggest that nearly 100 percent of dreams involve emotions (Merritt et al., 1994; Sikka et al., 2014). It also does not seem that emotions are merely transient occurrences in dreams: Sikka et al. (2014) found that the average dream involves seven different emotion categories. While this is not quite the same as saying that emotions occur throughout dreams, and while the results are preliminary, let us assume that emotions really are a phenomenological characteristic that is maximally stable across dreaming. Would we thereby have a viable candidate for identifying and empirically grounding the phenomenal core of dreaming? And could a model of dreaming be introduced according to which dreaming is essentially emotional experience during sleep? I think there are two main reasons for concern.

First, even assuming that one could defend the argument that emotions are necessary for dreaming, an obvious objection arises to the claim that they are sufficient. Isolated emotional experiences in sleep are the cause of a sleep-related phenomenon that is commonly taken to be distinct from dreaming: sleep terrors. Sleep terrors (also known as night terrors or *pavor nocturnus*) are episodes during which subjects awaken in a state of intense panic, often accompanied by screaming, bolting out of bed, sweating, panting, and so on. Sleep terrors are most common in children but may persist into adulthood and, unless associated with other sleep behaviors such as sleepwalking, are not dangerous. Importantly, according to Schenck (2007, p. 144), sleep terrors are not just different but "completely distinct" from nightmares. Whereas nightmares are vivid and emotionally intense dreams causing awakening, which mostly occur during REM sleep and in the second half of the night, sleep terrors are not proper creatures of sleep. Sleep terrors are behavioral episodes that take place in a dissociated state between sleep and wakefulness and typically arise from the deep stages of NREM sleep in the first half of the night. Also, whereas nightmares are often vividly recalled, subjects experiencing sleep terrors often remember either nothing or only a vague episode of fright preceding the episode. They also often forget the entire episode, including fear behavior, if asked about it the next morning. Consequently it seems that sleep terrors are preceded not by full-fledged, temporally extended and narratively complex dreams but, if at all, by short and isolated emotional experiences during sleep (Schenck, 2007, p. 148). This is also supported by the observation that night terrors are often triggered by a subtle stimulus, such as the sound of a door opening, which immediately and apparently without any buildup leads to a strong emotional reaction. But if this is correct, then isolated emotional experiences during sleep are not sufficient for full-fledged dreaming; if they were, the distinction between dreams and night terrors would be obliterated. This would be at odds with the goal of working toward a clear conceptual model of dreaming. So even if it turns out that the frequency of emotions in dreams, as seems to be suggested by newer studies, was severely underestimated by earlier studies (for details, see Sikka et al., 2014), and even if it turns out that dreams are consistently accompanied by emotions, emotions would not be a promising candidate for the phenomenal core of dreaming.

The second and related cause for concern is that identifying emotions as the phenomenal core of dreaming does not translate into an explanatory model of dreaming. In searching for the phenomenal core of dreaming, we are, after all, searching not just for any kind of maximally invariant phenomenal property of dreaming but for something that will yield the conceptual glue, if you will, to hold different types of dreams together. Emotion does not seem to be a promising candidate for this job, not only because it fails to honor the accepted distinction between dreams and sleep terrors, but also because there does not seem to be any deep, intrinsic connection between the concept of dreaming and that of emotions. Indeed, many researchers have quite willingly assumed that emotions are not necessary for dreaming and have been surprised by the discovery that emotions are more prevalent in dreams than previously thought. Of course, it is always possible that the folk-psychological concept of dreaming is too vague to be used as a guide for developing a new, empirically informed model. Yet all other things being equal, folk-psychological plausibility can serve as a first indication of whether a new conceptual model can help make sense of the target phenomenon and can be used as a successor, for instance to the folk-psychological concept of dreaming. For the emotional character of dreaming, this apparently is not the case; if it were, new research results on the prevalence of emotions in dreams would not be as surprising as they are.

Rather than focusing on specific types of dream imagery, on the cognitive and mnemonic characteristics of dreaming, dream bizarreness, or emotions, I propose that the phenomenal core of dreaming is closely related to the phenomenology of self-experience in dreams. But because the analysis over the last chapters has shown phenomenal selfhood to be highly variable in dreams, in the degree both of phenomenal-functional embodiment and of cogitative selfhood, and because it has also become clear that phenomenal-functional embodiment and cogitative selfhood can be lost altogether, neither dimension appears to be strictly necessary for dreaming. Yet underlying this variation in phenomenal selfhood, ranging from phenomenal disembodiment and indeterminacy with respect to the body to strong phenomenal embodiment, and from doxastic indeterminacy to strong forms of doxastic situatedness and cogitative selfhood, dreams consistently seem to be characterized by at least a basic sense of self. This suggests that questions about the phenomenal core of dreaming can be transformed into questions about the phenomenal core of phenomenal selfhood in dreams.

Considering phenomenal selfhood as a candidate for the invariant phenomenal core of dreaming is promising for several reasons. To begin with, it captures the intuition, widespread in folk psychology and, as we have seen, in philosophy, that dreaming involves the experience of being present in a world. This intuition gains some support from scientific studies of dreaming: studies consistently show that a self is present in a majority of dreams. While Strauch and Meier (1992, p. 112) found that as many as 10.6 percent of dreams are described as selfless, Occhionero et al. (2005) found that only 1 percent of REM reports (and nearly 9 percent of NREM reports) lack any reference to the self (tellingly, they give hypnagogic imagery, not dreaming, as an example of selfless dreams). Right at the outset, a possible objection to the claim that phenomenal selfhood is the phenomenal core of dreaming arises from occasional reports of selfless dreaming. Because we are aiming to identify the necessary conditions for dreaming, even such rare cases of selfless dreams are relevant. Even if it turns out, however, that a substantial proportion of dreams (and not just of sleep onset experiences or sleep thinking) are consistently described as selfless, one can still ask whether the key intuition that dreaming involves the experience of presence in a world can be preserved while accommodating such reports of selfless dreams. My suggestion is that this can be done by refocusing attention from the phenomenology of selfhood to the experienced relation between the phenomenal dream self and the dream world.

The property I have in mind is the immersive character of dreaming. We can begin to unpack this notion by saying that immersion is tied to the phenomenology of self-location. Self-location is still intimately related to the phenomenology of selfhood (in fact I will argue that it is the simplest form of phenomenal selfhood). Yet, at least when self-location is decoupled from the experience of being a phenomenally embodied self and cognitive agent, it is different enough from standard waking self-experience to possibly lead subjects to retrospectively describe the episodes in question as selfless (I return to the problem of selfless dreams in section 11.3.2).

For now, note that self-location can be understood in a spatial sense and in a temporal sense: on the phenomenological level of description, it refers to the experience of being located *here*, at a specific point in a larger spatial expanse, and *now*, at a specific moment within a duration or succession of unfolding moments. Both properties are relational: the phenomenal *here* and *now* refer to the sense of being located at a specific point within a broader spatial and temporal reference frame and relative to other persons, objects, or events in this reference frame. Spatial and temporal self-location situate us within space, but also within time; they serve as the center of the experienced world. The central idea of the present model is that dreams are immersive in that they essentially involve the experience of spatiotemporal situatedness, thus explaining the commonplace intuition that dreaming involves the phenomenology of presence in much the same way as standard waking experiences do. From now on, I will use the concepts of spatiotemporal self-location, spatiotemporal situatedness, and immersion interchangeably to refer to this phenomenal property.

Spatiotemporal situatedness is a promising candidate for identifying the phenomenal core of dreaming for a number of reasons. To begin with, it suggests a unified and state-independent account of conscious experience, both in dreaming and in wakefulness. Much has been made of the perspectival nature of consciousness: conscious experience always seems to be tied to a perspective; it is inherently subjective (see Metzinger, 2003, pp. 156–162). Spatiotemporal self-location is a highly invariant property of phenomenal experience: though the phenomenal *here* and *now* as such are constantly shifting, they consistently form the center of the phenomenal world. Spatiotemporal situatedness can also be described as a particular kind of representational content, representing the spatial and temporal relation between the phenomenal self and the experienced world.[2]

This basic idea has a rich philosophical history. It is closely related, for instance, to the Kantian idea that experience is structured by our pure intuitions of space and time. The temporal structure of experience is also of central importance for philosophers working in the classical phenomenological tradition of Husserl, including neurophenomenology (see Thompson, 2007, chap. 11). It is also related to the claim, popular in contemporary philosophy of mind, that the phenomenology of *here* and *now* is crucial for conscious experience. For instance, Revonsuo argues that "phenomenal space is a prerequisite of any qualities at all to spring into existence" (Revonsuo, 2006, p. 170); individual qualities are necessarily spatially located somewhere, and several different qualities are necessarily located relative to each other. He also argues that temporality plays a similarly central role for experience (p. 172).

Because spatiotemporal self-location might be a way of making sense of what it means to say that consciousness is centered on a single and unified perspective, it might also be particularly relevant for understanding the emergence of phenomenal selfhood. The experience of being in a world only makes sense if we assume that someone, a phenomenal self, is experienced as being located in this world. Phenomenal selfhood requires the experience of a self–world boundary; phenomenal selves are always situated selves, located at some point in the experienced world and distinguished from other persons or objects in it. They are also always firmly located in the present and bear a particular relation to remembered past and anticipated future selves (for a defense of the claim that minimal selfhood results from the temporal structure of consciousness, see Kiverstein, 2009). Spatiotemporal situatedness thus is a plausible candidate not only for identifying the phenomenal core of dreaming but also for understanding conscious experience and phenomenal selfhood as such.

Moreover, focusing on spatiotemporal self-location suggests a possible way of distinguishing dreaming and wakefulness. The basic idea is that the transition from wakefulness to dreaming involves a shift in spatiotemporal self-location, such that dreams occur when the phenomenal *here* and *now* is located no longer relative to a largely veridical, perceptual environment but rather relative to an alternative, largely internally generated environment, the so-called dream world, that is at best weakly constrained by external stimuli and hence hallucinatory. This shift in spatiotemporal self-location is typically not detectable on the level of phenomenal experience. With the exception of lucid dreams, we remain oblivious, while dreaming, to the fact that we are now located relative to a hallucinatory, rather than a perceptual, environment. The difference between dreaming and wakefulness becomes visible only when we look beyond the phenomenological level of description to the unique phenomenal-functional configuration underlying dreams. Here dreams are clearly distinguished from wakefulness by the comparatively weaker degree of external stimulus correlation. Moreover, the failure of this difference to show up on the phenomenological level is not at all surprising. Indeed, in the current model, it is to be expected: if spatiotemporal situatedness is the phenomenal core of dreaming, and if this basic phenomenal property is maximally invariant in dreaming and standard wake states, one would not expect the difference between dreaming and wakefulness to be easily detected while dreaming—to the contrary, the immersive character of dreams would be expected to counteract the ability to realize that this is just a dream. Cognitive corruption would thus be expected to follow on the heels of spatiotemporal self-location, and we have seen that this is exactly what happens, even in lucid dreams (cf. chap. 10).

A similar point can be made about the emotional character of dreaming. Again, the key idea is that spatiotemporal self-location is a relative property: it allows us to experience our location relative to other persons and objects around us. The present model thus predicts that spatiotemporal self-location will be closely connected to emotions, which are themselves, after all, a fast and immediate way of gauging, perhaps via the perception of bodily changes (cf. sec. 8.2.2), events in the environment and determining our reactions to them. If dreams are essentially immersive experiences, it thus seems unsurprising that they will also be highly emotional: assuming that the feeling of presence is connected to our emotional susceptibility to ongoing events, the current model would predict that dreams should be almost consistently characterized by strong emotions—and this is precisely what newer studies suggest.

The supposed connection between presence and emotions is also supported by studies investigating presence in immersive virtual reality. Riva and colleagues (2007) found evidence that presence and emotions influence each other, with the feeling of presence being stronger in emotionally loaded virtual environments (presented via head-mounted displays), and emotions also being influenced by the degree to which participants felt present in virtual environments. A similar connection between presence and emotions would seem to exist in dreams, with nightmares being a particularly clear example.

Departing from this brief overview, we can ask a number of more specific questions. What exactly is the relationship between spatiotemporal situatedness and the phenomenology of selfhood in dreams? In particular, is it possible for spatiotemporal situatedness to occur in isolation, independently of phenomenal embodiment and cogitative selfhood? What, in other words, is the minimal form of immersion occurring in dreams?

Conceptually, it is possible to describe self-location independently of phenomenal embodiment and cogitative selfhood; there is, at least, no obvious conceptual reason why the phenomenology of *here* and *now* should require the presence of bodily and cogitative experience. More importantly for present purposes, my analysis of the bodily and cogitative dream self (in chaps. 7 and 9) suggests that such dissociations are not just a logical and purely hypothetical possibility, but that they actually occur. Phenomenal disembodiment and phenomenal indeterminacy with respect to the body, coupled with doxastic indeterminacy and a lack of cogitative activity, suggest that there are cases in which self-experience is reduced to the experience of being an unextended point of awareness or a disembodied entity. Here, or so I would argue, we are approaching a pure form of spatiotemporal self-location requiring nothing more than a phenomenal *here* and a phenomenal *now* and arising independently of modality-specific forms of imagery. Because self-location is a relative property, amodal imagery might then characterize not just the sense of self-location but also the broader spatiotemporal reference frame within which the dreamer feels located.[3]

Again, there are several reasons for thinking that pure spatiotemporal situatedness arising independently of any modality-specific imagery is a phenomenal actuality, something that is occasionally experienced in sleep and can be reported after awakening. As LaBerge and DeGracia (2000) note, lucid dreamers sometimes report so-called minimal perceptual environments, in which most or even all sensory imagery is absent. Though this phenomenon has not, to my knowledge, been studied systematically in the laboratory, one does find anecdotal reports of dark periods in lucid dreams, imageless or nonrepresentational lucid dreams, or reports of lucid dreams involving the experience of being in a void (Magallón, 1991; Bogzaran, 2003; Hurd, 2008). Johnson (2014, p. 63) suggests that the combination of meditation and lucidity might be particularly useful for further investigating such experiences. She notes that in such dreams,

> we generally have a minimal body-sense, perhaps experiencing ourselves as a "dot of consciousness." Sensory perceptions tend to be minimized to an impression of light or darkness, and our thought processes are reduced to a state of nonthought where we might describe ourselves after the event as having existed as "pure awareness." A spatial sense often seems present as an impression of existing in "infinite space" . . . or simply as having no further sense of separate existence. (Johnson, 2014, p. 63)

Though such dreams involving minimal environments are rarely reported from nonlucid dreams, one might speculate that if they occurred in nonlucid dreams, they would easily be forgotten; in fact, the present model predicts, as will become clear in the next section, that they are on the verge of reportability. It is also possible that more detailed and systematic analyses of dream reports would reveal that even such minimal dream environments still consistently involve visual imagery of abstract geometric forms, and that even imageless lucid dreams still involve subtle vestibular sensations, such as the sensation of floating. Yet even rare reports of transient periods of purely amodal self-location would suggest that spatiotemporal situatedness cannot be reduced to a modality-specific form of imagery.

A related point is that dreams, as noted earlier, are often characterized by "just knowing" (Kahn et al., 2000, 2002). Consequently it seems possible that one could "just know" that one was dreaming of a particular environment without having corresponding modality-specific imagery. Again, the concept of phenomenal indeterminacy might be useful in this context. If dreams are indeed characterized by widespread phenomenal indeterminacy (see chaps. 5, 7), and if we allow that indeterminacy might spread over the entire phenomenal field in dreams, then perhaps completely amodal dreams could be redescribed as involving spatiotemporal self-location in a phenomenally indeterminate reference frame. Indeterminacy blindness, coupled with just knowing, might explain why we might nonetheless describe, after awakening, having dreamt of being in a specific environment while failing to notice that this environment was not, for instance, visually represented, or even that modality-specific imagery was lacking altogether.

This characterization seems particularly apt for describing the dreams of congenitally blind subjects. Take the example of a dream reported by a congenitally blind woman, describing how in her dream, she entered a room and was aware of its rectangular shape, as well as of the location of certain objects in it. This experience of the spatial layout of the room was independent of any modality-specific information and certainly was not of a visual nature (Kerr, 1993). Rather, it involved amodal spatial imagery, plus a sense of just knowing the shape of the room. But if this is the case, then a similar form of amodal spatial imagery might underlie, at least occasionally, the dreams of sighted subjects.

The claim that amodal spatial imagery, together with spatiotemporal self-location, might be a basic form of imagery underlying more complex forms of dreaming also makes sense in view of theoretical approaches to imagery in cognitive science. For instance, amodal spatial imagery plays a central role in the emulation theory proposed by Grush, who suggests that it may be "a core form of neurocognitive representation" (Grush, 2004, p. 394). The emulation theory posits an amodal environment emulator that uses purely spatial imagery to generate forward models that can be run online to generate expectations of sensory feedback during perception, as well as offline, for instance, during mental imagery or even dreaming. While amodal spatial imagery will often be accompanied by modality-specific imagery, Grush emphasizes that the amodal spatial emulator is dissociable from, and in a sense prior to, modality-specific emulators. The amodal environment emulator structures the subject's environment: it provides an egocentric reference frame in which objects are located relative to the self. If we combine this idea with the ISTH model of dreaming, the intriguing suggestion arises that amodal dreams occur when an amodal environment emulator operates independently of modality-specific imagery.

Farah and colleagues (1988) provide further evidence for the possible dissociation between spatial and visual representational systems for imagery. They describe the case of a patient who was severely impaired in visual imagery tasks while performing normally on spatial imagery tasks. They argue that this presents evidence for the existence of two functionally independent and neurologically dissociable subsystems of imagistic representation. The ISTH model of dreaming predicts that similar dissociations at least occasionally occur in the dreams of healthy subjects.

Anecdotal evidence for the occurrence of amodal dreaming also comes from the Buddhist tradition of dream and sleep yoga. According to this tradition, lucid dreaming is an intermediate step on the way toward maintaining awareness throughout dreamless sleep (Varela, 1997; Rinpoche, 1998; Norbu, 2002). Contrary to the Western literature on lucid dreaming, which emphasizes active participation and dream control, the underlying idea in dream yoga is to cultivate passive observation to overcome the dualistic separation of the dream self and dream world altogether. In this state, the sleeping subject is said to experience the "clear light of sleep," which is seen as a preliminary step toward the "clear light of death"; this type of experience is considered one of the highest, and most difficult, achievements of yoga. The experience of the "clear light of sleep" requires the dissolution not only of dream imagery but also of the self. Such clear-light dreams, as they are also sometimes called, are supposed to be devoid of content: "There is no subjective dreamer or dream ego, nor any self in a dualistic relationship with the dream or the dream content" (Rinpoche, 1998, p. 64). In this view, enlightenment is the end of all dreams; all that remains is the "luminous fundamental nature that is absolute reality" (Rinpoche, 1998, p. 33; see also Gillespie, 1988a; Varela, 1997, p. 108). Whether this type of state, or a subset thereof, still involves the experience of spatiotemporal self-location is a further question; if so, these experiences would count as minimal forms of dreaming in the sense introduced here. If not, we might rather want to say that lucid dreaming and amodal imagery precede or perhaps even enable a dissolution of dual awareness and of phenomenal selfhood that would itself count as experience during dreamless sleep (for discussion, see Thompson, 2014, 2015; Windt, 2015).

Reports of this state are similar to those sometimes described by Western lucid dreamers. Gillespie (1988b) describes that his experiences of what he calls dreamless sleep lacked imagery, emotions, and bodily sensations, preserving only the quality of expansiveness. He notes that they often ended in nothingness, and awareness itself simply dissolved, leaving him with no subsequent memory. Again, this is what the present theory would predict: if amodal spatiotemporal situatedness is the simplest form of dreaming, then its disappearance will also mark the transition to dreamless sleep and will hover on the borders of reportability. Perhaps what is described as the experience of *dreamless* sleep here is actually a form of imageless dreaming, in which dreaming is reduced to its minimal ingredients. Or perhaps, at least some of these states are even simpler than minimal dreaming by involving, for instance, a dissociation between spatial and temporal forms of self-location.

An intriguing though at this point highly speculative hypothesis is that a similar interpretation could be applied to white dreams, as well as to cases in which subjective judgments of sleep and objective (EEG and behavioral) markers of sleep come apart. As discussed in section 4.4, at least two different interpretations of white dreams, or the impression of having dreamt without remembering anything particular about one's dream (Noreika et al., 2009), are available. One, white dream reports might refer to perceptually vivid and narratively complex dreams, where everything specific about the dream has been forgotten. In this case, white dreaming would be an artifact of poor dream recall. Two, white dreaming might be a form of minimal dreaming, involving just enough phenomenal experience to retrospectively be recalled and described as dreaming, but lacking any specific forms of imagery or narrative. In this case, white dreaming would be a genuine sleep phenomenon rather than an artifact of poor dream recall. If this interpretation is correct, then at least a subset of white dreams might involve pure, imageless immersion or situatedness. This view gains some support from the finding that white dreams are particularly frequent during slow-wave sleep. Noreika et al. (2009) found that awakenings from stages 2 and 3 NREM sleep were followed by roughly equal rates of dream reports, white dream reports, and reports of nondreaming. This finding fits in well with the assumption that at least a subgroup of white dream reports might describe an intermediate phenomenon between nondreaming and dreaming, namely pure ISTH. It is also consistent with the conservative approach, as it takes reports of white dreaming at face value rather than suggesting that they are an artifact of poor dream recall. A similar phenomenon might underlie cases in which subjects, following awakenings mostly (but not exclusively) from the early stages of NREM sleep, deny having been asleep (Webb, 1975; Tanaka et al., 1996, p. 731). In patients suffering from subjective insomnia, this kind of subjective sleep perception is a cause of suffering (Mercer et al., 2002; Greene, 2008; Bianchi et al., 2013; cf. chap. 2).

For now, it is sufficient to note that based on anecdotal reports, there seems to be at least a superficial resemblance between the awareness of sleep cultivated in the tradition of dream yoga, so-called imageless lucid dreams, sleep perception (e.g., in subjective insomnia), and perhaps a subgroup of white dreams. A highly speculative hypothesis would be that by investigating reports from these states in a systematic manner, perhaps by increasing the expressive granularity of reports with the help of training or specific questionnaires, it might be possible to identify phenomenological similarities. A possible result could be that at least a subgroup of these experiences involve pure spatiotemporal situatedness of the type described here: there would be a continued sense of presence in an abstract, amodally experienced spatial expanse, and depending on the cultural and personal context, this early precursor to dreaming could be interpreted as the awareness of dreamless sleep, as imageless lucid dreaming, as white dreaming, or as wakefulness rather than sleep (as in subjective insomnia). In all four cases, increased attention to sleep—driven either by the attempt to cultivate lucid awareness of sleep and dreaming or by worries about the perceived inability to sleep—would increase the likelihood of experiencing and remembering such states. If this hypothesis were confirmed by future findings, we would have evidence for linking subjective sleep perception to minimal dreaming. This would be highly relevant for the project of identifying the neural basis of conscious experience and might even have therapeutic consequences for treating subjective insomnia. And it might also lead to the introduction of fine-grained conceptual distinctions, perhaps suggesting intermediate stages or even dissociations between the spatial and temporal aspects of self-location.

If experiences occurring on the border of sleep can count as dreams, this raises the question of whether sleep, in the current model, is a necessary condition for dreaming. One candidate for ISTH occurring beyond the borders of sleep is the so-called anesthesia dream (Munglani & Jones, 1992; Errando et al., 2008; Noreika et al., 2011). In their study of subjective experience during anesthesia, which they also call anesthesia dreams, Noreika et al. (2011) found evidence for a wide variety of hallucinatory experiences, with participants reporting subjective experiences in 60 percent of sessions. Reports could involve laboratory references or be wholly unrelated to the laboratory situation, and like sleep dreams, they could involve both positive and negative affect, as well as discontinuous jumps in the narrative. Subjects also occasionally reported feelings of having left their bodies, resembling both hypnagogia and certain (pre-)lucid dreams. Importantly, even reports involving laboratory references were often unrelated to the actual occurrences in the laboratory, thus resembling false awakenings. This suggests not only that a subgroup of anesthesia experiences involve ISTH—justifying their classification as dreams—but also that anesthesia experiences are of a similarly heterogeneous nature as is the case for sleep dreams. According to the current theory, sleep dreams and anesthesia dreams should equally count as dreaming insofar as they converge on ISTH as their common denominator. Importantly, this does not count for any experience during anesthesia. In particular, instances of external stimulus perception during anesthesia, so-called anesthesia awareness (Noreika et al., 2011, p. 328), should not count as an instance of dreaming in the sense proposed here. However, it would seem that anesthesia dreams are much more frequent than instances of anesthesia awareness, with dreams occurring in 6 to 53 percent and genuine awareness in fewer than 1 percent of general anesthesia cases.

Because the ISTH model defines dreaming via its unique phenomenal-functional configuration, it does not categorically exclude experiences occurring beyond the borders of sleep as dreams. Rather, the model remains neutral regarding this matter. It does, however, give rise to the following prediction. If immersive spatiotemporal hallucination really picks out the unified and distinctive phenomenal-functional core of dreaming, then one would expect experiences involving ISTH to show some degree of overlap on the neurophysiological level of description even when they occur beyond the borders of sleep as commonly defined. The idea is that ISTH might serve as a guide for identifying underlying similarities in the neural correlates, as well. In keeping with this hypothesis, Noreika et al. (2011) suggest that anesthesia dreams may be related to the occurrence of covert, REM-like processes and that the comparison of sleep dreams and anesthesia dreams may lead to the emergence of "an exciting new endeavor towards integration of anesthesia, sleep and consciousness research" (Noreika et al., 2011, p. 333). If this is correct, then Nielsen's (2000a) covert REM sleep hypothesis, originally conceived of as a hypothesis about the neural basis of dreams occurring during NREM sleep, might provide a unified account of anesthesia dreams, as well. Moreover, if the ISTH model really is a first step toward identifying the phenomenal-functional similarities between dreams occurring in sleep and in anesthesia—and perhaps even toward identifying common underlying neural mechanisms—then it at least seems conceivable that the concepts of sleep and dreaming could once more change in tandem. In chapter 2, we saw that this was frequently the case in the history of sleep and dream research: scientific discoveries on dreaming were closely linked to changing conceptions of sleep, and historically there is no such thing as a static and unchanging conception of sleep. The idea, then, is not that the ISTH model should decouple the concept of dreaming from the concept of sleep; rather, the model suggests a specific research strategy for first identifying the phenomenal-functional core of dreaming and then investigating the possibility of extending the concept of dreaming to cover dream-like experience occurring beyond sleep, at least as commonly defined. Such an extension will be warranted if it is theoretically fruitful, allowing for the detection of an overlap between a related family of experiences occurring predominately, but not exclusively, in sleep. For this reason, the ISTH model is also closely related, in spirit, to Nielsen's covert REM sleep hypothesis:[4] rather than focusing on the stereotyped features and conventional scoring criteria of their respective target phenomena, both hypotheses try to isolate their simplest forms. In doing so, they operate on different levels of description, but the idea is that it might ultimately be possible to integrate both approaches into a unified, multilevel framework.

At the same time, the inherent openness of the ISTH model also has the advantage of being able to accommodate the variability of dreaming and to solve a number of puzzles about dreaming. As discussed in sections 3.4 and 3.5, dream-enactment behavior in RBD patients and signal-verified lucid control dreams raise concerns about the generalizability of results to standard dreaming, because both types of dreams are distinguished from standard dreams along a number of phenomenological dimensions. At least for lucid dreams, it has been proposed that they might better be conceptualized as occurring in a hybrid state between dreaming and wakefulness (Voss et al., 2009) rather than being regarded as proper sleep dreams. Because the ISTH model focuses on the phenomenal core of dreaming and does not make dreaming contingent on the stereotyped characteristics of a majority of dreams, it affords a straightforward answer: it picks out what is common to many different types of dreaming, namely, ISTH, without thereby denying the existence of robust phenomenological differences between them. For the same reason, the model can accommodate the occurrence of dreams in different sleep stages, such as NREM sleep or sleep onset, rather than being restricted to REM sleep dreams. This both fits the folk-psychological concept of dreaming as well as empirical findings, both of which suggest that dreams can occur in all stages of sleep (see chap. 2).

Finally, note that this openness of the concept of dreaming is not simply a case of conceptual messiness: it is exactly the same openness in virtue of which the concept of dreaming can accommodate the inherent variability, or what is often called the multiplicity (Hunt, 1989), of dreaming. So far, the variability of dreaming—regarding both its phenomenology and its sleep-stage correlates—has foiled any attempt to advance a simple and uniform definition of dreaming. So the openness of the concept of dreaming proposed by the ISTH model appears to be grounded in the multiplicity of the target phenomenon itself.

However, a problem remains. To be turned into an interesting and theoretically fruitful conceptual tool, rather than a truism about conscious experience, the ISTH model will have to show which conscious states lack immersive spatiotemporal experience in the intended sense. Saying that dreams are distinguished from standard wake states because in dreams, spatiotemporal situatedness is experienced relative to what is in fact a hallucinatory reference frame is not quite enough, because there exist in-between cases that do not fall neatly on either side of the conceptual divide between hallucination and veridical perception. Sleep–wake transitions offer a reservoir of examples that blur this distinction: in fact, I argue that they offer evidence for a gradual shift in experienced self-location from perceptual to hallucinatory reference frames. The problem here is not that the difference is not available on the level of phenomenal experience: as argued earlier, this is standardly the case for nonlucid dreams. Rather, the problem is that the ISTH model, as introduced so far, suggests a schematic and dichotomous account according to which spatiotemporal self-location is relative either to a perceptual or to a hallucinatory reference frame and where only the second type of experience counts as dreaming. If the model ended here, it would fail to accommodate cases to which this dichotomous characterization does not apply and thus would ultimately be too schematic to be empirically plausible. In the next section, I argue that by carefully analyzing the trajectory from waking perception to full-fledged dreaming during sleep–wake transitions, we can supplement the ISTH model with a conceptual framework for describing intermediate states. Sleep onset experiences should be properly described, in this view, as *oneiragogic* experiences, or as experiences leading not just into sleep but into dreams. If the present theory is correct, they are a core target for future research.

11.2 Oneiragogia: Sleep–Wake Transitions and the Gradual Relocation of the Self

To bring out the problem for the ISTH model that is raised by experiences during sleep–wake transitions, let us consider the following experience reported by Ernst Mach.

> I awaken and lie quietly with my eyes closed. In front of me, I see, in great detail, my blanket with all its tiny folds, my hands quietly resting on top of it. If I open my eyes, the room is either completely dark, or it is light, but the blanket and hands are in a completely different position than they appeared to be. This is a particularly static and persistent phantasm, which I have never observed under different conditions. (Mach, 1919, p. 167; translation mine)

Is this experience a dream, or is it a sleep-related illusion? A first problem, of course, is that the report itself leaves much to be desired. For starters, it is not really a report of an individual experience at all but rather a general description of a type of experience that Mach has, apparently, had frequently in a similar manner. But let us take this description, for the sake of argument, at face value. In this case, we might begin by saying that the experience clearly does not involve a simple case of perception: because Mach's eyes are closed throughout the episode, he is not actually *seeing* his hands, but rather, the visual imagery experienced during the episode is hallucinatory. It is less clear, however, what to say about his bodily experience. The experienced position of his hands does not correspond to their actual position (as he readily notices when he opens his eyes), but what about the rest of his body? From the brief description quoted here, it is unclear whether the episode involves a postural illusion restricted to the position of the hands, or whether the felt position of the whole body is illusory, as would be the case if he experienced himself as lying in a different position from his actual one. Indeed, it is not even clear from his description whether this counts as an episode of full-phenomenal embodiment, for instance, in that he not only sees but also feels the position of his hands, as well as vividly experiences his entire body. Again for the sake of argument, let us imagine that we could fill in the details by saying that during the episode, Mach additionally feels the sheets on his body, feels his back gently sinking into the mattress, and feels the gentle rise and fall of his chest while breathing. A large proportion of his bodily experience, or so we will assume, would now be perceptual, whereas only certain aspects—the position of his hands—would be illusory. But let us further imagine that his visual imagery during the experience fails to correspond to his actual sleeping environment. Perhaps Mach is actually on the cusp of awakening in a hotel, but what he sees, in great detail, is his customary bedroom, complete with his favorite bedding. What should we now say about the spatiotemporal reference frame during the episode? Is this a case of hallucinatory or illusory spatiotemporal self-location, or perhaps one of largely perceptual self-location, constrained by his actual bodily position, but combined with hallucinatory visual imagery and the illusory mislocation of his hands? What should we say about this type of experience?

I think the challenge is to describe such experiences in a systematic manner while resisting the temptation to give a clear-cut, simple answer. Indeed, I think the type of hypothetical experience described here is truly transitional in that it resists classification as a dream or waking experience and as involving self-location in a hallucinatory, illusory, or perceptual spatiotemporal reference frame. Rather, we are apparently dealing with multiple overlapping reference frames: these include at least one that determines the coordinates of bodily position and is partly veridical and partly illusory; a visual one that is generated independently of contemporaneous visual inputs (remember that Mach's eyes are closed) and thus is hallucinatory; and one that involves amodal spatial imagery about which, at present, we don't know what to say at all. It is even less clear what to say about the temporal reference frame. It might make a difference, for instance, whether Mach is aware, during the episode, of the actual date, or whether, to vary the episode's description, what he sees in the experience is his childhood bedroom, in which he has not slept for decades. Rather than legislating an answer, I think one should do something else, namely, try to describe, in as much detail as possible, the gradual trajectory from cases in which we are fairly confident that they describe standard waking perception to those in which we are equally confident that they describe full-fledged dreaming.

But before I do so, I want to point out that the situation is further complicated by the fact that the terminology used for describing sleep onset experiences and sleep–wake transitions is almost as unclear as is the case for dreaming. A first point is that the term *hypnagogia*, literally "leading into sleep," is sometimes used to refer to a particular type of experience (e.g., by Maury [1848], who originally coined the term) and sometimes to refer to experiences occurring during a particular state, namely, the transitional state between sleep and wakefulness (Mavromatis, 1987, p. 3; Tanaka et al., 1996, 1997; for an early review including the historical literature, see Schacter, 1976). This is problematic because it is commonly acknowledged that full-fledged dreaming can also take place during sleep–wake transitions and may be related to REM-related processes occurring during sleep onset (Nielsen, 2000a; Takeuchi et al., 2001). If one takes hypnagogia in the phenomenological sense, there will be instances in which hypnagogic experience shades into full-fledged dreaming during sleep onset. This appears to be the sense intended by Mavromatis when he writes that "in hypnagogia a person may observe his imagery turning into a full-blown dream without a break in consciousness" (Mavromatis, 1987, p. 89). If, however, one takes hypnagogia to refer more generally to experiences occurring during sleep–wake transitions, then even if full-fledged dreaming occurs at sleep onset, it will count as hypnagogia. This latter reading of hypnagogia appears to underlie Mavromatis's strategy of discussing "the phenomenological similarities between some hypnagogic experiences and sleepdreams" (p. 89) and his claim "that the former can be comparable in all respects to the latter" (p. 89).

The failure to separate these two readings of hypnagogia renders the questions of how to distinguish dreaming from hypnagogia and of whether sleep onset supports the same type of experience as REM sleep utterly intractable. The problem is further exacerbated because, even moving beyond terminological disputes, the experiences occurring during sleep–wake transitions are almost as variable and heterogeneous as dreams, ranging from isolated imagery (Mavromatis, 1987; Stickgold et al., 2000) to dreamlike experiences (Nielsen, Stenstrom, et al., 2005; Stenstrom et al., 2012). As is the case for dreams, visual imagery is the most frequent type of imagery experienced during sleep onset, followed by auditory imagery and vestibular sensations, for instance, of floating or falling. Only rarely are all three modalities described in a single report, and most reports describe unimodal or bimodal imagery (Germain & Nielsen, 2001). Mavromatis cites a number of first-person reports likening hypnagogic imagery to seeing pictures, a succession of forms on a cinematograph screen (Mavromatis, 1987, pp. 14–15), or even animated or moving miniatures, as if seen on a magic lantern (p. 30). Sleep onset imagery is also typically described as being projected into the space in front of one's eyes or as being externalized, rather than as occurring "in the head," as is sometimes said to be the case for daydreams (pp. 29, 43). Subjects also often report seeing faces during the hypnagogic state, and these may develop into groups of animated figures moving through entire street scenes: "Groups of people stand about—women veiled and men clothed all in white and bright colours. Then through the archway rushes a tumultuous mob of people, whose features, despite the numbers are all painfully discernible" (p. 17). Subjects may also describe a progression from simple to more complex imagery, noting that faces "seem to come up out of the darkness, as a mist, and rapidly develop into sharp delineation, assuming roundness, vividness, and living reality. Then they fade off only to give place to others, which succeed with surprising rapidity and in enormous multitude" (p. 17). Others describe geometric imagery, such as a chessboard pattern stretched over the visual surface (Gillespie, 1997).

From these reports, one might conclude that the imagery experienced during sleep onset lacks the immersive character of dreaming. Perhaps sleep onset imagery is more akin to looking at a picture or watching a movie. If one takes this analogy seriously, one might want to say that while there is an experienced spatial relation between the subject and the image, it is not immersive: in looking at a picture or watching a movie, there is a spatial relation between the viewer and the physical properties of the picture, that is, the canvas of a painting or the movie screen, but one does not thereby feel located *in* the depicted scene or spatially related to the depicted persons or objects. The difference between nonimmersive sleep onset imagery and immersive dreaming would then be comparable to the difference between watching a conventional movie on a screen and an immersive virtual reality experience in which one can move through and interact with the virtual environment. Conrad (1968, p. 61) formulates the key intuition particularly clearly, stating that the transition from reverie to dreaming occurs when all contact to the external world is lost and the "there" of the imagined world turns into "here." He also takes this transition to occur at a single instant rather than gradually.

This description would not, however, do justice to the variability of sleep onset imagery. In their analysis of the sleep onset mentation reports from a single trained subject, Stenstrom et al. (2012) found that a majority described not just visual imagery but three-dimensional environments experienced from an internal, first-person perspective. There was almost always movement, mostly in the environment and in about half of the reports involving illusory own-body movement. Emotions were present in almost 70 percent of reports. Importantly, more than 80 percent of reports were judged to involve scene construction, with objects being integrated into a three-dimensional environment centered on the dream self. Only 13 percent, or four reports, failed to meet the criteria for scene construction: these reports involved either an environment without any objects (i.e., traveling through space) or objects without an environment (i.e., cubes floating in darkness or isolated objects). Rarely, there was a progression from simple objects to an integrated scene. A majority of reports, however, fulfilled the criteria for ISTH: to the extent that they involved the sense of presence in a hallucinatory environment, a scene that is centered on a first-person perspective (including objectless environments, reminiscent of imageless lucid dreams), they count as dreams. What this suggests is that sleep onset supports a range of experiences, at least some of which are fully immersive dreams. Yet, recognizing this variability, though important, at best gives us a rough idea of two extreme end points, where one involves isolated imagery and the other fully immersive dreaming.

Before analyzing in more detail the intermediate points on the trajectory from wakefulness to full-fledged dreaming, I should point out that part of the variability detected in reports of sleep onset imagery may be explained by the fact that different studies have used different methodologies. While some groups have focused on daytime naps (Nielsen, 1991–1992, 1995; Kussé et al., 2012), others have investigated imagery during nighttime sleep onset (Stickgold et al., 2000; Wamsley, Perry, et al., 2010). There is also the familiar difference between home (e.g., Wamsley, Perry, et al., 2010) and laboratory (e.g., Kussé et al., 2012; Stenstrom et al., 2012) studies. Sample sizes are variable, with a number of studies being conducted with specially trained single subjects (Stenstrom et al., 2012) or even using systematic self-observation performed by the researcher (Nielsen, 1991–1992, 1995). Aside from sleeping in different environments and different awakening methods, different sleeping postures have been used to sample sleep onset imagery. While subjects in most studies have slept in the supine position (Stickgold et al., 2000; Wamsley, Perry, et al., 2010; Kussé et al., 2012; Stenstrom et al., 2012), Nielsen (1991–1992, 1995) used an upright napping procedure. Here, awakening was either provoked by the images themselves or caused by muscle jerks associated with the onset of sleep-related atonia. Some studies have also investigated the influence of presleep stimuli, such as playing the computer game Tetris (Stickgold et al., 2000; Kussé et al., 2012) or the arcade game Alpine Racer (Wamsley, Perry, et al., 2010) on sleep onset imagery.

Finally, as discussed in chapter 2, characterizing the imagery by its occurrence during sleep onset is overly simplified, as well. Nine EEG substages reaching from wakefulness to stage 2 NREM sleep have been distinguished (Hori et al., 1994; Tanaka et al., 1996, 1997), and the rates of mentation recall are not equally spread across them but rather concentrated in substages 4 and 5. According to Stenstrom et al. (2012), awakenings from these substages are not only particularly likely to yield dreamlike reports (with a recall rate of just over 80 percent, equaling that of REM sleep awakenings) but also particularly variable regarding whether participants judge themselves to have been asleep, with approximately one-half of participants judging themselves to have been awake and approximately one-fifth being unsure whether they were awake or asleep. This suggests that more fine-grained phenomenological descriptions will ultimately have to be complemented by more fine-grained sleep-stage scoring procedures to make real progress toward understanding sleep onset imagery.

What about the intermediate points on the trajectory from isolated sleep onset imagery to full-fledged dreaming? The most systematic description, to my knowledge, comes from Nielsen's (1991–1992) analysis of nearly eighty reports collected using the self-observation, upright napping procedure described earlier. Here (and in Nielsen, 1995), he describes a fairly linear progression from the moment of closing his eyes to the moment of awakening. The first phase is characterized by fleeting imagery, consisting of brief, mostly unimodal visual or kinesthetic impressions (such as brief flashes of visual scenery) and sleep onset feelings (such as heat sensations, thickening or swelling of body parts, and falling sensations) localized in the body or in "apparent space" outside the body. Aside from brief lapses, awareness of surroundings is mostly intact. Imagery during this period is typically extremely brief and difficult to recall or describe in detail. The second period is characterized by "phenomenal silence": it involves a longer lapse of awareness, probably arising during a brief period of microsleep that subjectively lasts for an indeterminate period. The third stage, the "preimage context," involves the impression that a hallucinatory event, which, however, cannot be subsequently remembered, occurred before the hypnagogic image. As is the case for white dreams, it is unclear whether this results from forgetting an actual experience, or whether this stage involves cognitive processes just below the threshold of conscious experience and reportability. Finally, the fourth stage involves what Nielsen calls fully formed "hypnagogic images." He describes these as brief but dynamic images, resembling one- or two-second video clips. They are hallucinatory and experienced as sufficiently realistic to prompt reflex responses such as limb jerks. A further feature is their coherence: "An image does not appear to evolve from a series of more elementary constituents, but rather to erupt into consciousness from the silent period" (Nielsen, 1991–1992, p. 362). Fully formed images are also more easily recalled and described than fleeting images.

I want to suggest that these four phases provide evidence for the claim that the sleep onset period is accompanied by a gradual shift in self-location from a largely perceptual to a largely hallucinatory spatiotemporal reference frame. We should not imagine this transition as immediate or even as simple; rather, Nielsen's four stages suggest that we can distinguish different dimensions. This suggests why Mach's report, quoted at the beginning of this section, is puzzling: the reason is that during sleep–wake transitions, there can be multiple, partially overlapping reference frames, possibly even constrained to different degrees by external and bodily stimuli. Speaking of dreams as immersive spatiotemporal hallucinations thus is an idealization: sleep–wake transitions show that a more differentiated account is needed for describing the processes underlying this replacement of perceptual, largely externally constrained reference frames by hallucinatory, largely internally generated ones. I think it is a helpful idealization insofar as in full-fledged dreams, this replacement is typically complete: dreams involve the feeling of presence in what is now a hallucinatory spatiotemporal reference frame. However, this should not deflect attention from the fact that the underlying processes are in fact complex, and this replacement does not always occur instantaneously and in concert, but rather in a gradual and dissociated manner.

Here my central claim is that sleep onset is a particularly valuable research target precisely because it provides evidence for multiple dissociations and mutually overlapping reference frames, thus shedding light on the complexity of cognitive processes that otherwise (e.g., in contrasting only full-fledged dreams and standard wake states) appear to be a simple, all-or-nothing affair. For this reason, I want to suggest a new conceptual tool for describing experiences during sleep–wake transitions.[5] From now on, I will use the term *oneiragogia* to describe experiences occurring during sleep onset; by contrast, I will reserve the term *hypnagogia* for describing the state in which these experiences occur. Oneiragogic experiences, in this view, are not quite the same as dreams; they are experiences *leading* into dreams. The same is true for *hypnagogia*, which is derived from the Greek terms *hypnos* (sleep) and *agogeus* (guide or leader): it is a state leading into sleep, marking the transition between sleep and wakefulness. Similarly, *oneiragogia*, from the Greek word for dreams, *oneiros*, describes a transitional experience, a prequel or perhaps even a bridge, if you will, from wakefulness to dreaming. While oneiragogia culminates in dreamlike experiences, its earlier stages are not yet immersive spatiotemporal hallucinations. Yet their classification as oneiragogia seems warranted because despite their heterogeneity, they involve a linear progression toward dreamlike experience and systematically shed light on the cognitive processes, or so I claim, involved in dream formation and in particular on the trajectory underlying the shift in self-location from a perceptual to a hallucinatory (or dreamlike) reference frame.

From the available evidence, it is already possible to distinguish at least five different dimensions for describing this process. These are (1) the gradual emergence of an integrated visuospatial scene, beginning with isolated imagery and building up to an immersive hallucinatory environment; (2) changes in phenomenal-functional embodiment, leading up to the weak kind of phenomenal-functional embodiment often seen in dreams; (3) the gradual substitution of the temporal reference frame, leading from brief and isolated to prolonged and narratively organized episodes; (4) the integration and reorganization of waking-memory sources; and (5) the integration with autobiographical memory, or what might be called the historicity of the episodes, closely linked to the ability to recall them after awakening. I discuss these five dimensions in turn. Note, however, that I do not take them to be exclusive or to proceed in a linear manner; to the contrary, this constitutes a very first pass at a more detailed description, and in most cases, changes in different dimensions will occur simultaneously and in concert with each other.

Regarding (1) the emergence of an integrated visuospatial scene, there seems to be a development from isolated, static, and extremely brief imagery to the emergence of a more dynamic, prolonged, and immersive scene. Earlier I suggested that the contrast between nonimmersive sleep onset imagery and immersive dreaming can be compared metaphorically to the difference between watching a conventional movie and an immersive virtual reality experience. So far we have seen that this should not be regarded as a contrast between sleep onset imagery and dreaming; rather, the contrast describes the end points seen within oneiragogic experience itself. Yet, between these end points, there is room for further degrees of spatial situatedness involving more fine-grained distinctions. An important step toward the emergence of an immersive environment is not just the integration of individual objects into a coherent scene. It is at least as important that the location of perceived objects be not wholly observer-relative, but also subject to environmental constraints. For oneiragogic experiences, this is not always the case. For instance, Müller observes that dream images persisting into wakefulness often maintain their location in the visual field, moving in synchrony with his eyes. In this, they are similar to afterimages, though he claims they obscure part of the perceptual environment more completely (Müller, 1826/1951, p. 119; for a number of examples describing imagery ranging from transparent to opaque and completely occluding physical objects, see Green & McCreery, 1975, pp. 150–155). The ability to scan visual imagery, by contrast, involves a stronger degree of spatial and contextual embeddedness; the image's location is now not just observer-relative but also has a location in the environment. For instance, the German poet Isolde Kurz relates an experience in which she saw a ghostly phantom tree, as she calls it, standing next to her bed. She tried to obliterate the image by rubbing her eyes and turning in the other direction, but when she turned back to check, the phantom tree was still there. Despite its translucent appearance, she remarks that it could not possibly have been the shadow of a real tree (Kurz, 1919, p. 106; for similar examples, see Green & McCreery, 1975).

It seems likely that these types of phenomena result from an overlap of egocentric, or observer-relative, and allocentric reference frames, with the former representing subject-object and the latter object-object or object-environment locations. There is growing consensus that distinct systems with different neural correlates exist, but also that spatial imagery and memory involve both egocentric and allocentric representations (Klatzky, 1998; Zacks & Michelon, 2005; Zacks & Tversky, 2005; Burgess, 2006, 2008). A highly speculative hypothesis is that during oneiragogia, the emergence of an immersive scene involves the alignment of both types of representations. Also, in this view, the instability in visual imagery and spatial discontinuity in dreams can be redescribed as resulting from an incomplete alignment of egocentric and allocentric spatial processing. A tantalizing question for the future would be whether certain types of dream bizarreness—such as discontinuous changes in the scenery and the instability of objects—tend to concern their observer-relative locations or their location relative to other objects. Such findings could then be used to investigate whether disturbances are more likely to arise on the level of egocentric or allocentric spatial representations in dreams. A careful prediction from the current theory would be that disturbances in allocentric spatial representations should be more common in dreams: if spatiotemporal self-location really is the core of dream experience, then one would expect egocentric reference frames in dreams to have priority over allocentric ones. Indeed, it would seem that oneiragogic experience is often egocentric:

> Out of the whirl of uncouth figures to which I was accustomed, a big stout man, in a grey suit and Homburg hat, emerged, and came up (apparently) so close to me, staring pointedly at me, that I expected him to touch me. The movement and the figure were both so aggressive that I mentioned it to my family the next morning. (Mavromatis, 1987, p. 16)

At least as important as the dimension of visual imagery are (2) gradual changes in phenomenal-functional embodiment. Even the earliest stage of oneiragogia often involves sensations of tiredness located either inside or outside the body. Again, we can begin by describing the end points of this process, which range from standard bodily awareness to illusory full-body movement. According to van Eeden, the loss of own-body perception is crucial for the transition to what he calls initial dreams:

> In hypnagogic hallucinations we have visions, but we have full bodily perception. In the initial dream type I see and feel as in any other dream. I have a nearly complete recollection of day-life, I know that I am asleep and where I am sleeping, but all perceptions of the physical body, inner and outer, visceral or peripheral, are entirely absent. Usually I have the sensation of floating or flying, and I observe with perfect clearness that the feeling of fatigue, the discomfort of bodily overstrain, has vanished. I feel fresh and vigorous; I can move and float in all directions; yet I know that my body is at the same time dead tired and fast asleep. (van Eeden, 1913)

As is the case for visual scene construction, intermediate stages can be distinguished. Moreover, changes in phenomenal-functional embodiment will often happen in concert with changes in visual imagery.

> With the appearance of scenery and interiors I acquire a new faculty—that of travel. Though still aware that some aspect of myself remains in the body, I go on exploring expeditions through houses, streets and country lanes, or wander through the aisles of vast cathedrals. I meet people on these journeys, but if they (or for that matter "I") can be said to be there at all, I am certainly invisible to them. I remember coming face to face with a crowd of laughing children who, to my disappointment, simply walked through me. At any rate, my sense of being present in these scenes is so vivid that I have to remind myself that I can walk into the houses unseen, and that I need not stop to open doors, but can pass through them. (Mavromatis, 1987, p. 23)

Alongside subtle changes in bodily experience, one might also expect oneiragogia to be accompanied by a progression from passive observation to active participation. Indeed, a recent study suggests that REM dreams are more likely to involve active participation on the part of the dreamer, whereas sleep onset experiences are more likely to involve passive observation (Speth et al., 2013). Here the suggestion is that active participation tracks increased immersion: the greater the sense of presence in the dream, the more the phenomenal self will interact with the virtual environment.

It is at least as interesting to try to match changes in phenomenal-functional embodiment to changes in visuospatial and temporal aspects of the experience. Sartre recounts that during sleep onset, he would typically experience a gradual weakening of bodily sensations and motor abilities, including a loss of muscle tone, and this would happen in concert with uncertainty regarding spatial orientation and time:

> The sense of one's body is extremely confused, and the contact with the bedsheets and mattress is felt even more vaguely. The position of the body in space is very poorly determined. Orientation is subject to typical disturbances. Time perception is uncertain. (Sartre, 1940/2005, p. 86; translation mine)

A tantalizing possibility is that changes in bodily experience might not just occur in concert with visual scene construction but themselves prompt the emergence of a hallucinatory spatial reference frame. Recall that according to Nielsen (1991–1992), sleep onset feelings may be localized within the body or in "apparent space" outside the body, almost as if the interior of the body itself had become the immersive space. Indeed, he reports having observed

> countless instances of immersive, quasi-perceptual imagery that is exclusively body-oriented, e.g., images of hallucinatory changes occurring within the chest or other parts of the body. The image has that hallucinatory *feel* to it, but an external setting is indeterminate. Rather, it seems that the setting is the (now illusory) body itself within which focal events transpire. (Nielsen, personal communication, September 2012)

A speculative hypothesis is that in these cases, altered body perception may itself provide a spatial context for other objects and events. Indeed, Nielsen (1991–1992, 1995) investigated the systematic correlation between phasic neuromuscular events (such as limb jerks and head tilts) and sleep onset imagery. Neuromuscular events accompanied more than half of fully formed images. Importantly, they were differentially associated with two different types of experienced movements. Self-movements were much more likely to be accompanied by neuromuscular events than nonself (i.e., observed) movements. The type of neuromuscular event also made a difference, with leg jerks matching illusory movements in 80 percent and head tilts only in roughly 20 percent of all cases. An example of a match is given by the following report:

> I was standing at the top of a set of stairs. I had spilled some brightly colored candies all around me on the floor and down the stairs. I reached out behind me with my right foot and swept one of the candies down the stairs with my toes. On awakening, my right thigh muscles jerked slightly without displacing my leg. (Nielsen, 1991–1992, p. 359)

A tentative and highly speculative conclusion might be that the process of reorienting in a hallucinatory space, as well as the generation of visual imagery, is modulated by more or less direct forms of own-body perception. If so, then at least in some cases, real-body changes may be reflected not just on the level of experienced body movements but also on the level of visuospatial imagery. Metaphorically, we might say that at least in certain cases, the immersive hallucinatory space that is the hallmark of dreaming is a reflection of the physical body. As Nielsen (1995) points out, this fits in well with Silberer's (1919; see sec. 9.5) claim that experience during sleep onset is autosymbolic, symbolizing preceding mental activity and bodily state.

The deep connection between bodily and spatial imagery is confirmed by recent work on vestibular contributions to self-consciousness. In a recent review article, Pfeiffer and colleagues (2014) argue that vestibular processing, by coding head position in three-dimensional space, provides a spatial reference for bodily self-consciousness more generally (see also Lenggenhager & Lopez, 2015; Pfeiffer, 2015). They discuss evidence showing that vestibular processing contributes to and constantly updates spatial representations of the position of the whole body and its movements through extrapersonal space. The vestibular system may thus make a central contribution to self-location and the first-person perspective as the two central spatial aspects of bodily self-consciousness (Lopez et al., 2008). The same may be true for volumetric body representations, for instance, the perceived shape of body parts (Longo & Haggard, 2010).

Pfeiffer and colleagues (2014) also discuss evidence suggesting that spatial imagery is modulated by vestibular signals. In particular, they argue that egocentric spatial imagery of the type involving whole-body perspective transformations depends on vestibular processing; for instance, egocentric (but not allocentric) imagery is disturbed in patients with vestibular loss and can be systematically disturbed via vestibular stimulation in healthy subjects (Lenggenhager et al., 2008). All of this fits in well with my claim that ambiguous vestibular signals, for instance, in the supine position, may lie at the root of movement sensations in dreams (see sec. 8.4). Given that both egocentric spatial imagery and self-location are closely tied to vestibular processing, the immersive character of dreaming and of oneiragogic experience may likewise be deeply rooted in the vestibular system.

Finally, it might also be possible to connect social imagery in dreams and oneiragogia to vestibular processing and real-body stimuli. Indeed, spatial and vestibular processing has been suggested to contribute to social cognition (Lenggenhager & Lopez, 2015; Pfeiffer, 2015). Nielsen (1991–1992) notes that in addition to the linear progression of different stages of imagery during sleep onset, several unexpected patterns were observed. One involved self–other confusion, in which illusory movements observed in another character also seemed to involve the self, as if arising from a sudden shift in perspective. The second involved illusory movements that were in opposition to the actual body movement. These occasionally manifested on the level of self-movement but occasionally also concerned movement observed in nonself characters:

> Someone in front of me is doubled over toward me, praying. Someone else reaches around from behind this person and quickly lifts him into an upright position. At the same time I feel my head nodding slightly forward and it awakens me. (Nielsen, 1991–1992, p. 360)

This externalization of real-body changes to nonself characters is already familiar from full-fledged dreaming (Nielsen, 1993; see sec. 8.2.2). Again, the idea is that perhaps nonself characters in dreams and oneiragogia are a reflection of the sleeping body. Nielsen (2007) has made a similar proposal in relation to felt presence, which he argues is a variant of normal social imagery that, under certain conditions such as sleep paralysis, takes the form of spatial hallucination. He also, tellingly, links felt presence to phantom limb experiences.

A similar account may be available for nonself characters experienced at sleep onset or during dreams; indeed, one of the reasons why dreams are rife with social imagery might be the unique phenomenal-functional configuration underlying dreams. This suggests yet another way in which dreams might be weakly phenomenally-functionally embodied states, namely, in that real-body stimuli not only modulate bodily experience in dreams but are also the origin of nonself dream characters. Very preliminary support for the link between own-body perception and social imagery comes from the finding that low interoceptive sensitivity leads to a blurring of self–other distinctions, for instance, in a mirror recognition task (Tajadura-Jimenez & Tsakiris, 2013). Perhaps this effect of weakened own-body perception on the blurring of self–other distinctions and social imagery is even more pronounced during sleep onset and in dreams. In our highly speculative and preliminary study of self-tickling and being tickled by a nonself dream character during lucid dreams, we found that both were rated as similarly ticklish as self-tickling during wakefulness (Windt et al., 2014). We interpreted this as preliminary evidence for the claim that in the special case of lucid control dreams, sensory attenuation for self-produced tickles spreads to tickles produced by others. A possible interpretation is that sensory attenuation for other-produced tickles results from a blurring of self-other distinctions: at least in lucid control dreams, being tickled by another feels much like tickling oneself. Yet, it is noteworthy that the self-other distinction did not collapse completely, with there still being a clear experiential distinction between the self and nonself dream characters.

Note how these observations on the close interaction between changes in phenomenal-functional embodiment and spatial and perhaps even social imagery in oneiragogia and dreams give rise to a conceptual problem. If it is indeed possible to single out real-body inputs and vestibular processing as the conditions for the emergence of an immersive spatial reference frame, as well as perhaps of social imagery, might the ISTH model be a misnomer? In particular, might dreams better be described as immersive spatiotemporal illusions?

I don't think this necessarily follows. Recall that in section 8.4, I remarked that the distinction between illusions and hallucinations is a matter of degrees. It can be drawn,

I suggested, by assessing the appropriateness of the stimulus source. Experiences occasioned by an external stimulus count as hallucinations, in this view, if they are highly idiosyncratic and unrelated to an appropriate stimulus, whereas experiences that are more straightforwardly and predictably related to external stimulation are better described as illusions. So an intuitive way of responding to this objection would be to say that the immersive space experienced in dreams is generally too far removed from real-body inputs to count as a misperception of the body. The concept of illusion would require, in this view, that the illusory experience represent the same type of object, roughly, as its cause, or the object of which it is supposedly a misperception. A space, however, is not the same type of object as a body; hence even if dreamed spaces are constrained by own-body perception, they count as hallucinations, and the same would be true for nonself dream characters. An immersive space, in this view, is simply too different from the body to count as a misperception of the body.

A slightly different and I think more interesting way of responding to this objection would be to allow that in certain cases, immersive spatiotemporal imagery may indeed be considered illusory; this will be the case, for instance, if the subject undergoing the experience, as indicated by Nielsen, has the impression that the experienced space is a reflection of the body. If we take such reports seriously, then at least a subgroup of nonveridical immersive spatiotemporal experiences are illusions, not hallucinations. Note how this immediately gives rise to a prediction. If it turns out that such cases of illusory spatiotemporal imagery should arise more frequently during the early stages of oneiragogia than in full-fledged dreams, we could say that oneiragogic experience involves a gradual decoupling of spatiotemporal imagery from bodily experience. In the early stages, the spatial aspects of experience may still have a distinctly bodily flavor, whereas in full-fledged dreaming the connection between the scenery in which the dream is unfolding and the sleeping body is more idiosyncratic and perhaps even weakened to the point of being unrecognizable. Moreover, even if a subgroup of immersive spatiotemporal experiences can be linked to own-body perception, there may be other cases where they rely on internal brain activity alone. At least for a subgroup of cases, we might describe the stages of oneiragogia as a trajectory moving toward a gradual decoupling of spatiotemporal imagery from real-body inputs. If so, there is no contradiction in saying that while in some cases, spatial imagery is a form of illusory own-body perception, in full-fledged dreams, it is hallucinatory.

Moving to (3) the emergence of a hallucinatory temporal reference frame, it is remarkable that all four stages distinguished by Nielsen (1991–1992) have a distinct temporal component. This is already true for the basic distinction between fleeting and fully formed images: the latter involve the integration of objects not just into a spatial scene but also into a temporal context, characterized by a sense of duration and temporal dynamics. Temporality plays an even greater role for the two intermediate stages. Recall that phenomenal silence is described as involving a lapse of attention lasting for an indeterminate period, suggesting a departure of experienced subjective time from actual clock time. Nielsen tellingly describes this "as a silent background against which the hypnagogic scenario abruptly appears" (1991–1992, p. 361). The preimage context is once more of a temporal nature; more precisely, it involves the impression of a preceding but unrecalled hallucinatory event. Both stages contribute to the emergence of a temporal context: the fully formed image is experienced not as created wholly anew but as preceded by unrecalled hallucinatory events and an indeterminate larger "silent" context. In full-fledged dreams, the formation of an alternative temporal reference frame is complete: dreams are often, after all, described as involving a sense of duration and temporal progression.

Again, intermediate steps are possible. Noreika et al. (2009, 2010) found that subjects awakened from NREM sleep reported static dreams that lacked not only narrative progression but also movement sensations. The reports often described the experience of being present in a static scene, for instance, quietly sitting on a bench, with nothing else happening (Valdas Noreika, personal communication, July 2010; see also Noreika, 2014, p. 53). Although these dreams lacked the animated quality of more complex dreams, there was still a sense of duration, and according to subjective estimates, these simple dreams lasted from thirty seconds to one minute. This suggests that a basic sense of temporal progression can arise even when narrative progression and temporal dynamics are lacking. There is also already a phenomenal *now* that functions as the center of the unfolding temporal scene. Although dreams are often unstable and involve discontinuous jumps in dream narration, the unfolding narrative is still centered around a chain of phenomenal *nows*. What distinguishes even fleeting images from simple, static forms of dreaming is that in dreams the experience of duration is linked to the emergence of a temporal scene and, once more, to the experience of occupying a specific location—*now*—within this scene. This immersive quality is lacking in fleeting imagery, which is often described as having an instantaneous, snapshotlike quality (Mavromatis, 1987; Nielsen, 1991–1992, 1995). Another way of saying this is that the fleeting images that mark the first stages of oneiragogia lack temporal extension. Certainly they have the quality of happening *now*, but they lack a broader temporal structure or context, and hence there is no sense of temporal self-location *within* this structure. They are momentary flashes lacking embeddedness in a temporal scene or context.

The idea that dreams unfold not just in their own spatial but in their own temporal reference frame, unconstrained by objective clock time and real-life constraints on the duration of events, is not new. As Globus writes, "The dream has its own experienced time—an authentic time continuously created *de novo*—notwithstanding what the clocks of waking show" (Globus, 1987, p. 77). The most extreme version of this claim is the view that dreams are instantaneous memory insertions (as suggested by Dennett's 1976 cassette theory of dreaming; see sec. 1.3.4); the paradigmatic example is

Maury's (1861) dream of the French Revolution, with its seeming narrative buildup toward his execution at the guillotine, apparently caused by the bed's headboard falling onto his neck. Ample anecdotal evidence points to dreams apparently building up to an external stimulus. Yet, as Bossard (1987, p. 82) points out, even exceptional dream reports such as Maury's guillotine dream do not warrant the conclusion that dreams lack temporal extension or that dream narratives systematically build up toward the culminating event before it happens. They only show that dreams can flexibly incorporate external stimuli into an ongoing narrative. Moreover, external stimuli themselves may continue over a prolonged time span and may occasionally even lend direction to the dream narrative without being incorporated throughout the dream.

> Then I hear a doorbell or something ring and I go downstairs and find my old friend Joe at the door. I start a long discussion with him, we walk outside. Finally I wake up and realize my alarm has been ringing for quite a while. (Hartmann, 1998, p. 53)

We have also seen that evidence from signal-verified lucid dreams contradicts the claim that dreams lack duration. Erlacher et al. (2014) found that actions performed in lucid dreams such as walking or gymnastics actually took longer than in wakefulness, though the relative duration of parts of the task was preserved. All of this suggests that while dreams have temporal structure, this structure is not constrained in any simple or direct manner by external stimuli.

A further interesting difference between dreaming and wakefulness is that in dreams, our *subjective* temporal reference points are attenuated. As Bossard notes (1987, p. 83), the dream self is typically focused on the experienced *now* and only rarely is consciously concerned with past and future. This focus on the present, with at best a vague sense of the past and future, is especially obvious in dreams involving undetected jumps in dream narration. Such discontinuities, which can perhaps be likened to movies, may contribute to the anecdotal discrepancy between dream time and objective time. There might even be cases in which the succession of dream events and the seeming narrative buildup to an artful denouement are a retrospective construction resulting from the telling of dreams rather than from the actual sequential structure of dream experience during sleep. If dream events can be molded into a sensible narrative in the telling of the dream, then perhaps the fact that the dream itself contained numerous jumps, going back and forth from what will later be one point in the dream narrative to another, might be easily overlooked. Of course, the present theory would predict that if dreams are in fact less continuous than is suggested in dream reports, this trend should be reversed in reports gathered in improved reporting conditions (cf. chap. 4). And occasionally it is: studies of dream bizarreness show that for dreams, narrative continuity is not nearly as smooth as for waking experiences, as evidenced by frequent discontinuities in dream narration (Hobson, 1988; Revonsuo & Salmivalli, 1995).

Perhaps these remembered gaps are just a small portion of the more numerous skips and cracks in time experience in dreams.

The cognitive structure of time experience—our memories of the past and worries and anticipations about the future—may also be lacking, at least in some dreams. This might be the case in dreams in which one has at best a vague sense of the past or future, or even misremembers or forgets important waking life events. Consequently at least some dreams may provide an example of time experience independently of the objective constraints on the duration and succession of events, as well as of the cognitive and mnemonic factors of time experience, such as forming explicit beliefs about the past and future.

All of this suggests that dreams are experiences in which the temporal structure of experience is profoundly altered. While both dreaming and wakefulness involve a "window of presence" (Metzinger, 2003, pp. 126, 251), this window of presence becomes so predominant in dreams as to obscure, metaphorically speaking, the view of the past and future. Still, this is not to say that dreaming involves the experience of unconnected and momentary flashes of imagery, or a discrete succession of experiential *nows*. Though dreams often lack explicit cognitive reference to past and future, they still involve temporal situatedness: an experiential *now* within a broader temporal frame, though this frame is less structured than is the case for standard wake states. Perhaps the concept of phenomenal indeterminacy can meaningfully be applied to the temporal reference frame, as well: its simplest form may be the stages of phenomenal silence and the preimage context described by Nielsen as taking place during sleep onset and apparently accompanied by the impression that an indeterminate amount of time has passed. But even full-fledged dreams often involve vague insinuations of the past or present, such as typical dream themes of being late, being about to miss one's train, and so on. The vague sense of pastness or of the future may also be tied to doxastic indeterminacy. Recall that in Globus's (1987) dream of being at an academic meeting, he had a vague and undefined sense of his academic past but did not, judging from the dream report, have any particular beliefs about it (see sec. 10.3.2).

The claim that dreaming does not simply involve discrete experiential *nows* is borne out by general work on time experience. This idea dates back at least to James's (1890/1950a) concept of a "specious" or temporally extended present and has been confirmed by experimental work on time perception. Both the experience of simultaneity and what one experiences as the present moment take place within an extended window of presence; likewise, in certain conditions, subjects can identify events as nonsimultaneous without being able to specify which event came first (Pöppel, 2000). What subjectively appears as a simple, unstructured single moment is in fact extended and temporally stretched, as well as a highly complex construction. The upshot is that we never experience discrete moments but rather live within a "window of presence" within which temporal succession and the sharp boundary between past and present are eroded (see also Metzinger, 2009). While memory is central to our ability not just to recall the past (Pöppel, 2000, pp. 100–109) but also to anticipate and prepare for the future, dreaming appears to diminish our ability to cogitatively situate ourselves with respect to past or future events. But even without these cogitative aspects of duration, there is still an experiential *now*, integrated into an ongoing flow of events. Dreaming, as Globus writes, still involves "a sequence of variously stretched now durations" (Globus, 1987, p. 75) and a predominate concern for the present, rather than for the future or the past (p. 75). Compared to wakefulness, time experience in dreams has lost much of its structure, but it is not turned into a staccato of unconnected present moments. By contrast, the fleeting images during the first stages of oneiragogia may be a case in which the temporal embeddedness of individual moments is reduced to a minimum.

Experimental evidence also points to the connection between spatial and temporal relatedness and the self. Several studies have shown that thinking about time is closely related to spatial thinking, though the converse does not seem to be true (Boroditsky & Ramscar, 2002; Casasanto & Boroditsky, 2008). Temporal illusions can be induced by varying the spatial relationship between an observed object and the observer. For instance, a stimulus moving toward the observer is perceived as having a longer duration than a static object or one moving away from the observer, though the objective duration of the stimuli is the same (van Wassenhove et al., 2008). In an fMRI study, Wittman et al. (2010) replicated this finding to investigate the neural correlates of time perception. They found that stimuli that induced the illusion (i.e., that were subjectively experienced as having a longer duration) were correlated with increased activity in the posterior cingulate cortex. This region is generally thought to be involved not just in time experience but also in the processing of self-related stimuli (Kelley et al., 2002). Imaging studies of REM sleep have found this region to be comparatively deactivated during REM sleep—a finding that would fit in well with the observation that the cognitive structuring of time via memory and beliefs is often attenuated in dreams. However, another region implicated in time experience, the anterior cingulate cortex, is more strongly activated in REM sleep (Dang-Vu et al., 2005). Taken together, these findings lend at least initial plausibility to the claim that self-related processing, spatial relatedness, and time experience are intimately connected in both dreams and wakefulness.

A next step toward understanding the unique phenomenal-functional configuration of dreaming is to inquire into the relationship between spatiotemporal situatedness and the mnemonic sources of dreaming (4). Here a first general observation is that in dreams, spatiotemporal self-location is relative to a reference frame that is largely made up of fragments of waking-memory sources. Going back to Gregory's (1923) claim, it would seem that dreams often involve instances of cryptamnesia, or memories that are experienced not *as* memories but as occurring now. Much research has been dedicated to the memory sources of dreaming, but here a few brief remarks should suffice.

To begin with, it seems that while dream content draws from waking-memory sources and changes in response to ongoing concerns, researchers are still too far from identifying a straightforward pattern to say how exactly this process takes place. As discussed in chapter 6, the debate surrounding the continuity hypothesis is not close to being resolved, and progress might ultimately consist in moving beyond the dichotomous debate on continuity versus discontinuity between dreams and waking life. This inability to find a straightforward pattern for the integration of waking-memory sources is, however, in itself interesting, because it suggests that the situation is more complex than commonly assumed. A tempting idea is that the trajectory from wakefulness via oneiragogia to full-fledged dreaming might involve a reorganization and interweaving of increasingly distal memory sources. I will return to this issue in more detail in chapter 12; for now, note that this seems likely for a number of reasons. One, research suggests that episodic memories are almost never replayed in dreams, whereas fragmented memory sources are much more frequent (Fosse et al., 2003; Nielsen & Stenstrom, 2005). Two, evidence from studies investigating the dream-lag effect suggests that daytime events are most likely to be integrated into dreams immediately after the event and after a five- to seven-day interval but are less likely to be dreamed of in the intervening interval (Nielsen & Powell, 1992; Blagrove, Fouquet, et al., 2011; Blagrove, Henley-Einion, et al., 2011). Three, this may be related to a transition in memory processing from the hippocampus to the neocortex after about one week (Nielsen & Stenstrom, 2005). In their review of the memory sources of dreaming, Nielsen and Stenstrom (2005) explicitly suggest that the hippocampus is involved in the binding of disparate spatial and temporal memories into a coherent, immersive scene, thus sustaining "the moment-to-moment illusion that dreaming is taking place from the first-person perspective ('here') and in the subjective present ('now')" (Nielsen & Stenstrom, 2005, p. 1287). This suggests that dreaming somewhat paradoxically involves the phenomenology of spatiotemporal situatedness relative to disparate memory sources integrated into a more or less coherent scene and narrative context, where these memories are no longer recognized as memories.

Note that if this characterization is even superficially correct, one would predict that the emergence of an immersive spatiotemporal reference frame during oneiragogic experience might be linked to the integration of ever-more-disparate and abstract memory sources throughout the course of the night. This is highly speculative, but some evidence does point in this direction. Several studies have shown that when presleep stimuli are integrated into sleep onset imagery, the incorporation is often of an abstract nature. For instance, in two studies, reports of Tetris-related imagery in participants who had played Tetris before sleep tended to involve accurate representations of falling (or rotating) Tetris blocks but typically left out nonessential elements such as the computer and keyboard (Stickgold et al., 2000; Kussé et al., 2012). The same was true in a study that investigated the incorporation of imagery related to the arcade game Alpine Racer; again, the reports involved stereotyped descriptions of being involved in a race, skiing downhill, and so on, but did not describe the virtual skis, game control, or the arcade machine itself (Wamsley, Perry, et al., 2010). Oneiragogia seems to have involved a shift in self-location to the environment depicted by the game while leaving out the game itself. It is also noteworthy that game-related imagery grew less frequent, less veridical, and more abstract with increased sleep duration. For instance, whereas participants reported flashes of skiing or racing initially, later imagery was semantically related but temporally remote: "I was picturing stacking wood this time. . . . I felt like I was doing it at . . . a ski resort that I had been to before, like five years ago maybe" (Wamsley, Perry, et al., 2010, p. 67). A similar pattern was observed in Stenstrom et al.'s (2012) study of sleep onset imagery. Presleep memory sources were found only in reports from the first awakenings, whereas memory sources reaching back for several years appeared only in reports from later-night awakenings. Moreover, later-night reports did not just integrate older memories but also integrated memories spanning an increasingly large time frame, connecting recent memories with ever-more-remote ones. A tempting conclusion would be that the emergence of an immersive spatiotemporal scene during oneiragogia is related to the integration of recent with increasingly remote memory sources.

Note that the examples discussed here are also tempting for another reason. In particular, note that both Tetris and Alpine Racer involve visuospatial imagery. Yet they also differ in an important respect: Tetris is played by manipulating images on a screen, whereas Alpine Racer has players stand on a pair of skis and control the onscreen character by using actual leg movements and weight shifts. As Wamsley and colleagues note, "Alpine Racer challenges participants with an intense, interactive, multisensory learning experience played out in an engaging 3D environment" (Wamsley, Perry, et al., 2010, p. 66). The incorporation rate of 24 percent by far exceeded that of other studies using presleep stimuli, including the Tetris study, which had an incorporation rate of 7 percent. This tells us two things: one, sleep onset and dreaming do not integrate presleep imagery in any direct and straightforward manner. Two, imagery of the type that is similar to dreaming—that is, involving immersive visuospatial tasks—is nonetheless particularly likely to find its way into dreaming.

This also leads to an important methodological observation. If the studies just cited point in the right direction, then attempts to study the memory sources of dreaming by relying on presleep stimulation only may be too shortsighted to make the actual temporal structure of memory integration visible. Rather, if dreams really integrate remote memories sometimes reaching back several years, then research is once more thrown back on trusting subjects to identify the memory sources of their dreams themselves. This nicely echoes the main tenets of antiskepticism about dream reporting developed in chapter 4. As Nielsen and Stenstrom put it:

> Future progress may ultimately depend upon whether subjects can be taught to identify and report the non-obvious minutiae of their dreams and memories with sufficient accuracy that the dreamwork mechanisms connecting the two can be discerned. In an era of high-resolution brain imaging, similarly high-resolution reports of dream imagery may be needed. (Nielsen & Stenstrom, 2005, p. 1289)

This also ties in with the last dimension that I introduced earlier to describe the trajectory through oneiragogia to full-fledged dreaming: (5) the integration with autobiographical memory. Recall that according to Nielsen (1991–1992), the progression through the stages of sleep onset mentation is also one toward greater reportability. Whereas fleeting imagery is characterized by lapses of awareness and difficulty recalling and describing experiences, phenomenal silence and the preimage context involve the impression of an indeterminate preceding period and then of an unrecalled preceding hallucinatory episode. Nielsen notes that by contrast, fully formed images are more easily recalled and described. We might then say that with the emergence of a spatiotemporal reference frame and perhaps even of a narrative context, there is also a progression in the degree of integration in autobiographical memory and experienced historicity, or the degree to which the episode is experienced as really having occurred. In the early stages of oneiragogia, experiences are so fleeting that they can barely be captured by retrospective recall; they are fleeting not just because of their brevity but because retrospectively they are not really graspable as our own experiences at all. By contrast, in vivid nightmares, the properties of historicity and integration into autobiographical memory are much more strongly developed, sometimes to the point of being a cause of daytime anxiety. Finally, retrospective dream deception forms the end point on the continuum of experienced historicity because here what was in fact a dream is mistaken for a memory of a waking event. This integration of dreams into a coherent autobiographical context takes us back to the epistemologist's familiar nightmare: here the distinction between dreaming and wakefulness has been blurred even retrospectively, and this highest level of integration with autobiographical memory even after awakening is clearly maladaptive. For a philosophical theory of dreaming, this point is important, because if dreaming can indeed be described as contextualizing and binding disparate memory sources and internally generated imagery into a centered spatiotemporal reference frame, this function is best fulfilled by dreams that fall short of achieving this highest degree of autobiographical integration.

In sum, my discussion of the different dimensions of oneiragogia lends substance to the claim that dreams are immersive spatiotemporal hallucinations and helps redescribe sleep–wake transitions as involving gradual shifts along a number of dimensions. We now have a new conceptual model of dreaming according to which dreams integrate distal memory sources into a centered model of reality. With this model in hand, we can now also define a number of contrast conditions and reconsider the relationship between dreams, waking hallucinations, full-body illusions, virtual reality, and imagination. But before we do so, we must consider several objections to the current model.

11.3 Objections to the Immersive-Spatiotemporal-Hallucination Model of Dreaming

11.3.1 Is the Immersive-Spatiotemporal-Hallucination Model of Dreaming Different from Existing Theories? From Spatiotemporal Self-Location to Minimal Phenomenal Selfhood

The first objection to the ISTH model is that it merely restates the familiar conception of dreaming as the experience of a self in a world in slightly less colloquial terms. For instance, Revonsuo uses examples from dream and virtual reality research to show that the most general, but also the most fundamental, feature of conscious experience is the feeling of "being directly present or bodily immersed in the center of a world—the *world-for-me*—and being there surrounded by meaningful objects" (Revonsuo, 2006, p. 109). In this view, "to be conscious is to have the *sense of presence* in a world" (p. 126). Similarly, Metzinger argues that the simplest form not just of dreaming but of conscious experience is "the *appearance of a reality*, here and now" (Metzinger, 2003, p. 605). And Nielsen notes that "dreaming consists of *a convincing simulation of waking reality experience*" (Nielsen, 2010c, p. 595). So what, if anything, distinguishes the ISTH model from these existing claims about dreaming and about consciousness more generally?

Immersive spatiotemporal experience picks out one specific aspect—the core aspect, I believe—of what it means to experience oneself as a self in a world, while avoiding many of the additional background assumptions this essentially metaphorical description implies. The widespread practice of describing dreaming as involving the experience of a self in a world is not theoretically innocent. Rather, as seen over the last chapters, this initially metaphorical description introduces wake-state bias into theoretical descriptions of dreaming. Dreaming is then too easily modeled on the way we standardly experience (or think we experience) ourselves, including our bodies, in wakefulness. While the question of whether to describe dreaming as involving world experience or, more technically, as ISTH, is a terminological decision, I think it is an important one. Describing dreaming as ISTH helps formulate the key intuition of the present theory, namely, that while dreams and wakefulness are both characterized by the phenomenology of spatiotemporal situatedness, there are systematic differences in the phenomenology of selfhood between both states.

According to the ISTH model, dreaming and standard wake states converge on immersive spatiotemporal experience, and though this is an important and probably central ingredient of conscious experience, this is not the same as saying that dreaming replicates our waking experience of being a self in a world in all its subtlety. Indeed, one reason for avoiding such conclusions is that it is unclear what exactly such a general comparative statement about the phenomenology of dreaming and wakefulness would imply. According to the present theory, dreaming is interesting precisely because of the ways in which it fails to replicate standard waking experience: dreams typically involve a self, but this self is not phenomenally embodied in a strong sense, nor is it consistently a cogitative self, or a thinker of thoughts. Even more interesting are those rare cases of dreaming in which such commonplace characteristics of standard wakefulness are lacking completely, that is, dreams in which there is a phenomenally disembodied self or in which self-experience is phenomenally indeterminate with respect to the dream body, or dreams that lack the occurrence of any cogitations whatsoever and in which the dream self is best described as doxastically indeterminate. These dreams, I propose, approach pure spatiotemporal situatedness: an experience characterized by nothing other than the sense of self-location in a spatiotemporal reference frame. The choice of terminology involved in formulating the ISTH model is intended to reflect the fact that in dreams, the phenomenology of selfhood is typically quite different from that of standard waking experience.

Moreover, in the ISTH model, it is also not quite correct to say that dreams consistently involve world experience. Dreaming, according to the ISTH model, is characterized by the substitution of a largely veridical, spatiotemporal reference frame with a largely internally generated, hallucinatory one. This hallucinatory reference frame is often quite different from the actual environment of the sleeping subject not only in the specific surroundings, persons, and objects it represents but also in the way it is experienced. Reports of floating through darkness, the void, or geometrical shapes describe highly nonstandard environments that are distinctly different from the way we experience and interact with the world in wakefulness. Again, the ISTH model allows that dreaming and wakefulness share the common ingredient of spatiotemporal self-location. Yet self-location, for instance, in an amodally experienced spatial expanse is quite different from what we typically mean by world experience. The departure of the ISTH model from the familiar terminology of world experience aims to highlight this distinction.

This point is not just a matter of terminological convention but may also be theoretically fruitful. In particular, exactly because dreaming and wakefulness converge on the common ground of spatiotemporal situatedness, focusing on this aspect as their common denominator allows us to abstract all those dimensions that are not strictly necessary for the emergence of self- and world experience. For this reason, the present theory also claims that dreams are an important candidate for identifying the minimal conditions of phenomenal selfhood and that the theoretical discussion of dreaming can enrich and extend existing theoretical work on self-consciousness. In dreams, spatiotemporal situatedness is sufficient for the most basic form of self-experience, as well as for the retrospective description of these states as having involved a self.

Here is a single example. Drawing from research on OBEs and full-body illusions (see chap. 12 for details), Blanke and Metzinger argue that phenomenal selfhood minimally involves "a passive, multisensory and globalized experience of 'owning' a body," and this sense of "'global ownership'—functionally defined as availability of an integrated, transparent and global representation of the spatiotemporally situated body—is the simplest form of self-consciousness" (Blanke & Metzinger, 2009, p. 12). They contrast this position with earlier claims according to which selfhood essentially consists in the ability to control one's thoughts and bodily actions, their main point being that bodily ownership is sufficient for minimal phenomenal selfhood, even independently of agency. This simpler conception says that minimal phenomenal selfhood still comprises three components: (1) self-identification, or identification with the content of a global body representation; (2) self-location, resulting in the phenomenology of "here" and "now"; and (3) the weak first-person perspective, which they describe as a "purely geometrical feature of an egocentric model of reality, most often referred to as a visuospatial model. A weak 1PP [first-person perspective] includes a spatial frame of reference, plus a global body representation, with a perspective originating within this body representation" (p. 7).

The ISTH model extends this argument in two important respects. The model's main claim is that while bodily ownership characterizes standard wake states, examples of phenomenal disembodiment and bodily indeterminacy in dreams show that bodily experience is not strictly necessary for phenomenal selfhood. In this view, Blanke and Metzinger's definition of minimal phenomenal selfhood equivocates two distinct properties that in dreams can come apart: namely, bodily ownership, or identification with a global body representation, and spatiotemporal self-location. Contra Blanke and Metzinger, the simplest form of being or having a self cannot be described as involving "an integrated and global representation of the spatiotemporally situated body" (Blanke & Metzinger, 2009, p. 12); phenomenally disembodied forms of spatiotemporal situatedness exist, and they offer a simpler form of phenomenal selfhood than that suggested by Blanke and Metzinger.

Recently, Metzinger (2013c) has taken up this criticism, originally developed in Windt (2010), and pointed out that it has the consequence of collapsing the distinction between self-identification and self-location. Because examples of phenomenal disembodiment in dreams, but also in OBEs, show that body ownership is not necessary for self-location, we now have a new candidate for the simplest unit of identification. What we identify with, both during the ongoing dream and retrospectively in dream reports, is not a global body representation but rather the phenomenal *here* and *now*. Spatiotemporal self-location is sufficient for minimal phenomenal selfhood because it grounds subjective experience by offering a locus of identification.

Immersive spatiotemporal reference frames of the type seen in dreams are *centered* reference frames, and the experience of being or having a self emerges at exactly the moment at which a specific position within a broader spatial and temporal expanse becomes available for identification.

A similar analysis applies to the distinction, proposed by Blanke and Metzinger, between the first-person perspective and self-location. One way of describing centered spatiotemporal reference frames is to say that they are organized around an internal, spatiotemporal first-person perspective. They are centered not just on a particular spatiotemporal self-location but also because they provide a position from which this reference frame is experienced. Recall that spatiotemporal self-location is a relative property: it is a way of experiencing one's location relative to other persons and objects, but also to past and future moments. For this reason, spatiotemporal self-location is also the origin of the spatiotemporal first-person perspective. This perspective is not necessarily visual or auditory, though of course it can be. Even cases of amodal spatiotemporal situatedness involve an extremely basic form of a first-person perspective because they represent the spatiotemporal relation between the phenomenal self and the broader spatiotemporal expanse relative to which it feels located. A preliminary conclusion is that spatiotemporal situatedness grounds both self-identification and the first-person perspective by collapsing them into a single target property.[6]

Again, this is not just a conceptual point, but is linked to an empirical prediction, namely, that more complex forms of self-experience involving bodily ownership, or identification with a body model, and modality-specific variants of the first person perspective (e.g., the point of origin of visual and auditory experience) should be distinct, in the current model, from self-location. It should, for instance, be possible to devise experimental manipulations targeting bodily ownership but not self-location, and they should have distinct neural correlates (for a review of recent evidence that this is the case, see Serino et al., 2013).[7] If this is correct, then bodily self-consciousness will comprise distinct and dissociable components. Yet, a surprising conclusion is that the simplest form of self-consciousness is not inherently bodily at all—or at least not on the phenomenal level of experience. Phenomenally disembodied and indeterminate dreams are, in this view, not the only, but perhaps the clearest and most frequent example for a dissociation between bodily experience and minimal phenomenal selfhood.

11.3.2 Do Selfless Dreams Exist?

A second potential objection to the ISTH model targets the claim that dreaming essentially involves self-experience, even in the very basic form associated with pure self-location. True, a majority of dream reports describe the presence of a self—but recall that a small percentage of dreams are also described as selfless. Studies investigating self-representation in dreams have found that dreams are at least occasionally described as selfless (Occhionero et al., 2005), and lucid dreams are also sometimes described in this manner (LaBerge & DeGracia, 2000). So it would seem that based on the empirical evidence, and taking reports of selfless dreams at face value, the ISTH model does not capture the phenomenal core of dreaming. Perhaps some dreams are not immersive in the sense of involving self-location within the dream; or perhaps dreams can be centered on an internal perspective without thereby involving a self. This would threaten my proposal that phenomenal selfhood can arise independently of phenomenal embodiment. It would then also no longer be correct to describe such dreams as involving a sense of *self*-location: there would be no phenomenal self to feel localized *in* the dream in the first place. Immersion or situatedness, as I have been using the term, refers to the sense of self-location within a spatiotemporal reference frame. If some dreams do not involve a self, then immersion or situatedness—which is essentially a relational property—can no longer be instantiated. If either of the relata is missing—i.e. that is, the subject who is immersed or the reference frame in which the subject is immersed—the notion of immersion is nonsensical.

Revonsuo uses the example of observer dreams to argue against the view that conscious experience necessarily has the structure of a self in a world. This stands in apparent contradiction to his claim that "the dream world seems to have at least one permanent inhabitant: the dreamer herself" (Revonsuo, 2006, p. 82). While he takes the presence of a self to characterize a majority of dreams, he does not take it to be necessary. Indeed, he argues that observer dreams involve "seeing the dream world from a detached, disembodied third-person perspective" (p. 184), much as would be the case for a 3-D movie. In such dreams, "the point of view . . . still constitutes the *center* of the virtual space" (p. 184), but only "as an empty point from whose perspective the visuospatial world is constructed" (p. 184). In such cases, conscious experience has lost its egocentricity, and indeed "the ego (the virtual self) need not be in the center of the reference frame, or indeed present at all" (p. 184). According to Revonsuo, this has important consequences, showing that "the experience of phenomenal objects is not a *relation* obtaining between the self . . . and objects" (p. 185), "the virtual body or virtual self is entirely dissociable from virtual presence" (p. 185), and neither the phenomenology of embodiment nor *any* representation of the self is a necessary condition of consciousness.

There are different ways of spelling out the difference between our views. One way is to say that the difference is of a merely terminological nature: where I take self-location to be sufficient for phenomenal selfhood, Revonsuo takes phenomenal embodiment to be necessary. We both agree that dreams can be immersive without thereby involving a phenomenally embodied self; we merely disagree about whether to say that some form of phenomenal selfhood is still involved even in phenomenally disembodied dreams. A slightly different interpretation would say that the difference between our views is of an empirical nature. If evidence suggests that dream reports describe instances of disembodied *self*hood, this would support my theory; by contrast, reports of selfless dreams would support Revonsuo's.

Taking the second reading first, things are rarely quite so simple. In support of my view, recall that observer dreams are sometimes described as involving a sense of embodiment (Occhionero et al., 2005) and indeed may be quite variable in this respect. Consequently, observer dreams *as such* should be confused with neither disembodied nor strongly embodied dreams (see secs. 7.3.2 and 7.3.3). Moreover, the absence of a *bodily* dream self in phenomenally disembodied dreams should not be confused with the absence of a self as such. Even in reports of phenomenally disembodied dreams, dreamers still sometimes state that they were present as a disembodied entity, thus placing themselves in the dream (see sec. 7.3.2). However, this strategy will not yield a clean solution: Revonsuo's reading would still be supported by reports of selfless dreams. Here antiskepticism would seem to demand that we take reports of such dreams seriously: if even a small percentage of dreams are described as selfless, then selfless dreams exist. Spatiotemporal self-location could then no longer be claimed to constitute the phenomenal core of dreaming. Given that the ISTH model aspires to characterize the simplest form of dreaming, any such counterexamples, however rare, would be genuinely damaging to its key claims.

A first step toward defending the ISTH model against this objection is to point out that it is not at all clear what is meant by selfless experience. Throughout the book, it has become clear that folk-psychological concepts such as dreaming and the self are sorely underdefined. An important aim of the ISTH model has been to replace the folk-psychological concept of dreaming with a technical one that aspires to empirical and folk-psychological plausibility but will still involve some degree of conceptual legislation. So it stands to reason that a similar process of conceptual refinement may apply to the concept of phenomenal selfhood. If this is true, then Revonsuo's selfless dreams may well, in the ISTH model, be described as instances of minimal phenomenal selfhood. The difference between our views would now be a terminological one only, but would fail to point to any interesting and deep phenomenological differences. Perhaps reports of selfless dreams only describe experiences that were selfless in a qualified sense, that is, by lacking the phenomenology of embodied selfhood or that of being a cognitive agent. They would not thereby count as selfless in the strict technical sense used in the ISTH model; in particular, they would still involve a sense of self-location. Note that this translates into an empirical prediction: dreams described as selfless, according to the present theory, either will still be immersive (and thus qualify as involving minimal phenomenal selfhood) or will not count as proper dreams.

This terminologically motivated response to the problem of selfless dreams is, however, still overly simplified. Even if at least a subgroup of purportedly selfless dreams can be understood as selfless only in a qualified sense, this is not yet an argument for the stronger claim that dreaming *necessarily* involves a self. Can anything more be said in support of this stronger claim?

To begin with, note that nothing commits the present theory to the claim that there is a sharp cutoff point for distinguishing selfless experience from minimal phenomenal selfhood. To the contrary, one of the aims of the ISTH model is to capture the transition from wakefulness via oneiragogia to full-fledged dreaming, and the same is true for the gradual emergence of phenomenal selfhood. This is why, in fact, I have spent so much time pointing out that phenomenal selfhood is not a simple construct but a highly complex process involving different and potentially dissociable dimensions. It would make no sense, in this view, to demand that the simplest form of phenomenal selfhood be an all-or-nothing affair or that there must be an obvious criterion for distinguishing between selfless experience and phenomenal selfhood. In this view, terminological discrepancies, like the one between my theory and Revonsuo's, are to be expected: indeed, if spatiotemporal self-location really is a candidate for the simplest form of phenomenal selfhood, then one might expect there to be some disagreement about whether it forms part of phenomenal selfhood proper or itself is just below the threshold. Our disagreement over the appropriate terminology may be an indicator that we are on the right track. One strategy, then, might be to resist the demand for a clear-cut answer: what is important in the present theory is that self-location marks a crucial point on the trajectory toward the emergence of a phenomenal self in a more interesting sense, and this is something about which Revonsuo and I can agree.

Yet I think something more can be said in defense of my suggestion that spatiotemporal self-location be considered a form of minimal phenomenal selfhood. By applying the concept of phenomenal selfhood to experiences hovering on the border of what it means to be or have a self, rather than reserving if for clear-cut cases (such as phenomenally embodied selfhood, as suggested by Revonsuo), we actually enhance the precision of our conceptual tool kit. In particular, this terminological convention helps describe the trajectory toward more complex forms of selfhood while also alerting us to the common denominator of different forms of phenomenal selfhood, namely, spatiotemporal self-location. If this is correct, then my proposal has genuine explanatory value: by applying the concept of phenomenal selfhood to a broader range of cases, it enables a more unified perspective on the target phenomenon.[8]

The question, then, is whether the proposed framework can make sense of purportedly selfless experiences. I will discuss a single example to show there is reason to think that it can. Note that this example is similar to the "clear light of sleep" experiences from the tradition of dream yoga described in section 11.1. Similar cases of selfless experiences or dissolution or transcendence of the self are also reported to occur in meditative states and drug-induced states (Vaitl et al., 2005; Studerus et al., 2010; Beauregard, 2011; Berkovich-Ohana et al., 2013). For now, we will focus on a lucid dream reported by Clare Johnson on March 19, 1995:

> I am suspended in space—*dream space*, I think. There is nothing here, just millions of greyish dots and I am one of the dots, there's no dream-body anymore, I'm just a dot [of] pure consciousness suspended. A feeling of great peace comes over me and a sense of gentle, infinite expansion. It's as if everything and nothing are the same thing and there is a sense of effortless belonging. As the sense of expansion increases I am no longer a single dot of consciousness; all the dots are me and I am them. There's no "I" or "them." We are one. There's just a blissful sense of timelessness and oneness and a merging with the light. After an indefinable length of time, I start to feel the weight of my body in bed, and settle back into it, tingling all over. (Clare Johnson, unpublished dream report)

This dream report nicely illustrates my point that the transition from minimal phenomenal selfhood to selfless experience can be gradual. The beginning of the report still describes a case of disembodied spatiotemporal self-location, involving a sense of identifying with an extensionless point, a dot, in a larger space. This is followed by a sense of expansion, in the process of which the sense of temporal self-location dissolves in tandem with the sense of spatial location. Simultaneously, the sense of identifying with any specific point and any remaining sense of distinction between the self and the environment are lost. What we see here closely resembles the gradual emergence of an immersive spatiotemporal reference frame that characterizes the stages of oneiragogia, only in reverse. Moreover, the transition from minimal phenomenal selfhood to the dissolution of the self, if we take the report seriously, approaches the limits of reportability: with the loss of a sense of self-location, there is also a sense of timelessness, finally followed by awakening after an undefined period of time. This is exactly what would be predicted by the current theory: if dreams are immersive spatiotemporal hallucinations, and if dreams are for methodological reasons reportable experiences, then experiences that no longer count as immersive and as involving minimal phenomenal selfhood should be located on the borders of reportability. As long as we resist the demand for a clear cutoff point between minimal phenomenal selfhood and selfless experience, as well as between dreaming and nondreaming, I think this report nicely fits in with the current model.

Moreover, the conceptual tool kit offered by the ISTH model can be used to explain even the moment at which spatiotemporal self-location is lost. While the ISTH model says that minimal phenomenal selfhood still involves a sense of identification with an extensionless point, Metzinger (2013c) points out that there might also be something like a maximal unit of identification. This might be the case, he argues, for instances of "pure consciousness," in which the unit of identification is

> the most *general* phenomenal property available for identification at all: Philosophers might call it the global "unity of consciousness," or phenomenality *per se*, or awareness *as such*, namely the singular, integrated, all-pervading quality of conscious*ness* characterizing the current totality of experiential contents, as it is given in every single moment of experience. (Metzinger, 2013c, p. 5)

Such experiences can no longer be described as involving a self, as the distinction between self and environment has apparently been lost. They no longer involve a centered reference frame or a subjective perspective; the unit of identification is no longer distinct from the spatial reference frame. For this reason, I think that even the conditions for minimal phenomenal selfhood are no longer met. This interpretation thus suggests that their description as selfless experiences be taken seriously. Yet the conceptual repertoire used to describe this type of experience is not wholly discontinuous. Rather, it still makes sense to speak of a unit of identification—though this unit of identification is now maximally abstract and undifferentiated. Importantly, doing so can help describe the trajectory from minimal phenomenal selfhood to selfless states in continuous terms.

I think that describing pure spatiotemporal self-location as involving a minimal sense of self is also attractive for another reason. This is that it acknowledges the intuition, common in philosophy, that there is a deep and perhaps even conceptual connection between conscious experience and phenomenal selfhood. According to the present theory, the existence of selfless experience is not ruled out on the basis of conceptual considerations alone; indeed, anecdotal evidence supports its existence, and so it should be regarded as an empirical possibility. To say otherwise would be to commit the armchair fallacy. Yet even cases of nonstandard experience, as in phenomenally disembodied dreams, still count as involving a phenomenal self. This is a nod to the long-standing philosophical tradition of regarding conscious experience as inherently subjective and thus ensures that the present theory is compatible with existing philosophical theories.

At least since Descartes, it has been a conceptual truism that selfless experience cannot exist: as Descartes (1641/1901) argues in the *Second Meditation*, there can be no experience without a thinker, and having any experience whatsoever ensures him of his own existence. Generally a tight conceptual connection exists between the concepts of experience and self-consciousness; and indeed, the notions of experience, subjective experience, and phenomenal consciousness are often used synonymously. If, for instance, conscious experience has something to do with presence—in the sense of perceptual presence—then it necessarily involves a self in the minimal sense of a locus of experience. Presence requires the appearance of something as present to *someone*; consequently, experiencing an object as present requires an experiential subject, as well. This inherently perspectival and subjective character is precisely why conscious experience, to many, is such a profound challenge for an objective science of consciousness. Nagel's (1986) famous *view from nowhere* is an attempt to express the need for a wholly objective description of the world that would still comprise subjective viewpoints in it. Yet Nagel's view from nowhere involves a contradiction. A view from nowhere would no longer be a *view* in the phenomenal sense, nor would there be anyone who *had* this view. Viewpoints are inherently perspectival; the *view from nowhere* is a misnomer because it suggests that there could be a place, nowhere, from which this view could be had (cf. Metzinger, 2011).

Moving beyond these general observations, note that much contemporary work focusing on the phenomenal self and specifically on the concept of minimal phenomenal selfhood holds that conscious experience is necessarily and intrinsically subjective. I will give three short examples.

Metzinger (2011) discusses different variants of the "no-self alternative" or of antirealism about the self. Ontological antirealism about the self denies that selves are substances, in the sense of entities existing independently of the brain and the body. Methodological antirealism about the self is the position that the scientific investigation of self-consciousness is not committed to the assumption that selves actually exist. And semantic antirealism denies that indexical expressions such as "I" refer to an ontologically fundamental entity. Note, however, that among these different variants of antirealism about the self, one is missing: there is no variant of phenomenological antirealism about the self, and this absence is inextricably connected to, and indeed explained by, Metzinger's self-model theory of subjectivity. According to Metzinger, the *no-self alternative*, no matter in what reading, is profoundly counterintuitive. It "describes an impossible scenario, because it cannot be *actively* simulated and embedded—the property of cognitive agency itself would have to be dissolved. . . . For beings like us, it is therefore not a phenomenologically possible world" (Metzinger, 2011, p. 289). For Metzinger, the inability to imagine or phenomenally simulate a selfless world is not an accidental failure in our imaginative capacities but is rooted in our evolutionary history. For beings like us, antirealism about the self is inherently counterintuitive because believing in the necessity of our own existence proved to be adaptive. Realism about the self thus is implicit in the structure of phenomenal experience, and we cannot even imagine what it would be like to not be or have a self. The theory predicts that so-called selfless experiences, if they exist at all, will be exceedingly rare and highly unstable.

Similarly, in discussing the minimal subject, Strawson notes that if you consider a subject of experience and strip away, in thought, everything other than the ongoing experience, the subject remains:

> You can't get rid of the subject of experience, in taking a portion of experience or experiential what-it's-likeness and stripping away everything other than the existence of that experience. Concretely occurring experience can't possibly exist without a subject of experience existing. If you strip away the subject, you haven't got experience any more. (Strawson, 2011, p. 253)

Again, we find the idea that conscious experience and the existence of a phenomenal self are inextricably entwined: "Experience is experiencing: whatever remains if experience remains, something that is correctly called a subject must remain. . . . I'll call the subject that must remain when everything but experience has been stripped away the 'minimal' subject" (Strawson, 2011, p. 254).

Finally, Zahavi notes that subjectivity is not a specific experiential content and consequently is not something that could be stripped away from the experience. Rather, "the self is the very subjectivity of experience and not something that exists independently of the experiential flow" (Zahavi, 2011, p. 328). Zahavi concludes that the subjectivity of experience is inescapable: "There is subjectivity of experience and a minimal sense of self, not only when I realize that I am perceiving a candle, but whenever there is perspectival ownership, whenever there is first-personal presence or manifestation of experience" (p. 334). Moreover, he argues that "it is this pre-reflective sense of self which provides the experiential grounding for any subsequent self-ascription, reflective appropriation, and thematic self-identification" (p. 334).

Despite the differences between these three authors, they converge on a single point: conscious experiences and at least a minimal form of phenomenal selfhood or subjectivity are inextricably linked, either as a matter of nomological necessity (for instance, because, as predicted by Metzinger's self-model theory, the phenomenology of selfhood was evolutionarily advantageous and we cannot deliberately think ourselves out of it) or as a matter of conceptual necessity. The ISTH model preserves this intuition by suggesting a new candidate for empirically grounding reports of being or having a self. Moreover, spatiotemporal situatedness is also, according to the current theory, a precursor to more complex forms of self-consciousness. Dreams, as the most common example of minimal phenomenal selfhood, help strip phenomenal selfhood down to its core, thus identifying aspects commonly taken to characterize self-consciousness, such as phenomenal embodiment and cognitive agency, as contingent.

Before moving on to the final objection, I want to point out that a similar idea was already formulated by Freud, though in a different context. In *A Metapsychological Supplement to the Theory of Dreams* from 1917 (Strachey, 1957, pp. 217–236), Freud writes:

> We are not in the habit of devoting much thought to the fact that every night human beings lay aside the wrappings in which they have enveloped their skin, as well as anything which they may use as a supplement to their bodily organs (so far as they have succeeded in making good those organs' deficiencies by substitutes), for instance, their spectacles, their false hair and teeth, and so on. We may add that when they go to sleep they carry out an entirely analogous undressing of their minds and lay aside most of their psychical acquisitions. Thus on both counts they approach remarkably close to the situation in which they began life. (Strachey, 1957, p. 222)

Although Freud was more interested in the psychological implications of this undressing of our minds, a similar point is emphasized by the ISTH model: dreaming may be a state in which we set aside not just the wrappings enveloping our skins but also the functional constraints imposed by our physical bodies on bodily experience. At least occasionally, this may result in the loss of bodily experience altogether. Yet even in these cases, dreams still retain their subjective flavor. The ISTH model tries to do justice to the central intuition that conscious experience, including dreams, is tied to the presence of at least a minimal phenomenal self without turning this conceptual point into a matter of phenomenological necessity. While the ISTH model offers a simpler and drastically reduced model of what minimal phenomenal selfhood consists in, it preserves the commonplace intuition that experience is inherently perspectival and subjective. I think that this essentially conservative motivation is an argument for expanding the concept of phenomenal selfhood to include instances of pure spatiotemporal situatedness, rather than calling them selfless.

11.3.3 Is Cognitive Immersion an Alternative to the Immersive-Spatiotemporal-Hallucination Model of Dreaming?

A final objection to the ISTH model is that it targets the wrong type of immersion. The imagination view, as discussed earlier (cf. chap. 6), denies that dreaming involves the phenomenology of presence and of phenomenal selfhood in a robust phenomenological sense. Yet the imagination view must explain why we seem to believe in the reality of our dreams while they are occurring, and why we sometimes even seem to be emotionally involved in them. The imagination view has two different ways of responding to this challenge, and both have to do with the idea that our sense of cognitive and emotional involvement in dreams can be modeled on the type of fictional immersion experienced while reading a novel, watching a movie, or daydreaming. Because both versions of the imagination view rely on exclusively cognitive mechanisms to explain the phenomenology of dreaming, I call this the *cognitive immersion model* of dreaming.

The first version of the cognitive immersion model posits that one's sense of presence in dreams is exactly the same as when one is lost in a novel, movie, or daydream. In this view, immersion in dreams is of the purely cognitive type. It does not involve a sense of self-location in the dream, nor does it involve actual beliefs or emotions about the ongoing dream (Ichikawa, 2009; cf. chap. 6). I have already argued (in chaps. 6 and 9) that this type of wholesale argument against the occurrence of real beliefs and emotions in dreams is empirically implausible. For present purposes, a weaker version of the cognitive immersion model is more interesting.

This second version, proposed by McGinn, maintains that dreaming involves imaginative (rather than quasi-perceptual) experience, but allows that dreams invite genuine beliefs. Like Ichikawa, McGinn holds that fictional immersion of the type experienced in wakefulness merely mimics real belief. Yet because he also thinks that dream beliefs and emotions are real instances of their kind, fictional immersion alone cannot explain dream belief.

McGinn argues that we need two additional factors to explain dream belief. The first is that during sleep, any competing percepts that could distract attention from the unfolding images are lacking.

> The prime obstacle to full fictional immersion in the waking state is therefore absent. It is as if you could read a novel without having to look at marks on a page, or experience a movie without having to gaze at a screen. In such a condition your mind would make direct unmediated contact with the events and objects of the story. (McGinn, 2004, p. 106)

This view echoes the familiar Aristotelian theme that the clearest dream images unfold in the silence of sleep. The second factor, according to McGinn, is the extreme suggestibility that characterizes the dream state, which is comparable to hypnosis. This interplay of fictional immersion in the absence of distracting percepts and the condition of extreme suggestibility lifts dream belief to the level of real belief. This is analogous to McGinn's account of delusion formation, according to which "the delusions of madness . . . are traceable to a disturbance of the imagination" (McGinn, 2004, p. 114).

Elsewhere McGinn compares dreaming to the experience of watching a movie, predicting that "movies arouse in the viewer the same kinds of psychological mechanisms and processes that characterize the dreaming state" (McGinn, 2005b, pp. 101–102), for instance, by attenuating the capacity for critical thinking. He even suggests that "movies chemically alter the brain in the direction of its dreaming mode" (p. 134), thus producing a state of semisomnolence (p. 134). This state does not, however, enable dreams to involve the phenomenology of presence. As is the case for watching a movie, we may identify with a dream character or project ourselves to the position of the camera; yet according to McGinn, dream and movie objects are not experienced as having a particular location relative to the self (p. 101). The reason why we are nonetheless deceived by our dreams is that the dreamer, in McGinn's words, is a doxastic patsy (McGinn, 2004, pp. 98–99) who no longer grasps the difference between fiction and reality. This last point gives rise to a further prediction, namely, that in the cognitive immersion model, "You cannot be both immersed in a fiction and consciously thinking 'this is only fiction.' Your attention cannot be both on the fictional work *qua* fictional artifact *and* on the fictional world depicted" (p. 111).

Let us grant, for the sake of argument, that paying attention to a fictional work qua fictional artifact—say, thinking about the literary style of an author, the performance of an actor in a movie, or even deliberately plotting one's ongoing daydream—works against cognitive immersion. Even so, it is not clear that the same is true for dreaming. Fully lucid dreams are a particularly clear example. In the following dream report, it seems that the cognitive realization that this is a dream even leads to an increase in the dream's experienced vividness. This is exactly the opposite of what the cognitive immersion model would predict:

> I dreamed that I was standing on the pavement outside my home. . . . I was about to enter the house when, on glancing casually at [the pavement] stones, my attention became riveted by a passing strange phenomenon, so extraordinary that I could not believe my eyes—they had seemingly all changed their position in the night, and the long sides were parallel to the curb! Then the solution flashed upon me: though this glorious summer morning seemed as real as real could

be, I was *dreaming*! With the realization of this fact, the quality of the dream changed in a manner very difficult to convey to one who has not had the experience. Instantly, the vividness of life increased a hundred-fold. Never had the sea and sky and trees shone with such glamorous beauty; even the commonplace houses seemed alive and mystically beautiful. Never had I felt so absolutely well, so clear-brained, so inexpressibly *free*! The sensation was exquisite beyond words; but it lasted only a few minutes and I awoke. (Fox, 1962/1993, pp. 32–33)

This report is no exception, and a similar increase in vividness associated with the onset of lucidity is often reported in the literature on lucid dreaming (LaBerge, 1985; LaBerge & Rheingold, 1990; Brooks & Vogelsong, 1999). In the cognitive immersion model, this type of experience, coupled with lucid insight, should be impossible.

But there is a deeper problem with the cognitive immersion model. We have seen that while dreams can be belief inviting, they are not typically or consistently so (see chap. 10). The cognitive immersion model, by aiming to explain dream deception, misidentifies the explanandum: a majority of dreams are not, as assumed by McGinn, deceptive in any interesting sense. If dream deception (in the sense of believing in the reality of dream events) is a measure of cognitive immersion, then we are in fact only rarely cognitively immersed in our dreams. Consequently the sense in which we feel present in our dreams cannot be reduced to cognitive immersion. In fact, according to the ISTH model, the opposite is true: spatiotemporal self-location is a more basic type of immersion, and while it may well be necessary for cognitive forms of immersion and Cartesian dream deception, it is not sufficient. In dreams, doxastic situatedness does not always follow on the heels of spatiotemporal situatedness; in fact, fully lucid dreams are a good example for saying that these two kinds of situatedness can be almost completely dissociated in the dream state.

Yet I want to propose that in a slightly different reading, the cognitive immersion model can meaningfully be applied to dreams, after all. To see why, recall that in chapter 9, I argued that dreams are weakly cogitative states in that in most dreams, dream cogitation is only weakly differentiated from unintentional, automatic, and largely imagistic dream cognition. The sense in which we are cognitively immersed in our dreams, in this view, is not that we falsely believe in their reality but that we typically fail to form any beliefs about them whatsoever. Dreams involve the creation of centered spatiotemporal reference frames, but only rarely and intermittently do they create a comparable doxastic reference frame or a cognitive first-person perspective. This failure is intimately tied to the cognitive processes underlying dream formation. Recall that in the ISTH model, dreams can be described as integrating distal memory sources and internally generated spatiotemporal imagery into a centered model of reality. A speculative working hypothesis is that the computational processes underlying the generation of spatiotemporal self-location in dreams are closely linked to the widespread absence of cogitative activity and doxastic situatedness. In the dream state, quasi-propositional, conscious thought is systematically overridden by lower-level spatiotemporal cognitive processing that is experienced not as self-generated but as generated externally. Indeed, this might be why both fully lucid dreams and strong cases of dream deception, both of which require conscious cogitation and doxastic situatedness, are comparatively rare and highly unstable. Clearly the computational load would be considerably enhanced if dreams additionally required the formation of a full-fledged cognitive first-person perspective. For this reason, cognitive immersion, in a majority of dreams, does not refer to the fact that we form particular beliefs about our dreams; rather, it refers to the fact that due to the nature of cognitive processing underlying dream formation, the emergence of a cogitative self is largely prevented in the first place. If this is correct, we can now see that cognitive immersion is not a genuine alternative to the type of spatiotemporal situatedness proposed by the ISTH model; rather, the ISTH model explains why cognitive immersion takes the form that it does, at least in a majority of dreams.

11.4 Conclusions

This chapter suggests a new answer to the conceptualization problem, or the question of how exactly to describe conscious experience during the dream state. Dreams, according to the ISTH model, are centered reality models, characterized by the integration of distal memory sources and largely internally generated imagery into an immersive spatiotemporal reference frame. The model proceeds from the assumption that a conceptual model of dreaming should abstract from all the stereotyped but contingent phenomenal properties of a majority of dreams to identify the phenomenal core of dreaming, or the common denominator underlying different types of dreams. According to the present model, this phenomenal core consists in immersive spatiotemporal experience.

This purely phenomenological characterization is attractive because it suggests a unified account not just of dreaming but of conscious experience more generally, across standard and altered states of wakefulness and sleep. This is not yet, however, a distinctive account of dreaming: though dreams and standard wake states often do differ on the phenomenological level of description, they converge on spatiotemporal situatedness as their common phenomenal core. This convergence explains the widespread intuition that dreams involve the phenomenology of presence and why we rarely realize, while dreaming, that this is only a dream. The present model thus tries to preserve widespread intuitions about dreaming, both from folk psychology and from philosophy, while translating them into the more technical terminology of ISTH.

To obtain an account of what distinguishes dreaming from standard wake states, it is necessary to look beyond the phenomenology of dreaming. Here the criterion for distinguishing dreams from standard wake states is that dreaming involves a replacement of a largely veridical perceptual reference frame with a largely internally generated, hallucinatory one. This shift can be described more precisely by discerning a number of dimensions. The example of oneiragogia—or of experiences leading into dreams, occurring during sleep–wake transitions—is particularly helpful for teasing apart these different dimensions. Here the shift from perceptual to hallucinatory spatiotemporal situatedness is seen to result from the interplay of complex cognitive processes describable along at least five different dimensions: (1) the gradual emergence of an integrated, hallucinatory visuospatial scene; (2) a gradual weakening of phenomenal-functional embodiment; (3) the gradual substitution of a temporal reference frame; (4) the integration and reorganization of waking-memory sources; and (5) the integration with autobiographical memory.

Several important consequences for the concept of dreaming follow from the ISTH model and the analysis of oneiragogic experiences. The first is that the concept of dreaming can be extended to cover immersive spatiotemporal hallucinations occurring outside of sleep as commonly defined, including sleep–wake transitions, but also anesthesia dreams. The ISTH model thus offers a more precise characterization of the target phenomenon in phenomenal-functional terms, as well as a research strategy for investigating the neural basis of different forms of ISTH. Importantly, because of its focus on self-location, it also suggests a revised account of minimal phenomenal selfhood. If spatiotemporal self-location really is sufficient for the simplest form of self-experience, as proposed here, this suggests that dreaming can extend and enrich existing philosophical work on self-consciousness, as well as experimental work on full-body illusions and OBEs commonly taken to be relevant in this context. The surprising conclusion is that the focus of contemporary theories on *bodily* forms of self-consciousness mistakes state-dependent constraints on subjective experience that are characteristic of standard wake states for necessary conditions of phenomenal selfhood. By contrast, because phenomenal-functional embodiment in dreams is generally weakened and sometimes, in phenomenally disembodied and phenomenally indeterminate dreams, practically obliterated, dreams show that in healthy subjects, self-consciousness can arise independently of the phenomenology of embodiment. This is an important step toward identifying the state-independent necessary and sufficient conditions for phenomenal selfhood across sleep–wake states.

The next natural question is how the contrastive analysis (Windt & Noreika, 2011) of dreaming and standard and altered wake states can lead to the formulation of new questions for future research, but also to an enriched understanding of how to integrate dreaming into the broader framework of concepts used to describe conscious experience and the self. The next chapter discusses three examples, with the aim of showing how relocating the concept of dreaming on the map of concepts used for the description of standard and altered wake states can lead not just to a new understanding of dreaming but also to a new perspective on some core aspects of conscious experience.

12 Relocating Dreams on the Conceptual Map: Consequences and Perspectives for Future Research

12.0 Introduction

Throughout the book, I have been concerned with developing a conceptual framework for describing conscious experience in the dream state in its own terms. I have repeatedly emphasized the importance of avoiding premature comparisons between dreaming and wakefulness, which often, in the history of philosophical theorizing about dreaming, have led to oversimplified characterizations of dreaming tinged by similarly oversimplified characterizations of wakefulness. Now that a new conceptual model of dreams as immersive spatiotemporal hallucinations is in place, we can take the next step toward solving the conceptualization problem, or the problem of where to locate the concept of dreaming on the map of concepts used to describe standard and altered wake states. We have already seen that an empirically informed philosophical theory of dreaming requires a departure from some theoretical conceptions of dreaming popular in the contemporary literature. So it stands to reason that the revised framework for describing dreams developed in this book will not only enable us to describe the relationship between dreams and standard and altered wake states more precisely but also require us to relocate the concept of dreaming in several important respects. This is what the present chapter sets out to do.

First, however, it makes sense to take a step back and take a bird's-eye view on the conceptual territory covered so far. Recall that chapters 1 through 4 were concerned with the foundational epistemological and methodological problem underlying any theoretical or empirical investigation of dreaming: the trustworthiness of dream reports. Based on an analysis of different variants of philosophical skepticism about dream reports (chap. 1), an overview of historical changes in the concepts of sleep and dreaming (chap. 2), and a reconstruction of the background assumptions implicit in scientific dream research (chap. 3), I proposed antiskepticism as a constructive position on the trustworthiness of dream reports (chap. 4). Antiskepticism makes the following main assumptions:

- Dreams are reportable experiences (or phenomenal states) occurring during sleep.
- Dream reports, at least when gathered under ideal reporting conditions, are transparent in the sense of being trustworthy sources of evidence about the occurrence and phenomenal character of experience during sleep.
- Scientific dream research is constrained, for methodological reasons, by the space of reportable dreams.
- The conservative approach takes dream reports maximally seriously and simultaneously strives to improve the conditions under which dreams are reported so as to eventually identify and approximate ideal reporting conditions and optimize the relative transparency of dream reports.

Using antiskepticism as a point of departure, I then gradually worked toward a positive characterization of dreaming. Again, my strategy was to set out from assumptions from the history of philosophical theorizing about dreaming, confront them with a critical discussion of scientific research results and psychological dream theories, and propose a revised, conceptually sound, and empirically plausible view at the end of each chapter. From this, the following picture emerged:

- Dreaming does not fall soundly on either side of the putative conceptual divide between quasi-perceptual experience (or experiences that share the phenomenology of waking perception, i.e., hallucinations and illusions; see chap. 5) and imaginative experience or daydreaming (see chap. 6). Attempts to model the phenomenology of dreaming exclusively on either waking perception or imagination are oversimplified.
- Both for conceptual reasons and in view of research findings, it might be fruitful to consider dreaming as an intensified form of mind wandering that blurs the borders between quasi-perceptual and imaginative experience as commonly conceived (chap. 6); I return to this idea in section 12.3.
- Dreams are weakly phenomenally embodied states in which bodily experience is characterized by widespread phenomenal indeterminacy and disturbances in multisensory integration; at the same time, bodily experience in dreams is highly variable, ranging from phenomenal indeterminacy and phenomenal disembodiment to weak and strong phenomenal embodiment (chap. 7).
- Dreams are weakly functionally embodied states in which the processing of real-body inputs and production of motor outputs are attenuated, but not completely absent; consequently, bodily experiences occurring in dreams are best described as involving illusory own-body perception (chap. 8). Moreover, there continues to be a systematic relationship between phenomenal and functional embodiment, such that the unique pattern of bodily experience in dreams can be explained by appealing to alterations in the functional connection to the sleeping body and the continued processing of real-body inputs.
- Dreams are weakly cogitative states in which conscious cogitative activity (e.g., thinking, self-reflection, and metacognition) is highly unstable and typically only

weakly differentiated from unintentional, automatic, and largely imagistic dream cognition (chap. 9).

- Dreams are characterized by widespread doxastic indeterminacy in that dreamers typically form no beliefs, among other things, about the nature of their current conscious state. For this reason, a majority of dreams provide no foothold for classical Cartesian dream skepticism (chap. 10). Yet dreams can be genuinely belief inviting, and different forms of concurrent and retrospective dream deception can be distinguished. Of these, cognitive corruption is perhaps the most disturbing from a theoretical perspective, because it undermines our ability to recognize genuinely rational thought. By contrast, retrospective dream deception, in which dreams are misremembered as waking experiences or invite beliefs beyond the borders of sleep, are a source of practical and clinical concern, at least in certain subject groups.
- Dreams can be compared to wake-state delusions on a case-by-case basis but should not generally be described as delusional; a nosological analysis of insight in different types of lucid dreams may open new perspectives for future research (chap. 10).
- Dreams are immersive spatiotemporal hallucinations (chap. 11) in which spatiotemporal self-location is relative to a largely internally generated spatiotemporal reference frame rather than one that is perceptual and largely veridical, as in standard wake states.
- Oneiragogic experiences, or experiences leading into dreams, provide evidence for a gradual shift in spatiotemporal self-location toward a hallucinatory spatiotemporal reference frame along a number of different dimensions (chap. 11).
- Spatiotemporal self-location is a core target for future research and a candidate for (re)defining minimal phenomenal selfhood (chap. 11).

This cursory list shows that the taxonomy developed throughout the book has already led to a relocation of the concept of dreaming in a number of respects. It has already led, for instance, to a much more precise understanding of the relationship between dreaming and dream reports and of the relationship between phenomenal selfhood, phenomenal embodiment, and the physical body. It has also led to a new perspective on the relationship between dreaming and the phenomenology of being a thinking (or cogitative) self and rational agent, as well as to a reassessment of the problem of dream belief and of what it means to say that dreams are deceptive or even delusional experiences. The ISTH model of dreaming suggests that dreaming may be a key protagonist in the attempt to isolate the target property of minimal phenomenal selfhood. Finally, thanks to the new technical concept of oneiragogia, the relationship between full-fledged dreaming and sleep onset experiences can now be described in more precise terms.

I now want to look in more detail at those contrast conditions that I take to be most promising for future research and that follow most directly from the ideas developed so far. These are other instances of nonveridical spatiotemporal experience aside from dreaming, as in full-body illusions and felt presence, immersive virtual reality and vection, and mind wandering. The chapter's final section sketches the consequences for the questions of why we dream and why we dream the way we do.

12.1 Dreams, Amodal Agent Models, and Illusory Self-Location

Given that dreams, in the current theory, are immersive spatiotemporal hallucinations, this immediately gives rise to the question of whether nonimmersive spatiotemporal hallucinations exist. A good candidate occurring during both wakefulness and sleep–wake transitions is felt presence.[1] In *The Varieties of Religious Experience*, William James describes these experiences as involving "a sense of reality, a feeling of objective presence, a perception of what we may call 'something there,' more deep and more general than any of the special and particular 'senses'" (James, 1902/2003, p. 51). This sense of reality has an amodal yet a distinctly felt quality: according to James, it is a *feeling* of objective presence rather than a belief or judgment, and he describes it as both more basic than and independent of imagery in any of the sensory modalities.

In the newer literature, felt presence is described as involving the sense of a presence located at a more or less determinate position in the proximity of the self. The presence is typically felt in an amodal manner, though it may also be accompanied by visual, tactile, or auditory sensations (e.g., seeing a shadow on the wall, hearing footsteps, or feeling its touch). A further characteristic is that the presence is typically experienced as an animate being, an agent with specific intentions toward the dreamer, ranging from friendly or neutral observation to threatening and strongly menacing (for details, see Cheyne, 2001, 2012; Cheyne & Girard, 2007a, 2007b; Nielsen, 2007; Solomonova et al., 2011). Felt presence is closely associated with sleep–wake transitions and sleep paralysis but may also occur during epileptic seizures or arise spontaneously during periods of heightened stress and emotional arousal (e.g., bereavement, accidents, postpartum, prolonged isolation, etc.; see Nielsen, 2007). It can also result from lesions, especially in the temporoparietal areas, and can be induced through electrical stimulation (Arzy et al., 2006).

There is some debate as to whether felt presence should count as hallucinatory. Nielsen describes felt presence as involving hallucinations of a purely spatial nature (Nielsen, 2007), whereas Cheyne and Girard (2007a, 2007b, 2009) argue that felt presence is itself prehallucinatory and merely enables the formation of modality-specific (visual and auditory) hallucinations. The key question is whether the concept of amodal hallucination makes sense. The concept of hallucination is indeed often tied to specifically sensory imagery; for instance, de Boismont writes that "sensible objects are the exclusive materials of hallucinations" (1860, p. xii). Yet there is no principled objection against allowing hallucinations to be potentially independent of any modality-specific information.[2] What we perceive, arguably, are not sensory qualities but objects.

Felt presence is an interesting case in which an object can be felt to be present though it is not perceived in any of the sensory modalities. As highlighted earlier, this feeling of presence differs from belief; moreover, the object of the experience is highly specific: typically it is an agent with intentions directed at the subject of experience. Like perceptual objects, felt presence also involves a sense of spatial relatedness. This is nicely captured by James:

> It often happens that a hallucination is imperfectly developed: the person affected will feel a "presence" in the room, definitely localized, facing in one particular way, real in the most emphatic sense of the word . . . and yet neither seen, heard, touched, nor cognized in any of the usual "sensible" ways. (James, 1902/2003, p. 51)

In a study investigating the spatial characteristics of sleep paralysis experiences, Cheyne and Girard (2004) found that felt presence, as well as visual and auditory variants (or intruder experiences), and vestibulo-motor experiences were mostly experienced as located in external space (as opposed, for instance, to voices occasionally reported to be heard inside the head). Intruders were typically experienced as close to the subject, often within touching distance; by contrast, vestibulo-motor experiences often covered a greater distance, with flying experiences sometimes even involving excursions beyond the bedroom. As illustrated most strongly by threatening variants, felt presence is also clearly experienced as happening *now*.

The debate about whether to describe felt presence as hallucinatory is complemented by differing models of the cognitive processes underlying felt presence. Cheyne and Girard (2007a, 2007b) propose a model according to which sleep paralysis leads to an activation of the threat-activated vigilance system, with greater activation leading to the feeling of a presence that, in turn, later functions as an attractor for modality-specific (visual, auditory, and tactile) hallucinations. By contrast, Nielsen (2007) argues that felt presence is better described as the product of a social imagery generator. Distorted own-body perception during sleep–wake transitions may count among the causally enabling conditions for activating social imagery, but as he points out, felt presence can also occur in a variety of conditions that are quite independent of sleep. For this reason, he argues that felt presences during sleep paralysis "are variants of a much more generic capacity for generation of realistic social imagery" (Nielsen, 2007, p. 978), or animistic experiences. He also emphasizes that these presences are not consistently experienced as threatening. In certain individuals, the combination of sleep paralysis and social imagery may trigger fear, thus creating the experience of a menacing presence. Indeed, one study found that reports of sleep paralysis involving felt presence are associated with heightened scores on a social anxiety measure (Solomonova et al., 2011). Yet this threatening character is not, according to Nielsen, intrinsic to felt presence. His model thus is better adapted to explaining why felt presences can also be experienced as good-natured or neutral entities.

For present purposes, two points are of particular interest. The first is that in both accounts, felt presence is a precursor to, and indeed seems to function as an attractor for, modality-specific hallucinations. In Nielsen's (2007) model, felt presence may additionally drive emotional reactions, whereas according to Cheyne and Girard (2007a, 2007b), the feeling of being threatened precedes the felt presence. The point of disagreement is merely whether the concept of hallucination should be reserved for sensory experiences (Cheyne & Girard, 2007b) or whether it can meaningfully be applied to amodal spatial experience, as well. This is in part a terminological decision, but note that the ISTH model suggests a particular answer. This is that the specificity of felt presence—with respect to both its spatiotemporal characteristics and its contents—taken together with the felt quality of the experience suggest that felt presence is hallucinatory. This is especially true because amodal variants of felt presence may isolate an ingredient that underlies different modality-specific hallucinations and is crucial for explaining their unique phenomenological characteristics.

This leads us to the second point, which is that felt presence is not exhausted by its spatiotemporal character. Rather, it is additionally characterized by its content, typically involving a seemingly agentive, intentional character experienced as distinct from the self. This, in fact, is one of the reasons why Cheyne and Girard (2007b) object to its characterization as a spatial hallucination, which they argue cannot capture the unique content of the episodes. Yet note that something very similar is true for dreaming. I have argued that the phenomenal core of dreaming is best described as immersive spatiotemporal hallucination—yet this does not mean that dreaming is exhausted by its spatiotemporal characteristics. Rather, what turns ISTH into an interesting research target is that in the present model, it is also associated with minimal phenomenal selfhood. It only makes sense to describe dreams as immersive if we assume that their unique spatiotemporal characteristics are also intimately connected to the phenomenology of selfhood. The same is apparently true for felt presence: felt presence is not exhausted by its (nonimmersive) spatiotemporal character but additionally involves the (amodal) experience of an agent other than the self.

This suggests that a deep similarity exists between both phenomena, a similarity that can be brought to light by using the conceptual tools offered by the ISTH model of dreaming. The idea is that where imageless or amodal dreams are immersive spatiotemporal hallucinations characterized by minimal phenomenal selfhood, amodal variants of felt presence are nonimmersive spatiotemporal hallucinations characterized by the presence of a nonself entity, often with specific intentions toward the self. Whereas felt presence locates a nonself agent relative to the self, dreams relocate the self relative to a hallucinatory scene or environment. For both dreams and felt presence, altered own-body perception may play an important role by providing the causally enabling conditions for the episodes to arise.

It is also noteworthy that both phenomena, while presenting a deeply idiosyncratic departure from intersubjective reality, nonetheless typically have a distinctively social flavor. Recall that dreams often include a number of further nonself characters (the average dream has nearly four nonself characters, to be precise), and that these nonself dream characters are often identified on the basis of just knowing, rather than by their visual appearance (Kahn et al., 2002). In fact, social imagery is described in over 95 percent of adults' dream reports, and social interactions are actually more frequent in dream reports than in randomly timed reports of waking experiences (McNamara et al., 2005; for discussion and further references, see Nielsen & Lara-Carrasco, 2007).[3] Nonself dream characters are also experienced as highly realistic. Bizarreness is comparatively rare for nonself dream characters (Foulkes & Schmidt, 1983; Kahn et al., 2000), and they are often the target of theory of mind attributions, suggesting that they are experienced as having a mind of their own (McNamara, McLaren, Kowalczyk, & Pace-Schott, 2007). As perhaps most clearly illustrated by nightmares, social interactions in dreams are also often experienced as emotionally engaging (Kahn et al., 2002). The social characteristics of dreaming have recently led Revonsuo and colleagues (2015) to formulate the Social Simulation Theory of dreaming, which is intended not just as a descriptive claim about the prevalence of social imagery in dreams, but as a hypothesis about its evolutionary function, for instance by training and enhancing social skills and bonds. Independently of this evolutionary claim, it is telling that the inherently social flavor of dreaming leaks out beyond the borders of sleep as well, with nonself characters experienced, in the guise of felt presence, during sleep onset often being the first step towards a gradual shift in relocation to the dream world. We might even say that while felt presence involves the activation of an *amodal agent model*, which subjectively *feels* like a mind-independent entity because of its spatiotemporal characteristics and apparent intentions, dreams minimally involve the activation of an *amodal self-model*,[4] which functions as the center of a spatiotemporal environment and frequently also as the center of a social environment, in which the self is experienced as interacting with other nonself agents.

The central difference between dreams and felt presence, in this view, can be described by the different degrees to which they involve a substitution of the spatiotemporal reference frame. Again, felt presence experiences are variable in this respect. Subjects often describe being able to open their eyes and look around the room during the episodes (Cheyne & Girard, 2004). This might suggest that felt presence, or a subgroup thereof, involves an integration of an isolated and hallucinatory amodal agent model into a largely veridical perceptual reference frame. If this analysis is correct, then ISTH can be understood as lifting the more circumscribed spatiotemporal hallucinations of felt presence to a global level. The amodal type of hallucination that characterizes felt presence may then be a precursor not just to modality-specific hallucinations but also to *immersive* spatiotemporal hallucinations in dreams. Just as immersive spatiotemporal hallucination can be considered a minimal form of dreaming, so nonimmersive spatiotemporal hallucination can be considered a minimal form of hallucination. It can be rendered more complex by involving modality-specific imagery, or by involving a substitution not just of a small part but of the entire spatiotemporal reference frame, as in dreams.

An alternative explanation would be to say that episodes of felt presence in which visual perception remains intact are characterized by an incomplete integration or misalignment of overlapping spatiotemporal and visual reference frames characterized by different degrees of external-stimulus correlation. This suggestion is supported by certain observations on visual hallucinations. Green and McCreery (1975, p. 39) note that in apparitional experiences, which typically involve, again, the experience of seeing a nonself agent, the experienced spatial location of the agent may be dissociated from the place from which it is seen. Because visual hallucinations are not constrained by the physical position of the eyes, so-called apparitions may be experienced as occupying physically impossible positions, as if subjects were able to see something or someone standing behind them through the back of their heads. Although these experiences have a distinctly visual quality, the perceived location of the hallucinatory agent is not completely visual; rather, the felt location takes the form of a spatial hallucination, added to which is a visual hallucination of the object that is *seen* to be in a different location from its "felt" location. Similarly, in their recent comparison of dreams and hallucinations, Collerton and Perry note that in dementia patients, hallucinations are typically "set against an existing background as one part of a perceived scene" (Collerton & Perry, 2011, p. 1017). They give the following example:

> Every night I would see a man and a young child standing in the corner of the room staring at me. . . . It was really queer. They would move but not come any closer to me and didn't say anything. . . . They both had on old fashioned clothing, like Victorian style with cloaks on. (Collerton & Perry, 2011, p. 1017)

This impression of the man and child moving toward the subject without coming closer might result from a misalignment of visual and spatial reference frames, perhaps constrained to different degrees by external sensory input.

It is also noteworthy that Collerton and Perry emphasize that hallucinations in patients with Charles Bonnet syndrome are more dreamlike than those in dementia patients in that they often involve scenic properties and have a "surround" or "full-screen" quality. This may be facilitated by the fact that due to a visual deficit, the processing of external visual stimuli is largely absent in Charles Bonnet patients. Consider the following example:

> There was a huge, vast canyon. I had to step over the edge to get to my seat. I have seen large churches with gravestones inside and everything was larger than life, but doors appeared shorter.

> The rooms just carried on, and I have to walk through things to get around—like a bath or a wall. (Collerton & Perry, 2011, p. 1017)

Note that the foregoing account is highly similar to dream reports such as the following:

> I was in a hospital corridor, it was the department where I work as an intern and I saw many white gowns, a lot of people who were walking, then from the entrance an enormous person entered, really big—a gigantic person, who filled the whole doorway. (Collerton & Perry, 2011, p. 1017)

Collerton and Perry note that Charles Bonnet hallucinations are distinguished from dreams in that they lack strong emotions, loss of insight, and thought disorder. Yet both seem to converge, at least superficially, on ISTH, suggesting that in dreams and reports of "surround" visual hallucinations, as sometimes reported by Charles Bonnett patients, there is a stronger degree of integration or alignment of hallucinatory visual and spatial reference frames than is the case in circumscribed hallucinations occurring before the background of ongoing perception.

Yet there might be a further difference between visual hallucinations with a surround quality, as occasionally reported by Charles Bonnett patients, and dreams. In particular, it is unclear from the report just cited to which degree the episode involved *illusory* movement sensations, as well, rather than real ones. It is thus unclear whether we should say that in these episodes, self-location is relative to a hallucinatory visual reference frame, whether it is still constrained by the actual position of the body, or perhaps some mixture thereof. As in the example of Mach awakening to see his hands in a different position from their actual one (cf. sec. 11.2), it is unclear, in other words, whether scenic visual hallucinations alone suffice for immersive spatiotemporal *hallucination* of the type experienced in dreams. The crucial ingredient that may be missing from such scenic visual hallucinations may be the factor of weak-phenomenal-functional embodiment that characterizes dreams.

The next step, then, is to ask whether analogues of this factor exist in wakefulness. Specifically, what is the relationship between ISTH and illusory self-location in waking subjects? There is now a wealth of literature on both spontaneously occurring and experimentally induced full-body illusions. Among the first group, three types of full-body illusions are potentially relevant (for details, see Blanke & Arzy, 2005; Blanke & Mohr, 2005; Blanke et al., 2005; Heydrich & Blanke, 2013; Blanke, 2012). Autoscopy is characterized by seeing a double of oneself while self-location remains unaltered and tied to the physical body; in heautoscopy, the sense of self-location either alternates between the visual double and the physical body or is simultaneously located in both; and in out-of-body experiences (OBEs), one has a sense of having left one's physical body and often of observing it from an elevated perspective.

It may be tempting to think that a continuous relationship holds between these different full-body illusions, such that the transition from autoscopy via heautoscopy to

OBEs is related to a gradual shift in self-location. In this view, autoscopy might appear as a visual variant of felt presence involving the self, rather than a nonself agent. This type of view is suggested by Occhionero and Cicogna (2011) in their review of autoscopic phenomena and own-body experiences in dreams. They argue that different types of self-representation in dreams, ranging from observer dreams, to fully embodied, active self-participation, to a double representation of the self, may be understood along a continuum. They write:

> A key point in this continuum, which could constitute a link with autoscopic phenomena, is the alteration of the perspective from which the dreamer observes the dream scene. Indeed, total or partial self body images in dreams . . . are similar to out-of-body experiences because the dreamer sees his own body in another spatial location. . . . Double representation of Self is similar to the autoscopic hallucination of double location, and can also involve oscillations and uncertainty, as in heautoscopy, relative to perception, feelings and emotions. (Occhionero & Cicogna, 2011, p. 1014)

However, despite the clear surface similarities between autoscopy, heautoscopy, and OBEs, the attempt to place them on a continuum might be premature. Heydrich and Blanke (2013) argue that autoscopic hallucinations and heautoscopy have distinct neuroanatomical substrates, with heautoscopy being linked to sensorimotor (but not visual) deficits and the opposite being true for autoscopy. If this is correct, then heautoscopy cannot be modeled as an intermediate state between autoscopy and OBEs.

Moreover, it is not clear that OBEs consistently involve the experience of *seeing* one's body from the outside. Cheyne and Girard (2009) have argued that anomalous vestibulo-motor experiences, rather than altered visual imagery, may be the decisive factor in bringing about OBEs. They propose that anomalous vestibulo-motor experiences may give rise to out-of-body feelings, and these feelings in turn may be precursors to a particular form of autoscopic experience associated with out-of-body autoscopy, or the experience of seeing oneself from the outside as in the classical OBE. This would fit in well with the finding that in the hypnagogic state, vestibulo-motor hallucinations may precede full-fledged OBEs.

Researchers have long noted the similarity between OBEs, lucid dreams, and false awakenings (cf. sec. 8.2.2.2; see Irwin, 1988; Blackmore, 1988; Levitan & LaBerge, 1991; Green and McCreery, 1994; LaBerge & DeGracia, 2000). Sleep paralysis is often characterized by vestibulo-motor experiences, ranging from feelings of floating or falling to full-blown OBEs (Cheyne, 2003, 2005; cf. sec. 8.2.2). A connection with lucidity is implied by the fact that OBEs frequently involve some degree of awareness of undergoing an altered state, namely, of having left one's body. The overlap with false awakenings is suggested by the seemingly realistic quality of OBEs. For instance, subjects may believe that they have actually left their physical body and are seeing it from above, or that they are able to travel through their houses (LaBerge, 1985; see sec. 10.3.4). Despite the seemingly realistic quality of visual experience, subjects may later realize that what they saw during the OBE state was not quite accurate: they may discover, for instance, that the bedding they saw from their floating position above the bed was not the same, or that they saw themselves as having a different hairdo (Devinsky et al., 1989). At least a subgroup of OBEs consequently appear to involve not just altered self-location and an alteration in the visual perspective but also hallucinatory visual experience. For others, visual imagery might still be constrained more strongly by ongoing perception and short-term memory of what the subject's current environment looks like. It is also interesting to note that as in dreams, phenomenal embodiment during OBEs varies greatly; subjects may feel that they have a double body much like their actual one, but may also feel embodied in a spatial volume or feel that they are an unextended point in space (Wolfradt, 2000/2001; Alvarado, 2000; for discussion, see Metzinger, 2013c).

The ISTH model offers a framework for spelling out this surface similarity in more precise terms. In particular, certain OBEs might emerge as a subclass of ISTH additionally involving the experience of having left one's body and frequently of seeing it from an elevated perspective (Blanke & Mohr, 2005). As is the case for full-fledged dreaming, the exact similarity between OBEs and full-fledged dreams can be determined on a case-by-case basis along the five dimensions that I used in chapter 11 to describe the progression through oneiragogic experience. Their classification as full-fledged dreams will depend not just on shifts in spatiotemporal self-location but also on the integration of waking-memory sources and spontaneously generated imagery, integration into autobiographic memory, and changes in phenomenal-functional embodiment. It is also noteworthy that the impression of having left one's body or even of seeing oneself from the outside, as in OBEs, is importantly different from the amodal agent models that characterize felt presence: the representation of the physical body, in OBEs, is experienced as a deviant model of the self. The intersubjective character of felt presence and of many dreams thus is not characteristic of OBEs.

For now, we can note that in both OBEs and dreams, vestibulo-motor experiences are abundant, and insofar as they involve whole-body movements, they involve progressive changes in experienced self-location within and relative to a broader environment. If, while lying still, you feel that you are floating or moving but are aware that the actual position of your body is not changing, a natural interpretation would be that you are slowly slipping out of and moving away from your physical body; in a sense, this is the experience of gradually becoming disembodied (stereotyped "exit strategies" of leaving the body during the OBE state are described by Muldoon & Carrington, 1929; and Waelti, 1983). As we have seen, ambiguous signals from the otoliths in the inner ear (Schwabe and Blanke, 2008; cf. sec. 8.4) might be the culprit, especially in situations in which the processing of external sensory and bodily input is already weakened, as is the case for sleep onset. The phenomenology of slowly leaving one's body suggests that the transition from veridical self-location in a perceptual environment to nonveridical self-location in a hallucinatory environment can itself be experienced.

OBEs are the special case in which the changes in the degree of functional embodiment are reflected in the experienced shift in subjective self-location and the experience of losing touch with the physical body. The present model also predicts that progressive alterations in experienced self-location, associated with vestibulo-motor experiences and facilitated by ambiguous vestibular signals, are more basic than specific modal hallucinations and perhaps drive their generation. This would mean that feeling that one has left one's body precedes seeing it from the outside, as predicted by Cheyne and Girard (2009).

Spatiotemporal hallucinations as the crucial factor linking dreams to sleep paralysis, felt presence, and OBEs are also plausible with respect to their neural correlates. Both felt presence and OBEs have been linked to lesions in the temporoparietal area and have been experimentally induced through electrical stimulation of the temporoparietal junction (Arzy et al., 2006; Blanke & Arzy, 2005; Blanke & Mohr, 2005; Blanke et al., 2005; de Ridder et al., 2007; Blanke, 2012; for more on full-body illusions, see Lenggenhager, 2009). All of this fits in well with findings on dreaming. A recent study conducted by Mota-Rolim et al. (2010) found preliminary evidence that lucid dreams in which the dreamer described observing the dream scene from a third-person perspective were associated with increased gamma activity in the right parieto-temporal region. This is in keeping with the hypothesis that the comparative deactivation of the parietal cortex in REM sleep may be associated with a lack of distinction between the first- and third-person perspective in dreams (Dang-Vu et al., 2007, p. 100). The fundamental importance of spatial imagery for dreaming is also supported by Solms's (1997, 2000) theory of dreaming, which claims that activation in the parieto-temporo-occipital (PTO) junction is necessary not only for visual imagery in dreams but for dreaming per se (cf. sec. 2.2). All of this would suggest that the temporoparietal region is related to changes in experienced self-location across the sleep–wake cycle and thus, in the current model, makes a crucial contribution in bringing about all these altered states.

While the suggestion of an integrated theoretical framework for describing dreams and OBEs is tempting, it should be noted that the proposal of a systematic contrastive analysis of dreams and OBEs is hampered by methodological limitations. OBEs are rare, and their occurrence is near-impossible to predict or control experimentally. Moreover, existing data on the phenomenology of OBEs have been gathered either from anecdotal reports or from general questionnaires. The situation is not improved by the fact that a majority of questionnaires use only yes/no questions without gathering detailed reports of the individual episodes (Alvarado, 2000). Answers to such questionnaires are consequently significantly removed from the ideal conditions of dream reporting outlined in chapter 4. It is not even clear that in the present theory, they count as proper experience reports at all. This raises serious concerns about the trustworthiness of such essentially anecdotal accounts.

Researchers have attempted to circumvent this problem by devising ways of reproducing the core components of OBEs in healthy subjects in a carefully controlled manner. The details of the experimental setup vary but typically involve the use of multisensory conflict and virtual reality technology, with the goal of inducing shifts in experienced self-location toward, and identification with, a virtual avatar (Lenggenhager et al., 2007; Ehrsson, 2007; see also Ionta et al., 2011). In a now classic experiment, Lenggenhager et al. (2007) attempted to create a full-body variant of the rubber-hand illusion (Botvinick & Cohen, 1998; Tsakiris & Haggard, 2005; Slater et al., 2009). Participants were shown, through a head-mounted display, an image of themselves that was taped from behind, creating the illusion of seeing themselves from behind. They were then stroked on their backs while synchronously or asynchronously seeing the strokes on the back of the avatar in front of them. Often they reported that they identified with the virtual body and felt as if it was their own. After the experiment, they were blindfolded and displaced and then asked to return to the position at which they had stood during the experiment. The position to which they returned was localized toward the avatar and the degree of displacement was associated with the reported strength of the illusion.

Aside from visuo-tactile stimulation, as in the original experiment, researchers have also investigated the contribution of interoception (e.g., by investigating how visual information about participants' heartbeat influences the illusion; cf. Aspell et al., 2013) and of vestibular processing (e.g., by creating visuo-vestibular conflict about the direction of gravity; cf. Pfeiffer et al., 2013) to self-identification and self-location. In an extreme version of the experiment, experimenters were able to induce a body swap, enabling healthy subjects to feel that either another person's body or an artificial human body was their own (Petkova & Ehrsson, 2008; Petkova et al., 2011). Aside from subjective reports, skin conductance responses such as increased sweating in response to physical threats to the virtual body provided objective measures for the strength of the illusion. The most impressive finding was that participants undergoing the illusion of owning another person's or a virtual body could face their own body and shake hands with it without thereby interrupting the illusion.

Recently, Ionta et al. (2011) used robotic technology to induce visuo-tactile and visuo-vestibular conflicts, which resulted in predictable changes in experienced self-location in healthy subjects. In keeping with findings on the neural basis of spontaneous OBEs, they suggest that "multisensory integration at the TPJ [temporoparietal junction] reflects one of the most fundamental subjective feelings of humans: the feeling of being an entity localized at a position in space and perceiving the world from this position and perspective" (Ionta et al., 2011, p. 363).

The theoretical relevance of these experiments has long been recognized. It was this type of experiment, together with an analysis of spontaneously occurring full-body illusions, that led Blanke and Metzinger (2009) to suggest that self-location, self-identification or body ownership, and the first-person perspective are the conditions for minimal phenomenal selfhood (see sec. 11.3.1). Researchers have now also begun to investigate the exact relationship between these three factors in more detail, with a recent study finding evidence that body ownership and self-location are indeed distinct and suggesting that spatial perception may be the basis of body-part and whole-body perception (Dobricki et al., 2013).

Yet given my critique in chapter 11 that dreams provide evidence for an even simpler account of minimal phenomenal selfhood, in which self-identification and the first-person perspective collapse into self-location and in which bodily experience can be lost altogether, this should also alert us to a danger. In particular, the main methodological strength of experimental work on full-body illusions—namely, the focus on carefully controlled, experimentally induced effects in healthy, awake subjects—is also its central weakness. Insofar as such studies are taken to investigate the basis of minimal phenomenal selfhood as such, or the relationship between different dimensions of bodily self-consciousness, they are limited by the fact that their results are state dependent. Because these experiments focus on standard wake states, they introduce a confound between a specific type of bodily self-consciousness that is tightly constrained by real-body inputs and self-consciousness as such. And while the ease with which multisensory conflict can be used to induce full-body illusions, even in standard wake states, has been a source of surprise, revealing the "flimsiness of . . . bodily self-awareness" (Hohwy, 2010, p. 139), it is significant that these illusions are still more tightly constrained by real-body inputs than is the case for dreaming. Indeed, this is the condition of possibility for these experimental manipulations, which, after all, exploit tactile or vestibular cues to induce multisensory conflict. By contrast, in dreams, the real-body constraints on bodily experience are even weaker, occasionally even offering evidence for a dissociation between bodily experience and phenomenal selfhood.

It is also important to note that this comparatively stronger form of phenomenal-functional embodiment very likely limits the strength of the observed effects considerably. In their original study, Lenggenhager et al. (2007, p. 1098) emphasize that the "present illusion was neither associated with overt disembodiment nor with a change in visuospatial perspective," thus at best approximating certain aspects of illusory self-location in OBEs and heautoscopy. In particular, such experimental setups, while focusing on the induction of identification with and mislocalization toward a virtual body, do not achieve phenomenal disembodiment, or the loss of ownership of the real body. The factor of disownership of the real body has only begun to be investigated, with researchers finding that increased identification with the virtual body may come at the cost of reduced ownership, as measured by changed skin conductance rates, for the real body (Guterstam & Ehrsson, 2012). Yet this is at best an early precursor to the form of body disownership and weak-phenomenal-functional embodiment seen in spontaneous OBEs and certain dreams.

For these reasons, I suggest that dream research offers a possibility to extend research on full-body illusions beyond the state-dependent constraints characterizing waking experience. Because dreams are much more frequent than spontaneous OBEs, scientific dream research offers a much better infrastructure for using timed awakenings and specific reporting methods for gathering and analyzing dream reports. This infrastructure can be used, among other things, to investigate the target properties of pure, amodal ISTH and of minimal phenomenal selfhood in a more systematic manner.[5]

Finally, note that this affinity between scientific dream research and work on full-body illusions is also surprising for another reason. In particular, it suggests that the end points of the spectrum of phenomenal-functional embodiment are not as far removed from each other as has previously been assumed. While the ease with which full-body illusions can be induced suggests that bodily experience, even in standard wake states, is less tightly constrained by real-body inputs than previously thought, my description of dreams as weakly phenomenally-functionally embodied states suggests that, once more contrary to common opinion, even in dreams, bodily experience is still constrained by the physical body. This is not to deny that an interesting difference between dreams and standard wake states exists. Rather, it suggests that the phenomenal-functional profile of dreaming is not so dissimilar from that of standard wake states as to render an integrated, continuous account impossible. This is an important metatheoretical insight with potentially fruitful methodological implications.

12.2 Immersive Virtual Reality, Substitutional Reference Frames, and Vection: The Mirror Image of Dreaming?

We have seen that felt presence and spontaneously occurring OBEs, as well as to a weaker extent experimentally induced full-body illusions, can be described as involving two important components of full-fledged dreams: felt presence is characterized by nonimmersive spatiotemporal hallucinations taking the form of amodal agent models, whereas OBEs and other full-body illusions involve illusory self-location such that one feels located outside one's physical body. Dreaming lifts both factors to a more global level, involving immersive spatiotemporal hallucinations and amodal self-models (or minimal phenomenal selfhood), as well as, often, a number of nonself agent models. Unlike OBEs, this illusory shift in self-location to a position outside the physical body is often no longer reflected on the level of phenomenal experience. In most dreams, we do not *feel* located outside our bodies, but simply present in the experienced environment.

By focusing on nonimmersive spatiotemporal hallucinations and illusory self-location, however, we have diverted attention away from the fact that the phenomenology of dreaming is not just about illusory self-location but about presence in a world. Spatiotemporal self-location is a relational property and as such is inextricably linked to the generation of centered reference frames. Even the simplest forms of dreaming involving amodal self-models or pure spatiotemporal situatedness cannot be understood independently of the environment model at the center of which is the phenomenal dream self. The next plausible step is to compare dreams to the closest approximation in wakefulness, namely, to immersive virtual reality (VR) systems, which are increasingly being employed in interdisciplinary consciousness research. As has often been pointed out, there is a sense in which dreaming simply is a virtual reality: in dreams, we find ourselves interacting with a world that is experienced as real but is nonetheless largely independent of our actual surroundings.

The connection between the metaphorical description of dreaming *as* a virtual reality and the comparison between dreaming *and* virtual reality systems is particularly obvious in Revonsuo's (2006) work on dreaming. Revonsuo uses the analogy between dreaming and VR to argue that consciousness itself is nothing other than world simulation: we are never in direct experiential contact with the external world, and in this sense, conscious experience is a form of *inner* presence. This suggests that the technical concepts of immersion and presence, developed in the context of virtual reality research, can also meaningfully be applied to dreaming (Revonsuo, 2006, chap. 6). Similarly, Metzinger notes that the sense of presence achieved by dreams not only mirrors the sense of presence experienced in standard wake states but is also "exactly what modern designers of virtual realities are trying to achieve" (Metzinger, 2009, p. 137). Dreams and VR technologies are thus granted a similar role in illustrating the simulational nature of consciousness itself.

The comparison between dreaming and VR is also discussed by researchers from the field of virtual reality research itself. Blascovich and Bailenson (2011) begin their popular introduction to the topic with a chapter titled "Dream Machines," arguing that video games or novel VR systems simply extend our natural affinity to engage in alternate realities: "Stray thoughts, dreams, chemical influences, or religious beliefs all point to an inherent human need and desire to transcend reality 'as we know it.' There is nothing inhuman, in other words, about the virtual" (Blascovich & Bailenson, 2011, p. 23). Similarly, Moller and Barbera propose that "from an evolutionary perspective, dreaming can be understood as the most archetypal form of virtual reality" (Moller & Barbera, 2006, p. 98) and argue that the study of dreaming may prove theoretically fruitful for understanding presence in the context of VR research:

> Dreaming, firstly, may be regarded as a natural experiment in presence; in fact the "gold standard" by which other immersive environments may be compared. Secondly dreaming provides a unique window into the processes that determine how the brain organizes information (external and internal). (Moller & Barbera, 2006, p. 105)

And Ijsselsteijn and Riva note:

> The feeling of "being there," or presence, is not intrinsically bound to any specific type of technology—it is a product of the mind. In normal, daily life we are seldomly aware of our sense

of presence in the world. It may take something like an altered psychological state (e.g., a dream or hallucination), a leap of the imagination . . . or a mediated perception (through, e.g., cinema) to make us become aware of this "default" experience. (Ijsselsteijn and Riva, 2003, p. 5)

The first step, then, is to compare the concepts of presence and immersion as understood in the ISTH model to the way these concepts are used in VR research. In the ISTH model, presence and immersion are used synonymously to refer to the sense of being located within a spatiotemporal reference frame, or the sense of spatiotemporal situatedness. By contrast, Sanchez-Vives and Slater, in a now classic paper on presence in VR systems, define immersion as "the technical capability of the system to deliver a surrounding and convincing environment with which the participant can interact" (Sanchez-Vives & Slater, 2005, p. 333). In this view, the technological concept of immersion is complemented by the subjective experience of presence:

The degree of immersion is an objective property of a system that, in principle, can be measured independently of the human experience that it engenders. Presence is the human response to the system, and the meaning of presence has been formulated in many ways. (Sanchez-Vives & Slater, 2005, p. 333)

In the most common view, "presence is the sense of being in a VE [virtual environment] rather than the place in which the participant's body is actually located" (Sanchez-Vives & Slater, 2005, p. 333). Alternatively, presence may also be understood as the ability to interact with the virtual environment.[6]

Here we see a clear surface similarity to dreaming. Both dreaming and VR involve a sense of self-location in what is in fact a virtual environment. Yet once one tries to spell out this similarity in more precise terms, one encounters the conceptual problem of what exactly to say about the virtual environments generated by VR systems. To begin with, they are clearly not perceptual in the classical sense of putting participants in contact with largely veridical reference frames. Yet they are also not hallucinatory, in the sense of relying largely on internal stimulus generation. Like dreams, VR systems involve the substitution of perceptual reference frames; yet they do so in an importantly different manner. Sanchez-Vives and Slater explain:

Presence occurs when there is successful substitution of real sensory data by computer-generated sensory data, and when people can engage in normal motor actions to carry out tasks and to exercise some degree of control over their environment. By "successful," we mean that the person responds to the virtual stimuli as if they were real. (Sanchez-Vives & Slater, 2005, p. 338)

To capture the difference between VR and standard perceptual engagement with the world, as well as between VR and the largely internally generated reference frames that characterize dreaming, let us say that VR involves *substitutional reference frames*, where substitution is achieved by providing a technologically engineered alternative stimulus source that exploits ongoing perception and bodily experience in awake participants.

There are two main ways of doing this (Sanchez-Vives & Slater, 2005). The first is the so-called Cave, consisting essentially of four white screens (front, left, right, and floor), onto which 3-D images are projected. Participants wear 3-D glasses, as well as tracking devices, enabling the seen environment to change in concert with body movements. The second is the use of head-mounted displays, ideally blocking surrounding light and using carefully aligned images, displayed to the right and left eye, to create the illusion of a 3-D environment; again, tracking devices adjust the images to the participants' head and body movements. Even this brief description should alert us to an important point. VR systems at best involve *modality-specific* substitutional reference frames, typically by providing technologically generated visual and auditory input, and bringing them into alignment with largely unaltered sensations of movement, position, and touch with the goal of simulating realistic sensorimotor interaction (Slater, 2009). For head-mounted displays, this typically raises the additional problem of creating a visual representation of the body, as well, and the illusion is strongest when participants have not only a visual body representation (or avatar) but also one that moves in synchrony with their own movements (Slater et al., 2010). This is apparently more important than sensory realism, for instance regarding visual resolution or multisensory integration, which do not seem to be necessary for presence, though they may strengthen the illusion (Sanchez-Vives & Slater, 2005). This brief description shows why the contrast between immersion as an objective property of a VR system and presence as the human response to the system makes sense for VR research, but not for dreaming. In dreaming, immersive spatiotemporal reference frames are largely internally generated, and thus there is no experience-independent entity of which immersion could be a measurable, objective property. Rather, for dreams, presence and immersion are experiential properties only.

The importance of bringing the physical body into alignment with the virtual environment is particularly obvious in Slater's (2009) distinction of two separate illusions involved in VR experience. These are the place illusion, that is, the experience of presence or of "being there," located in a real place; and the plausibility illusion, or the illusion that the depicted scenario is actually occurring. Slater explains that the place illusion

> occurs as a function of the range of normal SCs [sensorimotor contingencies] that are possible. But we have also defined immersion in terms of the range of SCs that are possible, and an immersion hierarchy formed by the extent to which one system can be used to simulate another. . . . By definition if a person perceives the virtual world making use of motor actions to perceive *in the same way* as perceiving the real world, but on the other hand knows that is a virtual reality, then this must give rise to PI [place illusion]. (Slater, 2009, p. 3551)

By contrast, the plausibility illusion depends, according to Slater, on the sense that "events in the virtual environment over which you have no direct control refer directly to you" (Slater, 2009, p. 3553), thus enhancing the plausibility of the experienced events.

This brief description should once more alert us to two important differences from dreaming. The first is that dreaming relaxes the computational requirements on generating the place and plausibility illusions because in dreams, as weakly cogitative states, conscious thinking is only weakly and intermittently differentiated from spontaneous, imagistic cognitive processing. As argued in section 11.3.3, most dreams are characterized by doxastic indeterminacy and low-level cognitive immersion. With the exception of lucid dreams, there is no sense, however, in which the dreamer knows that the ongoing dream is not real. There is thus nothing to counteract the analogues of the place and plausibility illusions in most dreams.

There is, however, also an interesting parallel between the coexistence of the place and plausibility illusions in VR and C-lucid dreams involving cognitive insight into the nature of the ongoing experience (cf. sec. 9.3.3). This parallel gives rise to an intriguing prediction. Given that lucidity is prone to lucidity lapses, one might expect there to be a similar effect, if only to a weaker extent, in VR, in which beliefs and judgments are captured by the strength of the illusion. Do such cases of cognitive corruption exist in VR?

A first and extremely preliminary indication in this direction was a study conducted by Suzuki et al. (2012). The researchers aimed to create a substitutional reality system, in which participants would not just have the illusion of being present in an environment they knew to be virtual, but which would undermine their ability to distinguish between live and prerecorded scenes, as presented via a head-mounted display and headphones. The experiment had five stages. First, unbeknownst to each participant, participant and experimenter were filmed as they entered the lab. The participant was then fitted with a head-mounted display and headphones and, after an initial covert switch from a live to a prerecorded scene, interacted naturally with an experimenter. Once the illusion that the events were actually occurring had been established, participants were shown a recording of themselves entering the room (the so-called doppelgänger condition), thereby providing a vivid demonstration that the scene was not, despite appearances, live. Next, in the so-called fake live condition, participants were once more induced to interact with a prerecorded version of an experimenter. Surprisingly, 70 percent did not realize that this last scene was prerecorded but judged it to be real. This was followed by a genuine live scene, often associated with considerable uncertainty and confusion on the part of the participant, reminiscent of the confusion one might feel after a realistic false awakening.

What was surprising was not that participants initially failed to detect the covert switch from live to prerecorded scene, but that even when they had been presented with a strong reason for doubting the veridicality of what they were seeing—namely, the doppelgänger condition—they still continued to mistake recorded scenes for live events. On one level, one might say that the achievement of the experiment was primarily impressive for technological reasons; the seamless integration of real and prerecorded scenes, both enabling a similar degree of interaction, is clearly a sign of progress. But for present purposes, the more interesting point is that the seemingly realistic character of the experiment increased participants' gullibility—even in the face of strong reasons for doubting the veridicality of what they were seeing. This is certainly not the same as cognitive corruption in the dream state, but it may be a first careful step in this direction. Indeed, one of the goals of the study, according to the authors, was to develop an efficient way of studying metacognitive functions and psychiatric disorders.

Some evidence also suggests that even when participants are aware of interacting with a virtual environment, the place and plausibility illusions are not entirely doxastically inert. Indeed, the ability of illusions to influence emotions and behavior in subtle ways is part of what makes this line of research so interesting. Generally, social interactions in VR elicit behavioral and emotional responses similar to those one would expect in real life. For instance, the amount of anxiety exhibited by shy men interacting with a virtual woman is positively associated with the amount of anxiety they would exhibit if approached by a woman in real life. Similarly, participants with a fear of public speaking respond with heightened anxiety when asked to speak in front of a virtual audience, even when the representational and behavioral fidelity is low. Finally, participants confronted with a virtual pit will edge their way around it rather than walk over it (for discussion and further references, see Slater, 2009). As Slater puts it, "Since you are as real as can be, and this external sensed world appears to be addressing you, the reality of that external world is itself enhanced" (Slater, 2009, p. 3553).

Note that these examples bear a striking resemblance to the lucidity lapses discussed in chapter 9. There and in chapter 10, it became clear that cognitive corruption is not restricted to nonlucid dreams but can also occur before a background of genuine lucid insight. Recall that these lucidity lapses often arise with respect to other dream characters or to actions that would be threatening or painful in waking life. For instance, an otherwise lucid dreamer may be reluctant to jump off a cliff, cut herself with a knife, or engage in embarrassing or socially inappropriate behavior. Social imagery and potentially dangerous actions thus create a strong bias toward believing in the reality of what one is experiencing, and this is apparently the case both for lucid dreams and VR. Here are two examples from lucid dreaming:

> I am lying in an unfamiliar way on my stomach, next to various people who are sitting around and talking. Various improbable events which have just occurred have caused me to wonder for some time whether I might be dreaming or not. However, I am too embarrassed to try jumping up in the air in the presence of other people, so I decide to observe everything closely and with circumspection, so as not to regret later missing an opportunity. (Green & McCreery, 1994, p. 25)

> I know I'm dreaming as I fly about with R and others. I encourage R to try to remember this experience [not lucid enough to realize I'm talking with a dream character]. We hover in front of

a striking glass picture of pale green hues, with flower designs embossed into its surface. I tell R that lucid dreams are even more easy to recall than non-lucids. (Levitan, 1994)

It is tempting to assume that virtual reality can be used to create a similar bias in reasoning, as well; at least there is no obvious reason why conscious reasoning should be thought to be completely quarantined from the effects of the place and plausibility illusions in wakefulness. One would not expect the effect to be anywhere as pronounced as is the case for dreaming; yet future research could attempt to investigate the parallels between cognitive corruption in dreams and in VR in more detail. So while cognitive insight distinguishes VR from nonlucid dreams, interesting parallels in particular to cognitive corruption and lucidity lapses may exist.

The second difference between VR and dreaming concerns the relevance of realistic sensorimotor interactions for the place and plausibility illusions. While this is an important condition for VR, dreaming relaxes the computational requirements on simulating a realistic, wakelike interaction with the objects and persons encountered in the dream. This is achieved, one, by widespread indeterminacy in bodily experience (weak phenomenal embodiment), and two, by the fact that in dreams, the functional constraints on bodily experience are weakened. By contrast, the example of VR shows that stronger forms of phenomenal-functional embodiment place stronger constraints on the simulation of realistic sensorimotor engagement with the virtual environment, as well.

Yet there may still be interesting parallels. A particularly intriguing possibility is that VR setups involving a strong sense not just of presence in the virtual environment but also of identification with a virtual avatar might weaken the functional constraints on bodily experience, thus inching toward what we see, in a much more pronounced manner, in dreams. Murray and Gordon (2001) compared bodily experiences of subjects in immersive VR environments lacking a visual body representation to those of subjects using a desktop VR environment. They found that the desktop group experienced higher levels of skin, muscular, and cardiovascular awareness (e.g., awareness of itching, muscle tension, sweaty palms), while the immersive group experienced higher levels, among other things, of body boundary ambiguity (e.g., sensations of prickling, tingling, or numbness, or a swelling of certain body parts). Because the experiences described by the immersive group at least bear a surface similarity to oneiragogia, and because the experiences described by the desktop group are exactly what is missing in dreams, this lends very preliminary support to the claim that immersive VR may facilitate, though to a much weaker degree, the partial disconnection from the physical body involved in full-fledged dreaming.

But there is more. Newer studies have shown that ownership for a virtual arm in immersive virtual reality is associated with a decrease in temperature sensitivity in the real arm (Llobera et al., 2013) and a significant increase in the threshold for thermal pain (Martini et al., 2014; for details on embodiment in VR, see Kilteni, Groten, &

Slater, 2012). Participants can also experience ownership for an unrealistic limb, for instance, an unnaturally long arm (Kilteni, Normand, et al., 2012). Although the illusion declined with increasing length, the subjective illusion was confirmed by heightened skin conductance rates when the virtual arm was threatened by bending the virtual finger to an unnatural position. This shows that under certain conditions, prior knowledge about one's body can be overridden, for instance, by initially presenting participants with a normally sized arm and gradually lengthening it to an unnatural length. Another study found that a threat to a virtual hand elicited activation of the motor cortex, much as one would expect in response to a perceived threat to the real arm (González-Franco et al., 2014).

By varying the degree of realism of the avatar, immersive VR can also be used to investigate the conditions for body ownership in more detail. A recent study by Maselli and Slater (2013) found that when the virtual body as seen from the first-person perspective is highly realistic, both in its visual appearance and in its spatial overlap with the position of the real body, additional visuo-tactile or sensorimotor stimulation is not necessary for the illusion. By contrast, when spatial overlap or visual realism is low, multisensory cues can be used to boost or suppress the illusion. Bringing the virtual body into alignment with the physical body weakens the constraints on body ownership. This pattern is to be expected based on the fact that in dreams, self-consciousness and bodily experience arise even when multisensory integration is disturbed (chap. 7). Multisensory integration may only be necessary when trying to induce the experience of ownership for an unrealistic virtual body or body part. Moreover, dreams suggest that the suppression of own-body perception may be a further step, and in such cases, what is needed is not so much alignment of the experienced and physical body as the lack of an experienced discrepancy between them.

Researchers are also taking first steps toward completely bypassing the physical body to create the illusion of presence in a virtual environment and ownership for a virtual body (Friedman, Leeb, Dikovsky, et al., 2007; Friedman, Leeb, Guger, et al., 2007). In a cavelike environment, participants were instructed to attempt to control the movements of a virtual avatar via brain-computer interface, that is, only using their thoughts. Imagined leg and arm movements could be used to produce similar movements in the avatar. Three participants reported different degrees of presence and of projection into the virtual body. The participant for whom the illusion was the strongest tellingly switched from referring to the avatar as "she" to describing it as "I" after a few minutes. Whether this involved the phenomenology of embodiment in the avatar and a corresponding decrease of functional embodiment is at present unclear, but the parallel to dreaming is obvious.

But there is another and much more old-fashioned way of creating illusory shifts in self-location without any direct body stimulation whatsoever. Vection refers to self-motion illusions, primarily induced by moving visual stimuli (Riecke, 2011). The term is derived from the Latin *vectio* or *vehere* (to carry). Vection thus can metaphorically be described as the experience of being carried or lifted outside one's actual location and of moving while actually remaining still.

Vection is not restricted to experimental situations but occurs spontaneously outside the laboratory, as well. An example is the familiar experience of sitting on a train while observing an adjacent train. You feel your own train begin to move out of the station and slowly accelerate. Then at some point you notice that it was the adjacent train that was moving all along; the sensation that your own train was moving in the opposite direction was illusory. Because vection does not require direct bodily stimulation, it can be used in 3-D movies or theme park rides to create the illusion of movement while remaining stationary, even for large groups of subjects.

The experimental investigation of vection dates back at least to the nineteenth century. Mach (1919) notes that while standing on a bridge and watching the water flow beneath it, we usually feel at rest and that the water is moving. But if we keep our eyes fixed on the water for a prolonged period of time, the entire bridge, including the spectator and the environment, begin to move toward the water, which now appears still. Mach even built a contraption using a large piece of leather moving over a reel, which allowed him to induce the illusion and switch between rest and apparent self-motion at will (Mach, 1919, pp. 117–118). Vection can also, however, be used to generate the illusion of self-movement through virtual environments, without having "to go through the effort of allowing for large-scale physical locomotion" (Riecke, 2011, p. 150) and the simulation of realistic sensorimotor contingencies. Riecke also notes that vection is not restricted to visual stimulation but may also result from auditory, tactile, biomechanical, or direct galvanic current stimulation of the vestibular system. Again there are objective measures. For instance, Lepecq et al. (1993) asked participants to memorize the location of target objects and found that when asked to point at the targets, participants would modify their responses after a period of vection.

Generally, experiments inducing vection exploit the fact that the perception of movement and of stability are inextricably linked. Subjective movement is experienced relative to an apparently stable environment; and because we are naturally biased toward the expectation of environmental stability, we are prone to mistakenly experience ourselves as moving when a sufficiently large portion of the environment, such as an adjacent train or a river beneath us, moves. A majority of studies have focused on the physical stimulus properties involved in vection, such as the influence of stimulus size, stimulus movement, depth, and viewing angle on illusory self-movement. An interesting finding is that realistic stimuli, such as street scenes, have a stronger influence on vection than abstract points. However, Riecke et al. (2006) also found that cognitive factors such as attention and awareness of the real surroundings modulate the strength of the illusion. Priming theme park visitors to expect actual motion may thus increase the strength of the illusion.[7]

Illusory self-movement induced independently of bodily stimulation can be compared to dreaming in several respects. First, vection arises when environmental stimuli produce illusory movement and associated changes in experienced self-location, thus loosening the degree to which subjective self-location is constrained by the actual location of the physical body. Moreover, while both VR and vection can involve modality-specific substitutional reference frames, vection does not require such a substitution. Rather, the misinterpretation of environmental movement as self-motion is sufficient. Second, vection provides an alternative model for the generation of movement sensations in dreams. This means that we now have two different ways of explaining the frequency of movement sensations in dreams. One is that as in vection, self-movement sensations in dreams result from seen environmental movement. This fits in well with the notorious instability of dream imagery: given our tendency to assume large-scale environmental stability, unstable visual imagery might create the illusion of self-movement, especially under conditions in which the processing of real-body inputs is already attenuated. The other is that illusory movement sensations, such as sensations of weightlessness, floating, or falling caused by the loosening of the functional connection with the sleeping body, are prior to visual imagery in dreams. This would be the exact reverse of vection: the generation of visual imagery would be driven by movement sensations, rather than vice versa.

Clearly the difference between the two models is an empirical question. They also might not be mutually exclusive, and perhaps there are subgroups of dreams better explained by either one or the other. For now, note that there are a number of reasons for thinking that the second model is more attractive. We have already seen that both for OBEs and oneiragogia, altered own-body perception is likely a precursor to visual imagery, possibly even with sleep onset feelings giving way to an immersive yet distinctly bodily space (Nielsen, 1991–1992; cf. sec. 11.2). We have also seen that in SP nightmares, visual hallucinations of old hags seem to be triggered by perceived paralysis and irregular breathing (cf. sec. 8.2.2). Finally, recall that the vestibular system has been suggested to play a role in the generation REMs and motor inhibition (cf. sec. 8.3). A particularly important suggestion in this context is that PGO waves during REM sleep are equivalent to the orienting response, or an instinctive adjustment to unexpected stimuli (Kuiken, 2014; Morrison, 2014). REM-sleep and dreaming might simulate this process, and as Kuiken (2014; see also Kuiken & Nielsen, 1996) suggests, bizarre discontinuities in dreams may result from orienting activity in response to internally generated stimuli that are processed as if they were externally generated and unexpected. This fits in well with the finding that eye movements during REM sleep are similar to those associated with head movements (rather than a stationary head position), as might be conducted in response to an unexpected stimulus. Herman notes:

> A constant property of vivid REM dreaming is the sensation of orientation in the dreamt space, or the hallucinated impression that the dreamer is physically present in the dreamt scene, in three-dimensional space. The dreamer is directionally oriented and aware of a spatial relationship to the persons or objects present in the dream. The sensation of orientation is made possible by cortical-subcortical mechanisms, including the vestibular system. The associated eye movements are related to reflexive and orienting responses to the dreamt surround via feedback loop connecting cortex to vestibular nuclei. *It is postulated that the sense of orientation is the phenomenological equivalent in dreaming to the continuous vestibular activation during REM sleep.* (Herman, 2000, p. 950)

A tentative and highly speculative hypothesis might be that in dreams, visual imagery is tailored to fit, as it were, vestibular orienting activity. The situation would then be exactly the opposite of vection: environmental imagery would be driven by vestibular orienting activity, not vice versa. An interesting and somewhat paradoxical consequence would be that weak-phenomenal-functional embodiment might be complemented by a tighter coupling, as compared to wakefulness, of the residual processing of bodily stimuli and the production of environmental imagery.

If anything like this model turns out to be correct, this would suggest that attempts to describe dreams as virtual reality and perhaps even model them on immersive VR systems and movement illusions in wakefulness are in an important sense misleading. While interesting parallels exist, we might say that dreaming and wakefulness present similar solutions to diametrically opposed computational problems. Immersive VR shows how in wakefulness, self-location and body ownership are modulated by bodily and environmental stimuli; wakefulness, in this view, is dedicated to orienting and localizing the (bodily) self in the world, and what we experience as our body is driven, in part, by externally generated sensory input. By contrast, in dreams, the situation is essentially reversed, involving the generation of spatiotemporal reference frames in response to internally generated vestibular sensations and altered own-body perception. In greatly simplified terms, we might then say that while wakefulness is about being in and locating oneself in the world, dreaming is about creating a world centered on the self. Although this might seem paradoxical, there may then be a sense in which in wakefulness, bodily self-consciousness, and self-location are driven by the world, whereas in dreams, the experienced world is modulated by the body.

12.3 Wandering Minds, Predictive Brains, and Relocated Selves

Could immersive VR be rendered more dreamlike by creating a causal feedback loop with the body? Could immersive VR at least approximate the body-driven generation of centered world models that characterizes dreams? In the late 1990s, a researcher and an artist collaborated to create an immersive virtual environment, Osmose, that gave participants the sensation of floating through a surreal environment (Davies & Harrison, 1996). Rather than having participants navigate the virtual environment with a joystick, sensors measured the tilt of spine and breathing rate. Tilting the spine led to changes in direction, inhaling to a gentle upward movement and exhaling to moving downward. Participants sometimes described the experience as simultaneously involving a sense of embodiment and disembodiment, as well as, often, strong emotions to the point of euphoria and a temporary suspension of thought and reasoning. A similar attempt, aptly called SuperDreamCity (Friedman, Suji, & Slater, 2007), was to develop an immersive virtual environment that responded to electrodermal activity, ultimately using body feedback to determine the switch between positively and negatively toned scenes. While the study demonstrated the technical feasibility of such a closed-loop system, the authors note that this was still far removed from creating a subjectively meaningful experience for the participants.

Leaving technicalities to the side, the challenge of creating a subjectively meaningful virtual environment modulated by physiological changes points to a central problem not only of VR systems but also of the present model of dreaming. By emphasizing that dreaming is, at its core, immersive spatiotemporal hallucination and investigating its relationship to full-body illusions, immersive VR, and illusory movement, we seem to have moved away from the cognitive sources of dreaming. It is time to take up the lesson from chapters 5 and 6, according to which dreams fall soundly on neither side of the putative divide between quasi-perceptual and imaginative experience. Rather, dreams combine a core aspect of standard wakefulness—namely, the robust sense of spatiotemporal situatedness—with a variety of experiences ranging from modality-specific (especially visual) hallucinations, illusory own-body perception, and widespread phenomenal indeterminacy of the kind that characterizes waking imagination. I still owe an account of the cognitive sources of dreaming and of how dreams acquire their apparent and subjective meaning. Discussing this will take us to some new considerations on the functions of dreaming.

Hunter states the problem in particularly compelling terms. Describing a conversation in dreams, he writes:

> Sometimes there was just his voice, sometimes a wistful or a puzzled expression and nothing more, sometimes just a gesture of his hands as he spoke. The fragmentary nature of his embodiment in the dream was not due, as it sometimes is in waking life, to my having closed my eyes here, focused my attention on his hands there, and so on. There *was* nothing more, at a certain point, than hands gesturing; and yet while there was not at that moment a human body along with them, there was not the absence of a body either: they were not seen as hands moving eerily before my eyes as if to gesture. Had I not remembered the dream precisely I would with no sense of wonderment, have described this point in the dream simply as his having made an impatient gesture with his hands as he said this. (Hunter, 1976, p. 126)

The experience is richer, Hunter seems to be saying, than what is visually represented. At the same time, he notes that "there is not a hole where the face would have been, or a body without a head" (Hunter, 1976, p. 126). Phenomenal indeterminacy is pervasive in dreams, but this indeterminacy is discretely glossed over by indeterminacy blindness. The puzzle is how dreams nonetheless acquire their specificity on the level of dream content. To answer this question, Hunter somewhat fancifully envisions the idea of a

> rather unusual raconteur: a very inventive fellow who not only can compose elaborate stories as he goes along, but sometimes does this for his own gratification, and when he does does not need to articulate the story he is composing, as he would of course have to do if he were entertaining someone else with it. He knows how this story runs and he could lay it out in rich detail, but when alone does not go to that trouble. Let us imagine further that to enrich his imaginative endeavors he has invented a device with which, as a story unfolds, he can illustrate its features with representations, of woodland scenes, screeching brakes, wistful smiles—many of them schematically or impressionistically done. (Hunter, 1976, pp. 128–129)

Sudden discontinuities would not puzzle the raconteur; rather, they "would all meld together with the story he could, but did not, articulate, and leave him in a position afterwards to recount the tale without in any way guessing, or interpreting, what had happened" (Hunter, 1976, p. 129).

The key idea is that having an ability is importantly different from having an experience, and more specifically that being able to narrate dreams in a storylike form is importantly different from saying that dreams themselves are stories. This is an important distinction; what remains to be seen is whether we can turn this somewhat whimsical sketch into an empirically plausible account of dreaming.

The first natural question to ask is whether dreams can really be compared to stories. A number of studies have investigated the storylike narrative structure of dreams (Cipolli et al., 1998; Kilroe, 2000; Nielsen et al., 2001; Cariola, 2008). The strongest challenge to the analogy between dreams and stories comes from Allan Hobson's claim that dreams are the product of random, brain-stem-driven signal generation (e.g., Hobson et al., 2000). In addition, experiments with dream splicing, in which independent judges failed to distinguish genuine dream reports from ones that had been cut up and merged with others, challenge the view that dream reports have narrative structure and internal coherence (Hobson, 1999; Hobson & Kahn, 2007).

To this, one might object that the inability of judges to reliably identify artificially merged dream reports does not show that dream reports have no narrative structure. Dream reports may be too fragmented to count as canonical stories, for instance, by exhibiting signs of planning and of dramatic progression from beginning to end, but might still have narrative structure comparable to waking experience reports (Montangero, 2012b). They might also resemble canonical stories or works of fiction in another manner. Schwartz (2004) found that the pattern of word frequency in dream reports more strongly resembled that of novels or theater plays than that of waking experience reports, which more strongly resembled newspaper articles. Kramer (2007, p. 158) remarks that dreams are more like improvisation in that they typically evolve in a non-goal-directed manner. Similarly, Hunt (1989) allows that dreams can develop into genuine dramatic stories, but thinks that they typically fall short. "Specifically, they are not very bizarre and much of their bizarreness (especially confusional clouding) merely disorganizes and introduces discontinuity into the dream plot, which then never gets resolved—at least during the dream" (Hunt, 1989, p. 176). He concludes that "most dreams remain truncated, mundane, and inconclusive, lived out in immediate action sequences, and lack even a hint of narrative voice—these are not stories. Rather, they are proto-stories or quasi-plots" (p. 177). In this view,

> Dreaming may indeed be an Ur-form for some very fundamental cognitive capacities, in particular the narrative story as sequential configuration fusing concordance and discordance, the transformation of one situation into an expression or metaphor for another, and the metaphysical questioning of the reality of our lives—the idea, known to all peoples in all times, that in some basic sense life *is* like a dream. (Hunt, 1989, p. 214)

The idea, as States puts it, is that "whatever fictions do with our experience was done first in the dream" (States, 1997, p. 1). This view may have some historical plausibility, as some of the earliest documented written stories relate dreams (Schreier Rupprecht, 2007). This connection is still evident in contemporary literature, where to this day, dreams play an important role in the writing of fiction (Schrage-Früh, 2012).

Combining these different ideas, we might say that the question is not whether dreams, as reflected in dream reports, themselves have narrative or even storylike structure but whether the process of dream generation provides evidence of a storytelling capacity, perhaps actualized to different degrees or only intermittently. It is immediately apparent that in this weaker formulation, the storytelling capacity reflected in dreams should be highly similar to that of spontaneous mental activity in waking imagination and mind wandering.[8] Recent research on dreaming and waking mind wandering, understood in a general sense as spontaneous, largely stimulus- and task-independent thought, supports this view.

Based on a review of prefrontal cortical activity during REM sleep, Pace-Schott suggests that "dream plots may emerge in a manner analogous to spontaneous confabulation with the dreamer's uncritical acceptance of emerging imagery, along with a willingness to accept and explain, however illogically, bizarre occurrences and gaps in knowledge" (Pace-Schott, 2007, p. 141). Once more, we find the common theme that dreaming weaves disparate mnemonic sources together in such a way as to create an immersive spatiotemporal reference frame, experienced, for the time being, as real (Pace-Schott, 2013). Pace-Schott attributes this to the default-mode network, which is both implicated in dreaming and known to be involved in the simulation of future and re-creation of past events (Schacter et al., 2007; Schacter, 2012), and argues that dreaming is the product of "a 'hardwired' tendency to represent reality in the form of narrative—a 'story-telling' instinct or module" (Pace-Schott, 2013, p. 1).

Note that this fits in well with a number of points made so far. It fits in well, for instance, with Gerrans's (2013a, 2014) claim that dreams arise from default-mode network activity that in the absence of higher-level cortical supervision turns hyperassociative, as well as with Domhoff's (2011b) and Fox and colleagues' (2013) suggestion that dreaming is perhaps the quintessential form of mind wandering. It also fits in well with my earlier observation that one way of assessing the transition from the early stages of oneiragogia to full-fledged dreams is to ask about the degree of integration with autobiographical memory, which determines both the reportability of the experiences in question and their experienced historicity (cf. sec. 11.2). Reportability and narrative structure could thus be closely related to each other.

If dreaming is a form of mind wandering that reflects a storytelling capacity, the next question is whether insights from research on mind wandering in wakefulness can be interestingly compared to dream content. Klinger (2013) recently suggested that much of the content of waking thought is driven by goal commitments and that the same is true for dreaming. For instance, auditory cues related to subjects' personal goals are more likely to be integrated into dreams than neutral words, and presleep instructions to dream about particular topics are most effective when related to personal goals.

While this portrays dreaming as an efficiently goal-oriented type of mental activity, other research on the continuity of presleep thought and dreams paints a less optimistic picture. It has long been known that suppressed thoughts have a way of returning in wakefulness. In the white-bear effect, when people are asked not to think of a white bear, they invariably do—and apparently they do so more frequently than people asked to think about a white bear (Wegner et al., 1987; Wegner & Zanakos, 1994). This effect may be even more pronounced during sleep onset and in dreams (Schmidt & Gendolla, 2008). In one study, subjects were found to frequently dream about persons they had been primed to think about before sleep (Wegner et al., 2004). The effect was, again, stronger for subjects who had been asked *not* to think about the target person than for subjects who had been asked to think about that person.

This rebound effect may be particularly pronounced for subjects with a tendency to suppress unwanted thoughts (Taylor & Bryant, 2007). Moreover, Schmidt and van der Linden (2009, 2013; see also Schmidt et al., 2011) have suggested that feelings of regret, shame, and guilt may be particularly frequent during the presleep period and are associated with disturbed sleep and insomnia. One reason could be that for many people, the moments leading up to sleep are the first time they are able to review the day's events; another could be that as control over mental processes is lost during drowsiness, unwanted thoughts return. Given the relation between regrets at sleep onset and insomnia, dreaming of unwanted thoughts and regrets may be better than getting no sleep at all.

Schmidt and Gendolla (2008) found that at sleep onset, suppressed thoughts were particularly likely to occur in the context of thoughts that had previously been used as distractors. One participant who had talked about her parents' garden to suppress white-bear thoughts described hypnagogic imagery of

> a picture of cherries with a text column alongside, which resembled a recipe in a magazine she had read in the afternoon as she explained in the morning interview. Then, a vision of a myriad of small leaves pervaded her mind, abruptly followed by an image of a fern in a pot on the white, round table in her parents' garden. At this point, the picture of the face of a white bear intruded upon her. (Schmidt & Gendolla, 2008, p. 721)

Wegner et al. (2004) link this type of phenomenon to the deactivation of the prefrontal areas in REM sleep, suggesting that the weakening of executive control makes the return of suppressed thoughts particularly likely in dreams.[9] At the same time, the dream rebound effect also highlights the responsiveness of dreaming to waking thought, though perhaps in a highly selective manner.

The general theory in the background is that thought control is driven by ironic processes (Wegner, 1994, 2009): an operating process that directs conscious attention to thoughts that are consistent with the intended state, and a monitoring process aimed at detecting attentional failures and deviations. Under conditions of cognitive load or reduced capacity, this process can ironically produce the unwanted mental states and lead us to focus on unwanted thoughts. As Wegner (1997) puts it:

> Consciousness may be prone to wander when we try to hold it in place while we check to make sure it has not moved. In this checking, and in particular when our conscious mental energies are taxed, we inadvertently draw our minds toward precisely where they least intend to go. Our voluntary control becomes the culprit in our involuntary mind wandering. (Wegner, 1997, p. 312)

These findings on thought suppression and thought control are of considerable theoretical relevance. Based on his review of empirical findings on mind wandering, Metzinger (2013a) has gone so far as to suggest that our view of ourselves as cognitive agents able to deliberately control our conscious thoughts is a myth. He claims that mental autonomy is an exception rather than the rule: the prevalence of uncontrolled thought—that is, waking mind wandering and nocturnal dreams—shows that conscious thought is largely a subpersonal process. A somewhat more benevolent reading, based on Wegner's ironic process theory, would be to say that mind wandering is not the opposite of conscious thought control but a natural consequence of the attempt to monitor one's cognitive activity. Thought control and mind wandering would then be two sides of the same coin.

Either way, while these empirical findings on thought suppression nicely highlight how waking thought can be continuous with sleep onset and dreams, their ability to explain the contents of our dreams and their narrative structure is limited. For one thing, dream content cannot be reduced to suppressed thoughts; clearly the range of dream contents and the emotional tone of dreaming are more variable than would be predicted by the claim that dreaming, as the paradigm of an uncontrolled cognitive process, involves a return of suppressed thoughts only. In addition, with the exception of lucid dreams, it does not seem that we try to deliberately control our dreams,[10] suggesting that the ironic monitoring of deviations from intended thoughts cannot be the only explanation of dream contents, at least not when this is understood as a conscious form of monitoring. So while it would seem that research on mind wandering and thought suppression can fruitfully be connected to dream research, this at best provides the outlines of a very partial answer to the question of how dreams get their apparent meaning and narrative coherence.

At this point, it will be helpful to take yet another step back and ask whether findings about thought control and thought suppression in wakefulness can be used to understand, if not the contents, then the form of conscious cognitive processing in dreams. This brings us back to the claim that dreams are weakly cogitative states, in which conscious thought is driven and often corrupted by low-level, imagistic cognition. With the exception of lucid control dreams—and even these do not, as we have seen, always result in the expected outcome—only those types of cognitive processing that belong to consciously experienced thought, or cogitation proper, are candidates for thought control in dreams. At the same time, it has long been noted that the transition from conscious thoughts to dream imagery is often highly fluid. Gregory, discussing dreams, as well as visual and verbal images, went so far as to describe visualization as an instinct, noting that "images may invade the mind as the impulse to flight may invade it when fear arises" (Gregory, 1922, p. 328). He also remarked that "many dreams appear to transcribe thoughts directly into pictures by discarding the verbal expression and releasing the inhibited images" (p. 333). Consider the following dream report:

> Looking down the valley that we were descending into, I saw that the path lay across it like a white ribbon and then turned off to the right towards the distant village. "It looks just like a road marked out upon a map," I thought, "or like a railway map." As I looked again I saw that there was running down the length of the valley a railroad track which I had not seen before and which crossed the path that we were following. We came nearer, and I saw that no gates guarded the crossing of the roads. "What a dangerous sort of level crossing," I thought, "like some of those unguarded railway lines in America that we read of. There would be no warning whatever of a train coming from behind the hills except the sound of it." I listened, and then I began to hear, far away, the roar of a train; it came round the hill and round a sudden sharp curve at the foot of the valley, rushing towards us very fast. I stood clear of the line and watched it come and go. (Arnold-Forster, 1921, pp. 133–134)

Similarly, Mach (1919, p. 207) notes that even when he was on the verge of noticing that he was dreaming, his dreams often had a way of appeasing him and laying his worries to rest. At a time when he was concerned with questions of space, he dreamt that he was taking a walk in the forest. He noticed that the perspective of the trees was displaced and realized that he must be dreaming. The distortions, however, were corrected immediately—as if the dream imagery and his conscious thoughts were driven by the same source.

And, of course, they are. Because of their vivid phenomenology, the occasional convergence of dream imagery and conscious thought may seem almost eerie, even to the point of leading some to attribute dream narratives to intelligent design or an intentional dream producer (e.g. McGinn, 2004; sec. 6.1.2). But this approaches the problem from exactly the wrong direction. The process by which memory sources and spontaneous imagery are integrated into immersive spatiotemporal scenes, in dreams, is not intentional. Intentional, conscious thought is only the tip of the iceberg of cognitive activity, and in dreams it is often corrupted by lower-level, automatic cognitive processing. This does not, however, prevent lower-level dream cognition from being malleable and responsive to conscious thought. One might plausibly imagine, then, a mutual reinforcement such that the dreamer's current expectations, thoughts, and emotions partly constrain subsequent dream imagery, which in turn reinforces or modulates subsequent expectations, thoughts, and emotions, and so on. The challenge, I think, is not to model dreaming entirely as a process of low-level, autonomous signal generation, or to overemphasize the top-down, higher-level influences on dreaming, perhaps even to the point of regarding dreaming as the product of unconscious authorship. Rather, one should aim for an account that acknowledges both lower-level imagistic and bodily as well as higher-level cognitive contributions to dreaming.

This brings us back to Pace-Schott's suggestion that "dream hallucinosis itself may generate low-level narrative coherence by associative processes in which images evoke related images . . . that are successfully woven together by this putative tendency to organize experience as a story" (Pace-Schott, 2013, p. 2). Here he refers to an earlier study investigating bizarre transformations in dreams. Notably, Rittenhouse and colleagues (1994) found evidence of associational constraints on such transformations, such that while characters would occasionally transform into other characters and inanimate objects into other inanimate objects, characters would not typically transform into inanimate objects or vice versa. Bizarre transformation thus appeared to be remarkably coherent—providing evidence, according to the authors, for an interplay between random, bottom-up signal generation and top-down coherence (cf. Hobson, 1988; Hobson et al., 2000).

Recent years have seen the development of a powerful account dedicated precisely to describing this interplay. Put in greatly simplified terms, the basic idea of predictive processing accounts is that the brain does one single thing: it is continuously involved in the business of hypothesis testing by using predictions to optimize its models of its next expected states and minimize prediction errors (Clark, 2013b; Hohwy, 2013b). Despite the use of personal-level concepts such as prediction and hypothesis testing, this is assumed to be a largely subpersonal process involving unconscious Bayesian inference. Part of what makes predictive processing accounts attractive for present purposes, however, is that they may be able not only to offer a new account of information processing in the brain, for instance, during perception, but also to do justice to the phenomenology of perception (Madary, 2012, 2015). Another reason is their broad scope. The audacious yet compelling idea is that hypothesis-testing "is meant to explain perception and action and everything mental in between" (Hohwy, 2013b, p. 1). Predictive processing accounts thus aim to offer nothing less than a grand, unifying theory of the mind (Hohwy, 2015; for discussion, see Harkness, 2015). In this view,

> Perception, action, and attention are but three different ways of doing the very same thing. All three ways must be balanced carefully with each other in order to get the world right. The unity of conscious perception, the nature of the self, and our knowledge of our private mental world is at heart grounded in our attempts to optimize predictions about our ongoing sensory input. (Hohwy, 2013b, p. 2)

This way of thinking about the mind has deep philosophical implications for our relation to the world, the nature of perception as an active process of hypothesis testing, and the relationship between perception, action, and cognition, which are taken to be deeply continuous (Clark, 2013a, 2013b, 2015). Perception, rather than being regarded as passive and driven by sensory stimuli, is considered as the outcome of a delicate dance, as Clark (2013a) puts it, between top-down predictions and bottom-up signals, with active inference, or exploration of the environment, and the attempt to predict sensory stimuli even before they arrive playing an important role (Clark, 2015). This leads to a more cognitive reading of perception as "what happens in the 'meeting' of top-down and bottom-up processing in the brain" (Hohwy, 2010, p. 135).

Because they acknowledge the role of top-down predictive models in perception, predictive processing accounts also suggest a deep continuity between perception, illusions, hallucinations, and imagination—as well as dreams (Seth, 2014). Clark argues that

> Systems that know how to perceive an object as a cat are thus systems that, ipso facto, are able to use a top-down cascade to bring about the kinds of activity pattern that would be characteristic of the presence of a cat. . . . Perceivers like us, if this is correct, are inevitably potential dreamers and imaginers too. Moreover, they are beings who, in dreaming and imagining, are deploying many of the very same strategies and resources used in ordinary perception. (Clark, 2013a, p. 764)

Leaving details to the side, it should nonetheless be clear that this lends substance to my claim—originally departing from the phenomenological level of description—that dreaming blurs the borders between perception and imagination and that in dreams, lower-level cognitive processing is incompletely distinguished from conscious thought. The idea is that while predictive processing frameworks generally suggest that the divide between perception on the one hand and imagination and belief on the other is not nearly as great or as unbridgeable as traditionally assumed, this distinction is even further weakened under the unique phenomenal-functional conditions that characterize dreaming.

In particular, predictive processing accounts can elegantly explain several aspects of dreaming that are often likened to fiction and imagination. In their recent discussion of dreaming and REM sleep within a predictive processing framework, Hobson and Friston suggest that "the mechanics of predictive coding provide a compelling three-way link between the physiology of PGO waves, rapid eye movements and the fictive percepts of dreaming" (Hobson & Friston, 2012, p. 97; see also Hobson & Friston, 2014; Friston, 2014). In particular, the saccades involved in visual search in wakefulness have been likened to experiments, aimed at gathering data to test internally generated predictions (Friston et al., 2012). This also fits well with the observation that REMs resemble eye movements observed during head movement rather than during rest (Herman, 2000). Applied to rapid eye movements during REM sleep, such an account could possibly support a revised version of the original scanning hypothesis. The point here would be that rapid eye movements are not so much involved in passively scanning as in actively testing hypotheses about the environment. This process, in turn, may contribute to the generation of the experienced persons and objects in dreams and thus may indirectly drive the unfolding dream narrative.

Moreover, if this account of REMs as involving active hypothesis testing in dreams is correct, we would also expect the process of generating and updating predictive models to be disturbed in dreams, or at least to depart from waking hypothesis testing in specific ways. Predictive processing accounts suggest that because dreams are largely unconstrained by external stimuli and hence by prediction errors, representational accuracy, for instance, of visual dream imagery, is lost (cf. Hobson & Friston, 2012). This loss of accuracy and the resulting inability to update predictive models against a stable stimulus source may result in familiar forms of dream bizarreness. Fletcher and Frith (2008, p. 52) suggest that "the dream state arises from disruptions in hierarchical Bayesian processing, such that sensory firing is not constrained by top-down prior information and inferences are accepted without question owing to an attenuation of the prediction-error signal from lower to higher levels," thus suggesting why bizarre occurrences in dreams are often met with uncritical acceptance.

What is more surprising, in this view, is that dreams are not more bizarre than they are: dreaming, after all, is a state in which the process of hypothesis testing, including perception, or the use of sensory stimuli for the optimization of predictive models, and active inference, or the attempt to selectively alter sensory inputs to attain maximal fit with internal models, is simulated internally. Thus, both the models and the stimuli they are tested against have to be generated largely offline (see Windt et al., 2014). This is a sophisticated computational achievement, and it is remarkable that dreams are, with the exception of slight deviations, fairly realistic in their overall organization by involving mostly coherent world models, centered on a single self and including objects and largely realistic nonself dream characters. If predictive processing accounts are on the right track, then dreaming might indeed be a state in which hypothesis testing and prediction error minimization can be rehearsed and optimized (Clark, 2013a, 2013b; Hohwy, 2013; Hobson & Friston, 2012, 2014); and the degree to which this process functions smoothly is impressive. Moreover, this way of thinking about dreaming also nicely explains why dream imagery should often be responsive to higher-level expectations within the dream, and why unintentional imagistic cognition (on the level of experienced dream content and simulated sensory inputs), and conscious thoughts and expectations (on the level of dream cogitation), should be incompletely differentiated. The increased malleability and weak differentiation between unintentional, imagistic cognition and conscious thoughts and expectations might be particularly apt for promoting the ability of dreaming to rehearse and improve hypothesis testing.

Predictive processing accounts also have the resources to explain presence and phenomenal selfhood in dreams. The central claim of the ISTH model, according to which the phenomenal core of dreaming and minimal phenomenal selfhood converge, is not at all surprising in such accounts. The most basic prior driving phenomenal selfhood, as well as self-other distinctions, in dreams may be the assumption that at any given moment, there should not only be a self, but also that there should be no more than a single self at a time (Windt et al., 2014). As Limanowski and Blankenburg (2013) note, the continuous presence of a phenomenal self, at least in minimal forms, follows quite naturally from the fact that systems operating under such a model constantly expect, and hence predict, their own existence (for a general discussion of self-representation in the context of predictive processing, see also Apps & Tsakiris, 2014; Moutoussis, 2014). This fits in nicely with Metzinger's (2013c) claim that the unit of identification—for minimal phenomenal selfhood, an unextended point—will be the maximally invariant input source detected by the system at any given moment. In standard wake states, this is the physical body, but under conditions of weak-phenomenal-functional embodiment, as in dreams, this does not necessarily apply. Even in wakefulness, simple experimental manipulations, after all, are sufficient to cause bodily illusions. In the absence of real-body input, the unit of identification may even be reduced to an unextended point. The fact that phenomenal selfhood is not lost together with phenomenal embodiment in dreams reflects our strong natural bias for assuming experienced world-models to be centered on a self—even if the usual ways of experiencing this self are lost. More generally, the analysis of dreaming may help isolate the most basic prior expectations and patterns of predictions underlying experience. In a sense, dreams, after all, are perhaps the state in which these predictions are maximally unconstrained by external inputs. Rather than being adjusted and optimized in response to sensory inputs, we might even say that dreaming is the special case in which (simulated) sensory inputs are adjusted to fit our predictions. It stands to reason that dreaming should strongly reflect our most deeply ingrained prior expectations.[11]

Returning to the question posed at the beginning of this section, predictive processing elegantly circumvents the problem of how to account for the storylike structure of dreaming without ascribing it to the work of a homuncular author or dream producer. All the brain does, according to these accounts, is hypothesis testing, and it does so at different levels reaching from conscious thought to unintentional imagistic processing. In dreams, the relationship between these different levels is somewhat tighter than in wakefulness, and this might explain why dream imagery and dream thinking sometimes seemingly emerge in concert, occasionally giving rise to narratives resembling carefully crafted stories. Note how this might also explain, quite elegantly, Hunter's puzzle of how the fragmented images that populate our dreams can merge into a seemingly smooth though barely enunciated story. My cursory discussion of predictive processing accounts has enabled us to do away with the raconteur. Dreaming, as noted by Dennett, is "a process of narrative production, of detail accumulation, with no authorial intentions or plans at all—an illusion with no illusionist" (Dennett, 1991, p. 11); it involves a process of continuous question-and-answer, where the protagonists are neither especially critical nor particularly attentive to detail. Dreamers have, as Dennett puts it, poor epistemic appetites and are quickly satisfied by even the sketchiest, most indeterminate of answers. Indeterminate predictions may thus produce sketchy story lines as well, while failing to alert the dreamer to this very sketchiness. Moreover, because, as we have seen, low-level imagistic processing and higher-level conscious thought ultimately involve the same type of hypothesis testing, the alleged single-mindedness of dreams is to be expected.

Unlike Dennett's (1976) cassette theory, however, which has dreams scripted and stored away in a library of undreamed dream cassettes long before they are ever inserted into memory, we can now see how the presentation process, or the conscious experience of dreaming, and the composition process are inextricably bound together (cf. sec. 1.3.4). Dreams do not, after all, behave like movies or novels, unfolding independently of the dreamer's thoughts, expectations, and reactions, but rather are modulated online, in response to these—or at least this is an initially plausible working hypothesis. In the account presented here, dreams fluidly develop in response to predictive models at different levels. These may include unconscious visual search, perhaps in the form of rapid eye movements, unconscious priors and bias in the interpretation of ambiguous vestibular stimuli, emotions, or even conscious expectations, fears, or wonderings. Examples such as the dream reported by Arnold-Forster, in which a conscious expectation is smoothly transformed into dream imagery, are just a particularly obvious example. And though predictive processing accounts make room for such top-down cognitive, expectation-based contributions to perception, as well, the effect should clearly be much more pronounced for dreaming. A prediction for future research is that dreaming involves a tighter coupling of top-down and bottom-up processing than waking perception. This coupling might well lead to the illusion of just knowing, in a dream, that one is dreaming of a particular person—though there is no sufficiently detailed visual imagery to match.

Predictive processing accounts might thus also explain why phenomenal indeterminacy is even more pervasive in dreams than in standard perception. If predictive models are more concerned with the gist than with the exact perceptual details, then phenomenal indeterminacy would be the natural result (Madary, 2012; Seth, 2014). Given the indeterminacy of predictions about expected sensory input, this lack would not be easily noticed. Moreover, given that in dreams, detailed and stable sensory input is missing and the optimization of predictive models is consequently disturbed, one would expect the resolution of these models likewise to be low. Combined with more specific conscious expectations, this might account for the strange combination of specificity in content and phenomenal indeterminacy in dreams, but also in the amodal agent models that characterize felt presence. Something similar might be said about our frequent ability, after awakening, to formulate a dream in storylike terms that itself did not yet have a fully developed narrative structure. In dream reporting, as in dreaming itself, specificity in content may outrun detail in experience.

If this is on the right track, then it might be possible to introduce a new metaphor for dreaming. Instead of likening dreams to stories, we might say that dreaming is a process of mental improvisation, arising from predictions generated on the fly, under compromised conditions of hypothesis testing. This is occasionally sufficient to result in coherent and genuinely storylike narratives but sometimes, as in free jazz, also results in seemingly bizarre and dissonant dreams. The tendency to interpret dreams, which is a continued attempt to make sense of them even after awakening, may be rooted in the cognitive processes underlying their very generation (see Pace-Schott, 2013).[12]

To conclude, we can now see that dreams are in fact centered reference frames in more ways than one. They are centered on a spatiotemporally situated self, as the point of maximal invariance within the dream, and they are driven by predictions that in turn are generated in response to the ongoing dream. Some of these predictions reflect our most deeply ingrained and likely unconscious prior expectations, whereas others reflect the dreamer's fears, concerns, preferences, and personality—possibly explaining why dreams lend themselves to interpretation and why paying attention to one's dreams has long been thought to be a source of personal insight, for instance, in psychotherapy (Hill et al., 1993; Kramer, 2007, pp. 117–119; Edwards et al., 2013) and creativity (Hobson & Wohl, 2005; Schredl & Erlacher, 2007; Stumbrys & Daniels, 2010). Nothing in the present account requires that dreams are intrinsically meaningful, perhaps in the form of an unconscious message to the dreamer. The account proposed here might nonetheless help explain why the most deeply meaningful dreams—which are likely also the most memorable ones—are intimately connected to our own bodily states, our joys, fears, memories, and expectations. Because the dream world is created on the fly, in response to bodily inputs, but also to memories and expectations, dreams may achieve a degree of coupling with the phenomenal self that is not reached in wakefulness. If anything like this account is correct, this might also show why attempts to create virtual realities that respond to participants' bodily states fall short of dreaming, and why they will likely continue to do so even if researchers manage to optimize bodily feedback loops. Unless they additionally manage to tap into participants' emotions, memories, and ongoing thoughts, this level of integration will likely not be reached. This suggests that the specific ways in which dreams interweave memories and expectations with the generation of a spatiotemporal reference frame centered on the phenomenal self and often driven by residual own-body perception are deeply interconnected. Predictive processing accounts may provide the conceptual glue for holding these two aspects of mind wandering and bodily based self-location together.

12.4 Reflections on the Functions of Dreaming

Does the account outlined in the previous section give us any indication of why we dream? The question of why we dream is as controversial as it is important; indeed, it looms over much of the theoretical discussion of dreaming. Here a brief survey of the most prominent theories and main theoretical challenges will have to suffice.

A first important distinction is the one between the functions of sleep and different sleep stages on the one hand and the functions of dreaming on the other. REM sleep has been proposed to fulfill a number of phylogenetic and ontogenetic functions, for instance regarding thermoregulation and the development of cortical structures in birds and mammals (Jouvet, 1999; for details on the phylogeny and ontogeny of sleep, see Lesku et al., 2009). Further well-documented hypothetical functions include neurotransmitter repletion, the reconstruction and maintenance of little-used brain circuits, the structural development of the brain in early developmental phases, as well as the preparation of a repertoire of reflexive or instinctive behaviors (Hobson, 1988, 2009). Many studies have shown that sleep facilitates memory consolidation (Rauchs et al., 2005; Diekelmann et al., 2009; Diekelmann, 2014; Stickgold & Walker, 2005; Walker, 2005, 2009a, 2009b, 2009c). Declarative memory consolidation in particular has been associated with slow-wave sleep (Plihal & Born, 1999; Born & Wagner, 2004; Gais & Born, 2004), whereas motor learning may be enhanced by REM sleep (Laureys et al., 2002; Walker et al., 2005).

By contrast, only a few studies have directly investigated the relationship between dream content and memory consolidation in sleep (for an overview, see Nielsen & Stenstrom, 2005; Wamsley & Stickgold, 2011). Fosse et al. (2003) found that though a majority of dreams integrate waking-memory sources, only 1 to 2 percent of dreams involve an episodic replay of waking memories, suggesting a functional dissociation between episodic memory and REM sleep/dreaming. The replay of episodic memories may be most pronounced in recurrent dreams, such as those experienced by patients with post-traumatic stress disorder (PTSD; see Hartmann, 2001; Zadra, 2001). Such dreams, however, can be considered as dysfunctional because of their distressing effects on waking levels of well-being and are one of the main symptoms of PTSD (Nielsen & Germain, 2000).

More generally, it has been suggested that the associative and bizarre nature of dreams may strengthen weak associative connections (Stickgold, 1998) or may serve a selective processing function, such as the integration of novel features into existing cognitive memory networks (Schwartz, 2003). A new and ambitious theory by Llewellyn (2013) suggests that dreams have the function of elaboratively encoding episodic memories, much in the manner of ancient art-of-memory techniques (for my own critique of this view, see Windt, 2013a). Generally, however, the relation between memory processing during sleep and changes in dream content has not been well investigated, and these results are preliminary. Thus it is possible that the different sleep stages play a role in memory consolidation but do so independently of systematic changes in dream content.[13]

A second group of theories on the functions of dreaming links dreaming to emotional processing (Perogamvros et al., 2013). These theories typically focus on nightmares and stressful dreams (e.g., Levin & Nielsen, 2007, 2009; Nielsen & Lara-Carrasco, 2007; Lara-Carrasco et al., 2009). For instance, Hartmann proposes that dreams help integrate unresolved emotional concerns and provide the ideal conditions for "making connections in a safe place" (Hartmann, 1998, p. 129), pointing out that this function is most obvious in nightmares and distressing dreams. Similarly, Cartwright (2001, 2010), who studied the dream reports of recent divorcees, suggests that dreams can aid individual adjustment during depression. The involvement of dreaming in emotional processing is strengthened by the finding that changes in mood from evening to morning are often related to dream content (Kramer, 2007, pp. 122–128).

More general functional theories of dreaming include problem solving (Barrett, 2007; but see Montangero, 1993), costly signaling (McNamara, Harris, & Kookoolis, 2007), psychological adaption (for a review, see Valli & Revonsuo, 2007), and the simulation of imagery abilities and social and emotional skills (Germain et al., 2000; Franklin & Zyphur, 2005; Revonsuo et al., 2015).

Rather than concentrating on memory consolidation and emotional processing, another group of theories ascribes a future-oriented function to dreaming. Based on the high prevalence of negative emotions and threatening dream content, the threat simulation theory suggests that the evolutionary function of dreaming lies in the simulation of ancestral threats. The theory proposes that the rehearsal of threatening events and avoidance skills in dreams has an immediate adaptive value by enhancing the individual's chances of survival (see Revonsuo, 2000; Revonsuo & Valli, 2000; Valli, 2008). A number of studies have confirmed that waking exposure to threatening events enhances threatening or aggressive dream content (for details and further references, see Valli, 2008). The theory, however, has been criticized for focusing too strongly on threatening dream content (Germain et al., 2000), and Nielsen and Germain (2000) suggest that dreaming may fulfill a more general simulational function, for instance, by simulating social and emotional skills.

A much more general function is suggested by Hobson's (2009) protoconsciousness theory of REM sleep, which claims that REM sleep, even before birth, contributes to the development of brain regions that are crucial for our ability to perceive and interact with the world (for a discussion of how this relates to the threat simulation theory, see Revonsuo, 2014; for a discussion of how the theory is related to the ISTH model, see Windt, 2014a). The theory draws from findings on mammalian REM sleep, its tie to thermoregulation, as well as its prevalence in early developmental stages. Hobson argues that even before the emergence of consciousness, REM sleep plays an important role in fetal development by preparing the brain for the integrative function of consciousness.

> Originally dreamless, protoconscious REM sleep could therefore provide a virtual world model, complete with an emergent imaginary agent (the protoself) that moves (via fixed action patterns) through a fictive space (the internally engendered environment) and experiences strong emotion as it does so. (Hobson, 2009, p. 808)

From this brief quotation, it is clear that the theory does not necessarily ascribe a function to the conscious experience of dreaming itself but rather focuses on protoconscious REM sleep. Yet Hobson suggests that dream bizarreness may play an important role in this context:

> What we may need to navigate our waking world is an infinite set of charts from which we may draw the one best suited to an equally infinite set of real-life possibilities. To this end, it makes sense to tie even remotely connected scraps of experience together to give our library of plots enough verisimilitude to be useful. (Hobson, 2009, pp. 807–808)

Finally, the philosopher Owen Flanagan (1995, 2000) has proposed that dreams are mere epiphenomena of the periodic activation of the brain during REM sleep, claiming that dreams are "serendipitous accompaniments of what sleep is for" (Flanagan, 2000, p. 115). At the same time, Flanagan believes that dreams serve a derivative psychological function and thus can be truly meaningful.

A central problem for theories on the functions of dreaming is how to explain the fact that a majority of dreams are forgotten, raising the question of how dreams can fulfill their hypothetical function independently of recall. Crick and Mitchinson (1983) famously proposed that REM sleep "erases" or deletes surplus information and unnecessary memories, suggesting that enhanced dream recall might actually undermine the effectiveness of the memory-erasing system. Concerns about dream recall also raise a familiar methodological worry. Dream reporting is a cultural phenomenon, depending, among other things, on the ability to remember and communicate one's dreams and to conceptualize the difference between dreaming and wakefulness. These practices, in turn, may influence the frequency and content of dreaming itself (Hacking, 2001). To the extent that the concept of dreaming, and hence dreaming as an object of empirical investigation, is inextricably linked to the practice of dream reporting, dreaming is not an exclusively biological but also a cultural phenomenon, and as such it is not clear that it requires an explanation in terms of its adaptive value or evolutionary function at all. The attempt to identify an evolutionary as opposed to a cultural function of dreaming may thus be misguided. As is the case for theories of dreaming, theories on the functions of dreaming are constrained by the space of reportable dreams.

Some researchers even question the relation between REM sleep and memory consolidation itself. Vertes and Eastman (2000) cite impressive evidence supporting the claim that a loss of REM sleep, caused by selective REM sleep deprivation or a cessation of REM sleep following pontine lesions, does not impair cognitive functioning. William Dement, a pioneer of dream research who conducted early and largely unsuccessful sleep deprivation studies aiming to show that a prolonged loss of REM sleep/dreaming would render subjects psychotic (Dement, 1960), has also suggested that sleep itself may not necessarily fulfill a single vital function (Dement, 1999/2001). Relatedly, even if the different stages of sleep have a function, it is not clear why dreaming should be thought to contribute to this function rather than simply being correlated or continuous with, for instance, waking cognitive skills or emotional adaption. This is especially true because dreaming can be lost selectively, apparently without any detrimental effects on cognitive functioning in wakefulness (see sec. 3.6; Solms, 1997, 2000).

For all these reasons, I propose that the question of why we dream can be rendered more tractable by transforming it into a slightly different question, namely, the question of why we dream the way we do. In fact, I would submit that we have already made considerable progress on answering this new question. Before briefly recapitulating why this is the case, I want to add one further ingredient to the mix.

A question that is almost as interesting as the question of why we dream the way we do is why we *don't* dream differently, or why dreams typically lack certain phenomenological characteristics. Indeed, this second question is crucial for formulating a satisfying answer to the first. In an excellent and well-researched article on dreaming, Symons (1993) combines a highly differentiated analysis of dream experience with an evolutionary hypothesis. Although he does not offer a positive account of why we dream, he argues that considering the stuff dreams are *not* made of, to quote his paper's title, is an important step toward understanding the target phenomenon that a functional hypothesis should account for. He allows that dreaming involves visual and movement hallucinations similar to waking percepts, but denies that the same is true of other forms of imagery, arguing that rare types of imagery, but also auditory imagery, are conjured rather than hallucinated in dreams.

> In dreams we often eat and drink, but we rarely taste or smell; we sustain physical insults, but we feel no pain; we walk and run, but we do so effortlessly. Such differences between dream and wake-state sensory experience exist, according to the vigilance hypothesis, because natural selection has disfavored the occurrence during sleep of hallucinations that compromise vigilance. (Symons, 1993, p. 207)

While the *vigilance hypothesis* does not deny that rare sensations occasionally occur in dreams, it predicts that dreams systematically lack sensations that are critical to vigilance: "During sleep the brain does not hallucinate sensations that it could easily confound with sensations of exogenous origin, because to do so would compromise vigilance" (Symons, 1993, p. 182). Because most mammals sleep in the dark, with eyes closed and in an immobile position, having vivid visual and motor hallucinations during sleep poses no threat.

This view has several problems. A first is whether it really offers a satisfying account of auditory imagery in dreams, which according to Symons is conjured (or imaginative) rather than hallucinated. Dreamed speech, in this view, involves the mechanisms of speech production rather than those of auditory perception and is "telegraphic or cloudy because it is only abstract phonological or phonetic information" (Symons, 1993, p. 201). While this account fits in well with phenomenal indeterminacy in dreams, it is, according to antiskepticism about dream reporting, an empirical question how frequently this type of indeterminacy affects auditory dream imagery, and the difference between hallucinatory and conjured auditory imagery should be detectable in dream reports.

A second and more serious problem concerns instances of external-stimulus incorporation in dreams. My characterization of dreams as weakly phenomenally-functionally embodied states is contradicted by the vigilance hypothesis, according to which one would expect that auditory and especially tactile stimulation, as a more immediate threat to the body, should lead to awakening. Yet, as we have seen, tactile stimulation appears to be more likely to be incorporated in dreams than auditory stimulation (cf. chap. 8). A possible explanation might be that when sleeping in a safe environment, we are less likely to be awakened by such stimuli. Our ability to sleep through customary stimuli, such as the movements, touches, and sounds produced by sleeping partners, while still (at least usually) waking up in response to the alarm clock, may suggest that we maintain a degree of vigilance toward external stimuli throughout sleep, and one that is sensitive to their meaning. Note that this would make good functional sense. Perhaps the distorted incorporation of external and particularly bodily stimuli in dreams is a trade-off between preserved sleep on the one hand and preserved sensitivity to external and potentially dangerous stimuli on the other. Such stimuli would only lead to awakening if they were particularly strong, unexpected, or salient (explaining, perhaps, why mothers selectively awaken to their own babies' crying within hours after giving birth). Whether this line of argument offers a plausible defense of Symons's vigilance hypothesis is, at least in principle, open to empirical investigation.

Finally, other rare dream sensations may neither be likely to be incorporated into dreams nor lead to awakening. One study found that contrary to the common belief that threatening smells (such as smoke) lead to awakening, participants did not reliably respond to odors presented during sleep (Carskadon & Herz, 2004). If further substantiated, this finding would threaten the vigilance hypothesis. Despite these worries, Symons's vigilance hypothesis places, perhaps with some cosmetic adjustments, an important constraint on functional theories of dreaming, namely, to account for those aspects of waking experience that are typically missing in dreams.

To return to the initial question: why do we dream the way we do, and not differently? Rather than offering a final answer, my strategy will be to translate this question into a number of successor questions. I will offer no more than some highly speculative remarks on each of these. Yet my hope is that this deflationary account will offer some new ways of thinking about the question of why we dream at all.

- *If dreaming has a function, why are dreams only remembered if special steps are taken to retain them?*

This question points to a central problem for any theory on the functions of dreaming. As pointed out earlier, if dreaming had a function, it would have to have little if anything to do with actual instances of dream recall. Not only is spontaneous dream recall variable and unreliable, even in adults. In addition, in beings who are unable to distinguish dream recall from waking memories, as may well be the case for nonhuman animals and small children, dream recall would be outright dangerous. The examples of retrospective dream deception discussed in chapter 10 show that dream–reality confusion after awakening poses a genuine threat to the well-being of those afflicted. This suggests that the function (if any) of conscious experience during sleep should be separable from the ability to recall dreams—at least if this function, as one would expect of an evolutionary theory, is supposed to characterize the dreams of nonhuman animals and small children, and not just of human adults.

Formulating the question about the function of dreaming in this way, however, leads to a complication: the present theory defines dreams, in part, as reportable experiences during sleep. This reduces the question of whether dreaming, in the absence of dream recall, has an evolutionary function to absurdity: it is not clear how reportability in the absence of actual dream recall could fulfill any function of its own, and spontaneous dream recall is too unstable and too strongly tied to the abilities of adult humans to have a convincing evolutionary function. More importantly, asking about the function of unrecalled dreams in humans who are capable, in principle, of dream recall and dreaming reporting is different from asking about the function of dreams in creatures that are in principle incapable of reporting their dreams. Perhaps we can at least extrapolate from some of the most general characteristics of human dreams to conscious experience in the sleep of nonhuman mammals; but if we can, this type of experience will still likely be very different from human dreaming, and we will not be able to say much about its functions. Questions about why we dream the way we do, in the context of the present theory, are always questions about *reportable* dreams in human subjects; and even our own ability to report dreams is likely too evolutionarily recent and unstable to fulfill a function of its own. So while we may be able to make sense of some of the phenomenal-functional characteristics of dreams by placing them in the context of evolutionary considerations, we should not expect a clear-cut evolutionary answer to the question of why dream at all. Questions about the functions of dreaming are, in the context of the present theory, constrained by the space of reportable dreams, and these constraints are quite possibly such that they prevent a theory on the evolutionary functions of dreaming.

• *Why do dreams take the form of immersive spatiotemporal hallucinations during sleep? Or, put in slightly simplified terms, why is dreaming consistently characterized by a phenomenal here and now, or by minimal phenomenal selfhood?*

This structural feature of dream experience should tell us something about the central importance, for beings who continuously and dynamically interact with the world, of having centered reality models in which uncertainty as to one's current location—both in spatial and in temporal terms—is minimal. True, full-body illusions show that self-location can be disturbed and manipulated in special experimental setups in healthy, awake subjects. There may even be situations in which self-location is distributed between two points, for instance, between a virtual environment and the actual location of the physical body; and perhaps these situations can even be described as involving bilocation (Wissmath et al., 2011; Furlanetto et al., 2013). However, even if such cases of bilocation exist, these are hardly cases against which our ancestors would have needed to protect themselves. Rather, having a rough-and-ready but highly stable sense of spatiotemporal self-location was apparently enough. The phenomenology of dreaming reflects this very basic computational challenge.

Two more points seem relevant. The first is that we have a bias toward assuming self-movement rather than environmental movement. This bias may lie at the root of frequent movement sensations in dreams, and it also becomes evident in vection, or movement illusions in wakefulness, though in a slightly different manner. Precisely because typically we are dynamically moving through space, it is of vital importance that we only feel located at a single point in space at a given time. Moreover, in standard situations, the position at which we feel located should move along with our physical bodies.

The second point is that while dreams consistently involve a sense of self-location, the locations in which we feel located in dreams are typically not familiar. Dream locations are actually unfamiliar in roughly two-thirds of all cases, and familiar settings may be less frequent than other episodic memory elements (Fosse et al. 2003). Dream locations are also the most frequently discontinuous element in dreams (for discussion and further references, see Nielsen, 2013). This might suggest that the environment relative to which one feels located in dreams is simply not important, or that the sense of self-location in dreams is a by-product of brain activation during sleep. A bolder answer would be that dreaming is a cognitive workout, in which self-location must be generated with respect to continuously shifting and unfamiliar environments. As Diekelmann and Born (2014, p. 120) put it, "the reason humans produce REM sleep dreams is not to provide a more or less correct but sparse copy of the past, but to synthesize, out of randomness, an '*a priori* model of the world,' to create new associations to build new hypotheses that may eventually serve to control the waking future." To achieve this function, the integration of disparate or even randomly generated stimuli might be enough; in fact, integrating disparate memory sources and random stimuli into a coherent world model might be particularly challenging. There might be a deeper reason, then, for the unfamiliar and discontinuous nature of dream settings.

What about temporal self-location? An intriguing but highly speculative point is that just as we (or rather our ancestors) are continuously engaged in bodily movement, so are our minds continuously engaged in mental time travel. Mind wandering makes up roughly 50 percent of waking mental activity (Schooler et al., 2011) and is mostly oriented toward the future or the past (Corballis, 2013). Although voluntary mental time travel exists, a significant part of it is involuntary (Berntsen & Jacobsen, 2008). Having a maximally invariant temporal position—the experienced *now*—provides a crucial backdrop in the face of this constant cycling through remembered past and anticipated future events (cf. Droege, 2009). In this context, it is noteworthy that the temporal context in dreams is almost as diverse as the spatial one. Given that dreams draw richly from memory sources, it is surprising that they are still so strongly focused on the present moment. Again, this could mean that in dreams, accurate temporal self-location is simply not important enough to get right; but it could also mean that endowing distal and interwoven memory sources with the phenomenology of *now* contributes to the function, if any, of dreaming.

- *Why are bodily experiences, with the exception of movement sensations, systematically underrepresented in dreams? And why are such bodily experiences as do occur in dreams still weakly constrained by real-body inputs?*

Again, at least two answers are available. The first is that detailed bodily experiences (with the exception of movement sensations) are just not the kind of experience that it is necessary or beneficial to generate independently of real-body inputs. A slightly stronger reading is that for beings like us, decoupling the robust phenomenology of being an embodied subject from the physical body altogether would be positively detrimental; the physical body, after all, is the most reliable and stable anchor for our interaction with the world. Either way, it is interesting to note that even in dreams, the intimate connection between bodily experience and the physical body is not completely severed, but the physical body continues to modulate ongoing experience with varying degrees of subtlety. This should tell us something about the robustness of the connection between bodily experience and its real-body basis.

A recent study provides an interesting and again unexpected perspective on this problem. Honma and colleagues (2014) investigated whether repeated exposure to the rubber-hand illusion would have a learning effect on participants. They found, among other things, that proprioceptive drift was significantly attenuated after sleep, leading to the intriguing and somewhat paradoxical suggestion that "sleep resets the proprioceptive sensation closer to reality and maintains awareness that the RH [rubber hand] is an alien limb. . . . Hence, sleep does not seem to haphazardly enhance memories. . . . Rather, it seems to resist incongruent multisensory integration" (Honma et al., 2014, p. 7).

Given that the rubber-hand illusion is induced by multisensory conflict, and given that multisensory integration is itself disturbed in dreams (see chap. 7), this once more leads to the claim that in dreams, experienced self-location is more tightly constrained by internally generated signals than by external, multisensory stimuli. This might be why changes in experienced self-location or body-part location, as in the rubber-hand illusion, are attenuated after sleep. Whether there is any association with dream content would, of course, be a fascinating question for future research. Yet, for now, an intriguing though highly speculative hypothesis is that sleep, and perhaps dreaming, is closely related to the bodily (as opposed to external or multisensory) sources of experienced self-location.

- *Why is dreaming driven by low-level, imagistic cognition, and why are conscious thought and reasoning so unreliable in dreams?*

Dreams are dominated by visuospatial imagery, movement sensations, emotions, and egocentric reference frames. The interesting point is that in dreams, these are not experienced *as* cognitive processes but for the most part are experienced as real. Again, a weak answer would be to say that under the particular conditions of sleep, there is no harm in mistaking cognitive processes for external reality. A more audacious answer would start from the claim that dreams offer the clearest example of vivid visuospatial imagery. Mental simulation involving vivid visuospatial imagery is clearly a powerful cognitive tool and greatly enhances our behavioral flexibility. This is not to say that dreaming is necessary for the maintenance of waking imagination; rather, dreaming is the most vivid and impressive expression of one of our central mental capacities.

Again, two points are relevant. The first is that the present theory emphasizes the importance of spatial over visual imagery, which, at least in dreams, is claimed to be the most basic type of imagery. This sits nicely with findings according to which amodal (or supramodal) spatial imagery is one of our most important representational capacities (see chap. 11) and possibly even forms the basis of reasoning (Knauff, 2013).

Second, it is not quite correct, as we have seen, that dreaming is driven by low-level cognition in the absence of conscious thought; rather, imagistic cognition and conscious cogitation mutually influence each other, and both make an important contribution to the process of dream formation. This is illustrated by the recent finding that in patients with autoactivation deficit, characterized by a loss of self-initiated behavior and self-reported thought (or mental emptiness), at least a simple form of dreaming is preserved (Leu-Semenescu et al., 2013). Compared to healthy controls, patients' dream reports were shorter, as well as less complex, frequent, bizarre, and emotional. Still, this finding powerfully illustrates that while dreaming can apparently be preserved in the absence of intentional mental activity and top-down control, many of its key characteristics are lost.

Turning to the second part of the question, a simple answer would be that if dreaming had a function, conscious thought and reasoning would be irrelevant to this function. Yet, as we just saw, this is somewhat too simple. A slightly more interesting answer would be to say that if the function of dreaming consisted, in part, in generating environments experienced as realistic, the added computational load required to do this in the face of preserved critical reasoning and reality testing might be too high. This is especially true because dream deception itself is not adaptive. Maximally realistic dreams with belief-inviting effects would shoot beyond the mark, leading to pathological and potentially dangerous dream–reality confusion (Wamsley et al., 2014; chap. 10).

Another way of answering the question might be to say that in dreams, the lack of control over conscious thought itself is functional. Perhaps dreams are an ideal playground for prioritizing automatic, non-goal-directed, hyperassociative processing over conscious thought, and perhaps this complements more controlled thought processes in wakefulness. Note how this idea is closely related to the metaphorical description of dreams as involving mental improvisation. This speculative hypothesis once more suggests that in the future, it might be fruitful to compare the debate about the functions of dreaming with the debate about the functions of daydreaming. While much research has been dedicated to investigating the detrimental effects of mind wandering on ongoing task performance and mood, positive functions are only beginning to be investigated and might include future planning, creativity, attentional cycling between multiple goal states, and dishabituation, which is thought to facilitate refreshed attention to ongoing tasks (for discussion and further references, see Schooler et al., 2011;

Smallwood & Andrews-Hanna, 2013). Given that dreaming and waking mind wandering are closely related, one would expect these two theoretical debates to be able to fruitfully inform each other.

- *Why do dreams integrate distal memory sources as well as ongoing concerns?*

A humble answer is that just because dreams are made of memories, memory might not be what dreaming is *for* (Blagrove et al., 2013). A bolder answer might appeal to the fact that dreams bind increasingly distal memories into a single reference frame experienced as happening *now* (see sec. 11.2). If dreams are, as suggested by Hobson's protoconsciousness theory, as much about anticipating the future as they are about remembering the past, this ought to be expected. This fits in well with the finding that memory is crucial, among other things, for the ability to anticipate future events (Schacter et al., 2007; Schacter, 2012).

Moreover, recall that the types of memory sources that so far seem the most likely to be integrated into dreams center, once more, on spatial imagery tasks. Experimental findings suggest that hippocampal activity during sleep is closely related to memory processing (Saletin & Walker, 2012), and evidence points to a replay of hippocampal activity associated with spatial tasks during sleep in rats (Skaggs & McNaughton, 1996) and possibly in humans (Peigneux et al., 2004). Generally, the hippocampus, associated both with memory processing and with spatial cognition, is thought to be implicated in dreaming. Nielsen and Stenstrom (2005) even attribute the *here* and *now* of dream experience to hippocampal activity. Evidence also suggests that performance on virtual navigation tasks improves following dreams (Wamsley, Perry, et al., 2010; Wamsley, Tucker, et al. 2010; see also Wamsley & Stickgold, 2009, 2010, 2011). Schredl and Erlacher (2010) found evidence for a correlation between performance on a mirror tracing task and bizarre, long, intense, and negatively toned dreams; however, the effect was not related to specific incorporations of the task in dreams. Oudiette et al. (2011) studied the effects of learning a serial reaction time task on sleep in healthy subjects and RBD patients. Several of the RBD patients manifested complex behaviors during sleep, and in one patient, hand movements that were highly similar (as rated by independent judges) to those of waking subjects performing the task from memory were observed. The authors note that this is "the first direct and unambiguous demonstration of overt behavioral replay of a recently learned skill during human sleep" (Oudiette et al., 2011, p. 5). To what extent this was related to dream content, however, is unclear, and the authors note that only seven out of fifty-seven subjects reported any dreams. Two of these reported dreams that were related to the task.

Aside from emphasizing the importance of spatial processing in dreams, one might additionally want to claim that dreaming is a form of self-consolidation, for instance, by integrating disparate memory sources with each other and centering them on the self (Horton et al., 2009). A similar self-maintenance function has been suggested for mind wandering (Metzinger, 2013a). At least the spatial aspects of dreaming, however, seem to point to a more basic, generic function, which is unspecific to the dreamer's personality. This brings us back to my claim, at the end of the last section, that dreaming involves the generation of reference frames that are centered on the self in two distinct but intimately related ways, namely, by the robust phenomenology of *here* and *now*, and in a more metaphorical sense in which dreams are driven by waking-memory sources, concerns, and even ongoing cogitative activity such as thoughts and expectations about the unfolding dream.

The emerging picture is that dreaming reflects, or perhaps even subserves, two apparently distinct computational needs. The first is the need, for beings interacting with a dynamic and constantly changing world, to have a stable sense of self-location with respect to other agents in this world. The second is that the ability to engage in mental simulation—in planning, remembering, or daydreaming—requires the ability to partially dissociate from the real world and the physical body to engage in the imaginary one. It is noteworthy that even in wakefulness, vivid mental simulation is accompanied by an attenuation of external stimuli (Kam & Handy, 2013).[14] Yet this is nowhere near what is seen in dreaming. Ongoing input from the physical body ensures that even when lost in deliberate mental simulation and spontaneous mind wandering, we are still safely anchored in the body. Moreover, unlike in dreams, we often experience our daydreams from the outside, rather than imagining that we are located in them. In one study, nearly half of participants experienced their daydreams from a third-person rather than from a first-person perspective (Christian et al., 2013). By contrast, passive observation is much less frequent, as we have seen, in dreams. Both in the phenomenological and in the functional sense, self-location during wakefulness is more securely anchored in the real world and in the physical body than in the imaginary one. One might even say that this is a condition for our ability to engage in mind wandering and to safely travel imaginary worlds. By contrast, in the quiet of sleep, this type of continued anchoring retreats into the background, allowing us to more fully enter the world of dreams.

Metzinger has illustrated a related point with the help of the world-zero hypothesis (Metzinger, 2003, pp. 61–62). In a system using competing world models, that is, navigating the real world while lost in imaginary ones, one of these world models has to be indexed as the real one. The same will be true for the self. As soon as we acquire the ability to create imaginary selves, we need a self-zero that will act as "the functional anchor and reference basis for concomitant phenomenal self-simulations" (p. 285). Ongoing bodily experience goes a long way toward explaining why in imagining and daydreaming, we always already know we are not quite going to get what we are imagining (cf. Ryle, 1949/2000, p. 268). The important point is that it is only this contrast between alternative world models and self-models that introduces the need to index one of them as real.

Dreams, in this context, are the most vivid reflection of our mental capacity to simulate imaginary worlds, as well as imaginary selves in them; at the same time, in doing so, they consistently respect the need for a continuous and maximally stable sense of self-location in this imaginary world. What is unique about dreams is that they fulfill both needs at the same time, within a single spatiotemporal reference frame. As external and real-body inputs are weakened during sleep onset, the distinction between the real and the imaginary world collapses. Hippocampal activation may offer the common neural basis for both the spatiotemporal and the mnemonic properties of dreaming. If it is really true that a substantial portion of wakefulness, as indicated by research on mind wandering, involves mental simulations that are only weakly constrained by ongoing perception and thus is characterized by simultaneously overlapping spatiotemporal reference frames with different degrees of experienced realness, then there might even be a sense in which dreaming offers a glimpse of a more simple and unified form of consciousness experience: dreams might then be the rare exception in which we feel robustly present in a world that is ultimately the product of our own making and intimately centered on the self. Dreaming may also be the rare case in which spatiotemporal self-location and vivid mental simulation in the form of mind wandering coincide, rather than pulling us in different directions. Even if dreaming does not fulfill a specific evolutionary function, different aspects of the phenomenology of dreaming can still make good functional sense. Their occurrence in sleep may not be necessary for the maintenance of the corresponding waking capacities, but we might still say that the types of conscious experiences that we do have in sleep make functional sense and are, for beings like us, part of the story.

Even if this insight does not necessarily amount to a positive answer about the functions of dreaming, it certainly has an important metatheoretical function: it tells us something about the possible contribution of a philosophical theory of dreaming to our theoretical understanding of the core capacities of our minds. Within the framework of concepts used for describing the mind, dreaming straddles the seemingly distinct positions between embodied perceptual presence and mind wandering or imagination. Dreaming might even help us rethink the seemingly dichotomous distinction between them. Perhaps the phenomenology of presence is not so much there to anchor us in the real world and prevent us from losing ourselves in mind wandering. Instead, for minds continuously engaged in mental activity that is partially decoupled from the real world and from ongoing perception, the sense of presence might actually alleviate the need to constantly attend to the perceptual world, functioning as a rapid entry point only when our immediate and deliberate reaction is required. This is a subtle difference, but one that shifts our view of ourselves as embodied agents constantly engaging in perceptual interaction with the world to one that views us as continuously engaged in mental simulations that are only weakly constrained by the ongoing flow of perceptual and bodily experience. This latter reading is in keeping with recent attempts to describe mind wandering not as an irritating and defective deviation from goal- and task-oriented thought but as an integral, entirely normal, and perhaps even beneficial form of mental activity (Smallwood & Andrews-Hanna, 2013). Dreaming may simply be the fullest expression of this natural tendency: here external and bodily constraints on conscious experience reach their low point, while at the same time we enjoy a much more robust sense of presence in an imaginary world than we ever do in wakefulness. What is new about this idea is that losing oneself in imaginary worlds might not counteract the feeling of presence in the real world but help explain why the phenomenology of presence is so important in the first place. If the theoretical analysis of dreaming ever so gently nudges us toward a new perspective on these questions, it might enrich our understanding of conscious experience considerably.

12.5 Conclusions

We have come a long way. Most of the book has been concerned with defining the outlines of a conceptual framework for describing conscious experience in dreams. After the first four chapters clarified the methodological background assumptions of dream research and the epistemic status of dream reports, chapters 5 through 10 were dedicated to capturing the phenomenology of a majority of dreams, as well as the characteristic variability of dreaming. Then, by using the conceptual tools of immersive spatiotemporal hallucination and oneiragogia, chapter 11 delineated the borders of the target phenomenon and developed the conceptual resources for describing the trajectory from wakefulness via sleep onset imagery to full-fledged dreaming. This final chapter investigated how this conceptual map of dreaming fits into the taxonomy of conscious experience more generally, thus taking yet another step toward solving the conceptualization problem of dreaming.

This does not, of course, mean that I have fully or conclusively answered the problem of how to integrate dreaming into a broader theory of consciousness and the self. I hope, however, to have laid the groundwork for doing so and to have offered some new ways of thinking about this problem along the way. In particular, we can begin to see the outlines of how a theoretical account of dreaming might fit into a broader theory of cognitive processing. The emerging idea is that perception is about situating oneself in the world, whereas dreaming is about centering a world of one's own making on oneself. Dreaming thus simultaneously addresses two distinct yet intimately related computational needs: the need for maintaining spatiotemporal self-location as a maximally robust and stable phenomenal-functional property, and the need for mentally simulating, especially through visual and spatial imagery, imaginary worlds. Because dreaming involves the robust phenomenology of spatiotemporal situatedness relative to a reference frame created largely out of disparate memories and spontaneously generated visuospatial imagery, dreaming genuinely blurs the borders between mind wandering and standard waking perception. Recent predictive processing accounts might provide the theoretical context for spelling out this claim in more detail.

I suggest that my analysis of the contrast conditions discussed in this final chapter—ranging from felt presence and full-body illusions, to virtual reality and vection, to mind wandering—is not just an exercise in remapping the concept of dreaming. Rather, it is an invitation to consider new ways of contrasting dreaming and wakefulness that, aside from their metatheoretical interest, might also eventually lead to new research questions for interdisciplinary consciousness research. Recall that a central motivation of the book has been not just to develop an empirically informed philosophical account of dreaming but also to offer an account that is empirically informative, for instance, by suggesting how scientific dream research can contribute to interdisciplinary consciousness research. If dreaming can indeed contribute to the investigation of topics such as self-consciousness, presence, and mind wandering, this would be an important step toward turning dreaming into a respectable and productive target for interdisciplinary consciousness research. And if the conceptual framework I have developed here can help show how this can be done, one of the book's most important goals will have been reached.

Notes

General Introduction: The Conceptualization Problem of Dreaming

1. I use the term *experimental philosophy* here in the broad sense of using experiments to gather philosophically relevant data, rather than in the narrow sense of using questionnaires to probe the intuitions of laypeople. See Prinz (2008) and Alexander (2012) for details and further references; for my own take on experimental philosophy of mind, see Metzinger and Windt (2014).

2. A condensed version of this argument can be found in Windt (2013b).

3. Chap. 11 and part of chap. 12 build on an earlier publication (Windt, 2010) but offer a considerably expanded and revised discussion of my earlier views. An accessible, German-language introduction is Windt (2014b).

1 Dream Skepticism, Skepticism about Dreaming, and the Problem of Dream Experience

1. I return to the epistemological problem of whether dreams are deceptive experiences in chapter 10, where I use dream reports and findings from dream research to develop different scenarios of actual dream deception.

2. The dream example has been a staple of skeptical argumentation since antiquity. It appears in Plato's *Theaetetus* (1996, 157e–158e), as well as in Pyrrhonian skepticism (Diogenes Laertius, 1943; Sextus Empiricus, 1987), and was discussed by Cicero (1994, 88), Augustine (1950, 11.25), and Montaigne (1966), just to name a few prominent examples.

3. In his review of philosophical publications on dreaming since 1960, Dreisbach (2000, p. 37) found that more than half were about Cartesian dream skepticism.

4. Note that Descartes's argument crucially depends on the credibility of his example: only if it is truly representative of a best-case scenario of acquiring sensory-based beliefs about the world (cf. Stroud, 1984, p. 20), and only if the exact same type of experience could occur in a dream, does it support the claim that all (and not just some) sensory-based beliefs are dubitable.

5. This commits Descartes to the view that the indistinguishability between dreaming and wakefulness is bidirectional and that the two cannot be told apart in dreaming or during wakefulness. A weaker assumption would be to say that the indistinguishability between dreaming and wakefulness is unidirectional: granting that we are unable to distinguish between dreaming and wakefulness while *dreaming*, we might still be perfectly able to do so while awake. Indeed, this type of unidirectional indistinguishability underlies Descartes's own solution to the dream problem in the *Sixth Meditation*. For now, note that within the context of the dream problem as it is posed in the *First Meditation*, this weaker analysis would not improve the epistemic situation of the Cartesian inquirer: even if in wakefulness he can, as a matter of fact, distinguish dreaming from wakefulness, he would still have to know, by independent means, that he is now awake to trust his (apparent) ability to make the distinction. After all, if he were now dreaming, phenomenologically all things being equal, he would be mistaken. For the Cartesian inquirer, the problem is not whether he can distinguish dreaming and wakefulness while awake but whether he can do so *independently* of whether he is now dreaming or awake. Apparently he cannot, and thus he remains vulnerable to the threat of deception.

6. This is not to say, however, that Descartes assumes that he is caught in a lifelong dream: if he were always dreaming, it would make no sense to assert the indistinguishability between dreaming and wakefulness in the first place. Moreover, the stronger assumption that he is caught in a lifelong dream is not necessary for establishing the damaging potential of the dream example. To establish this, it is enough that at any given moment, he is unable to make the distinction.

7. From now on, I use "experience," "conscious experience," "phenomenal experience," and "phenomenal consciousness" interchangeably. Following Nagel (1974), phenomenal experience refers to *what it is like* to have an experience for the subject of that experience.

8. Epistemic transparency is intended in a nontechnical, metaphorical sense here; in particular, it should not be confused with the thesis that the mind is transparent to itself (for a general discussion and critique of this view, see Carruthers, 2011, chap. 2). Transparency is a visual metaphor, intending to capture the intuition that dream reports reveal, in as much detail as possible, what it is like to dream—they are the closest that researchers can come to "watching the sleeping mind" (Cartwright, 2010, p. 17). Similarly, Wamsley notes that subjective reports of dreaming, but also of mind wandering, "provide a valuable window into the cognitive processes occurring during offline states" (Wamsley, 2013, p. 5). Transparency is also meant to reflect that dream reports are not identical with, but better conceived of as separate from dreaming. Finally, transparency reflects the historical situation that the theoretical problems raised by dream reporting were nearly invisible in the philosophical discussion before Malcolm's (1956, 1959/1962) skeptical argument about dream reporting.

9. The dependence of Descartes's argument on first-person familiarity with his own dreams is closely related to the first-person narrative style of the *Meditations*. This style is crucial to the argumentative force of the *Meditations*, with Descartes implicitly assuming that his claims will be shared by anyone willing to follow the Cartesian meditator on his path from doubt through the different stages of skepticism to certainty. As Frankfurt (1970/2008, p. 5) notes, Descartes begins the *Meditations* on the basis of commonsense arguments and gradually works toward a more

refined philosophical position, thus reflecting the intellectual development he hopes to elicit in his readers. This journey is also one from naive to more critical phenomenologist, and when Descartes returns to the dream example in the *Sixth Meditation*, his description of dreaming has undergone an important change. The underlying assumption that the reader will share Descartes's intuitions about dreaming is both a literary device and connected to the structure of his theoretical argument. Not only does the first-person, present-tense narration of the dream example and ensuing doubt make the example much more poignant, but it is also crucial to Descartes's gradual resolution of his doubts, beginning with the *cogito* of the *Second Meditation* (Williams, 1978, chap. 3).

10. Of course, these are ultimately empirical hypotheses about the psychological effects of the dream example, but they do have some initial plausibility.

11. In an article titled "Descartes' Skepticism of the Senses," Bouwsma (1977) developed an anti-skeptical argument based on the conceptual analysis of the sentences "I am awake" and "I am asleep," arguing that Descartes's use of these expressions deviates from their standard meaning to the point of rendering them meaningless. Bouwsma argues that although Descartes has ruled out any reference to the body to determine his current state, he nonetheless "continues . . . to use language which is significant only in terms of the distinctions which he has abandoned. The questions 'Am I awake?' [and] 'Am I asleep?' are questions about bodies. Since he has ruled out bodies as within the range of the application of these terms, his language is now meaningless" (Bouwsma, 1977, p. 61). In her 1953 essay "Sleeping and Waking," Margaret Macdonald (1977) argued that discourses about waking and dreaming are governed by different criteria and that questions about the location of dreams in space and time are absurd (p. 67). For a critical discussion of her argument, see Yost and Kalish (1977).

12. Accordingly, my discussion of Malcolm's critics in the next section is necessarily selective and focuses on the lines of criticism that I take to be most relevant in the present context.

13. Some even suggest that Malcolm's analysis of dreaming can be seen as a mere elaboration of Wittgenstein's remarks on dreaming (Putnam, 1986, p. 304; Chihara & Fodor, 1977, p. 191). Indeed, there are several striking parallels. According to Wittgenstein, dream reports cannot be true or false in the sense that they agree or disagree with previous dream experience; they can only be truthful in the sense that they are delivered sincerely. To introduce a distinction between the truth and the truthfulness of dream reports, according to Wittgenstein, would be to introduce a new criterion of the concept of dreaming and thus to change its meaning (Wittgenstein, 1953, pp. 222–223; for discussion, see Child, 2007). Consequently both authors reject the possibility of misremembering dreams and emphasize the logical connection between dreaming and dream reporting. Both also reject the view that we can judge, while dreaming, that we are dreaming and that we can make any judgments whatsoever during sleep. Wittgenstein notes that even if we audibly say, "I am dreaming," while dreaming—as could happen in sleep talking—this does not count as a judgment, as it does not involve real thinking. At best, it involves *dreaming* that one said one was dreaming. Similarly, in stating that I am currently daydreaming, it is not clear that these words are a judgment about the daydream rather than part of it. In this view, dreaming, daydreaming, and anesthesia, just as making statements or judgments in a play or novel, belong

to a different language game: their meaning differs from their standard sense, which is only applicable to standard wake states (Wittgenstein, 1967, pars. 396–397).

14. A recent application of this view can be found in Bennett and Hacker's (2003) sweeping critique of cognitive neuroscience and philosophy of mind. They do not offer an extended analysis of dreaming, only remarking that "to dream is not to have any experiences, although it may involve dreaming that one has experiences (if one dreams about oneself, and dreams that one perceives, does and undergoes various things). However, to dream that one is watching a football match, enjoying a party, etc. is not to have any experience at all—it is only to dream that one is having such experiences" (Bennett & Hacker, 2003, p. 247). This is a nice contemporary example of how Wittgenstein's criterial view of language leads to Malcolmian-style skepticism about dreaming.

15. Malcolm himself would profoundly disagree with this reading: in his view, he is not uncovering contingent truths about the usage of a particular term but revealing deep and necessary truths about the concept of dreaming. This view of concepts, however, is itself extremely problematic, as will become clear in the next section.

16. Of course, you might insist that if dreams are not "experiences," then there is no "phenomenology" of dreaming for conceptual reasons and the conceptual link between "experience" and "phenomenology" is too tight to be severed. But this again would be a merely conceptual point. Losing some or all of the terms commonly used to describe the subjective or phenomenal character of dreaming would certainly be inconvenient, as it would require the introduction of new technical terms and a corresponding reformulation of the argument presented here (and I would even go so far as to say that this inconvenience could count *against* a purely conceptual argument, at least unless it was outweighed by other merits). But it would not change the deeper and more interesting point, namely, that mere conceptual arguments of this sort have no consequences for what it is like to dream. Importantly, this holds whether this what-it's-likeness is called phenomenology, experience, et cetera, or not. I would like to thank Sascha Fink for raising this objection.

17. The armchair fallacy is related to Dennett's "philosophers' syndrome," or "mistaking a failure of imagination for an insight into necessity" (Dennett, 1991, p. 401). While both result from an exaggerated confidence in apparently necessary truths, the armchair fallacy results not from a failure in exercising imaginative abilities but from a disregard for the phenomenal character of actual experience, coupled with a tendency to derive conclusions about phenomenal experience from conceptual analysis alone.

18. This is also why Malcolm thinks that the introduction of new criteria for the occurrence of dreams is something far more serious than the adoption of a new convention or the stipulation of a new usage (Malcolm, 1959/1962, p. 76); it is the introduction of an entirely new concept: "The desire to know more about dreaming should not lead scientists into transforming the concept in such a way that their subsequent discoveries do not pertain to *dreaming*" (p. 82).

19. The intimate connection between verificationism and logical behaviorism is aptly illustrated by Dennett (1979, p. 251).

20. Revonsuo (2006), who regards dreaming as consciousness in its purest form, decoupled from sensory input and motor output, is a counterexample. See chapter 5 for discussion.

21. Similarly, it is certainly no coincidence that Hacking (1975, pp. 103–112) devotes an entire chapter of *Why Does Language Matter to Philosophy?* to Malcolm's analysis of dreaming. According to Hacking, Malcolm's analysis of dreaming is important because it is a particularly strong form of verificationism, and because his insistence on criteria for verification and the determination of a sentence's meaning "leads us on to matters vastly more serious than dreaming" (Hacking, 1975, p. 112). Again, the debate about dreaming appears as a proxy war for larger issues.

22. Similarly, Schröder discusses the example of a victim of a burglary tied and gagged in such a manner as to be unable to give any behavioral or verbal indication of entertaining any thoughts or having any feelings. And though it is logically impossible for a person "to be at the same time completely unable to move at all or make any noise and still be able to express thoughts and feelings" (Schröder, 1997, p. 29), it would be absurd to assume that this state of inertia rendered him unable to have such thoughts or feelings.

23. This suggestion foreshadowed a series of experiments conducted by Arkin and colleagues; see Arkin et al. (1966, 1970a, 1970b); and section 3.4 for discussion.

24. For a more concessive approach, suggesting how a scientific theory of dreaming could be compatible with the criterial view defended by Malcolm, see Landesman (1977). He regards Malcolm's claim that dream reports are, as a matter of logical necessity, the criterion of dreaming as the result of stipulation, relative to an arbitrary description. But he notes that even if we accept Malcolm's arbitrary definition of dreams as dependent on the criterion of dream reports, this would not render scientific dream research impossible. Telling a dream can be the criterion of dreaming even if dreams in fact consist of brain processes.

> If a dream is a brain process, then relatively to any purely neurological description of it[,] it is logically independent of the dream impression and the dream narration. But if we say that dreams are only those brain processes which cause dream-impressions, then if a person awakes without a dream-impression, he can be sure he did not dream. And if a dream is any brain process which causes a dream-impression, then if a person awakes with a dream-impression he can be sure that he did dream. (Landesman, 1977, p. 315)

Landesman notes that "this identification would be contingent and hence liable to empirical refutation" (p. 313); he does not claim that it is confirmed, only that it is intelligible. But if his analysis is correct, then even if one accepts the criterial analysis of dreaming, this would not be enough to reject scientific dream research as irrelevant. This concessive approach to the criterial view of dreaming thus is particularly damaging in that it suggests that Malcolm's view is untenably strong even by his own standards.

25. Elsewhere Malcolm weakens his claim about the inferential nature of dream recall by acknowledging that typically one knows one dreamt without any process of inference. Dream recall is only inferential in the weaker sense that "one should always be able to give grounds, upon demand, for saying it was a dream and not reality" (Malcolm, 1959, p. 47). This suggests that in claiming that dream recall is inferential, he is using the term *inferential* not in the

psychological sense of consciously engaging in inference but in a justificatory sense. Does this help rebut Ayer's and Chihara's criticisms? Unfortunately, Malcolm does not explain what he means by "giving grounds" in this context. It seems implausible, however, that people actually do justify their claims of having dreamt by anything other than their impression that they did, or indeed that any further justification is ever demanded of them. Typically we do not justify our dream reports by appealing to the bizarre and improbable occurrences in our dreams (indeed, not all dreams are bizarre; see chap. 10 for a discussion of realistic false awakenings): we just note that we dreamt. It does not seem, then, that Malcolm's weaker account of the inferential nature of dream recall can allay the worries raised by his critics.

26. Similarly, Brown (1957) points out that the persistence of anxiety after violent nightmares is a particularly good example of the continuity between dreaming and wakefulness. He notes that people sometimes have to be comforted and reassured after awakening, and children, in particular, have to be taught the difference between dream fear and real fear. Again, this familiar phenomenon would seem to be at odds with Malcolm's claims about the sharp distinction between real and dreamed sensations, thoughts, and emotions.

27. Actually, a theory very similar to Dennett's cassette theory was championed by the nineteenth-century dream researcher Alfred Maury (1861). He reported having a long and elaborate dream about the French Revolution, which slowly built up toward his own execution at the guillotine. At the moment the blade fell, he woke up, only to find that the headboard of his bed had come loose and fallen on his neck. Inspired by the apparent buildup of the dream to this dramatic climax, he surmised that the entire dream must have been triggered in a single instant by the impact of the headboard. Since then, theorists have toyed with the possibility of dreams occurring instantaneously or being constructed at the moment of awakening rather than being temporally extended experiences during sleep. This view was also defended by Binz (1878), Goblot (1896), and Hall (1981), who argued for the so-called Goblot hypothesis based on empirical evidence. For a discussion of the older literature, see Freud (1899/2003, pp. 39–47); for a discussion from the perspective of contemporary dream research, see Kramer (2007, pp. 22–24).

Dennett's cassette theory also has a precedent in philosophy. Using a less technological metaphor, Gregory argued that certain dreams are "psychical explosions" occurring at the moment of awakening:

> It is plain that when a stimulus simultaneously wakes the sleeper and originates a dream this dream passes rapidly during the moment of waking. The dream then represents a sort of psychical explosion. The unconsciousness of sleep corresponds to complete mental relaxation, and the dream occurs as the mind springs forward from relaxation into activity. (Gregory, 1916, p. 198)

Such dreams, according to Gregory, do not properly occur during sleep at all.

28. Dennett joins ranks with many other of Malcolm's critics, then, in denying the alleged uniqueness of dreaming and dream-related sentences. But whereas other authors took this to indicate that dream reports could, after all, be corroborated, for Dennett, the outcome is the opposite: for him, dreaming is just a particularly clear example of why the distinction between experience and memory trace is *generally* untenable.

29. It is noteworthy that in her reply to Dennett, Emmett construes these broad implications of the cassette theory as counting against Dennett, writing that "against Dennett I will argue that his sceptical refusal to accept dream recall as evidence that dreams occur during sleep would effectively undermine any appeal to memory as a source of evidence about the past" (Emmett, 1978, p. 455). Dennett (1979) happily rises to the challenge, because in a certain sense, this is exactly what he set out to do.

30. A third possibility is that the question of whether dreams fall inside or outside the boundaries of experience is itself misposed. Perhaps the concept of experience itself does not point to a natural kind:

> In the end, the concept of experience may not prove to differentiate any one thing of sufficient theoretical interest to warrant time spent in determining its boundaries. Were this to occur, the received view of dreams, like the lay view of experience in general, would not be so much disproved as rendered obsolete. (Dennett, 1976, p. 171)

31. This suspicion is strengthened by a case discussed by Schröder. He invites us to consider the case of a tribe called the Hyperboreans, who "conceive of waking-remembrances as not being preceded by a dream-experience during sleep" (Schröder, 1997, p. 16); the very idea, according to Schröder, strikes them as preposterous. Furthermore, they fully acknowledge all empirical findings on REM sleep and dreaming known to us. It is just that what we take to be the neural correlates of dreaming, they consider to be the neural correlates of unconscious memory-trace production during sleep. Schröder argues that even if these were the same neural mechanisms that are responsible for waking memory—even, that is, if Dennett's question about whether dream recall is the product of nonstandard memory insertion were answered negatively—this would not disconfirm the Hyperboreans' view. "No empirical discovery could settle the question of whether we do experience dreams while asleep or whether this is only our impression on awakening. . . . It is a merely grammatical question" (Schröder, 1997, p. 17). The parallel to Dennett's thought experiment is striking; for present purposes, however, its moral is simply that the contrast between the cassette theory and the received view is *neither* clearly empirical (as claimed by Dennett) *nor* merely grammatical (as claimed by Schröder); if it were, such opposing views on two deeply similar formulations of the same problem would not be possible.

32. This reading is further supported by Dennett's (1993) discussion of qualia inversion in "Quining Qualia." Here Dennett invites us to consider the case of "gradual post-operative recovery," in which

> we have somehow "surgically inverted" Chase's taste bud connections in the standard imaginary way: post-operatively, sugar tastes salty, salt tastes sour, etc. But suppose further—and this is as realistic a supposition as its denial—that Chase has subsequently compensated—as revealed by his behavior. He now *says* that the sugary substance we place on his tongue is sweet, and no longer favors gravy on his ice cream. Let us suppose the compensation is so thorough that on all behavioral and verbal tests his performance is indistinguishable from that of normal subjects—and from his own pre-surgical performance. (Dennett, 1993, p. 394)

Now, is the compensatory adjustment pre- or post-qualia? That is, is it only Chase's reactions that have changed, or have his taste experiences changed along with them? Though it might seem that empirical research could help decide the question, Dennett notes that even if the change could be located in an adjustment in the memory-accessing process, two conflicting interpretations would still be available:

> (I) Chase's current qualia are still abnormal, but thanks to the revision in his memory-accessing process, he has in effect adjusted his memories of how things used to taste, so he no longer notices any anomaly.
> (II) The memory-comparison step occurs just prior to the qualia phase in taste perception; thanks to the revision, it now *yields* the same old qualia for the same stimulation. (Dennett, 1993, p. 395)

As Dennett notes:

> The physiological facts will not in themselves shed any light on where in the stream of physiological process twixt tasting and telling to draw the line at which the putative qualia appear as properties of that phase of the process. (Dennett, 1993, p. 394)

Similarly, if advanced dream researchers managed to tamper with dream reports, it would be impossible to determine whether the tampering had affected the process of dream production, presentation, or recall. Not only is the subject unable to tell the difference, but so are the researchers who made the change in the first place. Dennett's conclusion in "Quining Qualia" is that the philosophical concept of qualia is not amenable to empirical research and ought to be abandoned. This, however, is no longer an empirical matter, but the fault lies with the concept itself. Similarly, the distinction between dream presentation and recall turns out, for Dennett, to elude empirical investigation, and for principled reasons. Despite appearances to the contrary, the cassette theory is not an empirical alternative to the received view but, if successful, reveals a flaw in the concept of experience itself.

2 A Short Introduction to Empirical Dream Research: History, Methodology, and Changing Theoretical Conceptions

1. Note, however, that in the *American Academy of Sleep Medicine Manual for Scoring Sleep and Associated Events*, published in 2007 (Iber et al., 2007), the distinction between stages 3 and 4 NREM sleep has been dropped.

2. I am grateful to Tore Nielsen for emphasizing this point and drawing my attention to the relevant studies.

3. Again, the problem in the background is what to make of these subjective sleep reports. Assuming that sleep is defined by behavioral and EEG criteria, subjects who deny that they slept can easily be mistaken. This is not, however, a problem for the transparency assumption, which makes no claims about the transparency of subjective sleep reports—in the absence of further information, I hesitate to call them *dream* reports—with respect to *sleep stage.* The deeper problem is whether sleep reports are to be taken at face value insofar as they describe the presence, say, of rudimentary awareness and conscious experience, and it is this latter kind of transparency that is implicit in the transparency assumption. The occurrence of phenomenal experience during behavioral and EEG sleep could easily lead subjects to claim that they were not, subjectively, asleep. Unless one wants to claim that sleep, as defined by behavioral and EEG criteria, is strictly incompatible with the occurrence of phenomenal experience—which would be to commit the armchair fallacy—I do not see on what basis such subjective sleep reports can be

denied. To the contrary, if one wants to investigate the question of how subjective sleep maps onto behavioral and EEG criteria of sleep, then gathering sleep reports—under the assumption of transparency with respect to the occurrence of *phenomenal experience* during sleep—seems the only way to go.

Note also that this question is not of merely theoretical interest. The existence of a mismatch between subjective sleep perception and objective measures of sleep in insomniacs, though well-known, continues to be poorly understood (Bianchi et al., 2013). For patients who suffer from sleep loss, being told that they in fact had perfectly normal sleep during a night in the sleep laboratory, while helpful to some, is perceived by others as an infuriating instance of not being taken seriously. As aptly pointed out by Greene, the gap between subjective and objective insomnia highlights the degree to which sleep resists uniform conceptualizations in a particularly drastic manner: what is defined as sleep according to EEG measurements simply may not *feel* like sleep to an insomniac and thus may be the cause of very real suffering. See Greene (2008, pp. 270–278) for discussion; see also Mercer et al. (2002) and Harvey & Tang (2012).

4. At thirty weeks, the human fetus spends nearly twenty-four hours a day in a preliminary form of REM sleep. Jouvet (1999, p. 58) calls this immature form of REM sleep "active sleep" because no variations in cortical activity can be detected at this early developmental stage. The question of whether dreaming is present at early stages of development is a matter of controversy. Some authors argue that the neurophysiological features of REM sleep precede the appearance of dream consciousness (Jouvet, 1999, p. 59; Foulkes, 1999). Others argue that the question of whether infants and animals dream is moot (Hobson, 2002, p. 73), or even suggest that prenatal REM sleep may involve protoconscious experience (Hobson, 2009; for details, see sec. 5.2.4). The answer will hinge on what one means by the concept of dreaming. Even if one wants to allow that REM sleep in neonates or animals is accompanied by phenomenal experience—and this is an open question—one could argue that this will be too different from the dreams experienced and reported by children and adults to pass under the name of dreaming.

5. Aside from the conceptual problem of how to define dreaming, methodological factors may also influence results on the similarities versus differences between NREM and REM dreams. Nielsen, Kuiken, Hoffmann, and Moffitt (2001) used a story grammar tool to parse dream reports into their constituent components and derive measures of story coherence. They found that REM reports were more frequently organized in a storylike manner than reports from stage 2 NREM sleep. However, they found no difference between the two types of reports when they were scored for the co-occurrence of different types of dream content, such as characters, actions, and scenes. This finding suggests that different positions regarding the REM—NREM debate might in part be artifacts of the different methodologies employed in analyzing and comparing dream reports from different sleep stages. Again, the debate on the similarities and differences between REM and NREM reports may require a more complex, multilevel answer than assumed by early researchers.

6. Schopenhauer defended an internal-stimulus hypothesis (Hobson, 1988, pp. 45–46) of dream generation almost half a century earlier, arguing that "there is nothing left but to assume that it [the brain] receives some purely physiological stimulus from the inside of the organism" (Schopenhauer 1851/1988, p. 234; translation mine; see sec. 5.1.3 for details).

7. Note that these different classifications of REM sleep also depend in part on whether one understands them in terms of external stimulus processing or internal brain activation. Wakelike levels of brain activity commend REM sleep as a light stage of sleep, whereas external stimulus processing and muscle tone, which are lowest in REM sleep, commend it as a deeper stage of sleep than NREM sleep. Indeed, as noted by Debru (2014), Dement and Kleitman embraced the former view, whereas Jouvet embraced the latter.

8. Nielsen (2010a) performed a database search on sleep- and dream-related publications since the late 1940s to compare the development of the fields of sleep and dream research. He found that the two fields not only differ as to the number of publications per year—with sleep-related publications outnumbering those on dreaming more than tenfold—but also underwent different patterns of development. Whereas sleep-related publications increased linearly since the late 1940s, dream-related publications peaked in the 1960s, followed by a twenty-five-year period of stagnation. Since the beginning of the twenty-first century, evidence points to a renewed vitality of dream research and an increase in the publication of dream-related articles.

9. This finding would have confirmed the psychoanalytic claim that dreaming has psychological value, but was contradicted by the fact that antidepressant drugs proved to be beneficial despite leading to REM suppression. This suggested the possibility that dreaming might not be essential to well-being and psychological health, after all. The relationship between dreaming and psychological well-being continues to be investigated by contemporary dream researchers (e.g., Hartmann, 1998; Cartwright, 2010). For details on the effects of sleepiness and sleep deprivation on alertness, behavior, cognitive functioning, and mood, see Minkel and Dinges (2009); Minkel et al. (2009); Drummond and McKenna (2009).

10. For a detailed discussion and critique from a theory of science perspective, see Dawson and Conduit (2011).

11. It may also be applicable to disorders of consciousness such as coma, vegetative state, or minimally conscious state (Laureys, 2014).

12. Solms admits, however, that this evidence only indirectly refutes the brain-stem hypothesis of dream generation proposed by Hobson and does not prove that dreaming can be preserved independently of the brain-stem control mechanism of REM sleep (Solms, 2000, p. 845). Moreover, Hobson and colleagues (2000) do not assert that the brain stem is sufficient to trigger dreaming, only that it is necessary.

13. I am greatly indebted to Tore Nielsen for drawing my attention to this point.

3 The Methodological Background Assumptions of Scientific Dream Research

1. This description of a typical sleep laboratory setup is based on my visits to Tore Nielsen's Dream and Nightmare Laboratory in Montreal and to the dream laboratory used by Antti Revonsuo's Consciousness Research Group in Turku, Finland. In Turku, I had the opportunity to spend a night as a participant in a pilot study conducted as part of an interdisciplinary project I was participating in within the European Platform for Life Sciences, Mind Sciences, and the

Humanities sponsored by the VolkswagenStiftung. This experience alerted me to the difficulties involved in dream research, both for researchers and for participants. It is a very strange feeling indeed to be asked to go to sleep in a laboratory bed, with wires attached to your head and well knowing that researchers—typically strangers, though in my case friends and colleagues—are going to observe your every movement throughout the night. Knowing that others are losing sleep to study yours places high demands on the participant, especially if upon being awakened you are unable to remember any dreams. This, unfortunately, happened to me: I was so uncomfortable that I first could not go to sleep and then following most awakenings sincerely denied having been asleep, even though the polysomnographic measurements said otherwise. Certainly my discomfort was exacerbated by the strange mixture of personal and professional interest and social dynamics and by the pressure I put on myself to be a "good" subject. But though my experience in Turku certainly was not representative, I do want to emphasize that sleep and dream research generally takes a much higher toll on both researchers and participants than most experiments conducted with waking participants, where hundreds of reports, via button press, can be collected in a matter of hours. I have also heard several researchers say, in conversation, that they were no longer willing to conduct laboratory research on dreaming because it was simply too taxing in the mid- to long term. It is important, then, to realize that this in itself is a tremendous obstacle toward conducting large and prolonged laboratory studies of dreaming.

2. Noreika et al. (2009) used an early night serial awakenings paradigm, waking their subjects repeatedly after brief periods of NREM sleep. They found that an average of eight reports from NREM sleep could be collected during a single three-hour session. A similar paradigm was used by Siclari et al. (2013, 2014). In their study, awakenings were conducted every fifteen to thirty minutes, and an average of eighteen reports were collected from each participant per night. These studies suggest that serial awakenings paradigms may be useful to maximize the number of dream reports gathered in laboratory studies.

3. According to Goodenough and Koulack's (1976) classical arousal-retrieval hypothesis, insufficient arousal can prevent the dream from being transferred into long-term memory and may also impede retrieval from short-term memory, thus interfering with the subject's ability to report a dream (see also Goodenough, 1991; Kramer, 2007, pp. 46–48). Sleep inertia, sometimes described as a transitional state of lowered arousal after awakening involving a carryover of cognitive processes characterizing sleep into wakefulness (Tassi & Muzet, 2000), or distractions upon awakening may even have a larger influence on recall than temporal distance from the dream (see also Schredl, Wittmann, Ciric, & Götz, 2003). Moving upon awakening may also render dream recall more difficult, and it is sometimes recommended that people interested in improving dream recall should remain in their sleeping position after awakening (LaBerge & Rheingold, 1990).

4. Instead of asking participants to report their dreams, researchers sometimes ask them to report whatever was going through their minds before awakening (e.g., Noreika et al., 2009). Otherwise participants might fail to report certain experiences because they do not consider them to be dreams.

5. This careful formulation underscores the importance of distinguishing between what is in fact reported and what could in principle have been reported. Obviously dreams are easily forgotten if

no special steps are taken to retain them upon awakening; and even dreams that are remembered may not be reported for a variety of reasons. So an important challenge for scientific dream research is to optimize the conditions under which dream reports are elicited, thus ensuring that participants actually do report anything that they can remember and minimizing the danger of forgetting dreams altogether. The underlying intuition behind the reportability assumption, then, is that ideal reporting conditions minimize the gap between what subjects do and what they could have reported. By the same token, an inability to report previous experience even given ideal conditions indicates that nothing was, in fact, experienced. See chapter 4 for details.

6. There are, of course, exceptions. As Chihara (1977) notes, there is the familiar phenomenon of remembering a dream, or additional aspects of a dream, later in the day, and often this seems to be due to a priming-like effect, that is, to an experience that suddenly triggers the belated recall of the dream. Unfortunately, however, dream content is so variable and unpredictable that it is impossible to systematically improve dream recall by priming subjects for the relevant memories (though Arkin et al.'s experiments with sleep talkers might be an exception; for discussion, see Arkin et al., 1966, 1970a, 1970b; and sec. 3.4). Moreover, it is impossible to investigate the influence of temporal distance on dream recall directly. Trindler and Kramer (1971) tried to investigate the extent to which temporal distance from the dream, but also position (early versus late-night dreams), salience, and dramatic intensity, influence dream reporting by comparing dream reports elicited throughout the night to reports of the same dreams given in the morning, following several hours of sleep (for discussion, see Kramer, 2007, pp. 41–42; for a similar study, see Baekeland & Lasky, 1968). But this type of comparison only works for dreams that have already been recalled and reported in the first place—which might arguably have a considerable influence on the subject's ability to report the dream several hours later, as well. Also, for studies investigating the sleep-stage or neural correlates of dreaming, such instances of belated dream recall are not helpful, because they would no longer allow researchers to relate the belated dream report to the corresponding sleep stage. So again the principle of temporal proximity is mainly intended as a pragmatic, methodological constraint.

7. Note that Aserinsky and Kleitman (1953, p. 273) in fact describe "27 interrogations *during* ocular motility" (italics mine)—a formulation that, assuming the subjects were not actually questioned while they were still asleep, underscores the importance of temporal proximity in a particularly emphatic manner.

8. This brief quotation, as pointed out to me by Tore Nielsen, also immediately raises a problem for the reportability assumption as formulated earlier. In particular, the reportability assumption seems to assume a dichotomous distinction between reports of dreaming and of nondreaming; but, of course, the situation is not so simple. In particular, there are cases in which subjects report the feeling of having dreamt but are unable to describe any details. In one study investigating reports from NREM sleep, more than a third of reports described such white dreams, and the percentage of white dream reports was actually higher than that of reports of dreaming and of nondreaming, respectively (Noreika et al., 2009). Don't such white dream reports contradict the reportability assumption, because they describe the inability to report experiences? To the contrary, and in keeping with the transparency assumption, I suggest that reports of white dreaming should be taken at face value. If subjects claim to have had a detailed and vivid dream that they

are now unable to describe, then the safest assumption is that they really had a detailed and vivid dream but have forgotten all or most of it. Indeed, the reportability assumption should not be understood as demanding that all details of an experience can be reported upon awakening; it says only that the occurrence of experience during sleep can be reported. And this is exactly what subjects reporting white dreams are doing: they are reporting the occurrence of experience during sleep. A stronger reading of the reportability assumption would be to say that perhaps in such cases, the reporting conditions were suboptimal—perhaps the details of the now forgotten dream could have been reported under other conditions. Even so, the main point of the reportability assumption is not that everything that is experienced in sleep can in fact be reported but that, principally, assuming the existence of unreportable experiences during sleep does not make sense. White dream reports can be said to fulfill this condition to the extent that subjects report having experienced a dream—and this, for present purposes, is enough.

However, another interpretation of white dream reports is possible: perhaps the vague impression of having dreamt, coupled with an inability to remember or report any further details, is simply a report of indeterminate and hence indescribable experiences (cf. my discussion of phenomenal indeterminacy in chap. 7). If such a report is taken at face value, then perhaps the inability to report any specific experiences is not due to a failure in recall but reflects the phenomenal character of the experience itself. Conscious experience, in most accounts, is not an all-or-nothing affair (cf. Metzinger, 2003), so why not assume that the same is true for sleep experiences? To my knowledge, this possibility has not been investigated—but given the high percentage of white dream reports in certain sleep stages, why not allow that at least a subgroup of white dreams might fall somewhere on a continuum between dreaming and nondreaming? In fact, the Noreika et al. (2009) study scored dream reports on a scale distinguishing different degrees of complexity (ranging from isolated percepts to detailed narratives), finding that most reports from NREM sleep were scored as having a low degree of complexity; perhaps, then, the high percentage of white dream reports in their study indicates experiences just below the cutoff line of what can be reported and hence scored as a dream. Of course, all of this is highly speculative—but it might suggest, at least, that paying more attention to the exact wording of white dream reports and asking whether different types of white dream reports can be distinguished (i.e., the feeling of having had a detailed and complex dream that is now forgotten, versus the feeling of having had vague and hence indescribable experiences) might in itself be a promising and theoretically fruitful research project.

9. I am thankful to Valdas Noreika for pointing this out to me.

10. All of this also raises the question of how to individuate a dream. At least three possibilities come to mind: First, what counts as a single dream could be determined by the duration of the corresponding sleep phase, with sleep-stage transitions marking the beginning and end of individual dreams. But sleep-stage scoring is a matter of convention, so this seems rather unsatisfying and arbitrary. Second, proponents of the hypothesis that we dream continuously throughout sleep (cf. chap. 4) could argue that reportedly dreamless portions of sleep are merely forgotten fragments of the same dream—a possible consequence being that we not only dream throughout sleep but also have one long dream per night. This, however, seems to contradict the transparency assumption. Third, especially for early morning REM periods, which can last for up to fifty

minutes (Hobson et al., 2000), one could also argue that several dreams can be experienced even within a single REM period—possibly interrupted by brief periods of dreamless sleep or by a particularly stark cut in the dream narrative. This would seem to be the consequence of taking the transparency assumption seriously. Indeed, my own intuition is that one dream is distinguished from another if there seems, upon awakening, to be a memory gap and the impression of time having passed between the two. But this is just an intuition, and exactly what people mean when they describe "last night's dream" may not be the same for everyone. Perhaps the concept of dreaming is simply underdefined with respect to matters of individuation, and one should simply leave it at that; or perhaps one could conduct an empirical study to determine what people actually mean when they are describing a "single" dream (and whether what they mean is consistent within and across subjects); or yet again, researchers could try to say more clearly what they mean by "one" dream in a given study and hope that their subjects henceforth follow suit. To some, the very question of how to individuate a dream and distinguish it from another may appear absurd and purely theoretical; but a deeper problem lurks in the background: if it is impossible to say what "one" dream is, and to say clearly when it begins, ends, or merges into another, then one also cannot study the narrative structure, internal coherence versus bizarreness, duration, or sleep-stage and neural correlates of individual dreams. This would make the project of dream research much more messy and its target more elusive than commonly assumed. So once more, this shows how closely methodological challenges and conceptual problems about sleep and dreaming are intertwined.

11. For instance, researchers have recently been using retrospective reports to investigate anesthesia experience (e.g., Errando et al., 2008; Noreika et al., 2011), and determining whether patients in a vegetative state maintain residual perception and cognition continues to be an important (though notoriously difficult) research goal. Important advances have been made with patients in a minimally conscious state, showing that they are able to voluntarily engage in specific mental-imagery tasks in response to instructions, thus suggesting that conscious experience is preserved in some of these patients (Owen et al., 2006; Laureys et al., 2009; Gosseries et al., 2011). In all these cases, researchers are acknowledging the possibility of conscious experience in the absence of wakelike levels of brain activation. Comparing coma, anesthesia, vegetative state, and minimally conscious states to the different stages of sleep may ultimately be a helpful strategy for understanding the neural correlates of consciousness (see Laureys et al., 2009).

12. A more general problem is that the temporal resolution of PET is poorer than that of EEG, meaning that fast changes may not be detectable using this methodology. See Baars and Gage (2010, pp. 568–570).

13. Nofzinger et al. (1997) suggest that part of this methodological problem can be avoided through the use of FDG PET methodology, which allows subjects to sleep in the comparatively natural environment of the sleep laboratory rather than in the noisy scanner.

14. However, Kussé et al. (2010) also consider the difficulty of obtaining reliable dream reports to be one of the main limitations in this field.

15. This may be especially helpful in studying children's dreams, as described by Mallon (2002). Similarly, Cartwright (2010, p. 19) notes that children who are unable to report dreams verbally

may nonetheless be able to provide dream drawings. Consequently, the type of dream report used in a given study may influence not only the extent to which different aspects of the dream are reported but also the overall number of reported dreams.

16. This is what Valdas Noreika suggested doing in an unpublished study we conducted within the European Platform for Life Sciences, Mind Sciences, and the Humanities.

17. I borrow this term from Sascha Fink.

18. Studies asking subjects to report their most recent or earliest remembered dreams are an exception in that they target specific dreams while violating the principle of temporal proximity. Therefore, in my view, they fall somewhere between dream reporting and merely stating opinions about dreaming. At best, they are informative about particularly memorable dreams, but it is also possible that the dreams reported in such studies are strongly biased by, and saturated with, personal views about dreaming.

19. Note that this is another way in which opinions about dreaming differ from dream reports: the trustworthiness of opinions about dreaming is something for which independent evidence is to be had, and this independent evidence is provided, unsurprisingly, by dream reports.

20. Animals exhibit similar forms of specific, presumably oneiric or dream-related, activity. Michel Jouvet and colleagues managed to obliterate REM-sleep-related muscular atonia in cats by causing lesions in the area of the brain stem (Jouvet & François, 1959; Jouvet, 1999, p. 82). Shortly after the onset of REM sleep, the cats opened their eyes and raised their heads, "looking" upward and apparently watching an imaginary object. They then exhibited stereotyped behavioral patterns, behaving as if they were stalking mice, fighting imaginary enemies, or grooming themselves. While behaviorally awake, Jouvet's cats apparently remained oblivious to their environment and did not respond to visual stimuli even though their eyes were open. For instance, they would not react when offered a piece of meat, even if they were hungry. A deeper question, however, is whether this type of behavior can properly be considered as *dream*-enactment behavior. Before one can even gesture toward an answer, two points would have to be clarified. First, in the absence of a report, the association with phenomenal experience is unclear; and before one could generalize from human dream-enactment behavior to the existence of such behavior in animals, more would have to be known about the exact association between sleep behavior in humans and dream experience. Second, even if the question of experience associated with sleep behavior in animals could be answered affirmatively, there would be a separate question as to whether this type of experience should be regarded as *dreaming*. Perhaps such experiences in cats, if present, are simply too different from human dreams to be regarded as dreaming—or perhaps, there are good reasons to think that the structure of sleep experience in animals is indeed relevantly similar to that in humans. The answer will depend, in part, on how to best characterize human dreams.

21. A similar increase in aggressive dreams and dreams of animals has been observed in patients with Parkinson's disease; see Bugalho and Paiva (2011); Paiva et al. (2011). Patients with both Parkinson's disease and RBD may also be particularly prone to waking hallucinations; see Manni et al. (2011).

22. More examples can be found in Arkin et al. (1966, 1970a). A similar case of a close but not exact correspondence from REM sleep is the imperfect match between a peacock tree described in sleep speech and a butterfly tree described in the retrospective report; see Arkin et al. (1970a, p. 384); there are many more.

23. It is also a matter of philosophical controversy whether seemingly rational dream thoughts and dream beliefs, as occasionally occur in lucid dreams, count as actual thoughts and beliefs; see section 9.1 for details.

24. Although these experiments show that there is a correspondence between EOG measures of eye movements and gaze shifts within the dream, this does not mean that *all* eye saccades measured during REM sleep dreaming are related to visual exploration and intentional scanning. Research has shown that in REM sleep, the eye saccades are driven, to a considerable degree, by low-level, automatic control mechanisms related to reflexive orienting (Herman, 2000, p. 950). A related point is that the ability to control one's eye movements during a lucid dream is actually quite limited: if dreamers fix their gaze on something within the dream, this disrupts their rapid eye movements and causes awakening. In the literature on lucid dreaming, this is a well-known technique for ending a dream, resulting in either actual or false awakening (LaBerge & Rheingold, 1990, pp. 146–147; Tholey & Utecht, 1995/2000, pp. 59–60).

25. By contrast, the examples of dream enactment discussed in the previous section should generally be regarded as behaviors rather than as actions, because even within the dream state, they likely lack the type of intentionality and goal orientation that most would agree is required for action in an interesting sense. But what about hypothetical lucid RBD-enactment dreams? Imagine the case of an RBD patient prone to lucid dreaming who not only realizes that he is currently dreaming but also decides to perform a little experiment in his dream—say, jumping up and down and flapping his arms to see if he can fly. I think, in this case, one could consider such behavior an action insofar as it resulted from critical reflection and volitional control on the part of the dreamer. By contrast, the outward execution of the respective movements would not be intentional—let's just assume, for the sake of argument, that our lucid RBD dreamer was not aware of actually jumping up and down and flapping his arms but believed himself to be doing so only in his lucid dream (or alternatively that he was not lucid enough to realize he was actually enacting his dream movements with his physical body). As in most cases of RBD-enactment dreams, this behavior would also be vastly inappropriate to the sleeping subject's actual environment. But this leads to the somewhat strange consequence that lucid RBD dream-enactment behavior, insofar as it was intentionally directed at events within the dream, could count as action, but insofar as the behavior took place in the real world, with potentially detrimental effects to our imaginary dreamer (and his bed partner), the externally observable counterparts of these actions would be mere behaviors. By contrast, if our lucid RBD patient knew he was outwardly enacting his dream movements, we might be more prone to consider them as actions in the real world, as well. This, in turn, also leads to an important insight about the concept of action: what turns a given behavior into an action is not just its intentionality but the fact that it is intentional *relative to a given context or behavioral space*; and as the RBD example shows, a behavior that is intentional in one context, that is, the internal behavioral space of the dream world, does not necessarily fulfill the requirements for being an action in another context, such as the

real-world environment, in which the externally observable counterparts of these dream actions are unfolding.

Though the example, so far, is purely hypothetical, the questions it raises are not. Sleep- (and dream-) related violence raises not just theoretical but also legal problems (cf. Mahowald et al., 2005), and the same is true for the assessment of actions performed in virtual realities or video games and their (nonintentional) real-world consequences (cf. Lastowka, 2010; I am thankful to Michael Madary for pointing this out to me). Moreover, the new field of robot ethics deals with a similar set of questions. Metzinger (2013b, p. 260) envisions a case of "anarchic robot syndrome," in which a man is lying in a scanner, controlling a distant robot and feeling fully embodied in it. Through the robot's eyes, the man sees his ex-wife's new husband enter the room—and before he can inhibit the motor imagery arising in association with a violent fantasy, the robot has responded by killing the ex-wife's new husband. Metzinger uses this example to introduce the principle of veto autonomy: in such cases, moral responsibility depends on the ability to inhibit the outward execution of purely mental actions. The example of lucid RBD-enactment dreams, by contrast, suggests that in addition, a number of more basic conditions must be fulfilled; plausible candidates for assessing responsibility for sleep behavior could be (1) the ability to control one's behavior within the dream, (2) realizing that one was now dreaming (as in a lucid dream), and (3) realizing that the actions performed within one's lucid dream were being enacted by the sleeping body. I cannot fully resolve the question of moral responsibility in dreams here but only want to point out that the existence of dream-enactment behavior arising from lucid dreams may offer a new perspective on this historical problem: even if one assumes, as most philosophers since Augustine (400/1991) have, that nonlucid dream actions (including, presumably, nonlucid enactment dreams) do not fulfill the conditions for moral responsibility (cf. Driver, 2007), the situation might be different for a subgroup of overt actions arising from lucid dreams. For an anecdotal description of an RBD patient using lucid enactment dreams to have conversations with his wife, see Mahowald et al. (2011).

26. Given the relationship between the complexity of dreams and retrospective estimates of dream duration suggested by previous studies (cf. sec. 3.3), it would be interesting to know whether the lucid dreamers in this study would also have estimated that their dream actions took longer than they would have in wakefulness, and whether this effect would have been related to the varying degrees of complexity of the specific tasks (i.e., whether it would have been strongest for the gymnastics routine).

27. Pagel (2003) argues that a small percentage of the general population (0.38 percent) may also qualify as nondreamers (and not just poor dream recallers), independently of lesions, any signs of cognitive impairment, or changes in sleep architecture. If correct, this would suggest that there is a group of subjects who cannot properly be said to have lost the ability to dream, as they have never been able to remember any dreamlike experiences at all. Though Pagel investigated these subjects in the laboratory, however, they were only asked for mentation reports following morning awakenings, so his results should be treated with caution.

28. A nice correspondence between dreaming and wakefulness was observed in a case study published by Doricchi at al. (2007). They investigated changes in dreaming and eye movements during REM sleep in a single patient suffering from unilateral spatial neglect. They found that a

particular type of eye movement involving alternating rightward slow and leftward fast phases selectively disappeared both during wakefulness and during REM sleep; this directional asymmetry fit in well with dream reports of watching a train traveling in a leftward direction and observing the landscape flowing leftward while looking out of the window of a moving car. Because of the selective loss of this type of eye movement during REM sleep, this study also provided evidence of morphologically distinct and dissociable types of REMs, leading the authors to suggest a new classification system. Moreover, based on the comparison of eye movements during REM sleep and wakefulness in their patient, the authors suggest that "scanning" the dream scene is not equivalent to self-paced visual exploration in wakefulness, but dreaming and wakefulness share phylogenetically older, reflexive oculomotor mechanisms.

29. It is noteworthy, in this context, that in Sophie Schwartz's work, the centrality of dream reports for understanding not just the phenomenology but also the neural correlates of dreaming takes on a very special form: she dedicated her dissertation to the analysis of her own dream diary, thus following the tradition of many early dream researchers (Schwartz, 1999; cf. chap. 2). An important goal of her project was to use dream reports "to explore oneiric reality" (p. 17; my translation), or the reality of dream experience, itself. In Schwartz (2004), she presents data from the analysis of 1,770 of her own dream reports, collected over fifteen months. The validity of this approach is supported by the results from the analysis of a large corpus of dream reports from one hundred students. This demonstrates that for dream research, the use of the researcher's own dream reports may be not only courageous but also scientifically viable.

30. Two particularly obvious and well-known examples of dream-related science fiction fueling philosophical speculation are Grau's (2005) *Philosophers Explore "The Matrix,"* containing a number of excellent articles by well-known philosophers, and Johnson's (2012) *"Inception" and Philosophy*.

31. For a similar study using high density EEG during REM sleep to predict associated dream contents such as faces, places, movement, and speech, see Siclari et al. (2014).

4 Antiskepticism about Dreaming and Dream Reporting: From Default Assumption to Theoretical Justification

1. It is also close to Peirce's (1958, 5.189) standard definition of abduction:

(1) The surprising fact, *C*, is observed.
(2) But if *A* were true, *C* would be a matter of course.
(3) Hence there is reason to suspect that *A* is true.

2. A related challenge, as Lipton (2004, 2007) points out, is that in some cases, explanatory virtues may fail to map onto truth. Truth may not be the best explanation of the predictive success of a theory, and so on for the other explanatory virtues listed here. Thus, using essentially pragmatic and explanatory considerations as a guide to likeliness and truth may require an equally pragmatic account of truth.

3. This is not to say that using IBE as a response to skepticism is without problems or to rule out alternative views. A prominent alternative is Bayesian confirmation theory. Whereas IBE emphasizes explanatory considerations, Bayesianism focuses on the probability of hypotheses and the degrees of belief assigned to them both initially and conditional on subsequent evidence. Roughly, IBE considers a hypothesis as true if it explains the available evidence better than another (or, ideally, than any other); by contrast, Bayesianism says that a hypothesis is confirmed if its prior probability, conditional on a given piece of evidence, is greater than its prior unconditional probability (for a general introduction to Bayesian epistemology, see Talbott, 2011). In a Bayesian approach, explanatory considerations of the type raised by IBE might even be wholly irrelevant to confirmation. I cannot resolve this issue or even begin to go into the details here, but I do want to point out that several authors have argued that IBE and Bayesianism may be compatible and that IBE may even be a much-needed supplement to Bayesianism (Lipton, 2004; Weisberg, 2009; for discussion and further references, see Douven, 2011). For instance, Lipton suggests that the explanatory considerations raised by IBE "might be part of the way people 'realise' the Bayesian transformations" (Lipton, cited in Ladyman, 2005, p. 357; see also the other contributions there).

I think there are nonetheless reasons for thinking that IBE is particularly, though perhaps not uniquely, well poised to offer the kind of framework needed in the present context. One is that its relevance for skeptical arguments has been more thoroughly explored, and thus the framework provided by IBE provides a particularly clear connection to both classical dream skepticism and, as I hope to show, skepticism about dreaming. Of course, there are exceptions; for a recent proposal on how a Bayesian framework might be applied to external world skepticism, see Douven, 2009, who also discusses its advantages over IBE. Even so, I suspect that IBE may be better tailored to skepticism about dreaming for another reason, as well. As will become clear in the next section, skepticism about dreaming sometimes explicitly appeals to considerations of explanatory goodness, and even when it does not, it can fruitfully be reconstructed in this manner. Accordingly, assessing skepticism about dreaming within an IBE framework enables me to meet it on its own terms.

4. In *Our Knowledge of the External World,* Russell argues that the other-minds problem can also be solved by IBE. He notes that our belief in other minds gains some support from the argument from analogy. However:

> At the same time, it is a hypothesis which systematizes a vast body of facts and never leads to any consequences which there is reason to think false. There is therefore nothing to be said against its truth, and good reason to use it as a working hypothesis. When once it is admitted, it enables us to extend our knowledge of the sensible world by testimony. . . . In actual fact, whatever we may try to think as philosophers, we cannot help believing in the minds of other people, so that the question whether our belief is justified has a merely speculative interest. And if it is justified, then there is no further difficulty of principle in the vast extension of our knowledge, beyond our own private data, which we find in science and common sense. (Russell, 1914/2009, pp. 77–78)

Believing in the minds of other people renders our sensory impressions of them less surprising than, say, the hypothesis that they are zombies; it enables predictions regarding their future behavior, and it is productive by enabling an expansion of our knowledge. It is also a lovely explanation in virtue of its simplicity—that is, by allowing the analogy to our own minds—and by being instinctive, or intuitively lovely.

5. Note that at least initially, *reporting* cannot yet be used as a success word, but only in a way that is neutral as to whether this counts as genuine or only seeming reporting behavior. If it turned out that an alternative theory—say, the cassette theory—better explained dream reporting behavior than the received view, then what we call dream reports would not be reports of dreams understood as conscious experiences during sleep at all.

6. One might think that it would be enough to insert the experiential assumption here, but it is not: logically, dreams could be experiences *without* dream reports transparently showing this to be the case. In fact, this is precisely the possibility envisioned by Dennett (cf. sec. 1.3.4), namely, that even if dreams are experiences, this fact is revealed not by dream reports but by scientific findings on brain activation during sleep. By contrast, the transparency assumption already implicitly contains the experiential assumption: if dream reports are transparent with respect to previous experiences during sleep, then denying that dreams are experiences during sleep makes no sense. For this reason, I focus on the transparency assumption here and throughout this section.

7. This is not to deny that in other cases, IBE may be a psychologically plausible model of why we come to adopt a hypothesis in the first place. As Lipton notes:

> Inference to the Best Explanation seems itself to be a relatively lovely explanation of our inductive practices. It gives a natural description of familiar aspects of our inferential procedures. The simplest reason for this is that we are often aware that we are inferring an explanation of the evidence; but there is more to it. We are also often aware of making an inferential choice between competing explanations. (Lipton, 2004, p. 64)

My only contention is that in the cases discussed here—skepticism about the external world and skepticism about dreaming—we are, at least initially, naive realists. In fact, I think that the intuitive loveliness of the realist view is what makes it so hard to entertain skepticism as a live option. As Descartes notes at the end of the *First Meditation*, preventing himself from slipping back into his former, habitual beliefs about the external world and his own existence requires effort. Moreover, the claim that the transparency assumption of dream reporting and the experiential hypothesis of dreaming are intuitive is not to be confused with the claim that the occurrence of dreams in REM sleep, for instance, is intuitive—as we have seen, the precise localization of dreams in sleep and different sleep phases has changed considerably throughout the history of theorizing about dreaming and continues to be affected by new findings, for instance, on lucid dreams (see chaps. 2 and 3).

8. This strategy of acknowledging the pull of intuitions while treating them with due caution is also, I think, in keeping with Lipton's (2004, p. 206) own remarks on intuitions.

9. By contrast, note that it is not at all clear that IBE can be meaningfully applied to Malcolm's purely conceptually motivated type of skepticism about dreaming—or that it needs to be. Recall that his analysis of dreaming was in fact a war of proxy on more general issues concerning verification, the meaning and acquisition of concepts, and the relationship between philosophy and scientific research (cf. section 1.3). Malcolm's analysis of dreaming was a reaction to what he perceived as an untenable form of competition, with observational data gathered in sleep laboratories threatening to undermine the authority of conceptual analysis. Dreaming as a heretofore scientifically unexplored phenomenon, the most private (after all, according to Malcolm, *logically*

private) of all experiences, was the ideal bastion from which to conduct this war of proxy. Consequently, Malcolm did not suggest an alternative explanation of dream reporting to be evaluated on the basis of its explanatory goodness—he legislated one. Hence, in order to assess Malcolm's theory on its own terms, explanatory considerations may be inadequate; instead, the battle over Malcolm's purely conceptually motivated brand of skepticism is best fought on the level of meaning, concepts, verification, and philosophical methodology, and it is on this level that its weaknesses become most apparent.

10. Note, however, that this would not apply to versions of the cassette theory in which the composition of dream narratives occurs smoothly and in a linear, beginning-to-end fashion throughout REM sleep, but only to those that posit an instantaneous composition process at the moment of awakening or something similar (cf. Dennett, 1991, p. 158). As Dennett later seems to prefer the former version of the cassette theory (cf. Dennett, 1991, p. 160), it would seem that he is once more left without a satisfying account of precognitive dreams.

11. Rather than backing this suspicion up with an independent argument, I want to draw readers' attention to a thought experiment that is similar to Dennett's cassette theory but predates it by five years. Hunter (1971) imagines a situation in which the "neurological correlate" of waking experience occurs during sleep and in which the same mechanism that leads to the production of waking memories also leads to the production of dream recall—but importantly, *without* giving rise to experience during sleep. He then compares the explanatory loveliness of the received view to that of this rival theory and assesses their respective advantages and disadvantages. He concludes that the "counter-argument so far makes it moderately clear that the neurological correlate thesis does not account for the known facts or reasonable suppositions, while the common sense view does" (Hunter, 1971, p. 74). Hunter notes that the commonsense view better explains a number of observations about dreaming that otherwise remain mysterious—for instance, why we react to dreams, after awakening, with delight or fear, or why we naturally and noninferentially use the same language to describe dream and waking events. And importantly, this ability to offer a better explanation suggests that the received view is also theoretically justified.

12. A related problem is that even if we hit on a particularly lovely explanation, there is no guarantee that its loveliness will be matched by the loveliness of the actual world. As Lipton puts it, "Why should we suppose that ours is the loveliest of all possible worlds?" (Lipton, 2000, p. 11; for discussion, see Lipton, 2004).

13. There is, of course, always another alternative. Aside from arguing that a rival theory that better explains dream reporting might yet come along—and this possibility can never be completely excluded—the determined skeptic might dig his heels in and reject IBE as a viable anti-skeptical strategy altogether. The skeptic might continue to insist on independent evidence confirming the transparency assumption and reject IBE as too weak and unreliable. The consequences of this move are nicely illustrated by Squires (1973), who pursues something very much like this strategy. Unlike Malcolm, Squires thinks that the received view of dreaming cannot be ruled out based on a priori conceptual considerations: he thinks it can, and indeed should, be ruled out *after* investigation (Squires, 1973, p. 248). Note that Squires does not argue that the

received view is empirically false; rather, he argues that it is best rejected because an examination of empirical findings shows that no independent evidence of dream experience and the trustworthiness of dream reports is forthcoming. In this way, his is an intermediate position between Malcolm's purely analytic, armchair approach and Dennett's reconstruction of the problem of dream experience as an open empirical question. Squires writes:

> The attraction of the inner event account is that it offers an explanation of our waking impressions as recollections of our night life. The alternative, it may seem, is to leave our impressions unexplained. But unless there is independent confirmation of the framework in which an explanation operates, it is an intellectual encumbrance. (Squires, 1973, p. 249)

I find this passage interesting because it suggests that Squires and I can actually agree on a number of relevant points. We agree, for instance, that the received view is lovelier than the alternative in that it offers an explanation of dream reporting behavior, and that without it, this behavior would seem mysterious. We also agree that no independent confirmation of this view is forthcoming. Unless the transparency of dream reports is assumed from the outset, the problem is insoluble. However, the conclusions we draw from this assessment are diametrically opposed: while for Squires, explanatory loveliness without independent confirmation is insufficient to support the received view, for me, the demand for independent confirmation is misleading, and explanatory virtue is the best that an account of dream reporting can aspire to. Squires concludes on an almost poetic note:

> But in the dark hours of the morning, when the perfumes of the night are in one's nostrils or the flesh creeps at the thought of terrors past, how absurd it seems to deny that something is remembered! It is the magic of this primitive certainty that sustains the scotched myths; not laboratory evidence, reflection on unusual occurrences or even grammatical illusions. In the cold light of day, perhaps we should also wake up to the fact that this spellbinding certainty only shows what it shows: that we sometimes wake up with apparent memories. (Squires, 1973, p. 259)

To which I would respond that dreaming is quite possibly a case where primitive certainty (or intuitive loveliness) withstands the cold light of day (or rather: an evaluation of its explanatory loveliness). What, then, is Squires's positive account of dream reporting? In a later paper, he writes:

> Dreams (what we relate, the content of our false recollections) do not occur in sleep (or at any other time), although they are formed in sleep. So if dreaming refers to dream formation then we dream in sleep, just as we digest our food then. But if dreaming is meant to be a "conscious process," like reflecting or visualizing or hallucinating, then we do not dream in sleep or indeed at all. (Squires, 1995, p. 91)

His end point, then, is close to the cassette theory's vision of a nonphenomenal composition process of dream narratives; just that Squires, unlike Dennett, who remains neutral on this matter, prefers this theory. At the least, I hope to have shown that this is an increasingly costly and unattractive preference. Perhaps settling for a preliminary explanation is still better than being left with no explanation at all.

14. Note that a further ambiguity concerns the scope of Schwitzgebel's skeptical argument. While all his specific arguments focus on dream color, he sometimes seems to suggest that he distrusts not only beliefs about dream color but beliefs about the phenomenology of dreaming in general. It is unclear, however, whether doubts about opinions about dream color provide a sufficient reason for doubting opinions about dreaming in general. For present purposes, I focus on

the more limited claim; to support the broader skeptical claim about dreaming, an independent argument would be needed.

15. Indeed, elsewhere he suggests that even concurrent judgments about the phenomenology of dreaming can be erroneous:

> But in dreams we make badly incoherent judgments, or at least very stupid ones. . . . Indeed, it seems to me not entirely preposterous to suppose that we have no color experiences at all in our sleep, or have them only rarely, and that our judgments about the colors of dream objects are . . . purely creative fiction unsupported by any distinctive phenomenology. If so, the corresponding judgments about the coloration of our *experiences* of those dream objects will be equally unsupported. (Schwitzgebel, 2011c, p. 124)

16. Note that the situation is similar to Descartes's predicament in the *First Meditation*: to undermine his sensory-based beliefs in general, he has to show not only that he is occasionally deceived by the senses (as in familiar cases of sensory illusions), but that even in a best-case scenario of sensory perception, deception cannot be ruled out (see sec. 1.1).

17. Confidence in such hypothetical findings is bolstered by a recent study showing that visual imagery is constrained by features of visual experience in a highly specific manner. The study investigated participants' ability to imagine individuals known either through black-and-white or color media, finding that visual images of individuals from the black-and-white media era, such as Albert Einstein, were also experienced as less colored (Pearson & Hollings, 2013). This suggests that mental imagery is responsive to waking experience in a highly selective manner that is specific to the objects or persons that are imagined.

18. I think a similarly constructive reading can be applied to Rosen's (2013) *narrative fabrication thesis* of dreaming. She reviews Malcolm's and Dennett's skeptical arguments and concludes that their rejection of dream experience fails. However, while rejecting skepticism about dreaming, she makes an empirically informed case for skepticism about dream reporting. In doing so, she oscillates between a stronger and a weaker skeptical argument. In the stronger version, dream reports are *generally* untrustworthy because their accuracy cannot be verified:

> Although it may be possible for some dreams to be accurately reported, given our current technology and methods it is very difficult to discern whether any particular dream report is an accurate representation of the dream content, and there is no fail-safe method to ensure accuracy. (Rosen, 2013, p. 1)

As we have seen, however, the demand for conclusive verification or for a fail-safe method for ensuring the accuracy of dream reports is too strong. Thus I think a weaker reading of Rosen's argument is more constructive. In this view:

> Dream reports are often confabulated so that the narrative reported is not accurate of the dream experience itself. The main reasons this occurs . . . are that dreams can be bizarre which leads to rationalization of their content, and memory of dreams is poor, which can cause confabulation of dream events. (Rosen, 2013, p. 5)

While Rosen thinks that "we cannot know to what extent a dream has been rationalized even if the dreamer is woken directly from REM sleep" (Rosen, 2013, p. 7), she allows that "certain conditions will increase the probability of an accurate report" (p. 7). For instance, she allows that lucid dream reports may be "more accurate than nonlucid dream reports due to increased memory and rational capacities" (p. 11). She also suggests that the narrative fabrication thesis could be verified, for instance, by using sophisticated signaling in lucid dreams or through advanced neuroimaging technology. While I think that regarding such possibilities as *verifying* the accuracy of dream reports is problematic, I fully agree that they may be used for improving the conditions of dream reporting.

In keeping with this weaker reading, much of the evidence she uses is derived from comparing different types of dream reports in terms of their *relative* trustworthiness (e.g., dream reports given in the morning as compared to those given following REM sleep awakenings, as investigated by Foulkes, 1999; the effect of rehearsal on dream recall, as investigated by Horton, 2011a, 2011b; or the contrast between dream reports and sleep talking, as investigated by Arkin, 1966; Arkin et al., 1966, 1970a, 1970b; cf. sec. 3.4). To defend the more general claim that even reports given immediately after awakening might be fabricated, she argues that similar rationalization processes affect waking-memory reports, and memory is always reconstructive rather than directly replaying remembered events. However, even here a more moderate reading is possible: if dream reports are indeed similar to waking reports in being affected by rationalization, for instance, then this can once more help determine (and henceforth avoid) the conditions under which such rationalization is particularly likely to occur; moreover, if memory is always reconstructive, this does not thereby show, as she rightly points out, that all memories are false, and such a comparative analysis can once more be used for improving reporting conditions.

Finally, Rosen points out, quite correctly, that dreamers may knowingly rationalize or censor their dream reports or may admit to being unable to recall their dreams in detail. But again, this is not a principled problem for the transparency assumption: while there is a very real problem of how to distinguish sincere and willfully fabricated dream reports, the problem is a practical one, not one that raises principled reasons for distrusting dream reports.

In sum, to the extent that Rosen succeeds in casting doubt on the *relative* trustworthiness of dream reports, as I think she does quite convincingly, she fails in defending a stronger and principled form of skepticism about dream reporting. Her central claim, then, is best read as saying that as a matter of empirical fact, we often fabricate and rationalize our dream reports, sometimes deliberately, sometimes not. But this is not at all the same as casting principled doubt on dream reporting under ideal reporting conditions.

19. Of course, subtle differences between these positions exist. For instance, Squires (1973, 1995) thinks that the question of dream experience should be ruled out not based on purely conceptual considerations but because scientific dream research is unable to produce the kind of evidence that could resolve the question. So while his starting point is different from Malcolm's, his end point is similar: dream reports are the product of *apparent* memories, and it is misguided to think that dreams are experiences during sleep or that their temporal location and duration during sleep could be discovered. Similarly, McFee argues that because dream reports are the only basis for the attribution of dreams, they are not memory reports, nor can dreams be given an exact temporal location and duration during sleep. While he explicitly acknowledges the centrality of dreaming for philosophy of mind, noting that "accounts of the mind will be inadequate to the degree that they cannot accommodate dreaming" (McFee, 1994, p. 95), he argues that taking a commonsense view on dream reporting should "combat an account of the mind that places undue stress on its physical underpinnings" (p. 113). Finally, Schröder (1997) follows Malcolm in claiming that the location of dreams during sleep is determined by grammar and is not

an empirical fact, but claims that this is not enough to rule out the experiential assumption. According to Schröder, most dreams, indeed, are not experiences; the language game used to describe dreams is typically too different from that used to describe waking experiences. Yet violent nightmares, or dreams to which the dreamer attributes a particular significance, may be exceptions and count as genuine experiences. In either case, it is up to the dreamer which language game to choose when reporting a dream: where this language game "merges into one of dealing with real experiences, the experiences of a dream may indeed be said to attain the status of real experiences" (Schröder, 1997). In this view, empirical facts, for instance, about REM sleep, say nothing about the occurrence or temporal location of dreams during sleep.

20. Again, this is an oversimplification. In particular, the proponent of empirically informed skepticism about dream reporting of the type defended by Schwitzgebel and Rosen has, as I see it, two options: the first (depicted in table 4.1) is to insist on principled skepticism about dream reporting, with the result of collapsing into a Dennettian position according to which dream reports are irrelevant to scientific dream research. Yet here the ideal of providing *empirical* reasons for doubting dream reports, even under ideal reporting conditions, is questionable (cf. sec. 4.3). The second option is to defend a more moderate and constructive version of skepticism about dream reporting, based on empirical evidence about the *relative* transparency of different types of reports. At this point, however, empirically informed skepticism collapses into antiskepticism about dreaming, with all the consequences.

21. Again, a number of practical issues exist. For instance, Domhoff notes that due to motivational factors and high dropout rates, the number of dreams reported in brief dream diaries is typically low and varies strongly between subjects. He also notes that requiring students to report a certain number of dreams increases the risk of willful fabrication (Domhoff, 2003, p. 50). By contrast, Domhoff does not take these issues to be problematic for long-term personal diaries, arguing that it is not so much spontaneous dream recall he is worried about as whether brief dream diaries really reflect the space of spontaneously recalled dreams.

22. Studies might even suggest that morning recall is more detailed and trustworthy than reports following experimental awakenings. Montangero et al. (2003) found that morning dream reports often contain new information compared to dream reports following REM sleep awakenings. Their finding is interesting because they also investigated the accuracy of subjects' reports of dreamlike films. After being awakened from REM sleep, subjects were shown a short video and then asked to describe it, once immediately afterward and once in the morning after several hours of sleep. Again, the morning reports (following memory cues) often yielded additional details, and 86 percent of this new information was accurate. The authors tentatively conclude that this supports the accuracy of dream and waking reports and might even indicate a hypermnesic effect for morning recall.

23. Using adaptation nights, as often done in sleep laboratory studies, might also be considered as "training" subjects to sleep in an unfamiliar environment, at least in a very general sense.

24. This type of approach can also backfire or produce unclear results. In a study investigating dreams in congenitally blind subjects, Bertolo et al. (2003) asked blind and sighted subjects to draw their dreams upon awakening. They found that the drawings of blind and sighted subjects

(produced with eyes closed) were similar. Moreover, they found a significant negative correlation between the presumably visual content of dreams and EEG alpha power. Their conclusion that congenitally blind subjects experience visual dream content contradicts the widely accepted view that if sight is lost before the age of five to seven years, visual dreaming is lost, as well (Kerr & Domhoff, 2004; for critical discussion, see Lopes da Silva, 2003). Does the conservative approach have anything to say on this matter? Note that caution is warranted for several reasons. First, congenitally blind subjects certainly have spatial imagery, and it is not at all inconceivable that they could experience nonvisual spatial imagery (as well as auditory and tactile imagery and movement sensations) in their dreams, enabling them to both draw and verbally describe dream events in a manner that could suggest, to sighted researchers, that they experienced visual imagery as well. Second, blind subjects often describe their experience in visual terms; an impressive example is Helen Keller, who, though blind and deaf from the age of nineteen months, claimed to be able to imagine what it is like to see and hear, and also stated that she occasionally saw a light in her dreams (which otherwise consisted of nonvisual imagery; see Keller, 1908/2003). Third, as Kerr and Domhoff (2004) point out, the attenuation of alpha power may be related not just to visual imagery but also to imagery in other sensory modalities. Consequently it may well be that the use of dream drawings, in this case, was more misleading than helpful.

Consider a case study reported by Zeman et al. (2010) of a patient who reported abruptly losing the ability to generate visual images in wakefulness as well as visual dreaming. Intriguingly, he performed normally on a number of standard tests of visuospatial imagery, had no trouble finding his way around familiar environments, was able to draw pictures from memory, and "could give detailed descriptions of scenes and landmarks around Edinburgh, including visual details: when asked how he could do this he commented: 'I can remember visual details, but I can't see them. . . . I can't explain that. . . . From time to time I do miss being able to see'" (Zeman et al., 2010, p. 147). fMRI revealed a different pattern of brain activation during imagery tasks compared to healthy controls, suggesting that the patient used a different cognitive strategy to solve them. Perhaps, in such cases, a verbal report is needed to decide whether a drawing is indeed a report of *visual* experience or of nonvisual spatial imagery. And for congenitally blind subjects, to whom the contrast between visual and nonvisual spatial imagery is not available, there may ultimately be no unequivocal answer. Here the most careful and conservative strategy is probably to withhold judgment.

25. Note that this strategy of arguing from inter- and intrasubjective variation to possible error is similar to the standardized types of arguments, called tropes, traditionally used by Pyrrhonian skeptics. Here, between- and within-subject variability was used to question the reliability of sensory perception, but also of moral beliefs. This argumentative strategy is already described in detail by Diogenes Laertius (1943) and Sextus Empiricus (1987). Schwitzgebel's approach, however, is novel in applying this type of argument to introspection and our knowledge of ongoing phenomenal experience. In doing so, he reverses the classical Pyrrhonian strategy: where the Pyrrhonian accepts genuine variability in experience and indeed takes this variability to present an epistemological problem for our knowledge of the external world, Schwitzgebel essentially argues that variability in judgments or opinions about experience undermines introspective knowledge of the inner world. At the same time, his willingness to supplement his arguments

with "confessional epistemology—sincere expression of my own degree of confidence or doubt on the matters at hand" (Schwitzgebel, 2013b, p. 1163) and his aim of inducing personal, not just theoretical, doubt in his readers (2011c, p. ix) have an almost classical flavor: *ataraxia*, the goal of the Pyrrhonian, is, after all, a state of mind, in which withholding judgment finally leads to peace and mental quietude.

26. While for Locke, the example of unconscious, dreamless sleep shows that psychological continuity is necessary for personal identity, Johnstone (1973) suggests that periods of unconsciousness during sleep may be necessary for our ability to think of ourselves as conscious at all.

> A person who recognized no gaps in his experience, then, could have no idea what it would mean to be unconscious. It must be stressed, however, that no interruption of consciousness would *necessitate* the framing of the hypothesis of an interruption. Even if there are gaps in a person's experience, he does not *have* to interpret them as intervals in which he has lost consciousness. I am arguing, then, that the occurrence of a gap in experience is a necessary, not a sufficient, condition for a person's forming the idea of unconsciousness. (Johnstone, 1973, 76)

By the same token, only the occasional loss of consciousness can make us aware of being conscious. In this view, Descartes's *cogito* argument would not be possible if continuous thinking were indeed essential to his being; without occasional gaps in experience, he could not even recognize this basic feature of himself. Likewise, one could propose that the contrast between periods of unconsciousness and consciousness during sleep are essential to our ability to realize that we sometimes enjoy conscious experience during sleep, or dream.

27. For an imaginative conversation between two fictive proponents of these different historical views, see Tononi (2012).

28. Sebástian (2014) recently defended the claim that dreams are a real-world example of conscious experience arising independently of cognitive access. This is so, he claims, because (nonlucid) dreams are experiences in which the prefrontal areas as the most plausible candidate for enabling cognitive access are selectively deactivated. What this does not explain, however, is why we are nonetheless frequently able to report even nonlucid dreams.

29. This distinction might also be interestingly related to the question of whether we mistake detail in the world for detail in experience. For instance, the existence of a blind spot, color blindness in the periphery of the visual field, and inattentional and change blindness suggest that we do not have uniformly detailed, sharply focused, quasi-photographic experience over the entire visual field. If so, then (as some have argued) the visual world may be a grand illusion, in that we take visual experience to be more detailed than it is. However, as Noë (2002) points out, it is not clear whose illusion this is supposed to be. The snapshot view of visual experience may be a theorist's illusion, or it might only show that uninformed convictions about attention and perception are untrustworthy; if so, then the grand illusion argument itself might turn out to be a grand illusion (Cohen, 2002). In my terminology, it would at best cast doubt on untutored opinions about experience, and investigating inattentional blindness or the blind spot would be a way of improving the reporting conditions needed to make progress on how detailed our visual experience really is.

30. A question that I have been avoiding so far, and which is beyond the scope of this book, is how to proceed in cases in which no reports are to be had from certain subject groups, even in

principle. Here other types of evidence may still compel researchers to ascribe experience, as is the case for animals (for discussion, see Allen, 2011), infants, or adult patients in vegetative states (Owen et al., 2006). In these cases, where report-based consciousness research is not straightforwardly possible, nothing I have been saying about the importance of the transparency and reportability assumptions applies: in particular, a failure to report conscious experience obviously cannot be used to infer an absence of experience. If a subject is generally unable to give experience reports in any of the ways open to healthy adult humans, one should obviously be extremely lenient in what one considers a report or evidence of conscious experience independently of reports. Saying anything else would be morally reckless.

Vegetative patients, for instance, can respond to researchers' instructions by engaging in specific imagery tasks, and perhaps this qualifies as a type of reporting behavior. Owen et al. write that their patient

> retained the ability to understand spoken commands and to respond to them through her brain activity, rather than through speech or movement. Moreover, her decision to cooperate with the authors by imagining particular tasks when asked to do so represents a clear act of intention, which confirmed beyond any doubt that she was consciously aware of herself and her surroundings. Of course, negative findings in such patients cannot be used as evidence for lack of awareness, because false negative findings in functional neuroimaging studies are common, even in healthy volunteers. (Owen et al., 2006, p. 1402)

Again, it seems there are two questions: one, what counts as a report; and two, what counts as report-independent evidence of conscious experience in subjects unable, for whatever reason, to engage in any reporting activity. Yet, Owen and colleagues' remark that their patient's cooperation was a clear act of intention suggests that she was, indeed, using imagery tasks as a form of experience report. Consequently, reporting behavior might be broadened to include purely mental forms of behavior.

Can anything constructive be said about whether animals dream? While animal models play an important role in sleep research, their contribution to dream research is unclear (cf. Dawson & Conduit, 2011, pp. 141–142). For instance, one study showed that hippocampal activity measured while rats performed specific behavioral tasks (running along a circular track) was replayed during REM sleep with a similar, though slightly longer, duration (Louie & Wilson, 2001). Yet this in itself is not sufficient to license the inference that this pattern was associated with spatial phenomenal experience rather than with nonconscious processing (for discussion, see Revonsuo, 2006, pp. 324–327). Moreover, one might allow that animals have phenomenal experience during sleep while denying that this qualifies as dreaming (in the sense of dreams investigated by scientific dream research); this will depend on what exactly one means by the term, as well as on the type of animal one is investigating. It might seem clear, at least, that animals don't report their dreams. But while this typically seems to be true, this might not be a principled limitation. Some animals, such as great apes, can give behavioral or even linguistic reports under appropriate conditions (for details, see Seth et al., 2005), and Dawson and Conduit (2011, p. 142) mention a study involving rhesus monkeys who had been trained to press a bar in response to visual stimuli and purportedly did so in REM sleep as well. Whether this counts as a concurrent dream report of visual imagery, however, is a different question. The same is true for the orienting behavior observed in cats following the obliteration of REM-sleep-related muscular atonia through experimentally induced lesions to the brain stem (Jouvet & François, 1959; Jouvet, 1999, p. 82;

cf. sec. 3.4). Again, it is not clear that this counts as behavioral evidence of phenomenal experience during sleep or even of dreaming. Given these theoretical uncertainties, it would seem that the most cautious conclusion is that animal models cannot, at present, contribute to dream research proper.

31. Researchers have also done interesting work comparing objective and subjective measures of consciousness. Prominent candidates are confidence ratings, postdecision wagering, and the perceptual awareness scale (Dienes & Seth, 2010; Overgaard et al., 2010; Sandberg et al., 2010; Irvine, 2013). The reliability of these different measures is still controversial; however, preliminary evidence suggests that the perceptual awareness scale, which is a purely introspective measure, may be the clearest measure of subjective experience. As Sandberg et al. (2010) note, this might be because only the perceptual awareness scale asks participants to report experiential qualities (e.g., the clarity of visual experience), whereas more "objective" measures ask subjects to assess their degree of confidence or wager on their decision (for a similar assessment, see Overgaard et al., 2010). This is still very much an ongoing debate; yet this view fits in nicely with my emphasis on reports as intentional and sincere descriptions of experience. However, if more objective measures such as confidence ratings were further substantiated as measures of conscious experience, perhaps they could be used to improve reporting conditions. As Sandberg et al. note, "Different measures tap into slightly different aspects of what it means to be aware of a particular state of affairs, but the measures also interact in subtle ways with stimulus difficulty and performance" (Sandberg et al., 2010, p. 1077). Again, this is mainly an empirical question; a careful prediction is that the reliability of such putatively objective measures will depend, at least in part, on their ability to match up with more subjective measures.

32. It is interesting to note that Schwitzgebel collaborated with Russell Hurlburt in a joint project designed to resolve their disagreement regarding first-person methods in general and descriptive experience sampling in particular. They conducted a number of prolonged interviews with a naive subject and used these interviews as the starting point for a theoretical discussion, with Hurlburt acting as proponent and Schwitzgebel as skeptic. In his final assessment of the joint experiment, Schwitzgebel's position is characteristically ambivalent. He remarks that descriptive experience sampling should be placed on an equal footing with other introspective methods (Hurlburt & Schwitzgebel, 2007, p. 227); yet he notes that he continues to be tempted by both radical skepticism and cautious hope that progress may be possible (p. 250).

33. Note that when I first introduced the reportability assumption, I did so in the context of my discussion of laboratory research on dreaming in section 3.1. The formulation provided here in table 4.2 is more neutral and thus better able to accommodate the possibility that laboratory awakenings may not provide ideal reporting conditions after all. Perhaps only a subgroup of timed, laboratory awakenings does so; or perhaps timed awakenings are best conducted outside the laboratory in the home environment; or perhaps some other, superior method of collecting dream reports will yet turn up.

5 Dreaming as Quasi-Perceptual Experience: The Traditional View

1. An important omission, for instance, is medieval philosophy. Medieval philosophers such as Augustine and Aquinas were primarily interested in the ethical question of whether we are morally responsible for our dreams (Dreisbach, 2000), and as their interest in dreams thus points in a different direction than my aims in this chapter, I do not discuss them here. Augustine, however, also held that bodily experiences in dreams are real experiences, and he can thus be regarded as a proponent of the quasi-perceptual view. I discuss this topic in more detail in chapter 8.

2. Note that Metzinger's use of transparency differs significantly from the way I use this term in my discussion of the epistemic status of dream reports (cf. sec. 4.4). Within his theory, transparency is a property of phenomenal states: a mental state is transparent when it is experienced not as constructed or simulational but rather as immediately given, as real. Metzinger's use of the concept also differs from its use in the philosophy of perception, where transparency is typically used to describe the openness of perception, that is, the naive realistic experience of being in direct contact with the objects of perception (for discussion and further references, see Crane, 2011). Metzinger's use of transparency is broader, because it refers not only to perceptual experience but to phenomenal experience more generally, including the phenomenology of self-experience. Moreover, while transparency, in the philosophy of perception, refers to a property of perceptual experience, Metzinger uses it to refer to the transparency of underlying processing stages: when transparency is lost, earlier processing stages become available to attention (cf. Metzinger, 2003, pp. 163–179).

3. A similar claim is defended by McGinn (2004, p. 80) and Ichikawa (2009), who both argue that if dreams were percepts, they would wake us up.

4. There is also a growing body of evidence suggesting that sensory cues, such as odors that were previously associated with a learning task, can be used to trigger memory reactivation during sleep, and this may even accelerate sleep-related memory consolidation. For a review and discussion of the literature, see Diekelmann (2014). Generally, an association with dreaming is tempting, but still highly speculative; for discussion, see Diekelmann and Born (2014).

6 Dreaming as Imaginative Experience: The Rival View

1. Note that a similar problem underlies disjunctivism. Disjunctivism is one of several possible ways of solving the problem of perception (Soteriou, 2009; Crane, 2011; I offer a brief discussion in Windt, 2011). Because illusions and hallucinations, understood in a philosophical sense, are subjectively indistinguishable from genuine instances of perception, they present a problem for the commonsense view of perception as putting us in direct contact with the objects of perception. Hallucinations, in particular, show that the "openness of perception" is a phenomenal property that can turn out to be false on the third-person epistemological level of description, because hallucinations arise independently of mind-independent objects. The problem of

perception is the problem of how to reconcile the seeming openness of perception with the possibility of subjectively indistinguishable hallucinations and illusions.

Disjunctivism tries to uphold the commonsense, naive realist view of perception by claiming that indistinguishability is all that hallucinations and genuine perception have in common, where indistinguishability is understood as a purely epistemological problem, but fails to point to any deeper phenomenological similarity (for a critical discussion, see Siegel, 2008). The most basic description of such experiences is disjunctive: they are either hallucinatory or perceptual but have nothing more basic in common. Specifically, the phenomenal character of perception is determined by the mind-independent objects of perception, and hallucinations fail to have this phenomenal character. For the same reason, disjunctivism is also, like the imagination view of dreaming, sometimes considered as a response to skepticism about sensory-based knowledge (Macpherson, 2013).

Most disjunctivists do not give a positive account of hallucinations (but for an exception, as well as for an empirically informed defense of this position, see Fish, 2008, 2009), and some even deny that hallucinations have phenomenal character at all (e.g., Martin, 2009; for discussion, see Nudds, 2009). To the extent that disjunctivism derives from a purely conceptually motivated rejection of the common-kind assumption, and to the extent that the proposed disjunctive characterization of perception and hallucination is regarded as an argument for denying that hallucinations have phenomenal character, this position commits the armchair fallacy: even if we admit the key disjunctivist claim that hallucinations and veridical perceptions are fundamentally different kinds of mental states—whatever that is taken to mean—this in itself would have no necessary phenomenological consequences.

2. This assumption is by no means obvious. A brief analysis of Ryle's discussion of imagination will show why. To begin with, Ryle rejects the thesis, which he mainly attributes to Hume (cf. sec. 5.1.2), that imagining and perceiving are gradually distinguished by their different degrees of vivacity. According to Ryle, Hume's (1739/2000) "mistake was to suppose that 'seeing' [i.e., visualizing or seeing before the mind's eyes] is a species of seeing, or that 'perception' is the name of a genus of which there are two species, namely impressions and ghosts or echoes of impressions [i.e., Humean ideas]. There are no such ghosts, and if there were, they would merely be extra impressions; and they would belong to seeing, not to 'seeing'" (Ryle, 1949/2000, p. 250). The fact that we sometimes fail to realize that we are "not seeing, but only 'seeing,' as in dreams, delirium, extreme thirst, hypnosis and conjuring-shows, does not in any degree tend to obliterate the distinction between the concept of seeing and that of 'seeing,' any more than the fact that it is often difficult to tell an authentic from a forged signature tends to obliterate the distinction between the concept of a person signing his own name and that of someone else forging it" (Ryle, 1949/2000, p. 246).

For Ryle, seeing, as all perceptual concepts, is an achievement word; it can only be applied to those cases in which the object of perception was actually present and accurately perceived.

> Just as a person cannot win a race unsuccessfully, or solve an anagram incorrectly, since "win" means "race victoriously" and "solve" means "rearrange correctly," so a person cannot detect mistakenly, or see incorrectly. To say that he has detected something means that he is not mistaken, and to say that he sees, in its dominant sense, means that he is not at fault. It is not that the perceiver has used a procedure which prevented him from

going wrong or set a Faculty to work which is fettered to infallibility, but that the perception verb employed itself connotes that he did not go wrong. (Ryle, 1949/2000, p. 238)

This rules out hallucinations as instances of perception and shows that it is absurd to assume that perception could be deceptive. Yet note that Ryle is not concerned here with denying that we can fail to distinguish imagined from actual seeing; he is using the concept of perception as an achievement word only and in doing so recognizes that this does not have any consequences for our ability to distinguish, for instance, dreams from genuine perception on the phenomenological level of description. From a purely third-person epistemological perspective, perceiving and imagining can be distinguished sharply even if cases exist in which this distinction is blurred (perhaps even systematically) on the level of phenomenal experience.

3. One example that is often discussed in this context is the "light switch experiment," in which lucid dreamers try to switch on a light in a dark room, that is, to create a visual environment within a nonvisual dream. Some authors state that lucid dreamers can easily generate visual imagery in this way (Levitan & LaBerge, 1993). Others state that large-scale changes of the dream environment, such as creating a visual dream scene in a nonvisual dream or changing the environment at will, can at best be achieved after a certain time lag, with results that often do not correspond exactly to the dreamer's intentions (Worsley, 1988; Gillespie, 1991; Green & McCreery, 1994). Likewise, suppressing imagery within a lucid dream seems hard to achieve. For instance, if lucid dreamers close their eyes, they will experience darkness, but visual imagery will reappear after some time. This technique can be used to achieve a scene change but apparently does not eliminate visual imagery for more than a few seconds (Brooks & Vogelsong, 1999, pp. 90–91).

4. Do such instances of recognition by just knowing also occur in wakefulness? Recognition by just knowing, likely closely related to emotional reactions, occurs in neuropsychological syndromes such as the Fregoli and Capgras syndrome, where subjects either misidentify strangers or believe relatives to have been replaced by impostors (Schwartz & Maquet, 2002). Perhaps this aspect of just knowing is particularly prominent in dreams; or perhaps it is only particularly obvious, because in dreams, it often appears independently of the perceptual cues that are available in wakefulness. Researching this in more detail might be promising, and existing work on the feeling of knowing (see Koriat, 2000; Metcalfe, 2000) would seem to be relevant to this question.

5. A related question is whether anything new can be learned from mental images or dreams. This possibility is rejected by Wittgenstein (1953, 1967) and Sartre (1940/2005), as well as by Ryle (1949/2000), their general point being that a mental image could contain nothing that was not already known to the subject (for a detailed critique of this thesis in relation to Sartre and Wittgenstein, see Stock, 2007). At the same time, the question of whether imagination—either by forming visual images or by propositional imagination, as in imagining *that* or supposing *that*—can provide new information is also central to the epistemological discussion of whether imagination is a guide to genuine insight (Gendler & Hawthorne, 2002; Byrne, 2007).

It is empirically well documented that subjects use mental imagery to answer specific questions (Kosslyn, 1994). Asked about the shape of an object, for example, a German shepherd's ears, many try to answer by imagining a German shepherd's ears. Thompson et al. (2008) have suggested that previously unnoticed properties of objects can be detected by forming visual images of them, supporting the view that "the same representations and processes are used in both visual perception and visual mental imagery" (Thompson et al., 2008, p. 1031). Mental imagery may also be interestingly related to perceptual learning (for a recent review, see Mast et al., 2012), and like dreams, it may enable unexpected recall of previously forgotten memories (Sheehan, 1972). Generally, the process of imagining may provide access to information that would otherwise not have been explicitly available to conscious reasoning. It may also provide insights about things that are not contained or directly entailed by the image, for instance, about one's emotional reactions (Stock, 2007).

The view that something new can be learned from mental images requires neither that an image is actually observed nor that the imagined object itself is observed. Generally, the image view of imagination is fraught with problems because it appears to require a homuncular observer and thus results in an infinite regress (for discussion, see Kind, 2001, 2005; Byrne, 2010). Acknowledging these points, however, does not contradict the description of mental imagery as involving the phenomenology of observing. In this purely phenomenological sense, waking mental imagery may very well involve an observational attitude, and this may be enough for generating new insights. A careful and intermediate position would concede that both waking imagination and dreams depend on information internal to the system, but only through the formation of the image does this information become explicitly available to the subject. The ensuing knowledge is new in the sense that the process of imagining makes this information available for use in further reasoning. If knowledge has something to do with the phenomenology of knowing, with knowing that you know and with having conscious access to this knowledge, it is possible that imagery and dreams can make available forms of knowledge that were either forgotten or of which one was never consciously aware in the first place. Perhaps this might even be a first step toward understanding why dreams often seem to be a source of insight (cf. Edwards et al., 2013).

6. Foulkes's argument that the generativity and narrative structure of dreams requires the assumption of an underlying grammar is also reminiscent of Fodor's language of thought hypothesis (Fodor, 1975; Aydede, 2010). Fodor's central claim is that the productivity and systematicity of language, which are crucial to our ability to generate and understand an infinite number of sentences, require the assumption that thought itself has a combinatorial syntax and semantics. This, in turn, can only be explained through the existence of a language of thought. Foulkes's claim, then, is that dreaming, as a form of thinking, is an expression of the language of thought.

7. The debate on continuity versus discontinuity is also reflected in different explanations of disorder in dreams. Several studies have focused on the investigation of dream bizarreness, of which different types can be distinguished (Hobson, 1988; see also Revonsuo & Salmivalli, 1995; Revonsuo & Tarkko, 2002). These are incongruity (e.g., of an object appearing in an incongruous context, such as a dream in which Barack Obama is sitting in my kitchen), discontinuity (or sudden changes, e.g., Obama suddenly changing into George W. Bush), and vagueness (e.g., my uncertainty as to whether I dreamt about Barack Obama or George W. Bush). A problem related to the study of bizarreness, however, is that for independent judges to decide whether a given

dream element is bizarre, they may require additional information about the dreamer. For instance, raters might fail to detect that the presence of my grandmother (who has been dead for years) in my kitchen is bizarre. They may also erroneously think a feature is bizarre though it is not, for instance, if, unbeknownst to them, Obama had actually visited me and sat in my kitchen to discuss an urgent matter of state.

While this is a practical problem, a more principled question is which level of description is relevant for the explanation of bizarre dream features. States (1997, 2000), for instance, suggests that dream bizarreness is not so much a corruption of thought as a way of thinking in its own right, and argues that it contributes to the metaphoric quality of dreams. By contrast, Hobson (1988; Porte & Hobson, 1986a; Mamelak & Hobson, 1989; Kahn et al., 2000, 2002) argues that bizarreness results from the chaotic input sources of dreams, thus reflecting a breakdown of the higher-level cognitive and synthetic aspects of dream formation: "The trouble arises at or near the sensori-motor level. . . . The cognitive domain is dominated by an inability to describe and/or a vain attempt to explain a host of perceptual and motoric abnormalities" (Hobson, 1988, p. 267). While this suggests that bizarreness is linked to chaotic PGO activity and changes in neurochemical balance during REM sleep, Schwartz and Maquet (2002) suggest that dream bizarreness can be linked to neuropsychological syndromes. For instance, a dream in which I identify a dream character resembling George W. Bush as my grandmother is similar to the Fregoli syndrome, in which subjects misidentify strangers as close friends or family members, though there is no obvious physical resemblance. Schwartz and Maquet hypothesize that the occurrence of Fregoli-like phenomena in dreams is related to regional changes in activation patterns similar to the lesion topography that causes the disorder in wakefulness. Finally, Revonsuo (1999; Revonsuo & Salmivalli, 1995; Revonsuo & Tarkko, 2002) links dream bizarreness to binding errors, arguing that bizarreness should be analyzed at the phenomenological and cognitive levels rather than being related to neurochemical balance or changes in regional activation patterns. Revonsuo and Tarkko hypothesize that "the probability of binding errors in different dream features depends on the degree of modularity in the processing that produces a feature" (Revonsuo & Tarkko, 2002, pp. 14–15). The question, then, is not only whether dreaming is an ordered or random process but also whether disorder and bizarreness reflect a form of (meaningful) thinking, whether they can be used to understand the cognitive binding mechanisms involved in waking consciousness, or whether they reflect changes in neurochemical binding or regional changes in brain activation patterns similar to lesion-induced neuropsychological syndromes. It is interesting to note that the tension between lower-level, bottom-up and higher-level, top-down cognitive explanations exists not only in attempts to explain the alleged continuity of dreaming and wakefulness but also in theoretical accounts of bizarre dream elements that most clearly depart from waking experience.

8. Cartwright (2010) points out, however, that the paucity of small children's dream reports may be an artifact of Foulkes's use of verbal dream reports. She suggests that dream drawings may be a better way to probe small children's dreams.

7 Are Dreams Subjective Experiences (I)? Phenomenal Selfhood and Bodily Experiences in Dreams

1. This is not to be confused with the claim that fully embodied selfhood necessarily accompanies dreaming: Revonsuo (2006, p. 184; for discussion, see sec. 11.3.2) explicitly allows that the phenomenology of embodied selfhood can be lost in dreams.

2. In Windt et al. (2014), we investigated an alternative and perhaps slightly less disturbing version of the pinching test: the question of whether it is possible to tickle oneself and be tickled by another in a lucid dream. Our results suggested that at least in lucid dreams, both self-tickling and being tickled by another fall short of waking ticklish sensations. While we found these findings quite interesting from a predictive processing perspective, we are not confident, however, that the tickling test can provide a satisfying answer to the skeptic.

3. Note that this is also different from the claim that an implicit, tacit, and prereflective form of bodily experience pervades sensory experience (for discussion, cf. Legrand, 2006; Thompson, 2007, chap. 9). Your visual experience of this book, for instance, has the book as its perceptual object; the body's role in mediating this experience (i.e., the eye saccades allowing you to scan the page, or the postural adjustments made while reading) does not necessarily enter into experience but still structures it in subtle yet important ways. But even if it is true that the body is always at least implicitly or prereflectively given in perceptual experience, one can still distinguish cases in which the body is invisible, as it were, on the level of phenomenal experience from cases in which the experience of being a *bodily* self takes on a distinct and explicit quality. It is the latter cases that I am interested in here. For now, I will focus on cases in which the body is not just an (invisible) accompaniment of perception but enters into experience in its own right. This is important, I think, to distinguish bodily experience from nonbodily experience (the slightly inflationary alternative is to claim that all experience is bodily on the phenomenological level of description), and to focus on bodily experiences that are detectable in dream reports.

4. Many other fine-grained questions about bodily experience are not addressed here, for instance, concerning the basis of body ownership (or *myness*), the *spatiality* of bodily experience, the question of whether emotions are necessarily linked to bodily sensations (such as elevated heartbeat), whether perceptual exploration always has a distinctly bodily flavor, and whether bodily experience itself should be modeled on perception. See de Vignemont (2011) for discussion and further references.

5. This distinctly social flavor of dreaming leads to an important computational problem: while dreams are commonly populated with a number of dream characters, only one of these is typically identified as the self. I discuss this problem in Windt (2014b), and in Windt et al. (2014); for an excellent review of studies on social imagery in dreams, see Revonsuo et al. (2015).

6. I thank Tore Nielsen for raising this objection.

7. Note that this weaker reading would still be compatible with Saurat and colleagues' proposal that "the human brain has an inner walking program, which is either genetic or develops via

mirror neurons" (Saurat et al., 2011, p. 1431), thus enabling paraplegics to experience types of movement in dreams that they are not familiar with from wakefulness.

8. The researchers also found that the specificity of dreams about babies or children decreased in late as compared to early third-trimester dreams, a finding that might be related to a decrease in spontaneous fetal movements and associated nighttime microarousals during this period of pregnancy.

9. As Coleridge, in whose work sleep, dreams, and specifically painful dreams were a recurring theme (for discussion, see Ford, 1999), noted, pain may also take on a distinctive quality in dreams:

> This astonishing multiplication of Pain into itself, in dreams,—I do not understand it. This Evening sleeping I—for the first time I recollect, had a most intolerable sense of Pain as Pain, without affright or disgustful Ideas—a sense of an excruciating patience-mocking Rheumatism in my right arm . . . the astonishing difference in the degrees of the Pain felt or supposed when asleep & when awaked. (Coleridge, 1957–1990, entry 2838)

10. A possible explanation, especially for bodily experiences occurring in the context of lucid control dreams, might be that sensory attenuation for the consequences of self-generated movements prevents them from reaching their customary waking vividness (cf. Windt et al., 2014).

11. Importantly, he found that participants' existing presuppositions concerning the richness or thinness (Schwitzgebel, 2007) or abundance or sparseness (Schwitzgebel, 2011c) of experience were only weakly correlated with their experience reports. He considers this as preliminary evidence that their reports were not strongly driven by existing biases.

12. Note that I am not here committing to descriptionalism about the representational format of mental imagery (for a general introduction, see Kind, 2005), nor to any specific position on the existence of fictional objects (Sainsbury, 2010), but am only using this as a rough analogy. Tye (1991), for instance, has argued that indeterminacy can be accommodated by both descriptionalist and depictive approaches to the representational format of mental imagery.

13. An interesting question would be whether phenomenal disembodiment is more frequent in lucid dreams, as would be predicted by the present theory. Perhaps the added attentional resources and cognitive insight that accompany the transition to lucid dreams turn the inconspicuous absence of the body into an experienced absence.

14. I thank Thomas Metzinger for suggesting this term to me.

15. If dreams were phenomenally indeterminate with respect to bodily experience, one might expect, after all, dreamers to be uncertain as to whether bodily experiences occurred in their dream.

16. Note how this also gives rise to the semantic question of where to draw the cutoff line between bodily and disembodied or even *a*bodily experience (cf. Metzinger, 2013c). My suggestion is to answer this pragmatically, in view of how such dreams are actually reported. Indeed, dreamers seem to be quite willing to describe examples of the type given above in bodily terms, whereas the experience of being a point in space is often supplemented by the remark that the

dreamer had no body. If this observation is borne out by further research, I suggest using this as a guide for introducing an empirically informed cutoff line between bodily and disembodied, or even phenomenally indeterminate, experience.

17. This claim is not restricted, of course, to dreaming. Hohwy (2013b, p. 232) reports a variant of the rubber-hand illusion in which subjects reported feeling a touch that was located in thin air. A touch felt, then, also does not imply a corresponding visual representation of the relevant body part.

18. Heydrich et al. (2010) recently reported two fascinating cases of epileptic patients experiencing rare forms of bodily distortions and altered bodily experience, including numbness and a sense of disownership for particular body parts, during seizures. These experiences did not, however, spread evenly over the whole body. Instead certain body parts—for instance, the neck, trunk, and lower extremities in one of the patients—were felt to be *less* present than others and disconnected from the rest of the body. This shows that the experienced realness of the body need not be uniform but can vary across different parts of the body.

19. I am grateful to Tore Nielsen for pointing this out to me.

20. Clare Johnson has since told me that "it was sensation rather than vision that suggested my double position to me, just like closing your eyes as you sit reading this and feeling yourself seated, and yet bizarrely also being able to feel yourself lying down across the room" (personal communication, January 2014). Yet there could, of course, be cases in which there additionally is a double visual representation of the body, as well.

21. If phantom limb patients frequently dream of their missing limbs, the opposite is apparently true of BIID patients. Brugger (2012) notes that following desired amputation, they dream of themselves as after amputation more quickly than is the case for traumatic amputees—a finding that supports the view of BIID as a "negative phantom," or a loss of phenomenal embodiment for a limb that is still present. Similarly, Giummarra et al. (2012) report the case of a thirty-four-year-old male desiring a loss of control over the lower body who has been dreaming of himself as paraplegic since the age of ten. Apparently dreams reflect not so much objective properties of the physical body as the way the body is experienced in wakefulness. Yet they do so without exactly replicating waking bodily experience.

22. It should be noted, however, that in another study, in which she was asked in structured interviews to give vividness ratings of her waking phantoms, the ratings showed variation over different body parts and were generally somewhat lower, at around 3 to 5 on a scale from 0 to 6, than in our study (Brugger et al. 2000). This difference may be due to the reporting method; for practical reasons, we were only able to give her a short and very general questionnaire, and perhaps our method was too coarse-grained to obtain accurate results. Alternatively, the intensity of her phantoms may have changed with growing meditation experience, as she herself indicated. Unfortunately, it is no longer possible to resolve this question; yet, I would like to point that our study, unlike the study conducted by Brugger and colleagues, asked her to compare the vividness of her phantoms not only in dream and wake states, but also to the vividness of her physical body. These comparative ratings might be informative even if the absolute ratings for individual

body parts we obtained were not as precise as those obtained in the study conducted by Brugger and colleagues.

8 Are Dreams Disembodied Experiences? The Role of the Body and of the Brain in Shaping Bodily Experience in Dreams

1. For the sake of completeness, note that the independence hypothesis does not entail that dreams typically or even potentially *replicate* wakelike phenomenal embodiment, as predicted by the bodily duplicate hypothesis. While the two hypotheses are indeed linked in the examples discussed in this section, the independence hypothesis need only claim that any bodily experiences that do arise in dreams do so independently of real-body inputs. Strictly speaking, this is compatible with saying that dreams involve weak phenomenal embodiment; such claims would, however, owe an explanation of why exactly phenomenal embodiment varies in dreams before a background of uniform functional disembodiment.

2. Note, however, that this does not sit well with Descartes's endorsement of the retirement view, according to which we dream throughout sleep but are only able to recall our dreams as "long as the mind is joined to the body" (Descartes, 1984, II.247), at least in the limited sense of leaving its imprint on the brain. In this view, in deep sleep, the brain is unsuited to receive these traces, resulting in the failure to recall any dreams upon awakening (cf. sec. 4.5). Descartes's remark that his dreams frequently replicate the wakelike experience of being a bodily agent, as in his famous dream of sitting by the fire, consequently can only refer to dreams that he was able to recall upon awakening, and hence dreams in which his soul was still joined, as it were, to the body. This would suggest, however, that his account of dream recall requires that at least the subgroup of reportable dreams does provide him with a primitive certainty of having a body: to recall a given dream, the dream must have left its traces on his brain. At least in this limited sense, reportable dreams should, by his own lights, reassure him of his own material existence.

To this idea, Descartes might object that his skeptical argument refers only to dreams in which this rudimentary link to the physical body is obliterated; perhaps he still experiences himself as a wakelike bodily agent in (unreportable) dreams in which the mind has retired from the body altogether. Or perhaps he simply cannot tell the difference between retirement dreams and those in which there continues to be a functional link to the sleeping body. These responses, however, undercut the plausibility of Descartes's account in another manner. To the extent that his skeptical argument owes its credibility to its appeal to actual, real-world instances of dream recall, it cannot be informed by dreams in which the link between experience and the brain has been severed completely. For this reason, Descartes's skeptical argument is actually inconsistent with his own account of dream recall.

3. The independence hypothesis also has repercussions for the moral status of dream actions. Augustine, for instance, thought that while the body experienced in dreams is not real, pleasure and pain sensations felt in dreams are real instances of their kind: "For when we suffer sadness (tristia patimur), even if only in our dreams, although it is only the likeness of bodily limbs which acts, and not bodily limbs themselves, still there is not just a likeness of suffering, but real suffering; joy is felt in the same way" (quoted in Sirridge, 2005, p. 229).

As Sirridge (2005) points out, this position on dream suffering creates a tension with Augustine's position that we are not morally responsible for our dream actions, such as committing adultery in dreams. For even if in dreams we are not acting with our real bodies or truly deciding to act in a certain manner, our dream actions are nonetheless felt or experienced as if they were real—and for Augustine, this in itself is a cause of moral concern. For details and further references, see Windt, in press.

4. For the same reason, any strong version of what one might call the functional-embodiment hypothesis, according to which a direct causal relationship exists between bodily experience in dreams and the sleeping body, is not a viable alternative and so will not be discussed here.

5. By contrast, Bastuji et al. (2008) found no evidence of direct incorporation of pain in dreams following thermal laser pulses to the hand during sleep. However, this may be because subjects were not asked to complete a questionnaire about their dreams until they woke up the next morning. The time lag between any associated dreams and the morning reports may have led to forgetting.

6. As muscular atonia also characterizes NREM sleep, especially following REM sleep deprivation (Nielsen 2014), a similar explanation may apply to NREM dreams as well.

7. Likewise, in a study of 625 quadriplegic men, Comarr et al. (1983) found that although 84 percent reported having sex dreams before injury, only 56 percent reported having sex dreams after the injury. So while there is some suggestion that sex dreams may be influenced by waking functionality, this finding does not sit well with the Freudian theory of wish fulfillment.

8. I am grateful to Tore Nielsen and Antonio Damasio for urging me to include at least a brief discussion of emotion in this chapter.

9. An early version of this view was proposed by Rignano, who suggested that in sleep, we enter into an emotional state "of a purely somatic origin, that afterwards calls up in the dream the images that are in harmony with it: we do not feel any terror because we have committed a crime, but we dream we have committed a crime because a visceral disturbance of some kind reflects itself in us under the form of a state of anxiety" (Rignano, 1920, pp. 316–317).

10. It is well known, for instance, that vestibular simulation accelerates sleep onset and has a calming effect on sleep. Because vestibular pathways are linked to the amygdala, rocking may both affect sleep regulation and explain why it is pleasant and relaxing to be rocked to sleep (Bayer et al., 2011). An exciting question is whether this would also be reflected in the associated dream content and dream emotions.

11. Dream enactment may also suggest an association with enactive theories of consciousness and perception. The enactive approach, as introduced by Varela et al. (1993, p. 205; see also Thompson, 2007, pp. 13–15), suggests that cognition no longer be seen "as problem solving on the basis of representations; instead, cognition in its most encompassing sense consists in the enactment or bringing forth of a world by a viable history of structural coupling." Without my going into the details, it should be clear that any attempt to apply the concept of enactment to dreaming is only loosely and at best metaphorically related to enactment in this sense. Even if

dreams are the experience of a world, and the actions of the dream self bring it about (e.g., if dream plot construction is modulated by the reactions of the dream self), this is quite different from the technical sense in which an autopoietic system, in generating and maintaining itself, transforms its environment into a meaningful cognitive domain. The extent to which the enactment of a dream world depends on a "viable history of structural coupling" is also, without a detailed account spelling out this claim, metaphorical at best. This is not to deny that an enactive approach could meaningfully be applied to dreaming in general and the relationship between dream actions, real-body movements, and dream content in particular, but only to point out that the concept of enactment as used in the debate on embodiment and enactivism cannot be straightforwardly applied to dreaming. Dream enactment, as used in the scientific literature to refer to dream-related behaviors, involves a much more theoretically innocent and neutral concept of enactment.

12. Note that the same type of argument can be applied to the investigation of the motor output blockade. To investigate the extent to which the outward execution of internal dream behavior is blocked, one has to correlate retrospective descriptions of such behavior with objective measurements of movement, muscular twitches, or cardiovascular responses in the sleeping body. The main instances of doing so come from cases in which motor inhibition fails, such as RBD, or is incomplete, as in other instances of dream-related behavior and lucid dreams. Because it is nearly impossible to study the exact relationship between dream movement and real-body changes in nonlucid dreams for methodological reasons (though the correlation between nightmares and periodic leg movements may offer a counterexample), it is plausible that the real-body effects of lucid dream movements do not so much present evidence of a failure of the output blockade as shed light on its normal functioning. The frequent occurrence of micromovements during REM sleep further supports this claim. If this is correct, then the motor output blockade is incomplete to begin with, and moreover, the only available evidence for its investigation consists in instances in which it is incomplete.

This analysis is further strengthened by the following considerations. Dream-related behavior, either during REM sleep as in RBD or during NREM sleep as in sleepwalking, is typically regarded as the outward enactment of internally experienced behavior (e.g., Revonsuo, 2006; Metzinger, 2003). However, in RBD, it seems to be not just the motor control system that goes back online—in many cases, the apparent specificity of dream-related behavior seems to owe itself to a partial restoration of external information processing, as well. If RBD were only an issue of outwardly enacting dream behavior relative to an exclusively internally generated dream environment, the disorder would, in a way, be less dangerous. However, common reports of RBD patients attacking their wives during RBD episodes suggest that within their dream, they are in fact acting on a mixture of internally and externally generated inputs. For instance, in the stereotyped RBD episode of a patient attacking his wife, all the while dreaming of defending himself against an attacker, the dream attacker may actually be a distorted perception of the real-world wife intruding on her husband's dream. If RBD enactment in fact results from a temporary partial breakdown of both the input and the output blockade, this would explain not only the tendency of RBD patients to attack an actual target—rather than groping into thin air—but also, possibly, the occurrence of more intense dreams leading to such enactment episodes in the first place. The reason for the heightened intensity and threatening character of RBD dreams could be that these patients are acting, within their dreams, on a distorted perception of external-world stimuli intruding on the dream. It seems plausible that such stimuli, within the dream, would take on a disturbing and particularly threatening character, hence accounting for the nightmarish emotional tone of associated dreams (Fantini, Corona, Clerici, & Ferini-Strambi, 2005; Schenck, 2005). Though speculative, this analysis would seem to offer a plausible explanation for both the change in the phenomenology of RBD dreams and the neurofunctional alterations that lead to their outward execution. Consequently it might be incorrect to say that RBD dreams reveal dreams to unfold in an *exclusively internal* behavioral space (contra Windt and Metzinger, 2007); and since RBD dreams are some of the best evidence for saying that dreams give rise to such an internal behavioral space, this assumption itself now appears to be tarnished. Perhaps the most vivid dream environments are those that no longer wholly belong to the world of dreams but rather integrate real-world stimuli, as well.

13. A similar argument has been defended by Thompson and Cosmelli (2011) with relation to brain-in-a-vat thought experiments. Their main aim is to question the empirical plausibility of the thought experiment, arguing that "any adequately functional 'vat' will be a surrogate body . . . sufficiently like ours in its functional organization" (Thompson & Cosmelli, 2011, p. 172), though not in its material composition. They acknowledge that this does not undermine the conceivability or possibility in principle of an envatted brain, but argue that "such conceivability or possibility in principle tells us virtually nothing of interest with regard to what concerns us here, namely, the explanatory framework of the neuroscience of consciousness in relation to embodied cognitive science" (p. 174). Relatedly, they note that "the mere truth of internal supervenience provides no support for internalist explanation" (p. 176).

14. Similarly, Thomas Metzinger has argued that "the available empirical evidence from research on dreams and hallucinations generally, but also from whole-body illusions . . . certainly makes it overwhelmingly plausible that the phenomenal experience of embodiment could continue" in a brain in a vat (Metzinger, 2006, p. 5). However, in Metzinger (2014) he explicitly acknowledges that the case for saying that dreams are functionally disembodied states is weaker than he originally assumed.

15. To be clear on this point, the disagreement between Revonsuo and Noë is not about bodily experience but about perceptual experience or even consciousness as such. There is also some disagreement about whether Noë's argument should really be read as defending a metaphysical claim about the supervenience base of experience (Block, 2005) or merely an explanatory one (Thompson & Cosmelli, 2011); in the explanatory reading, the disagreement between Noë and Revonsuo would be narrowed, though not obliterated. Revonsuo, for instance, thinks that dreaming is an explanatory and research model of conscious experience (cf. Revonsuo, 2006, chap. 21). This means that he not only takes the phenomenology of dreaming to be wholly dependent only on internal brain activity but also thinks that it is sufficient to appeal to brain activity to understand why we dream the way we do, and that the same is ultimately true for conscious experience in general. By contrast, Noë is at least committed to the claim that any explanation of perception (though not of dreaming) will have to appeal to its bodily (and not just brain) basis.

16. In its strongest reading, this would only be true during death or coma.

17. Indeed, increasing evidence indicates that in appropriate experimental setups, real-body stimulation can be exploited to induce highly nonstandard bodily experiences in healthy subjects. These include feeling touch on a cardboard box or even in thin air, feeling touch in two locations while being stimulated in only one, feeling embodied in a small wooden doll, and so on. Apparently, bodily experience is not restricted to standard bodily shapes and perhaps does not even require location in a material object at all. See Hohwy and Paton (2010) and Hohwy (2013b) for discussion and further examples.

18. Note that the same point applies to phantom limb experience. According to philosophical usage, a phantom limb might count as a hallucination because the limb itself no longer exists. The same might even be true for psychological usage, depending, once more, on how exactly the limits of appropriate stimulus sources and idiosyncratic perception are spelled out. By contrast, my account offers a precise way of describing phantom limbs as illusory rather than as hallucinatory. A prominent theory suggests that phantom limb sensations can be explained by appealing to the remapping of cortical body-part representations. In this process, tactile stimulation, say on parts of the face, systematically causes sensations of touch in the phantom hand, a finding that fits in well with the adjacency of the cortical face and hand areas (Ramachandran & Hirstein, 1998). Importantly, in this framework, (referred) phantom limb sensations can be understood by appealing to real-body stimulation as a causally enabling condition, and real-body stimulation forms an important part of any research strategy for their systematic investigation. Indeed, had researchers not investigated the effects of facial stimulation on phantom limb sensations, the connection would have remained invisible. This is enough, I would claim, for conceptualizing such sensations as illusory.

19. The body-brain-body problem has at least two ancestors and yet differs slightly from both. Metzinger (2006, 2014) introduces a conceptual framework distinguishing three different kinds of embodiment. First-order embodiment explains the emergence of intelligent behavior out of "dynamical, self-organizing interactions between the environment and a purely physical, reactive system that does not possess anything like a central processor or 'software' and no explicit computation" (Metzinger, 2006, p. 2), as in biorobotics. Second-order embodiment refers to systems that additionally possess a single, coherent representation of themselves as embodied agents, and whose behavior can be understood by describing them as representational systems. Intelligent robots, but also primitive animals and possibly sleepwalkers and patients during epileptic absence seizures are examples. Systems with third-order embodiment additionally experience themselves as embodied agents and possess a phenomenal self-model. Metzinger explicitly notes that third-order embodiment is episodically realized in humans in standard wake states and dreams. He relates the question of how third-order embodiment is grounded in second- and first-order embodiment to the "self-grounding problem," or the problem of how phenomenal selfhood is grounded in low-level physical dynamics and how this might be described computationally.

The body-brain-body problem, while maintaining Metzinger's three-level embodiment structure, is more limited in scope, by focusing on the special case of how dynamic, causal coupling between the brains and bodies of biological organisms brings about bodily experience. Metzinger's broader question, which extends to artificial systems and robotics, is a natural extension and generalized version of the body-brain-body problem, whereas the more modest scope of the body-brain-body problem itself is aimed at maximum empirical plausibility, in bodily experience in general and bodily experience in the dreams of humans in particular. The body-brain-body problem also should not, at least at this point, be understood as analogous to the self-grounding problem. It can indeed contribute to the solution of a particular variant of the self-grounding problem, namely, the question of how *embodied* selfhood is grounded in lower-level dynamics. However, the present theory is not committed to the claim that phenomenal selfhood can be reduced to (phenomenally) embodied selfhood, and rather suggests that there may even be empirical dissociations between the two (cf. chap. 11). Finally, the body-brain-body problem emphasizes that varying degrees of functional embodiment will depend on the degree of causal coupling between the physical body and the brain; by contrast, in Metzinger's (2014) framework, it additionally makes sense to ask about the possibility of bodily experience in the absence of body representation, or second-order embodiment. The body-brain-body problem sets out from the commonsense assumption that bodily experience is at least grounded in brain activity; it assumes that a meaningful relationship holds between brain activity and bodily experience, and merely adds that in addition, a meaningful relationship may also hold between the physical body and the brain. Again, this perspective is more limited and humble, as well as more soundly focused on real-world cases of biological organisms.

The second ancestor of the body-brain-body problem is Thompson's (2007, pp. 235–237, 244–245) body-body problem, which refers to the relationship between the lived body and the material or physical body. Thompson explicitly introduces this as a variant and special case of the mind—body problem. He also emphasizes that the body-body problem should not be confused with the claim that there are two bodies; this is first and foremost a phenomenological distinction and is emphatically not intended to introduce an ontological duality between the lived body and the physical body. By integrating classical philosophical phenomenology in the tradition of Husserl and Merleau-Ponty with biology and dynamic systems theory, Thompson presents a sweeping and highly ambitious account, first, of how a physical system can turn into a living, autopoietic system, and second—and this is the target of the body-body problem—of how a living organism can also be described as having a lived body, understood as the locus of feeling and intentionality. This narrows the intractable gap between mind and body, as Thompson notes, to the relationship between two types of embodiment.

Again, the body-brain-body problem is more modest and much more limited, and it has a different starting point. Thompson's aim is, first, to present a biologically plausible and philosophically sound theory of life and, second, to use this account to investigate the origin of the mind. By contrast, the present chapter only has the aim of understanding a highly specific kind of phenomenology—bodily experiences in dreams—in functional terms. The functional-disembodiment hypothesis as the predominant view in the literature on dreaming also provided a theoretical starting point that was diametrically opposed to Thompson's: the assumption that the sleeping body is completely irrelevant, both causally and explanatorily, to bodily experiences in dreams. The weak-functional-embodiment hypothesis takes a first step toward a rapprochement between bodily experience in dreams and the sleeping body; and given the contemporary intellectual climate in thinking about dreams, it is indeed surprising that its result turns out to be located anywhere in the vicinity of Thompson's body-body problem at all. Yet it is also importantly different. Because of its particular theoretical context, the body-brain-body problem places a much stronger emphasis on the distinction between bodily experience and the sleeping body; indeed, I think dreams are an excellent example of why this distinction is crucial to any understanding of bodily experience. The gap between the phenomenal and the physical body is consequently larger than it might seem in Thompson's theory. The body-brain-body problem emphasizes that the phenomenally embodied self is first and foremost embrained, and only via the causal coupling between the brain and the body can it also be said, at least derivatively, to be functionally embodied. What is surprising from the perspective of the current theory is that even in dreams, this weak and indirect type of functional embodiment persists rather than being wholly obliterated. The idea is that ultimately this gap can be bridged by experimentally investigating and computationally modeling the causal interactions between brain and body, with the goal of mapping changes in bodily experience onto changes in causal coupling between the brain and the physical body.

9 Are Dreams Subjective Experiences (II)? The Phenomenology of Thinking and the Problem of Dream Belief

1. The main point here is that dreaming offers general grounds for doubting all beliefs about the world and about oneself; no subgroup of beliefs of this kind is protected. They could all occur in a dream.

2. If one endorses strict dispositionalism about beliefs, one may hold that such occurrences are manifestations of beliefs, that is, of a dispositional structure, but are not themselves beliefs. Describing such occurrences as beliefs would then be a category mistake. I do not want to enter into an extended discussion here of the relationship between occurrent beliefs and beliefhood (see Schwitzgebel, 2011a) but wish only to point out that this essentially terminological question does not substantially affect the analysis offered here. In particular, saying that occurrences of believing are not themselves beliefs but manifestations of beliefs would only shift the debate to the question of whether occurrences in dreams count as manifestations of real beliefs and are relevant for assessing whether the dispositional profile of a given belief was fulfilled.

3. A similar argument might be applied to the supposed disconnection between beliefhood and perception. In the quasi-perceptual view (which Sosa endorses, though Ichikawa does not), Laura may have acquired the dream belief of being in London based on her quasi-perceptual visual experience of dream-London. Only if one already assumes, for independent reasons, the imagination view to be correct, as does Ichikawa, would the connection between quasi-perception and beliefhood be severed. Moreover, if, as I have argued, the dichotomy between the quasi-perceptual and the imagination view is overstated, then appealing to the connection between beliefs and perception would once more offer no easy way of objecting to the doxastic status of dream belief.

4. For this reason, the claim is quite different from considering dreaming itself as a form of thinking, as is done by Foulkes (1978) or Freud (1899/2003; cf. sec. 6.2.1). Cogitating, for Descartes, means experiencing these cogitations *as* cogitations. By contrast, saying that dreaming is a form of thinking does not require that it *feel* like thinking. I return to this issue in section 9.3.1.

5. To say that engaging in cogitations has a phenomenology is also not to say that cogitating in general, or engaging in specific cogitations such as thinking or doubting in particular, has a proprietary, distinctive, or individuative phenomenology that is independent of the accompanying auditory or visual imagery, emotions, and so on. This more specific claim is the object of much recent philosophical debate (see Horgan & Tienson, 2002; Pitt, 2004; Bayne & Montague, 2011). Here, I am just making the general claim, accepted by both sides in the contemporary debate about cognitive phenomenology, that it feels like something to cogitate, and that cogitating is associated with the experience of being a cogitative self, or a thinker of thoughts.

6. Williams (1978, p. 78) notes that Descartes's use of the Latin *cogitare* and *cogitation* is in fact wider than the English *thinking* and *thought*, and argues that Descartes ultimately takes it refer to any conscious state whatsoever. My use of the term *cogitation* is closer to its original signification, which is related to thinking, imagining, planning, or intending. This corresponds to Descartes's introduction of the term *cogitation*, as shown hereafter; only later does he broaden its meaning. I take cogitation to refer to the act of engaging in various types of thinking, as opposed to largely automatic and unintentional cognitive processes.

7. This is not to say that one cannot be wrong about what one is thinking, or about the content of one's thoughts. As famously noted by Putnam (1981), if I were a brain-in-a-vat, I could not think I am a brain-in-a-vat; unbeknownst to me, I would be thinking in vat-English, and my brain-in-a-vat thoughts would no longer refer to real brains in real vats. But crucially, I would still be thinking.

Note also that Putnam's argument cannot straightforwardly be applied to dreaming. The brain-in-a-vat has no history of causal interaction with the kinds of things it takes itself to be thinking about. The dreamer, however, does. Presumably, a merely temporary causal disconnection would not deprive the dreamer of the ability to think meaningful thoughts in standard English. Granted, we may have the impression of thinking meaningful gibberish in dreams as well, and Freud gives many examples of bizarre word creations in dreams (Freud, 1899/2003); but this is a problem related to cognition and memory, not to reference. A dreamer stuck in an endless dream would, admittedly, dream in dream-English, just as the envatted brain thinks in vat-English—but then, this would no longer be a real-world instance of dreaming in the sense we are interested in here. So Putnam-style arguments concerning reference do not directly affect the occurrence of thoughts in dreams or their meaning.

8. A similar claim can be found in Crowther's remark that "the phenomenal character of the wakeful condition is marked by capacitation with respect to the agent's capacities for autonomous and self-determined mental activity; activity that constitutively involves immediate knowledge of experience and why it unfolds as it does. The dreamer is one for whom such knowledge is inaccessible through the obtaining of a state of incapacitation with respect to mental activity" (Crowther, in press, p. 29). The *capacitation thesis*, then, says that the availability of cognitive capacities leaves its phenomenal mark on the associated experiences.

9. A similar distinction should be applied to theories that view dreaming as a form of problem solving or creative thought. As Blagrove notes, "The cognitive psychological theories of thinking usually applied to dreams (imagination, analogical reasoning, and creativity) each have, in

waking life, conscious or reflexive goal directed action control as an important component, and so are largely inapplicable" to dreams (Blagrove, 1996, p. 124). If Blagrove is correct, then attempts to understand dreaming through the application of psychological terms used to characterize waking thought will not only fail to elucidate dreaming but may even impede a deeper theoretical understanding.

10. Elsewhere, Hobson has defended the claim that lucid dreams are distinguished from nonlucid ones because nonlucid dreams only involve primary consciousness and lucid dreams additionally involve secondary consciousness (e.g., Hobson & Voss, 2011; for a critical discussion, see Kuiken, 2010; Voss et al., 2013). This is, roughly, the empirical equivalent to the no-cogitation hypothesis occasionally defended in the philosophical literature.

11. By contrast, Metzinger (2003, p. 36) uses the concept of introspection to refer not only to "inner perception" and consciously experienced cognitive self-reference but also to external attention and consciously experienced cognitive reference to nonself objects. Metacognition, in the way I use it, is reserved for the type of introspection that involves consciously experienced cognitive self-reference, or the experience of thinking about one's cognitive processes *as* one's cognitive processes.

12. Similarly, Reuter (2011) argues that at least some forms of imagination are introspective to the extent that the imagined object is posited as nonexistent and that subjects, in reporting imaginative experience, report on their imaginary state rather than on the imagined objects. He also notes that for a similar reason, lucid dreams can be regarded as introspective.

13. It is also not clear that the failure to recognize hallucinations as such always involves a breakdown in the ability to engage in rational thought. One-factor accounts of delusions argue that at least some wake-state delusions may arise from altered experience rather than involving a violation of rational thought (for discussion, see Bortolotti, 2010). If this is correct, then we may never be good at realizing whether we are hallucinating or perceiving; it may be a question that typically we do not dwell on at all. I return to this issue in chapter 10.

14. Metzinger (2003, p. 538; see also Windt & Metzinger, 2007) additionally assumes that this involves the ability to direct attention at the underlying construction processes or vehicle properties. However, there is a logical difference between being able to attend to the fact that one's current experience is constructed (as in being aware that one is now dreaming) and being able to attend to the underlying construction processes themselves. Here I only make the weaker assumption that A-lucidity involves awareness of ongoing experience *as* constructed. I am grateful to Kevin Reuter for pointing this distinction out to me.

15. For a similar reason, the present account may explain the gullibility characterizing much of dream thought. If the feeling of familiarity for dream characters and places can be understood as the deliverance of misinterpreted metacognitive monitoring, the same might be true for bizarre occurrences in dreams more generally. The reason such occurrences are rarely questioned or even met with surprise in dreams may simply be that they are experienced as vaguely familiar and expected. And indeed, they are: the only difference is that it is not the content of dream imagery but its source that is familiar. Again, a misinterpreted awareness of the self-generated nature of

bizarre dream events on the level of dream cognition may lead to a lack of critical thinking about their occurrence on the level of dream cogitation.

16. Similarly, it has been suggested that hallucinations arise from a misattribution of mental events to external sources, or defective source monitoring (for discussion and further references, see Larøi et al., 2005; Aleman & Larøi, 2008, pp. 118–122). To this, the present account adds that the feeling of familiarity of certain hallucinations and dream contents may reflect residual (but misattributed) metacognitive awareness of their internal source resulting when the attribution to an external source is incomplete.

17. The idea that dream imagery is a symbolic and associative expression of underlying, unconscious thought may seem reminiscent of the Freudian distinction between latent and manifest dream content. There is, however, a crucial difference. Where Freudian dream theory posits the existence of inherently meaningful underlying thoughts, translated according to certain rules into manifest dream imagery, the present theory says only that dreaming is a form of imagistic cognitive processing driven by largely unconscious, lower-level processes. As in Jackendoff's theory, the feeling of meaningfulness—both during the dream and after awakening—is first and foremost a phenomenological characterization and should not be confused with actual, inherent meaning. Indeed, in dreams, the feeling of meaningfulness can be a free rider; examples are the occasional puniness of dream discoveries, inventions, and insights, felt to be deeply meaningful within the dream, or mystic experiences in lucid dreams that the dreamer, upon awakening, has no tendency to regard as profound revelations (cf. chap. 10).

18. The claim that beliefs do not occur in isolation but rather require a whole network of interrelated beliefs is sometimes also used to deny the existence of beliefs in nonhuman animals (cf. Davidson, 2001).

10 Are Dreams Deceptive Experiences? Deception, Delusion, and Insight

1. Dreams have played a leading role in skeptical arguments since antiquity (for prominent examples, see Plato, 1996; Diogenes Laertius, 1943; Sextus Empiricus, 1987; Cicero, 1994; Augustine, 1950; de Montaigne, 1966). Here I focus on Descartes's version of the problem because it offers its most poignant formulation; cf. chapter 1.

2. Note that this is markedly different from Plato's position in the *Theaetetus*, where he claimed that because we dream throughout sleep, we spend as much time being deceived by our dreams as we do correctly believing that we are awake (Plato, 1996, 158d).

3. It is interesting to note that Western philosophy and Indian philosophy, including the tradition of dream yoga, hold starkly contrasting views of dreaming. Whereas Western philosophers commonly consider dreaming as an obstacle to knowledge, the dream state, in the Indian tradition, holds the promise of true knowledge and is seen as an ideal practicing ground for obtaining insight. According to Western philosophy, one must awaken from dreams to know, whereas Indian philosophy teaches one to awaken *into* the dream—both during sleep and during wakefulness (see Rinpoche, 1998; Varela, 1997; Norbu, 2002; for discussion, see O'Flaherty, 1984; Young, 1999). For an excellent discussion of dreaming and sleep integrating neuroscientific evidence and

Indian philosophy, see Thompson (2014). For a discussion of the famous butterfly dream described in the Zhuangzi, often regarded as the Chinese equivalent of dream skepticism, see Moeller (2004, esp. pp. 44–55), who argues that this reading is in fact a projection of largely Western ideas on the original text.

However, a parallel also runs between the notion of lucid wakefulness and the ideal of true knowledge in Western philosophy. In the *Republic*, Plato observes that we can only have true knowledge if we awaken from the dream world; as long as we reside in the dream world, as most people do, we can never conceive of things in themselves and are limited to fallible beliefs and opinions. "Isn't dreaming precisely the state, whether one is asleep or awake, of taking something to be the real thing, when it is actually only a likeness? Yes, that's what I'd say dreaming is, he said" (Plato, 1994, 476c). Here wakefulness itself is seen as a continued dream, and the same is true in Indian philosophy, where dreams show waking life itself to be an illusion. This theme resurfaces in contemporary philosophy of mind, where it is sometimes claimed that even waking consciousness is virtual and dreamlike (Metzinger, 2003, 2009; Revonsuo, 2006).

4. Another way of putting this is to say that doxastic situatedness enables one to experience oneself *as* a *self in the act of believing, affirming*, and so on. This is what Metzinger calls the "phenomenal model of the intentionality relation." He describes it as a conscious mental model whose content "is an ongoing, episodic subject–object relation" (Metzinger, 2003, p. 411). It involves not only experienced directedness but experiencing the *type* of directedness, the type of attitude that one currently has. Doxastic situatedness as I use the term is just the special case in which the phenomenal model of the intentionality relation involves the experience of having a doxastic attitude, of believing or affirming that *p*.

In more recent writings, Metzinger (2013a, 2013b) has introduced the concept of an *epistemic agent model*. He suggests that the property of experiencing oneself as an epistemic agent capable of knowledge may be crucial for understanding the transition from a simple form of phenomenal selfhood to more interesting and robust forms of self-consciousness. Dreams, or so I suggest in this section, provide the optimal conditions for investigating this transition, precisely because they abound with instances in which the phenomenal-functional property of epistemic agency is only weakly expressed or even missing altogether.

5. Note that this does not contradict my claim in chapters 5 and 6 that the contrast between the quasi-perceptual and the imagination view of dreaming is largely overstated and unhelpful. All that is required for deception from perceptual realism is that some dreams indeed have a realistic, quasi-perceptual character, and that this character may be the cause of dream deception. In particular, I am not here claiming that this is true for a majority or even many dreams. I am only claiming that a subgroup of deceptive dreams are deceptive in virtue of their realistic phenomenal character—and as I also argue that interestingly deceptive dreams themselves are quite rare, this will likely only be true for a small number of dreams.

6. Similarly, subjects experiencing SP night-mares often fear that these frightening but mostly harmless experiences are a sign of physical or mental illness. And shockingly, according to one study, 50 percent of psychiatrists misdiagnose SP-related hallucinations as a psychotic disorder (for details and further references, see Adler, 2011, pp. 27–28, 70–73).

7. Mazzoni and Loftus (1996) suggest that in special laboratory conditions, subtle suggestions can lead subjects to misattribute dream memories to wakefulness, and argue that this raises uncomfortable questions about suggestibility in psychotherapy, where dream memories might be suggested to be related to childhood trauma. Payne and colleagues (2009), investigating subjects' ability to recall word lists before and after sleep, suggest that SWS may preferentially enhance false memories, preserving the meaning or gist of learned information at the cost of details. Their intriguing conclusion is that sleep might facilitate systematic memory distortion. Whether this is in any way related to dreaming is unclear; from a philosophical perspective, however, it is interesting to note that if this turned out to be correct, then epistemological concerns about dream deception or sleep-related false memories may run contrary to the evolutionary functions of dreaming. Finally, one recent study suggests that subjects with high dream awareness and lucidity more frequently make types of memory errors consistent with impaired reality monitoring than is the case for controls (Corlett et al., 2014). A highly speculative conclusion would be that cognitive corruption might indeed extend into wakefulness.

8. There are, of course, exceptions. As Schröder (1997, p. 38) points out, the truly pressing question in thinking about dream belief is to understand why we occasionally ascribe meaning to dream beliefs and come to treat them as real beliefs even after awakening. To the extent that we do, he notes, there is no reason to set dream beliefs and dream thoughts apart from waking beliefs and thoughts. Similarly, Flanagan (1995, 2000) notes that even if dreams are not intrinsically meaningful, the question of why we are sometimes willing to treat dreams as meaningful experiences upon awakening is philosophically relevant.

9. Note that the answer is directly relevant to the problem of dream deception, as well as to the philosophical analysis of dream lucidity. Hunter notes that in worrying about the indistinguishability of dreaming and wakefulness, it might be necessary to consider *who* is trying to distinguish the two states. The idea is that if the dreamer is not a sufficiently sensitive observer, the job may not be done reliably; in this case, the indistinguishability between dreaming and wakefulness would not be particularly worrisome. He writes: "We might have some confidence in the waking Descartes' assurance that he could find no difference, but it might be unsafe to rely on the judgment about this, of someone in Descartes' dreams, even the person cast as himself. We cannot assume that such discernment as the waking Descartes may have is of course carried over into his dreams" (Hunter, 1983, p. 84).

The question, then, is whether dream-Descartes is really Descartes. Hunter is skeptical that the question can be answered affirmatively even in cases that are unlike Rosen and Sutton's vicarious dreams: "Was it not Descartes who, in Descartes' dream, sat by the fire in a study very like Descartes' and hoped that a servant would bring him some tea? Well, it was not Henry VIII, but if it was an average dream, neither were the biographical facts anywhere near sufficient to establish that it was Descartes" (Hunter, 1983, p. 85).

Bernhard (2005) raises a similar concern about lucid dreaming. Who exactly is the subject, he asks, who realizes, while dreaming, that this is a dream? He argues that ascribing the judgment "I assert that I am dreaming" to the lucid dream self, the protagonist of the ongoing dream, equivocates between two different uses of "I." The subject performing the judgment, in this view, is not the same as the dreaming subject; rather, lucid dreaming involves the development of an incomplete third-person perspective within the dream and hence a split between the judging and dreaming subject. If this analysis is correct, it would mean that the dream self cannot properly become lucid: lucidity involves a split between two selves.

10. For a general account of narrative approaches to consciousness, see Fireman et al. (2003).

11. Might this present a problem for my antiskeptical account of dream reporting defended in chapter 4 and in Windt (2013b)? Specifically, don't skeptical worries occasioned by retrospective dream deception support, contra my own argument, skepticism about the trustworthiness of first-person reports? I don't think this follows. Note that while retrospective dream deception has us deceived about the sources of our memories in dream or in waking experience, antiskepticism emphasizes the trustworthiness of dream recall and dream reports with respect to the contents of experience. Even if I miscategorize, upon awakening, my foregoing experience as, say, an OBE or even as a real occurrence rather than as a dream, this does not mean that I thereby also misdescribe the phenomenal experiences involved or that my report is rendered useless for investigating the phenomenology of the respective experience. Moreover, even mistakes regarding the sources of experience can be avoided by improving reporting conditions; for narcoleptics confusing dreams and waking experiences, for instance, special steps, such as sleeping in a laboratory, might be needed to identify the root of the problem. Finally, note that the account of real-world scenarios of dream deception developed here is committed to the transparency assumption, because it leans heavily on the trustworthiness of dream reports. Dream skepticism thus presumes antiskepticism about dream reporting.

12. An interesting question that, for reasons of space, I cannot discuss here is whether a relationship exists between imagination inflation, where subjects mistake imagined events for childhood memories (Garry et al., 1996), and retrospective dream deception, perhaps such that subjects who are particularly prone to the imagination inflation also more likely to experience retrospective dream deception.

13. Note also that other forms of delusions such as thought insertion or thought control, which are frequent in schizophrenia (for a philosophical discussion, see Stephens & Graham, 2000), almost never seem to occur in dreams. To the best of my knowledge, dream reports rarely describe that dreamers lacked the phenomenology of agency for their own thoughts, suggesting that dream thinking does not resemble the phenomenology of thought insertion that forms part of the positive symptoms of schizophrenia (for a philosophical discussion of thought insertion, see Stephens & Graham, 2000). A possible explanation is that cogitative processes lacking the phenomenology of ownership and agency in dreams are externalized and experienced as speech by other dream characters. This seems particularly plausible for cases in which dream speech does not involve hearing but has a telepathic, thoughtlike quality (see sec. 6.1.3).

14. There may also be an interesting parallel between A-lucid dreams and depersonalization and derealization, in which perceptual experience is often described as taking on an unreal, dreamlike quality. Subjects experiencing derealization are not delusional—that is, they know that what they are experiencing is real—but this realistic character is no longer written into experience. For

details on depersonalization and derealization, see Simeon and Abugel (2006); for a philosophical discussion, see Ratcliffe (2008, pp. 165–186).

15. It has also been suggested that dreams are similar to anosognostic states seen in patients with blindness denial, denial for hemiplegia, and so on. Dieguez and colleagues (2007) note that neurological patients may maintain insight into their predicament but can also be indifferent or outright endorse false beliefs about their symptoms. They argue that indifference toward, or unawareness of, bodily defects should not be considered as delusional, but the term should be reserved for outright false beliefs—and this is exactly what I have argued for dreaming.

11 From Oneiragogia to Full-Fledged Dreaming: The Immersive-Spatiotemporal-Hallucination Model of Dreaming

1. This chapter is a substantially revised and expanded version of ideas developed in Windt (2010), where I first introduced the ISTH model of dreaming.

2. Spatiotemporal situatedness is another special case of what Metzinger calls the *phenomenal model of the intentionality relation* (Metzinger, 2003, pp. 411–420), which he describes as involving experienced direction and depicting "a certain relationship as currently holding between the system, as transparently represented to itself, and an object component" (p. 411). This relationship may be of very different types, ranging from the low-level and preconceptual experience of spatiotemporal relatedness to higher-level cognitive relations, as in consciously experienced thought and occurrent belief (or what I call doxastic situatednes). The claim of the ISTH model is that a specific type of such a model is required to understand the experience of dreaming: in dreams, we always have a phenomenal model of our spatiotemporal relation to the world.

3. By amodal imagery, I mean imagery that is not modality specific, namely, that is dissociable from and possibly more basic than imagery in any of the standard sensory modalities (i.e., visual, auditory, gustatory, olfactory, tactile, and proprioceptive experience). One could also argue that spatial imagery is in fact a type of modal imagery corresponding to some sort of "spatial sense," much as is sometimes claimed to be the case for proprioception. I take this to be a largely terminological question and want to remain neutral on this matter.

4. I am grateful to Tore Nielsen for emphasizing this point.

5. In keeping with the literature, I focus on sleep onset imagery here. Of course, a complete account would have to describe the process of emerging from a dream upon awakening in equal detail.

6. We can use this insight to connect the ISTH model to further accounts of conscious experience and phenomenal selfhood from contemporary philosophy of mind. For instance, I agree partially with Thompson's claim that "if something appears perspectivally, then the subject to whom it appears must be spatially related to it" (Thompson, 2007, p. 248). It is only his further claim, according to which "to be spatially related to something requires that one be embodied" (p. 248), that I object to—at least insofar as this is taken to be a phenomenological claim. Thompson continues:

> The body functions as the "zero point," "null point of orientation," or absolute indexical "here" in relation to which things appear perspectivally. . . . In this way among others, perceptual experience involves a non-object-directed and implicit awareness of one's lived body, an intransitive and prereflective bodily self-awareness. (Thompson, 2007, p. 249)

While I think that standard wake states are indeed characterized by a close connection between phenomenal selfhood, spatial relatedness, and the phenomenology of embodiment, my discussion of dreaming shows that this is not necessarily the case. Rather, as my analysis has shown, phenomenal embodiment, or the lived body, is not necessary for spatial relatedness, and there can be a sense in which spatial self-location is independent of bodily experience altogether.

7. Another intriguing possibility is that in standard cases of embodied self-consciousness, the locus of identification may spread to a spatially expanded position or even be ambiguous between different parts of the body. A recent study found that when asked to indicate the exact position of self-location in their own bodies, healthy subjects indicated two distinct regions, the upper face and upper torso (Alsmith & Longo, 2014). Perhaps embodied self-consciousness is associated with increased indeterminacy in the locus of self-location, or perhaps the phenomenology of embodiment enables subjects to feel located in more places than one, with self-location spreading with varying degrees of intensity over the body. Either way, this fits in well with the claim that the phenomenology of occupying an extended volume is an important step toward phenomenal embodiment, and also with the observation that this seems to be lost in dreams in which the dreamer identifies with an extensionless point, or a maximally determinate locus of self-location.

8. One might also worry about the coherence of reporting selfless experiences. Metzinger argues that first-person reports of selfless states, including reports of lucid dreams involving the dissolution of the phenomenal self, involve a performative self-contradiction (Metzinger, 2003, pp. 539, 563–567). In a selfless state in the strict sense of the term, there would no longer be an experiential subject whose memories of experiencing such a state could later be reported. So even if a selfless experience could be had, it would be no one's experience, and (so the argument goes) no one could report it. Dream reports of selfless experiences, if taken literally to refer to the absence of any phenomenal self, are thus to be rejected as inherently contradictory. While this argument alerts us to an important grammatical peculiarity of first-person reports of selfless states, it should not, however, be taken to present a substantial objection against the occurrence of selfless experience. It highlights a grammatical inconsistency in reports using the first-person pronoun to refer to the occurrence of selfless experiences (e.g., "I had no self"). But this should not be confused with a phenomenological fact: one could still have the robust phenomenology of lacking a self, and this might be true even if such an experience cannot be described without conceptual contradiction.

12 Relocating Dreams on the Conceptual Map: Consequences and Perspectives for Future Research

1. An earlier version of this section was published in Windt (2010) but has since undergone considerable revision and expansion.

2. Note, however, that allowing for the occurrence of amodal hallucinations does not commit one to a particular view on amodal perception. For the debate about amodal perception and the possible contribution of imagination, see Nanay (2009a, 2010) and Briscoe (2011).

3. For this reason, dreaming also gives rise to a particular computational problem. Because dreams involve multiple dream characters, but typically only a single self at a given time, this raises the problem of which of these dream characters is the locus of self-identification. The present theory predicts that this will have to do, to a considerable degree, with spatiotemporal self-location, though additional factors (such as narrative aspects) may also play a role. A related prediction is that abrupt changes or discontinuities in self-location might lead to discontinuous jumps in identification with different dream characters; I briefly return to this idea in note 5.

4. The terminology of agent models and self-models is inspired by Metzinger's self-model theory of subjectivity; see Metzinger (2003, 2009).

5. This is not to deny, of course, that dream research can equally profit from the rigorous and carefully controlled experimental manipulations used in current research on full-body illusions. Quite to the contrary, the idea is that these two fields should mutually inform each other. In particular, the systematic comparison between dreams and full-body illusions might be used to formulate hypotheses about the cognitive processes underlying dream phenomena that so far are incompletely understood. To give a single example, consider vicarious dreams (Rosen & Sutton, 2013), in which the dreamer identifies with dream characters other than herself, or even successively identifies with different dream characters. Revonsuo reports the following dream: "The person in the dream that so far had been me, now was suddenly my classmate J. Somehow I became (physically) detached from myself and I noticed that I was not me but him. This was accompanied by a funny feeling" (Revonsuo, 2000, pp. 213–214). Based on the similarity of this dream with the shift in identification to an avatar, as in Lenggenhager and colleagues' (2007) original study, as well as based on the ISTH model, we can now say that the "funny feeling" accompanying this dream is a shift in self-location. A speculative hypothesis would be that at least a subgroup of vicarious dreams involve sudden and discontinuous shifts in experienced self-location that lead, on the level of the dream narrative, to the impression of suddenly having become someone else, of having turned into another dream character.

6. Note that presence is sometimes also discussed in relation to nonimmersive computer games (Gaggioli et al., 2003), and theorists sometimes claim that computer games, like movies, novels, or daydreams, involve a kind of cognitive or fictional immersion (for discussion, see Therrien, 2009). Ijsselsteijn and Riva (2003) distinguish different types of presence, namely, *physical presence*, or the sense of being physically located in mediated space, *social presence*, or the feeling of social interaction, and *copresence*, or the sense of being together in a shared space, and consider not just VR but also cinema, television, and paintings as instances of physical presence. Gackenbach and Rosie compare presence in dreams to presence in video games, which they in turn consider to be a "VR like experience" (Gackenbach & Rosie, 2011, p. 104). They found that items related to cognitive immersion (such as losing track of time or the realistic quality of visual and auditory imagery) did not differ significantly between dreams and computer games. Three other

items, however, received significantly higher ratings for dreams than for video games: the extent to which the dream or video game caused real feelings and emotions, the extent to which the participant felt she was present in the dream, and the extent to which events seemed to be actually happening. This is exactly what would be predicted by the ISTH model, but also by work on immersive VR systems. In keeping with this hypothesis, Slater writes that whereas immersive VR systems create the effortless and robust feeling of presence, "In the case of a desktop system the situation is quite different; the feeling reported as 'being there' if it comes at all is after much greater exposure, requires deliberate attention and is not automatic—it is not simply a function of how the perceptual system normally works, but is something that essentially needs to be learned, and may be regarded as more complex" (Slater, 2009, p. 3552). In immersive VR, cognitive forms of immersion are driven by lower-level forms of presence, not vice versa. The debate on presence in VR and desktop computer games is thus roughly analogous to the debate on cognitive immersion in dreams (see sec. 11.3.3).

7. Research on vection also provides evidence for the connection between the spatial and the temporal aspect of self-location. One study investigated the influence of movement illusions on the direction of mental time travel. Miles, Karpinska, et al. (2010) found that illusory forward movements induced mental time travel toward the future, whereas illusory backward movements were related to daydreaming about the past. Another study (Miles, Nind, & Macrae, 2010) found a correlation between the temporal direction of mental time travel and shifts in posture, that is, leaning forward or backward. Might sleeping posture and the direction of dream movement likewise be related to dreams of the past or future?

8. It might seem that in making this point, I am conflating the properties of dream reports with those of dreaming. This difficulty is circumvented, however, by the present model's claim that dreams are reportable experiences (chaps. 3–4). Moreover, because dream reports are taken to transparently reflect preceding dream experience, the current theory assumes that the reportability of dreams will be intimately associated with their narrative structure.

9. The observation that sleep onset is associated with diminished control over conscious thought processes also fits in well with the finding that sleep perception, or the ability to notice that one has been asleep, is associated not so much with the attenuation of perception as with the loss of control over conscious thought (Yang et al., 2010).

10. It is interesting to note, however, that findings on thought suppression nicely complement the finding that while lucid dream control is a learnable skill, the ability to successfully control one's dreams is unstable even in practiced lucid dreamers and often produces unintended effects (Stumbrys et al., 2012).

11. Another potentially relevant proposal is Seth's (2013; see also Seth, 2015) claim that emotions arise from probabilistic, predictive models of the causes of interoceptive signals (see sec. 8.2.2.3) and are intimately linked to embodied selfhood and the experience of body ownership. Seth and colleagues explain that "presence depends on a match between *informative* interoceptive signals and top-down predictions arising from a dynamically evolving brain–body–world interaction" (Seth et al., 2012, p. 3). This sits well with my description of dreams as weakly

phenomenally-functionally embodied states, according to which immersion and illusory movement in dreams are often grounded in distorted own-body perception. Moreover, Seth and colleagues' (2012) claim that the anterior insular cortex is the brain region most strongly associated with interoceptive predictive processing and hence is crucial to presence fits in well with the finding that during REM sleep, activation in the amygdala covaries with activation in the right anterior insular cortex (Dang-Vu et al., 2010), once more suggesting a connection between presence and emotions in dreams.

12. An alternative account is that dreaming results from an essentially mindless process of self-organization, comparable to laser beams, termite nests, or flight patterns in birds. Kahn argues that self-organization can explain why the contents of dreams are organized in a coherent manner, claiming that a "self-organization model predicts that a central story or theme will emerge from the interaction of elements. . . . This idea here is that self-organization is a process by which the interacting elements compete with each other until a dominant one succeeds, in a sense, enslaving all other elements" (Kahn, 2013, p. 5; cf. Haken, 1983). We might add that for dreams, the phenomenal self and emotional character of dreams will typically function as an attractor, molding disparate images into a common story line, typically involving the self. The self-organization model does not explain, however, why we dream of particular contents. Due to their more distinctly cognitive flavor, predictive processing accounts might offer a smoother explanation of the occasional continuity of dreams with waking concerns.

13. As Wamsley and Stickgold (2011) note, the analysis of dream reports may also, however, be of methodological importance for investigating the replay of memories during sleep. Neuroimaging can only show "that the same memory systems engaged at learning are doing something during sleep, but does not demonstrate the 'reactivation' of a particular memory" (p. 10). By contrast, because of the transparency of dream reports, the systematic study of the memory sources of dreaming might "provide a useful window into consolidation-related activities of the sleeping brain" (p. 11). This adds an interesting methodological twist, suggesting that the study of dreaming may be crucial to the study of memory-consolidation during sleep after all.

14. Likewise, imaginary movement is typically not completely imaginary but rather involves small muscle twitches. For instance, imaginatively engaging in sports activities leads to an increase in heart rate and ventilation. A large body of research indicates that imagined and actual movements also rely on largely the same neural correlates. This means that motor imagery is never completely offline and that there will typically be degrees of decoupling between real and imaginary movement (Guillot & Collet, 2005; Demougeot et al., 2009).

References

Adler, S. R. (2011). *Sleep paralysis: Night-mares, nocebos, and the mind-body connection: Studies in medical anthropology*. New Brunswick, NJ: Rutgers University Press.

Adriani, G. (1892). Fisiologica psicologica del sonniloquio. *Annals of Neurology*, 10, 299–308.

Agargun, M., Boysan, M., & Hanoglu, L. (2004). Sleeping position, dream emotions, and subjective sleep quality. *Sleep and Hypnosis*, *6*(1), 8–13.

Aleman, A., Böcker, K. B., & de Haan, E. H. (2001). Hallucinatory predisposition and vividness of auditory imagery: Self-report and behavioral indices. *Perceptual and Motor Skills*, *93*(1), 268–274.

Aleman, A., & De Haan, E. (1998). On redefining hallucination. *American Journal of Orthopsychiatry*, *68*, 656–658.

Aleman, A., & Larøi, F. (2008). *Hallucinations: The science of idiosyncratic perception*. Washington, DC: American Psychological Association.

Aleman, A., & Vercammen, A. (2012). Functional neuroimaging of hallucinations. In J. D. Blom & I. E. C. Sommer (Eds.), *Hallucinations: Research and practice* (pp. 267–282). New York: Springer.

Alessandria, M., Vetrugno, R., Cortelli, P., & Montagna, P. (2011). Normal body scheme and absent phantom limb experience in amputees while dreaming. *Consciousness and Cognition*, *20*(4), 1831–1834.

Alexander, J. (2012). *Experimental philosophy: An introduction*. Cambridge: Polity Press.

Allen, C. (2011). Animal consciousness. In E. N. Zalta (Ed.), *The Stanford encyclopedia of philosophy (winter 2011 edition)*. Retrieved from http://plato.stanford.edu/archives/win2011/entries/consciousness-animal.

Allen, P., & Modinos, G. (2012). Structural neuroimaging in psychotic patients with auditory verbal hallucinations. In J. D. Blom & I. E. C. Sommer (Eds.), *Hallucinations: Research and practice* (pp. 251–266). New York: Springer.

Alsmith, A. (2013). Eric Schwitzgebel: Perplexities of consciousness. *Phenomenology and the Cognitive Sciences*, *13*(3), 497–501.

Alsmith, A. J. T., & Longo, M. R. (2014). Where exactly am I? Self-location judgements distribute between head and torso. *Consciousness and Cognition, 24*, 70–74.

Alvarado, C. (2000). Out-of-body experiences. In S. J. Lynn, S. Krippner, & E. Cardeña (Eds.), *Variety of anomalous experience* (pp. 183–218). Washington, DC: American Psychology Association.

Amador, X. F., & David, A. S. (2004). *Insight and psychosis: Awareness of illness in schizophrenia and related disorders*. Oxford: Oxford University Press.

Amador, X. F., Flaum, M., Andreasen, N. C., Strauss, D. H., Yale, S. A., Clark, S. C., et al. (1994). Awareness of illness in schizophrenia and schizoaffective and mood disorders. *Archives of General Psychiatry, 51*(10), 826–836.

Amador, X. F., & Kronengold, H. (2004). Understanding and assessing insight. In X. F. Amador & A. S. David (Eds.), *Insight and psychosis: Awareness of illness in schizophrenia and related disorders* (2nd ed., pp. 3–30). Oxford: Oxford University Press.

Amador, X. F., Strauss, D. H., Yale, S. A., & Gormann, J. M. (1991). Awareness of illness in schizophrenia. *Schizophrenia Bulletin, 17*(1), 113–132.

Andrews-Hanna, J. R. (2012). The brain's default network and its adaptive role in internal mentation. *Neuroscientist: A review journal bringing neurobiology, neurology, and psychiatry, 18*(3), 251–270.

Andrews-Hanna, J. R., Reidler, J. S., Sepulcre, J., Poulin, R., & Buckner, R. L. (2010). Functional-anatomic fractionation of the brain's default network. *Neuron, 65*(4), 550–562.

Antrobus, J. S. (1990). The neurocognition of sleep mentation: rapid eye movements, visual imagery, and dreaming. In R. Bootzin, J. Kihlstrom, & D. Schacter (Eds.), *Sleep and cognition* (pp. 3–24). Washington, DC: American Psychological Association.

Antrobus, J. S., Fein, G., Jordan, L., Ellman, S. J., & Arkin, A. M. (1991). Measurement and design in research on sleep reports. In S. Ellman & J. S. Antrobus (Eds.), *The mind in sleep: Psychology and psychophysiology* (pp. 83–121). New York: John Wiley and Sons.

Antrobus, J. S., Hartwig, P., Rosa, D., & Fein, G. (1987). Brightness and clarity of REM and NREM imagery: Photo response scale. *Sleep Research, 16*, 240.

Antrobus, J. S., & Singer, J. L. (1964). Eye movements accompanying daydreaming, visual imagery, and thought suppression. *Journal of Abnormal and Social Psychology, 69*, 244–252.

Antrobus, J. S., & Wamsley, E. J. (2009). REM/NREM differences in dream content. In R. Stickgold & M. P. Walker (Eds.), *The neuroscience of sleep* (pp. 310–315). London: Elsevier Academic Press.

Apps, M. A. J., & Tsakiris, M. (2014). The free-energy self: a predictive coding account of self-recognition. *Neuroscience and Biobehavioral Reviews, 41*, 85–97. doi:10.1016/j.neubiorev.2013.01.029.

Aristotle. (350 BCE/1994–2009a). *On dreams* (J. I. Beare, Trans.). Retrieved from http://classics.mit.edu/Aristotle/dreams.html.

Aristotle. (350 BCE/1994–2009b). *On the soul* (J. A. Smith, Trans.). Retrieved from http://classics.mit.edu/Aristotle/soul.html.

Aristotle. (350 BCE/1994–2009c). *The history of animals* (D'Arcy Wentworth Thompson, Trans.). Retrieved from http://classics.mit.edu/Aristotle/history_anim.html.

Aristotle. (350 BCE/1997). *Kleine naturwissenschaftliche Schriften (Parva naturalia)* (Eugen Dönt, Ed. & Trans.). Stuttgart: Reclam.

Arkin, A. M. (1966). Sleep-talking: A review. *Journal of Nervous and Mental Disease, 143*(2), 101–122.

Arkin, A., & Antrobus, J. (1991). The effects of external stimuli applied prior to and during sleep on sleep experience. In S. Ellman & J. S. Antrobus (Eds.), *The mind in sleep: Psychology and psychophysiology* (pp. 265–307). New York: John Wiley and Sons.

Arkin, A. M., Hastey, J. M., & Reiser, M. F. (1966). Post-hypnotically stimulated sleep-talking. *Journal of Nervous and Mental Disease, 142*(4), 293–309.

Arkin, A. M., Toth, M. F., Baker, J., & Hastey, J. M. (1970). The degree of concordance between the content of sleep talking and mentation recalled in wakefulness. *Journal of Nervous and Mental Disease, 151*(6), 375–393.

Arkin, A. M., Toth, M. F., Baker, J., & Hastey, J. M. (1970). The frequency of sleep talking in the laboratory among chronic sleep talkers and good dream recallers. *Journal of Nervous and Mental Disease, 151*(6), 369–374.

Arnold-Forster, M. (1921). *Studies in dreams*. London: George Allen & Unwin.

Arnulf, I. (2011). The "scanning hypothesis" of rapid eye movements during REM sleep: A review of the evidence. *Archives Italiennes de Biologie, 149*(4), 367–382. doi:10.4449/aib.v149i4.1246.

Arnulf, I. (2012). REM sleep behavior disorder: Motor manifestations and pathophysiology. *Movement Disorders: Official Journal of the Movement Disorder Society, 27*(6), 677–689. doi:10.1002/mds.24957.

Artemidorus. (1975). *The interpretation of dreams* (R. J. White, Trans.). Park Ridge, NJ: Noyes Press.

Arzi, A., Shedlesky, L., Ben-Shaul, M., Nasser, K., Oksenberg, A., Hairston, I. S., et al. (2012). Humans can learn new information during sleep. *Nature Neuroscience, 15*(10), 1460–1465.

Arzy, S., Seeck, M., Ortigue, S., Spinelli, L., & Blanke, O. (2006). Induction of an illusory shadow person. *Nature, 443*(7109), 287.

Aserinsky, E., Joan, A. L., Mack, M. E., Tzankoff, S. P., & Hurn, E. (1985). Comparison of eye motion in wakefulness and REM sleep. *Psychophysiology, 22*(1), 1–10.

Aserinsky, E., & Kleitman, N. (1953). Regularly occurring periods of eye motility and concurrent phenomena during sleep. *Sleep Medicine Reviews, 118*, 273–274.

Aserinsky, E., & Kleitman, N. (1955). Two types of ocular motility occurring in sleep. *Journal of Applied Physiology, 8*, 1–10.

Aspell, J. E., Heydrich, L., Marillier, G., Lavanchy, T., Herbelin, B., & Blanke, O. (2013). Turning body and self inside out: Visualized heartbeats alter bodily self-consciousness and tactile perception. *Psychological Science, 24*(12), 2445–2453. doi:10.1177/0956797613498395.

Aspell, J., Lenggenhager, B., & Blanke, O. (2011). Multisensory perception and bodily self-consciousness: From out-of body to inside-body experience. In M. M. Murray, & M. T. Wallace (Eds.), *The neural bases of multisensory processes* (pp. 467–481). Boca Raton: CRC Press.

Atance, C. M. (2008). Future thinking in young children. *Current Directions in Psychological Science, 17*(4), 295–298.

Attarian, H. P., Schenck, C. H., & Mahowald, M. W. (2000). Presumed REM sleep behavior disorder arising from cataplexy and wakeful dreaming. *Sleep Medicine, 1*(2), 131–133.

Augustine. (400/1991). *Confessions* (H. Chadwick, Trans.). Oxford: Oxford University Press.

Augustine. (1950). *Against the academics* (J. J. O'Meara, Trans.). Westminster, MD: Newman Press.

Augustine. (1997). *The nature and origin of the soul: The works of Saint Augustine: A translation for the 21st century*. Part 1, Vol. 23. (R. Teske, Ed. & Trans.). New York: New City Press.

Austin, J. L. (1962). *Sense and sensibilia: Reconstructed from the manuscript notes by G. J. Warnock*. Oxford: Oxford University Press.

Aydede, M. (2010). The language of thought hypothesis. In E. N. Zalta (Ed.), *The Stanford encyclopedia of philosophy (fall 2010 edition)*. Retrieved from http://plato.stanford.edu/archives/fall2010/entries/language-thought.

Ayer, A. J. (1946/1952). *Language, truth, and logic*. New York: Dover.

Ayer, A. J. (1977). Professor Malcolm on dreams. In C. E. M. Dunlop (Ed.), *Philosophical essays on dreaming* (pp. 127–148). Ithaca: Cornell University Press. (Reprinted from *Journal of Philosophy* 57 [1960], pp. 517–535.)

Baars, B. B., & Gage, N. M. (2010). *Cognition, brain, and consciousness: Introduction to cognitive neuroscience* (2nd ed.). Burlington, MA: Elsevier Academic Press.

Baekeland, F., & Lasky, R. (1968). The morning recall of rapid eye movement period reports given earlier in the night. *Journal of Nervous and Mental Disease, 147*(6), 570–579.

Baillet, A. [1691] (1946). *La vie de Monsieur Descartes*. Paris: La Table Ronde.

Baker, L. R. (1998). The first-person perspective: A test for naturalism. *American Philosophical Quarterly, 35*(4), 327–348.

Baker, L. R. (2013). *Naturalism and the first-person perspective*. Oxford: Oxford University Press.

Baldridge, B. J. (1966). Physical concomitants of dreaming and the effect of stimulation on dreams. *Ohio State Medical Journal, 62*, 1272–1275.

Baldridge, B. J., Whitman, R., & Kramer, M. (1965). The concurrence of fine muscle activity and rapid eye movements during sleep. *Psychosomatic Medicine, 27*, 19–26.

Barbera, J. (2008). Sleep and dreaming in Greek and Roman philosophy. *Sleep Medicine, 9*, 906–910.

Barrett, D. (1991). Flying dreams and lucidity: An empirical study of their relationship. *Dreaming, 1*(2), 129–134.

Barrett, D. (2007). An evolutionary theory of dreams and problem-solving. In D. Barrett & P. McNamara (Eds.), *The new science of dreaming: Vol. 4. Cultural and theoretical perspectives* (pp. 133–154). Westport: Praeger Perspectives.

Bassetti, C., Vella, S., Donati, F., Wielepp, P., & Weder, B. (2000). SPECT during sleepwalking. *Lancet, 356*(9228), 484–485.

Bastuji, H., Mazza, S., Perchet, C., Frot, M., Mauguiere, F., Magnin, M., et al. (2012). Filtering the reality: Functional dissociation of lateral and medial pain systems during sleep in humans. *Human Brain Mapping, 33*(11), 2638–2649. doi:10.1002/hbm.21390.

Bastuji, H., Perchet, C., Legrain, V., Montes, C., & Garcia-Larrea, L. (2008). Laser evoked responses to painful stimulation persist during sleep and predict subsequent arousals. *Pain, 137*(3), 589–599.

Bayer, L., Constantinescu, I., Perrig, S., Vienne, J., Vidal, P.-P., Muhlethaler, M., & Schwartz, S. (2011). Rocking synchronizes brain waves during a short nap. *Current Biology: CB, 21*(12), R461–462. doi:10.1016/j.cub.2011.05.012.

Bayne, T. (2012). *The unity of consciousness*. Oxford: Oxford University Press.

Bayne, T. (2015). Introspective insecurity. In T. Metzinger & J. M. Windt (Eds.), *Open MIND: 3(T)*. Frankfurt am Main: MIND Group. doi: 10.15502/9783958570214.

Bayne, T., & Montague, M. (Eds.). (2011). *Cognitive phenomenology*. Oxford: Oxford University Press.

Bayne, T. J., & Pacherie, E. (2005). In defence of the doxastic conception of delusions. *Mind and Language, 20*(2), 163–188.

Bayne, T., & Spener, M. (2010). Introspective humility. *Philosophical Issues, 20*, 1–22.

Beare, J. I. (1906). *Greek theories of elementary cognition from Alcmaeon to Aristotle*. Oxford: Clarendon Press.

Beauregard, M. (2011). Transcendent experiences and brain mechanisms. In E. Cardeña & M. Winkelman (Eds.), *Altering consciousness: Multidisciplinary perspectives: Vol. 2. Biological and psychological perspectives* (pp. 63–84). Santa Barbara, CA: Praeger.

Bennett, M. R., & Hacker, P. M. S. (2003). *The philosophical foundations of neuroscience*. Malden, MA: Blackwell.

Berger, H. (1929). Über das Elektroenkephalogram des Menschen. *Archiv für Psychiatrie und Nervenkrankheiten, 87*, 527–570.

Berger, R. J. (1963). Experimental modification of dream content by meaningful verbal stimuli. *British Journal of Psychiatry, 109*, 722–740.

Bergson, H. (1914/2007). *Dreams* (Edward E. Slosson, Trans.). Retrieved from http://www.gutenberg.org/ebooks/20842.

Berkeley, G. (1710). *A treatise concerning the principles of human knowledge*. Dublin: Aharon Rhames. Retrieved from http://galenet.galegroup.com/servlet/ECCO?c=1&stp=Author&ste=11&af=BN&ae=T077986&tiPG=1&dd=0&dc=flc&docNum=CW118263402&vrsn=1.0&srchtp=a&d4=0.33&n=10&SU=0LRF&locID=dfg_ecco.

Berkovich-Ohana, A., Dor-Ziderman, Y., Glicksohn, J., & Goldstein, A. (2013). Alterations in the sense of time, space, and body in the mindfulness-trained brain: A neurophenomenologically-guided MEG study. *Frontiers in Psychology, 4*, 912. doi:10.3389/fpsyg.2013.00912.

Bernhard, P. (2005). Luzide Träume und analytische Philosophie. *Zeitschrift für Anomalistik, 5*, 101–114.

Berntsen, D., & Jacobsen, A. S. (2008). Involuntary (spontaneous) mental time travel into the past and future. *Consciousness and Cognition, 17*(4), 1093–1104. doi:10.1016/j.concog.2008.03.001.

Berrios, G. E., & Marková, I. S. (2004). Insight in the psychoses: A conceptual history. In X. F. Amador & A. S. David (Eds.), *Insight and psychosis: Awareness of illness in schizophrenia and related disorders* (2nd ed., pp. 31–50). Oxford: Oxford University Press.

Bertolo, H., Paiva, T., Pessoa, L., Mestre, T., Marques, R., & Santos, R. (2003). Visual dream content, graphical representation and EEG alpha activity in congenitally blind subjects. *Cognitive Brain Research, 15*, 277–284.

Bianchi, M. T., Williams, K. L., McKinney, S., & Ellenbogen, J. M. (2013). The subjective–objective mismatch in sleep perception among those with insomnia and sleep apnea. *Journal of Sleep Research, 22*(5), 557–568. doi:10.1111/jsr.12046.

Bigelsen, J., & Schupak, C. (2011). Compulsive fantasy: Proposed evidence of an under-reported syndrome through a systematic study of 90 self-identified non-normative fantasizers. *Consciousness and Cognition, 20*(4), 1634–1648. doi:10.1016/j.concog.2011.08.013.

Binz, C. (1878). *Über den Traum*. Bonn: Adolph Marcus.

Bischof, M., & Bassetti, C. F. (2004). Total dream loss: A distinct neuropsychological dysfunction after bilateral PCA stroke. *Annals of Neurology, 56*, 583–586.

Blackmore, S. (1983). *Beyond the body: An investigation of out-of-the-body experiences*. London: Paladin.

Blackmore, S. (1987). Where am I? Perspectives in imagery and the out-of-body experience. *Journal of Mental Imagery, 11*(2), 53–66.

Blackmore, S. (1988). A theory of lucid dreams and OBEs. In J. Gackenbach & S. LaBerge (Eds.), *Conscious mind, sleeping brain: Perspectives on lucid dreaming* (pp. 373–387). New York: Plenum Press.

Blackmore, S. (1996). *In search of the light: The adventures of a parapsychologist.* New York: Prometheus.

Blagrove, M. (1996). Problems with the cognitive psychological modeling of dreaming. *Journal of Mind and Behavior, 17*, 99–134.

Blagrove, M. (2007). Dreaming and personality. In D. Barrett & P. McNamara (Eds.), *The new science of dreaming: Vol. 2. Content, recall, personality correlates* (pp. 115–158). Westport: Praeger Perspectives.

Blagrove, M., Farmer, L., & Williams, E. (2004). The relationship of nightmare frequency and nightmare distress to well-being. *Journal of Sleep Research, 13*, 129–136.

Blagrove, M., Fouquet, N. C., Henley-Einion, J. A., Pace-Schott, E. F., Davies, A. C., Neuschaffer, J. L., et al. (2011). Assessing the dream-lag effect for REM and NREM stage 2 dreams. *PLoS ONE, 6*(10), e26708. doi:10.1371/journal.pone.0026708.

Blagrove, M., Henley-Einion, J., Barnett, A., Edwards, D., & Heidi Seage, C. (2011). A replication of the 5–7 day dream-lag effect with comparison of dreams to future events as control for baseline matching. *Consciousness and Cognition, 20*(2), 384–391.

Blagrove, M., & Pace-Schott, E. F. (2010). Trait and neurobiological correlates of individual differences in dream recall and dream content. *International Review of Neurobiology, 92*, 155–180.

Blagrove, M., Ruby, P., & Eichenlaub, J.-B. (2013). Dreams are made of memories, but maybe not for memory. *Behavioral and Brain Sciences, 36*(06), 609–610. doi:10.1017/S0140525X13001222.

Blanke, O. (2012). Multisensory brain mechanisms of bodily self-consciousness. *Nature Reviews: Neuroscience, 13*, 556–571. doi:10.1038/nrn3292.

Blanke, O., & Arzy, S. (2005). The out-of-body experience: Disturbed self-processing at the temporo-parietal junction. *Neuroscientist, 11*(1), 16–24. doi:10.1177/1073858404270885.

Blanke, O., & Metzinger, T. (2009). Full-body illusions and minimal phenomenal selfhood. *Trends in Cognitive Sciences, 13*(1), 7–13.

Blanke, O., & Mohr, C. (2005). Out-of-body experience, heautoscopy, and autoscopic hallucination of neurological origin: Implications for neurocognitive mechanisms of corporeal awareness and self-consciousness. *Brain Research Reviews, 50*(1), 184–199.

Blanke, O., Mohr, C., Michel, C. M., Pascual-Leone, A., Brugger, P., & Seeck, M. (2005). Linking out-of-body experience and self-processing to mental own-body imagery at the temporoparietal junction. *Journal of Neuroscience, 25*(3), 550–557.

Blanke, O., Morgenthaler, F. D., Brugger, P., & Overney, L. S. (2009). Preliminary evidence for a fronto-parietal dysfunction in able-bodied participants with a desire for limb amputation. *Journal of Neuropsychology, 3*(Pt 2), 181–200. doi:10.1348/174866408X318653.

Blascovich, J., & Bailenson, J. (2011). *Infinite reality: Avatars, eternal life, new worlds, and the dawn of the virtual revolution*. New York: Harper Collins.

Blischke, K., & Erlacher, D. (2007). How sleep enhances motor learning—a review. *Journal of Human Kinetics, 17*, 3–14.

Blischke, K., Erlacher, D., Kresin, H., Brueckner, S., & Malangré, A. (2008). Benefits of sleep in motor learning—prospects and limitations. *Journal of Human Kinetics, 20*, 23–36.

Block, N. (1981). Introduction: What is the issue? In N. Block (Ed.), *Imagery* (pp. 1–18). Cambridge, MA: MIT Press.

Block, N. (2005). Review of Alva Noë: Action in perception. *Journal of Philosophy, 102*(5), 259–272.

Block, N. (2007). Consciousness, accessibility, and the mesh between psychology and neuroscience. *Behavioral and Brain Sciences, 30*, 499–548.

Block, N. (2011). Perceptual consciousness overflows cognitive access. *Trends in Cognitive Sciences, 15*(12), 567–575. doi:10.1016/j.tics.2011.11.001.

Blom, J. D. (2010). *A dictionary of hallucinations*. New York: Springer.

Blom, J. D., & Sommer, I. E. C. (2012). Hallucinations of bodily sensation. In J. D. Blom & I. E. C. Sommer (Eds.), *Hallucinations: Research and practice* (pp. 157–170). New York: Springer.

Blumberg, M. S. (2010). Beyond dreams: Do sleep-related movements contribute to brain development? *Frontiers in Neurology, 1*, 140. doi:10.3389/fneur.2010.00140.

Blumberg, M. S., Coleman, C. M., Gerth, A. I., & McMurray, B. (2013). Spatiotemporal structure of REM sleep twitching reveals developmental origins of motor synergies. *Current Biology: CB, 23*(21), 2100–2109. doi:10.1016/j.cub.2013.08.055.

Boeve, B. F. (2010). REM sleep behavior disorder: Updated review of the core features, the REM sleep behavior disorder-neurodegenerative disease association, evolving concepts, controversies, and future directions. *Annals of the New York Academy of Sciences, 1184*, 15–54. doi:10.1111/j.1749-6632.2009.05115.x.

Bogzaran, F. (2003). Lucid art and hyperspace lucidity. *Dreaming, 13*(1), 29–42. doi:10.1023/A:1022186217703.

Born, J., & Wagner, U. (2004). Memory consolidation during sleep: Role of cortisol feedback. *Annals of the New York Academy of Sciences, 1032*, 198–201.

Boroditsky, L., & Ramscar, M. (2002). The roles of body and mind in abstract thought. *Psychological Science, 13*(2), 185–189.

Borst, G., & Kosslyn, S. M. (2008). Visual mental imagery and visual perception: Structural equivalence revealed by scanning processes. *Memory & Cognition, 36*(4), 849–862. doi:10.3758/MC.36.4.849.

Bortolotti, L. (2005). Delusions and the background of rationality. *Mind and Language, 20*(2), 189–208.

Bortolotti, L. (2009). *Delusions and other irrational beliefs*. Oxford: Oxford University Press.

Bortolotti, L. (2010). Delusion. In E. N. Zalta (Ed.), *The Stanford encyclopedia of philosophy (fall 2010 edition)*. Retrieved from http://plato.stanford.edu/archives/fall2010/entries/delusion.

Bossard, R. (1987). *Traumpsychologie: Wachen—Schlafen—Träumen*. Frankfurt am Main: Fischer Taschenbuch.

Botterill, G. (2008). The internal problem of dreaming: Detection and epistemic risk. *International Journal of Philosophical Studies, 16*(2), 139–160. doi:10.1080/09672550701740252.

Botvinick, M., & Cohen, J. (1998). Rubber hands "feel" touch that eyes see. *Nature, 391*(6669), 756. doi:10.1038/35784.

Bouwsma, O. K. (1977). Descartes' skepticism of the senses. In C. E. M. Dunlop (Ed.), *Philosophical essays on dreaming* (pp. 52–63). Ithaca: Cornell University Press. (Reprinted from *Mind* 54 [1945], pp. 313–322.)

Bradley, F. H. (1894). On the failure of movement in dream. *Mind, 3*(11), 373–377.

Braun, A. R. (2014). Your PET studies demonstrate major differences between REM sleep and waking. How do you conceive of these differences in relation to theories of waking and dreaming consciousness? In N. Tranquillo (Ed.), *Dream consciousness: Allan Hobson's new approach to the brain and its mind* (pp. 95–100). New York: Springer.

Braun, A. R., Balkin, T. J., Wesensten, N. J., Carson, R. E., Varga, M., Baldwin, P., et al. (1997). Regional cerebral blood flow throughout the sleep-wake cycle: An H2(15)O PET study. *Brain, 120*(7), 1173–1197. doi:10.1093/brain/120.7.1173.

Braun, A. R., Balkin, T. J., Wesensten, N. J., Gwadry, F., Carson, R. E., Varga, M., et al. (1998). Dissociated pattern of activity in visual cortices and their projections during human rapid eye movement sleep. *Science, 279*(2), 91–95.

Braz, J., Vázquez, P.-P., & Pereira, J. M. (Eds.). *GRAPP 2007: Proceedings of the Second International Conference on Comupter Graphics Theory and Applications*, Vol. AS/IE. Barcelona: INSTICC.

Breger, L., Hunter, I., & Lane, R. W. (1971). *The effect of stress on dreams*. New York: International Universities Press.

Brentano, F. (1874). *Psychologie vom empirischen Standpunkt*. Leipzig: Duncker & Humblot.

Breuer, J., & Freud, S. (1893/2003). *Studien über Hysterie*. Frankfurt am Main: Fischer Taschenbuch.

Briesen, J. (2008). Skepticism, externalism, and inference to the best explanation. *Abstracta, 4*(1), 5–26.

Briscoe, R. (2011). Mental imagery and the varieties of amodal perception. *Pacific Philosophical Quarterly, 92*(2), 153–173.

Brooks, J. E., & Vogelsong, J. A. (1999). *The conscious exploration of dreaming: Discovering how we create and control our dreams.* Bloomington: First Books Library.

Broughton, J. (2002). *Descartes's method of doubt.* Princeton: Princeton University Press.

Broughton, J. (2005). Dreamers and madmen. In C. Mercer & E. O'Neill (Eds.), *The history of early modern philosophy* (pp. 1–23). Oxford: Oxford University Press.

Broughton, J. (2006). Impressions and ideas. In S. Traiger (Ed.), *The Blackwell guide to Hume's Treatise* (pp. 43–57). Malden, MA: Blackwell.

Brown, R. (1957). Sound sleep and sound scepticism. *Australasian Journal of Philosophy, 35*(May), 47–53.

Browne, A. (1977). Descartes's dreams. *Journal of the Warburg and Courtauld Institutes, 40*, 256–273.

Brüggen, A. S. (2015). Imagination and experience—A commentary on Jérôme Dokic and Margherita Arcangeli. In T. Metzinger & J. M. Windt (Eds.), *Open MIND: 11(C).* Frankfurt am Main: MIND Group. doi: 10.15502/9783958570481.

Brugger, P. (2006). From phantom limb to phantom body: Varieties of extracorporeal awareness. In G. Knoblich, I. M. Thornton, M. Grosjean, & M. Shiffrar (Eds.), *Human body perception from the inside out* (pp. 171–210). Oxford: Oxford University Press.

Brugger, P. (2008). The phantom limb in dreams. *Consciousness and Cognition, 17*(4), 1272–1278. doi:10.1016/j.concog.2008.01.005.

Brugger, P. (2012). Phantom limb, phantom body, phantom self: A phenomenology of "body hallucinations." In J. D. Blom & I. E. C. Sommer (Eds.), *Hallucinations: Research and practice* (pp. 203–218). New York: Springer.

Brugger, P., Kollias, S. S., Müri, R. M., Crelier, G., Hepp-Reymond, M.-C., & Regard, M. (2000). Beyond re-membering: Phantom sensations of congenitally absent limbs. *Proceedings of the National Academy of Sciences of the United States of America, 97*(11), 6167–6172. doi:10.1073/pnas.100510697.

Bryant, R. A., Wyzenbeek, M., & Weinstein, J. (2011). Dream rebound of suppressed emotional thoughts: The influence of cognitive load. *Consciousness and Cognition, 20*(3), 515–522. doi:10.1016/j.concog.2010.11.004.

Buckner, R. L., Andrews-Hanna, J. R., & Schacter, D. L. (2008). The brain's default network: Anatomy, function, and relevance to disease. *Annals of the New York Academy of Sciences, 1124*, 1–38. doi:10.1196/annals.1440.011.

Bugalho, P., & Paiva, T. (2011). Dream features in the early stages of Parkinson's disease. *Journal of Neural Transmission, 118*(11), 1613–1619. doi:10.1007/s00702-011-0679-5.

Bulkeley, K., Broughton, B., Sanchez, A., & Stiller, J. (2005). Earliest remembered dreams. *Dreaming, 15*(3), 205–222.

Burgess, N. (2006). Spatial memory: How egocentric and allocentric combine. *Trends in Cognitive Sciences, 10*(12), 551–557.

Burgess, N. (2008). Spatial cognition and the brain. *Annals of the New York Academy of Sciences, 1124*, 77–97. doi:10.1196/annals.1440.002.

Burks, A. W. (1946). Peirce's theory of abduction. *Philosophy of Science, 13*(4), 301–306.

Burroughs, W. S. (1995). *My education: A book of dreams*. London: Picador.

Buzzi, G. (2011). False awakenings in light of the dream protoconsciousness theory: A study in lucid dreamers. *International Journal of Dream Research, 4*(2). Retrieved from http://archiv.ub.uni-heidelberg.de/ojs/index.php/IJoDR/article/view/9085.

Byrne, A. (2007). Possibility and imagination. *Philosophical Perspectives, 21*, 125–144.

Byrne, A. (2010). Recollection, perception, imagination. *Philosophical Studies, 48*, 15–26.

Caldwell, R. L. (1977). Malcolm and the criterion of sleep. In C. E. M. Dunlop (Ed.), *Philosophical essays on dreaming* (pp. 157–173). Ithaca: Cornell University Press. (Reprinted from *Australasian Journal of Philosophy* 43 [1965], pp. 339–352.)

Calkins, M. W. (1893). Statistics of dreams. *American Journal of Psychology, 5*, 311–343.

Callwood, J. (1990). *The sleepwalker: The trial that made Canadian legal history*. Toronto: Seal Books.

Campbell, J. (1999). Immunity to error through misidentification and the meaning of a referring term. *Philosophical Topics, 26*(1–2), 89–104.

Canfield, J. V. (1977). Judgments in sleep. In C. E. M. Dunlop (Ed.), *Philosophical essays on dreaming* (pp. 149–156). Ithaca: Cornell University Press. (Reprinted from *Philosophical Review* 70 [1961], pp. 224–230.)

Cariola, L. A. (2008). A structural and functional analysis of dream narratives. *Dreaming, 18*(1), 16–26.

Carruthers, P. (2011). *The opacity of mind: An integrative theory of self-knowledge*. Oxford: Oxford University Press.

Carskadon, M. A., & Herz, R. S. (2004). Minimal olfactory perception during sleep: Why odor alarms will not work for humans. *Sleep, 27*(3), 402–405.

Cartwright, R. (1990). A network model of dreams. In R. Bootzin, J. Kihlstrom, & D. Schacter (Eds.), *Sleep and cognition* (pp. 179–189). Washington, DC: American Psychological Association.

Cartwright, R. (2001). Dreams and adaptation to divorce. In D. Barrett (Ed.), *Trauma and dreams* (pp. 179–185). Cambridge, MA: Harvard University Press.

Cartwright, R. D. (2010). *The twenty-four hour mind: The role of sleep and dreaming in our emotional lives*. Oxford: Oxford University Press.

Casagrande, M., & Cortini, P. (2008). Spoken and written dream communication: Differences and methodological aspects. *Consciousness and Cognition, 17*(1), 145–158. doi:10.1016/j.concog.2007.11.002.

Casali, A. G., Gosseries, O., Rosanova, M., Boly, M., Sarasso, S., Casali, K. R., et al. (2013). A theoretically based index of consciousness independent of sensory processing and behavior. *Science Translational Medicine, 5*(198), 198ra105. doi:10.1126/scitranslmed.3006294.

Casasanto, D., & Boroditsky, L. (2008). Time in the mind: Using space to think about time. *Cognition, 106*(2), 579–593. doi:10.1016/j.cognition.2007.03.004.

Chalmers, D. (2000). What is a neural correlate of consciousness? In T. Metzinger (Ed.), *Neural correlates of consciousness: Empirical and conceptual questions* (pp. 17–39). Cambridge, MA: MIT Press.

Charcot, J. M. (1883). Un cas de suppression brusque et isolée de la vision mentale des signes et des objets (formes et couleurs). *Le Progres Medical, 11*, 568–571.

Charland-Verville, V., & Laureys, S. (2014). How does your formulation of lesion-induced states of diminished consciousness fit with AIM? Do you suppose that brain stem damage affects activation (A) and modulation (M)? In N. Tranquillo (Ed.), *Dream consciousness. Allan Hobson's new approach to the brain and its mind* (pp. 101–108). New York: Springer.

Chellappa, S. L., & Cajochen, C. (2013). Ultradian and circadian modulation of dream recall: EEG correlates and age effects. *International Journal of Psychophysiology, 89*(2), 165–170. doi:10.1016/j.ijpsycho.2013.03.006.

Chellappa, S. L., Munch, M., Blatter, K., Knoblauch, V., & Cajochen, C. (2009). Does the circadian modulation of dream recall modify with age? *Sleep, 32*(9), 1201–1209.

Cherpillod, C., Bugermeister, J. J., Dubois, C., & Castellanos, G. (1965). Quelques particularités du sommeil paradoxal de l'homme. *Schweizer Archiv für Neurologie, Neurochirurgie und Psychiatrie, 96*(1), 141–153.

Cheyne, J. A. (2001). The ominous numinous: Sensed presence and "other" hallucinations. *Journal of Consciousness Studies, 8*(5–7), 133–150.

Cheyne, J. A. (2003). Sleep paralysis and the structure of waking-nightmare hallucinations. *Dreaming, 13*, 163–179.

Cheyne, J. A. (2005). Sleep paralysis episode frequency and number, types, and structure of associated hallucinations. *Journal of Sleep Research, 14*, 319–324.

Cheyne, J. A. (2012). Sensed presences. In J. D. Blom & I. E. C. Sommer (Eds.), *Hallucinations: Research and practice* (pp. 219–234). New York: Springer.

Cheyne, J. A., & Girard, T. A. (2004). Spatial characteristics of hallucinations associated with sleep paralysis. *Cognitive Neuropsychiatry*, *9*(4), 281–300. doi:10.1080/13546800344000264.

Cheyne, J. A., & Girard, T. A. (2007a). The nature and varieties of felt presence experiences: A reply to Nielsen. *Consciousness and Cognition*, *16*, 984–991.

Cheyne, J. A., & Girard, T. A. (2007b). Paranoid delusions and threatening hallucinations: A prospective study of sleep paralysis experiences. *Consciousness and Cognition*, *16*(4), 959–974. doi:10.1016/j.concog.2007.01.002.

Cheyne, J. A., & Girard, T. A. (2009). The body unbound: Vestibular-motor hallucinations and out-of-body experiences. *Cortex*, *45*(2), 201–215. doi:10.1016/j.cortex.2007.05.002.

Cheyne, J. A., Rueffer, S. D., & Newby-Clark, I. R. (1999). Hypnagogic and hypnopompic hallucinations during sleep paralysis: Neurological and cultural construction of the night-mare. *Consciousness and Cognition*, *8*, 319–337.

Chihara, C. S. (1977). What dreams are made on. In C. E. M. Dunlop (Ed.), *Philosophical essays on dreaming* (pp. 250–264). Ithaca: Cornell University Press. (Reprinted from *Theoria* 31 [1965], pp. 145–158.)

Chihara, C. S., & Fodor, J. A. (1977). Operationalism and ordinary language: A critique of Wittgenstein. In C. E. M. Dunlop (Ed.), *Philosophical essays on dreaming* (pp. 174–204). Ithaca: Cornell University Press. (Reprinted from *American Philosophical Quarterly* 2 [1965], pp. 281–295.)

Child, W. (2007). Dreaming, calculating, thinking: Wittgenstein and anti-realism about the past. *Philosophical Quarterly*, *57*(227), 252–272.

Chokroverty, S., & Sahota, P. (Eds.). (2011). *Acute and emergent events in sleep disorders*. New York: Oxford University Press.

Chomsky, N. (1957). *Syntactic structures*. The Hague: Mouton.

Chow, H. M., Horovitz, S. G., Carr, W. S., Picchioni, D., Coddington, N., Fukunaga, M., et al. (2013). Rhythmic alternating patterns of brain activity distinguish rapid eye movement sleep from other states of consciousness. *Proceedings of the National Academy of Sciences of the United States of America*, *110*(25), 10300–10305. doi:10.1073/pnas.1217691110.

Christian, B. M., Miles, L. K., Parkinson, C., & Macrae, C. N. (2013). Visual perspective and the characteristics of mind wandering. *Frontiers in Psychology*, *4*, 699. doi:10.3389/fpsyg.2013.00699.

Churchland, P. S. (1988). Reduction and the neurobiological basis of consciousness. In A. Marcel & E. Bisiach (Eds.), *Consciousness in contemporary science* (pp. 273–304). Oxford: Oxford University Press.

Churchland, P. S. (2005). A neurophilosophical slant on consciousness research. *Progress in Brain Research*, *149*, 285–293. doi:10.1016/S0079-6123(05)49020-2.

Cicero, M. T. (1994). *Academica* (H. Rackham, Trans.). Loeb Classical Library. Cambridge, MA: Harvard University Press.

Cichy, R. M., Heinzle, J., & Haynes, J.-D. (2012). Imagery and perception share cortical representations of content and location. *Cerebral Cortex, 22*(2), 372–380. doi:10.1093/cercor/bhr106.

Cicogna, P., & Bosinelli, M. (2001). Consciousness during dreams. *Consciousness and Cognition, 10*, 26–41.

Cicogna, P., Natale, V., Occhionero, M., & Bosinelli, M. (2000). Slow wave and REM sleep mentation. *Sleep Research Online, 3*, 67–72.

Cipolli, C., Bolzani, R., & Tuozzi, G. (1998). Story-like organization of dream experience in different periods of REM sleep. *Journal of Sleep Research, 7*, 13–19.

Clark, A. (2005). The twisted matrix: Dream, simulation, or hybrid? In C. Grau (Ed.), *Philosophers explore The Matrix* (pp. 177–197). Oxford: Oxford University Press.

Clark, A. (2013a). Dreaming the whole cat: Generative models, predictive processing, and the enactivist conception of perceptual experience. *Mind, 121*(483), 753–771. doi:10.1093/mind/fzs106.

Clark, A. (2013b). Whatever next? Predictive brains, situated agents, and the future of cognitive science. *Behavioral and Brain Sciences, 36*(03), 181–204. doi:10.1017/S0140525X12000477.

Clark, A. (2015). Embodied prediction. In T. Metzinger & J. M. Windt (Eds.), *Open MIND: 7(T)*. Frankfurt am Main: MIND Group. doi: 10.15502/9783958570115.

Clarke, D. M. (2006). *Descartes: A biography*. New York: Cambridge University Press.

Cohen, J. (2002). The grand grand illusion illusion. In A. Noë (Ed.), *Is the visual world a grand illusion?* (pp. 141–157). Thorverton: Imprint Academic. (Special issue, *Journal of Consciousness Studies*, Vol. 9, no. 5–6.)

Cohen, M. A., & Dennett, D. C. (2011). Consciousness cannot be separated from function. *Trends in Cognitive Sciences, 15*(8), 358–364.

Cole, J. R. (1992). *The Olympian dreams and youthful rebellion of René Descartes*. Urbana: University of Illinois Press.

Cole, J. (1998). *About face*. Cambridge, MA: MIT Press.

Coleridge, S. T. (1957–1990). *Notebooks of S. T. Coleridge* (Kathleen Coburn, Ed.). New York: Routledge & Kegan Paul.

Collerton, D., Dudley, R., & Mosimann, U. P. (2012). Visual hallucinations. In J. D. Blom & I. E. C. Sommer (Eds.), *Hallucinations: Research and practice* (pp. 75–90). New York: Springer.

Collerton, D., & Perry, E. (2011). Dreaming and hallucinations—continuity or discontinuity? Perspectives from dementia with Lewy bodies. *Consciousness and Cognition, 20*(4), 1016–1020.

Coltheart, M. (2007). Cognitive neuropsychiatry and delusional belief. *Quarterly Journal of Experimental Psychology, 60*(8), 1041–1062.

Coltheart, M., Langdon, R., & McKay, R. (2011). Delusional belief. *Annual Review of Psychology, 62*(1), 271–298. doi:10.1146/annurev.psych.121208.131622.

Comarr, A. E., Cressy, J. M., & Letch, M. (1983). Sleep dreams of sex among traumatic paraplegics and quadriplegics. *Sexuality and Disability, 6*(1), 25–29.

Conrad, T. (1968). *Zur Wesenslehre des psychischen Lebens und Erlebens: Phenomenologica*. Den Haag: Martinus Nijhoff.

Corballis, M. C. (2013). Wandering tales: Evolutionary origins of mental time travel and language. *Frontiers in Psychology, 4, 485*. doi:10.3389/fpsyg.2013.00485.

Corlett, P. R., Canavan, S. V., Nahum, L., Appah, F., & Morgan, P. T. (2014). Dreams, reality and memory: Confabulations in lucid dreamers implicate reality-monitoring dysfunction in dream consciousness. *Cognitive Neuropsychiatry, 19*(6), 540–553. doi:10.1080/13546805.2014.932685.

Cottingham, J. (Ed.). (2005). *The Cambridge companion to Descartes*. Cambridge: Cambridge University Press.

Crane, T. (2011). The problem of perception. In E. N. Zalta (Ed.), *The Stanford encyclopedia of philosophy 2011*. Retrieved from http://plato.stanford.edu/archives/spr2011/entries/perception-problem.

Crick, F., & Mitchinson, G. (1983). The function of dream sleep. *Nature, 304*, 111–114.

Crowther, T. (in press). Experience, dreaming and the phenomenology of wakeful consciousness. In F. Dorsch, F. Macpherson, & M. Nida-Rumelin (Eds.), *Phenomenal presence*. Oxford: Oxford University Press. Retrieved from http://publications.heythrop.ac.uk/2009.

Crugnola, C. R., Maggiolini, A., Caprin, C., De Martini, C., & Giudici, F. (2008). Dream content of 10- to 11-year-old preadolescent boys and girls. *Dreaming, 18*(3), 201–216.

Curley, E. (1977). Dreaming and conceptual revision. In C. E. M. Dunlop (Ed.), *Philosophical essays on dreaming* (pp. 317–346). Ithaca: Cornell University Press. (Reprinted from *Australasian Journal of Philosophy* 53 [1975], pp. 119–141.)

Currie, G. (2000). Imagination, delusion and hallucination. *Mind and Language, 15*(1), 168–183.

Currie, G., & Ravenscroft, I. (2002). *Recreative minds: Imagination in philosophy and psychology*. Oxford: Clarendon Press.

Cvetkovic, D., & Cosic, I. (Eds.). (2011). *States of consciousness: Experimental insights into meditation, waking, sleep and dreams*. Frontiers collection. Heidelberg: Springer.

Czisch, M., Wehrle, R., Kaufmann, C., Wetter, T. C., Holsboer, F., Pollmacher, T., et al. (2004). Functional MRI during sleep: BOLD signal decreases and their electrophysiological correlates. *European Journal of Neuroscience, 20*(2), 566–574. doi:10.1111/j.1460-9568.2004.03518.x.

D'Agostino, A., Castelnovo, A., & Scarone, S. (2013). Dreaming and the neurobiology of self: Recent advances and implications for psychiatry. *Frontiers in Psychology, 4*, 680. doi:10.3389/fpsyg.2013.00680.

D'Agostino, A., Manni, R., Limosani, I., Terzaghi, M., Cavallotti, S., & Scarone, S. (2012). Challenging the myth of REM sleep behavior disorder: No evidence of heightened aggressiveness in dreams. *Sleep Medicine, 13*(6), 714–719.

D'Agostino, A., & Scarone, S. (2011). From dreams to psychosis: A European Science Foundation exploratory workshop. [Editorial]. *Consciousness and Cognition, 20*(4), 985–986. doi:10.1016/j.concog.2011.06.011.

Dainton, B. (2011). *The phenomenal self.* Oxford: Oxford University Press.

Damasio, A. R. (1994). *Descartes' error: Emotion, reason, and the human brain.* New York: Penguin.

Damasio, A. R. (1999). *The feeling of what happens: Body and emotion in the making of consciousness* (1st ed.). New York: Harcourt.

Damasio, A. R. (2010). *Self comes to mind: Constructing the conscious brain* (1st ed.). New York: Vintage.

Damasio, A. (2014). Does your "feeling of what happens" definition of consciousness extend to dreaming? If so, how do you conceptualize internally generated FWHs? In N. Tranquillo (Ed.), *Dream consciousness: Allan Hobson's new approach to the brain and its mind* (pp. 111–112). New York: Springer.

Dang-Vu, T. T. (2012). Neuronal oscillations in sleep: Insights from functional neuroimaging. *Neuromolecular Medicine, 14*(3), 154–167. doi:10.1007/s12017-012-8166-1.

Dang-Vu, T. T., Desseilles, M., Albouy, G., Darsaud, A., Gais, S., Rauchs, G., et al. (2005). Dreaming: A neuroimaging view. *Schweizer Archiv für Neurologie und Psychiatrie, 156*(8), 415–425.

Dang-Vu, T. T., Desseilles, M., Peigneux, P., Laureys, S., & Maquet, P. (2009). PET activation patterns. In R. Stickgold & M. P. Walker (Eds.), *The neuroscience of sleep* (pp. 30–36). London: Elsevier Academic Press.

Dang-Vu, T. T., Schabus, M., Desseilles, M., Schwartz, S., & Maquet, P. (2007). Neuroimaging of sleep and dreaming. In D. Barrett & P. McNamara (Eds.), *The new science of dreaming: Biological aspects* (Vol. 1, pp. 95–114). Westport: Praeger Perspectives.

Dang-Vu, T. T., Schabus, M., Desseilles, M., Sterpenich, V., Bonjean, M., & Maquet, P. (2010). Functional neuroimaging insights into the physiology of human sleep. *Sleep, 33*(12), 1589–1603.

Darling, M., Hoffmann, R., Moffitt, A., & Purcell, S. (1993). The pattern of self-reflectiveness in dream reports. *Dreaming, 3*(1), 9–19.

David, A. S. (2004). The clinical importance of insight: An overview. In X. F. Amador & A. S. David (Eds.), *Insight and psychosis: Awareness of illness in schizophrenia and related disorders* (2nd ed., pp. 359–392). Oxford: Oxford University Press.

Davidson, D. (2001). Rational animals. In D. Davidson, *Subjective, intersubjective, objective* (pp. 95–106). Oxford University Press.

Davies, C., & Harrison, J. (1996). Osmose. *ACM SIGGRAPH Computer Graphics, 30*(4), 25–28. doi:10.1145/240806.240808.

Davies, M., Coltheart, M., Langdon, R., & Breen, N. (2001). Monothematic delusions: Towards a two-factor account. *Philosophy, Psychiatry, and Psychology, 8*(2), 133–158. doi:10.1353/ppp.2001.0007.

Davies, O. (2003). The nightmare experience, sleep paralysis, and witchcraft accusations. *Folklore, 114*(2), 181–203.

Dawson, J. L., & Conduit, R. (2011). The substrate that dreams are made on: An evaluation of current neurobiological theories of dreaming. In D. Cvetkovic & I. Cosic (Eds.), *States of consciousness: Experimental insights into meditation, waking, sleep and dreams* (pp. 133–156). Frontiers collection. New York: Springer.

de Boismont, A. B. (1860). *Hallucinations: A history and explanation of apparitions, visions, dreams, ecstasy, magnetism, and somnambulism* (R. T. Hulme, Trans.). Columbus, OH: Joseph Riley.

Debru, C. (2014). On the relationship between neuroscience and philosophy: The case of sleep and dreaming. In M. Galavotti, E. Nemeth, & F. Stadler (Eds.), *Vienna Circle Institute yearbook: European philosophy of science—philosophy of science in Europe and the Viennese heritage* (pp. 57–65). Springer International. Retrieved from http://dx.doi.org/10.1007/978-3-319-01899-7_4.

Debruyne, H., Portzky, M., Van den Eynde, F., & Audenaert, K. (2009). Cotard's syndrome: A review. *Current Psychiatry Reports, 11*(3), 197–202.

DeCicco, T. L., Lyons, T., Pannier, W., Wright, C., & Clarke, J. (2010). Exploring the dreams of women with breast cancer: Content and meaning of dreams. *International Journal of Dream Research, 3*(2), 104–110.

de Gennaro, L., Marzano, C., Cipolli, C., & Ferrara, M. (2012). How we remember the stuff that dreams are made of: Neurobiological approaches to the brain mechanisms of dream recall. *Behavioural Brain Research, 226*(2), 592–596. doi:10.1016/j.bbr.2011.10.017.

Dement, W. C. (1955). Dream recall and eye movements during sleep in schizophrenics and normal. *Journal of Nervous and Mental Disease, 122*, 263–269.

Dement, W. C. (1960). The effect of dream deprivation: The need for a certain amount of dreaming each night is suggested by recent experiments. *Science, 131*(3415), 1705–1707.

Dement, W. C. (1999/2001). *The promise of sleep: The scientific connection between health, happiness and a good night's sleep*. London: Pan.

Dement, W. C., & Kleitman, N. (1957a). Cyclic variations in EEG during sleep and their relation to eye movements, body motility, and dreaming. *Electroencephalography and Clinical Neurophysiology, 9*(4), 637–690.

Dement, W. C., & Kleitman, N. (1957b). The relation of eye movements during sleep to dream activity: An objective method for the study of dreaming. *Journal of Experimental Psychology, 53*, 89–97.

Dement, W. C., & Wolpert, E. (1958). The relation of eye movements, body motility, and external stimuli to dream content. *Journal of Experimental Psychology, 55*, 543–553.

de Montaigne, M. (1966). The apology for Raymond Sebond. In R. M. Popkin (Ed.), *Readings in the history of philosophy: The philosophy of the 16th and 17th centuries*. Toronto: Collier-Macmillan.

Demougeot, L., Normand, H., Denise, P., Papaxanthis, C., & Aleman, A. (2009). Discrete and effortful imagined movements do not specifically activate the autonomic nervous system. *PloS ONE, 4*(8), e6769. doi:10.1371/journal.pone.0006769.

Dennett, D. C. (1976). Are dreams experiences? *Philosophical Review, 85*(2), 151–171.

Dennett, D. C. (1979). The onus re experiences: A reply to Emmett. *Philosophical Studies, 35*(April), 315–318.

Dennett, D. C. (1981a). The nature of images and the introspective trap. In N. Block (Ed.), *Imagery* (pp. 51–62). Cambridge, MA: MIT Press.

Dennett, D. C. (1981b). True believers: The intentional strategy and why it works. In A. F. Heath (Ed.), *Scientific explanation: Papers based on Herbert Spencer lectures given in the University of Oxford*. Oxford: Clarendon.

Dennett, D. C. (1987). *The intentional stance*. Cambridge, MA: MIT Press.

Dennett, D. C. (1991). *Consciousness explained*. Boston: Little, Brown.

Dennett, D. C. (1993). Quining qualia. In A. I. Goldman (Ed.), *Readings in philosophy and cognitive science*. Cambridge, MA: MIT Press. (Originally published in Marcel, A. J., & Bisiach, E. [Eds.]. [1988]. *Consciousness in contemporary science* [pp. 381–414]. Oxford: Oxford University Press.)

Dennett, D. C. (1995). *Darwin's dangerous idea: Evolution and the meanings of life*. London: Penguin.

Dennett, D. (2003). Who's on first? Heterophenomenology explained. *Journal of Consciousness Studies, 10*(9–10), 19–30.

Dennett, D. C. (2007). Heterophenomenology reconsidered. *Phenomenology and the Cognitive Sciences, 6*(1), 247–270.

Dennett, D. C. (2013). *Intuition pumps and other tools for thinking*. New York: W. W. Norton.

Dennett, D. C. (2014). What is dreaming for, if anything? In N. Tranquillo (Ed.), *Dream consciousness. Allan Hobson's new approach to the brain and its mind* (pp. 113–118). New York: Springer.

Dennett, D. C. (2015). Why and how does consciousness seem the way it seems? In T. Metzinger & J. M. Windt (Eds.), *Open MIND: 10(T)*. Frankfurt am Main: MIND Group. doi:10.15502/9783958570245.

Dennett, D. C., & Akins, K. (2008). Multiple drafts model. *Scholarpedia, 3*(4), 4321. Retrieved from http://www.scholarpedia.org/article/Multiple_drafts_model.

Dennett, D. C., & Kinsbourne, M. (1992). Time and the observer: The where and when of consciousness in the brain. *Behavioral and Brain Sciences, 15*, 183–201.

de Ridder, D., van Laere, K., Dupont, P., Menovsky, T., & van de Heyning, P. (2007). Visualizing out-of-body experience in the brain. *New England Journal of Medicine, 357*(18), 1829–1833. doi:10.1056/NEJMoa070010.

de Saint-Denys, H. (1867). *Les rêves et les moyens de les diriger*. Paris: Amyot.

de Sanctis, S. (1901). *Die Träume: Medizinisch-psychologische Untersuchungen; Autorisierte und durch zahlreiche Nachträge des Verfassers erweiterte Übersetzung von Dr. O. Schmidt, nebst Einführung von Dr. P. J. Möbius*. Leipzig: Verlag von Carl Marhold.

Descartes, R. (1637/2008). *Discourse on the method of rightly conducting the reason, and seeking truth in the sciences*. Retrieved from http://www.gutenberg.org/files/59/59-h/59-h.htm.

Descartes, R. (1641/1901). *Descartes' Meditations* (John Veitch, Trans.). Retrieved from http://www.wright.edu/cola/descartes/mede.html.

Descartes, R. (1972). *Meditationen über die Grundlagen der Philosophie mit den sämtlichen Einwänden und Erwiderungen: Zum erstenmal vollständig* (A. Buchenau, Trans. & Ed.). Hamburg: Felix Meiner.

Descartes, R. (1984). *The philosophical writings of Descartes* (J. Cottingham, R. Stoothoff, & D. Murdoch, Trans. & Eds.). Cambridge: Cambridge University Press.

Desmedt, J. E., & Manil, J. (1970). Somatosensory evoked potentials of the normal human neonate in REM sleep, in slow wave sleep and in waking. *Electroencephalography and Clinical Neurophysiology, 29*, 113–126.

de Sousa, R. (2009). Epistemic feelings. *Mind and Matter, 7*(2), 139–161.

de Sousa, R. (2013). Emotion. In E. N. Zalta (Ed.), *The Stanford encyclopedia of philosophy (spring 2013 edition)*. Retrieved from http://plato.stanford.edu/archives/spr2013/entries/emotion.

Desseilles, M., Dang-Vu, T. T., Sterpenich, V., & Schwartz, S. (2011). Cognitive and emotional processes during dreaming: A neuroimaging view. *Consciousness and Cognition, 20*(4), 998–1008.

Destexhe, A., Hughes, S. W., Rudolph, M., & Crunelli, V. (2007). Are corticothalamic "up" states fragments of wakefulness? *Trends in Neurosciences, 30*(7), 334–342. doi:10.1016/j.tins.2007.04.006.

de Vignemont, F. (2011). Bodily awareness. In Edward N. Zalta (Ed.), *The Stanford encyclopedia of philosophy (fall 2011 edition)*. Retrieved from http://plato.stanford.edu/archives/fall2011/entries/bodily-awareness.

de Vignemont, F. (in press). A multimodal conception of bodily awareness. *Mind*.

Devinsky, O., Feldmann, E., Burrowes, K., & Bromfield, E. (1989). Autoscopic phenomena with seizures. *Archives of Neurology, 46*(10), 1080–1088.

Diaz, B. A., van der Sluis, S., Moens, S., Benjamins, J. S., Migliorati, F., Stoffers, D., et al. (2013). The Amsterdam Resting-State Questionnaire reveals multiple phenotypes of resting-state cognition. *Frontiers in Human Neuroscience, 7*, 446. doi:10.3389/fnhum.2013.00446.

Dieguez, S., Annoni, J.-M., & Godefroy, O. (2013). Asomatognosia: Disorders of the bodily self. In O. Godefroy (Ed.), *The behavioral and cognitive neurology of stroke* (pp. 170–192). Cambridge: Cambridge University Press.

Dieguez, S., & Blanke, O. (2011). Altered states of bodily consciousness. In E. Cardeña & M. Winkelman (Eds.), *Altering consciousness: Multidisciplinary perspectives: Vol. 2. Biological and psychological perspectives* (pp. 237–262). Santa Barbara, CA: Praeger.

Dieguez, S., Staub, F., & Bogousslavsky, J. (2007). Asomatognosia. In O. Godefroy & J. Bogousslavsky (Eds.), *The behavioral and cognitive neurology of stroke* (pp. 215–253). Cambridge: Cambridge University Press.

Diekelmann, S. (2014). Sleep for cognitive enhancement. *Frontiers in Systems Neuroscience, 8*, 46. doi:10.3389/fnsys.2014.00046.

Diekelmann, S., & Born, J. (2014). Your interest in sleep and memory posits a semantic learning function to NREM sleep. What are your views of associative memory enhancement by REM? In N. Tranquillo (Ed.), *Dream consciousness. Allan Hobson's new approach to the brain and its mind* (pp. 119–122). New York: Springer.

Diekelmann, S., Wilhelm, I., & Born, J. (2009). The whats and whens of sleep-dependent memory consolidation. *Sleep Medicine Reviews, 13*(5), 309–321. doi:10.1016/j.smrv.2008.08.002.

Dienes, Z., & Seth, A. (2010). Gambling on the unconscious: A comparison of wagering and confidence ratings as measures of awareness in an artificial grammar task. *Consciousness and Cognition, 19*(2), 674–681. doi:10.1016/j.concog.2009.09.009.

Diogenes Laertius. (1943). *Lives of eminent philosophers* (R. D. Hicks, Trans.). Cambridge, MA: Harvard University Press.

Dobricki, M., de La Rosa, S., & Tsakiris, M. (2013). The structure of conscious bodily self-perception during full-body illusions. *PLOS ONE, 8*(12), e83840. doi:10.1371/journal.pone.0083840.

Dokic, J. (2012). Seeds of self-knowledge: Noetic feelings and metacognition. In M. J. Beran, J. Brandl, J. Perner, & J. Proust (Eds.), *Foundations of metacognition* (pp. 302–321). Oxford: Oxford University Press.

Dokic, J. & Arcangeli, M. (2015). The heterogeneity of experiential imagination. In T. Metzinger & J. M. Windt (Eds.), *Open MIND: 11(T)*. Frankfurt am Main: MIND Group. doi:10.15502/9783958570085.

Domhoff, G. W. (1996). *Finding meaning in dreams*. New York: Plenum Press.

Domhoff, G. W. (1999). Drawing theoretical implications from descriptive empirical findings on dream content. *Dreaming, 9*, 201–210.

Domhoff, G. W. (2000). *Dreams and parapsychology*. Retrieved from http://www.dreamresearch.net/Library/domhoff_2000c.html.

Domhoff, G. W. (2001). A new neurocognitive theory of dreams. *Dreaming, 11*, 13–33.

Domhoff, G. W. (2003). *The scientific study of dreams: Neural networks, cognitive development, and content analysis*. Washington, DC: American Psychological Association.

Domhoff, G. W. (2005). Refocusing the neurocognitive approach to dreams: A critique of the Hobson versus Solms debate. *Dreaming, 15*, 3–20.

Domhoff, G. W. (2011a). Dreams are embodied simulations that dramatize conceptions and concerns: The continuity hypothesis in empirical, theoretical, and historical context. *International Journal of Dream Research, 4*(2), 50–62.

Domhoff, G. W. (2011b). The neural substrate for dreaming: Is it a subsystem of the default network? *Consciousness and Cognition, 20*(4), 1163–1174. doi:10.1016/j.concog.2011.03.001.

Domhoff, G. W., & Schneider, A. (2008a). Similarities and differences in dream content at the cross-cultural, gender, and individual levels. *Consciousness and Cognition, 17*(4), 1257–1265. doi:10.1016/j.concog.2008.08.005.

Domhoff, G. W., & Schneider, A. (2008b). Studying dream content using the archive and search engine on DreamBank.net. *Consciousness and Cognition, 17*(4), 1238–1247. doi:10.1016/j.concog.2008.06.010.

Doricchi, F., Iaria, G., Silvetti, M., Figliozzi, F., & Siegler, I. (2007). The "ways" we look at dreams: Evidence from unilateral spatial neglect (with an evolutionary account of dream bizarreness). *Experimental Brain Research, 178*(4), 450–461. doi:10.1007/s00221-006-0750-x.

Doricchi, F., & Violani, C. (1992). Dream recall in brain-damaged patients: a contribution to the neuropsychology of dreaming through a review of the literature. In J. S. Antrobus & M. Bertini (Eds.), *The neuropsychology of sleep and dreaming* (pp. 99–129). Hillsdale, NJ: Erlbaum.

Dorsch, F., Macpherson, F., & Nida-Rumelin, M. (Eds.). (in press). Phenomenal presence. In *Phenomenal presence*. Oxford: Oxford University Press.

Douven, I. (2009). Can the skepticism debate be resolved? *Synthese, 168*(1), 23–52. doi:10.1007/s11229-008-9311-0.

Douven, I. (2011). Abduction. In E. N. Zalta (Ed.), *The Stanford encyclopedia of philosophy (spring 2011 edition)*. Retrieved from http://plato.stanford.edu/cgi-bin/encyclopedia/archinfo.cgi?entry=abduction.

Dreisbach, C. (2000). Dreams in the history of philosophy. *Dreaming, 10*(1), 31–41.

Dresler, M., Eibl, L., Fischer, C. F. J., Wehrle, R., Spoormaker, V. I., Steiger, A., et al. (2014). Volitional components of consciousness vary across wakefulness, dreaming, and lucid dreaming. *Frontiers in Psychology, 4*, 987. doi:10.3389/fpsyg.2013.00987.

Dresler, M., Koch, S., Wehrle, R., Spoormaker, V. I., Holsboer, F., Steiger, A., et al. (2011). Dreamed movement elicits activation in the sensorimotor cortex. *Current Biology, 21*(21), 1833–1837.

Dresler, M., Koch, S., Wehrle, R., Spoormaker, V. I., Holsboer, F., Steiger, A., et al. (2012). Neural correlates of dream lucidity obtained from contrasting lucid versus non-lucid REM sleep: A combined EEG/fMRI case study. *Sleep, 35*(7), 1017–1020. doi:10.5665/sleep.1974.

Dresler, M., Spoormaker, V., Wehrle, R., & Czisch, M. (2014). Your imaging group has coined the term "dream imaging": Please summarize the concept in relation to dream theory. In N. Tranquillo (Ed.), *Dream consciousness: Allan Hobson's new approach to the brain and its mind* (pp. 123–129). New York: Springer.

Dresler, M., Wehrle, R., Koch, S., Sämann, P. G., Steiger, A., Obrig, H., et al. (2008). Dream imaging—how to read the sleeping brain. *Journal of Sleep Research, 17*(Suppl. 1), 92.

Dresler, M., Wehrle, R., Spoormaker, V., Holsboer, F., Steiger, A., Czisch, M., & Hobson, J. (2014). Neural correlates of insight in dreaming and psychosis. *Sleep Medicine Review*. doi: 10.1016/j.smrv.2014.06.004.

Dresler, M., Wehrle, R., Spoormaker, V. I., Koch, S., Holsboer, F., Steiger, A., et al. (2009). Neural correlates of consciousness—insights from sleep imaging. *8th Göttingen Meeting of the German Neuroscience Society*, T24–3C.Göttingen: Neurowissenschaftliche Gesellschaft. https://www.nwg-goettingen.de/2009/upload/file/Programm2009.pdf.

Driver, J. (2007). Dream immorality. *Philosophy, 82*(01), 5–22. doi:10.1017/S0031819107319013.

Droege, P. (2009). Now or never: How consciousness represents time. *Consciousness and Cognition, 18*(1), 78–90. doi:10.1016/j.concog.2008.10.006.

Drummond, S. P. A., & McKenna, B. S. (2009). Sleep deprivation and brain function. In R. Stickgold & M. P. Walker (Eds.), *The neuroscience of sleep* (pp. 249–253). London: Elsevier Academic Press.

Dunlop, C. E. M. (1977). Introduction. In C. E. M. Dunlop (Ed.), *Philosophical essays on dreaming* (pp. 13–51). Ithaca: Cornell University Press.

Duyn, J. H. (2012). EEG-fMRI methods for the study of brain networks during sleep. *Frontiers in Neurology, 3*, 100. doi:10.3389/fneur.2012.00100.

Edelman, G. M. (2003). Naturalizing consciousness: A theoretical framework. *Proceedings of the National Academy of Sciences of the United States of America, 100*(9), 5520–5524. doi:10.1073/pnas.0931349100.

Edwards, C. L., Ruby, P. M., Malinowski, J. E., Bennett, P. D., & Blagrove, M. T. (2013). Dreaming and insight. *Frontiers in Psychology, 4*, 979. doi:10.3389/fpsyg.2013.00979.

Ehrsson, H. H. (2007). The experimental induction of out-of-body experiences. *Science, 317*(5841), 1048. doi:10.1126/science.1142175.

Eichenlaub, J.-B., Bertrand, O., Morlet, D., & Ruby, P. (2014). Brain reactivity differentiates subjects with high and low dream recall frequencies during both sleep and wakefulness. *Cerebral Cortex, 24*(5), 1206–1215. doi:10.1093/cercor/bhs388.

Eisinger, K., & Schilder, P. (1929). Träume bei Labyrinthläsionen. *Monatsschrift für Psychiatrie und Neurologie, 73*(5–6), 314–327.

Ekirch, A. R. (2001). Sleep we have lost: Pre-industrial slumber in the British Isles. *American Historical Review, 106*(2), 343–387.

Ellis, H. (1911/1931). *The world of dreams*. London: Constable & Company.

Ellis, H. D., & Szulecka, T. K. (1996). The disguised lover: A case of Frégoli delusion. In P. W. Halligan & J. C. Marshall (Eds.), *Method in madness: Case studies in cognitive neuropsychiatry* (pp. 39–50). East Sussex: Psychology Press.

Ellman, S., & Antrobus, J. S. (Eds.). (1991). *The mind in sleep: Psychology and psychophysiology*. New York: John Wiley and Sons.

Emmett, K. (1978). Oneiric experiences. *Philosophical Studies, 34* (November), 445–450.

Erlacher, D. (2005). *Motorisches Lernen im luziden Traum: Phänomenologische und experimentelle Betrachtungen*. Retrieved from http://www.ub.uni-heidelberg.de/archiv/5896.

Erlacher, D. (2009). Recall of a specific word list in lucid dreams: An explorative online study. *International Journal of Dream Research, 2*(1), 37–40.

Erlacher, D. (2010). Mentales Training als Simulation. *Zeitschrift für Sportpsychologie, 17*(3), 69–77.

Erlacher, D., Schädlich, M., Stumbrys, T., & Schredl, M. (2014). Time for actions in lucid dreams: Effects of task modality, length, and complexity. *Frontiers in Psychology, 4*, 1013. doi:10.3389/fpsyg.2013.01013.

Erlacher, D., & Schredl, M. (2004). Time required for motor activity in lucid dreams. *Perceptual and Motor Skills, 99*(3 Pt 2), 1239–1242.

Erlacher, D., & Schredl, M. (2008a). Cardiovascular responses to dreamed physical exercise during REM lucid dreaming. *Dreaming, 18*(2), 112–121.

Erlacher, D., & Schredl, M. (2008b). Do REM (lucid) dreamed and executed actions share the same neural substrate? *International Journal of Dream Research, 1*(1), 7–14.

Erlacher, D., & Schredl, M. (2010). Frequency of sport dreams in athletes. *International Journal of Dream Research, 3*(1), 91–94.

Erlacher, D., Schredl, M., & LaBerge, S. (2003). Motor area activation during dreamed hand clenching: A pilot study on EEG alpha band. *Sleep and Hypnosis, 5*, 182–187.

Ermis, U., Krakow, K., & Voss, U. (2010). Arousal thresholds during human tonic and phasic REM sleep. *Journal of Sleep Research, 19*(3), 400–406. doi:10.1111/j.1365-2869.2010.00831.x.

Errando, C. L., Sigl, J. C., Robles, M., Calabuig, E., Garciá, J., Arocas, F., et al. (2008). Awareness with recall during general anaesthesia: A prospective observational evaluation of 4001 patients. *British Journal of Anaesthesia, 101*(2), 178–185.

Esposito, M. J., Nielsen, T. A., & Paquette, T. (2004). Reduced Alpha Power associated with the recall of mentation from Stage 2 and Stage REM sleep. *Psychophysiology, 41*, 288–297.

Evans, F. J. (1990). Behavioral responses during sleep. In R. Bootzin, J. Kihlstrom, & D. Schacter (Eds.), *Sleep and Cognition* (pp. 77–87). Washington, DC: American Psychological Association.

Evans, J. S. B. T. (2003). In two minds: Dual-process accounts of reasoning. *Trends in Cognitive Sciences, 7*(10), 454–459.

Evans, J. S. B. T., & Stanovich, K. E. (2013). Dual-process theories of higher cognition: Advancing the debate. *Perspectives on Psychological Science, 8*(3), 223–241. doi:10.1177/1745691612460685.

Fantini, M. L., Corona, A., Clerici, S., & Ferini-Strambi, L. (2005). Aggressive dream content without daytime aggressiveness in REM sleep behavior disorder. *Neurology, 65*(7), 1010–1015. doi:10.1212/01.wnl.0000179346.39655.e0.

Fantini, M. L., Ferini-Strambi, L., & Montplaisir, J. (2005). Idiopathic REM sleep behavior disorder: Toward a better nosologic definition. *Neurology, 64*(5), 780–786. doi:10.1212/01.WNL.0000152878.79429.00.

Farah, M. J., Hammond, K. M., Levine, D. N., & Calvanio, R. (1988). Visual and spatial mental imagery: Dissociable systems of representation. *Cognitive Psychology, 20*(4), 439–462. doi:10.1016/0010-0285(88)90012-6.

Feinberg, T. E., Venneri, A., Simone, A. M., Fan, Y., & Northoff, G. (2009). The neuroanatomy of asomatognosia and somatoparaphrenia. *Journal of Neurology, Neurosurgery, and Psychiatry, 81*(3), 276–281. doi:10.1136/jnnp.2009.188946.

Fell, J., Fernandez, G., Lutz, M. T., Kockelmann, E., Burr, W., Schaller, C., et al. (2006). Rhinal-hippocampal connectivity determines memory formation during sleep. *Brain, 129*(Pt 1), 108–114. doi:10.1093/brain/awh647.

Fenwick, P., Schatzmann, M., Worsley, A., Adams, J., Stone, S., & Backer, A. (1984). Lucid dreaming: Correspondence between dreamed and actual events in one subject during REM sleep. *Biological Psychology, 18*, 243–252.

Ferrè, E. R., Bottini, G., & Haggard, P. (2012). Vestibular inputs modulate somatosensory cortical processing. *Brain Structure and Function, 217*(4), 859–864. doi:10.1007/s00429-012-0404-7.

Ferrè, E. R., Bottini, G., Iannetti, G. D., & Haggard, P. (2013). The balance of feelings: Vestibular modulation of bodily sensations. *Cortex, 49*(3), 748–758. doi:10.1016/j.cortex.2012.01.012.

Ferrè, E. R., Vagnoni, E., & Haggard, P. (2013). Vestibular contributions to bodily awareness. *Neuropsychologia, 51*(8), 1445–1452. doi:10.1016/j.neuropsychologia.2013.04.006.

ffytche, D. H. (2007). Visual hallucinatory syndromes: Past, present, and future. *Dialogues in Clinical Neuroscience, 9*(2), 173–189.

ffytche, D. H. (2013). The hallucinating brain: Neurobiological insights into the nature of hallucinations. In F. Macpherson & D. Platchias (Eds.), *Hallucination: Philosophy and psychology* (pp. 45–64). Cambridge, MA: MIT Press.

ffytche, D. H. (2014). What does your theory of hallucinosis make of dream experience? Do both reveal the operation of internal image generator mechanisms in the brain? In N. Tranquillo (Ed.), *Dream consciousness: Allan Hobson's new approach to the brain and its mind* (pp. 133–136). New York: Springer.

ffytche, D. H., Blom, J. D., & Catani, M. (2010). Disorders of visual perception. *Journal of Neurology, Neurosurgery, and Psychiatry, 81*(11), 1280–1287. doi:10.1136/jnnp.2008.171348.

ffytche, D. H., Howard, R. J., Brammer, M. J., David, A., Woodruff, P., & Williams, S. (1998). The anatomy of conscious vision: An fMRI study of visual hallucinations. *Nature Neuroscience, 1*(8), 738–742. doi:10.1038/3738.

ffytche, D. H., & Zeki, S. (2011). The primary visual cortex, and feedback to it, are not necessary for conscious vision. *Brain, 134*(Pt 1), 247–257. doi:10.1093/brain/awq305.

Fireman, G. D., McVay, T. E., & Flanagan, O. (2003). Introduction. In G. D. Fireman, T. E. McVay, & O. Flanagan (Eds.), *Narrative and consciousness* (pp. 3–16). Oxford: Oxford University Press.

First, M. B. (2005). Desire for amputation of a limb: Paraphilia, psychosis, or a new type of identity disorder. *Psychological Medicine, 35*(6), 919–928.

Fish, W. (2008). Disjunctivism, indistinguishability, and the nature of hallucination. In A. Haddock & F. Macpherson (Eds.), *Disjunctivism: Perception, action, knowledge* (pp. 144–167). Oxford: Oxford University Press.

Fish, W. (2009). *Perception, hallucination, and illusion*. Oxford: Oxford University Press.

Fisher, C. (1966). Dreaming and sexuality. In R. Loewenstein, L. Newman, M. Schur, & A. Solnit (Eds.), *Psychoanalysis: A general psychology* (pp. 537–569). New York: International Universities Press.

Fitch, T., & Armitage, R. (1989). Variations in cognitive style among high and low frequency dream recallers. *Personality and Individual Differences, 10*(8), 869–875.

Flanagan, O. (1995). Deconstructing dreams: The spandrels of sleep. *Journal of Philosophy, 92*, 5–27.

Flanagan, O. (1996). *Self expressions: Mind, morals, and the meaning of life*. Oxford: Oxford University Press.

Flanagan, O. (2000). *Dreaming souls: Sleep, dreams, and the evolution of the conscious mind*. Oxford: Oxford University Press.

Flavell, J. H. (1976). Metacognitive aspects of problem solving. In L. B. Resnick (Ed.), *The nature of intelligence* (pp. 231–236). Hillsdale, NJ: Erlbaum.

Fletcher, P. C., & Frith, C. D. (2009). Perceiving is believing: A Bayesian approach to explaining the positive symptoms of schizophrenia. *Nature Reviews. Neuroscience, 10*(1), 48–58. doi:10.1038/nrn2536.

Floyd, J. A., Janisse, J. J., Jenuwine, E. S., & Ager, J. W. (2007). Changes in REM-sleep percentage over the adult lifespan. *Sleep, 30*(7), 829–836.

Fodor, J. A. (1975). *The language of thought*. Cambridge, MA: Harvard University Press.

Ford, J. (1999). Samuel Taylor Coleridge and the pains of sleep. *History Workshop Journal, 48*, 169–186.

Fosse, M. J., Fosse, R., Hobson, J. A., & Stickgold, R. J. (2003). Dreaming and episodic memory: A functional dissociation? *Journal of Cognitive Neuroscience, 15*, 1–9.

Fosse, R. (2000). REM mentation in narcoleptics and normals: An empirical test of two neurocognitive theories. *Consciousness and Cognition, 9*(4), 488–509. doi:10.1006/ccog.2000.0466.

Fosse, R., Stickgold, R. J., & Hobson, J. A. (2001). Reciprocal variation in thoughts and hallucinations. *Psychological Science, 12*(1), 30–36.

Fosse, R., Stickgold, R., & Hobson, J. A. (2004). Thinking and hallucinating: Reciprocal changes in sleep. *Psychophysiology, 41*(2), 298–305. doi:10.1111/j.1469-8986.2003.00146.x.

Foulkes, D. (1966). *The psychology of sleep*. New York: Scribner.

Foulkes, D. (1978). *A grammar of dreams*. New York: Basic Books.

Foulkes, D. (1993a). Children's dreaming. In C. Cavallero & D. Foulkes (Eds.), *Dreaming as cognition* (pp. 114–132). New York: Harvester Wheatsheaf.

Foulkes, D. (1993b). Dreaming and REM sleep. *Journal of Sleep Research, 2*, 199–202.

Foulkes, D. (1999). *Children's dreaming and the development of consciousness*. Cambridge, MA: Harvard University Press.

Foulkes, D., & Domhoff, G. W. (2014). Bottom-up or top-down in dream neuroscience? A top-down critique of two bottom-up studies. *Consciousness and Cognition, 27*, 168–171. doi:10.1016/j.concog.2014.05.002.

Foulkes, D., & Fleisher, S. (1975). Mental activity in relaxed wakefulness. *Journal of Abnormal Psychology, 84*(1), 66–75.

Foulkes, D., & Kerr, N. H. (1994). Point of view in nocturnal dreaming. *Perceptual and Motor Skills, 78*, 690.

Foulkes, D., & Schmidt, M. (1983). Temporal sequence and unit composition in dream reports from different stages of sleep. *Sleep, 6*, 265–280.

Fox, K. C. R., Nijeboer, S., Solomonova, E., Domhoff, G. W., & Christoff, K. (2013). Dreaming as mind wandering: Evidence from functional neuroimaging and first-person content reports. *Frontiers in Human Neuroscience, 7*, 412. doi:10.3389/fnhum.2013.00412.

Fox, K. C. R., Zakarauskas, P., Dixon, M., Ellamil, M., Thompson, E., & Christoff, K. (2012). Meditation experience predicts introspective accuracy. *PLoS ONE, 7*(9), e45370. doi:10.1371/journal.pone.0045370.

Fox, O. (1962/1993). *Astral projections: A record of out-of-the-body experiences*. New York: Citadel Press.

Frank, B., & Lorenzoni, E. (1989). Experiences of phantom limb sensations in dreams. *Psychopathology, 22*, 182–187.

Frankfurt, H. G. (1958). Peirce's notion of abduction. *Journal of Philosophy, 55*(14), 593–597.

Frankfurt, H. G. (1970/2008). *Demons, dreamers, and madmen: The defense of reason in Descartes's Meditations*. Princeton: Princeton University Press.

Franklin, M. S., & Zyphur, M. J. (2005). The role of dreams in the evolution of the human mind. *Evolutionary Psychology, 3*, 59–78.

Freud, S. (1899/2003). *Die Traumdeutung*. Frankfurt am Main: Fischer Taschenbuch.

Freud, S. (1940/2003). *Schriften über Träume und Traumdeutungen*. Frankfurt am Main: Fischer Taschenbuch.

Friedman, D., Leeb, R., Dikovsky, L., Reiner, M., Pfurtscheller, G., & Slater, M. (2007). Controlling a virtual body by thought in a highly-immersive virtual environment—a case study in using a brain-computer interface in a virtual-reality cave-like system. In J. Braz, P.-P. Vázquez, & J.M. Pereira (Eds.), *GRAPP 2007: Proceedings of the Second International Conference on Comupter Graphics Theory and Applications* (pp. 83–90), Vol. AS/IE. Barcelona: INSTICC.

Friedman, D., Leeb, R., Guger, C., Steed, A., Pfurtscheller, G., & Slater, M. (2007). Navigating virtual reality by thought: What is it like? *Presence, 16*(1), 100–110. doi:10.1162/pres.16.1.100.

Friedman, D., Suji, K., & Slater, M. (2007). SuperDreamCity: An immersive virtual reality experience that responds to electrodermal activity. *Affective Computing and Intelligent Interaction: Lecture Notes in Computer Science, 4738*, 570–581.

Friston, K. (2014). Please comment on the predictive tenet of the protoconsciousness hypothesis: Is this idea consistent with the Helmholtzian model of free energy that you are developing? In N. Tranquillo (Ed.), *Dream consciousness: Allan Hobson's new approach to the brain and its mind* (pp. 137–142). New York: Springer.

Friston, K., Adams, R. A., Perrinet, L., & Breakspear, M. (2012). Perceptions as hypotheses: Saccades as experiments. *Frontiers in Psychology, 3*, 151. doi:10.3389/fpsyg.2012.00151.

Froese, T., Gould, C., & Seth, A. K. (2011). Validating and calibrating first- and second-person methods in the science of consciousness. *Journal of Consciousness Studies, 18*(2), 38–64.

Fromm, E. (1957). *The forgotten language*. New York: Grove Press.

Fulford, B. (2004). Insight and delusion: From Jaspers to Kraeplin and back again via Austin. In X. F. Amador & A. S. David (Eds.), *Insight and psychosis: Awareness of illness in schizophrenia and related disorders* (2nd ed., pp. 51–78). Oxford: Oxford University Press.

Fuller, P., & Lu, J. (2009a). Dopamine. In R. Stickgold & M. P. Walker (Eds.), *The neuroscience of sleep* (pp. 125–130). London: Elsevier Academic Press.

Fuller, P., & Lu, J. (2009b). Hypothalamic regulation of sleep. In R. Stickgold & M. P. Walker (Eds.), *The neuroscience of sleep* (pp. 91–98). Burlington, MA: Elsevier Academic Press.

Funkhouser, A. T., & Schredl, M. (2010). The frequency of déjà vu (déjà rêve) and the effects of age, dream recall frequency, and personality factors. *International Journal of Dream Research, 3* (1). Retrieved from http://archiv.ub.uni-heidelberg.de/ojs/index.php/IJoDR/article/view/473.

Furlanetto, T., Bertone, C., & Becchio, C. (2013). The bilocated mind: New perspectives on self-localization and self-identification. *Frontiers in Human Neuroscience, 7,* 71. doi:10.3389/fnhum.2013.00071.

Gackenbach, J. (1988). The psychological content of lucid dreams versus nonlucid dreams. In J. Gackenbach & S. LaBerge (Eds.), *Conscious mind, sleeping brain: Perspectives on lucid dreaming* (pp. 181–219). New York: Plenum Press.

Gackenbach, J., & Rosie, M. (2011). Presence in video game play and nighttime dreams: An empirical inquiry. *International Journal of Dream Research, 4*(2), 98–109.

Gaggioli, A., Bassi, M., & Delle Fave, A. (2003). Quality of experience in virtual environments. In G. Riva, F. Davide, & W. Ijsselsteijn (Eds.), *Being there: Concepts, effects, and measurement of user presence in synthetic environments* (pp. 121–135). Amsterdam: IOS Press.

Gais, S., & Born, J. (2004). Low acetylcholine during slow-wave sleep is critical for declarative memory consolidation. *Proceedings of the National Academy of Sciences of the United States of America, 101,* 2140–2144.

Galavotti, M., Nemeth, E., & Stadler, F. (Eds.). (2014). *Vienna Circle Institute yearbook: European philosophy of science—philosophy of science in Europe and the Viennese heritage.* Springer International.

Galen. (1996). *On diagnosis in dreams* (Lee Pearcy, Trans.). Retrieved from http://www.ucl.ac.uk/~ucgajpd/medicina%20antiqua/tr_GalDreams.html.

Gallagher, S. (2000). Self-reference and schizophrenia: A cognitive model of immunity to error through misidentification. In D. Zahavi (Ed.), *Exploring the self: Philosophical and psychopathological perspectives on self-experience* (pp. 203–239). Amsterdam: John Benjamins.

Gallagher, S., & Cole, J. R. (1995). Body schema and body image in a deafferented subject. *Journal of Mind and Behavior, 16,* 369–390.

Garcia-Rill, E. (2009). Reticular activating system. In R. Stickgold & M. P. Walker (Eds.), *The neuroscience of sleep* (pp. 133–139). London: Elsevier Academic Press.

Garfield, P. (1988). *Women's bodies, women's dreams.* New York: Ballantine.

Garfield, P. (1995). *Creative dreaming: Plan and control your dreams to develop creativity, overcome fears, solve problems, and creative a better self.* New York: Simon & Schuster.

Garfield, P. (2001). Dreams in bereavement. In D. Barrett (Ed.), *Trauma and dreams* (pp. 186–211). Cambridge, MA: Harvard University Press.

Garry, M., Manning, C., Loftus, E., & Sherman, S. (1996). Imagination inflation: Imagining a childhood event inflates confidence that it occurred. *Psychonomic Bulletin and Review, 3*(2), 208–214. doi:10.3758/BF03212420.

Gavie, J., & Revonsuo, A. (2010). The future of lucid dreaming treatment. *International Journal of Dream Research, 3*, 13–15.

Gehring, P. (2008). *Traum und Wirklichkeit: Zur Geschichte einer Unterscheidung*. New York: Campus.

Gendler, T. S. (2010). Philosophical thought experiments, intuitions, and cognitive equilibrium. In T. S. Gendler (Ed.), *Intuition, imagination, and philosophical methodology* (pp. 116–132). Oxford: Oxford University Press.

Gendler, T. S. (2013). Imagination. In E. N. Zalta (Ed.), *The Stanford encyclopedia of philosophy (fall 2013 edition)*. Retrieved from http://plato.stanford.edu/archives/fall2013/entries/imagination.

Gendler, T., & Hawthorne, J. (2002). *Conceivability and possibility*. Oxford: Clarendon Press.

Germain, A., & Nielsen, T. A. (2001). EEG power associated with early sleep onset images differing in sensory content. *Sleep Research Online, 4*, 83–90.

Germain, A., & Nielsen, T. A. (2003a). Impact of imagery rehearsal treatment on distressing dreams, psychological distress, and sleep parameters in nightmare patients. *Behavioral Sleep Medicine, 1*(3), 140–154.

Germain, A., & Nielsen, T. A. (2003b). Sleep pathophysiology in posttraumatic stress disorder and idiopathic nightmare sufferers. *Biological Psychiatry, 54*, 1092–1098.

Germain, A., Nielsen, T. A., Zadra, A., & Montplaisir, J. (2000). The prevalence of typical dream themes challenges the specificity of the threat simulation theory. *Behavioral and Brain Sciences, 23*, 940–941.

Germain, A., & Zadra, A. (2009). Dreams and nightmares in PTSD. In R. Stickgold & M. P. Walker (Eds.), *The neuroscience of sleep* (pp. 316–322). London: Elsevier Academic Press.

Gerrans, P. (2002). A one-stage explanation of the Cotard delusion. *Philosophy, Psychiatry, and Psychology, 9*(1), 47–53.

Gerrans, P. (2012). Dream experience and a revisionist account of delusions of misidentification. *Consciousness and Cognition, 21*(1), 217–227. doi:10.1016/j.concog.2011.11.003.

Gerrans, P. (2013a). Delusional attitudes and default thinking. *Mind and Language, 28*(1), 83–102. doi:10.1111/mila.12010.

Gerrans, P. (2013b). Unraveling the mind. *Behavioral and Brain Sciences, 36*(03), 214–215. doi:10.1017/S0140525X1200235X.

Gerrans, P. (2014). Pathologies of hyperfamiliarity in dreams, delusions, and deja vu. *Frontiers in Psychology, 5*, 97. doi:10.3389/fpsyg.2014.00097.

Gerrans, P. (2015). All the self we need. In T. Metzinger & J. M. Windt (Eds.), *Open MIND: 15(T)*. Frankfurt am Main: MIND Group. doi: 10.15502/9783958570078.

Gert, B. (2007). Hobbes's psychology. In P. Springborg (Ed.), *The Cambridge companion to Hobbes's Leviathan* (pp. 157–174). Cambridge: Cambridge University Press.

Giguère, B., & LaBerge, S. (1995). To touch a dream: An experiment in touch, pain, and pleasure. *NightLight, 7*(1), 1–6, 11.

Gillespie, G. (1988a). Lucid dreams in Tibetan Buddhism. In J. Gackenbach & S. LaBerge (Eds.), *Conscious mind, sleeping brain: Perspectives on lucid dreaming* (pp. 27–35). New York: Plenum Press.

Gillespie, G. (1988b). Without a guru: An account of my lucid dreaming. In J. Gackenbach & S. LaBerge (Eds.), *Conscious mind, sleeping brain: Perspectives on lucid dreaming* (pp. 343–350). New York: Plenum Press.

Gillespie, G. (1991). Problems related to experimentation while dreaming lucidly. *Tenth Anniversary Issue of Lucidity Letter.* Retrieved from http://www.sawka.com/spiritwatch/problems.htm.

Gillespie, G. (1997). Hypnopompic imagery and visual dream experience. *Dreaming, 7*(3), 187–194.

Girard, T. A., & Cheyne, J. A. (2006). Timing of spontaneous sleep-paralysis episodes. *Journal of Sleep Research, 15*(2), 222–229. doi:10.1111/j.1365-2869.2006.00512.x.

Giummarra, M. J., Bradshaw, J. L., Hilti, L. M., Nicholls, M. E. R., & Brugger, P. (2012). Paralyzed by desire: A new type of body integrity identity disorder. *Cognitive and Behavioral Neurology, 25*(1), 34–41. doi:10.1097/WNN.0b013e318249865a.

Giummarra, M. J., Bradshaw, J. L., Nicholls, M. E. R., Hilti, L. M., & Brugger, P. (2011). Body integrity identity disorder: Deranged body processing, right fronto-parietal dysfunction, and phenomenological experience of body incongruity. *Neuropsychology Review, 21*(4), 320–333. doi:10.1007/s11065-011-9184-8.

Globus, G. G. (1987). *Dream life, wake life: The human condition through dreams.* New York: State University of New York Press.

Goblot. (1896). Sur le souvenir des rêves. *Revue philosophique, XLII.*

Gold, I., & Hohwy, J. (2000). Rationality and schizophrenic delusion. *Mind and Language, 15*(1), 146–167.

González-Franco, M., Peck, T., Rodríguez-Fornells, A., & Slater, M. (2014). A threat to a virtual hand elicits motor cortex activation. *Experimental Brain Research, 232*(3), 875–887. doi:10.1007/s00221-013-3800-1.

Goodenough, D. (1991). Dream recall: History and current status of the field. In S. Ellman & J. S. Antrobus (Eds.), *The mind in sleep: Psychology and psychophysiology* (pp. 143–171). New York: John Wiley and Sons.

Goodenough, D., & Koulack, D. (1976). Dream recall and dream recall failure: An arousal-retrieval model. *Psychological Bulletin, 83*, 975–984.

Gosseries, O., Vanhaudenhuyse, A., Bruno, M.-A., Demertzi, A., Schnakers, C., Boly, M., et al. (2011). Disorders of consciousness: Coma, vegetative, and minimally conscious states. In D. Cvetkovic & I. Cosic (Eds.), *States of consciousness: Experimental insights into meditation, waking, sleep, and dreams* (pp. 29–56). Frontiers collection. Heidelberg: Springer.

Gottesmann, C. (1999). Neurophysiological support of consciousness during waking and sleep. *Progress in Neurobiology, 59*, 469–508.

Gottesmann, C. (2002). The neurochemistry of waking and sleeping mental activity: The disinhibition-dopamine hypothesis. *Psychiatry and Clinical Neurosciences, 56*, 345–354.

Gottesmann, C. (2006). The dreaming sleep stage: A new neurobiological model of schizophrenia? *Neuroscience, 140*, 1105–1115.

Gottesmann, C. (2007). A neurobiological history of dreaming. In D. Barrett & P. McNamara (Eds.), *The new science of dreaming: Vol. 1. Biological aspects* (pp. 1–51). Westport: Praeger Perspectives.

Grau, C. (2005). Bad dreams, evil demons, and the experience machine: Philosophy and *The Matrix*. In C. Grau (Ed.), *Philosophers explore The Matrix*. Oxford: Oxford University Press.

Green, C. (1968). *Lucid dreams*. Oxford: Institute of Psychophysical Research.

Green, C., & McCreery, C. (1975). *Apparitions*. London: Hamish Hamilton.

Green, C., & McCreery, C. (1994). *Lucid dreaming: The paradox of consciousness during sleep*. London: Routledge.

Greene, G. (2008). *Insomniac*. Berkeley: University of California Press.

Gregory, J. C. (1916). Dreams as psychical explosions. *Mind, 25*(98), 193–205.

Gregory, J. C. (1918). The dream of "frustrated effort": A suggested explanation. *Mind, 27*(105), 125–128.

Gregory, J. C. (1922). Visual images, words, and dreams. *Mind, 31*(123), 321–344.

Gregory, J. C. (1923). Memory, forgetfulness, and mistakes of recognition in waking and dreaming. *Monist, 33*(1), 15–32. doi:10.5840/monist19233314.

Gruberger, M., Simon, E. B., Levkovitz, Y., Zangen, A., & Hendler, T. (2011). Towards a neuroscience of mind-wandering. *Frontiers in Human Neuroscience, 5*, 56. doi:10.3389/fnhum.2011.00056.

Grundmann, T. (2002). Die Struktur des skeptischen Traumarguments. *Grazer Philosophische Studien, 64*, 57–81.

Grush, R. (2004). The emulation theory of representation: Motor control, imagery, and perception. *Behavioral and Brain Sciences, 27*(3), 377–396.

Guillot, A., & Collet, C. (2005). Contribution from neurophysiological and psychological methods to the study of motor imagery. *Brain Research: Brain Research Reviews, 50*(2), 387–397. doi:10.1016/j.brainresrev.2005.09.004.

Gustafson, D. F., & Tapscott, B. L. (Eds.). (1979). *Body, mind, and method: Essays in honor of Virgil C. Aldrich*. Dordrecht: D. Reidel.

Guterstam, A., & Ehrsson, H. H. (2012). Disowning one's seen real body during an out-of-body illusion. *Consciousness and Cognition, 21*(2), 1037–1042. doi:10.1016/j.concog.2012.01.018.

Hacking, I. (1975). *Why does language matter to philosophy?* Cambridge: Cambridge University Press.

Hacking, I. (2001). Dreams in place. *Journal of Aesthetics and Art Criticism, 59*(3), 245–260.

Haken, H. (1983). *Advanced synergetics: Instability hierarchies of self-organizing systems and devices*. Springer series in synergetics (Vol. 20). Berlin: Springer Berlin Heidelberg.

Hall, C. (1953). A cognitive theory of dreams. *Journal of General Psychology, 49*, 273–282.

Hall, C. S. (1981). Do we dream during sleep? Evidence for the Globot hypothesis. *Perceptual and Motor Skills, 53*(1), 239–246. doi:10.2466/pms.1981.53.1.239.

Hall, C., & van de Castle, R. (1966). *The content analysis of dreams*. New York: Appleton-Century-Crofts.

Halligan, P. W., & Marshall, J. C. (Eds.). (1996). *Method in madness: Case studies in cognitive neuropsychiatry*. East Sussex: Psychology Press.

Harkness, D. L. (2015). From explanatory ambition to explanatory power—A commentary on Jakob Hohwy. In T. Metzinger & J. M. Windt (Eds.), *Open MIND: 19(C)*. Frankfurt am Main: MIND Group. doi: 10.15502/9783958570153.

Harman, G. H. (1965). The inference to the best explanation. *Philosophical Review, 74*(1), 88–95.

Harsh, J., & Badia, P. (1990). Stimulus control and sleep. In R. Bootzin, J. Kihlstrom, & D. Schacter (Eds.), *Sleep and cognition* (pp. 58–66). Washington, DC: American Psychological Association.

Hartmann, E. (1998). *Dreams and nightmares: The new theory on the origin and meaning of dreams*. Cambridge, MA: Perseus.

Hartmann, E. (2000). We do not dream of the 3 R's: Implications for the nature of dreaming mentation. *Dreaming, 10*(2), 103–110. doi:10.1023/A:1009400805830.

Hartmann, E. (2001). Who develops PTSD and who doesn't? In D. Barrett (Ed.), *Trauma and dreams* (pp. 100–113). Cambridge, MA: Harvard University Press.

Harvey, A. G., & Tang, N. (2012). (Mis)perception of sleep in insomnia: A puzzle and a resolution. *Psychological Bulletin, 138*(1), 77–101. doi:10.1037/a0025730.

Haskell, R. (1985a). Dreaming, cognition, and physical illness: Part I. *Journal of Medical Humanities and Bioethics, 6*(1), 46–56. doi:10.1007/BF01149838.

Haskell, R. (1985b). Dreaming, cognition, and physical illness: Part II. *Journal of Medical Humanities and Bioethics, 6*(2), 109–122. doi:10.1007/BF01142305.

Haug, O. & Jung, M. F. (2015). Are there counterexamples to the immunity principle? Some restrictions and clarifications—A commentary on Caleb Liang. In T. Metzinger & J. M. Windt (Eds.), *Open MIND: 24(C)*. Frankfurt am Main: MIND Group. doi: 10.15502/9783958570467.

Hawthorne, N. (1837). *Twice-told tales*. Retrieved from http://www.eldritchpress.org/nh/hmind.html.

Haynes, J. D. (2015). An information-based approach to consciousness: Mental state decoding. In T. Metzinger & J. M. Windt (Eds.), *Open MIND: 17(T)*. Frankfurt am Main: MIND Group. doi: 10.15502/9783958570276.

Haynes, J.-D., & Rees, G. (2006). Decoding mental states from brain activity in humans. *Nature Reviews: Neuroscience*, *7*(7), 523–534. doi:10.1038/nrn1931.

Hearne, K. (1978). *Lucid dreams: An electro-physiological and psychological study*. Unpublished doctoral thesis, University of Liverpool, England. Retrieved from http://www.keithhearne.com/phd-download.

Hering, J. (1947). Concerning image, idea, and dream. *Philosophy and Phenomenological Research*, *8*(2), 188–205.

Herman, J. (2000). Reflexive and orienting properties of REM sleep dreaming and eye movements. *Behavioral and Brain Sciences*, *23*(6), 950.

Heydrich, L., & Blanke, O. (2013). Distinct illusory own-body perceptions caused by damage to posterior insula and extrastriate cortex. *Brain: A journal of neurology*, *136*(Pt 3), 790–803. doi:10.1093/brain/aws364.

Heydrich, L., Dieguez, S., Grunwald, T., Seeck, M., & Blanke, O. (2010). Illusory own body perceptions: Case reports and relevance for bodily self-consciousness. *Consciousness and Cognition*, *19*(3), 702–710.

Heydrich, L., Dodds, T., Aspell, J., Herbelin, B., Buelthoff, H., Mohler, B., et al. (2013). Visual capture and the experience of having two bodies: Evidence from two different virtual reality techniques. *Frontiers in Psychology*, *4*, 946. doi:10.3389/fpsyg.2013.00946.

Hilgendorf, E., & Günther, J.-P. (Eds.). (2013). *Robotik und Gesetzgebung*. Baden-Baden: Nomos.

Hill, C. E., Diemer, R., Hess, S., Hillyer, A., & Seeman, R. (1993). Are the effects of dream interpretation on session quality, insight, and emotions due to the dream itself, to projection, or to the interpretation process? *Dreaming*, *3*, 269–280.

Hill, J. (2004a). Descartes' dreaming argument and why we might be sceptical of it. *Richmond Journal of Philosophy*, *8*, 1–9.

Hill, J. (2004b). The philosophy of sleep: The views of Descartes, Locke, and Leibniz. *Richmond Journal of Philosophy*, *6*, 20–26.

Hilti, L. M., & Brugger, P. (2010). Incarnation and animation: Physical versus representational deficits of body integrity. *Experimental Brain Research*, *204*(3), 315–326. doi:10.1007/s00221-009-2043-7.

Hilti, L. M., Hanggi, J., Vitacco, D. A., Kraemer, B., Palla, A., Luechinger, R., et al. (2013). The desire for healthy limb amputation: Structural brain correlates and clinical features of xenomelia. *Brain, 136*(1), 318–329. doi:10.1093/brain/aws316.

Hirsh, J. B., Mar, R. A., & Peterson, J. B. (2013). Personal narratives as the highest level of cognitive integration. *Behavioral and Brain Sciences, 36*(03), 216–217. doi:10.1017/S0140525X12002269.

Hishikawa, Y., & Shimizu, T. (1995). Physiology of REM sleep, cataplexy, and sleep paralysis. *Advances in Neurology, 67*, 245–271.

Hobbes, T. (1651/1985). *Leviathan* (C. B. Macpherson, Ed.). London: Penguin.

Hobbes, T., & Descartes, R. (1641/1975–1999). *Objections to Descartes' Meditations, with Descartes' Replies* (George MacDonald Ross, Trans.). Retrieved from http://www.earlymoderntexts.com/pdfs/descartes1642_2.pdf.

Hobson, J. A. (1988). *The dreaming brain*. New York: Basic Books.

Hobson, J. A. (1999). *Dreaming as delirium: How the brain goes out of its mind*. Cambridge, MA: MIT Press.

Hobson, J. A. (2002). *Dreaming: An introduction to the science of sleep*. Oxford: Oxford University Press.

Hobson, J. A. (2005). *Thirteen dreams Freud never had: The new mind science*. New York: Pi Press.

Hobson, J. A. (2009). REM sleep and dreaming: Towards a theory of protoconsciousness. *Nature Reviews: Neuroscience, 10*(November), 803–813.

Hobson, J. A. (2011). *Dream life: An experimental memoir*. Cambridge, MA: MIT Press.

Hobson, J. A. (2014). The William James Lectures on dream consciousness. In N. Tranquillo (Ed.), *Dream consciousness: Allan Hobson's new approach to the brain and its mind* (pp. 3–79). New York: Springer.

Hobson, J. A., & Friston, K. J. (2012). Waking and dreaming consciousness: Neurobiological and functional considerations. *Progress in Neurobiology, 98*(1), 82–98. doi:10.1016/j.pneurobio.2012.05.003.

Hobson, J. A., & Friston, K. J. (2014). Consciousness, dreams, and inference: The Cartesian theatre revisited. *Journal of Consciousness Studies, 21*(1–2), 6–32.

Hobson, J. A., & Kahn, D. (2007). Dream content: Individual and generic aspects. *Consciousness and Cognition, 16*, 850–858.

Hobson, J. A., & McCarley, R. W. (1977). The brain as a dreamstate generator: An activation-synthesis hypothesis of the dream process. *American Journal of Psychiatry, 134*, 1335–1348.

Hobson, J. A., & Pace-Schott, E. F. (2002). The cognitive neuroscience of sleep: Neuronal systems, consciousness, and learning. *Nature Reviews: Neuroscience, 3*(September), 679–693.

Hobson, J. A., Pace-Schott, E. F., & Stickgold, R. (2000). Dreaming and the brain: Toward a cognitive neuroscience of conscious states. *Behavioral and Brain Sciences, 23*, 793–842.

Hobson, J. A., Sangsanguan, S., Arantes, H., & Kahn, D. (2011). Dream logic—the inferential reasoning paradigm. *Dreaming, 21*(1), 1–15. doi:10.1037/a0022860.

Hobson, J. A., & Schredl, M. (2011). The continuity and discontinuity between waking and dreaming: A dialogue between Michael Schredl and Allan Hobson concerning the adequacy and completeness of these notions. *International Journal of Dream Research, 4*(1), 3–7.

Hobson, A., & Voss, U. (2011). A mind to go out of: Reflections on primary and secondary consciousness. *Consciousness and Cognition, 20*(4), 993–997. doi:10.1016/j.concog.2010.09.018.

Hobson, J. A., & Wohl, H. (2005). *From angels to neurones: Art and the new science of dreaming.* Parma: Mattioli.

Hodges, M., & Carter, W. R. (1969). Nelson on dreaming a pain. *Philosophical Studies, 20*(3), 43–46.

Hoelscher, T. J., Klinger, E., & Barta, S. G. (1981). Incorporation of concern- and nonconcern-related verbal stimuli into dream content. *Journal of Abnormal Psychology, 90*, 88–91.

Hoff, H. (1929). Zusammenhang von Vestibularfunktion, Schlafstellung und Traumleben. *Monatsschrift für Psychiatrie und Neurologie, 71*(5–6), 366–372.

Hoff, H., & Plötzl, O. (1937). Über die labyrinthären Beziehungen von Flugsensationen und Flugträumen. *Monatsschrift für Psychiatrie und Neurologie, 97*(4), 193–211.

Hohwy, J. (2010). The hypothesis testing brain: Some philosophical applications. In *Proceedings of the 9th Conference of the Australasian Society for Cognitive Science* (pp. 135–144). Macquarie Centre for Cognitive Science.

Hohwy, J. (2011). Phenomenal variability and introspective reliability. *Mind and Language, 26*(3), 261–286.

Hohwy, J. (2013a). Delusions, illusions and inference under uncertainty. *Mind and Language, 28*(1), 57–71. doi:10.1111/mila.12008.

Hohwy, J. (2013b). *The predictive mind.* Oxford: Oxford University Press.

Hohwy, J. (2015). The neural organ explains the mind. In T. Metzinger & J. M. Windt (Eds.), *Open MIND: 19(T)*. Frankfurt am Main: MIND Group. doi: 10.15502/9783958570016.

Hohwy, J., & Paton, B. (2010). Explaining away the body: Experiences of supernaturally caused touch and touch on non-hand objects within the rubber hand illusion. *PLoS ONE, 5*(2), e9416.

Hohwy, J., & Rajan, V. (2012). Delusions as forensically disturbing perceptual inferences. *Neuroethics, 5*(1), 5–11.

Holzinger, B. (1997). *Der luzide Traum: Phänomenologie und Physiologie.* Vienna: WUV-Universitätsverlag.

Holzinger, B., LaBerge, S., & Levitan, L. (2006). Psychophysiological correlates of lucid dreaming. *Dreaming, 16*(2), 88–95. doi:10.1037/1053-0797.16.2.88.

Honma, M., Yoshiike, T., Ikeda, H., Kim, Y., Kuriyama, K., & Hamed, S. B. (2014). Sleep dissolves illusion: Sleep withstands learning of visuo-tactile-proprioceptive integration induced by repeated days of rubber hand illusion training. *PLOS ONE, 9*(1), e85734. doi:10.1371/journal.pone.0085734.

Hopkins, R. (2012). What Perky did not show. *Analysis, 72*(3), 431–439. doi:10.1093/analys/ans063.

Horgan, T. E., & Tienson, J. L. (2002). The intentionality of phenomenology and the phenomenology of intentionality. In D. J. Chalmers (Ed.), *Philosophy of mind: Classical and contemporary readings*. New York: Oxford University Press.

Hori, T., Hayashi, M., & Morikawa, T. (1994). Topographical EEG changes and the hypnagogic experience. In R. D. Ogilvie & J. R. Harsh (Eds.), *Sleep onset: Normal and abnormal processes* (pp. 237–253). Washington, DC: American Psychological Association.

Horikawa, T., Tamaki, M., Miyawaki, Y., & Kamitani, Y. (2013a). Neural decoding of visual imagery during sleep. *Science, 340*(6132), 639–642. doi:10.1126/science.1234330.

Horikawa, T., Tamaki, M., Miyawaki, Y., & Kamitani, Y. (2013b). *Supplementary materials for neural decoding of visual imagery during sleep*. Retrieved from http://www.sciencemag.org/content/340/6132/639/suppl/DC1.

Horowitz, M. J. (1972). Image formation: Clinical observations and a cognitive model. In P. W. Sheehan (Ed.), *The function and nature of imagery* (pp. 282–311). New York: Academic Press.

Horton, C. L. (2011a). Recall and recognition of dreams and waking events: A diary paradigm. *International Journal of Dream Research, 4* (1). Retrieved from http://archiv.ub.uni-heidelberg.de/ojs/index.php/IJoDR/article/view/6176.

Horton, C. L. (2011b). Rehearsal of dreams and waking events similarly improves the quality but not the quantity of autobiographical recall. *Dreaming, 21*(3), 181–196. doi:10.1037/a0024860.

Horton, C. L., Moulin, C. J., & Conway, M. A. (2009). The self and dreams during a period of transition. *Consciousness and Cognition, 18*(3), 710–717. doi:10.1016/j.concog.2009.07.001.

Hoss, R. J. (2010). Content analysis on the potential significance of color in dreams: A preliminary investigation. *International Journal of Dream Research, 3*(1), 80–90. doi:10.11588/ijodr.2010.1.508.

Howard, R. J., ffytche, D. H., Barnes, J., McKeefry, D., Ha, Y., Woodruff, P. W., et al. (1998). The functional anatomy of imagining and perceiving colour. *Neuroreport, 9*(6), 1019–1023.

Huber, R., Ghilardi, M. F., Massimini, M., & Tononi, G. (2004). Local sleep and learning. *Nature, 430*(6995), 78–81. doi:10.1038/nature02663.

Hufford, D. J. (1982). *The terror that comes in the night: An experience-centered study of supernatural assault traditions*. Philadelphia: University of Pennsylvania Press.

Hughes, J. D. (2000). Dream interpretation in ancient civilizations. *Dreaming, 10*(1), 7–18.

Hume, D. (1739/2000). *A treatise of human nature* (D. F. Norton & M. J. Norton, Eds.). Oxford: Oxford University Press.

Hume, D. (1748/2004). *An enquiry concerning human understanding*. Mineola, NY: Dover.

Hunt, H. T. (1989). *The multiplicity of dreams: Memory, imagination, and consciousness*. New Haven: Yale University Press.

Hunt, H. T., & Ogilvie, R. D. (1988). Lucid dreams in their natural series: Phenomenological and psychophysiological findings in relation to meditative states. In J. Gackenbach & S. LaBerge (Eds.), *Conscious mind, sleeping brain: Perspectives on lucid dreaming* (pp. 389–417). New York: Plenum Press.

Hunter, J. F. M. (1971). Some questions about dreaming. *Mind, 80*(January), 70–92.

Hunter, J. F. M. (1976). A puzzle about dreaming. *Analysis, 36*(3), 126. doi:10.2307/3327834.

Hunter, J. F. M. (1983). The difference between dreaming and being awake. *Mind, 92*(January), 80–93.

Hurd, R. (2008). *Exploring the void in lucid dreaming*. Retrieved from http://dreamstudies.org/2010/05/13/exploring-the-void-in-lucid-dreaming.

Hurd, R., & Bulkeley, K. (2014). *Lucid dreaming: New perspectives on consciousness in sleep. Two vols.* Santa Barbara: Praeger.

Hurlburt, R. T., & Akhter, S. A. (2006). The descriptive experience sampling method. *Phenomenology and the Cognitive Sciences, 5*(3–4), 271–301.

Hurlburt, R. T., & Schwitzgebel, E. (2007). *Describing inner experience? Proponent meets skeptic*. Cambridge, MA: MIT Press.

Husserl, E. (1966). *Analysen zur passiven Synthesis: Aus Vorlesungs- und Forschungsmanuskripten, 1918–1926* (M. Fleischer, Ed.). Den Haag: Martinus Nijhoff.

Husserl, E. (2006). *Phantasie und Bildbewußtsein*. Herausgegeben und eingeleitet von Eduard Marbach. Hamburg: Felix Meiner.

Hustvedt, S. (2006). *A Plea for Eros*. London: Sceptre.

Iber, C., Ancoli-Israel, S., Chesson, A. L., & Quan, S. F. (2007). *The AASM manual for the scoring of sleep and associated events: Rules, terminology, and technical specifications*. Westchester, IL: American Academy of Sleep Medicine.

Ichikawa, J. (2008). Scepticism and the imagination model of dreaming. *Philosophical Quarterly, 58*(232), 519–527. doi:10.1111/j.1467-9213.2007.546.x.

Ichikawa, J. (2009). Dreaming and imagination. *Mind and Language, 24*(1), 103–121.

Ijsselsteijn, W., & Riva, G. (2003). Being there: The experience of presence in mediated environments. In G. Riva, F. Davide, & W. Ijsselsteijn (Eds.), *Being there: Concepts, effects, and measurement of user presence in synthetic environments* (pp. 3–16). Amsterdam: IOS Press.

Ionta, S., Heydrich, L., Lenggenhager, B., Mouthon, M., Fornari, E., Chapuis, D., et al. (2011). Multisensory mechanisms in temporo-parietal cortex support self-location and first-person perspective. *Neuron, 70*(2), 363–374.

Irvine, E. (2013). *Consciousness as a scientific concept: A philosophy of science perspective: Studies in brain and mind* (Vol. 5). New York: Springer.

Irwin, H. J. (1988). Out-of-the-body experiences and dream lucidity: Empirical perspectives. In J. Gackenbach & S. LaBerge (Eds.), *Conscious mind, sleeping brain: Perspectives on lucid dreaming* (pp. 353–371). New York: Plenum Press.

Jack, A., & Roepstorff, A. (2003). Trusting the subject? The use of introspective evidence in cognitive science. In *Trusting the subject?* Exeter: Imprint Academic. (Special issue, *Journal of Consciousness Studies*, 10 [9–10].)

Jackendoff, R. S. (2012). *A user's guide to thought and meaning*. New York: Oxford University Press.

Jackson, J. D., Weinstein, Y., & Balota, D. A. (2013). Can mind-wandering be timeless? Atemporal focus and aging in mind-wandering paradigms. *Frontiers in Psychology, 4*, 742. doi:10.3389/fpsyg.2013.00742.

Jacobson, A. (1930). Electrical measurements of neuromuscular states during mental activities. III. Visual imagination and recollection. *American Journal of Physiology, 95*, 694–702.

Jakobson, A.J., Conduit, R., & Fitzgerald, P. B. (2012). Investigation of visual dream reports after transcranial direct current stimulation (tDCS) during REM sleep. *International Journal of Dream Research, 5*(1). Retrieved from http://archiv.ub.uni-heidelberg.de/ojs/index.php/IJoDR/article/view/9272.

Jakobson, A. J., Fitzgerald, P. B., & Conduit, R. (2012a). Induction of visual dream reports after transcranial direct current stimulation (tDCS) during Stage 2 sleep. *Journal of Sleep Research, 21*(4), 369–379. doi:10.1111/j.1365-2869.2011.00994.x.

Jakobson, A. J., Fitzgerald, P. B., & Conduit, R. (2012b). Investigation of dream reports after transcranial direct current stimulation (tDCS) during slow wave sleep (SWS). *Sleep and Biological Rhythms, 10*(3), 169–178. doi:10.1111/j.1479-8425.2012.00538.x.

James, W. (1890/1950a). *The principles of psychology: Volume one*. New York: Dover.

James, W. (1890/1950b). *The principles of psychology: Volume two*. New York: Dover.

James, W. (1902/2003). *The varieties of religious experience: A study in human nature*. With a new introduction by Peter J. Gomes. New York: Penguin.

James, W. (1907/1981). *Pragmatism* (Bruce Kuklick, Ed.). Indianapolis: Hackett Publishing.

Jaspers, K. (1948). *Allgemeine Psychopathologie* (5th ed.). Berlin: Springer.

Jaynes, J. (1976). *The origin of consciousness in the breakdown of the bicameral mind*. Boston, MA: Houghton Mifflin.

Jezower, I. (1928/1985). *Das Buch der Träume*. Frankfurt am Main: Ullstein.

Johnson, C.R. (2005). DreamSpeak: An interview Clare Johnson, PhD (R. Waggoner & L. Gillis, Eds.). *Lucid Dream Exchange, 34*, 2–7. Retrieved from http://www.dreaminglucid.com/dreamspeak/DreamSpeak%2034%20Clare%20Johnson,%20PhD.pdf.

Johnson, C. R. (2014). Magic, meditation, and the void: Creative dimensions of lucid dreaming. In R. Hurd & K. Bulkeley (Eds.), *Lucid dreaming: New perspectives on consciousness in sleep*. Santa Barbara: Praeger.

Johnson, D. (2012). *Inception and philosophy: Because it's never just a dream*. Blackwell philosophy and pop culture series. Hoboken, NJ: Wiley.

Johnstone, H. W. (1973). Toward a philosophy of sleep. *Philosophy and Phenomenological Research, 34*(1), 73–81.

Jones, B. E. (2009a). Acetylcholine. In R. Stickgold & M. P. Walker (Eds.), *The neuroscience of sleep* (pp. 121–124). London: Elsevier Academic Press.

Jones, B. E. (2009b). Locus coeruleus and raphe nucleus. In R. Stickgold & M. P. Walker (Eds.), *The neuroscience of sleep* (pp. 166–170). London: Elsevier Academic Press.

Jones, E. (1951). *On the nightmare*. New York: Liveright.

Jones, E. (1999). The phenomenology of abnormal belief: A philosophical and psychiatric inquiry. *Philosophy, Psychiatry, and Psychology, 6*(1), 1–16.

Jouvet, M. (1962). Recherches sur les structures nerveuses et les mécanismes responsables des différentes phases du sommeil physiologiques. *Archives Italiennes de Biologie, 100*, 125–206.

Jouvet, M. (1999). *The paradox of sleep: The story of dreaming* (L. Garey, Trans.). Cambridge, MA: MIT Press.

Jouvet, M., & Michel, F. (1959). Correlations electromyographiques du sommeil chez le chat décortiqué et mésencéphalique chronique. *Comptes Rendus de la Société Biologie, 153*, 422–425.

Kahan, T. (1994). Lucid dreaming as metacognition: implications for cognitive science. *Consciousness and Cognition, 3*, 246–264.

Kahan, T. (2001). Consciousness in dreaming: A metacognitive approach. In K. Bulkeley (Ed.), *Dreams: A reader on the religious, cultural, and psychological dimensions of dreaming* (pp. 333–360). New York: Palgrave.

Kahan, T., & LaBerge, S. (1996). Cognition and metacognition in dreaming and waking: Comparisons of first- and third-person ratings. *Dreaming, 6*(4), 235–249.

Kahan, T. L., & LaBerge, S. P. (2011). Dreaming and waking: Similarities and differences revisited. *Consciousness and Cognition, 20*(3), 494–514. doi:10.1016/j.concog.2010.09.002.

Kahan, T. L., LaBerge, S., Levitan, L., & Zimbardo, P. (1997). Similarities and differences between dreaming and waking cognition: An exploratory study. *Consciousness and Cognition, 6*(1), 132–147. doi:10.1006/ccog.1996.0274.

Kahan, T. L., & Sullivan, K. T. (2012). Assessing metacognitive skills in waking and sleep: A psychometric analysis of the Metacognitive, Affective, Cognitive Experience (MACE) questionnaire. *Beyond the Comparator Model, 21*(1), 340–352. doi:10.1016/j.concog.2011.11.005.

Kahn, D. (2007). Metacognition, recognition, and reflection while dreaming. In D. Barrett & P. McNamara (Eds.), *The New Science of Dreaming: Vol. 2. Content, Recall, Personality Correlates* (pp. 245–268). Westport: Praeger Perspectives.

Kahn, D. (2012). Response to comments on "Dream logic: The inferential reasoning paradigm." *Dreaming, 22*(1), 78–81. doi:10.1037/a0026142.

Kahn, D. (2013). Brain basis of self: Self-organization and lessons from dreaming. *Frontiers in Psychology, 4*, 408. doi:10.3389/fpsyg.2013.00408.

Kahn, D., & Gover, T. (2010). Consciousness in dreams. *International Review of Neurobiology, 92*, 181–195.

Kahn, D., & Hobson, J. A. (2005). State-dependent thinking: A comparison of waking and dreaming thought. *Consciousness and Cognition, 14*(3), 429–438. doi:10.1016/j.concog.2004.10.005.

Kahn, D., Pace-Schott, E. F., & Hobson, J. A. (1997). Consciousness in waking and dreaming: The roles of neuronal oscillation and neuromodulation in determining similarities and differences. *Neuroscience, 78*, 13–38.

Kahn, D., Pace-Schott, E. F., & Hobson, J. A. (2002). Emotion and cognition: Feeling and character identification in dreaming. *Consciousness and Cognition, 11*, 34–50.

Kahn, D., Stickgold, R., Pace-Schott, E. F., & Hobson, J. A. (2000). Dreaming and waking consciousness: A character recognition study. *Journal of Sleep Research, 9*, 317–325.

Kam, J. W. Y., & Handy, T. C. (2013). The neurocognitive consequences of the wandering mind: A mechanistic account of sensory-motor decoupling. *Frontiers in Psychology, 4*, 725. doi:10.3389/fpsyg.2013.00725.

Kant, I. (1766/2002). *Träume eines Geistersehers, erläutert durch Träume der Metaphysik*. Stuttgart: Reclam.

Kant, I. (1833). *Anthropologie in pragmatischer Hinsicht: Mit einem Vorwort von I. F. Herbart (Vierte Original-Ausgabe)*. Leipzig: Verlag von Immanuel Müller.

Kant, I. (1988). *Vorkritische Schriften bis 1768* (6th ed.). *Suhrkamp-Taschenbuch Wissenschaft (Vol. 187)*. Frankfurt am Main: Suhrkamp.

Kantor, J. (1970). Pinching and dreaming. *Philosophical Studies, 21*(1–2), 28–32.

Karim, A. A. (2010). Transcranial cortex stimulation as a novel approach for probing the neurobiology of dreams: Clinical and neuroethical implications. *International Journal of Dream Research, 3* (1). Retrieved from http://archiv.ub.uni-heidelberg.de/ojs/index.php/IJoDR/article/view/593.

Karlsson, L. (2001). *Dream experience: Phenomenology meets cognitive neuroscience.* Lund: Lund University.

Keefer, M. H. (1996). The dreamer's path: Descartes and the sixteenth century. *Renaissance Quarterly, 49*(1), 30–76.

Keelin. (1995). Adventures with the NovaDreamer. *NightLight, 7.* Retrieved from http://www.lucidity.com/NL71.KeelinNova.html.

Keller, H. (1908/2003). *The world I live in* (R. Shattuck, Ed.). New York: Review Books.

Kelley, W. M., Macrae, C. N., Wyland, C. L., Caglar, S., Inati, S., & Heatherton, T. F. (2002). Finding the self? An event-related fMRI study. *Journal of Cognitive Neuroscience, 14*(5), 785–794.

Kelly, P. H. (1998). Defective inhibition of dream event memory formation: A hypothesized mechanism in the onset and progression of symptoms of schizophrenia. *Brain Research Bulletin, 46*(3), 189–197.

Kerouac, J. (1961/2001). *Book of dreams.* San Francisco: City Lights.

Kerr, N. H. (1993). Mental imagery, dreams, and perception. In C. Cavallero & D. Foulkes (Eds.), *Dreaming as cognition* (pp. 18–37). New York: Harvester Wheatsheaf.

Kerr, N., & Domhoff, G. (2004). Do the blind literally "see" in their dreams? A critique of a recent claim that they do. *Dreaming, 14*(4), 230–233.

Khalsa, S. S., Rudrauf, D., Damasio, A. R., Davidson, R. J., Lutz, A., & Tranel, D. (2008). Interoceptive awareness in experienced meditators. *Psychophysiology, 45*(4), 671–677. doi:10.1111/j.1469-8986.2008.00666.

Kilroe, P. (2000). The dream as text, the dream as narrative. *Dreaming, 10,* 125–137.

Kilteni, K., Groten, R., & Slater, M. (2012). The sense of embodiment in virtual reality. *Presence, 21*(4), 373–387. doi:10.1162/PRES_a_00124.

Kilteni, K., Normand, J.-M., Sanchez-Vives, M. V., Slater, M., & Tsakiris, M. (2012). Extending body space in immersive virtual reality: A very long arm illusion. *PLOS ONE, 7*(7), e40867. doi:10.1371/journal.pone.0040867.

Kim, J.-J. (Ed.). (2011). *Virtual reality.* Rijeka, Croatia: InTech.

Kind, A. (2001). Putting the image back in imagination. *Philosophy and Phenomenological Research, 62*(1), 85–109.

Kind, A. (2005). Imagery and imagination. In J. Fieser & B. Dowden (Eds.), *Internet encyclopedia of philosophy.* Retrieved from http://www.iep.utm.edu/imagery.

King, D. B., & DeCicco, T. L. (2007). The relationship between dream content and physical health, mood, and self-construal. *Dreaming, 17*(3), 127–139.

Kirwan, C. B., Ashby, S. R., & Nash, M. I. (2014). Remembering and imagining differentially engage the hippocampus: A multivariate fMRI investigation. *Cognitive Neuroscience*, 1–9. doi: 10.1080/17588928.2014.933203.

Kiverstein, J. D. (2009). Minimal sense of self, temporality and the brain. *Psyche, 15*(1), 59–74.

Klatzky, R. (1998). Allocentric and egocentric spatial representations: Definitions, distinctions, and interconnections. In C. Freksa, C. Habel, & K. Wender (Eds.), *Spatial cognition* (pp. 1–17). Lecture notes in computer science. Berlin: Springer Berlin Heidelberg.

Klinger, E. (2013). Goal commitments and the content of thoughts and dreams: Basic principles. *Frontiers in Psychology, 4*, 415. doi:10.3389/fpsyg.2013.00415.

Knauff, M. (2013). *Space to reason: A spatial theory of human thought.* Cambridge, MA: MIT Press.

Knobe, J. M., & Nichols, S. (Eds.). (2008). *Experimental philosophy.* Oxford: Oxford University Press.

Knoblich, G., Thornton, I. M., Grosjean, M., & Shiffrar, M. (Eds.). (2006). *Human body perception from the inside out.* Oxford: Oxford University Press.

Knoth, I. S., & Schredl, M. (2011). Physical pain, mental pain, and malaise in dreams. *International Journal of Dream Research, 4*(1), 17–23.

Knudsen, S., Gammeltoft, S., & Jennum, P. J. (2010). Rapid eye movement sleep behaviour disorder in patients with narcolepsy is associated with hypocretin-1 deficiency. *Brain, 133*(2), 568–579. doi:10.1093/brain/awp320.

Knudson, R. M. (2001). Significant dreams: Bizarre or beautiful? *Dreaming, 11*(4), 167–177.

Knudson, R. M., Adame, A. L., & Finocan, G. M. (2006). Significant dreams: Repositioning the self narrative. *Dreaming, 16*(3), 215–222.

Koch, C., & Davis, J. (Eds.). (1994). *Large-scale neuronal theories of the brain.* Cambridge, MA: MIT Press.

Koivisto, M., Mantyla, T., & Silvanto, J. (2010). The role of early visual cortex (V1/V2) in conscious and unconscious visual perception. *NeuroImage, 51*(2), 828–834. doi:10.1016/j.neuroimage.2010.02.042.

Koivisto, M., Railo, H., & Salminen-Vaparanta, N. (2011). Transcranial magnetic stimulation of early visual cortex interferes with subjective visual awareness and objective forced-choice performance. *Consciousness and Cognition, 20*(2), 288–298. doi:10.1016/j.concog.2010.09.001.

Kouider, S., Andrillon, T., Barbosa, L. S., Goupil, L., & Bekinschtein, T. A. (2014). Inducing task-relevant responses to speech in the sleeping brain. *Current Biology, 24*(18), 2208–2214. doi:10.1016/j.cub.2014.08.016.

Koriat, A. (2000). The feeling of knowing: Some metatheoretical implications for consciousness and control. *Consciousness and Cognition, 9*(2 Pt 1), 149–171. doi:10.1006/ccog.2000.0433.

Kosslyn, S. M. (1994). *Image and brain: The resolution of the imagery debate*. Cambridge, MA: MIT Press.

Kosslyn, S. M., Thompson, W. L., & Ganis, G. (2006). *The case for mental imagery*. Oxford: Oxford University Press.

Koulack, D. (1969). Effects of somatosensory stimulation on dream content. *Archives of General Psychiatry, 20*, 718–725.

Kozmová, M., & Wolman, R. N. (2006). Self-awareness in dreaming. *Dreaming, 16*(3), 196–214.

Kramer, M. (2007). *The dream experience: A systematic exploration*. New York: Routledge.

Kramer, M., Czaya, J., & Roth, T. (1974). The development of psychological content across the REM period. *Sleep Research, 3*, 121.

Kramer, M., Kinney, L., & Scharf, M. (1983). Dream incorporation and dream function. In W. Koella (Ed.), *Sleep 1982* (pp. 369–371). Basel: S. Karger.

Kramer, M., Roth, T., & Czaya, J. (1975). Dream development within a REM period. In W. Koella & P. Levin (Eds.), *Sleep 1974* (pp. 406–408). Basel: S. Karger.

Kriegel, U. (2007). The phenomenologically manifest. *Phenomenology and the Cognitive Sciences, 6*(1–2), 115–136.

Kriegel, U. (2013). A hesitant defense of introspection. *Philosophical Studies, 165*(3), 1165–1176. doi:10.1007/s11098-013-0148-0.

Krippner, S. (2007). Anomalous experiences in dreams. In D. Barrett & P. McNamara (Eds.), *The new science of dreaming: Vol. 2. Content, recall, personality correlates* (pp. 285–306). Westport: Praeger Perspectives.

Kroker, K. (2007). *The sleep of others and the transformations of sleep research*. Toronto: University of Toronto Press.

Kryger, M. H., Roth, T., & Dement, W. C. (2011). *Principles and practice of sleep medicine* (5th ed.). Philadelphia, PA: Saunders/Elsevier.

Kuiken, D. (2010). Primary and secondary consciousness during dreaming: Commentary on "The neurobiology of consciousness: Lucid dreaming wakes up" by J. Allan Hobson. *International Journal of Dream Research, 3*(1), 21–25.

Kuiken, D. (2014). Does Morrison's PGO wave/startle hypothesis help us explain such robust dream features as surprise and scene shift? In N. Tranquillo (Ed.), *Dream consciousness: Allan Hobson's new approach to the brain and its mind* (pp. 153–155). New York: Springer.

Kuiken, D., Busink, R., Dukewich, T., & Gendlin, E. T. (1996). Individual differences in orienting activity mediate feelings of realization in dreams II: Evidence from concurrent reports of movement inhibition. *Dreaming, 6*, 251–264.

Kuiken, D., & Nielsen, T. A. (1996). Individual differences in orienting activity mediate feeling realization in dreams: I. Evidence from retrospective reports of movement inhibition. *Dreaming, 6*, 201–217.

Kurz, I. (1919). *Traumland*. Stuttgart: Deutsche Verlagsanstalt.

Kussé, C., Muto, V., Mascetti, L., Matarazzo, L., Foret, A., Bourdiec, A. S.-L., et al. (2010). Neuroimaging of dreaming: State of the art and limitations. *International Review of Neurobiology, 92*, 87–99. doi:10.1016/S0074-7742(10)92005-9.

Kussé, C., Shaffii-LE Bourdiec, A., Schrouff, J., Matarazzo, L., & Maquet, P. (2012). Experience-dependent induction of hypnagogic images during daytime naps: A combined behavioural and EEG study. *Journal of Sleep Research, 21*(1), 10–20. doi:10.1111/j.1365-2869.2011.00939.x.

LaBerge, S. (1985). *Lucid dreaming: The power of being awake and aware in your dreams*. New York: Ballantine.

LaBerge, S. (1988a). Lucid dreaming in Western literature. In J. Gackenbach & S. LaBerge (Eds.), *Conscious mind, sleeping brain: Perspectives on lucid dreaming* (pp. 11–26). New York: Plenum Press.

LaBerge, S. (1988b). The Psychophysiology of Lucid Dreaming. In J. Gackenbach & S. LaBerge (Eds.), *Conscious mind, sleeping brain: Perspectives on lucid dreaming* (pp. 135–153). New York: Plenum Press.

LaBerge, S. (1990). Lucid dreaming: Psychophysiological studies of consciousness during REM sleep. In R. Bootzin, J. Kihlstrom, & D. Schacter (Eds.), *Sleep and cognition* (pp. 109–126). Washington, DC: American Psychological Association.

LaBerge, S. (2007). Lucid dreaming. In D. Barrett & P. McNamara (Eds.), *The new science of dreaming: Vol. 1. Biological aspects* (pp. 307–328). Westport: Praeger Perspectives.

LaBerge, S., & DeGracia, D. J. (2000). Varieties of lucid dreaming. In R. G. Kunzendorf & B. Wallace (Eds.), *Individual differences in conscious experience* (pp. 269–307). Amsterdam: John Benjamins.

LaBerge, S., & Dement, W. C. (1982). Voluntary control of respiration during lucid REM dreaming. *Sleep Research, 11*, 107.

LaBerge, S., Greenleaf, W., & Kedzierski, B. (1983). Physiological responses to dreamed sexual activity during REM sleep. *Psychophysiology, 19*, 454–455.

LaBerge, S., & Levitan, L. (1995). Validity established of dreamlight cues for eliciting lucid dreaming. *Dreaming, 5*, 159–168.

LaBerge, S., Nagel, L. E., Dement, W. C., & Zarcone, V. P. (1981). Lucid dreaming verified by volitional communication during REM sleep. *Perceptual and Motor Skills, 52*, 727–732.

LaBerge, S., Phillips, L., & Levitan, L. (1994). An hour of wakefulness before morning naps makes lucidity more likely. *NightLight, 6*(3). Retrieved from http://www.lucidity.com/NL63.RU.Naps.html.

LaBerge, S., & Rheingold, H. (1990). *Exploring the world of lucid dreaming*. New York: Ballantine.

LaBerge, S., & Zimbardo, P. (2000). *Smooth tracking eye-movements discriminate both dreaming and perception from imagination*. Paper presented at the Toward a Science of Consciousness Conference, Tucson, April 10–15. Retrieved from http://www.lucidity.com/Tucson2000abs.html.

Ladd, G. T. (1892). Contribution to the psychology of visual dreams. *Mind, 1*, 299–304.

Ladyman, J. (2005). Wouldn't it be lovely: Explanation and scientific realism. *Metascience, 14*(3), 331–361. doi:10.1007/s11016-005-3431-7.

Lamme, V. A. F., Supèr, H., Landman, R., Roelfsema, P. R., & Spekreijse, H. (2000). The role of primary visual cortex (V1) in visual awareness. *Vision Research, 40*(10–12), 1507–1521. doi:10.1016/S0042-6989(99)00243-6.

Lancee, J., Spoormaker, V. I., & van Den Bout, J. A. N. (2011). Long-term effectiveness of cognitive-behavioural self-help intervention for nightmares. *Journal of Sleep Research, 20*(3), 454–459. doi:10.1111/j.1365-2869.2010.00894.x.

Landesman, C. (1977). Dreams: Two types of explanations. In C. E. M. Dunlop (Ed.), *Philosophical essays on dreaming* (pp. 309–316). Ithaca: Cornell University Press. (Reprinted from *Philosophical Studies* 15 [1964], pp. 17–23.)

Landesman, C. (2002). *Skepticism: The central issues*. Malden, MA: Blackwell.

Lane, T., & Liang, C. (2011). Self-consciousness and immunity. *Journal of Philosophy, 108*(2), 78–99.

Lara-Carrasco, J., Nielsen, T. A., Solomonova, E., Levrier, K., & Popova, A. (2009). Overnight emotional adaptation to negative stimuli is altered by REM sleep deprivation and is correlated with intervening dream emotions. *Journal of Sleep Research, 18*, 178–187.

Lara-Carrasco, J., Simard, V., Saint-Onge, K., Lamoureux-Tremblay, V., & Nielsen, T. (2013). Maternal representations in the dreams of pregnant women: A prospective comparative study. *Frontiers in Psychology, 4*, 551. doi:10.3389/fpsyg.2013.00551.

Larøi, F., Barr, W. B., & Keefe, R. S. E. (2004). The neuropsychology of insight in psychiatric and neurological disorders. In X. F. Amador & A. S. David (Eds.), *Insight and psychosis: Awareness of illness in schizophrenia and related disorders* (2nd ed., pp. 119–156). Oxford: Oxford University Press.

Larøi, F., Collignon, O., & van der Linden, M. (2005). Source monitoring for actions in hallucination proneness. *Cognitive Neuropsychiatry, 10*(2), 105–123. doi:10.1080/13546800344000354.

Lastowka, F. G. (2010). *Virtual justice: The new laws of online worlds*. New Haven: Yale University Press.

Laureys, S., Boly, M., Moonen, G., & Maquet, P. (2009). Coma. In R. Stickgold & M. P. Walker (Eds.), *The neuroscience of sleep* (pp. 146–155). London: Elsevier Academic Press.

Laureys, S., Peigneux, P., Perrin, F., & Maquet, P. (2002). Sleep and motor skill learning. *Neuron, 35*, 5–7.

Leclair-Visonneau, L., Oudiette, D., Gaymard, B., Leu-Semenescu, S., & Arnulf, I. (2010). Do the eyes scan dream images during rapid eye movement sleep? Evidence from the rapid eye movement sleep behaviour disorder model. *Brain, 133*(6), 1737–1746. doi:10.1093/brain/awq110.

Legrand, D. (2006). The bodily self: The sensori-motor roots of pre-reflective self-consciousness. *Phenomenology and the Cognitive Sciences, 5*(1), 89–118.

Legrand, D. P. M. (2007). Pre-reflective self-consciousness: On being bodily in the world. *Janus Head, 9*(2), 493–519.

Leibniz, G. W. (1704/2005). *New essays on human understanding* (Jonathan Bennett, Ed.). Retrieved from http://www.earlymoderntexts.com/authors/leibniz.html.

Leibniz, G. W. (1880). *Die Philosophischen Schriften.* Berlin: Weidmannsche Buchhandlung.

Leibniz, G. W. (1956). *Philosophical papers and letters: A selection* (Leroy E. Loemker, Trans. & Ed.). Chicago: University of Chicago Press.

Leijenhorst, C. (2007). Sense and nonsense about sense: Hobbes and the Aristotelians on sense perception and imagination. In P. Springborg (Ed.), *The Cambridge companion to Hobbes's Leviathan* (pp. 82–108). Cambridge: Cambridge University Press.

Lenggenhager, B. (2009). *Multisensory mechanisms underlying bodily self-consciousness.* Unpublished doctoral dissertation presented to Faculté des Sciences de la Vie, EPFL, Lausanne.

Lenggenhager, B. & Lopez, C. (2015). Vestibular contributions to the sense of body, self, and others. In T. Metzinger & J. M. Windt (Eds.), *Open MIND: 23(T).* Frankfurt am Main: MIND Group. doi: 10.15502/9783958570023.

Lenggenhager, B., Lopez, C., & Blanke, O. (2008). Influence of galvanic vestibular stimulation on egocentric and object-based mental transformations. *Experimental Brain Research, 184*(2), 211–221. doi:10.1007/s00221-007-1095-9.

Lenggenhager, B., Tadi, T., Metzinger, T., & Blanke, O. (2007). Video ergo sum: Manipulating bodily self-consciousness. *Science, 317*(5841), 1096–1099. doi:10.1126/science.1143439.

Lepecq, J. C., Jouen, F., & Dubon, D. (1993). The effect of linear vection on manual aiming at memorized directions of stationary targets. *Perception, 22*(1), 49–60.

Lesku, J. A., Martinez-Gonzalez, D., & Rattenborg, N. C. (2009). Phylogeny and ontogeny of sleep. In R. Stickgold & M. P. Walker (Eds.), *The neuroscience of sleep* (pp. 61–69). London: Elsevier Academic Press.

Leslie, K., & Ogilvie, R. D. (1996). Vestibular dreams: The effect of rocking on dream mentation. *Dreaming, 6*(1), 1–16.

Leu-Semenescu, S., Uguccioni, G., Golmard, J.-L., Czernecki, V., Yelnik, J., Dubois, B., et al. (2013). Can we still dream when the mind is blank? Sleep and dream mentations in auto-activation deficit. *Brain, 136*(10), 3076–3084. doi:10.1093/brain/awt229.

Levin, R., & Nielsen, T. A. (2007). Disturbed dreaming, posttraumatic stress disorder, and affect distress: A review and neurocognitive model. *Psychological Bulletin, 133*(3), 482–528.

Levin, R., & Nielsen, T. A. (2009). Nightmares, bad dreams, and emotions dysregulation: A review and new neurocognitive model of dreaming. *Current Directions in Psychological Science, 18*(2), 84–88.

Levin, Y., & Aharon, I. (2011). What's on your mind? A brain scan won't tell. *Review of Philosophy and Psychology, 2*(4), 699–722. doi:10.1007/s13164-011-0063-2.

Levitan, L. (1994). A fool's guide to lucid dreaming. *NightLight, 6.* Retrieved from http://www.lucidity.com/NL62.FoolsGuide.html.

Levitan, L., & LaBerge, S. (1991). Other worlds: Out-of-body experiences and lucid dreams. *NightLight, 3.* Retrieved from http://www.lucidity.com/NL32.OBEandLD.html.

Levitan, L., & LaBerge, S. (1993). Testing the limits of dream control: The light and mirror experiment. *NightLight, 5.* Retrieved from http://www.lucidity.com/NL52.LightandMirror.html.

Levitan, C., Ward, P. B., Catts, S. V., & Hemsley, D. R. (1996). Predisposition toward auditory hallucinations: The utility of the Launay-Slade Hallucination Scale in psychiatric patients. *Personality and Individual Differences, 21,* 287–289.

Lewis, H. D. (1969). *The elusive mind: Based on the first series of the Gifford Lectures delivered in the University of Edinburgh, 1966–1968.* London: George Allen & Unwin.

Liang, C. (2015). Self-as-subject and experiential ownership. In T. Metzinger & J. M. Windt (Eds.), *Open MIND: 24(T).* Frankfurt am Main: MIND Group. doi: 10.15502/9783958570030.

Limanowski, J., & Blankenburg, F. (2013). Minimal self-models and the free energy principle. *Frontiers in Human Neuroscience, 7,* 547. doi:10.3389/fnhum.2013.00547.

Limosani, I., D'Agostino, A., Manzone, M. L., & Scarone, S. (2011a). The dreaming brain/mind, consciousness and psychosis. In *From Dreams to Psychosis: A European Science Foundation Exploratory Workshop. Consciousness and Cognition* (special issue), *20*(4), 987–992. doi:10.1016/j.concog.2010.11.014.

Limosani, I., D'Agostino, A., Manzone, M. L., & Scarone, S. (2011b). Bizarreness in dream reports and waking fantasies of psychotic schizophrenic and manic patients: Empirical evidences and theoretical consequences. *Psychiatry Research, 189*(2), 195–199. doi:10.1016/j.psychres.2011.02.023.

Lipton, P. (2000). Inference to the best explanation. In W. H. Newton-Smith (Ed.), *A companion to the philosophy of science* (pp. 184–193). Malden, MA: Blackwell.

Lipton, P. (2004). *Inference to the best explanation* (2nd ed.). International Library of Philosophy. London: Routledge.

Lipton, P. (2007). Précis of *Inference to the best explanation,* 2nd edition. *Philosophy and Phenomenological Research, 74*(2), 421–423. doi:10.1111/j.1933-1592.2007.00027.x.

Llewellyn, S. (2013). Such stuff as dreams are made on? Elaborative encoding, the ancient art of memory, and the hippocampus. *Behavioral and Brain Sciences, 36*(06), 589–607. doi:10.1017/S0140525X12003135.

Llinás, R., & Paré, D. (1991). Of dreaming and wakefulness. *Neuroscience, 44*, 521–535.

Llinás, R., & Ribary, U. (1994). Perception as an oneiric-like state modulated by the senses. In C. Koch & J. Davis (Eds.), *Large-Scale Neuronal Theories of the Brain* (pp. 111–124). Cambridge, MA: MIT Press.

Llobera, J., Sanchez-Vives, M. V., & Slater, M. (2013). The relationship between virtual body ownership and temperature sensitivity. *Journal of the Royal Society, Interface/the Royal Society, 10*(85), 20130300. doi:10.1098/rsif.2013.0300.

Locke, J. (1689/1997). *An essay concerning human understanding* (R. Woolhouse, Ed.). London: Penguin.

Loeb, L. E. (2005). The Cartesian circle. In J. Cottingham (Ed.), *The Cambridge companion to Descartes* (pp. 200–235). Cambridge: Cambridge University Press.

Loewenstein, R., Newman, L., Schur, M., & Solnit, A. (Eds.). (1966). *Psychoanalysis: A general psychology*. New York: International Universities Press.

Logothetis, N. K. (2008). What we can do and what we cannot do with fMRI. *Nature, 453*(7197), 869–878. doi:10.1038/nature06976.

Longo, M. R., Azanon, E., & Haggard, P. (2010). More than skin deep: Body representation beyond primary somatosensory cortex. *Neuropsychologia, 48*(3), 655–668. doi:10.1016/j.neuropsychologia.2009.08.022.

Longo, M. R., & Haggard, P. (2010). An implicit body representation underlying human position sense. *Proceedings of the National Academy of Sciences of the United States of America, 107*(26), 11727–11732. doi:10.1073/pnas.1003483107.

Longo, M. R., & Haggard, P. (2012). What is it like to have a body? *Current Directions in Psychological Science, 21*(2), 140–145. doi:10.1177/0963721411434982.

Longo, M. R., Schuur, F., Kammers, M. P. M., Tsakiris, M., & Haggard, P. (2008). What is embodiment? A psychometric approach. *Cognition, 107*(3), 978–998. doi:10.1016/j.cognition.2007.12.004.

Lopes da Silva, F. H. (2003). Visual dreams in the congenitally blind? *Trends in Cognitive Sciences, 7*(8), 328–330.

Lopez, C., Halje, P., & Blanke, O. (2008). Body ownership and embodiment: Vestibular and multisensory mechanisms. *Neurophysiologie Clinique, 38*(3), 149–161. doi:10.1016/j.neucli.2007.12.006.

Lortie-Lussier, M. (2000). The consistency and continuity hypotheses revisited through the dreams of women at two periods of their lives. *Dreaming, 10*, 67–76.

Louie, K., & Wilson, M. A. (2001). Temporally structured replay of awake hippocampal ensemble activity during rapid eye movement sleep. *Neuron, 29*(1), 145–156.

Lustenberger, C., & Huber, R. (2012). High density electroencephalography in sleep research: Potential, problems, future perspective. *Frontiers in Neurology, 3*, 77. doi:10.3389/fneur.2012.00077.

Lutz, A., Lachaux, J.-P., Martinerie, J., & Varela, F. J. (2002). Guiding the study of brain dynamics by using first-person data: Synchrony patterns correlate with ongoing conscious states during a simple visual task. *Proceedings of the National Academy of Sciences of the United States of America, 99*(3), 1586–1591. doi:10.1073/pnas.032658199.

Lutz, A., & Thompson, E. (2003). Neurophenomenology—integrating subjective experience and brain dynamics in the neuroscience of consciousness. *Journal of Consciousness Studies, 10*(9–10), 31–52.

Lycos, K. (1964). Aristotle and Plato on "appearing." *Mind, 73*(292), 496–514.

Lydic, R., & Baghdoyan, H. A. (Eds.). (1999). *Handbook of behavioral state control: Molecular and cellular mechanisms*. Boca Raton, FL: CRC Press.

Lynn, S. J., Krippner, S., & Cardeña, E. (Eds.). (2000). *Variety of anomalous experience*. Washington, DC: American Psychology Association.

Lyons, W. E. (1986). *The disappearance of introspection*. Cambridge, MA: MIT Press.

Macdonald, M. (1977). Sleeping and waking. In C. E. M. Dunlop (Ed.), *Philosophical essays on dreaming* (pp. 64–80). Ithaca: Cornell University Press. (Reprinted from *Mind* 72 [1953], pp. 202–215.)

MacFarlane, J. G., & Wilson, T. L. (2006). A relationship between nightmare content and somatic stimuli in a sleep-disordered population: A preliminary study. *Dreaming, 16*(1), 53–59.

Mach, E. (1919). *Die Analyse der Empfindungen und das Verhältnis des Physischen zum Psychischen: Achte Auflage*. Jena: Verlag von Gustav Fischer.

Mackonis, A. (2013). Inference to the best explanation, coherence and other explanatory virtues. *Synthese, 190*(6), 975–995.

Macnish, R. (1838). *The Philosophy of Sleep*. Glasgow: W. R. M'Phun.

Macpherson, F. (2013). The philosophy and psychology of hallucination: An introduction. In F. Macpherson & D. Platchias (Eds.), *Hallucination: Philosophy and psychology* (pp. 1–38). Cambridge, MA: MIT Press.

Madary, M. (2012a). How would the world look if it looked as if it were encoded as an intertwined set of probability density distributions? *Frontiers in Psychology, 3*, 419. doi:10.3389/fpsyg.2012.00419.

Madary, M. (2012b). Husserl on perceptual constancy. *European Journal of Philosophy, 20*(1), 145–165. doi:10.1111/j.1468-0378.2010.00405.x.

Madary, M. (2013). Anticipation and variation in visual content. *Philosophical Studies, 165*(2), 335–347. doi:10.1007/s11098-012-9926-3.

Madary, M. (2015). Extending the explanandum for predictive processing—A commentary on Andy Clark. In T. Metzinger & J. M. Windt (Eds.), *Open MIND: 7(C)*. Frankfurt am Main: MIND Group. doi:10.15502/9783958570313.

Madow, L., & Snow, L. H. (Eds.). (1970). *The psychodynamic implications of the physiological studies on dreams*. Springfield, IL: Charles S. Thomas.

Madsen, P. L., Holm, S., Vorstrup, S., Friberg, L., Lassen, N. A., & Wildschiødtz, G. (1991). Human regional cerebral blood flow during rapid-eye-movement sleep. *Journal of Cerebral Blood Flow and Metabolism: Official Journal of the International Society of Cerebral Blood Flow and Metabolism, 11*(3), 502–507. doi:10.1038/jcbfm.1991.94.

Madsen, P. L., & Vorstrup, S. (1991). Cerebral blood flow and metabolism during sleep. *Cerebrovascular and Brain Metabolism Reviews, 3*, 281–296.

Magallón, L. (1991). Awake in the dark: Imageless lucid dreaming. *Lucidity, 10*(1–2), 46–48. Retrieved from http://www.sawka.com/spiritwatch/awakein.htm.

Maggiolini, A., Persico, A., & Crippa, F. (2007). Gravity content in dreams. *Dreaming, 17*(2), 87–97.

Mahowald, M. W., & Schenck, C. H. (1999). Dissociated states of wakefulness and sleep. In R. Lydic & H. A. Baghdoyan (Eds.), *Handbook of behavioral state control: Molecular and cellular mechanisms*. Boca Raton, FL: CRC Press.

Mahowald, M. W., Schenck, C. H., & Cramer Bornemann, M. A. (2005). Sleep-related violence. *Current Neurology and Neuroscience Reports, 5*(2), 153–158.

Mahowald, M. W., Schenck, C. H., & Bornemann, M. A. C. (2011). Parasomnias and sleep forensics. In S. Chokroverty & P. Sahota (Eds.), *Acute and emergent events in sleep disorders* (pp. 93–129). New York: Oxford University Press.

Malamud, J. R. (1988). Learning to become fully lucid: A program for inner growth. In J. Gackenbach & S. LaBerge (Eds.), *Conscious mind, sleeping brain: Perspectives on lucid dreaming* (pp. 309–319). New York: Plenum Press.

Malcolm, N. (1956). Dreaming and skepticism. *Philosophical Review, 65*(1), 14–37.

Malcolm, N. (1958). *Ludwig Wittgenstein: A memoir*. Oxford: Oxford University Press.

Malcolm, N. (1959). Stern's dreaming. *Analysis, 19*(December), 47.

Malcolm, N. (1959/1962). *Dreaming*. London: Routledge & Kegan Paul.

Malcolm, N. (1961). Professor Ayer on dreaming. *Journal of Philosophy, 58*(11), 294–297.

Mallon, B. (2002). *Dream time with children: Learning to dream, dreaming to learn*. London: Jessica Kingsley.

Mamelak, A. N., & Hobson, J. A. (1989a). Dream bizarreness as the cognitive correlate of altered neuronal behavior in REM sleep. *Journal of Cognitive Neuroscience, 1*, 201–222.

Mamelak, A., & Hobson, J. A. (1989b). Nightcap: A home-based sleep monitoring system. *Sleep, 12*(2), 157–166.

Manni, R., Terzaghi, M., Ratti, P.-L., Repetto, A., Zangaglia, R., & Pacchetti, C. (2011). Hallucinations and REM sleep behaviour disorder in Parkinson's disease: Dream imagery intrusions and other hypotheses. *Consciousness and Cognition, 20*(4), 1021–1026. doi:10.1016/j.concog.2010.10.009.

Maquet, P., Dive, D., Salmon, E., Sadzot, B., Franco, G., Poirrier, R., et al. (1990). Cerebral glucose utilization during sleep-wake cycle in man determined by positron emission tomography and [18F]2-fluoro-2-deoxy-d-glucose method. *Brain Research, 513*(1), 136–143. doi:10.1016/0006-8993(90)91099-3.

Maquet, P., Laureys, S., Peigneux, P., Fuchs, S., Petiau, C., Phillips, C., et al. (2000). Experience-dependent changes in cerebral activation during human REM sleep. *Nature Neuroscience, 3*(8), 831–836. doi:10.1038/77744.

Maquet, P., Peters, J., Aerts, J., Delfiore, G., Degueldre, C., Luxen, A., et al. (1996). Functional neuroanatomy of human rapid-eye-movement sleep and dreaming. *Nature Reviews: Neuroscience, 383*(6596), 163–166.

Maquet, P., Ruby, P., Maudoux, A., Albouy, G., Sterpenich, V., Dang-Vu, T., et al. (2005). Human cognition during REM sleep and the activity profile within frontal and parietal cortices: A reappraisal of functional neuroimaging data. In Steven Laureys (Ed.), *Progress in brain research: The boundaries of consciousness: Neurobiology and neuropathology* (pp. 219–595). Amsterdam: Elsevier. Retrieved from http://www.sciencedirect.com/science/article/pii/S0079612305500165.

Margolis, E., & Laurence, S. (2011). Concepts. In E. N. Zalta (Ed.), *The Stanford encyclopedia of philosophy (fall 2011 edition)*. Retrieved from http://plato.stanford.edu/entries/concepts.

Marinelli, L., & Mayer, A. (Eds.). (2000). *Die Lesbarkeit der Träume: Zur Geschichte von Freuds Traumdeutung*. Frankfurt am Main: Fischer Taschenbuch.

Marinelli, L., & Mayer, A. (2002). *Träume nach Freud: Die "Traumdeutung" und die Geschichte der psychoanalytischen Bewegung*. Vienna: Turia & Kant.

Marková, I. S., & Berrios, G. E. (1994). Delusional misidentifications: Facts and fantasies. *Psychopathology, 27*, 136–143.

Marks, D. F. (1972). Individual differences in the vividness of visual imagery and their effect on function. In P. W. Sheehan (Ed.), *The function and nature of Imagery* (pp. 83–108). New York: Academic Press.

Martin, M. (2009). The reality of appearances. In A. Byrne & H. Logue (Eds.), *Disjunctivism: Contemporary readings* (pp. 91–116). Cambridge, MA: MIT Press.

Martini, M., Perez-Marcos, D., & Sanchez-Vives, M. (2014). Modulation of pain threshold by virtual body ownership. *European Journal of Pain*. doi:10.1002/j.1532-2149.2014.00451.x.

Marzano, C., Ferrara, M., Mauro, F., Moroni, F., Gorgoni, M., Tempesta, D., et al. (2011). Recalling and forgetting dreams: Theta and alpha oscillations during sleep predict subsequent dream recall. *Journal of Neuroscience, 31*(18), 6674–6683. doi:10.1523/JNEUROSCI.0412-11.2011.

Maselli, A., & Slater, M. (2013). The building blocks of the full body ownership illusion. *Frontiers in Human Neuroscience, 7*, 83. doi:10.3389/fnhum.2013.00083.

Massimini, M., Ferrarelli, F., Huber, R., Esser, S. K., Singh, H., & Tononi, G. (2005). Breakdown of cortical effective connectivity during sleep. *Science, 309*(5744), 2228–2232. doi:10.1126/science.1117256.

Massimini, M., Ferrarelli, F., Murphy, M., Huber, R., Riedner, B., Casarotto, S., & Tononi, G. (2010). Cortical reactivity and effective connectivity during REM sleep in humans. *Cognitive Neuroscience, 1*(3), 176–183. doi:10.1080/17588921003731578.

Massimini, M., Ferrarelli, F., Sarasso, S., & Tononi, G. (2012). Cortical mechanisms of loss of consciousness: Insight from TMS/EEG studies. *Archives Italiennes de Biologie, 150*(2–3), 44–55. doi:10.4449/aib.v150i2.1361.

Mast, F. W., Tartaglia, E. M., & Herzog, M. H. (2012). New percepts via mental imagery? *Frontiers in Psychology, 3*, 360. doi:10.3389/fpsyg.2012.00360.

Maury, A. (1848). Des hallucinations hypnagogiques ou des erreurs des sens dans l'état intermédiaire entre la veille et le sommeil. *Annales Médico-Psychologiques du Système Nerveux, 11*, 26–40.

Maury, L.-F. A. (1861). *Le sommeil et les rêves: Études psychologiques sur ces phénomènes et les divers états qui s'y rattachent*. Paris: Didier et Cie Libraires-Éditeurs.

Mavromatis, A. (1987). *Hypnagogia: The unique state of consciousness between wakefulness and sleep*. London: Routledge.

Mazzetti, M., Bellucci, C., Mattarozzi, K., Plazzi, G., Tuozzi, G., & Cipolli, C. (2009). REM-dreams recall in patients with narcolepsy-cataplexy. *Brain Research Bulletin, 81*(1), 133–140. doi:10.1016/j.brainresbull.2009.10.021.

Mazzoni, G. A., & Loftus, E. F. (1996). When dreams become reality. *Consciousness and Cognition, 5*(4), 442–462. doi:10.1006/ccog.1996.0027.

McCarley, R. W., & Hoffman, E. (1981). REM sleep dreams and the activation-synthesis hypothesis. *American Journal of Psychiatry, 138*(7), 904–912.

McFee, G. (1994). The surface grammar of dreaming. *Proceedings of the Aristotelian Society, 94*, 95–115.

McGeoch, P. D., Brang, D., Song, T., Lee, R. R., Huang, M., & Ramachandran, V. S. (2011). Xenomelia: A new right parietal lobe syndrome. *Journal of Neurology, Neurosurgery, and Psychiatry, 82*(12), 1314–1319. doi:10.1136/jnnp-2011-300224.

McGinn, C. (2004). *Mindsight: Image, dream, meaning*. Cambridge, MA: Harvard University Press.

McGinn, C. (2005a). The matrix of dreams. In C. Grau (Ed.), *Philosophers explore The Matrix* (pp. 62–70). Oxford: Oxford University Press.

McGinn, C. (2005b). *The power of movies: How screen and mind interact*. New York: Vintage.

McMillan, R., Kaufman, S. B., & Singer, J. L. (2013). Ode to positive constructive daydreaming. *Frontiers in Psychology*, *4*, 626. doi:10.3389/fpsyg.2013.00626.

McNamara, P., Harris, E., & Kookoolis, A. (2007). Costly signaling theory of dream recall and dream sharing. In D. Barrett & P. McNamara (Eds.), *The New Science of Dreaming: Vol. 3. Cultural and Theoretical Perspectives* (pp. 117–132). Westport: Praeger Perspectives.

McNamara, P., Johnson, P., McLaren, D., Harris, E., Beauharnais, C., & Auerbach, S. (2010). REM and NREM sleep mentation. *International Review of Neurobiology*, 69–86. Elsevier.

McNamara, P., McLaren, D., & Durso, K. (2007). Representation of the self in REM and NREM dreams. *Dreaming*, *17*(2), 113–126. doi:10.1037/1053-0797.17.2.113.

McNamara, P., McLaren, D., Kowalczyk, S., & Pace-Schott, E. F. (2007). "Theory of mind" in REM and NREM dreams. In D. Barrett & P. McNamara (Eds.), *The New Science of Dreaming: Vol. 2. Content, Recall, Personality Correlates* (pp. 201–220). Westport: Praeger Perspectives.

McNamara, P., McLaren, D., Smith, D., Brown, A., & Stickgold, R. (2005). A "Jekyll and Hyde" within: Aggressive versus friendly interactions in REM and non-REM dreams. *Psychological Science*, *16*(2), 130–136. doi:10.1111/j.0956-7976.2005.00793.

Meaidi, A., Jennum, P., Ptito, M., & Kupers, R. (2014). The sensory construction of dreams and nightmare frequency in congenitally blind and late blind participants. *Sleep Medicine*, *15*(5), 586–595. doi:10.1016/j.sleep.2013.12.008.

Meier, B. (1993). Speech and thinking in dreams. In C. Cavallero & D. Foulkes (Eds.), *Dreaming as cognition* (pp. 58–76). New York: Harvester Wheatsheaf.

Menary, R. (2007). *Cognitive integration: Attacking the bounds of cognition*. New York: Palgrave Macmillan.

Menary, R. (2010). Cognitive integration and the extended mind. In R. Menary (Ed.), *The extended mind* (pp. 227–244). Cambridge, MA: MIT Press.

Menary, R. (2015). Mathematical cognition—A case of enculturation. In T. Metzinger & J. M. Windt (Eds.), *Open MIND: 25(T)*. Frankfurt am Main: MIND Group. doi:10.15502/9783958570818.

Mercer, J. D., Bootzin, R. R., & Lack, L. C. (2002). Insomniacs' perception of wake instead of sleep. *Sleep*, *25*(5), 564–571.

Merritt, J. M., Stickgold, R., Pace-Schott, E., Williams, J., & Hobson, J. A. (1994). Emotion profiles in the dreams of men and women. *Consciousness and Cognition*, *3*, 46–60.

Metcalfe, J. (2000). Feelings and judgments of knowing: Is there a special noetic state? *Consciousness and Cognition, 9*(2), 178–186. doi:10.1006/ccog.2000.0451.

Metzinger, T. (2003). *Being no one: The self-model theory of subjectivity.* Cambridge, MA: MIT Press.

Metzinger, T. (2005). Out-of-body experiences as the origin of the concept of a soul. *Mind and Matter, 3*(1), 57–84.

Metzinger, T. (2006). Reply to Gallagher: Different conceptions of embodiment. *Psyche, 12*(4).

Metzinger, T. (2009). *The ego tunnel: The science of the mind and the myth of the self.* New York: Basic Books.

Metzinger, T. (2011). The no-self-alternative. In S. Gallagher (Ed.), *The Oxford handbook of the self* (pp. 279–296). Oxford: Oxford University Press.

Metzinger, T. K. (2013a). The myth of cognitive agency: Subpersonal thinking as a cyclically recurring loss of mental autonomy. *Frontiers in Psychology, 4*, 931. doi:10.3389/fpsyg.2013.00931.

Metzinger, T. (2013b). Two principles for robot ethics. In E. Hilgendorf & J.-P. Günther (Eds.), *Robotik und Gesetzgebung* (pp. 247–286). Baden-Baden: Nomos.

Metzinger, T. K. (2013c). Why are dreams interesting for philosophers? The example of minimal phenomenal selfhood, plus an agenda for future research. *Frontiers in Psychology, 4*, 746. doi:10.3389/fpsyg.2013.00746.

Metzinger, T. (2014a). First-order embodiment, second-order embodiment, third-order embodiment. In L. Shapiro (Ed.), *The Routledge handbook of embodied cognition.* New York: Routledge.

Metzinger, T. (2014b). What is the specific significance of dream research for philosophy of mind? In N. Tranquillo (Ed.), *Dream consciousness: Allan Hobson's new approach to the brain and its mind* (pp. 161–166). New York: Springer.

Metzinger, T., & Windt, J. M. (2014). Die phänomenale Signatur des Wissens: Experimentelle Philosophie des Geistes mit oder ohne Intuitionen? In T. Grundmann, J. Horvath, & J. Kipper (Eds.), *Experimentelle Philosophie* (pp. 279–321). Frankfurt am Main: Suhrkamp.

Metzinger, T. & Windt, J. M. (2015). What does it mean to have an open MIND? In T. Metzinger & J. M. Windt (Eds.), *Open MIND: 0(C).* Frankfurt am Main: MIND Group. doi:10.15502/9783958571044.

Meumann, E. (1907). Über Organempfindungsträume und eine merkwürdige Traumerinnerung. *Archiv fur die Gesamte Psychologie, 9*(1), 63.

Middleton, W. C. (1942). The frequency with which a group of unselected college students experience colored dreaming and colored hearing. *Journal of General Psychology, 27*, 221–229.

Mignot, E., & Lin, L. (2009). Narcolepsy. In R. Stickgold & M. P. Walker (Eds.), *The neuroscience of sleep* (pp. 270–277). London: Elsevier Academic Press.

Miles, L. K., Karpinska, K., Lumsden, J., Macrae, C. N., & Gilbert, S. (2010). The meandering mind: Vection and mental time travel. *PLoS ONE, 5*(5), e10825. doi:10.1371/journal.pone.0010825.

Miles, L. K., Nind, L. K., & Macrae, C. N. (2010). Moving through time. *Psychological Science, 21*(2), 222–223. doi:10.1177/0956797609359333.

Minkel, J. D., Banks, S., & Dinges, D. F. (2009). Behavioral change with sleep deprivation. In R. Stickgold & M. P. Walker (Eds.), *The neuroscience of sleep* (pp. 241–248). London: Elsevier Academic Press.

Minkel, J. D., & Dinges, D. F. (2009). Circadian rhythms in sleepiness, alertness, and performance. In R. Stickgold & M. P. Walker (Eds.), *The neuroscience of sleep* (pp. 183–188). London: Elsevier Academic Press.

Miyawaki, Y., Uchida, H., Yamashita, O., Sato, M., Morito, Y., Tanabe, H., et al. (2008). Visual image reconstruction from human brain activity using a combination of multiscale local image decoders. *Neuron, 60*(5), 915–929. Retrieved from http://linkinghub.elsevier.com/retrieve/pii/S0896627308009586.

Moeller, H.-G. (2004). *Daoism explained: From the dream of the butterfly to the fishnet allegory*. Ideas explained series (Vol. 1). Chicago: Open Court.

Moffitt, A. (1995). Dreaming: Functions and meaning. *Impuls, 3*, 18–31.

Moffitt, A., Hoffmann, R., Mullington, J., Purcell, S., Pigeau, R., & Wells, R. (1988). Dream psychology: Operating in the dark. In J. Gackenbach & S. LaBerge (Eds.), *Conscious mind, sleeping brain: Perspectives on lucid dreaming* (pp. 429–439). New York: Plenum Press.

Moiseeva, N. I. (1975). The characteristics of EEG activity and the subjective estimation of time during dreams of different structure. *Electroencephalography and Clinical Neurophysiology, 38*(6), 569–577.

Moller, H. J., & Barbera, J. (2006). Media presence, consciousness, and dreaming. In G. Riva, M. T. Anguera, B. K. Wiederhold, & F. Mantovani (Eds.), *From communication to presence: Cognition, emotions, and culture towards the ultimate communicative experience; Festschrift in honor of Luigi Anolli* (pp. 97–123). Amsterdam: IOS Press.

Monroe, W. S. (1905). Mental elements of dreams. *Journal of Philosophy, Psychology, and Scientific Methods, 2*(24), 650–652.

Montangero, J. (1993). Dream, problem-solving, and creativity. In C. Cavallero & D. Foulkes (Eds.), *Dreaming as cognition* (pp. 93–113). New York: Harvester Wheatsheaf.

Montangero, J. (2012a). Dream thought should be compared with waking world simulations: A comment on Hobson and colleagues' paper on dream logic. *Dreaming, 22*(1), 70–73. doi:10.1037/a0026140.

Montangero, J. (2012b). Dreams are narrative simulations of autobiographical episodes, not stories or scripts: A review. *Dreaming, 22*(3), 157–172. doi:10.1037/a0028978.

Montangero, J., Tihon Ivanyi, C., & de Saint-Hilaire, Z. (2003). Completeness and accuracy of morning reports after a recall cue: Comparison of dream and film reports. *Consciousness and Cognition, 12*(1), 49–62. doi:10.1016/S1053-8100(02)00029-6.

Morrison, A. P., Wells, A., & Nothard, S. (2000). Cognitive factors in predisposition to auditory and visual hallucinations. *British Journal of Clinical Psychology, 39*, 67–78.

Morrison, A. R. (2014). You have interpreted the PGO waves of REM sleep as activation of the startle network of the brain: What is your theory of the function of off-line startle and what impact, if any, does this activation have upon dreaming? In N. Tranquillo (Ed.), *Dream consciousness: Allan Hobson's new approach to the brain and its mind* (pp. 171–173). New York: Springer.

Morrison, A. R., & Pompeiano, O. (1970). Vestibular control of autonomic functions during the rapid eye movements of desynchronized sleep. *Archives Italiennes de Biologie, 108*, 154–180.

Mota-Rolim, S. A., Erlacher, D., Tort, A. B. L., Araujo, J. F., & Ribeiro, S. (2010). Different kinds of subjective experience during lucid dreaming may have different neural substrates. *International Journal of Dream Research, 3*(1), 33–35.

Mota-Rolim, S. A., Targino, Z. H., Souza, B. C., Blanco, W., Araujo, J. F., & Ribeiro, S. (2013). Dream characteristics in a Brazilian sample: An online survey focusing on lucid dreaming. *Frontiers in Human Neuroscience, 7*, 836. doi:10.3389/fnhum.2013.00836.

Moutoussis, M., Fearon, P., El-Deredy, W., Dolan, R. J., & Friston, K. J. (2014). Bayesian inferences about the self (and others): A review. *Consciousness and Cognition, 25*(100), 67–76. doi:10.1016/j.concog.2014.01.009.

Mulder, T., Hochstenbach, J., Dijkstra, P. U., & Geertzen, J. H. B. (2008). Born to adapt, but not in your dreams. *Consciousness and Cognition, 17*(4), 1266–1271. doi:10.1016/j.concog.2007.04.001.

Muldoon, S. J., & Carrington, H. (1929). *The projection of the astral body*. London: Psychic Book Club.

Müller, J. (1826/1951). *Über die phantastischen Gesichtserscheinungen*. Hannover: Schmorl & von Seefeld Nachf.

Munglani, R., & Jones, J. G. (1992). Altered consciousness: Pharmacology and phenomenology: Sleep and general anaesthesia as altered states of consciousness. *Journal of Psychopharmacology, 6*(3), 399–409. doi:10.1177/026988119100500442.

Murray, C. D., & Gordon, M. S. (2001). Changes in bodily awareness induced by immersive virtual reality. *Cyberpsychology and Behavior, 4*(3), 365–371.

Murzyn, E. (2008). Do we only dream in colour? A comparison of reported dream colour in younger and older adults with different experiences of black and white media. *Consciousness and Cognition, 17*, 1228–1237.

Murzyn, E. (2010). *Individual differences in visual memory, imagery style, and media experience and their effect on the visual qualities of dreams*. Unpublished doctoral dissertation, University of

Dundee, Dundee, United Kingdom. Retrieved from http://discovery.dundee.ac.uk/portal/files/1199055/Murzyn_phd_2010.pdf.

Murzyn, E. (2012). Memory for colour, visual imagery preferences and the reported colour of dreams. *International Journal of Dream Research, 5*(2). Retrieved from http://archiv.ub.uni-heidelberg.de/ojs/index.php/IJoDR/article/view/9179.

Muzur, A., Pace-Schott, E. F., & Hobson, J. A. (2002). The prefrontal cortex in sleep. *Trends in Cognitive Sciences, 6*, 475–481.

Mwenge, B., Brion, A., Uguccioni, G., & Arnulf, I. (2013). Sleepwalking: Long-term home video monitoring. *Sleep Medicine, 14*(11), 1226–1228. doi:10.1016/j.sleep.2013.04.027.

Nagel, T. (1959). Dreaming. *Analysis, 19*(5), 112–116.

Nagel, T. (1974). What is it like to be a bat? *Philosophical Review, 83*(October), 435–450.

Nagel, T. (1986). *The view from nowhere*. New York: Oxford University Press.

Nanay, B. (2009a). Four theories of amodal perception. In D. S. McNamara & J. G. Trafton (Eds.), *Proceedings of the 29th Annual Conference of the Cognitive Science Society (CogSci 2007)* (pp. 1331–1336). Hillsdale, NJ: Erlbaum.

Nanay, B. (2009b). How speckled is the hen? *Analysis, 69*(3), 499–502. doi:10.1093/analys/anp072.

Nanay, B. (2010). Perception and imagination: Amodal perception as mental imagery. *Philosophical Studies*, 150(2), 239–254.

Nanay, B. (2012). The philosophical implications of the Perky experiments: Reply to Hopkins. *Analysis, 72*(3), 439–443. doi:10.1093/analys/ans066.

Neider, M., Pace-Schott, E. F., Forselius, E., Pittman, B., & Morgan, P. T. (2011). Lucid dreaming and ventromedial versus dorsolateral prefrontal task performance. *Consciousness and Cognition, 20*(2), 234–244. doi:10.1016/j.concog.2010.08.001.

Nelson, J. O. (1966). Can one tell that he is awake by pinching himself? *Philosophical Studies, 17*(6), 81–84.

Nielsen, T. A. (1991). Reality dreams and their effects on spiritual belief: A revision of animism theory. In J. Gackenbach & A. A. Sheikh (Eds.), *Dream images: A call to mental arms* (pp. 233–264). Amityville: Baywood Publishing.

Nielsen, T. A. (1991–1992). A self-observational study of spontaneous hypnagogic imagery using the upright napping procedure. *Imagination, Cognition, and Personality, 11*(4), 353–366.

Nielsen, T. A. (1993). Changes in kinesthetic content of dreams following somatosensory stimulation of leg during REM sleep. *Dreaming, 3*, 99–113.

Nielsen, T. A. (1995). Describing and modeling hypnagogic imagery using a systematic self observation procedure. *Dreaming, 5*(2), 75–94.

Nielsen, T. A. (2000a). A review of mentation in REM and NREM sleep: "Covert" REM sleep as a possible reconciliation of two opposing models. *Behavioral and Brain Sciences, 23*(6), 851–866.

Nielsen, T. A. (2000b). Dream mentation production and narcolepsy: A critique. *Consciousness and Cognition, 9*(4), 510–513. doi:10.1006/ccog.2000.0467.

Nielsen, T. A. (2004). Chronobiological features of dream production. *Sleep Medicine Reviews, 8*, 403–424.

Nielsen, T. A. (2007). Felt presence: Paranoid delusion or hallucinatory social imagery? *Consciousness and Cognition, 16*(4), 975–983. doi:10.1016/j.concog.2007.02.002.

Nielsen, T. A. (2010a). The changing historical context of dream research. In M. Kryger, T. Roth, & W. C. Dement (Eds.), *Principles and practice of sleep medicine* (5th ed., pp. 561–562). New York: Elsevier.

Nielsen, T. A. (2010b). Disturbed dreaming as a factor in medical conditions. In M. Kryger, T. Roth, & W. C. Dement (Eds.), *Principles and practice of sleep medicine* (5th ed., pp. 1116–1127). New York: Elsevier.

Nielsen, T. A. (2010c). Dream analysis and classification: The reality simulation perspective. In M. Kryger, T. Roth, & W. C. Dement (Eds.), *Principles and practice of sleep medicine* (5th ed., pp. 595–603). New York: Elsevier.

Nielsen, T. A. (2010d). Nightmares associated with the eveningness chronotype. *Journal of Biological Rhythms, 25*(1), 56–62.

Nielsen, T. A. (2013). The method of loci (MoL) and memory consolidation: Dreaming is not MoL-like. *Behavioral and Brain Sciences, 36*(06), 624–625. doi:10.1017/S0140525X13001398.

Nielsen, T. A. (2014). What is the current status of your "Covert REM Process" theory, especially in the light of the new protoconsciousness hypothesis? In N. Tranquillo (Ed.), *Dream consciousness: Allan Hobson's new approach to the brain and its mind* (pp. 175–180). New York: Springer.

Nielsen, T. A., & Chénier, V. (1999). Variations in EEG coherence as an index of the affective content of dreams from REM sleep: Relationships with face imagery. *Brain and Cognition, 41*, 200–212.

Nielsen, T. A., Deslaurier, D., & Baylor, G. (1999). Emotions in dream and waking event reports. *Dreaming, 1*, 287–300. doi:10.1037/h0094340.

Nielsen, T. A., & Germain, A. (2000). Post-traumatic nightmares as a dysfunctional state. *Behavioral and Brain Sciences, 23*, 978–979.

Nielsen, T. A., Kuiken, D., Hoffmann, R., & Moffitt, A. (2001). REM and NREM sleep mentation differences: A question of story structure? *Sleep and Hypnosis, 3*(1), 9–17.

Nielsen, T. A., Kuiken, D., & McGregor, D. (1989). Effects of dream reflection on waking affect: Awareness of feeling, Rorschach movement, and facial EMG. *Sleep, 12*, 277–286.

Nielsen, T. A., Laberge, L., Paquet, J., Tremblay, R. E., Vitaro, F., & Montplaisir, J. (2000). Development of disturbing dreams during adolescence and their relation to anxiety symptoms. *Sleep, 23*, 1–10.

Nielsen, T. A., & Lara-Carrasco, J. (2007). Nightmares, dreaming, and emotion regulation: A review. In D. Barrett & P. McNamara (Eds.), *The new science of dreaming: Vol. 2. Content, recall, personality correlates* (pp. 253–284). Westport: Praeger Perspectives.

Nielsen, T. A., & Levin, R. (2009). Theories and correlates of nightmares. In R. Stickgold & M. P. Walker (Eds.), *The neuroscience of sleep* (pp. 323–329). London: Elsevier Academic Press.

Nielsen, T. A., Ouellet, L., & Zadra, A. L. (1995). Pressure stimulation during REM sleep alters dream limb activity and body bizarreness. *Sleep Research, 24*, 134.

Nielsen, T. A., & Paquette, T. (2007). Dream-associated behaviors affecting pregnant and postpartum women. *Sleep, 30*(9), 1162–1169.

Nielsen, T. A., Paquette, T., Solomonova, E., Lara-Carrasco, J., Colombo, R., & Lanfranchi, P. (2010). Changes in cardiac variability after REM sleep deprivation in recurrent nightmares. *Sleep, 33*(1), 113–122.

Nielsen, T. A., & Powell, R. A. (1992). The day-residue and dream-lag effects: A literature review and limited replication of two temporal effects in dream formation. *Dreaming, 2*, 67–77.

Nielsen, T. A., & Stenstrom, P. (2005). What are the memory sources of dreaming? *Nature, 437*(7063), 1286–1289. doi:10.1038/nature04288.

Nielsen, T. A., Stenstrom, P., & Levin, R. (2006). Nightmare frequency as a function of age, gender, and September 11, 2001: Findings from an Internet questionnaire. *Dreaming, 16*(3), 145–158.

Nielsen, T. A., Stenstrom, P., Takeuchi, T., Saucier, S., Lara-Carrasco, J., Solomonova, E., et al. (2005). Partial REM-sleep deprivation increases the dream-like quality of mentation from REM sleep and sleep onset. *Sleep, 28*(9), 1083–1089.

Nielsen, T. A., Svob, C., & Kuiken, D. (2009). Dream-enacting behaviors in a normal population. *Sleep, 32*(12), 1629–1636.

Nielsen, T. A., Zadra, A. L., Simard, V., Saucier, S., Stenstrom, P., Smith, C., et al. (2003). The typical dreams of Canadian university students. *Dreaming, 13*, 211–235.

Nightingale, S., Orgill, J. C., Ebrahim, I. O., de Lacy, S. F., Agrawal, S., & Williams, A. J. (2005). The association between narcolepsy and REM behavior disorder (RBD). *Sleep Medicine, 6*(3), 253–258. doi:10.1016/j.sleep.2004.11.007.

Nir, Y., & Tononi, G. (2010). Dreaming and the brain: From phenomenology to neurophysiology. *Trends in Cognitive Sciences, 14*(2), 88–100. doi:10.1016/j.tics.2009.12.001.

Nishimoto, S., Vu, A. T., Naselaris, T., Benjamini, Y., Yu, B., & Gallant, J. L. (2011). Reconstructing visual experiences from brain activity evoked by natural movies. *Current Biology, 21*(19), 1641–1646. doi:10.1016/j.cub.2011.08.031.

Nishino, S. (2009). Cataplexy. In R. Stickgold & M. P. Walker (Eds.), *The neuroscience of sleep* (pp. 278–284). London: Elsevier Academic Press.

Nobili, L., Ferrara, M., Moroni, F., Gennaro, L. d., Lo Russo, G., Campus, C., et al. (2011). Dissociated wake-like and sleep-like electro-cortical activity during sleep. *NeuroImage*, *58*(2), 612–619. doi:10.1016/j.neuroimage.2011.06.032.

Nobili, L., de Gennaro, L., Proserpio, P., Moroni, F., Sarasso, S., Pigorini, A., et al. (2012). Local aspects of sleep: Observations from intracerebral recordings in humans. *Progress in Brain Research*, *199*, 219–232. doi:10.1016/B978-0-444-59427-3.00013-7.

Noë, A. (2002). Is the visual world a grand illusion? In A. Noë (Ed.), *Is the visual world a grand illusion?* (pp. 1–12). Thorverton: Imprint Academic. (Special issue of *Journal of Consciousness Studies*, 9[5–6].)

Noë, A. (2004). *Action in perception*. Cambridge, MA: MIT Press.

Nofzinger, E. A., Mintun, M. A., Wiseman, M., Kupfer, D. J., & Moore, R. Y. (1997). Forebrain activation in REM sleep: An FDG PET study. *Brain Research*, *770*(1–2), 192–201. doi:10.1016/S0006-8993(97)00807-X.

Norbu, C. N. (2002). *Dream yoga and the practice of natural light* (Michael Katz, Ed.). Ithaca, NY: Snow Lion Publications.

Noreika, V. (2011). Dreaming and waking experiences in schizophrenia: How should the (dis) continuity hypotheses be approached empirically? *Consciousness and Cognition*, *20*(2), 349–352, discussion 353–354. doi:10.1016/j.concog.2010.06.004.

Noreika, V. (2014). *Alterations in the states and contents of consciousness: Empirical and theoretical aspects*. Doctoral thesis. Turku: Annales Universitatis Turkuensis.

Noreika, V. (2015). It's not just about the contents: Searching for a neural correlate of a state of consciousness—A commentary on Wolf Singer. In T. Metzinger & J. M. Windt (Eds.), *Open MIND: 36(C)*. Frankfurt am Main: MIND Group. doi:10.15502/9783958570504.

Noreika, V., Jylhänkangas, L., Móró, L., Valli, K., & Kaskinoro, K. (2011). Consciousness lost and found: Subjective experiences in an unresponsive state. *Brain and Cognition*, *77*, 327–334.

Noreika, V., Valli, K., Lahtela, H., & Revonsuo, A. (2009). Early-night serial awakenings as a new paradigm for studies on NREM dreaming. *International Journal of Psychophysiology*, *74*(1), 14–18.

Noreika, V., & Windt, J. M. (2008). On the importance of 19th century dream research: Progress in dream research between Aristotle's work on dreaming and the discovery of REM sleep. *Sleep Medicine*, *9*, 919.

Noreika, V., Windt, J. M., Arstila, V., Falter, C. M., Kiverstein, J. D., & Revonsuo, A. (2010). The subjective and objective duration of static NREM sleep dreams. *International Journal of Dream Research*, *3*(Suppl. 1), S6–S7.

Noreika, V., Windt, J. M., Lenggenhager, B., & Karim, A. A. (2010). New perspectives for the study of lucid dreaming: From brain stimulation to philosophical theories of self-consciousness;

Commentary on "The neurobiology of consciousness: Lucid dreaming wakes up" by J. Allan Hobson. *International Journal of Dream Research, 3*(1), 36–46.

Nottelmann, N. (Ed.). (2013). *New essays on belief.* New York: Palgrave Macmillan.

Nudds, M. (2009). Recent work in perception: Naive realism and its opponents. *Analysis, 69,* 334–346.

Occhionero, M., & Cicogna, P. C. (2011). Autoscopic phenomena and one's own body representation in dreams. In *From Dreams to Psychosis: A European Science Foundation Exploratory Workshop* (pp. 1009–1015). (Special issue, *Consciousness and Cognition, 20*[4].) doi:10.1016/j.concog.2011.01.004.

Occhionero, M., Cicogna, P., Natale, V., Esposito, M. J., & Bosinelli, M. (2005). Representation of self in SWS and REM dreams. *Sleep and Hypnosis, 7*(2), 77–83.

O'Flaherty, W. D. (1984). *Dreams, illusion, and other realities.* Chicago: University of Chicago Press.

Ogilvie, R., Vieira, K. P., & Small, R. J. (1991). *EEG activity during lucid dreaming.* Tenth anniversary issue of the Lucidity Letter. Retrieved from http://www.sawka.com/spiritwatch/eeg.htm.

Ogilvie, R. D., & Harsh, J. R. (Eds.). (1994). *Sleep onset: Normal and abnormal processes.* Washington, DC: American Psychological Association.

Okada, H., Matsuoka, K., & Hatakeyama, T. (2005). Individual differences in the range of sensory modalities experienced in dreams. *Dreaming, 15*(2), 106–115.

O'Regan, J. K., & Noë, A. (2001). A sensorimotor account of vision and visual consciousness. *Behavioral and Brain Sciences, 24*(5), 883–917.

O'Shaughnessy, B. (2000). *Consciousness and the world.* Oxford: Oxford University Press.

O'Shaughnessy, B. (2002). Dreaming. *Inquiry, 45,* 399–432.

Oudiette, D., Constantinescu, I., Leclair-Visonneau, L., Vidailhet, M., Schwartz, S., Arnulf, I., et al. (2011). Evidence for the re-enactment of a recently learned behavior during sleepwalking. *PLoS ONE, 6*(3), e18056. doi:10.1371/journal.pone.0018056.

Oudiette, D., Dealberto, M.-J., Uguccioni, G., Golmard, J.-L., Merino-Andreu, M., Tafti, M., et al. (2012). Dreaming without REM sleep. *Consciousness and Cognition, 21*(3), 1129–1140. doi:10.1016/j.concog.2012.04.010.

Oudiette, D., de Cock, V. C., Lavault, S., Leu, S., Vidailhet, M., & Arnulf, I. (2009). Nonviolent elaborate behaviors may also occur in REM sleep behavior disorder. *Neurology, 72*(6), 551–557. doi:10.1212/01.wnl.0000341936.78678.3a.

Oudiette, D., Leu, S., Pottier, M., Buzare, M. A., Brion, A., & Arnulf, I. (2009). Dreamlike mentations during sleepwalking and sleep terrors in adults. *Sleep, 32*(12), 1621–1627.

Ouspensky, P. D. [1931] (2005). *On the study of dreams and hypnotism.* Whitefish, MT: Kessinger Publishing.

Overgaard, M., Gallagher, S., & Ramsøy, T. Z. (2008). An integration of first-person methodologies in cognitive science. *Journal of Consciousness Studies, 15*(5), 100–120.

Overgaard, M., Timmermans, B., Sandberg, K., & Cleeremans, A. (2010). Optimizing subjective measures of consciousness. *Consciousness and Cognition, 19*(2), 682–684, discussion 685–686. doi:10.1016/j.concog.2009.12.018.

Owen, A. M., Coleman, M. R., Boly, M., Davis, M. H., Laureys, S., & Pickard, J. D. (2006). Detecting awareness in the vegetative state. *Science, 313*, 1402.

Pace-Schott, E. F. (2007). The frontal lobes and dreaming. In D. Barrett & P. McNamara (Eds.), *The new science of dreaming: Vol. 2. Content, recall, personality correlates* (pp. 115–154). Westport: Praeger Perspectives.

Pace-Schott, E. F. (2009). Sleep architecture. In R. Stickgold & M. P. Walker (Eds.), *The neuroscience of sleep* (pp. 11–17). London: Elsevier Academic Press.

Pace-Schott, E. F. (2013). Dreaming as a story-telling instinct. *Frontiers in Psychology, 4*, 159. doi:10.3389/fpsyg.2013.00159.

Pace-Schott, E. F., & Hobson, J. A. (2002). The neurobiology of sleep: Genetics, cellular physiology, and cortical networks. *Nature Reviews: Neuroscience, 3*(August), 591–605.

Pagel, J. F. (2003). Non-dreamers. *Sleep Medicine, 4*, 235–241.

Pagel, J. F., Blagrove, M., Levin, R., States, B., Stickgold, B., & White, S. (2001). Definitions of dream: A paradigm for comparing field descriptive specific studies of dream. *Dreaming, 11*, 195–202.

Paiva, T., Bugalho, P., & Bentes, C. (2011). Dreaming and cognition in patients with frontotemporal dysfunction. *From Dreams to Psychosis: A European Science Foundation Exploratory Workshop* (pp. 1027–1035). (Special issue, *Consciousness and Cognition, 20*[4]). doi:10.1016/j.concog.2011.06.008.

Palagini, L., Gemignani, A., Feinberg, I., Guazzelli, M., & Campbell, I. G. (2004). Mental activity after early afternoon nap awakenings in healthy subjects. *Brain Research Bulletin, 63*, 361–368.

Paqueron, X., Leguen, M., Rosenthal, D., Coriat, P., Willer, J. C., & Danziger, N. (2003). The phenomenology of body image distortions induced by regional anaesthesia. *Brain, 126*, 702–712.

Parker, J. D., & Blackmore, S. J. (2002). Comparing the content of sleep paralysis and dream reports. *Dreaming, 12*(1), 45–59.

Paul, F., & Erlacher, D. (2011). Lucid dream induction by visual and tactile stimulation. *International Journal of Dream Research, 4*(Supplement 1), S49–S50.

Paulsson, T., & Parker, A. (2006). The effects of a two-week reflection-intention training program on lucid dream recall. *Dreaming, 16*(1), 22–35.

Payne, J. D., Schacter, D. L., Propper, R. E., Huang, L.-W., Wamsley, E. J., Tucker, M. A., et al. (2009). The role of sleep in false memory formation. *Neurobiology of Learning and Memory, 92*(3), 327–334. doi:10.1016/j.nlm.2009.03.007.

Pears, D. F. (1977). Dreaming. In C. E. M. Dunlop (Ed.), *Philosophical essays on dreaming* (pp. 205–226). Ithaca: Cornell University Press. (Reprinted from *Mind* 70 [1961], pp. 145–163.)

Pearson, D. G., & Hollings, J. (2013). Einstein's jacket: Evidence for long-term perceptual specificity in mental imagery. *Consciousness and Cognition, 22*(1), 148–154. doi:10.1016/j.concog.2012.11.004.

Peigneux, P., Laureys, S., Fuchs, S., Collette, F., Perrin, F., Reggers, J., et al. (2004). Are spatial memories strengthened in the human hippocampus during slow wave sleep? *Neuron, 44*(3), 535–545. doi:10.1016/j.neuron.2004.10.007.

Peirce, C. S. (1958). *The collected works of Charles Sanders Peirce*. Cambridge, MA: Harvard University Press.

Pelayo, R., & Guilleminault, C. (2009). History of sleep research. In R. Stickgold & M. P. Walker (Eds.), *The neuroscience of sleep* (pp. 3–8). London: Elsevier Academic Press.

Perky, C. W. (1910). An experimental study of imagination. *American Journal of Psychology, 21*, 422–452.

Perogamvros, L. (2013). Consciousness and the *Invention of Morel*. *Frontiers in Human Neuroscience, 7*, 61. doi:10.3389/fnhum.2013.00061.

Perogamvros, L., Dang-Vu, T. T., Desseilles, M., & Schwartz, S. (2013). Sleep and dreaming are for important matters. *Frontiers in Psychology, 4*, 474. doi:10.3389/fpsyg.2013.00474.

Perogamvros, L., & Schwartz, S. (2012). The roles of the reward system in sleep and dreaming. *Neuroscience and Biobehavioral Reviews, 36*(8), 1934–1951. doi:10.1016/j.neubiorev.2012.05.010.

Perry, E., Ashton, H., & Young, A. (Eds.). (2002). *Neurochemistry of consciousness: Neurotransmitters in mind: Advances in consciousness research*. Amsterdam: John Benjamins.

Petkova, V. I., & Ehrsson, H. H. (2008). If I were you: Perceptual illusion of body swapping. *PLoS ONE, 3*(12), e3832. doi:10.1371/journal.pone.0003832.

Petkova, V. I., Khoshnevis, M., & Ehrsson, H. H. (2011). The perspective matters! Multisensory integration in egocentric reference frames determines full-body ownership. *Frontiers in Psychology, 2*, 35.

Pfeiffer, C. (2015). Multisensory spatial mechanisms of the bodily self and social cognition—A commentary on Vittorio Gallese and Valentina Cuccio. In T. Metzinger & J. M. Windt (Eds.), *Open MIND: 14(C)*. Frankfurt am Main: MIND Group. doi:10.15502/9783958570450.

Pfeiffer, C., Lopez, C., Schmutz, V., Duenas, J. A., Martuzzi, R., Blanke, O., et al. (2013). Multisensory origin of the subjective first-person perspective: Visual, tactile, and vestibular mechanisms. *PLOS ONE, 8*(4), e61751. doi:10.1371/journal.pone.0061751.

Pfeiffer, C., Serino, A., & Blanke, O. (2014). The vestibular system: A spatial reference for bodily self-consciousness. *Frontiers in Integrative Neuroscience, 8*, 31. doi:10.3389/fnint.2014.00031.

Pfurtscheller, G. (1992). Event-related synchronization (ERS): An electrophysiological correlate of cortical areas at rest. *Electroencephalography and Clinical Neurophysiology, 83*, 62–69.

Picard, F. (2013). State of belief, subjective certainty, and bliss as a product of cortical dysfunction. *Cortex, 49*(9), 2494–2500. doi:10.1016/j.cortex.2013.01.006.

Piccinini, G. (2003). Data from introspective reports: Upgrading from common sense to science. *Journal of Consciousness Studies, 10*(9–10), 141–156.

Piccinini, G. (2009). First-person data, publicity, and self-measurement. *Philosophers' Imprint, 9*(9), 1–16.

Pigeon, W. R., Campbell, C. E., Possemato, K., & Ouimette, P. (2013). Longitudinal relationships of insomnia, nightmares, and PTSD severity in recent combat veterans. *Journal of Psychosomatic Research, 75*(6), 546–550. doi:10.1016/j.jpsychores.2013.09.004.

Pitt, D. (2004). The phenomenology of cognition, or What is it like to think that P? *Philosophy and Phenomenological Research, 69*(1), 1–36.

Plato. (360 BCE/1994–2009). *Phaedrus* (Benjamin Jowett, Trans.). Retrieved from http://classics.mit.edu/Plato/phaedrus.html.

Plato. (360 BCE/2009). *Timaeus* (Benjamin Jowett, Trans.). Retrieved from http://classics.mit.edu/Plato/timaeus.html.

Plato. (1994). *Republic* (Robin Waterfield, Trans.). Oxford: Oxford University Press.

Plato. (1996). *Theaetetus* (H. N. Fowler, Trans.). Loeb Classical Library. Cambridge, MA: Harvard University Press.

Plihal, W., & Born, J. (1999). Memory consolidation in human sleep depends on inhibition of glucocorticoid release. *Neuroreport, 10*, 2741–2747.

Pompeiano, O., & Morrison, A. R. (1965). Vestibular influences during sleep. I. Abolition of the rapid eye movements of desynchronized sleep following vestibular lesions. *Archives Italiennes de Biologie, 103*, 569–595.

Pompeiano, O., & Morrison, A. R. (1966). Vestibular origin of the rapid eye movements during desynchronized sleep. *Experientia, 22*, 60–61.

Pöppel, E. (2000). *Grenzen des Bewußtseins: Wie kommen wir zur Zeit, und wie entsteht Wirklichkeit?* Frankfurt am Main: Insel.

Porte, H., & Hobson, J. A. (1986a). Fictive movement in sleep: REM vs. NREM sleep reports. *Sleep Research, 15*, 82.

Porte, H., & Hobson, J. (1986b). Bizarreness in REM and SWS reports. *Sleep Research, 15*, 81.

Postuma, R. B., Gagnon, J.-F., & Montplaisir, J. Y. (2012). REM sleep behavior disorder: From dreams to neurodegeneration. *Neurobiology of Disease, 46*(3), 553–558. doi:10.1016/j.nbd.2011.10.003.

Powell, R. A., & Nielsen, T. A. (1998). Was Anna O's black snake hallucination a sleep paralysis nightmare? Dreams, memories, and trauma. *Psychiatry and Clinical Neurosciences, 61*, 239–248.

Price, R. F., & Cohen, D. B. (1988). Lucid dream induction: An empirical evaluation. In J. Gackenbach & S. LaBerge (Eds.), *Conscious mind, sleeping brain: Perspectives on lucid dreaming* (pp. 105–134). New York: Plenum Press.

Princess Elisabeth of Bohemia & René Descartes. (2007). *The correspondence between Princess Elisabeth of Bohemia and René Descartes* (L. Shapiro, Trans. & Ed.). Chicago: University of Chicago Press.

Prinz, J. J. (2004). *Gut reactions: A perceptual theory of emotion.* Philosophy of mind series. Oxford: Oxford University Press.

Prinz, J. J. (2008). Empirical philosophy and experimental philosophy. In J. M. Knobe & S. Nichols (Eds.), *Experimental philosophy* (pp. 189–208). Oxford: Oxford University Press.

Prinz, J. (2012). Emotions: How many are there? In E. Margolis, R. Samuels, & S. P. Stich (Eds.), *The Oxford handbook of philosophy of cognitive science* (pp. 183–200). New York: Oxford University Press.

Prosser, S., & Recanati, F. (Eds.). (2012). *Immunity to error through misidentification: New essays.* Cambridge: Cambridge University Press.

Proust, J. (2007). Metacognition and metarepresentation: Is a self-directed theory of mind a precondition for metacognition? *Synthese, 159*(2), 271–295.

Proust, J. (2013). *Philosophy of metacognition: Mental agency and self-awareness.* Oxford: Oxford University Press.

Pseudo-Mayne. (1728/1983). *Über das Bewußtsein: Englisch-Deutsch; Übersetzt und mit Einleitung und Anmerkungen* (R. Brandt, Ed.). Hamburg: Meiner.

Punamäki, R.-L. (1999). The relationship of dream content and changes in daytime mood in traumatized vs. non-traumatized children. *Dreaming, 9*, 213–233.

Purcell, S., Moffitt, A., & Hoffmann, R. (1993). Waking, dreaming, and self-regulation. In A. Moffitt, M. Kramer, & R. Hoffmann (Eds.), *The functions of dreaming* (pp. 197–260). Albany: State University of New York Press.

Purcell, S., Mullington, J., Moffitt, A., Hoffmann, R., & Pigeau, R. (1986). Dream self-reflectiveness as a learned cognitive skill. *Sleep, 9*(3), 423–437.

Putnam, H. (1975). The meaning of "meaning." *Minnesota Studies in the Philosophy of Science, 7*, 131–193.

Putnam, H. (1981). Brains in a vat. In H. Putnam, *Reason, truth, and history*. Cambridge: Cambridge University Press.

Putnam, H. (1986). Dreaming and "depth grammar." In H. Putnam, *Philosophical papers: Vol. 2. Mind, language, and reality* (pp. 304–324). Cambridge: Cambridge University Press. (Reprinted from Butler, R. [Ed.]. [1962]. *Analytical philosophy first series*. Oxford: Basil Blackwell & Mott.)

Pylyshyn, Z. W. (2002). Mental imagery: In search of a theory. *Behavioral and Brain Sciences, 25*(2), 157–182.

Radden, J. (2011). *On delusion*. London: Routledge.

Radin, D. I. (2006). *Entangled minds: Extrasensory experiences in a quantum reality*. New York: Paraview Pocket Books.

Raehlmann, E., & Witkowski, L. (1877). Über atypische Augenbewegungen. *Archiv für Anatomie und Physiologie*, 454–471.

Raehlmann, E., & Witkowski, L. (1878). Über das Verhalten der Pupillen während des Schlafes nebst Bemerkungen zur Innervation der Iris. *Archiv für Anatomie und Physiologie*, 109–121.

Raffman, D. (1993). *Language, music, and mind*. Cambridge, MA: MIT Press.

Raffman, D. (1995). On the persistence of phenomenology. In T. Metzinger (Ed.), *Conscious experience*. Paderborn: Schöningh/Imprint Academic.

Raichle, M. E., MacLeod, A. M., Snyder, A. Z., Powers, W. J., Gusnard, D. A., & Shulman, G. L. (2001). Inaugural article: A default mode of brain function. *Proceedings of the National Academy of Sciences of the United States of America, 98*(2), 676–682. doi:10.1073/pnas.98.2.676.

Rainville, R. E. (1994). The role of dreams in the rehabilitation of the adventitiously blind. *Dreaming, 4*(3), 155–164. Retrieved from http://www.asdreams.org/journal/articles/4-3rainville.htm.

Ramachandran, V. S., & Hirstein, W. (1998). The perception of phantom limbs: The D. O. Hebb lecture. *Brain, 121*(9), 1603–1630.

Rassin, E., Merckelbach, H., & Spaan, V. (2001). When dreams become a royal road to confusion: Realistic dreams, dissociation, and fantasy proneness. *Journal of Nervous and Mental Disease, 189*(7), 478–481.

Ratcliffe, M. (2008). *Feelings of being: Phenomenology, psychiatry and the sense of reality*. Oxford: Oxford University Press.

Rauchs, G., Desgranges, B., Foret, J., & Eustache, F. (2005). The relationships between memory systems and sleep stages. *Journal of Sleep Research, 14*(2), 123–140.

Raymond, I., Nielsen, T. A., Lavigne, G., & Choinière, M. (2002). Incorporation of pain in dreams of hospitalized burn victims. *Sleep, 25*, 41–46.

Rechtschaffen, A. (1975). Scientific method in the study of altered states of consciousness with illustrations from sleep and dream research. In The Drug Abuse Council Inc. (Ed.), *Altered states of consciousness: Current views and research problems* (pp. 135–191). Washington, DC: Smithsonian Institution Press.

Rechtschaffen, A. (1978). The single-mindedness and isolation of dreams. *Sleep, 1*(1), 97–109.

Rechtschaffen, A. (1983). Visual dimensions and correlates of dream images. *Sleep Research, 12,* 189.

Rechtschaffen, A., & Buchignani, C. (1992). The visual appearance of dreams. In J. S. Antrobus & M. Bertini (Eds.), *The neuropsychology of sleep and dreaming* (pp. 143–156). Hillsdale, NJ: Erlbaum.

Rechtschaffen, A., & Foulkes, D. (1965). Effect of visual stimuli on dream content. *Perceptual and Motor Skills, 20,* 1149–1160.

Redshaw, J., & Suddendorf, T. (2013). Foresight beyond the very next event: Four-year-olds can link past and deferred future episodes. *Frontiers in Psychology, 4,* 404. doi:10.3389/fpsyg.2013.00404.

Reed, T. M. (1979). Dreams, scepticism, and waking life. In D. F. Gustafson & B. L. Tapscott (Eds.), *Body, mind, and method: Essays in honor of Virgil C. Aldrich* (pp. 37–64). Boston: D. Reidel.

Resnick, J., Stickgold, R., Rittenhouse, C. D., & Hobson, J. A. (1994). Self-representation and bizarreness in children's dream reports collected in the home setting. *Consciousness and Cognition, 3,* 30–45.

Resnick, L. B. (Ed.). (1976). *The nature of intelligence.* Hillsdale, NJ: Erlbaum.

Reuter, K. (2011). Is imagination introspective? *Philosophia, 39*(1), 31–38. doi:10.1007/s11406-010-9275-4.

Revonsuo, A. (1995). Consciousness, dreams, and virtual realities. *Philosophical Psychology, 8,* 35–58.

Revonsuo, A. (1999). Binding and the phenomenal unity of consciousness. *Consciousness and Cognition, 8,* 173–185.

Revonsuo, A. (2000). The reinterpretation of dreams: An evolutionary hypothesis of the function of dreaming. *Behavioral and Brain Sciences, 23,* 877–901.

Revonsuo, A. (2005). The self in dreams. In T. E. Feinberg & J. P. Keenan (Eds.), *The lost self: Pathologies of the brain and identity* (pp. 206–219). Oxford: Oxford University Press.

Revonsuo, A. (2006). *Inner presence: Consciousness as a biological phenomenon.* Cambridge, MA: MIT Press.

Revonsuo, A. (2014). The idea, championed by your group, that dreaming functions as threat avoidance would seem to be quite compatible with protoconsciousness theory. But is it really dreaming (as against REM) that performs that function? In other words, are you a dualist or a neutral monist? In N. Tranquillo (Ed.), *Dream consciousness: Allan Hobson's new approach to the brain and its mind* (pp. 191–194). New York: Springer.

Revonsuo, A., & Salmivalli, C. (1995). A content analysis of bizarre elements in dreams. *Dreaming, 5*, 169–187.

Revonsuo, A., & Tarkko, K. (2002). Binding in dreams: The bizarreness of dream images and the unity of consciousness. *Journal of Consciousness Studies, 9*, 3–24.

Revonsuo, A., Tuominen, J. & Valli, K. (2015). The avatars in the machine—Dreaming as a simulation of social reality. In T. Metzinger & J. M. Windt (Eds.), *Open MIND: 32(T)*. Frankfurt am Main: MIND Group. doi:10.15502/9783958570375.

Revonsuo, A., & Valli, K. (2000). Dreaming and consciousness: Testing the threat simulation theory of the function of dreaming. *Psyche, 6*. Retrieved from http://www.theassc.org/files/assc/2467.pdf.

Riecke, B. E. (2011). Compelling self-motion through virtual environments without actual self-motion—using self-motion illusions ("vection") to improve VR user experience. In J.-J. Kim (Ed.), *Virtual reality* (pp. 149–176). Rijeka, Croatia: InTech.

Riecke, B. E., Schulte-Pelkum, J., Avraamides, M. N., von der Heyde, M., & Bülthoff, H. H. (2006). Cognitive factors can influence self-motion perception (vection) in virtual reality. *ACM Transactions on Applied Perception, 3*(3), 194–216. doi:10.1145/1166087.1166091.

Rignano, E. (1920). A new theory of sleep and dreams. *Mind, 29*(115), 313–322.

Rinpoche, T. W. (1998). *The Tibetan yogas of dream and sleep* (M. Dahlby, Ed.). Ithaca, NY: Snow Lion Publications.

Rissling, M., & Ancoli-Israel, S. (2009). Sleep in aging. In R. Stickgold & M. P. Walker (Eds.), *The neuroscience of sleep* (pp. 78–84). London: Elsevier Academic Press.

Rittenhouse, C. D., Stickgold, R., & Hobson, J. A. (1994). Constraint on the transformation of characters, objects, and settings in dream reports. *Consciousness and Cognition, 3*(1), 100–113. doi:10.1006/ccog.1994.1007.

Riva, G., Davide, F., & Ijsselsteijn, W. (Eds.). (2003). *Being there: Concepts, effects, and measurement of user presence in synthetic environments*. Amsterdam: IOS Press.

Riva, G., Mantovani, F., Capideville, C. S., Preziosa, A., Morganti, F., Villani, D., Gaggioli, A., Botella, C., Alcaniz, M. (2007). Affective interactions using virtual reality: The link between presence and emotions. *Cyberpsychology and Behavior: The Impact of the Internet, Multimedia, and Virtual Reality on Behavior and Society, 10*(1), 45–56. doi:10.1089/cpb.2006.9993.

Robert, G., & Zadra, A. (2014). Thematic and content analysis of idiopathic nightmares and bad dreams. *Sleep*. doi:10.5665/sleep.3426.

Rosen, M. G. (2013). What I make up when I wake up: Anti-experience views and narrative fabrication of dreams. *Frontiers in Psychology, 4*, 514

Rosen, M., & Sutton, J. (2013). Self-representation and perspectives in dreams. *Philosophy Compass, 8*(11), 1041–1053. doi:10.1111/phc3.12082.

Ruby, P. (2011). Experimental research on dreaming: State of the art and neuropsychoanalytic perspectives. *Frontiers in Psychology, 2,* 286. doi:10.3389/fpsyg.2011.00286.

Ruby, P., Blochet, C., Eichenlaub, J.-B., Bertrand, O., Morlet, D., & Bidet-Caulet, A. (2013). Alpha reactivity to first names differs in subjects with high and low dream recall frequency. *Frontiers in Psychology, 4,* 419. doi:10.3389/fpsyg.2013.00419.

Russell, B. (1912/1999). *The problems of philosophy*. Mineola, NY: Dover.

Russell, B. (1914/2009). *Our knowledge of the external world as a field for scientific method in philosophy*. London: Routledge.

Russell, B. (1921/2008). *The analysis of mind*. Rockville: Arc Manor.

Russell, B. (1948/2009). *Human knowledge: Its scope and limits*. London: Routledge.

Ryle, G. (1949/2000). *The concept of mind*. With an introduction by Daniel C. Dennett. Chicago: University of Chicago Press.

Sacks, O. (1985). *The man who mistook his wife for a hat*. London: Picador.

Sacks, O. (1999a). *Awakenings*. New York: Vintage.

Sacks, O. (1999b). *Migraine* (revised and expanded ed.). New York: Vintage.

Sacks, O. (2001). Neurological dreams. In D. Barrett (Ed.), *Trauma and dreams* (pp. 212–216). Cambridge, MA: Harvard University Press.

Sacks, O. (2004). Nothingness. In R. L. Gregory (Ed.), *The Oxford companion to the mind* (2nd ed., pp. 674–675). Oxford: Oxford University Press.

Sainsbury, R. M. (1995). *Paradoxes* (2nd ed.). Cambridge: Cambridge University Press.

Sainsbury, R. M. (2010). *Fiction and fictionalism*. London: Routledge.

Saletin, J. M., & Walker, M. P. (2012). Nocturnal mnemonics: Sleep and hippocampal memory processing. *Frontiers in Neurology, 3,* 59. doi:10.3389/fneur.2012.00059.

Salminen-Vaparanta, N., Koivisto, M., Noreika, V., Vanni, S., & Revonsuo, A. (2012). Neuronavigated transcranial magnetic stimulation suggests that area V2 is necessary for visual awareness. *Neuropsychologia, 50*(7), 1621–1627. doi:10.1016/j.neuropsychologia.2012.03.015.

Sämann, P. G., Wehrle, R., Hoehn, D., Spoormaker, V. I., Peters, H., Tully, C., et al. (2011). Development of the brain's default mode network from wakefulness to slow wave sleep. *Cerebral Cortex, 21*(9), 2082–2093. doi:10.1093/cercor/bhq295.

Sanchez-Vives, M. V., & Slater, M. (2005). Opinion: From presence to consciousness through virtual reality. *Nature Reviews: Neuroscience, 6*(4), 332–339. doi:10.1038/nrn1651.

Sandberg, K., Timmermans, B., Overgaard, M., & Cleeremans, A. (2010). Measuring consciousness: Is one measure better than the other? *Consciousness and Cognition, 19*(4), 1069–1078. doi:10.1016/j.concog.2009.12.013.

Sartre, J.-P. (1940/2005). *L'imaginaire: Psychologie phénoménologique de l'imagination.* Paris: Gallimard.

Sartre, J.-P. (1957). *The transcendence of the ego: An existentialist theory of consciousness.* New York: Noonday Press.

Saurat, M.-T., Agbakou, M., Attigui, P., Golmard, J.-L., & Arnulf, I. (2011). Walking dreams in congenital and acquired paraplegia. *Consciousness and Cognition, 20*(4), 1425–1432. doi:10.1016/j.concog.2011.05.015.

Sauvageau, A., Nielsen, T. A., & Montplaisir, J. (1998). Effects of somatosensory stimulation on dream content in gymnasts and control participants: Evidence of vestibulomotor adaptation in REM sleep. *Dreaming, 8,* 125–134.

Sauvageau, A., Nielsen, T. A., Ouellet, L., & Montplaisir, J. (1993). Eye movement density and amplitude after somatosensory stimulation during REM sleep: Effects on gymnasts and controls. *Sleep Research, 22,* 133.

Scaglione, C., Vignatelli, L., Plazzi, G., Marchese, R., Negrotti, A., Rizzo, G., et al. (2005). REM sleep behaviour disorder in Parkinson's disease: A questionnaire-based study. *Neurological Sciences, 25*(6), 316–321. doi:10.1007/s10072-004-0364-7.

Scarone, S., Manzone, M. L., Gambini, O., Kantzas, I., Limosani, I., D'Agostino, A., et al. (2007). The dream as a model for psychosis: An experimental approach using bizarreness as a cognitive marker. *Schizophrenia Bulletin, 34*(3), 515–522. doi:10.1093/schbul/sbm116.

Schabus, M., Dang-Vu, T. T., Heib, D. P. J., Boly, M., Desseilles, M., Vandewalle, G., et al. (2012). The fate of incoming stimuli during NREM sleep is determined by spindles and the phase of the slow oscillation. *Frontiers in Neurology, 3,* 40. doi:10.3389/fneur.2012.00040.

Schacter, D. L. (1976). The hypnagogic state: A critical review of the literature. *Psychological Bulletin, 83*(3), 452–481.

Schacter, D. (2012). Constructive memory: Past and future. *Dialogues in Clinical Neuroscience, 14*(1), 7–18.

Schacter, D. L., Addis, D. R., & Buckner, R. L. (2007). Remembering the past to imagine the future: The prospective brain. *Nature Reviews: Neuroscience, 8*(9), 657–661. doi:10.1038/nrn2213.

Schädlich, M., & Erlacher, D. (2012). Applications of lucid dreams: An online study. *International Journal of Dream Research, 5*(2), 134–138. doi:10.11588/ijodr.2012.2.9505.

Schanfald, D., Pearlman, C., & Greenberg, R. (1985). The capacity of stroke patients to report dreams. *Cortex, 21*(2), 237–247.

Schatzman, M., Worsley, A., & Fenwick, P. (1988). Correspondence during lucid dreams between dreamed and actual events. In J. Gackenbach & S. LaBerge (Eds.), *Conscious mind, sleeping brain: Perspectives on lucid dreaming* (pp. 155–179). New York: Plenum Press.

Schechtman, M. (2011). The narrative self. In S. Gallagher (Ed.), *The Oxford handbook of the self* (pp. 394–417). Oxford: Oxford University Press.

Schenck, C. (2005). *Paradox lost: Midnight in the battleground of sleep and dreams*. Minneapolis: Extreme Nights.

Schenck, C. (2007). *Sleep: The mysteries, the problems, and the solutions*. New York: Avery.

Schenck, C. H., & Mahowald, M. W. (1996). REM sleep parasomnias. *Neurologic Clinics, 14*, 697–720.

Schenck, C. H., & Mahowald, M. W. (2002). REM sleep behavior disorder: Clinical, developmental, and neuroscience perspectives 16 years after its formal identification in *Sleep*. *Sleep, 25*(2), 120–138.

Scherner, K. A. (1861). *Das Leben des Traumes*. Berlin: Verlag von Heinrich Schindler.

Schmidt, R. E., & Gendolla, G. H. E. (2008). Dreaming of white bears: The return of the suppressed at sleep onset. *Consciousness and Cognition, 17*(3), 714–724. doi:10.1016/j.concog.2007.09.002.

Schmidt, R. E., Renaud, O., & van der Linden, M. (2011). Nocturnal regrets and insomnia in elderly people. *International Journal of Aging and Human Development, 73*(4), 371–393.

Schmidt, R. E., & van der Linden, M. (2009). The aftermath of rash action: Sleep-interfering counterfactual thoughts and emotions. *Emotion, 9*(4), 549–553. doi:10.1037/a0015856.

Schmidt, R., & van der Linden, M. (2013). Feeling too regretful to fall asleep: Experimental activation of regret delays sleep onset. *Cognitive Therapy and Research, 37*(4), 872–880. doi:10.1007/s10608-013-9532-5.

Schönhammer, R. (2004). *Fliegen, Fallen, Flüchten: Psychologie intensiver Träume*. Tübingen: Deutsche Gesellschaft für Verhaltenstherapie.

Schönhammer, R. (2005). "Typical dreams": Reflections of arousal. *Journal of Consciousness Studies, 12*(4–5), 18–37.

Schooler, J. W., Smallwood, J., Christoff, K., Handy, T. C., Reichle, E. D., & Sayette, M. A. (2011). Meta-awareness, perceptual decoupling, and the wandering mind. *Trends in Cognitive Sciences, 15*(7), 319–326. doi:10.1016/j.tics.2011.05.006.

Schopenhauer, A. (1847/1988). *Kleinere Schriften*. Zürich: Haffmans.

Schopenhauer, A. (1851/1988). *Parerga und Paralimpomena: Kleine philosophische Schriften* (Vol. 1). Zürich: Haffmans.

Schopenhauer, A. (1859/1988). *Die Welt als Wille und Vorstellung* (Vol. 1). Zürich: Haffmans.

Schrage-Früh, M. (2012). "The roots of art are in the dream": Dreams, literature, and evolution. In D. Vanderbeke & C. Gansel (Eds.), *Telling stories: Literature and evolution/evolution of literature*. Berlin: De Gruyter, 156-171.

Schredl, M. (1998). Dream content in patients with narcolepsy: Preliminary findings. *Dreaming, 8*(2), 103–107.

Schredl, M. (2003). Continuity between waking and dreaming: A proposal for a mathematical model. *Sleep and Hypnosis, 5*, 38–52.

Schredl, M. (2004). Reliability and stability of a dream recall frequency scale: Perceptual and Motor Skills. *Perceptual and Motor Skills, 98*(3c), 1422–1426. doi:10.2466/pms.98.3c.1422-1426.

Schredl, M. (2006). Factors affecting the continuity between waking and dreaming: Emotional intensity and emotional tone of the waking-life event. *Sleep and Hypnosis, 8*, 1–5.

Schredl, M. (2007). Gender differences in dreaming. In D. Barrett & P. McNamara (Eds.), *The new science of dreaming: Vol. 1. Biological aspects* (Vol. 1, pp. 29–48). Westport: Praeger Perspectives.

Schredl, M. (2008a). Laboratory references in dreams: Methodological problem and/or evidence for the continuity hypothesis of dreaming? *International Journal of Dream Research, 1*(1), 3–6.

Schredl, M. (2008b). Personality correlates of flying dreams. *Imagination, Cognition and Personality, 27*, 129–137.

Schredl, M. (2008c). *Traum*. München: Reinhardt/UTB.

Schredl, M. (2009). Dreams in patients with sleep disorders. *Sleep Medicine Reviews, 13*(3), 215–221. doi:10.1016/j.smrv.2008.06.002.

Schredl, M. (2010a). Characteristics and contents of dreams. *International Review of Neurobiology, 92*, 135–154. doi:10.1016/S0074-7742(10)92007-2.

Schredl, M. (2010b). History of dream research: The dissertation "Entstehung der Träume (Origin of dreams)" of Wilhelm Weygandt published in 1893. *International Journal of Dream Research, 3*(1), 95–97.

Schredl, M. (2011a). Frequency and nature of flying dreams in a long dream series. *International Journal of Dream Research, 4*(1), 31–34.

Schredl, M. (2011b). Toilet dreams: Incorporation of waking-life memories? *International Journal of Dream Research, 4*(1), 41–44.

Schredl, M. (2014). You emphasize the continuity between waking and dreaming. But what about continuity in the other direction, i.e. between dreaming and waking? And what about discontinuity? Do you deny its existence? In N. Tranquillo (Ed.), *Dream consciousness: Allan Hobson's new approach to the brain and its mind* (pp. 197–200). New York: Springer.

Schredl, M., Atanasova, D., Hörmann, K., Maurer, J. T., Hummel, T., & Stuck, B. A. (2009). Information processing during sleep: The effect of olfactory stimuli on dream content and dream emotions. *Journal of Sleep Research, 18*, 285–290.

Schredl, M., & Erlacher, D. (2004). Lucid dreaming frequency and personality. *Personality and Individual Differences, 37*(7), 1463–1473.

Schredl, M., & Erlacher, D. (2007). Self-reported effects of dreams on waking-life creativity: An empirical study. *Journal of Psychology, 141*(1), 35–46.

Schredl, M., & Erlacher, D. (2010). Is sleep-dependent memory consolidation of a visuo-motor task related to dream content? *International Journal of Dream Research, 3*(1), 74–79.

Schredl, M., & Erlacher, D. (2011a). Dream consciousness and sleep physiology. In D. Cvetkovic & I. Cosic (Eds.), *States of consciousness: Experimental insights into meditation, waking, sleep, and dreams* (pp. 93–108). Frontiers collection. Heidelberg: Springer.

Schredl, M., & Erlacher, D. (2011b). Frequency of lucid dreaming in a representative German sample. *Perceptual and Motor Skills, 112*(1), 104–108.

Schredl, M., Fuchedzhieva, A., Hämig, H., & Schindele, V. (2008). Do we think dreams are in black and white due to memory problems? *Dreaming, 18*(3), 175–180.

Schredl, M., & Fulda, S. (2005). Reliability and stability of dream recall frequency. *Dreaming, 15*(4), 240–244.

Schredl, M., Funkhouser, A., & Arn, N. (2005–2006). Dreams of truck drivers: A test of the continuity hypothesis of dreaming. *Imagination, Cognition, and Personality, 25*(2), 179–186.

Schredl, M., Henley-Einion, J., & Blagrove, M. (2012). Lucid dreaming in children: The UK library study. *International Journal of Dream Research, 5*(1). Retrieved from http://archiv.ub.uni-heidelberg.de/ojs/index.php/IJoDR/article/view/9274.

Schredl, M., & Hofmann, F. (2003). Continuity between waking activities and dream activities. *Consciousness and Cognition, 12*, 298–308.

Schredl, M., & Reinhard, I. (2008). Gender differences in dream recall: A meta-analysis. *Journal of Sleep Research, 17*(2), 125–131. doi:10.1111/j.1365-2869.2008.00626.x.

Schredl, M., & Schmitt, J. (2009). Dream recall frequency and nightmare frequency in patients with sleep disordered breathing. *Somnologie, 13*, 12–17.

Schredl, M., Schmitt, J., Hein, G., Schmoll, T., Eller, S., & Haaf, J. (2006). Nightmares and oxygen desaturations: Is sleep apnea related to heightened nightmare frequency? *Sleep and Breathing, 10*, 203–209.

Schredl, M., Wittmann, L., Ciric, P., & Götz, S. (2003). Factors of home dream recall: A structural equation model. *Journal of Sleep Research, 12*, 133–141.

Schreier Rupprecht, C. (2007). Dreaming, language, literature. In D. Barrett & P. McNamara (Eds.), *The new science of dreaming: Vol. 3. Cultural and theoretical perspectives* (pp. 1–34). Westport: Praeger Perspectives.

Schröder, S. (1997). The concept of dreaming: On three theses by Malcolm. *Philosophical Investigations, 20*(1), 15–38.

Schupak, C., & Rosenthal, J. (2009). Excessive daydreaming: A case history and discussion of mind wandering and high fantasy proneness. *Consciousness and Cognition, 18*(1), 290–292. doi:10.1016/j.concog.2008.10.002.

Schwabe, L., & Blanke, O. (2008). The vestibular component in out-of-body experiences: A computational approach. *Frontiers in Human Neuroscience, 2*, 17.

Schwartz, S. (1999). *Matière à rêver: Exploration statistique et neuropsychologique des phénomènes oniriques au travers des textes et des images de rêves*. Unpublished doctoral thesis, University of Lausanne, Switzerland.

Schwartz, S. (2000). A historical loop of one hundred years: Similarities between 19th century and contemporary dream research. *Dreaming, 10*, 55–66.

Schwartz, S. (2003). Are life episodes replayed during dreaming? *Trends in Cognitive Sciences, 7*, 325–327.

Schwartz, S. (2004). What dreaming can reveal about cognitive and brain functions during sleep: A lexico-statistical analysis of dream reports. *Psychologica Belgica, 44*(1–2), 5–42.

Schwartz, S., Dang-Vu, T. T., Ponz, A., Duhoux, S., & Maquet, P. (2005). Dreaming: A neuropsychological view. *Schweizer Archiv für Neurologie und Psychiatrie, 156*(8), 426–439.

Schwartz, S., & Maquet, P. (2002). Sleep imaging and the neuropsychological assessment of dreams. *Trends in Cognitive Sciences, 6*(1), 23–30.

Schwitzgebel, E. (2002a). A phenomenal, dispositional account of belief. *Noûs, 36*(2), 249–275.

Schwitzgebel, E. (2002b). How well do we know our own conscious experience? The case of visual imagery. In A. Noë (Ed.), *Is the visual world a grand illusion?* (pp. 35–53). Charlottesville: Imprint Academic.

Schwitzgebel, E. (2002c). Why did we think we dreamed in black and white? *Studies in History and Philosophy of Science, 33*, 649–660.

Schwitzgebel, E. (2003). Do people still report dreaming in black and white? An attempt to replicate a questionnaire from 1942. *Perceptual and Motor Skills, 96*, 25–29.

Schwitzgebel, E. (2007). Do you have constant tactile experience of your feet in your shoes? Or is experience limited to what's in attention? *Journal of Consciousness Studies, 14*(3), 5–35.

Schwitzgebel, E. (2011a). Belief. In E. N. Zalta (Ed.), *The Stanford encyclopedia of philosophy (winter 2011 edition)*. Retrieved from http://plato.stanford.edu/archives/win2011/entries/belief.

Schwitzgebel, E. (2011b). Mad belief? *Neuroethics, 5*(1), 13–17. doi:10.1007/s12152-011-9127-3.

Schwitzgebel, E. (2011c). *Perplexities of consciousness*. Cambridge, MA: MIT Press.

Schwitzgebel, E. (2013a). A dispositional approach to attitudes. In N. Nottelmann (Ed.), *New essays on belief*. New York: Palgrave Macmillan.

Schwitzgebel, E. (2013b). Précis: Perplexities of consciousness. *Philosophical Studies, 165*(3), 1161–1163. doi:10.1007/s11098-013-0149-z.

Schwitzgebel, E. (2013c). Reply to Kriegel, Smithies, and Spener. *Philosophical Studies, 165*(3), 1195–1206. doi:10.1007/s11098-013-0152-4.

Schwitzgebel, E., Huang, C., & Zhou, Y. (2006). Do we dream in color? Cultural variations and skepticism. *Dreaming, 16*(1), 36–42.

Searle, J. R. (1992). *The rediscovery of the mind*. Cambridge, MA: MIT Press.

Sebastián, M. Á. (2014). Dreams: An empirical way to settle the discussion between cognitive and non-cognitive theories of consciousness. *Synthese, 191*(2), 263–285. doi:10.1007/s11229-013-0385-y.

Sedda, A. (2011). Body integrity identity disorder: From a psychological to a neurological syndrome. *Neuropsychology Review, 21*(4), 334–336. doi:10.1007/s11065-011-9186-6.

Segal, S. J. (1972). Assimilation of a stimulus in the construction of an image: The Perky effect revisited. In P. W. Sheehan (Ed.), *The function and nature of imagery* (pp. 203–233). New York: Academic Press.

Serino, A., Alsmith, A., Costantini, M., Mandrigin, A., Tajadura-Jimenez, A., & Lopez, C. (2013). Bodily ownership and self-location: Components of bodily self-consciousness. *Consciousness and Cognition, 22*(4), 1239–1252. doi:10.1016/j.concog.2013.08.013.

Seth, A. K. (2013). Interoceptive inference, emotion, and the embodied self. *Trends in Cognitive Sciences, 17*(11), 565–573. doi:10.1016/j.tics.2013.09.007.

Seth, A. K. (2014). A predictive processing theory of sensorimotor contingencies: Explaining the puzzle of perceptual presence and its absence in synesthesia. *Cognitive neuroscience, 5*(2), 97–118. doi:10.1080/17588928.2013.877880.

Seth, A. K. (2015). The cybernetic Bayesian brain—From interoceptive inference to sensorimotor contingencies. In T. Metzinger & J. M. Windt (Eds.), *Open MIND: 35(T)*. Frankfurt am Main: MIND Group. doi:10.15502/9783958570108.

Seth, A. K., Baars, B. J., & Edelman, D. B. (2005). Criteria for consciousness in humans and other mammals. *Consciousness and Cognition, 14*(1), 119–139. doi:10.1016/j.concog.2004.08.006.

Seth, A. K., Suzuki, K., & Critchley, H. D. (2012). An interoceptive predictive coding model of conscious presence. *Frontiers in Psychology, 2*, 395. doi:10.3389/fpsyg.2011.00395.

Sevilla, J. C. (2004). *Wrestling with ghosts: A personal and scientific account of sleep paralysis*. Xlibris.

Sextus Empiricus. (1987). *Outlines of Pyrrhonism* (R. G. Bury, Trans.). Cambridge, MA: Harvard University Press.

Sheehan, P. W. (1972). A functional analysis of the role of visual imagery in unexpected recall. In P. W. Sheehan (Ed.), *The function and nature of imagery* (pp. 149–174). New York: Academic Press.

Shoemaker, S. (1968). Self-reference and self-awareness. *Journal of Philosophy, 65*(19), 555–567.

Siclari, F., Khatami, R., Urbaniok, F., Nobili, L., Mahowald, M. W., Schenck, C., et al. (2010). Violence in sleep. *Brain: A Journal of Neurology, 133*(Pt 12), 3494–3509. doi:10.1093/brain/awq296.

Siclari, F., LaRocque, J. J., Postle, B. R., & Tononi, G. (2013). Assessing sleep consciousness within subjects using a serial awakening paradigm. *Frontiers in Consciousness Research, 4*, 542.

Siclari, F., LaRocque, J. J., Bernardi, G., Postle, B. R., & Tononi, G. (2014). The neural correlates of consciousness in sleep: A no-task, within-state paradigm. http://biorxiv.org/content/early/2014/12/30/012443.

Siegel, J. (2002). *The neural control of sleep and waking*. New York: Springer.

Siegel, S. (2008). The epistemic conception of hallucination. In A. Haddock & F. Macpherson (Eds.), *Disjunctivism: Perception, action, and knowledge* (pp. 205–224). Oxford: Oxford University Press.

Siegler, F. A. (1977). Remembering dreams. In C. E. M. Dunlop (Ed.), *Philosophical essays on dreaming* (pp. 265–279). Ithaca: Cornell University Press. (Reprinted from *Philosophical Quarterly* 17 [1967], pp. 14–24.)

Sikka, P., Valli, K., Virta, T., & Revonsuo, A. (2014). I know how you felt last night, or do I? Self- and external ratings of emotions in REM sleep dreams. *Consciousness and Cognition, 25*, 51–66. doi:10.1016/j.concog.2014.01.011.

Silberer, H. (1919). *Der Traum: Einführung in die Traumpsychologie*. Stuttgart: Ferdinand Enke.

Simeon, D., & Abugel, J. (2006). *Feeling unreal: Depersonalization disorder and the loss of the self*. Oxford: Oxford University Press.

Singer, E. A. (1924). On pain in dreams. *Journal of Philosophy, 21*(22), 589–601.

Sirridge, M. (2005). Dream bodies and dream pains in Augustine's "De natura et origine animae." *Vivarium, 43*(2), 213–249.

Skaggs, W. E., & McNaughton, B. L. (1996). Replay of neuronal firing sequences in rat hippocampus during sleep following spatial experience. *Science, 271*(5257), 1870–1873.

Slade, P. D., & Bentall, R. P. (1988). *Sensory deception: A scientific analysis of hallucination*. London: Croom Helm.

Slater, M. (2009). Place illusion and plausibility can lead to realistic behaviour in immersive virtual environments. *Philosophical Transactions of the Royal Society of London, Series B: Biological Sciences, 364*(1535), 3549–3557. doi:10.1098/rstb.2009.0138.

Slater, M., Perez-Marcos, D., Ehrsson, H. H., & Sanchez-Vives, M. V. (2009). Inducing illusory ownership of a virtual body. *Frontiers in Neuroscience, 3*(2), 214–220.

Slater, M., Spanlang, B., & Corominas, D. (2010). Simulating virtual environments within virtual environments as the basis for a psychophysics of presence. *ACM Transactions on Graphics, 29*(4), 1. doi:10.1145/1778765.1778829.

Smallwood, J., & Andrews-Hanna, J. (2013). Not all minds that wander are lost: The importance of a balanced perspective on the mind-wandering state. *Frontiers in Psychology, 4*, 441. doi:10.3389/fpsyg.2013.00441.

Smallwood, J., Brown, K., Baird, B., & Schooler, J. W. (2012). Cooperation between the default mode network and the frontal-parietal network in the production of an internal train of thought. *Brain Research, 1428*, 60–70. doi:10.1016/j.brainres.2011.03.072.

Smallwood, J., Schooler, J. W., Turk, D. J., Cunningham, S. J., Burns, P., & Macrae, C. N. (2011). Self-reflection and the temporal focus of the wandering mind. *Consciousness and Cognition, 20*(4), 1120–1126. doi:10.1016/j.concog.2010.12.017.

Smithies, D. (2013). On the unreliability of introspection. *Philosophical Studies, 165*(3), 1177–1186. doi:10.1007/s11098-013-0150-6.

Snyder, F. (1970). The phenomenology of dreaming. In L. Madow & L. H. Snow (Eds.), *The psychodynamic implications of the physiological studies on dreams* (pp. 124–151). Springfield, IL: Charles S. Thomas.

Snyder, T. J., & Gackenbach, J. (1988). Individual differences associated with lucid dreaming. In J. Gackenbach & S. LaBerge (Eds.), *Conscious mind, sleeping brain: Perspectives on lucid dreaming* (pp. 221–259). New York: Plenum Press.

Solms, M. (1997). *The neuropsychology of dreams: A clinico-anatomical study*. Mahwah, NJ: Erlbaum.

Solms, M. (2000). Dreaming and REM sleep are controlled by different brain mechanisms. *Behavioral and Brain Sciences, 23*, 843–850.

Solms, M. (2002). Dreaming: Cholinergic and dopaminergic hypotheses. In E. Perry, H. Ashton, & A. Young (Eds.), *Neurochemistry of consciousness: Neurotransmitters in mind* (pp. 123–131). Advances in consciousness research. Amsterdam: John Benjamins.

Solms, M. (2011). Neurobiology and the neurological basis of dreaming. *Handbook of Clinical Neurology, 98*, 519–544. doi:10.1016/B978-0-444-52006-7.00034-4.

Solms, M. (2014). Do you agree that Freud's dream theory was erroneous with respect to (1) the genesis of dreaming (release of repressed infantile wishes) and (2) the bizarreness of dreams (disguise and censorship of unacceptable wishes)? In N. Tranquillo (Ed.), *Dream consciousness: Allan Hobson's new approach to the brain and its mind* (pp. 205–209). New York: Springer.

Solomonova, E., Fox, K. C. R., & Nielsen, T. (2014). Methodological considerations for the neurophenomenology of dreaming: commentary on Windt's "Reporting dream experience." *Frontiers in Human Neuroscience, 8*, 317. doi:10.3389/fnhum.2014.00317.

Solomonova, E., Frantova, E., & Nielsen, T. (2011). Felt presence: The uncanny encounters with the numinous other. *AI and Society, 26*(2), 171–178. doi:10.1007/s00146-010-0299-x.

Solomonova, E., Nielsen, T., Stenstrom, P., Simard, V., Frantova, E., & Donderi, D. (2008). Sensed presence as a correlate of sleep paralysis distress, social anxiety, and waking state social imagery. *Consciousness and Cognition, 17*(1), 49–63. doi:10.1016/j.concog.2007.04.007.

Sosa, E. (2007). *A virtue epistemology: Apt belief and reflective knowledge* (Vol. 1). Oxford: Clarendon Press.

Sosa, E., & Ichikawa, J. (2009). Dreaming: Philosophical perspectives. In T. Bayne, P. Wilken, & A. Cleeremans (Eds.), *Oxford companion to consciousness* (pp. 237–240). Oxford: Oxford University Press.

Soteriou, M. (2014). The disjunctive theory of perception. In E. N. Zalta (Ed.), *The Stanford encyclopedia of philosophy (summer 2014 edition)*.

Spanos, N. P., McNaulty, S. W. A., DuBreuil, S. C., Pires, M., & Burgess, M. F. (1995). The frequency and correlates of sleep paralysis in a university sample. *Journal of Research in Personality, 29*, 285–305.

Spener, M. (2013). Moderate scepticism about introspection. *Philosophical Studies, 165*(3), 1187–1194. doi:10.1007/s11098-013-0151-5.

Sperling, G. (1960). The information available in brief visual presentations. *Psychological Monographs, 74*(11), 1–29. doi:10.1037/h0093759.

Speth, J., Frenzel, C., & Voss, U. (2013). A differentiating empirical linguistic analysis of dreamer activity in reports of EEG-controlled REM-dreams and hypnagogic hallucinations. *Consciousness and Cognition, 22*(3), 1013–1021. doi:10.1016/j.concog.2013.07.003.

Spoormaker, V. I., Czisch, M., & Dresler, M. (2010). Lucid and non-lucid dreaming: Thinking in networks. *International Journal of Dream Research, 3*(1), 49–51.

Spoormaker, V. I., Gleiser, P., & Czisch, M. (2012). Frontoparietal connectivity and hierarchical structure of the brain's functional network during sleep. *Frontiers in Neurology, 3*, 80. doi:10.3389/fneur.2012.00080.

Sprenger, A., Lappe-Osthege, M., Talamo, S., Gais, S., Kimmig, H., & Helmchen, C. (2010). Eye movements during REM sleep and imagination of visual scenes. *Neuroreport, 21*(1), 45–49. doi:10.1097/WNR.0b013e32833370b2.

Springett, B. (2013). Philosophy of dreaming. *Internet Encyclopedia of Philosophy*. Retrieved from http://www.iep.utm.edu/dreaming.

Squires, R. (1973). The problem of dreams. *Philosophy, 48*(185), 245–259.

Squires, R. (1995). Dream time. *Proceedings of the Aristotelian Society, 95*, 83–91.

States, B. (1997). *Seeing in the dark: Reflections on dreams and dreaming*. New Haven: Yale University Press.

States, B. O. (2000). Dream bizarreness and inner thought. *Dreaming, 10*, 179–192.

Staunton, H., & O'Rourke, K. (2012). The creation of a topographical world and its contents in the dreams of the congenitally blind. *Dreaming, 22*(1), 53–57. doi:10.1037/a0026690.

Stawarczyk, D., Cassol, H., & D'Argembeau, A. (2013). Phenomenology of future-oriented mind-wandering episodes. *Frontiers in Psychology, 4*, 425. doi:10.3389/fpsyg.2013.00425.

Stazicker, J. (2011). Attention, visual consciousness, and indeterminacy. *Mind and Language, 26*(2), 156–184.

Stenstrom, P., Fox, K., Solomonova, E., & Nielsen, T. (2012). Mentation during sleep onset theta bursts in a trained participant: A role for NREM stage 1 sleep in memory processing? *International Journal of Dream Research, 5*(1). Retrieved from http://journals.ub.uni-heidelberg.de/index.php/IJoDR/article/view/9135.

Stephan, A., & Walter, S. (Eds.). (2013). *Handbuch Kognitionswissenschaft*. Stuttgart: Metzler.

Stephens, G. L., & Graham, G. (2000). *When self-consciousness breaks: Alien voices and inserted thoughts*. Cambridge, MA: MIT Press.

Steriade, M. (2009). Sleep oscillations and PGO waves. In R. Stickgold & M. P. Walker (Eds.), *The neuroscience of sleep* (pp. 22–29). London: Elsevier Academic Press.

Stickgold, R. (1998). Sleep: Off-line memory reprocessing. *Trends in Cognitive Sciences, 2*, 484–492.

Stickgold, R., Malia, A., Maguire, D., Roddenberry, D., & O'Connor, M. (2000). Replaying the game: Hypnagogic images in normals and amnesics. *Science, 290*(5490), 350–353.

Stickgold, R., Malia, A., Fosse, R., Propper, R., & Hobson, J. A. (2001). Brain-mind states: I. Longitudinal field study of sleep/wake factors influencing mentation report length. *Sleep, 24*(2), 171–179.

Stickgold, R., & Walker, M. P. (2005). Memory consolidation and reconsolidation: What is the role of sleep? *Trends in Cognitive Sciences, 28*, 408–415.

Stock, K. (2007). Sartre, Wittgenstein, and learning from imagination. In P. Goldie & E. Schellekens (Eds.), *Philosophy and conceptual art* (pp. 171–194). Oxford: Oxford University Press.

Stone, J. (1984). Dreaming and certainty. *Philosophical Studies, 45*(May), 353–368.

Strachey, J. (Ed. & Trans.). (1957). *The standard edition of the complete psychological works of Sigmund Freud: Volume 14 (1914–1916). On the history of the psycho-analytic movement, papers on metapsychology, and other works*. London: Hogarth Press and Institute of Psycho-analysis.

Strauch, I. (2005). REM dreaming in the transition from late childhood to adolescence: A longitudinal study. *Dreaming, 15*(3), 155–169.

Strauch, I., & Lederbogen, S. (1999). The home dreams and waking fantasies of boys and girls between ages 9 and 15: A longitudinal study. *Dreaming, 9*(2–3), 153–161.

Strauch, I., & Meier, B. (1992). *Den Träumen auf der Spur: Ergebnisse der experimentellen Traumforschung*. Bern: Huber.

Strauch, I., & Meier, B. (1996). *In search of dreams: Results of experimental dream research* (M. Ebon, Trans.). New York: State University of New York Press.

Strawson, G. (2011). The minimal subject. In S. Gallagher (Ed.), *The Oxford handbook of the self* (pp. 253–278). Oxford: Oxford University Press.

Strelen, J. (2006). *Akustisch evozierte Potentiale bei luziden Träumen*. Unpublished doctoral thesis, Johannes-Gutenberg University, Mainz, Germany.

Stroud, B. (1984). *The significance of philosophical scepticism*. Oxford: Clarendon Press.

Studerus, E., Gamma, A., Vollenweider, F. X., & Bell, V. (2010). Psychometric evaluation of the altered states of consciousness rating scale (OAV). *PLoS ONE, 5*(8), e12412. doi:10.1371/journal.pone.0012412.

Stumbrys, T., & Daniels, M. (2010). An exploratory study of creative problem solving in lucid dreams: Preliminary findings and methodological considerations. *International Journal of Dream Research, 3*(2), 121–129.

Stumbrys, T., & Erlacher, D. (2012). Lucid dreaming during NREM sleep: Two case reports. *International Journal of Dream Research, 5*(2). Retrieved from http://archiv.ub.uni-heidelberg.de/ojs/index.php/IJoDR/article/view/9483.

Stumbrys, T., Erlacher, D., & Johnson, M. &. Schredl, M. (2014). The phenomenology of lucid dreaming: An online survey. *American Journal of Psychology, 127*(2), 191–204.

Stumbrys, T., Erlacher, D., Schädlich, M., & Schredl, M. (2012). Induction of lucid dreams: A systematic review of evidence. *Consciousness and Cognition, 21*(3), 1456–1475. doi:10.1016/j.concog.2012.07.003.

Stumbrys, T., Erlacher, D., & Schmidt, S. (2011). Lucid dream mathematics: An explorative online study of arithmetic abilities of dream characters. *International Journal of Dream Research, 4*(1), 35–40.

Stumbrys, T., Erlacher, D., & Schredl, M. (2013). Testing the involvement of the prefrontal cortex in lucid dreaming: A tDCS study. *Consciousness and Cognition, 22*(4), 1214–1222. doi:10.1016/j.concog.2013.08.005.

Suzuki, H., Uchiyama, M., Tagaya, H., Ozaki, A., Kuriyama, K., Aritake, S., et al. (2004). Dreaming during non-rapid eye movement sleep in the absence of prior rapid eye movement sleep. *Sleep, 27*(8), 1486–1490.

Suzuki, K., Wakisaka, S., & Fujii, N. (2012). Substitutional reality system: A novel experimental platform for experiencing alternative reality. *Scientific Reports, 2*. doi:10.1038/srep00459.

Symons, D. (1993). The stuff that dreams aren't made of: Why wake-state and dream-state sensory experiences differ. *Cognition, 47*(3), 181–217.

Sze, J. A., Gyurak, A., Yuan, J. W., & Levenson, R. W. (2010). Coherence between emotional experience and physiology: Does body awareness training have an impact? *Emotion, 10*(6), 803–814. doi:10.1037/a0020146.

Szücs, A., Janszky, J., Holló, A., Migléczi, G., & Halász, P. (2003). Misleading hallucinations in unrecognized narcolepsy. *Acta Psychiatrica Scandinavica, 108*, 314–317.

Tajadura-Jimenez, A., & Tsakiris, M. (2013). Balancing the "inner" and the "outer" self: Interoceptive sensitivity modulates self–other boundaries. *Journal of Experimental Psychology, 143*(2), 736–744. doi:10.1037/a0033171.

Takeuchi, T., Miyasita, A., Inugami, M., & Yamamoto, Y. (2001). Intrinsic dreams are not produced without REM sleep mechanisms: Evidence through elicitation of sleep onset REM periods. *Journal of Sleep Research, 10*(1), 43–52.

Talbott, W. (2011). Bayesian epistemology. In E. N. Zalta (Ed.), *The Stanford encyclopedia of philosophy (summer 2011 edition)*. Retrieved from http://plato.stanford.edu/archives/sum2011/entries/epistemology-bayesian.

Tanaka, H., Hayashi, M., & Hori, T. (1996). Statistical features of hypnagogic EEG measured by a new scoring system. *Sleep, 19*(9), 731–738.

Tanaka, H., Hayashi, M., & Hori, T. (1997). Topographical characteristics and principal component structure of the hypnagogic EEG. *Sleep, 20*(7), 523–534.

Tarokh, L., & Carskadon, M. A. (2009). Sleep in adolescents. In R. Stickgold & M. P. Walker (Eds.), *The neuroscience of sleep* (pp. 70–77). London: Elsevier Academic Press.

Tassi, P., & Muzet, A. (2000). Sleep inertia. *Sleep Medicine Reviews, 4*(4), 341–353.

Taylor, F., & Bryant, R. A. (2007). The tendency to suppress, inhibiting thoughts, and dream rebound. *Behaviour Research and Therapy, 45*(1), 163–168. doi:10.1016/j.brat.2006.01.005.

Therrien, C. (2009). Making sense in ludic worlds. The idealization of immersive postures in movies and video games. In B. Atkins, H. Kennedy, & T. Krzywinska (Eds.), *Breaking new ground: Innovation in games, play, practice, and theory; Proceedings of the 2009 Digital Games Research Association Conference*. London: Brunel University.

Tholey, P. (1988). A model for lucidity training as a means of self-healing and psychological growth. In J. Gackenbach & S. LaBerge (Eds.), *Conscious mind, sleeping brain: Perspectives on lucid dreaming* (pp. 263–287). New York: Plenum Press.

Tholey, P., & Utecht, K. (1995/2000). *Schöpferisch Träumen: Wie Sie im Schlaf das Leben meistern; Der Klartraum als Lebenshilfe*. Eschborn bei Frankfurt am Main: Klotz.

Thomas, N. J. T. (2010). *The multidimensional spectrum of imagination: Images, dreams, hallucinations, and active, imaginative perception*. Unpublished draft, February 3, 2010. Retrieved from http://www.imagery-imagination.com/Spectrum.pdf.

Thomas, N. J. T. (2011). Mental imagery. In Edward N. Zalta (Ed.), *The Stanford encyclopedia of philosophy (winter 2011 edition)*. Retrieved from http://plato.stanford.edu/archives/win2011/entries/mental-imagery.

Thompson, E. (2007). *Mind in life: Biology, phenomenology, and the sciences of the mind*. Cambridge, MA: Harvard University Press.

Thompson, E. (2014). *Waking, dreaming, being: New light on the self and consciousness from neuroscience, meditation, and philosophy*. New York: Columbia University Press.

Thompson, E. (2015). Dreamless sleep, the embodied bind, and consciousness—The relevance of a classical Indian debate to cognitive science. In T. Metzinger & J. M. Windt (Eds.), *Open MIND: 37(T)*. Frankfurt am Main: MIND Group. doi: 10.15502/9783958570351.

Thompson, E., & Cosmelli, D. (2011). Brain in a vat or body in a world? Brainbound versus enactive views of experience. *Philosophical Topics, 39*(1), 163–180. doi:10.5840/philtopics201139119.

Thompson, W. L., Kosslyn, S. M., Hoffman, M. S., & van der Kooij, K. (2008). Inspecting visual mental images: Can people "see" implicit properties as easily in imagery and perception? *Memory and Cognition, 36*(5), 1024–1032. doi:10.3758/MC.36.5.1024.

Tiriac, A., Del Rio-Bermudez, C., & Blumberg, M. S. (2014). Self-generated movements with "unexpected" sensory consequences. *Current Biology, 24*(18), 2136–2141. doi:10.1016/j.cub.2014.07.053.

Tonay, V. (1990–1991). California women and their dreams: A historical and sub-cultural comparison of dream content. *Imagination, Cognition, and Personality, 10*, 83–97.

Tong, F. (2003). Primary visual cortex and visual awareness. *Nature Reviews: Neuroscience, 4*(3), 219–229.

Tononi, G. (2004). An information integration theory of consciousness. *BMC Neuroscience, 5*(1), 42. doi:10.1186/1471-2202-5-42.

Tononi, G. (2008). Consciousness as integrated information: A provisional manifesto. *Biological Bulletin, 215*(3), 216–242.

Tononi, G. (2012). *Phi: A voyage from the brain to the soul*. New York: Pantheon.

Tononi, G. (2014). How does your PHI formula deal with the evidence that consciousness is state dependent? More specifically, if PHI were higher in REM than in waking, would you conclude that dreaming was more conscious even than waking? In N. Tranquillo (Ed.), *Dream consciousness: Allan Hobson's new approach to the brain and its mind* (pp. 215–217). New York: Springer.

Tononi, G., & Massimini, M. (2008). Why does consciousness fade in early sleep? *Annals of the New York Academy of Sciences, 1129*, 330–334. doi:10.1196/annals.1417.024.

Traiger, S. (Ed.). (2006). *The Blackwell guide to Hume's Treatise*. Malden, MA: Blackwell.

Trindler, J., & Kramer, M. (1971). Dream recall. *American Journal of Psychiatry, 128*(3), 296–301.

Tsakiris, M., & Haggard, P. (2005). The rubber hand illusion revisited: Visuotactile integration and self-attribution. *Journal of Experimental Psychology: Human Perception and Performance, 31*(1), 80–91. doi:10.1037/0096-1523.31.1.80.

Tsuchiya, N., & Koch, C. (2008). Attention and consciousness. *Scholarpedia, 3*(5), 4173. Retrieved from http://www.scholarpedia.org/article/Attention_and_consciousness.

Tye, M. (1991). *The imagery debate*. Cambridge, MA: MIT Press.

Tye, M. (2009). A new look at the speckled hen. *Analysis, 69*(2), 258–263. doi:10.1093/analys/anp011.

Tye, M. (2010). Up close with the speckled hen. *Analysis, 70*(2), 283–286.

Tye, M. (2011). *Consciousness revisited: Materialism without phenomenal concepts*. Representation and mind series. Cambridge, MA: MIT Press.

Uga, V., Lemut, M. C., Zampi, C., Zilli, I., & Salzarulo, P. (2006). Music in dreams. *Consciousness and Cognition, 15*(2), 351–357. doi:10.1016/j.concog.2005.09.003.

Ullman, M., Krippner, S., & Vaughan, A. (1973). *Dream telepathy*. New York: Macmillan.

Ullman, M., Krippner, S., & Vaughan, A. (1973/2002). *Dream telepathy: Studies in consciousness*. Charlottesville, VA: Hampton Roads.

Underwood, E. (2013). Neuroscience: How to build a dream-reading machine. *Science, 340*(6128), 21. doi:10.1126/science.340.6128.21.

Uslar, D. von. (1964). *Der Traum als Welt: Untersuchungen zur Ontologie und Phänomenologie des Traums*. Pfullingen: Neske.

Vaitl, D., Birbaumer, N., Gruzelier, J., Jamieson, G. A., Kotchoubey, B., Kübler, A., et al. (2005). Psychobiology of altered states of consciousness. *Psychological Bulletin, 131*(1), 98–127. doi:10.1037/0033-2909.131.1.98.

Valberg, J. J. (2007). *Dream, death, and the self*. Princeton: Princeton University Press.

Vallar, G., & Ronchi, R. (2009). Somatoparaphrenia: A body delusion; A review of the neuropsychological literature. *Experimental Brain Research, 192*(3), 533–551.

Valli, K. (2008). *Threat simulation: The function of dreaming?* Doctoral thesis. Turku: Annales Universitatis Turkuensis.

Valli, K., & Revonsuo, A. (2007). Evolutionary psychological approaches to dream content. In D. Barrett & P. McNamara (Eds.), *The new science of dreaming: Vol. 3. Cultural and theoretical perspectives* (pp. 95–116). Westport: Praeger Perspectives.

Valli, K., Revonsuo, A., Pälkäs, O., Isamail, K. H., Ali, K. J., & Punamäki, R.-J. (2005). The threat simulation theory of the evolutionary function of dreaming: Evidence from dreams of traumatized children. *Consciousness and Cognition, 14*, 188–218.

van de Castle, R. L. (1994). *Our dreaming mind*. New York: Ballantine.

Vanderbeke, D., & Gansel, C. (Eds.). (in press). *Telling stories: Literature and evolution/evolution of literature*. Berlin: De Gruyter.

van Eeden, F. (1913). A Study of Dreams. *Proceedings of the Society for Psychical Research, 26*, 431–461. Retrieved from http://www.lucidity.com/vanEeden.html.

van Fraassen, B. C. (1980). *The scientific image*. Oxford: Clarendon Press.

van Lutterfeld, R., & Ford, J. M. (2012). Neurophysiological research: EEG and MEG. In J. D. Blom & I. E. C. Sommer (Eds.), *Hallucinations: Research and practice* (pp. 283–296). New York: Springer.

van Wassenhove, V., Buonomano, D. V., Shimojo, S., Shams, L., & Eagleman, D. (2008). Distortions of subjective time perception within and across senses. *PLoS ONE, 3*(1), e1437. doi:10.1371/journal.pone.0001437.

Varela, F. J. (1996). Neurophenomenology: A methodological remedy for the hard problem. *Journal of Consciousness Studies, 3*(4), 330–349.

Varela, F. J. (Ed.). (1997). *Sleeping, dreaming, and dying: An exploration of consciousness with the Dalai Lama*. Boston, MA: Wisdom Publications.

Varela, F. J., Rosch, E., & Thompson, E. (1993). *The embodied mind: Cognitive science and human experience*. Cambridge, MA: MIT Press.

Varela, F. J., & Shear, J. (2000). *The view from within: First-person approaches to the study of consciousness*. Bowling Green, OH: Imprint Academic.

Verdone, P. (1965). Temporal reference of manifest dream content. *Perceptual and Motor Skills, 20*(Supplement), 1253–1268.

Vertes, R. P., & Eastman, K. E. (2000). The case against memory consolidation in REM sleep. *Behavioral and Brain Sciences, 23*, 867–876.

Vetrugno, R., Arnulf, I., & Montagna, P. (2008). Disappearance of "phantom limb" and amputated arm usage during dreaming in REM sleep behaviour disorder. *Journal of Neurology, Neurosurgery, and Psychiatry, 79*, 481–483.

Vogel, J. (1990). Cartesian skepticism and inference to the best explanation. *Journal of Philosophy, 87*(11), 658–666.

Vold, J. M. (1910/1912). *Über den Traum* (Vols. 1 and 2). O. Klemm (Ed.). Leipzig: Johann Ambrosius Barth.

Voss, U. (2014). Please summarize your findings on the dream representation of disability in your studies of handicapped people. What bearing do these data have on protoconsciousness theory? In N. Tranquillo (Ed.), *Dream consciousness: Allan Hobson's new approach to the brain and its mind* (pp. 221–222). New York: Springer.

Voss, U., Frenzel, C., Koppehele-Gossel, J., & Hobson, A. (2012). Lucid dreaming: An age-dependent brain dissociation. *Journal of Sleep Research, 21*(6), 634–642. doi:10.1111/j.1365-2869.2012.01022.x.

Voss, U. & Hobson, A. (2015). What is the state-of-the-art on lucid dreaming?—Recent advances and questions for future research. In T. Metzinger & J. M. Windt (Eds.), *Open MIND: 38(T)*. Frankfurt am Main: MIND Group. doi:10.15502/9783958570306.

Voss, U., Holzmann, R., Hobson, A., Paulus, W., Koppehele-Gossel, J., Klimke, A., et al. (2014). Induction of self awareness in dreams through frontal low current stimulation of gamma activity. *Nature Neuroscience, 17*(6), 810–812. doi:10.1038/nn.3719.

Voss, U., Holzmann, R., Tuin, I., & Hobson, J. A. (2009). Lucid dreaming: A state of consciousness with features of both waking and non-lucid dreaming. *Sleep, 32*(9), 1191–1200.

Voss, U., Schermelleh-Engel, K., Windt, J. M., Frenzel, C., & Hobson, J. A. (2013). Measuring consciousness in dreams: The lucidity and consciousness in dreams scale. *Consciousness and Cognition, 22*, 8–21.

Voss, U., Tuin, I., Schermelleh-Engel, K., & Hobson, A. (2011). Waking and dreaming: Related but structurally independent. Dream reports of congenitally paraplegic and deaf-mute persons. *Consciousness and Cognition, 20*(3), 673–687. doi:10.1016/j.concog.2010.10.020.

Voss, U., & Voss, G. (2014). A neurobiological model of lucid dreaming. In R. Hurd & K. Bulkeley (Eds.), *Lucid dreaming: New perspectives on consciousness in sleep* (pp. 23–36). Santa Barbara, CA: Praeger.

Vyazovskiy, V. V., Olcese, U., Hanlon, E. C., Nir, Y., Cirelli, C., & Tononi, G. (2011). Local sleep in awake rats. *Nature, 472*(7344), 443–447. doi:10.1038/nature10009.

Wackermann, J., Pütz, P., Büchi, S., Strauch, I., & Lehmann, D. (2002). Brain electrical activity and subjective experience during altered states of consciousness: Ganzfeld and hypnagogic states. *International Journal of Psychophysiology, 46*, 123–146.

Waelti, E. (1983). *Der dritte Kreis des Wissens*. Interlaken: Ansata.

Walker, M. P. (2005). A refined model of sleep and the time course of memory formation. *Behavioral and Brain Sciences, 28*, 51–64.

Walker, M. P. (2009a). REM, dreams, and emotional brain homeostasis. *Frontiers in Neuroscience, 3*, 442–443.

Walker, M. P. (2009b). The role of sleep in cognition and emotion. *Annals of the New York Academy of Sciences, 1156*(1), 168–197. doi:10.1111/j.1749-6632.2009.04416.x.

Walker, M. P. (2009c). Sleep-dependent memory processing. In R. Stickgold & M. P. Walker (Eds.), *The neuroscience of sleep* (pp. 230–240). London: Elsevier Academic Press.

Walker, M. P., Stickgold, R., Alsop, D., Gaab, N., & Schlaug, G. (2005). Sleep-dependent motor memory plasticity in the human brain. *Neuroscience, 133*, 911–917.

Walker, M. P., & van der Helm, E. (2009). Overnight therapy? The role of sleep in emotional brain processing. *Psychological Bulletin, 135*(5), 731–748. doi:10.1037/a0016570.

Wamsley, E. J. (2013). Dreaming, waking conscious experience, and the resting brain: Report of subjective experience as a tool in the cognitive neurosciences. *Frontiers in Psychology, 4*, 637. doi:10.3389/fpsyg.2013.00637.

Wamsley, E., Donjacour, C. E., Scammell, T. E., Lammers, G. J., & Stickgold, R. (2014). Delusional confusion of dreaming and reality in narcolepsy. *Sleep*. doi:10.5665/sleep.3428.

Wamsley, E. J., Hirota, Y., Tucker, M. A., Smith, M. R., & Antrobus, J. S. (2007). Circadian and ultradian influences on dreaming: A dual rhythm model. *Brain Research Bulletin, 71*(4), 347–354. doi:10.1016/j.brainresbull.2006.09.021.

Wamsley, E. J., Perry, K., Djonlagic, I., Reaven, L. B., & Stickgold, R. (2010). Cognitive replay of visuomotor learning at sleep onset: Temporal dynamics and relationship to task performance. *Sleep, 33*(1), 59–68.

Wamsley, E. J., & Stickgold, R. (2009). Incorporation of waking events into dreams. In R. Stickgold & M. P. Walker (Eds.), *The neuroscience of sleep* (pp. 330–336). London: Elsevier Academic Press.

Wamsley, E. J., & Stickgold, R. (2010). Dreaming and offline memory processing. *Current Biology, 20*(23), R1010–R1013. doi:10.1016/j.cub.2010.10.045.

Wamsley, E. J., & Stickgold, R. (2011). Memory, sleep, and dreaming: Experiencing consolidation. *Sleep Medicine Clinics, 6*(1), 97–108.

Wamsley, E. J., Tucker, M., Payne, J. D., Benavides, J. A., & Stickgold, R. (2010). Dreaming of a learning task is associated with enhanced sleep-dependent memory consolidation. *Current Biology, 20*(9), 850–855. doi:10.1016/j.cub.2010.03.027.

Ward, A. F., & Wegner, D. M. (2013). Mind-blanking: When the mind goes away. *Frontiers in Psychology, 4*, 650. doi:10.3389/fpsyg.2013.00650.

Warren, J. (2009). *The head trip: Adventures on the wheel of consciousness*. Richmond: Oneworld.

Watson, D. (2003). To dream, perchance to remember: Individual differences in dream recall. *Personality and Individual Differences, 34*, 1271–1286.

Waxman, W. (1994). *Hume's theory of consciousness*. Cambridge: Cambridge University Press.

Webb, W. B. (1975). *Sleep: The gentle tyrant*. Englewood Cliffs, NJ: Prentice Hall.

Weed, S. C., Hallam, F. M., & Phinney, E. D. (1896). A study of dream-consciousness. *American Journal of Psychology, 7*, 405–411.

Wegner, D. M. (1994). Ironic processes of mental control. *Psychological Review, 101*(1), 34–52.

Wegner, D. M. (1997). Why the mind wanders. In J. D. Cohen & J. W. Schooler (Eds.), *Carnegie Mellon Symposia on Cognition: Scientific approaches to consciousness* (pp. 295–315). Mahwah, NJ: Erlbaum.

Wegner, D. M. (2009). How to think, say, or do precisely the worst thing for any occasion. *Science, 325*(5936), 48–50. doi:10.1126/science.1167346.

Wegner, D. M., Schneider, D. J., Carter, S. R., & White, T. L. (1987). Paradoxical effects of thought suppression. *Journal of Personality and Social Psychology, 53*(1), 5–13.

Wegner, D. M., Wenzlaff, R. M., & Kozak, M. (2004). Dream rebound: The return of suppressed thoughts in dreams. *Psychological Science, 15*(4), 232–236. doi:10.1111/j.0963-7214.2004.00657.x.

Wegner, D. M., & Zanakos, S. (1994). Chronic thought suppression. *Journal of Personality, 62*(4), 615–640. doi:10.1111/j.1467-6494.1994.tb00311.x.

Wehrle, R., Czisch, M., Kaufmann, C., Wetter, T. C., Holsboer, F., Auer, D. P., et al. (2005). Rapid eye movement–related brain activation in human sleep: A functional magnetic resonance imaging study. *Neuroreport, 16*(8). Retrieved from http://journals.lww.com/neuroreport/Fulltext/2005/05310/Rapid_eye_movement_related_brain_activation_in.15.aspx.

Wehrle, R., Kaufmann, C., Wetter, T. C., Holsboer, F., Auer, D. P., Pollmächer, T., et al. (2007). Functional microstates within human REM sleep: First evidence from fMRI of a thalamocortical network specific for phasic REM periods. *European Journal of Neuroscience, 25*(3), 863–871. doi:10.1111/j.1460-9568.2007.05314.x.

Weinstein, L., Schwartz, D., & Ellman, S. J. (1988). The development of scales to measure the experience of self-participation in sleep. *Sleep and Biological Rhythms, 11*(5), 437–447.

Weisberg, J. (2009). Locating IBE in the Bayesian framework. *Synthese, 167*(1), 125–143. doi:10.1007/s11229-008-9305-y.

Weygandt, W. (1893). *Entstehung der Träume*. Leipzig: Grübel & Sommerlatte.

White, A. R. (1990). *The language of imagination*. Oxford: Blackwell.

Wilbrand, H. (1887). *Die Seelenblindheit als Herderscheinung und ihre Beziehung zur Alexie und Agraphie*. Wiesbaden: Bergmann.

Wilde-Frenz, J., & Schulz, H. (1983). Rate and distribution of body movements during sleep in humans. *Perceptual and Motor Skills, 56*(1), 275–283.

Williams, B. (1978). *Descartes: The project of pure enquiry*. Middlesex: Pelican Books.

Williams, J., Merritt, J., Rittenhouse, C., & Hobson, J. A. (1992). Bizarreness in dreams and fantasies: Implications for the activation-synthesis hypothesis. *Consciousness and Cognition, 1*(2), 172–185. doi:10.1016/1053-8100(92)90059-J.

Windmann, S., Wehrmann, M., Calabrese, P., & Güntürkün, O. (2006). Role of the prefrontal cortex in attentional control over bistable vision. *Journal of Cognitive Neuroscience, 18*(3), 456–471.

Windt, J. M. (2010). The immersive spatiotemporal hallucination model of dreaming. *Phenomenology and the Cognitive Sciences, 9*, 295–316.

Windt, J. M. (2011). Altered consciousness in philosophy. In E. Cardeña & M. Winkelman (Eds.), *Altering consciousness: Multidisciplinary perspectives: Vol. 1. History, culture, and the humanities* (pp. 229–254). Santa Barbara, CA: Praeger.

Windt, J. M. (2013a). Minding the dream self: Perspectives from the analysis of self-experience in dreams. *Behavioral and Brain Sciences, 36*(06), 633. doi:10.1017/S0140525X13001477.

Windt, J. M. (2013b). Reporting dream experience: Why (not) to be skeptical about dream reports. *Frontiers in Human Neuroscience, 7*, 708. doi:10.3389/fnhum.2013.00708.

Windt, J. M. (2014a). How can the protoconsciousness hypothesis contribute to philosophical theories of consciousness and the self? In N. Tranquillo (Ed.), *Dream consciousness: Allan Hobson's new approach to the brain and its mind* (pp. 225–230). New York: Springer.

Windt, J. M. (2014b). Träume, Bewusstsein und das Selbst: Eine Analyse aus Sicht der Philosophie des Geistes. In H. Fink & R. Rosenzweig (Eds.), *Bewusstsein—Selbst—Ich: Die Hirnforschung und das Subjektive* (pp. 209–245). Münster: mentis.

Windt, J. M. (2015). Just in time—Dreamless sleep experience as pure subjective temporality: A commentary on Evan Thompson. In T. Metzinger & J. M. Windt (Eds.), *Open MIND: 37(T)*. Frankfurt am Main: MIND Group.

Windt, J. M. (in press). Dreams and dreaming. In E. N. Zalta (Ed.), *The Stanford encyclopedia of philosophy*.

Windt, J. M., Harkness, D. L., & Lenggenhager, B. (2014). Tickle me, I think I might be dreaming! Sensory attenuation, self-other distinction, and predictive processing in lucid dreams. *Frontiers in Human Neuroscience, 8*, 717. doi:10.3389/fnhum.2014.00717.

Windt, J. M., & Metzinger, T. (2007). The philosophy of dreaming and self-consciousness: What happens to the experiential subject during the dream state? In D. Barrett & P. McNamara (Eds.), *The new science of dreaming: Vol. 3. Cultural and theoretical perspectives* (pp. 193–248). Westport: Praeger Perspectives.

Windt, J. M., & Metzinger, T. (2012). The self in dreams. In D. Barrett (Ed.), *Encyclopedia of sleep and dreams: The evolution, function, nature, and mysteries of slumber* (pp. 575–577). Santa Barbara, CA: Greenwood.

Windt, J. M., & Noreika, V. (2011). How to integrate dreaming into a general theory of consciousness: A critical review of existing positions and suggestions for future research. *Consciousness and Cognition, 20*(4), 1091–1107.

Windt, J. M., Schredl, M., & Hobson, J. A. (2013). Träumen. In A. Stephan & S. Walter (Ed.), *Handbuch Kognitionswissenschaft* (pp. 452–459). Stuttgart: Metzler.

Winget, C., & Kramer, M. (1979). *Dimensions of dreams*. Gainesville: University Presses of Florida.

Wissmath, B., Weibel, D., Schmutz, J., & Mast, F. W. (2011). Being present in more than one place at a time? Patterns of mental self-localization. *Consciousness and Cognition, 20*(4), 1808–1815. doi:10.1016/j.concog.2011.05.008.

Wittgenstein, L. (1953). *Philosophical investigations* (G. E. M. Anscombe, Trans.). Oxford: Blackwell.

Wittgenstein, L. (1967). *Zettel* (G. E. M. Anscombe & G. H. von Wright, Eds.; G. E. M. Anscombe, Trans.). Berkeley: University of California Press.

Wittgenstein, L. (1977). *On certainty* (Repr. with corrections). Oxford: Blackwell.

Wittmann, M., van Wassenhove, V., Craig, A. D., & Paulus, M. P. (2010). The neural substrates of subjective time dilation. *Frontiers in Human Neuroscience, 4*(2), 1–9. doi:10.3389/neuro.09.002.2010.

Wolfradt, U. (2000/2001). Außerkörpererfahrungen (AKE) aus differentiell-psychologischer Perspektive. *Zeitschrift für Parapsychologie und Grenzgebiete der Psychologie, 42–43*, 65–108.

Wolfson, E. R. (2011). *A dream interpreted within a dream: Oneiropoiesis and the prism of imagination.* New York: Zone.

Wolman, R. N., & Kozmová, M. (2007). Last night I had the strangest dream: Varieties of rational thought processes in dream reports. *Consciousness and Cognition, 16*(4), 838–849. doi:10.1016/j.concog.2006.09.009.

Woods, R. L., & Greenhouse, H. B. (Eds.). (1974). *The new world of dreams.* New York: Macmillan.

Worsley, A. (1988). Personal experiences in lucid dreaming. In J. Gackenbach & S. LaBerge (Eds.), *Conscious mind, sleeping brain: Perspectives on lucid dreaming* (pp. 321–341). New York: Plenum Press.

Wundt, W. M. (1880). *Grundzüge der physiologischen Psychologie (2nd ed.).* Leipzig: Engelmann.

Yang, C.-M., Han, H.-Y., Yang, M.-H., Su, W.-C., & Lane, T. (2010). What subjective experiences determine the perception of falling asleep during sleep onset period? *Consciousness and Cognition, 19*(4), 1084–1092. doi:10.1016/j.concog.2009.12.017.

Yost, R. M., Jr., & Kalish, D. (1977). Miss Macdonald on sleeping and waking. In C. E. M. Dunlop (Ed.), *Philosophical essays on dreaming* (pp. 81–102). Ithaca: Cornell University Press. (Reprinted from *Philosophical Quarterly* 5 [1955], pp. 109–124.)

Young, S. (1999). *Dreaming in the lotus: Buddhist dream narrative, imagery, and practice.* Boston, MA: Wisdom Publications.

Yu, C. K.-C. (2008). Typical dreams experienced by Chinese people. *Dreaming, 18*(1), 1–10.

Yu, C. K.-C. (2009a). Delusions and the factor structure of typical dreams. *Dreaming, 19*(1), 42–54. doi:10.1037/a0014789.

Yu, C. K.-C. (2009b). Paranoia in dreams and the classification of typical dreams. *Dreaming, 19*(4), 255–272. doi:10.1037/a0017583.

Yu, C. K.-C. (2010). Recurrence of typical dreams and the instinctual and delusional predispositions of dreams. *Dreaming, 20*(4), 254–279.

Yu, C. K.-C. (2012). The effect of sleep position on dream experiences. *Dreaming, 22*(3), 212–221. doi:10.1037/a0029255.

Zacks, J. M., & Michelon, P. (2005). Transformations of visuospatial images. *Behavioral and Cognitive Neuroscience Reviews, 4*(2), 96–118.

Zacks, J. M., & Tversky, B. (2005). Multiple systems for spatial imagery: Transformations of objects and bodies. *Spatial Cognition and Computation, 5*(4), 271–306.

Zadra, A. L. (2001). Recurrent dreams: Their relation to life events. In D. Barrett (Ed.), *Trauma and dreams* (pp. 231–247). Cambridge, MA: Harvard University Press.

Zadra, A., & Donderi, D. C. (2000). Threat perceptions and avoidance in recurrent dreams. *Behavioral and Brain Sciences, 23*(6), 1017–1018.

Zadra, A., & Robert, G. (2012). Dream recall frequency: Impact of prospective measures and motivational factors. *Consciousness and Cognition, 21*(4), 1695–1702. doi:10.1016/j.concog.2012.08.011.

Zahavi, D. (2001). Unity of consciousness and the problem of self. In S. Gallagher (Ed.), *The Oxford handbook of the self* (pp. 316–335). Oxford: Oxford University Press.

Zeman, A. Z. J., Della Sala, S., Torrens, L. A., Gountouna, V.-E., McGonigle, D. J., & Logie, R. H. (2010). Loss of imagery phenomenology with intact visuo-spatial task performance: A case of "blind imagination." *Neuropsychologia, 48*(1), 145–155. doi:10.1016/j.neuropsychologia.2009.08.024.

Zikmund, V. (1972). Physiological correlates of visual imagery. In P. W. Sheehan (Ed.), *The function and nature of imagery* (pp. 355–387). New York: Academic Press.

Index

ss
ssl
n Street, 9th floor
42

nit.edu
2882

orized representative in the EU for product safety and compliance is

ess System Europe Oü, 16879218
ie tee 50,
21

uests@easproject.com
968 939

80262549912
D: 152925671

www.ingramcontent.com/pod-product-compliance
Ingram Content Group UK Ltd.
Pitfield, Milton Keynes, MK11 3LW, UK
UKHW062007041026
15591UKWH00017B/1218

9 780262 5499